Multimedia
Security
Handbook

INTERNET and COMMUNICATIONS

This new book series presents the latest research and technological developments in the field of Internet and multimedia systems and applications. We remain committed to publishing high-quality reference and technical books written by experts in the field.

If you are interested in writing, editing, or contributing to a volume in this series, or if you have suggestions for needed books, please contact Dr. Borko Furht at the following address:

Borko Furht, Ph.D.
Department Chairman and Professor
Computer Science and Engineering
Florida Atlantic University
777 Glades Road
Boca Raton, FL 33431 U.S.A.

E-mail: borko@cse.fau.edu

Multimedia Security Handbook

Editors-in-Chief and Authors

Borko Furht
Darko Kirovski

CRC Press
Taylor & Francis Group
Boca Raton London New York

CRC Press is an imprint of the
Taylor & Francis Group, an **informa** business

CRC Press
Taylor & Francis Group
6000 Broken Sound Parkway NW, Suite 300
Boca Raton, FL 33487-2742

© 2005 by Taylor & Francis Group, LLC
CRC Press is an imprint of Taylor & Francis Group, an Informa business

First issued in paperback 2019

No claim to original U.S. Government works

ISBN-13: 978-0-367-45423-4 (pbk)
ISBN-13: 978-0-8493-2773-5 (hbk)

Visit the Taylor & Francis Web site at
http://www.taylorandfrancis.com

and the CRC Press Web site at
http://www.crcpress.com

Library of Congress Card Number 2004055105

Library of Congress Cataloging-in-Publication Data

Multimedia security handbook / edited by Borko Furht, Darko Kirovski.
    p. cm. – (Internet and communications)
    Includes bibliographical references and index.
    ISBN 0-8493-2773-3 (alk. paper)
    1. Multimedia systems – Security measures – Handbooks, manuals, etc.
    2. Intellectual property – Handbooks, manuals, etc.   I. Furht, Borivoje.
    II. Kirovski, Darko.   III. Title.   IV. Series.

QA76.9.A25M85 2004
005.8–dc22   2004055105

Preface

Recent advances in digital communications and storage technologies have brought major changes for consumers. High-capacity hard disks and DVDs can store a large amount of audiovisual data. In addition, faster Internet connection speeds and the emerging high-bit-rate DSL connections provide sufficient bandwidth for entertainment networks. These improvements in computers and communication networks are radically changing the economics of intellectual property reproduction and distribution. Intellectual property owners must exploit new ways of reproducing, distributing, and marketing their intellectual property. However, a major problem with current digital distribution and storage technologies is the great threat of piracy.

The purpose of the *Multimedia Security Handbook* is to provide a comprehensive reference on advanced topics in this field. The handbook is intended both for researchers and practitioners in the field, and for scientists and engineers involved in designing and developing systems for the protection of digital multimedia content. The handbook can also be used as the textbook for graduate courses in the area of multimedia security.

The handbook addresses a variety of issues related to the protection of digital multimedia content, including audio, image, and video protection. The state-of-the art multimedia security technologies are presented, including protection architectures, multimedia encryption, watermarking, fingerprinting and authentication techniques, and various applications.

This handbook is comprised of 26 chapters divided into 6 parts. Part I, General Issues, introduces fundamental concepts applied in the protection of multimedia content and discusses the vulnerability of various protection schemes. Part II, Multimedia Encryption, includes chapters on audio, image, and video encryption techniques. These techniques deal with selective video encryption, which meet real-time requirements, chaos-based encryption, and techniques for protection of streaming media. Part III, Multimedia Watermarking, consists of chapters dealing with various watermarking techniques, including audio, image, and video watermarking. Current state-of-the art and future trends are addressed, including multidimensional, fragile, and robust watermarks. Part IV, Multimedia Data Hiding, Fingerprinting, and Authentication, includes chapters on various issues related to these techniques. The topics include fundamentals on lossless data hiding, digital media fingerprinting techniques, scalable and signature-based media authentication, and attacking such media protections schemes. Part V, Applications, includes chapters that describe applications of multimedia protection schemes. The topics, such as application taxonomy, digital rights management, and techniques for adult image filtering, are covered in this part.

We would like to thank the authors, who are world experts in the field, for their contributions of individual chapters to the handbook. Without their expertise and effort, this handbook would never come to fruition. CRC Press editors and staff also deserve our sincere recognition for their support throughout the project.

Borko Furht and Darko Kirovski

Editors-in-Chief and Authors

Borko Furht is a professor and chairman of the Department of Computer Science and Engineering at Florida Atlantic University (FAU) in Boca Raton, Florida. Before joining FAU, he was a vice-president of research and a senior director of development at Modcomp (Ft. Lauderdale), a computer company of Daimler Benz, Germany, and a professor at the University of Miami in Coral Gables, Florida. Professor Furht received Ph.D. degrees in electrical and computer engineering from the University of Belgrade. His current research is in multimedia systems, video coding and compression, video databases, wireless multimedia, and Internet computing. He is the author of numerous books and articles in the areas of multimedia, computer architecture, real-time computing, and operating systems. He is a founder and editor-in-chief of the *Journal of Multimedia Tools and Applications* (Kluwer Academic Publishers). He has received several technical and publishing awards and has consulted for many high-tech companies, including IBM, Hewlett-Packard, Xerox, General Electric, JPL, NASA, Honeywell and RCA. He has also served as a consultant to various colleges and universities. He has given many invited talks, keynote lectures, seminars, and tutorials.

Darko Kirovski received his Ph.D. degree in computer science from the University of California, Los Angeles, in 2001. Since April 2000, he has been a researcher at Microsoft Research. His research interests include certificates of authenticity, system security, multimedia processing, biometric identity authentication, and embedded system design and debugging. He has received the 1999 Microsoft Graduate Research Fellowship, the 2000 ACM/IEEE Design Automation Conference Graduate Scholarship, the 2001 ACM Outstanding Ph.D. Dissertation Award in Electronic Design Automation, and the Best Paper Award at the ACM Multimedia 2002.

List of Contributors

Adnan M. Alattar, Digimarc Corporation, 19801 SW 72nd Ave., Ste. 250, Tualatin, OR 97062

Rolf Blom, Ericsson Research, Communications Security Laboratory, Ericsson AB, Sweden

A. Briassouli, Beckman Institute, Department of Electrical and Computer Engineering, University of Illinois at Urbana-Champaign, Urbana, IL 61801

Elisabetta Carrara, Ericsson Research, Communications Security Laboratory, Ericsson AB, Sweden

Shih-Fu Chang, Department of Electrical Engineering, Columbia University, New York, New York 10027

Guanrong Chen, Department of Electronic Engineering, City University of Hong Kong, Kowloon Toon, Hong Kong SAR, China

Mohamed Daoudi, MIIRE Group, LIFL/INT ENIC-Telecom Lille1, Cité Scientifique, 59655 Villeneuve d'Ascq, France

Petros Daras, Information Processing Laboratory, Electrical and Computer Engineering Department, Aristotle University of Thessaloniki, 540 06 Thessaloniki, Greece

Edward J. Delp, Video and Image Processing Laboratory (*VIPER*), School of Electrical and Computer Engineering, Purdue University, West Lafayette, Indiana 47907

Chabane Djeraba, MIIRE Group, LIFL/INT ENIC-Telecom Lille1, Cité Scientifique, 59655 Villeneuve d'Ascq, France

Jean-Luc Dugelay, EURECOM, 2229 route des Cretes, Sophia Antipolis, France

Sabu Emmanuel, Nanyang Technological University, Singapore

Ahmet M. Eskicioglu, Department of Computer and Information Science, Brooklyn College, The City University of New York, Brooklyn, New York 11210

Borko Furht, Department of Computer Science and Engineering, Florida Atlantic University, Boca Raton, Florida 33431-0991

Jaap Haitsma, Philips Research Eindhoven, 5656AA Eindhoven, The Netherlands

Yu-Feng Hsu, Department of Electrical Engineering, National Taiwan University, Taipei, Taiwan 10617, Republic of China

Bo Hu, Electrical Engineering Department, Fudan University, Shanghai, China

Jiwu Huang, Sun Yat-Sen University, Guangzhou, China

Ebroul Izquierdo, Department of Electronic Engineering, Queen Mary, University of London, London, United Kingdom

Ton Kalker, Philips Research Eindhoven, 5656AA Eindhoven, The Netherlands

Xiangui Kang, Sun Yat-Sen University, Guangzhou, China

Mohan S. Kankanhalli, National University of Singapore, Singapore

Hyungshin Kim, Department of Computer Engineering, Chungnam National University, Korea

Darko Kirovski, Microsoft Research, One Microsoft Way, Redmond, WA

Deepa Kundur, Texas A&M University, Department of Electrical Engineering, 111D Zachry Engineering Center, 3128 TAMU, College Station, TX 77843-3128

Ken Levy, Digimarc, Tualatin, Oregon 97062

Shujun Li, Department of Electronic Engineering, City University of Hong Kong, Kowloon Toon, Hong Kong SAR, China

Fredrik Lindholm, Ericsson Research, Communications Security Laboratory, Ericsson AB, Sweden

Zheng Liu, C4 Technology, Inc., Tokyo 141-0021, Japan

Jeffrey Lotspiech, 2858 Hartwick Pines Dr., Henderson, NV 89052

Ya-Wen Lu, Department of Electrical Engineering, National Taiwan University, Taipei, Taiwan 10617, Republic of China

William Luh, Texas A&M University, Department of Electrical Engineering, 111D Zachry Engineering Center, 3128 TAMU, College Station, TX 77843-3128

Mohamed F. Mansour, Texas Instruments Inc., 12500 TI Blvd, MS8634, Dallas, Texas 75243

Edin Muharemagic, Department of Computer Science and Engineering, Florida Atlantic University, Boca Raton, Florida 33431-0991

Mats Näslund, Ericsson Research, Communications Security Laboratory, Ericsson AB, Sweden

Karl Norrman, Ericsson Research, Communications Security Laboratory, Ericsson AB, Sweden

Soo-Chang Pei, Department of Electrical Engineering, National Taiwan University, Taipei, Taiwan 10617, Republic of China

Tony Rodriguez, Digimarc, Tualatin, Oregon 97062

Yun Q. Shi, New Jersey Institute of Technology, Newark, New Jersey

D. Simitopoulos, Informatics and Telematics Institute, 1st km Thermi-Panorama Road, Thermi-Thessaloniki 57 001, Greece

Dimitrios Skraparlis, Toshiba Research Europe Limited, Telecommunications Research Laboratory, 32, Queen Square, Bristol BS1 4ND, U.K.

Daniel Socek, Department of Computer Science and Engineering, Florida Atlantic University, Boca Raton, Florida 33431-0991

Michael G. Strintzis, Informatics and Telematics Institute, 1st km Thermi-Panorama Road, 570 01, Thermi-Thessaloniki, Greece

Wei Su, U.S. Army CECOM, I2WD, AMSEL-RD-IW-IR, Fort Monmouth, New Jersey

Qibin Sun, Institute for Infocomm Research, Singapore 119613

Ahmed H. Tewfik, Department of Electrical and Computer Engineering, University of Minnesota, 200 Union Street, S.E., Minneapolis, MN 55455

Christophe Tombelle, MIIRE Group, LIFL/INT ENIC-Telecom Lille1, Cité Scientifique, 59655 Villeneuve d'Ascq, France

Dimitrios Tzovaras, Informatics and Telematics Institute, 1st km Thermi-Panorama Road, 570 01, Thermi-Thessaloniki, Greece

Guorong Xuan, Tongji University, Shanghai, China

Heather Yu, Panasonic Digital Networking Laboratory, Princeton, New Jersey 08540

Dimitrios Zarpalas, Informatics and Telematics Institute, 1st km Thermi-Panorama Road, 570 01, Thermi-Thessaloniki, Greece

Huicheng Zheng, MIIRE Group, LIFL/INT ENIC-Telecom Lille1, Cité Scientifique, 59655 Villeneuve d'Ascq, France

Xuan Zheng, Department of Electrical and Computer Engineering, University of Virginia, Charlottesville, Virginia 22904-4743

Contents

Contents

Part I
General Issues

1

Protection of Multimedia Content in Distribution Networks

Ahmet M. Eskicioglu and Edward J. Delp

INTRODUCTION

In recent years, advances in digital technologies have created significant changes in the way we reproduce, distribute, and market intellectual property (IP). Digital media can now be exploited by IP owners to develop new and innovative business models for their products and services. The lowered cost of reproduction, storage, and distribution, however, also invites much motivation for large-scale commercial infringement. In a world where piracy is a growing potential threat, the rights of the IP owners can be protected using three complementary weapons: technology, legislation, and business models. Because of the diversity of IP (ranging from e-books to songs and movies) created by copyright industries, no single solution is applicable to the protection of multimedia products in distribution networks.

Intellectual property is created as a result of intellectual activities in the industrial, scientific, literary, and artistic fields [1]. It is divided into two general categories:

- Industrial property: This includes inventions (patents), trademarks, industrial designs, and geographic indications of source. A *patent* is an exclusive right granted for an invention, which is a product or a process that provides either a new way of doing something or a new technical solution to a problem. A *trademark* is a distinctive sign that identifies certain goods or services as those produced or provided by a specific person or enterprise. It provides protection to the owner by ensuring the exclusive right to use it to identify goods or services or to authorize another to use it in return for payment. An *industrial design* is the ornamental or aesthetic aspect of an article. The design may consist of three-dimensional features (such as the shape or surface of an article) or of two-dimensional features (such as patterns, lines, or color). A *geographical indication* is a sign used on goods that have a specific geographical origin and possess qualities or a reputation that are due to that place of origin. In general, a geographical indication is associated with the name of the place of origin of the goods. Typically, agricultural products have qualities that derive from their place of production identified by specific local factors such as climate and soil.

- Copyright: This includes literary and artistic works such as novels, poems and plays, films, musical works, artistic works such as drawings, paintings, photographs, and sculptures, and architectural designs. As many creative works protected by copyright require mass distribution, communication, and financial investment for their dissemination (e.g., publications, sound recordings, and films), creators usually sell the rights to their works to individuals or companies with a potential to market the works in return for payment. Copyright and its related rights are essential to human creativity, providing creators incentives in the form of recognition and fair economic rewards and assurance that their works can be disseminated without fear of unauthorized copying or piracy.

To understand the increasing importance of copyrighted content protection, we should apprehend the essential difference between old and new technologies for distribution and storage. Prior to the development of digital technologies, content was created, distributed, stored, and displayed by analog means. The popular video cassette recorders (VCRs) of the 1980s introduced a revolutionary way of viewing audiovisual (A/V) content but, ironically, allowed unauthorized copying, risking the investments made in IP. The inherent characteristics of analog recording, however, prevented piracy efforts from reaching alarming proportions. If a taped content is copied on a VCR, the visual quality of the new (i.e., the first generation) copy is relatively reduced. Further generational copies result in noticeably less quality, decreasing the commercial value of the content. Today, reasonably efficient analog copy protection methods exist and have recently been made mandatory in consumer electronics devices to further discourage illegal analog copying.

With the advent of digital technologies, new tools have emerged for making perfect copies of the original content. We will briefly review digital representation of data to reveal why generational copies do not lose their quality. A text, an image, or a video is represented as a stream of bits (0s and 1s) that can be conveniently stored on magnetic or optical media. Because digital recording is a process whereby each bit in the source stream is read and copied to the new medium, an exact replica of the content is obtained. Such a capability becomes even more threatening with the ever-increasing availability of the Internet, an immense and boundless digital distribution mechanism. Protection of digital multimedia content therefore appears to be a crucial problem for which immediate solutions are needed.

Recent inventions in digital communications and storage technologies have resulted in a number of major changes in the distribution of multimedia content to consumers:

- Magnetic and optical storage capacity is much higher today. Even the basic configuration of personal computers comes with 40 GB of magnetic hard disk storage. Although a DVD (digital versatile disk) is the same physical size as a CD, it has a much higher optical storage capacity for audiovisual data. Depending on the type of DVD, the capacity ranges between 4 and 17 GB (2–8 h of video).

- The speed of the Internet connection has grown rapidly in recent years. Currently, cable modems and Asymmetric Digital Subscriber Line (ADSL) are the two technologies that dominate the industry. The emerging VDSL (very high bit-rate DSL) connection with speeds up to 52 Mbps will provide sufficient bandwidth for entertainment networks.

End-to-end security is the most critical requirement for the creation of new digital markets where copyrighted content is a major product. In this chapter, we present an overview of copyright and copyright industries and examine how the technological, legal, and business solutions help maintain the incentive to supply the lifeblood of the markets.

WHAT IS COPYRIGHT?

To guide the discussion into the proper context, we will begin with the definition of "copyright" and summarize the important aspects of the copyright law. Copyright is a *form of protection provided by the laws of the United States (title 17, U.S. Code) to the authors of "original works of authorship," including literary, dramatic, musical, artistic, and certain other intellectual works* [2]. Although copyright literally means "right to copy," the term is now used to cover a number of exclusive rights granted to the authors for the protection of their work. According to Section 106 of the 1976 Copyright Act [3], the owner of copyright is given the exclusive right to do, and to authorize others to do, any of the following:

- To *reproduce the copyrighted work* in copies or phonorecords
- To prepare *derivative works* based on the copyrighted work
- To *distribute copies or phonorecords* of the copyrighted work to the public by sale or other transfer of ownership, or by rental, lease, or lending
- To *perform the copyrighted work publicly*, in the case of literary, musical, dramatic, and choreographic works, pantomimes, and motion pictures and other audiovisual works
- To *display the copyrighted work publicly*, in the case of literary, musical, dramatic, and choreographic works, pantomimes, and pictorial, graphic, or sculptural works, including the individual images of a motion picture or other audiovisual work
- To *perform the copyrighted work publicly* by means of a digital audio transmission, in the case of sound recordings

It is illegal to violate the rights provided by the copyright law to the owner of the copyright. There are, however, limitations on these rights as established in several sections of the 1976 Copyright Act. One important limitation, the doctrine of "fair use," has been the subject of a major discussion on content protection. Section 107 states that the use of a copyrighted work by reproduction in copies or phonorecords or by any other means specified by the law, for purposes such as criticism, comment, news reporting, teaching (including multiple copies for classroom use), scholarship, or research, is not an infringement of copyright. In any particular case, the following criteria, among others, may be considered in determining whether fair use applies or not:

1. The purpose and character of the use, including whether such use is of a commercial nature or is for nonprofit educational purposes
2. The nature of the copyrighted work
3. The amount and substantiality of the portion used in relation to the copyrighted work as a whole
4. The effect of the use upon the potential market for, or value of, the copyrighted work

For copyright protection, the original work of authorship needs to be fixed in a tangible medium of expression from which it can be perceived, reproduced, or otherwise communicated, either directly or with the aid of a machine or device. This language incorporates three fundamental concepts [4] of the law: fixation, originality, and expression. *Fixation* (i.e., the act of rendering a creation in some tangible form) may be achieved in a number of ways depending on the category of the work. *Originality* is a necessary (but not a sufficient) condition for a work produced by the human mind to be copyrightable. Scientific discoveries, for example, are not copyrightable, as they are regarded as the common property of all people (however, an inventor may apply for a patent, which is another form of protection). Finally, it is the *expression* of an idea, and not idea itself, that is copyrightable. Ideas, like facts, are in the public domain without a need for protection. Nevertheless, the separation of an idea from an expression is not always clear and can only be studied on a case-by-case basis. When the three basic requirements of fixation, originality, and expression are met, the law provides a highly broad protection. Table 1.1 summarizes the range of copyrightable works.

It is interesting to note that copyright is secured as soon as the work is created by the author in some fixed form. No action, including publication and registration, is needed in the Copyright Office. *Publication* is the distribution of copies or phonorecords of a work to the public by sale or other transfer of ownership or by rental, lease, or lending. *Registration* is a legal process to create a public record of the basic facts of a particular copyright. Although neither publication nor registration is a requirement for protection, they provide certain advantages to the copyright owner.

The copyright law has different clauses for the protection of published and unpublished works. All unpublished works are subject to protection, regardless of the nationality or domicile of the author. The published works are protected if certain conditions are met regarding the type of work, citizenship, residency, and publication date and place.

TABLE 1.1. Main categories of copyrightable and not copyrightable items

Copyrightable	Not Copyrightable
Literary works	Works that have not been fixed in a tangible form of expression
Musical works (including any accompanying words)	Titles, names, short phrases, and slogans; familiar symbols or designs; mere variations of typographic ornamentation, lettering, or coloring; mere listings of ingredients or contents
Dramatic works (including any accompanying music)	
Pantomimes and choreographic works	
Pictorial, graphic, and sculptural works	
Motion pictures and other audiovisual works	Ideas, procedures, methods, systems, processes, concepts, principles, discoveries, or devices, as distinguished from a description, explanation, or illustration,
Sound recordings	
Architectural works	Works consisting entirely of information that is common property and containing no original authorship.

International copyright laws do not exist for the protection of works throughout the entire world. The national laws of individual countries may include different measures to prevent unauthorized use of copyrighted works. Fortunately, many countries offer protection to foreign works under certain conditions through membership in international treaties and conventions. Two important international conventions are the Berne Convention and the Universal Copyright Convention [3].

A work created on or after January 1, 1978 is given copyright protection that endures 70 years after the author's death. If more than one author is involved in the creation, the term ends 70 years after the last surviving author's death. For works predating January 1, 1978, the duration of copyright depends on whether the work was published or registered by that date.

A law enacted by the U.S. Congress in 1870 centralized the copyright system in the Library of Congress. Today, the U.S. Copyright Office is a major service unit of the Library, providing services to the Congress and other institutions in the United States and abroad. It administers the copyright law, creates and maintains public records, and serves as a resource to the domestic and international copyright communities. Table 1.2 lists some of the copyright milestones in the United States for the last two centuries [5–7].

TABLE 1.2. Notable dates in the U.S. history of copyright

Date	Event
May 31, 1790	First copyright law, derived from the English copyright law (Statute of Anne) and common law, enacted under the new constitution
April 29, 1802	Prints added to protected works
February 3, 1831	First general revision of the copyright law
August 18, 1856	Dramatic compositions added to protected works
March 3, 1865	Photographs added to protected works
July 8, 1870	Second general revision of the copyright law
January 6, 1897	Music protected against unauthorized public performance
July 1, 1909	Third general revision of the copyright law
August 24, 1912	Motion pictures, previously registered as photographs, added to classes of protected works
July 30, 1947	Copyright law codified as Title 17 of the U.S. Code
October 19, 1976	Fourth general revision of the copyright law
December 12, 1980	Copyright law amended regarding computer programs
March 1, 1989	United States joined the Berne Convention
December 1, 1990	Copyright protection extended to architectural works
October 28, 1992	Digital Audio Home Recording Act required serial copy management systems in digital audio recorders
October 28, 1998	The Digital Millennium Copyright Law (DCMA) signed into law

U.S. COPYRIGHT INDUSTRIES

The primary domestic source of marketable content is the U.S. copyright industries [8], which produce and distribute materials protected by national and international copyright laws. The products include the following categories:

1. All types of computer software (including business applications and entertainment software)
2. Motion pictures, TV programs, home videocassettes, and DVDs
3. Music, records, audio cassettes, audio DVDs and CDs
4. Textbooks, tradebooks, and other publications (both in print and electronic media)

Depending on the type of activity, the U.S. copyright industries can be studied in two groups: "core" and "total." The core industries are those that create copyrighted works as their primary product. The total copyright industries include the core industries and portions of many other industries that create, distribute, or depend on copyrighted works. Examples are retail trade (with sales of video, audio, books, and software) and the toy industry.

The International Intellectual Property Alliance (IIPA) [8] is a private-sector coalition that represents the U.S. copyright-based industries in bilateral and multilateral efforts to improve international protection of copyrighted materials. Formed in 1984, IIPA is comprised of six trade associations, each representing a different section of the U.S. copyright industry. The member associations are:

Association of American Publishers (AAP): the principal trade association of the book publishing industry

American Film Marketing Association (AFMA): a trade association whose members produce, distribute, and license the international rights to independent English language films, TV programs, and home videos

Business Software Alliance (BSA): an international organization representing leading commercial software industry and its hardware partners

Entertainment Software Association (ESA): the U.S. association of the companies publishing interactive games for video game consoles, handheld devices, personal computers, and the Internet

Motion Picture Association of America (MPAA): the MPAA, along with its international counterpart the Motion Picture Association (MPA), serve as the voice and advocate of seven of the largest producers and distributors of filmed entertainment

Recording Industry Association of America (RIAA): a trade association that represents companies that create, manufacture, and/or distribute approximately 90% of all legitimate sound recordings in the United States.

"Copyright Industries in the U.S. Economy: The 2002 Report," which updates eight prior studies, details the importance of the copyright industries to the U.S. economy based on three economic indicators: value added to GDP, share of national employment, and revenues generated from foreign sales and exports. This report gives an indication of the significance of the copyright industries to the U.S. economy:

- In 2001, the U.S. core copyright industries accounted for 5.24% ($535.1 billion) of the U.S. Gross Domestic Product (GDP). Between 1977 and 2001, their share of the GDP grew more than twice as fast as the remainder of the U.S. economy (7% vs. 3%).
- Between 1977 and 2001, employment in the U.S. core copyright industries grew from 1.6% (1.5 million workers) to 3.5% (4.7 million workers) of the U.S. workforce. Average annual employment growth

was more than three times as fast as the remainder of the U.S. economy (5% vs. 1.5%).

- In 2001, the U.S. core copyright industries estimated foreign sales and exports was $88.97 billion, leading all major industry sectors (chemical and allied products; motor vehicles, equipment and parts; aircraft and aircraft parts; electronic components and accessories; computers and peripherals).

Special 301, an annual review, requires the U.S. Trade Representative (USTR) to identify those countries that deny adequate and effective protection for intellectual property rights or deny fair and equitable market access for persons who rely on intellectual property protection. It was created by the U.S. Congress when it passed the Omnibus Trade and Competitive Act of 1988, which amended the Trade Act of 1974. According to IIPA's 2003 Special 301 Report on Global Copyright Protection and Enforcement, the U.S. copyright industries suffered estimated trade losses due to piracy of nearly $9.2 billion in 2002 as a result of the deficiencies in the copyright regimes of 56 countries. The losses for the five copyright-based industry sectors are given in Table 1.3. In USTR 2003 "Special 301" Decisions on Intellectual Property, this data is updated for 49 countries to be almost $9.8 billion. The annual losses due to piracy of U.S. copyrighted materials around the world are estimated to be $20–22 billion (not including Internet piracy) [8].

A major study titled "The Digital Dilemma — Intellectual Property in the Information Age" [9] was initiated by the Computer Science and Telecommunications Board (CSTB) to assess issues related to the nature, evolution, and use of the Internet and other networks and to the generation, distribution, and protection of content accessed through networks. The study committee convened by CSTB included experts from industry, academia, and the library and information science community. The work was carried out through the expert deliberations of the committee and by soliciting input and discussion from a wide range

TABLE 1.3.   Estimated trade losses due to copyright piracy in selected 56 countries in 2002 (in millions of U.S. dollars)

Industry	Estimated losses
Motion pictures	1322.3
Records and music	2142.3
Business software applications	3539.0
Entertainment software	1690.0
Books	514.5
Total	**9208.1**

of institutions and individuals. An important contribution of the study is a detailed review of the mechanisms for protecting IP. After a careful analysis of the technical tools and business models, the report concludes that

> There is great diversity in the kinds of digital intellectual property, business models, legal mechanisms, and technical protection services possible, making a one-size-fits-all solution too rigid. Currently, a wide variety of new models and mechanisms are being created, tried out, and in some cases discarded, at a furious pace. This process should be supported and encouraged, to allow all parties to find models and mechanisms well suited to their needs.

In the above study, the reliability of the figures attempting to measure the size of the economic impact of piracy was found questionable for two reasons:

- The IIPA reports may imply that the copyrights industries' contribution to the GDP depends on copyright policy and protection measures, pointing to a need for greater levels of protection in the digital world. However, from an economics viewpoint, the specific relation between the level of protection and revenue of a business in the copyright industries is not clear.
- The accuracy of the estimates of the cost of piracy is problematic.

At any rate, there is evidence of illegal copying, and we need to have a better understanding of its complex economic and social implications.

TECHNICAL SOLUTIONS

Figure 1.1 shows a "universe" of digital content distribution systems with five primary means of delivery to consumers: satellite, cable, terrestrial, Internet, and prerecorded media (optical and magnetic).

This universe is commonly used for distributing and storing entertainment content that is protected by copyright. The basic requirements for end-to-end security from the source to the final destination include:

- *Secure distribution of content and access keys*: In secure multimedia content distribution, the audiovisual stream is compressed, packetized, and encrypted. Encryption is the process of transforming an intelligible message (called the plaintext) into a representation (called the ciphertext) that cannot be understood by unauthorized parties [10,11]. In *symmetric key* ciphers, the enciphering and deciphering keys are the same or can be easily determined from each other. In *asymmetric (public) key* cipher, the enciphering and deciphering keys differ in such a way that at least one key is computationally infeasible to determine from the other.

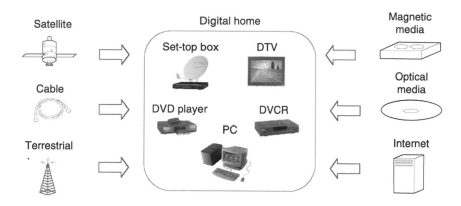

Figure 1.1. The universe of digital systems.

Symmetric key ciphers are commonly used for the protection of content, whereas the decryption keys sent to the consumers may be protected by public key ciphers (the protection of the decryption keys is normally privately defined for security reasons).

- *Authentication of source and sink consumer devices in home networks*: In a digital home network, copyrighted content may be moved from one digital device to another. Before this transfer can take place, the source and sink devices are normally engaged in mutual authentication to provide evidence that they are manufactured with the licensed protection technology. If a device is not able to produce that evidence, depending on its role, it cannot receive or transmit content.

- *Association of digital rights with content*: For the association of the digital rights with multimedia content, two approaches have been proposed: using metadata or watermarks. In the last few years, rights expression languages (RELs) have become an essential component of digital rights management systems. RELs are a means of expressing the rights of a party to certain assets, and they serve as standardized exchange formats for rights expressions. In the market, there are competing proposals that need to be standardized:

    - Started in 2001, the Open Digital Rights Language (ODRL) Initiative is an international effort aimed at developing an open standard and promoting the language at numerous standards bodies. The ODRL specification supports an extensible language and vocabulary (data dictionary) for the expression of terms and conditions over any content, including permissions, constraints, obligations, conditions, and offers and

13

agreements with rights holders. The Open Mobile Alliance (OMA) has adopted the ODRL as the rights commerce standard for mobile content.

- The eXtensible rights Markup Language (XrML) is a general-purpose, XML-based specification grammar for expressing rights and conditions associated with digital content, services, or any digital resource. The goal of XrML is to expand the usefulness of digital content, resources, and web services to rights holders, technology developers, service providers, and users by providing a flexible, extensible, and interoperable industry standard language that is platform, media, and format independent. XrML was selected as the basis for MPEG-21 Rights Expression Language (MPEG-21 REL) which is a Final Draft International Standard. XrML is also the choice of the Open eBook Rights and Rules Working Group (RRWG) as the rights expression language for its Rights Grammar specification.

- *Renewability of content protection systems*: Most of the content protection systems define renewability as device revocation. If a licensed device's secret information (keys, etc.) is compromised by the hacker, he can use the same secret information in devices that do not use the licensed protection technology. When such pirated devices appear in the black market in large numbers and identified by law enforcement agencies, the technology provider adds the hacked device's ID to a revocation list that is distributed to all licensed devices with proper cryptographic methods. When the new revocation list reaches the authorized devices, any pirated device will fail the authentication process and be unable to process protected content.

There are three industries with vested interest in the digital content protection arena: motion picture, consumer electronics, and information technology. Table 1.4 lists some of the key players that represent companies ranging from content owners to device manufacturers and service providers.

In the last two decades, several protection systems have been proposed and implemented in commonly used digital distribution networks. These include:

- Conditional access (CA) systems for satellite, cable, and terrestrial distribution
- Digital Rights Management (DRM) systems (unicast-based and multicast-based) for Internet distribution
- Copy protection (CP) systems for distribution within digital home networks

TABLE 1.4. Key players in multimedia content protection

Player	Brief information
ATSC [12]	ATSC (Advanced Television Systems Committee) Inc. is an international, nonprofit organization developing voluntary standards for digital television. Currently, there are approximately 140 members representing the broadcast, broadcast equipment, motion picture, consumer electronics, computer, cable, satellite, and semiconductor industries. ATSC incorporated on January 3, 2002.
CEA [13]	CEA (Consumers Electronics Association) represents more than 1000 companies within the U.S. consumer technology industry. The Board of Directors meets three times annually to review the state of the industry and determine CEA strategic goals to assist the industry and members. The Board of Directors elects the Officers of CEA (Executive Board members, including division chairs) and EIA Board of Governors representatives. CEA produces the International CES, the world's largest consumer technology event.
CPTWG [14]	CPTWG (Copy Protection Technical Working Group) was formed in early 1996 with the initial focus of protecting linear motion picture content on DVD. Supported by the motion picture, consumer electronics, information technology, and computer software industries, the scope of CPTWG now covers a range of issues from digital watermarking to protection of digital television.
DVD Forum [15]	DVD Forum is an international association of hardware manufacturers, software firms, and other users of DVDs. Originally known as the DVD Consortium, the Forum was created in 1995 for the purpose of exchanging and disseminating ideas and information about the DVD format and its technical capabilities, improvements, and innovations. The 10 companies who founded the organization are Hitachi, Ltd., Matsushita Electric Industrial Co. Ltd., Mitsubishi Electric Corporation, Pioneer Electronic Corporation, Royal Philips Electronics N.V., Sony Corporation, Thomson, Time Warner Inc., Toshiba Corporation, and Victor Company of Japan, Ltd.

(Continued)

TABLE 1.4. Continued

Player	Brief information
SCTE [16]	SCTE (Society of Cable Telecommunications Engineers) is a nonprofit professional organization committed to advancing the careers of cable telecommunications professionals and serving the industry through excellence in professional development, information, and standards. Currently, SCTE has almost 15,000 members from the United States and 70 countries worldwide and offers a variety of programs and services for the industry's educational benefit.
MPAA [17]	MPAA (Motion Picture Association of America) and its international counterpart, the Motion Picture Association (MPA), represent the American motion picture, home video, and television industries, domestically through the MPAA and internationally through the MPA. Founded in 1922 as the trade association of the American film industry, the MPAA has broadened its mandate over the years to reflect the diversity of an expanding industry. The MPA was formed in 1945 in the aftermath of World War II to reestablish American films in the world market and to respond to the rising tide of protectionism resulting in barriers aimed at restricting the importation of American films.
RIAA [18]	RIAA (Recording Industries Association of America) is the trade group that represents the U.S. recording industry. Its mission is to foster a business and legal climate that supports and promotes the members' creative and financial vitality. The trade group's more than 350 member companies create, manufacture, and distribute approximately 90% of all legitimate sound recordings produced and sold in the United States.
IETF [19]	IETF (Internet Engineering Task Force) is a large, open international community of network designers, operators, vendors, and researchers concerned with the evolution of the Internet architecture and the smooth operation of the Internet. The IETF working groups are grouped into areas and are managed by Area Directors (ADs). The ADs are members of the Internet Engineering Steering Group (IESG). Architectural oversight is provided by the Internet Architecture Board, (IAB). The IAB and IESG are chartered by the Internet Society (ISOC) for these purposes. The General Area Director also serves as the chair of the IESG and of the IETF and is an ex-officio member of the IAB.

(Continued)

TABLE 1.4. Continued

Player	Brief information
MPEG [20]	MPEG (Moving Pictures Expert Group) is originally the name given to the group of experts that developed a family of international standards used for coding audiovisual information in a digital compressed format. Established in 1988, the MPEG Working Group (formally known as ISO/IEC JTC1/SC29/WG11) is part of JTC1, the Joint ISO/IEC Technical Committee on Information Technology. The MPEG family of standards includes MPEG-1, MPEG-2, and MPEG-4, formally known as ISO/IEC-11172, ISO/IEC-13818, and ISO/IEC-14496, respectively.
DVB [21]	DVB (Digital Video Broadcasting) Project is an industry-led consortium of over 300 broadcasters, manufacturers, network operators, software developers, regulatory bodies, and others in over 35 countries committed to designing global standards for the global delivery of digital television and data services. The General Assembly is the highest body in the DVB Project. The Steering Board sets the overall policy direction for the DVB Project and handles its coordination, priority setting and, management, aided by three Ad Hoc Groups on Rules & Procedures, Budget, and Regulatory issues. The DVB Project is divided in four main Modules, each covering a specific element of the work undertaken. The Commercial Module and Technical Module are the driving force behind the development of the DVB specifications, with the Intellectual Property Rights Module addressing IPR issues and the Promotion and Communications Module dealing with the promotion of DVB around the globe.

Many problems, some of which are controversial, are still open and challenge the motion picture, consumer electronics, and information technology industries.

In an end-to-end protection system, a fundamental problem is to determine whether the consumer is authorized to access the requested content. The traditional concept of controlling physical access to places (e.g., cities, buildings, rooms, highways) has been extended to the digital world in order to deal with information in binary form. A familiar example is the access control mechanism used in computer operating systems to manage data, programs, and other system resources. Such systems can be effective in "bounded" communities [9] (e.g., a corporation or a college campus), where the emphasis is placed on the original access to information rather than how the information is used once it is in the possession of the user. In contrast, the conditional access systems for digital content in "open" communities need to provide reliable services for long periods of time (up to several decades) and be capable of controlling the use of content after access.

We will look at three approaches for restricting access to content. The first approach has been used by the satellite and terrestrial broadcasters and cable operators in the last few decades (for both analog and digital content). The second approach is adopted by the developers of emerging technologies for protecting Internet content. In the third approach, we have a collection of copy protection systems for optical and magnetic storage and two major digital interfaces.

Security in CA Systems for Satellite, Cable, and Terrestrial Distribution

A CA system [22–30] allows access to services based on payment or other requirements such as identification, authorization, authentication, registration, or a combination of these. Using satellite, terrestrial, or cable transmissions, the service providers deliver different types of multimedia content ranging from free-access programs to services such as PayTV, Pay-Per-View, and Video-on-Demand.

Conditional access systems are developed by companies, commonly called the CA providers, that specialize in the protection of audio visual (A/V) signals and secure processing environments. A typical architecture of a CA system and its major components are shown in Figure 1.2. The common activities in this general model are:

1. Digital content (called an "event" or a "program") is compressed to minimize bandwidth requirements. MPEG2 is a well-known industry standard for coding A/V streams. More recent MPEG alternatives (MPEG4, MPEG7, and MPEG21) are being considered for new applications.

2. The program is sent to the CA head-end to be protected and packaged with entitlements indicating the access conditions.

3. The A/V stream is scrambled[1] and multiplexed with the entitlement messages. There are two types of entitlement message [31] associated with each program: The Entitlement Control Messages (ECMs) carry the decryption keys (called the "control words") and a short description of the program (number, title, date, time, price, rating, etc.) whereas the Entitlement Management Messages

---

[1]In the context of CA systems, *scrambling* is the process of content encryption. This term is inherited from the analog protection systems where the analog video was manipulated using methods such as line shuffling. It is now being used to distinguish the process from the protection of *descrambling* keys.

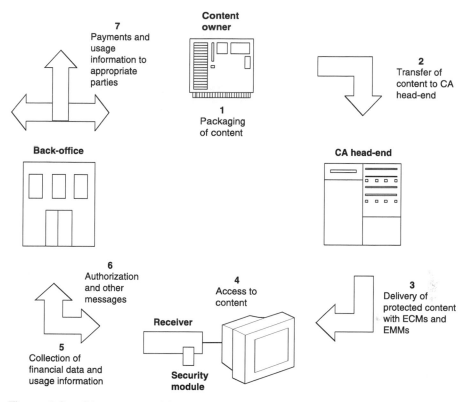

Figure 1.2. CA system architecture.

(EMMs) specify the authorization levels related to services. In most CA systems, the EMMs can also be sent via other means such as telephone networks. The services are usually encrypted using a symmetric cipher such as the Data Encryption Standard (DES) or any other public-domain or private algorithm. The lifetime and the length of the scrambling keys are two important system parameters. For security reasons, the protection of the ECMs is often privately defined by the CA providers, but public key cryptography and one-way functions are useful tools for secure key delivery.

4. If the customer has received authorization to watch the protected program,[2] the A/V stream is descrambled by the receiver (also called a "decoder"), and sent to the display unit for viewing. A removable security module (e.g., a smartcard) provides a safe environment for the processing of ECMs, EMMs, and other sensitive functions such as user authorization and temporary storage of purchase records.

---

[2]The program may come directly from the head-end or a local storage device. Protection of local storage (such as a hard disk) is a current research area.

5. The back office is an essential component of every CA system, handling billings and payments, transmission of EMMs, and interactive TV applications. A one-to-one link is established between the back office and the decoder (or the removable security module, if it exists) using a "return channel," which is basically a telephone connection via a modem. As with other details of the CA system, the security of this channel may be privately defined by the CA providers. At certain times, the back office collects the purchase history and other usage information for processing.

6. Authorizations (e.g., EMMs) and other messages (system and security updates, etc.) are delivered to the customer's receiver.

7. Payments and usage information are sent to the appropriate parties (content providers, service operators, CA providers, etc.).

In today's CA systems, the security module is assigned the critical task of recovering the descramling keys. These keys are then passed to the receiver for decrypting the A/V streams. The workload is, therefore, shared between the security module and its host. Recently, two separate standards have evolved to remove all of the security functionality from navigation devices. In the United States the National Renewable Security Standard (NRSS) [32] defines a renewable and replaceable security element for use in consumer electronics devices such as digital set-top boxes and digital TVs. In Europe, the DVB project has specified a standard for a common interface (CI) between a host device and a security module.

The CA systems currently in operation support several purchase methods, including subscription, pay-per-view, and impulsive pay-per-view. Other models are also being considered to provide more user convenience and to facilitate payments. One such model uses prepaid "cash cards" to store credits that may be obtained from authorized dealers or ATM-like machines.

Note that the model described in Figure 1.2 has been traditionally used to provide conditional access for viewing purposes. Recording control depends on the type of the signal output from the receiver. If the device has NTSC, PAL, or SECAM output (which is the case for some devices in the field today), protection can be provided by a Macrovision [33] system, which modifies the video in such a way that it does not appreciably distort the display quality of the video but results in noticeable degradation in recording. For higher-definition analog or digital outputs, however, the model is drastically changing, requiring solutions that are more complex and relatively more expensive.

The DVB Project has envisaged two basic CA approaches: "Simulcrypt" and "Multicrypt" [27,34].

- *Simulcrypt*: Each program is transmitted with the entitlement messages for multiple CA systems, enabling different CA decoders to receive and correctly descramble the program.
- *Multicrypt*: Each decoder is built with a common interface for multiple CA systems. Security modules from different CA system operators can be plugged into different slots in the same decoder to allow switching between CA systems.

These architectures can be used for satellite, cable, and terrestrial transmission of digital television. The ATSC [35] has adopted the Simulcrypt approach.

A recent major discussion item on the agenda for content owners and broadcasters has been the broadcast flag [36]. Also known as the ATSC flag, the broadcast flag is a sequence of digital bits sent with a television program that signals that the program must be protected from unauthorized redistribution. It is argued that implementation of this broadcast flag will allow digital TV (DTV) stations to obtain high-value content and assure consumers a continued source of attractive, free, over-the-air programming without limiting the consumer's ability to make personal copies. The suitability of the broadcast flag for protecting DTV content was evaluated by the Broadcast Protection Discussion Group (BPDG) that was comprised of a large number of content providers, television broadcasters, consumer electronics manufacturers, information technology companies, interested individuals, and consumer activists. The group completed its mission with the release of the BPDG Report. The broadcast flag can be successfully implemented once the suppliers of computer and electronics systems that receive broadcast television signals incorporate the technical requirements of the flag into their products. Undoubtedly, full implementation requires a legislative or regulatory mandate. In November 2003, the Federal Communications Commission (FCC) adopted a broadcast flag mandate rule, in spite of the objections of thousands of individuals and dozens of organizations [37].

Security in DRM Systems for Internet Distribution

DRM refers to the protection, distribution, modification and enforcement of the rights associated with the use of digital content. In general, the primary responsibilities of a DRM system are:

- Packaging of content
- Secure delivery and storage of content
- Prevention of unauthorized access
- Enforcement of usage rules
- Monitoring the use of content

Although such systems can, in principle, be deployed for any type of distribution media, the present discussions weigh heavily on the Internet.

The unprecedented explosion of the Internet has opened potentially limitless distribution channels for the electronic commerce of content. Selling goods directly to consumers over an open and public network, without the presence of a clerk at the point of sale, has significant advantages. It allows the businesses to expand their market reach, reduce operating costs, and enhance customer satisfaction by offering personalized experience. Although inspiring new business opportunities, this electronic delivery model raises challenging questions about the traditional models of ownership. The lessons learned from the MP3 phenomenon, combined with the lack of reliable payment mechanisms, have shown the need for protecting the ownership rights of copyrighted digital material.

A DRM system uses cryptography (symmetric key ciphers, public-key ciphers, and digital signatures) as the centerpiece for security-related functions, which generally include secure delivery of content, secure delivery of the content key and the usage rights, and client authentication.

Figure 1.3 shows the fundamentals of an electronic delivery system with DRM: a publisher, a server (streaming or Web), a client device, and a financial clearing house. The communication between the server and the customer is assumed to be unicast (i.e., point-to-point). Although details may vary among DRM systems, the following steps summarize typical activities in a DRM-supported e-commerce system:

1. The publisher packages the media file (i.e., the content) and encrypts it with a symmetric cipher. The package may include information about the content provider, retailer, or the Web address to contact for the rights.

2. The protected media file is placed on a server for downloading or streaming. It can be located with a search engine using the proper content index.

3. The customer requests the media file from the server.

4. The file is sent after the client device is authenticated. The customer may also be required to complete a purchase transaction. Authentication based on public-key certificates is commonly used for this purpose. Depending on the DRM system, the usage rules and the key to unlock the file may either be attached to the file or need to be separately obtained (e.g., in the form of a license) from the clearinghouse or any other registration server. The attachment or the license are protected in such a way that only

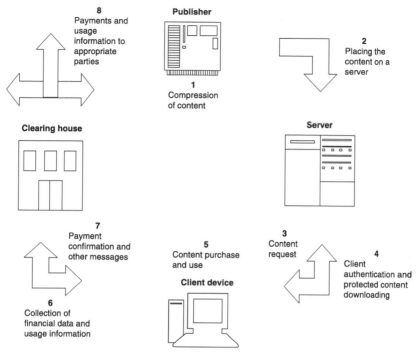

Figure 1.3. DRM system architecture.

the client is able to retrieve the information. Public-key ciphers are appropriately used here.

5. The customer purchases the content and uses it according to the rights and rules.

6. At certain times, the clearinghouse collects financial records and usage information from the clients.

7. Payment confirmation and other messages (system and security updates, etc.) are delivered to the client.

8. Payments and usage information are sent to the appropriate parties (content providers, publishers, distributors, authors, artists, etc.).

Renewability is achieved by upgrading DRM system components and preventing the compromised devices from receiving content. New security software may be released as a regular enhancement or in response to a threat or hack. Revocation lists allow the servers to refuse service to revoked clients (multimedia applications software can also be revoked if it can be authenticated by the DRM component that resides in the client).

A DRM enables the content owners to specify their own business models in managing the use of content. A wide range of sales models can be supported, including subscription, pay-per-use, and superdistribution. Time-limited reading of an e-book, multiple viewings of a movie, and transfer of a song to a portable music player are all possible scenarios.

Superdistribution is a relatively new concept for redistributing content across the Internet. It is a process that allows the consumers to forward a content that they have acquired to other consumers (friends, relatives, and associates) in the market. The content forwarded to a potential buyer cannot be accessed until the new rights are obtained. This approach has important advantages and drawbacks:

- It is an efficient way of using a DRM-supported e-commerce system because repeated content downloads are avoided.
- From an economic point of view, superdistribution may help widen the market penetration. Once a particular content is downloaded to a client, the customer, with sufficient encouragement, can act as an agent of the retailer with minimal effort and cost.
- From a security point of view, if the content, encrypted with a particular key, becomes available in the market in large quantities, it may increase the likelihood of the key being compromised. For increased security, different copies of a media file need to be encrypted with different keys. This, of course, requires new downloads from the server.

The unicast-based model in Figure 1.3 can be extended to multicast networks where the data can be efficiently delivered from a source to multiple receivers. In the past few years, there has been a substantial amount of research in group key management [38].

A number of organizations are in the process of standardizing Internet-based DRM systems for handling different types of content. The major efforts led by MPEG, Open eBook Forum, and IRTF are summarized in Table 1.5.

Security in Multicast Systems for Internet Distribution

The traditional mechanism to support multicast communications is IP multicast [44]. It uses the notion of a group of members identified with a given group address. When a sender sends a message to this group address, the network uses a multicast routing protocol to optimally replicate the message and forward copies to group members located throughout the network.

Although the Internet community began discussing architectural issues in the mid-1980s using Internet Engineering Task Force (IETF) Request for Comments (RFCs), significant activity in multicast IP did not

24

TABLE 1.5. DRM-related activities

Organization	Recent efforts
MPEG (Moving Picture Experts Group) [39] is a working group of ISO/IEC developing a family of standards used for coding audiovisual information in a digital compressed format.	MPEG-4: Latest compression standard designed specially for low-bandwidth (less than 1.5Mit/sec *bitrate*) video/audio encoding purposes. As a universal language for a range of multimedia applications, it will provide additional functionality such as bitrate scalability, object-based representation, and intellectual property management and protection. MPEG-4 IPMP (version 1) is a simple "hook" DRM architecture standardized in 1999. As each application may have different requirements for the protection of multimedia data, MPEG-4 allows the application developers to design domain-specific IPMP systems (IPMP-S). MPEG-4 standardizes only the MPEG-4 IPMP interface with IPMP Descriptors (IPMP-Ds) and IPMP Elementary Streams (IPMP-ES), providing a communication mechanism between IPMP systems and the MPEG-4 terminal. MPEG-7: Formally named "Multimedia Content Description Interface," the MPEG-7 standard provides a set of standardized tools to describe multimedia content. The main elements of the standard are description tools (Descriptors [D] and Description Schemes [DS]), a Description Definition Language (DDL) based on the XML Schema Language, and system tools. The DDL defines the syntax of the Description Tools and allows the creation of new Description Schemes and Descriptors as well as the extension and modification of existing Description Schemes. System tools enable the deployment of descriptions, supporting binary-coded representation for efficient storage and transmission, transmission mechanisms, multiplexing of descriptions, synchronization of descriptions with content, and management and protection of intellectual property in MPEG-7 descriptions.

(Continued)

25

TABLE 1.5. Continued

Organization	Recent efforts
	MPEG-21: MPEG-21 defines a normative open framework for multimedia delivery and consumption that can be used by content creators, producers, distributors, and service providers in the delivery and consumption chain. The framework is based on two essential concepts: the definition of a fundamental unit of distribution and transaction (the Digital Item) and the concept of Users interacting with Digital Items. Development of an interoperable framework for Intellectual Property Management and Protection (IPMP) is an ongoing effort that will become a part of the MPEG-21 standard.
	IPMP-X (Intellectual Property Management and Protection Extension) [40] is a DRM architecture that provides a normative framework to support many of the requirements of DRM solution (renewability, secure communications, verification of trust, granular and flexible governance at well-defined points in the processing chain, etc.). IPMP-X comes in two flavors: MPEG-2 IPMP-X (applicable to MPEG-2 based systems) and MPEG-4 IPMP-X (applicable to MPEG-4 based systems). The MPEG-4 IPMP extensions were standardized in 2002 as an extension to MPEG-4 IPMP "hooks." IPMP Tools are modules that perform IPMP functions such as authentication, decryption, and watermarking. In addition to specifying syntax to signal and trigger various IPMP Tools, IPMP-X specifies the architecture to plug the IPMP Tools seamlessly into IPMP-X terminal.
	MPEG LA, LLC [41] provides one-stop technology platform patent licensing with a portfolio of essential patents for the international digital video compression standard known as MPEG-2. In addition to MPEG-2, MPEG LA licenses portfolios of essential patents for the IEEE 1394 Standard, the DVB-T Standard, the MPEG-4 Visual Standard, and the MPEG-4 Systems Standard. In October 2003, MPEG LA, LLC, issued a call for patents that are essential to digital rights management technology (DRM) as described in DRM Reference Model v 1.0. The DRM Reference Model does not define a standard for interoperability among DRM devices, systems, or methods, or provide a specification of commercial products. It is an effort to provide users with convenient, fair, reasonable, nondiscriminatory access to a portfolio of essential worldwide patent rights under a single license. If the initial evaluation of the submitted patents is completed by the end of 2003, a joint patent license may become

Open eBook Forum (OeBF) [42] is an international trade and standards organization for the electronic publishing industries.

Its members consist of hardware and software companies, print and digital publishers, retailers, libraries, accessibility advocates, authors, and related organizations. The OeBF engages in standards and trade activities through the operation of Working Groups and Special Interest Groups. The Working Groups are authorized to produce official OeBF documents such as specifications and process documents (such as policies and procedures, position papers, etc.). In the current organization, there are five Working groups: Metadata & Identifiers WG, Publication Structure WG, Requirements WG, Rights & Rules WG, and Systems WG. The mission of the Rights and Rules Working Group is to create an open and commercially viable standard for interoperability of digital rights management (DRM) systems.

The Internet Research Task Force (IRTF) [43] is composed of a number of small Research Groups working on topics related to Internet protocols, applications, architecture, and technology.

Internet Digital Rights Management (IDRM) was an IRTF Research Group formed to research issue and technologies relating to Digital Rights Management (DRM) on the Internet. The IRTF is a sister organization of the Internet Engineering Task Force (IETF). There were three IRTF drafts, formally submitted through IDRM, that carried the IRTF title. The IDRM group is now closed.

occur until the creation of the Mbone in 1992. The Mbone is a set of multicast-enabled subnetworks connected by IP tunnels. Tunneling is a technique that allows multicast traffic to traverse parts of the network by encapsulating multicast datagrams within unicast datagrams.

In IPv4, multicast IP addresses are defined by Class D, which differs from Classes A, B, and C that are used for point-to-point communications. The multicast address space, assigned by the Internet Assigned Numbers Authority (IANA), covers the range (224.0.0.0 – 239.255.255.255). IPv6 has 128 bits of address space compared with 32 bits in IPv4.

The Internet Group Management Protocol (IGMP) defines a protocol for multicast-enabled hosts and routers to manage group membership information. Developed by the Defense Advance Research Projects Agency (DARPA), the Transmission Control Protocol/Internet Protocol (TCP/IP) connects networks designed by different vendors into a network of networks (i.e., the Internet). It has two transport layers for the applications: the Transport Control Protocol (TCP) and the User Datagram Protocol (UDP). Currently, UDP is the only protocol for IP multicast, providing minimal services such as port multiplexing and error detection. Any host can send a UDP packet to a multicast address, and the multicast routing mechanism will deliver the packet to all members of the multicast group. TCP provides a higher level of service with packet ordering, port multiplexing, and error-free data delivery. It is a *connection-oriented* protocol (unlike UDP, which is *connectionless*) and does not support multicast applications.

MSEC is a Working Group (WG) in the Internet Engineering Task Force (IETF). Its purpose is to "standardize protocols for securing group communication over internets, and in particular over the global Internet." The initial primary focus of the MSEC WG will be on scalable solutions for groups with a single source and a very large number of recipients. The standard will be developed with the assumption that each group has a single trusted entity (i.e., the Group Controller) that sets the security policy and controls the group membership. It will attempt to guarantee at least the following two basic security features:

- Only legitimate group members will have access to current group communication. (This includes groups with highly dynamic membership.)
- Legitimate group members will be able to authenticate the source and contents of the group communication. (This includes cases in which group members do not trust each other.)

We will look at the recent developments in key management, authentication, and watermarking for secure group communications in wired and wireless networks. The proposed methods provide solutions to address three different issues of secure multimedia data distribution:

- Controlling access to multimedia data among group members
- Assuring the identity of participating group members (senders or receivers)
- Providing copyright protection

Some of the challenging questions regarding these issues include the following:

- How does a group manager, if it exists, accept members to the group?
- How is the group key generated and distributed to members?
- How is multimedia data source authenticated by the receivers?
- How is the group key changed when a member joins or leaves a group?
- How does multicast multimedia data received by a member have a unique watermark?

Secure multicast communications in a computer network involves efficient packet delivery from one or more sources to a large group of receivers having the same security attributes. The four major issues of IP multicast security are [45]:

- *Multicast data confidentiality*: As the data traverses the public Internet, a mechanism is needed to prevent unauthorized access to data. Encryption is commonly used for data confidentiality.
- *Multicast group key management*: The security of the data packets is made possible using a group key shared by the members that belong to the group. This key needs to change every time a member joins (leaves) the group for backward access control (forward access control). In some applications, there is also a need to change the group key periodically. Encryption is commonly used to control access to the group key.
- *Multicast data source authentication*: An assurance of the identity of the data source is provided using cryptographic means. This type of authentication also includes evidence of data integrity. Digital signatures and Message Authentication Codes (MACs) are common authentication tools.
- *Multicast security policies*: The correct definition, implementation, and maintenance of policies governing the various mechanisms of multicast security is a critical factor. The two general categories are the policies governing group membership and the policies regarding security enforcement.

In multicast communications, a session is defined as the time period in which data are exchanged among the group members. The type of member participation characterizes the nature of a session. In a *one-to-many* application, data are multicast from a single source to multiple receivers. Pay-per-view, news feeds, and real-time delivery of stock market information are a few examples. A *many-to-many* application involves multiple senders and multiple receivers. Applications such as teleconferencing, white boarding, and interactive simulation allow each member of the multicast group to send data as part of group communications.

Wired Network Security
Key Management Schemes for Wired Networks. Many multicast key management schemes have been proposed in the last 10–15 years. Four classifications from the literature are:

1. *Nonscalable* and *scalable* schemes [46]. The scalable schemes are, in turn, divided into three groups: hierarchical key management (node-based and key-based), centralized flat key management, and distributed flat key management.
2. *Flat* schemes, *clustered* schemes, *tree-based* schemes, and other schemes [47].
3. *Centralized* schemes, *decentralized* schemes, and *distributed* schemes [48].
4. *Key tree-based* schemes, *contributory key agreement* schemes, *computational number theoretic* schemes, and *secure multicast framework* schemes [49].

It may be possible to have a new classification using two criteria: the entity who exercises the control and whether the scheme is scalable or not — *centralized group control, subgroup control, and member control*:

1. *Centralized group control*: A single entity controls all the members in the group. It is responsible for the generation, distribution, and replacement of the group key. Because the controlling server is the single point of failure, the entire group is affected as a result of a malfunction.
2. *Subgroup control*: The multicast group is divided into smaller subgroups, and each subgroup is assigned a different controller. Although decentralization substantially reduces the risk of total system failure, it relies on trusted servers, weakening the overall system security.
3. *Member control*: With no group or subgroup controllers, each member of the multicast group is trusted with access control and contributes to the generation of the group key.

Each of the above classes is further divided into *scalable* and *nonscalable* schemes. In the context of multicast key management, scalability refers to the ability to handle a larger group of members without considerable performance deterioration. A scalable scheme is able to manage a large group over a wide geographical area with highly dynamic membership. If the computation and communication costs at the sender increase linearly with the size of the multicast group, then the scheme is considered to be nonscalable. Table 1.6 lists the key management schemes according to the new criteria.

Hierarchical key distribution trees form an efficient group of proposals for scalable secure multicasting. They can be classified into two groups: *hierarchical key-based* schemes and *hierarchical node-based* schemes.

Table 1.6. Classification of key management schemes

	Scalable	Non-scalable
Centralized group control	Wong et al, 97 [50] Caronni et al, 98 [51] Balenson et al, 99 [52] Canetti et al, 99 [53] Chang et al, 99 [54] Wallner et al, 99 [55] Waldvogel et al, 99 [56] Banerjee and Bhattacharjee, 01 [57] Eskicioglu and Eskicioglu, 02 [58] Selcuk and Sidhu, 02 [59] Zhu and Setia, 03 [60] Huang and Mishra, 03 [61] Trappe et al, 03 [62]	Chiou and Chen, 89 [63] Gong and Shacham, 94 [64] Harney and Muckenhirn, 97 [65] Dunigan and Cao, 98 [66] Blundo et al, 98 [67] Poovendran et al, 98 [68] Chu et al, 99 [69] Wallner et al, 99 [55] Scheikl et al, 02 [70]
Subgroup control	Mittra, 97 [71] Dondeti et al, 99 [72] Molva and Pannetrat, 99 [73] Setia et al, 00 [74] Hardjono et al, 00 [75]	Ballardie, 96 [76] Briscoe, 99 [77]
Member control	Dondeti et al, 99 [78] Perrig, 99 [79] Waldvogel et al, 99 [56] Rodeh et al, 00 [80] Kim et al, 00 [81]	Boyd, 97 [82] Steiner et al, 97 [83] Becker and Willie, 98 [84]

A hierarchical key-based scheme assigns a set of keys to each member depending on the location of the member in the tree. Hierarchical node-based schemes define internal tree nodes that assume the role of subgroup managers in key distribution.

Among the schemes listed in Table 1.6, three hierarchical schemes, namely the Centralized Tree-Based Key Management (CTKM), Iolus and DEP, are compared through simulation using real-life multicast group membership traces [85]. The performance metrics used in this comparison are (1) the encryption cost at the sender and (2) encryption and decryption cost at the members and subgroup managers. It is shown that hierarchical node-based approaches perform better than hierarchical key-based approaches, in general. Furthermore, the performance gain of hierarchical node-based approaches increases with the multicast group size.

An Internet Draft generated by the MSEC WG presents a common architecture for MSEC group key management protocols that support a variety of application, transport, and internetwork security protocols. The document includes the framework and guidelines to allow for a modular and flexible design of group key management protocols in order to accommodate applications with diverse requirements [86].

Periodic Batch Rekeying. In spite of the efficiency of the tree-based scalable schemes for one-to-many applications, changing the group key after each join or leave (i.e., individual *rekeying*) has two major drawbacks:

- *Synchronization problem*: If the group is rekeyed after each join or leave, synchronization will be difficult to maintain because of the interdependencies among rekey messages and also between rekey and data messages. If the delay in rekey message delivery is high and the join or leave requests are frequent, a member may need to have memory space for a large number of rekey and data messages that cannot be decrypted.
- *Inefficiency*: For authentication, each rekey message may be digitally signed by the sender. Generation of digital signatures is a costly process in terms of computation and communication. A high rate of join or leave requests may result in a performance degradation.

One particular study attempts to minimize these problems with *periodic batch rekeying* [87]. In this approach, join or leave requests are collected during a rekey interval and are rekeyed in a batch. The out-of-sync problems are alleviated by delaying the use of a new group key until the next rekey interval. Batch processing also leads to a definite performance advantage. For example, if digital signatures are used for data source authentication, the number of signing operations for J join and L leave requests is reduced from $J+L$ to 1.

Periodic batch rekeying provides a trade-off between performance improvement and delayed group access control. A new member has to wait longer to join the group, and a leaving member can stay longer with the group. The period of the batch rekeying is, thus, a design parameter that can be adjusted according to security requirements. To accommodate different application needs, three modes of operation are suggested:

- *Periodic batch rekeying*: The key server processes both join and leave requests periodically in a batch.
- *Periodic batch leave rekeying*: The key server processes each join request immediately to reduce the delay for a new member to access group communications but processes leave requests in a batch.
- *Periodic batch join rekeying*: The key server processes each leave request immediately to reduce the exposure to members who have left but processes join requests in a batch.

A *marking algorithm* is proposed to update the key tree and to generate a rekey subtree at the end of each rekey interval with a collection of J join and L leave requests. A rekey subtree is formed using multiple paths corresponding to multiple requests. The objectives of the marking algorithm are to reduce the number of encrypted keys, to maintain the balance of the updated key tree, and to make it efficient for the users to identify the encrypted keys they need. To meet these objectives, the server uses the following steps:

1. Update the tree by processing join and leave requests in a batch. If $J \leq L$, J of the departed members with the smallest IDs are replaced with the J newly joined members. If $J > L$, L departed members are replaced with L of the newly joined members. For the insertion of the remaining $J-L$ new members, three strategies have been investigated [88,89].
2. Mark the key nodes with one of the following states: Unchanged, Join, Leave, and Replace.
3. Prune the tree to obtain the rekey subtree.
4. Traverse the rekey subtree, generate new keys, and construct the rekey message.

Balanced Key Trees. The efficiency of a tree-based key management scheme depends highly on how well the tree remains balanced. In this context, a tree is balanced if the difference between the distances from the root node to any two leaf nodes does not exceed 1 [90]. For a balanced binary tree with n leaves, the distance from the root to any leaf is $\log_2 n$. The issue of maintaining trees in a balanced manner is critical for any real implementation of a key management tree. Several techniques, based on the scheme described by Wallner et al., are introduced to maintain a balanced tree in the presence of arbitrary group membership updates [90]. Although we have complete control over how the tree is edited for new member additions, there is no way to predict the locations in the tree at which the deletions will occur. Hence, it is possible to imagine extreme cases leading to costs that have linear order in the size of the group. Two simple tree rebalancing schemes have been proposed to avoid this cost increase [90]. The first is a modification of the deletion algorithm; the other allows the tree to become imbalanced after a sequence of key updates and periodically invokes a tree rebalancing algorithm to bring the tree back to a balanced state.

A recent work presents the design and analysis of three scalable online algorithms for maintaining multicast key distribution trees [91]. To minimize worst-case costs and to have good average-case performance, there was a trade-off between tree structures with worst-case bounds and the restructuring costs required to maintain those trees. Simulations showed that the height-balanced algorithm performed better than the weight-balanced algorithm.

Authentication. In multicast architectures, group membership control, dictated by security policies, allows access to a secure multicast group. *Member authentication* involves methods ranging from the use of access control lists and capability certificates [92] to mutual authentication [93] between the sender and the receiver.

- *Access control lists*: The sender maintains a list of hosts who are either authorized to join the multicast group or excluded from it. When a host sends a join request, the sender checks its identity against the access control list to determine if membership is permitted. The maintenance of the list is an important issue, as the list may be changing dynamically based on new authorizations or exclusions.

- *Capability certificates*: Issued by a designated Certificate Authority, a capability certificate contains information about the identity of the host and the set of rights associated with the host. It is used to authenticate the user and to allow group membership.

- *Mutual authentication*: The sender and the host authenticate each other via cryptographic means. Symmetric or public-key schemes can be used for this purpose.

A challenging problem in secure group communications is *data source authentication* (i.e., providing assurance of the identity of the sender and

34

the integrity of the data). Depending on the type of multicast application and the computational resources available to the group members, three levels of data source authentication can be used [94]:

- *Group authentication*: Provides assurance that the packet was sent by a registered group member (a registered sender or a registered receiver)
- *Source authentication*: Provides assurance that the packet was sent by a registered sender (not by a registered receiver)
- *Individual sender authentication*: Provides assurance of the identity of the registered sender of the packet

In a naive approach, each data packet can be digitally signed by the sender. For group (source) authentication, all members, sender or receiver (all senders), can share a private key to generate the same signature on the packets. Individual sender authentication, however, requires each sender to have a unique private key. Although digital signature-based authentication per packet is desirable as a reliable tool, it exhibits a poor performance because of lengthy keys and computational overhead for signature generation and verification.

Recent research has led to more efficient authentication methods, including:

- *Multiple Message Authentication Codes (MACs)* [53]
- *Stream signing* [95]
- *Authentication tree-based signatures* [96]
- *Hybrid signatures* [97]
- *TESLA, EMSS, and BiBa* [98–100]
- *Augmented chain* [101]
- *Piggybacking* [102]
- *Multicast packet authentication with signature amortization* [103]
- *Multicast packet authentication* [104]

A Message Authentication Code (MAC) is a keyed hash function used for data source authentication in communication between two parties (sender and receiver). At the source, the message is input to a MAC algorithm that computes the MAC using a key K shared by both parties. The sender then appends the MAC to the message and sends the pair {message|MAC} to the receiver. In an analysis of the generalization of MACs to multicast communications, it is shown that a short and efficient collusion-resistant multicast MAC (MMAC) cannot be constructed without a new advance in digital signature design [105].

Watermarking. Watermarking (data hiding) [106–108] is the process of embedding data into a multimedia element such as image, audio or video. This embedded data can later be extracted from, or detected in, the multimedia for security purposes. A watermarking algorithm consists of the watermark structure, an embedding algorithm, and an extraction, or a detection, algorithm. Watermarks can be embedded in the pixel domain or a transform domain. In multimedia applications, embedded watermarks should be invisible, robust, and have a high capacity [109]. Invisibility refers to the degree of distortion introduced by the watermark and its affect on the viewers or listeners. Robustness is the resistance of an embedded watermark against intentional attacks and normal A/V processes such as noise, filtering (blurring, sharpening, etc.), resampling, scaling, rotation, cropping, and lossy compression. Capacity is the amount of data that can be represented by an embedded watermark. The approaches used in watermarking still images include least significant bit encoding, basic M-sequence, transform techniques, and image-adaptive techniques [110]. Because video watermarking possesses additional requirements, development of more sophisticated models for the encoding of video sequences is currently being investigated.

Typical uses of watermarks include *copyright protection* (*identification of the origin of content, tracing illegally distributed copies*) *and disabling unauthorized access to content.* Requirements and characteristics for the digital watermarks in these scenarios are different, in general. Identification of the origin of content requires the embedding of a single watermark into the content at the source of distribution. To trace illegal copies, a unique watermark is needed based on the location or identity of the recipient in the multimedia network. In both of these applications, watermark extraction or detection needs to take place only when there is a dispute regarding the ownership of content. For access control, the watermark should be checked in every authorized consumer device used to receive the content. Note that the cost of a watermarking system will depend on the intended use and may vary considerably.

The *copyright protection* problem in a multicast architecture raises a challenging issue. All receivers in a multicast group receive the same watermarked content. If a copy of this content is illegally distributed to the public, it may be difficult to find the parties responsible for this criminal act. Such a problem can be eliminated in a unicast environment by embedding a unique watermark for each receiver. To achieve uniqueness for multicast data, two distinct approaches are feasible:

1. Multiple copies of content, each with a different watermark, are created to allow the selection of appropriate packets in distribution.
2. A single copy of unwatermarked content is created to allow the insertion of appropriate watermarks in distribution.

The following proposals are variations of these two approaches:

- *A different version of video for each group member* [69]: For a given multicast video, the sender applies two different watermark functions to generate two different watermarked frames, $d_{i,w0}$ and $d_{i,w1}$, for every frame i in the stream. The designated group leader assigns a randomly generated bit stream to each group member. The length of the bit string is equal to the number of video frames in the stream. For the ith watermarked frame in stream j, $j = 0$, 1, a different key $K_{i,j}$ is used to encrypt it. The random bit stream determines whether the member will be given K_{i0} or K_{i1} for decryption. If there is only one leaking member, its identification is made possible with the collaboration of the sender who can read the watermarks to produce the bit stream and the group leader who has the bit streams of all members. The minimum length of the retrieved stream to guarantee a c-collusion detection, where c is the number of collaborators, is not known. An important drawback of the proposal is that it is not scalable and two copies of the video stream need to be watermarked, encrypted, and transmitted.

- *Distributed watermarking (watercasting)* [111]: For a multicast distribution tree with maximum depth d, the source generates a total of n differently watermarked copies of each packet such that $n \geq d$. Each group of n alternate packets is called a transmission group. On receiving a transmission group, a router forwards all but one of those packets to each downstream interface on which there are receivers. Each last hop router in the distribution tree will receive $n - d_r$ packets from each transmission group, where d_r is the depth of the route to this router. Exactly one of these packets will be forwarded onto the subnet with receivers. The goal of this filtering process is to provide a stream for each receiver with a unique sequence of watermarked packets. The information about the entire tree topology needs to be stored by the server to trace an illegal copy. A major potential problem with watercasting is the support required from the network routers. The network providers may not be willing to provide a security-related functionality unless video delivery is a promising business for them.

- *Watermarking with a hierarchy of intermediaries* [112]: WHIM Backbone (WHIM-BB) introduces a hierarchy of intermediaries into the network and forms an overlay network between them. Each intermediary has a unique ID used to define the path from the source to the intermediary on the overlay network. The Path ID is embedded into the content to identify the path it has traveled. Each intermediary embeds its portion of the Path ID into the content before it forwards the content through the network. A watermark embedded by a WHIM-BB identifies the domain of a receiver.

WHIM-Last Hop (WHIM-LH) allows the intermediaries to mark the content uniquely for any child receivers they may have. Multiple watermarks can be embedded using modified versions of existing algorithms. The above two "fingerprinting" schemes [69,111] require a certain number of video frames in order to deduce sufficient information about the recipient, whereas WMIN requires only one frame because the entire trace is embedded into each frame. A serious overhead for this scheme, however, is the hierarchy of intermediaries needed for creating and embedding the fingerprint.

Finally, the two techniques described below appear to be viable approaches for copyright protection and access control, respectively.

- *Hierarchical tagging and bulk tagging* [113]: Hierarchical tagging allows an artist to insert a different watermark for each of his distributors. Similarly, each distributor can insert a watermark for several subdistributors. This process can continue until the individual customers receive tagged content identifying the artist and all the distributors in the chain. In practice, however, more than a few layers of watermarks may reduce the visual quality to an unacceptable level. With bulk tagging, the distributor creates multiple, tagged versions of the data. The contents are hidden using cryptographic techniques and distributed as a single dataset. Each customer receives the same dataset, performs some preprocessing, and retrieves only the tagged data prepared for him. A simple approach is described to show the feasibility of bulk-tagging for images. It requires registration with the producer and the delivery of keys to decrypt the consumer's individually tagged copy. The preprocessing required by the client device creates a weakness in system security, as the individual tag is used for access control only. If the decryption keys are recovered for one consumer, the content would become available in-the-clear and there would be no trace to the illegal distributor.

Wireless Network Security. Key management in wireless networks is a more complicated problem because of the mobility of group members [114,115]. When a member joins or leaves a session, the group key needs to change for backward access control and forward access control. Because secure data cannot be communicated during the rekeying process, an important requirement for a key management scheme is to minimize the interruption in secure data communications. Mobility also allows the members to move to other networks without leaving the session. The existence of a member whose position changes with time adds another dimension of complexity to the design of rekeying algorithms.

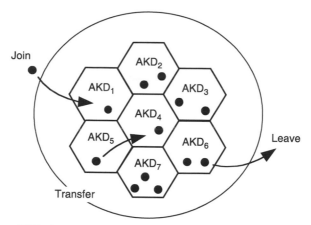

DKD: domain key distributor, AKD: area key distributor

Figure 1.4.   Mobility framework.

A common approach in designing a scalable multicast service is to use a hierarchical structure in group key distribution. The hierarchical key management schemes fall into two major groups [92]: *logical* hierarchy of keys and *physical* hierarchy of servers. These schemes divide the key management domain into smaller areas in order to distribute the processing workload. Members of the multicast group belong to a key distribution tree having a root at the sender. In *hierarchical key-based* schemes, the set of keys kept by a member is determined by the location of the member in the tree. In *hierarchical node-based* schemes, internal tree nodes assume the role of subgroup managers in key distribution. For mobile members, the latter approach is more appropriate.

Consider the mobility framework in Figure 1.4. All the members in the group belong to a "domain," denoted by the collection of pentagons, managed by a *Domain Key Distributor* (DKD). The domain is divided into several independent "areas," each managed by an *Area Key Distributor*. An area is defined in such a way that member movement within an area does not require any rekeying, and a join or leave is handled locally by an *intra-area* rekeying algorithm. When a member moves between the areas, *interarea* rekeying algorithms provide the coordination for the transfer of security relationships.

The DKD generates the *data encryption key* (DEK) for the session and distributes it to all AKDs. Each AKD is responsible for distributing the DEK to its members. As the distribution of the DEK has to be secure, it is protected by a local *key encryption key* (KEK). For intra-area rekeying, several approaches, including the hierarchical key-based schemes, can be used.

We will now summarize the three operations: join, leave, and transfer [116].

Joining the group via area i: When a member joins the group via area i, it sends a signaling message to AKD_i to notify AKD_i of its arrival. AKD_i creates a new KEK_i and securely distributes it to area i existing members and the new member. Once the new KEK_i is in place, the new DEK can be securely multicast among the AKDs and then from each AKD to area members.

Leaving the group via area i: When a member leaves the group via area i, all AKDs, j, for which the departing member holds a valid key KEK_j must be notified. A new KEK_j is created and securely distributed to remaining members for all areas, j, for which the departing member holds a valid key KEK_j. Once the new KEK_js are in place, the new DEK can be securely multicast among the AKDs and then from each AKD to area members.

Transfer from area i to area j: For member transfer from one area to another, three interarea rekeying algorithms have been defined:

- *Baseline rekeying (BR)*: The member first leaves the group via area i and then rejoins the group via area j. The data transmission is halted during the distribution of the KEKs and the DEK. In BR, when a member leaves the group, a notification is sent to its current AKD.
- *Immediate rekeying (IR)*: The member initiates a transfer by sending one notification to AKD_i and one notification to AKD_j. Area i performs a KEK_i rekey and area j performs a KEK_j rekey. The only KEK held by a group member is for the area in which it currently resides. Unlike the baseline algorithm, no DEK is generated and data transmission continues uninterrupted. In IR, when a member leaves the group, a notification is sent to its current AKD.
- *Delayed rekeying (DR)*: The member sends one notification to AKD_i and one notification to AKD_j. Area j performs a KEK_j rekey, but area i does not perform a KEK_i rekey. AKD_i adds the member to the Extra Key Owner List (EKOL). The EKOL is reset whenever a local rekey occurs. A member accumulates KEKs as it visits different areas. If the entering member has previously visited area j, no KEK_j rekey occurs for j. If the member is entering area j for the first time, a KEK_j rekey occurs for j. To limit the maximum amount of time that KEK_i can be held by a member outside area i, each AKD_i maintains a timer. At $t = T_i$ (a threshold value), the KEK_i is updated and the timer is set to zero. At this point, no group member outside of area i has a valid KEK_i. In DR, when a member leaves the group, a notification is sent to all the AKDs.

Two studies that compare the above algorithms show that delayed rekeying, with reduced communication load and rekeying rate, can improve the performance of key management [114,116]. The first study uses messaging overhead, the KEK rekey rate, and the number of KEKs held by a member as the performance metrics. The second study employs rekeying rates, mean number of extra keys, and percentage of off-line time to compare the interarea rekeying algorithms.

Security in Digital Home Networks

A digital home network (DHN) is a cluster of digital A/V devices, including set-top boxes, TVs, VCRs, DVD players, and general-purpose computing devices such as personal computers [117]. The problem of content protection in home networks has the following dimensions:

- Protection of content across digital interfaces
- Protection of content on storage media
- Management of rights associated with content

This problem, it is believed, turns out to be the most difficult problem to solve for a number of technical, legal, and economic reasons:

1. Private CA systems are, by definition, proprietary and can be defined and operated using the strictest possible security means and methods. In comparison, the protection systems needed for the devices and interfaces in home networks have to be developed with a consensus among the stakeholders, making the determination of requirements very difficult.

2. Renewability of a protection system needs to be properly defined and implemented. Important parameters are the cost, user convenience, and liabilities resulting from copyright infringements.

3. The new copyright legislation introduces controversial prohibitions subject to different interpretations.

4. The payer of the bill for the cost of protection in home networks is unclear. Several business models are under consideration.

Two groups of technologies arc believed to be useful in designing technical solutions: encryption based and watermark based. The potential value of these technologies has been the subject of prolonged discussions in the last decade. Because each group presents strengths and weaknesses, some are of the opinion that both types of solution should be implemented to increase the robustness to possible attacks.

Encryption and watermarking each provide a different "line of defense" in protecting content. The former, the *first line of defense*, makes the content unintelligible through a reversible mathematical transformation based on a secret key. The latter, the *second line of defense*, inserts data directly into the content at the expense of imperceptible degradation in quality. The theoretical level of security provided by encryption depends on the cipher strength and key length. Other factors such as tamper-resistant hardware or software also play an important role in implementations. Watermarking has several useful applications that dictate how and where the watermark is placed. For content protection purposes, the watermark should be embedded in such a way that it is imperceptible to the human eye and robust against attacks. Its real value comes from the fact that the information it represents remains with the content in both analog and digital domains.

The technical solutions developed in the last several years are listed in Table 1.7. They represent the major components of a comprehensive framework called the Content Protection System Architecture (CPSA) [118]. Using 11 axioms, this architecture describes how compliant devices manage copy control information (CCI), playback, and recording.

In the CSS system, each manufacturer is assigned a distinct master key. If a particular master key is compromised, it is replaced by a key that is used in the subsequent manufacturing of DVD players. With the new key assignment, future releases of DVDs cannot be played on the players manufactured with the compromised key. The CSS system was hacked in 1999 because a company neglected to protect its master key during the software implementation for a DVD drive. The software utility created by a group of hackers is commonly referred to as DeCSS. A computer that has the DeCSS utility can make illegal copies of decrypted DVD content on blank DVDs for distribution on the Internet.

In the rest of the protection systems, renewability is defined as device revocation as well. When a pirated device is found in the consumer market, its ID is added to the next version of the revocation list. Updated versions of the revocation lists are distributed on new prerecorded media or through external connections (Internet, cable, satellite, and terrestrial). Some examples are:

- DVD players can receive updates from newer releases of prerecorded DVDs or other compliant devices.
- Set-top boxes (digital cable transmission receivers or digital satellite broadcast receivers) can receive updates from content streams or other compliant devices.
- Digital TVs can receive updates from content streams or other compliant devices.
- Recording devices can receive updates from content streams, if they are equipped with a tuner or other compliant devices.

Table 1.7. Content protection in digital home networks

	Solution	What is protected?	Brief description
Optical media	CSS [119]	Video on DVD-ROM	CSS-protected video is decrypted during playback on the compliant DVD player or drive.
	CPPM [120]	Audio on DVD-ROM	CPPM-protected audio is decrypted during playback on the compliant DVD player or drive.
	CPRM [121]	Video or audio on DVD-R/RW/RAM	A/V content is re-encrypted before recording on a DVD recordable disc. During playback, the compliant player derives the decryption key.
	4C/Verance Watermark [122]	Audio on DVD-ROM	Inaudible watermarks are embedded into the audio content. The compliant playback or recording device detects the CCI represented by the watermark and responds accordingly.
	To be determined	Video on DVD-ROM/ R/RW/RAM	Invisible watermarks are embedded into the video content. The compliant playback or recording device detects the CCI represented by the watermark and responds accordingly. If a copy is authorized, the compliant recorder creates and embeds a new watermark to represent "no-more-copies."
Magnetic media	HDCP [123]	Video on digital tape	Similar in function to the Content Scrambling System (CSS).

(Continued)

Table 1.7. Continued

	Solution	What is protected?	Brief description
Digital interfaces	DTCP [124]	IEEE 1394 serial bus	The source device and the sink device authenticate each other, and establish shared secrets. A/V content is encrypted across the interface. The encryption key is renewed periodically.
	HDCP [125]	Digital Visual Interface (DVI) and High Definition Multimedia Interface (HDMI)	Video transmitter authenticates the receiver and establishes shared secrets with it. A/V content is encrypted across the interface. The encryption key is renewed frequently.

- Personal computers (PCs) can receive updates from Internet servers.

Each solution in Table 1.7 defines a means of associating the CCI with the digital content it protects. The CCI communicates the conditions under which a consumer is authorized to make a copy. An important subset of CCI is the two Copy Generation Management System (CGMS) bits for digital copy control: "11" (copy-never), "10" (copy-once), "01" (no-more-copies), and "00" (copy-free). The integrity of the CCI should be ensured to prevent unauthorized modification. The CCI can be associated with the content in two ways: (1) The CCI is included in a designated field in the A/V stream and (2) the CCI is embedded as a watermark into the A/V stream.

A CPRM-compliant recording device refuses to make a copy of content labeled as "copy-never" or "no-more-copies." It is authorized to create a copy of "copy-once" content and label the new copy as "no-more-copies." The DTCP carries the CGMS bits in the isochronous packet header defined by the interface specification. A sink device that receives content from a DTCP-protected interface is obliged to check the CGMS bits and respond accordingly. As the DVI is an interface between a content source and a display device, no CCI transmission is involved.

In addition to those listed in Table 1.7, private DRM systems may also be considered to be content protection solutions in home networks. However, interoperability of devices supporting different DRM systems is an unresolved issue today.

There are other efforts addressing security in home networking environments. Two notable projects are being discussed by the Video Electronics Standards Association (VESA) [126] and the Universal Plug and Play (UPnP) [127] Forum. VESA is an international nonprofit organization that develops and supports industrywide interface standards for the PC and other computing environments. The VESA and the Consumer Electronics Association (CEA) have entered a memo of understanding that allowed the CEA to assume all administration of the VESA Home Network Committee. UPnP Forum is an industry initiative designed to enable easy and robust connectivity among stand-alone devices and PCs from many different vendors.

In Europe, the work of the DVB Project is shared by its Modules and Working Groups. The Commercial Module discusses the commercial issues around a DVB work item, leading to a consensus embodied in a set of "Commercial Requirements" governing each and every DVB specification. Proving technical expertise, the Technical Module works according to requirements determined by the Commercial Module and delivers specifications for standards via the Steering Board to the recognized standards setting entities, notably the EBU/ETSI/CENELEC Joint Technical Committee. The IPR Module is responsible for making recommendations concerning the DVB's IPR policy, overseeing the functioning of the IPR policy, dealing with IPR related issues in other areas of the DVB's work, and making recommendations on antipiracy policies. The Promotions and Communications Module handles the external relations of the DVB Project. It is responsible for trade shows, conferences, DVB & Multimedia Home Platform (MHP) Web sites, and press releases. The Copy Protection Technologies (CPT) subgroup of the Technical Module was set up in March 2001 to develop a specification for a DVB Content Protection & Copy Management (CPCM) system based upon the Commercial Requirements produced by the Copy Protection subgroup of the Commercial Module and ratified by the Steering Board. The CPT subgroup is working to define a CPCM system to provide interoperable, end-to-end copy protection in a DVB environment [128].

Finally, we summarize in Table 1.8 the technical protection solutions for all the architectures we have discussed in this chapter.

LEGAL SOLUTIONS

Intellectual property plays an important role in all areas of science and technology as well as literature and the arts. The World Intellectual Property Organization (WIPO) is an intergovernmental organization responsible for the promotion of the protection of intellectual property throughout the world through cooperation among member States and for the administration of various multilateral treaties dealing with the legal

Table 1.8. Protection systems in the digital universe

Media protected		Protection type	Device authentication	Association of digital rights	Licensed technology	System renewability
Prerecorded media	Video on DVD-ROM	Encryption	Mutual between DVD drive and PC	Metadata	CSS [119]	Device revocation
	Audio on DVD-ROM	Encryption	Mutual between DVD drive and PC	Metadata	CPPM [120]	Device revocation
		Watermarking	na	Watermark	4C/Verance Watermark [122]	na
	Video or audio on DVD-R/RW/RAM	Encryption	Mutual between DVD drive and PC	Metadata	CPRM [121]	Device revocation
	Video on digital tape	Watermarking	na	Watermark	tbd	na
		Encryption	na	Metadata	High Definition Copy Protection (HDCP) [123]	Device revocation

Category	Technology	Encryption	Authentication	Metadata	Standard	Revocation
Digital interface	IEEE 1394	Encryption	Mutual between source and sink	Metadata	DTCP [124]	Device revocation
	Digital Visual Interface (DVI) and High Definition Multimedia Interface (HDMI)	Encryption	Mutual between source and sink	Metadata	HDCP [125]	Device revocation
	NRSS interface	Encryption	Mutual between host device and removable security device	Metadata	Open standards [129,130,131]	Service revocation
Broadcasting	Satellite transmission (Privately defined by service providers and CA vendors.)	Encryption	None	Metadata	Conditional access system [132,133] privately defined by service providers	Smartcard revocation
	Terrestrial transmission	Encryption	None	Metadata	Conditional access system [131] framework defined by ATSC	Smartcard revocation
Cable transmission		Encryption	None	Metadata	Conditional access system [134] privately defined by OpenCable	Smartcard revocation

(Continued)

Table 1.8. Continued

Media protected		Protection type	Device authentication	Association of digital rights	Licensed technology	System renewability
Internet	Unicast-based DRM systems are privately defined, and hence are not interoperable.	Encryption	Receiver	Metadata	DRM [135,136]	Software update
	Multicast-based DRM systems are yet to appear in the market.	Encryption	Sender and receiver (depends on the authentication type)	Metadata	Group key management [137]	tbd
	An Internet Draft defines a common architecture for MSEC group key management protocols that support a variety of application, transport and internetwork security protocols.				Watermarking proposals [38]	
	A few watermarking schemes have been proposed for multicast data.					

and administrative aspects of intellectual property. With headquarters in Geneva, Switzerland, it is 1 of the 16 specialized agencies of the United Nations system of organizations. WIPO has currently 179 member states, including the Unites States, China, and the Russian Federation.

The legal means of protecting copyrighted digital content can be classified into two categories:

- National laws (copyright laws and contract laws)
- International treaties and conventions

Two WIPO treaties — the WIPO Copyright Treaty and the WIPO Performances and Phonograms Treaty — obligate the member states to prohibit circumvention of technological measures used by copyright owners to protect their works and to prevent the removal or alteration of copyright management information.

The international conventions that have been signed for the worldwide protection of copyrighted works include [1,138]:

- The Berne Convention, formally the International Convention for the Protection of Literary and Artistic Works (1886)
- The Universal Copyright Convention (1952)
- Rome Convention for the Protection of Performers, Producers of Phonograms and Broadcasting Organizations (1961)
- The Geneva Convention for the Protection of Producers of Phonograms Against Unauthorized Duplication of Their Phonograms (1971)
- Brussels Convention Relating to the Distribution of Programme-Carrying Signals Transmitted by Satellite (1974)
- TRIPS Agreement (1995)

Since the end of the 1990s, we have seen important efforts to provide legal solutions regarding copyright protection and management of digital rights in the United States.

The most important legislative development in the recent years was the Digital Millennium Copyright Act (DCMA). Signed into a law on October 28, 1998, this Act implements the WIPO Copyright Treaty and the WIPO Performances and Phonograms Treaty. Section 103 of the DMCA amends Title 17 of the U.S. Code by adding a new chapter 12. Section 1201 makes it illegal to circumvent technological measures that prevent unauthorized access and copying, and Section 1202 introduces prohibitions to ensure the integrity of copyright management information. The DCMA has received earnest criticism with regard to the ambiguity and inconsistency in expressing the anticircumvention provisions [9,139]

The second major attempt by the U.S. Congress to strike a balance between the U.S. laws and technology came with the Consumer Broadband and Digital Television Promotion Act (CBDTPA). Introduced by Senator Fritz Hollings on March 21, 2002, the Act intends "to regulate interstate commerce in certain devices by providing for private sector development of technological protection measures to be implemented and enforced by Federal regulations to protect digital content and promote broadband as well as the transition to digital television, and for other purposes." In establishing open security system standards that will provide effective security for copyrighted works, the bill specifies the standard security technologies to be reliable, renewable, resistant to attacks, readily implementable, modular, applicable to multiple technology platforms, extensible, upgradable, not cost prohibitive, and based on open source code (software portions).

Recently, a draft legislation was introduced by Senator Sam Brownback. Known as the "Consumers, Schools, and Libraries Digital Rights Management Awareness Act of 2003," the bill seeks to "to provide for consumer, educational institution, and library awareness about digital rights management technologies included in the digital media products they purchase, and for other purposes." With several important provisions, the bill prevents copyright holders from compelling an Internet service provider (ISP) to disclose the names or other identifying information of its subscribers prior to the filing of a civil lawsuit, requires conspicuous labeling of all digital media products that limits consumer uses with access or redistribution restrictions, imposes strict limits on the Federal Communication Commission's ability to impose federal regulations on digital technologies, and preserves the right to donate digital media products to libraries and schools.

In every country, legally binding agreements between parties would also be effective in copyright protection. All technological measures, without any exception, must include IP (mostly protected by patents) to be licensable. Before a particular technology is implemented in A/V devices, the licensee signs an agreement with the owner of the technology agreeing with the terms and conditions of the license. Contract laws deal with the violations of the license agreements in the event of litigations.

BUSINESS MODELS

In addition to technical and legal means, owners of digital copyrighted content can also make use of new, creative ways to bring their works to the market. A good understanding of the complexity and cost of protection is probably a prerequisite to be on the right track. With the wide availability of digital A/V devices and networks in the near future, it will become increasingly difficult to control individual behavior and detect infringements of copyright. Would it then be possible to develop business models that are not closely tied to the inherent qualities of digital content? The current business models, old and new, for selling copyrighted works are summarized in Table 1.9 [9]. Some of these models are relevant for the marketing of digital copyrighted content. In general, the selection of a business model depends on a number of factors including:

- Type of content
- Duration of the economic value of content
- Fixation method
- Distribution channel
- Purchase mechanism
- Technology available for protection
- Extent of related legislation

A good case study to explore the opportunities in a digital market is superdistribution. Figure 1.5 shows the players in a DRM-supported e-commerce system: a content owner, a clearinghouse, several retailers, and many customers. Suppose that the media files requested by the customers are hosted by the content owner or the retailers, and the licenses are downloaded from the clearinghouse. The following are a few of the possible ideas for encouraging file sharing and creating a competitive market [140]:

Promotions: The media file can be integrated with one or more promotions for retail offerings. Examples are a bonus track or a concert ticket for the purchase of an album or a discount coupon for the local music store. Attractive promotions may result in more file sharing.

Packaged media with a unique retailer ID: During the packaging of the media file, a unique retailer ID is added. The file is shared among customers. When a customer in the distribution chain is directed to the clearinghouse to get his license, the clearinghouse credits the original retailer. This may be an incentive for a retailer to offer the best possible deal for a content.

Packaged media with a unique customer ID: During the packaging of the media file, a unique customer ID is added. The file is shared among customers. When a customer in the distribution chain is directed to the clearinghouse to get his license, the clearinghouse credits the customer who initiated the distribution. This may be a good motivation for a customer to be the "first" in the chain.

Table 1.9. Business models for copyrighted works

Traditional		
Type	**Examples**	**Relevance to copyright protection**
Models based on fees for products and services		
Single transaction purchase	Books, videos, CDs, photocopies	
Subscription purchase	Newsletter and journal subscriptions	
Single transaction license	Software	
Serial transaction license	Electronic subscription to a single title	High sensitivity to unauthorized use
Site license	Software for a whole company	
Payment per electronic use	Information resource paid per article	
Models relying on advertising		
Combined subscription and advertising	Web sites for newspapers	Low sensitivity to unauthorized access.
Advertising only	Web sites	Concern for reproduction and framing.
Models with free distribution		
Free distribution	Scholarly papers on preprint servers	
Free samples	Demo version of a software	Low concern for reproduction. Sensitivity for information integrity.
Free goods with purchases	Free browser software to increase traffic on an income-producing Web site	
Information in the public domain	Standards, regulations	
Recent		
Give away the product and sell an auxiliary product or service	Free distribution of music because it enhances the market for concerts, t-shirts, posters, etc.	Services or products not subject to replication difficulties of the digital content.
Give away the product and sell upgrades	Antivirus software	Products have short shelf life.

(Continued)

Table 1.9.   Continued

	Traditional	
Type	**Examples**	**Relevance to copyright protection**
Extreme customization of the product	Personalized CDs	No demand from other people.
Provide a large product in small pieces	Online databases	Difficulty in copying.
Give away digital content to increase the demand for the actual product	Full text of a book online to increase demand for hard copies	
Give away one piece of digital content to create a market for another	Adobe's Acrobat Reader	Need for protecting the actual product sold.
Allow free distribution of the product but request payment	Shareware	
Position the product for low-priced, mass market distribution	Microsoft XP	Cost of buying converges with cost of stealing.

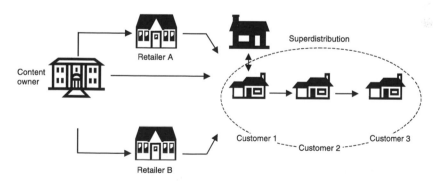

Figure 1.5.   E-commerce with superdistribution.

SUMMARY

We presented an overview of the complex problem of copyrighted multimedia content protection in digital distribution networks. After an introduction to copyright and copyright industries, we examined the

53

technical, legal, and business solutions for multimedia security in commonly used satellite, cable, terrestrial, and Internet-based architectures as well as digital home networks. Our analysis can be summarized as follows:

- End-to-end security is a key requirement for the growth of digital markets. Digital copyrighted content needs to be protected in every stage of its lifecycle in order to prevent piracy losses and encourage continued supply of products and services.

- The copyright industries, a segment of the U.S. economy with a high growth rate, are the primary source of marketable digital content. A better estimate of the economic impact of piracy on these industries is needed. Appropriate copyright policies can only be developed by understanding the complex social, cultural, legal, and ethical factors that influence the consumer behavior.

- Digital media offers certain advantages:

  - Perfect reproduction: Copies produced are indistinguishable from the original.

  - Reduced costs for storage and distribution: Because of efficient compression methods, high-quality content can be stored on lower-capacity media and transmitted through lower-bandwidth channels.

  - New business models: Content owners (artists, authors, musicians, etc.) can have direct access to consumers, specify and dynamically change their business rules, and regularly obtain customer information. Superdistribution appears to be a promising model for quick and inexpensive distribution of products and promotions.

- Irrespective of the model used, the following elements interact in a commerce cycle for the management of digital rights:

  - Content owner (or its agent): Packages content according to established business rules.

  - Content distributor: Makes content available to consumers through retail stores or other delivery channels.

  - Customer (with a compliant receiver): Purchases and consumes content according to usage rules.

  - Clearinghouse: Keeps track of financial information and collects usage information.

- Encryption and watermarking are two groups of technology used in developing technical solutions for the copy protection problem in DHNs [117,141–143]. Encryption, the first line of defense, is the most effective way to achieve multimedia security. Ciphers can be classified into two major categories: symmetric key ciphers and asymmetric key ciphers. Watermarking, the second line of defense, is the process of embedding a signal into multimedia content (images, video, audio, etc.). Depending on the purpose of the embedded watermark, there is an essential difference between the functionalities of the consumer electronics devices:

- *Copyright protection*: The open literature on watermarking has so far focused on copyright protection for which the receiver does not have to assume an active role in responding to the watermark. When a dispute arises regarding the ownership of content, the watermark needs to be detected or extracted by authorized entities such as the legal institutions.

- *Access control*: The use of watermarking for content protection has been the subject of prolonged discussions at the Copy Protection Technical Working Group (CPTWG) meetings in California in the last few years. The three industries (information technology, consumer electronics, and motion picture) supporting the CPTWG have agreed in principle to implement a watermarking system in DVD playback and recording devices. According to a set of principles, the playback and recording devices will detect and respond to watermarks representing the CGMS bits ("11" [copy-never], "10" [copy-once], "01" [no-more-copies], and "00" [copy-free]). If an unauthorized copy is detected, the playback device will prevent the playback of the copy and the recording device will refuse to make a next-generation copy. In spite of several years of research and testing, the Interim Board of Directors of the DVD Copy Control Association (DVD CCA) decided not to select a watermarking system for copy protection before ending its term in the summer of 2002.[3] The new board has inherited the task of determining the next steps in the selection process.

- In every type of content protection system based on secrets (conditional access, digital rights management [unicast and multicast], and copy protection), key management (i.e., generation, distribution, and maintenance of keys) is a critical issue. System renewability should be defined with the right balance among economic, social, and legal factors.

- The U.S. copyright law has been evolving in the last 200 years in response to technological developments. The recent obligations introduced by the DCMA will most likely need future revision for clarity and consistency. Additional legislative efforts are underway

---

[3]The DVD CCA is a not-for-profit corporation with responsibility for licensing CSS (Content Scramble System) to manufacturers of DVD hardware, disks, and related products.

both in the United States and in other countries to update national laws in response to technological developments.

- Worldwide coordination for copyright protection is a challenging task. In spite of international treaties, there are significant differences among countries, making it difficult to track national laws and enforcement policies.

- To complement the technical and legal solutions, content owners are also developing new business models for marketing digital multimedia content. The important factors used in selecting a business model include type of content, duration of the economic value of content, fixation method, distribution channel, purchase mechanism, technology available for protection, and extent of related legislation.

ACKNOWLEDGMENTS

The authors would like to acknowledge the permission granted by Springer-Verlag, Elsevier, the Institute of Electrical and Electronics Engineers (IEEE), the International Association of Science and Technology for Development (IASTED), and the International Society for Optical Engineering (SPIE) for partial use of the authors' following research material published by them:

- Eskicioglu, A.M. and Delp, E.J., Overview of Multimedia Content Protection in Consumer Electronics Devices, in Proceedings of SPIE Security and Watermarking of Multimedia Content II, 2000, San Jose, CA, Vol. 3971, pp. 246–263.

- Eskicioglu, A.M. and Delp, E.J., Overview of multimedia content protection in consumer electronics devices, *Signal Process.:Image Commn.*, 16(7), 681–699, 2001.

- Eskicioglu, A.M., Town J., and Delp, E.J., Security of Digital Entertainment Content from Creation to Consumption, in Proceedings of SPIE Applications of Digital Image Processing XXIV, San Diego, CA, 2001, Vol. 4472, pp. 187–211.

- Eskicioglu, A.M., Multimedia Security in Group Communications: Recent Progress in Wired and Wireless Networks, in Proceedings of the IASTED International Conference on Communications and Computer Networks, Cambridge, MA, 2002, pp. 125–133.

- Eskicioglu, A.M., Town, J., and Delp, E.J., Security of digital entertainment content from creation to consumption, *Signal Process.: Image Commn.*, 18(4), 237–262, 2003.

- Eskicioglu, A.M., Multimedia security in group communications: Recent progress in key management, authentication, and watermarking, *Multimedia Syst. J.*, 239–248, 2003.

- Eskicioglu, A.M., Protecting intellectual property in digital multimedia networks, *IEEE Computer*, 39–45, 2003.
- Lin, E.T., Eskicioglu, A.M., Lagendijk, R.L., and Delp, E.J., Advances in digital video content protection, *Proc. IEEE*, 2004.

REFERENCES

1. http://www.wipo.org.
2. http://www.loc.gov/copyright/circs/circ1.html.
3. http://www.loc.gov/copyright/title17/92chap1.html#106.
4. Strong, W.S., *The Copyright Book*, MIT Press, Cambridge, MA, 1999.
5. http://www.loc.gov/copyright/docs/circ1a.html.
6. http://arl.cni.org/info/frn/copy/timeline.html.
7. Goldstein, P., *Copyright's Highway*, Hill and Wang, 1994.
8. http://www.iipa.com.
9. National Research Council, *The Digital Dilemma: Intellectual Property in the Information Age*, National Academy Press, Washington, DC, 2000.
10. Menezes, J., van Oorschot, P.C., and Vanstone, S.A., *Handbook of Applied Cryptography*, CRC Press, Boca Raton, FL, 1997.
11. Schneier, B., *Applied Cryptography*, John Wiley & Sons, 1996.
12. Advanced Television Systems Committee, available at http://www.atsc.org.
13. Consumers Electronics Association, available at http://www.ce.org.
14. Copy Protection Technical Working Group, available at http://www.cptwg.org.
15. DVD Forum, available at http://www.dvdforum.org.
16. Society of Cable Telecommunications Engineers, available at http://www.scte.org.
17. Motion Picture Association of America, available at http://www.mpaa.org.
18. Recording Industries Association of America, available at http://www.riaa.org.
19. Internet Engineering Task Force, available at http://www.ietf.org/overview.html.
20. Moving Pictures Expert Group, available at http://mpeg.telecomitalialab.com.
21. Digital Video Broadcasting Project, available at http://www.dvb.org.
22. de Bruin, R. and Smits, J., *Digital Video Broadcasting: Technology, Standards and Regulations*, Artech House, 1999.
23. Benoit, H., *Digital Television: MPEG-1, MPEG-2 and Principles of the DVB System*, Arnold, London, 1997.
24. Guillou, L.C. and Giachetti, J.L., Encipherment and conditional access, *SMPTE J.*, 103(6), 398–406, 1994.
25. Mooij, W. Conditional Access Systems for Digital Television, International Broadcasting Convention, IEE Conference Publication, 397, 1994, pp. 489–491.
26. Macq, B.M. and Quisquater, J.J., Cryptology for digital TV broadcasting, *Proc. IEEE*, 83(6), 1995.
27. Rossi, G., Conditional access to television broadcast programs: Technical solutions, *ABU Tech.* Rev. 166, 3–12, September–October 1996.
28. Cutts, D., DVB conditional access, *Electron. Commn., Engi. J.*, 9(1), 21–27, 1997.
29. Mooij, W., Advances in Conditional Access Technology, International Broadcasting Convention, IEE Conference Publication, 447, 1997, pp. 461–464.
30. Eskicioglu, A.M., A Key Transport Protocol for Conditional Access Systems, in Proceedings of the SPIE Conference on Security and Watermarking of Multimedia Contents III, San Jose, CA, USA, 2001, pp. 139–148.
31. International Standard ISO-IEC 13818-1 Information technology — Generic coding of moving pictures and associated audio information: Systems, First Edition, 1996.
32. EIA-679B National Renewable Security Standard, September 1998.

33. http://www.macrovision.com.
34. EBU Project Group B/CA, Functional model of a conditional access system, *EBU Techn. Rev.* 266, Winter 1995–1996.
35. http://www.atsc.org/standards/a_70_with_amendment.pdf.
36. http://www.mpaa.org/Press/broadcast_flag_qa.htm.
37. Federal Communications Commission, Report and Order and Further Notice of Proposed Rulemaking, November 4, 2003.
38. Eskicioglu, A.M., Multimedia security in group communications: Recent progress in key management, authentication, and watermarking, *Multimedia Syst. J.*, 239–248, September 2003.
39. http://www.chiariglione.org/mpeg/faq/mp4-sys/sys-faq-ipmp-x.htm.
40. http://www.chiariglione.org/mpeg.
41. MPEG LA, LLC, available at http://www.mpegla.com.
42. Open eBook Forum, available at http://www.openebook.org.
43. http://www.irtf.org.
44. Miller, C.K., *Multicast Networking and Applications*, Addison-Wesley Longman, Reading, MA, 1999.
45. Hardjono, T. and Tsudik, G., IP Multicast security: Issues and directions, *Ann. Telecom*, 324–334, July–August 2000.
46. Dondeti, L.R., Mukherjee, S., and Samal, A., Survey and Comparison of Secure Group Communication Protocols, Technical Report, University of Nebraska–Lincoln, 1999.
47. Bruschi, D. and Rosti, E., Secure multicast in wireless networks of mobile hosts and protocols and issues, *ACM Baltzer MONET J.*, 7(6), 503–511, 2002.
48. Rafaeli, S. and Hutchison, D., A survey of key management for secure group communication, *ACM Computing Surv.*, 35(3), 309–329, 2003.
49. Chan, K.-C. and Chan, S.-H.G., Key management approaches to offer data confidentiality for secure multicast, *IEEE Network*, 30–39, September/October 2003.
50. Wong, C.K., Gouda, M.G., and Lam, S.S., Secure Group Communications Using Key Graphs, Technical Report TR-97-23, Department of Computer Sciences, The University of Texas at Austin, July 1997.
51. Caronni, G., Waldvogel, M., Sun, D., and Plattner, B., Efficient Security for Large and Dynamic Groups, Technical Report No. 41, Computer Engineering and Networks Laboratory, Swiss Federal Institute of Technology, February 1998.
52. Balenson, D., McGrew, D., and Sherman, A., Key Management for Large Dynamic Groups: One-Way Function Trees and Amortized Initialization, Internet Draft (work in progress), February 26, 1999.
53. Canetti, R., Garay, J., Itkis, G., Micciancio, D., Naor, M., and Pinkas, B., Multicast Security: A Taxonomy and Some Efficient Constructions, in Proceedings of IEEE INFOCOM, Vol. 2, New York, March 1999, pp. 708–716.
54. Chang, I., Engel, R., Kandlur, D., Pendakaris, D., and Saha, D., Key Management for Secure Internet Multicast Using Boolean Function Minimization Techniques, in Proceedings of IEEE INFOCOM, Vol. 2, New York, March 1999, pp. 689–698.
55. Wallner, D., Harder, E., and Agee, R., Key Management for Multicast: Issues and Architectures, RFC 2627, June 1999.
56. Waldvogel, M., Caronni, G., Sun, D., Weiler, N., and Plattner, B., The VersaKey framework: versatile group key management, *JSAC*, Special Issue on Middleware, 17(8), 1614–1631, 1999.
57. Banerjee, S., and Bhattacharjee, B., Scalable Secure Group Communication over IP Multicast, presented at International Conference on Network Protocols, Riverside, CA, November 10–14, 2001.
58. Eskicioglu, A.M., and Eskicioglu, M.R., Multicast Security Using Key Graphs and Secret Sharing, in Proceedings of the Joint International Conference on Wireless LANs and Home Networks and Networking, Atlanta, GA, August 26–29, 2002, pp. 228–241.

59. Selcuk, A.A., and Sidhu, D., Probabilistic optimization techniques for multicast key management, *Computer Networks*, 40(2), 219–234, 2002.
60. Zhu, S., and Setia, S., Performance Optimizations for Group Key Management Schemes, presented at 23rd International Conference on Distributed Computing Systems, Providence, RI, May 19–22, 2003.
61. Huang, J.-H., and Mishra, S., Mykil: A Highly Scalable Key Distribution Protocol for Large Group Multicast, presented at IEEE 2003 Global Communications Conference, San Francisco, CA, December 1–5, 2003.
62. Trappe, W., Song, J., Poovendran, R., and Liu, K.J.R., Key management and distribution for secure multimedia multicast, *IEEE Trans. Multimedia*, 5(4), 544–557, 2003.
63. Chiou, G.H., and Chen, W.T., Secure broadcast using the secure lock, *IEEE Trans. on Software Engi.*, 15(8), 929–934, 1989.
64. Gong, L., and Shacham, N., Elements of Trusted Multicasting, in Proceedings of the IEEE International Conference on Network Protocols, Boston, MA, October 1994, pp. 23–30.
65. Harney, H., and Muckenhirn, C., Group Key Management Protocol (GKMP) Architecture, RFC 2094, July 1997.
66. Dunigan, T., and Cao, C., *Group Key Management*, Oak Ridge National Laboratory, Mathematical Sciences Section, Computer Science and Mathematics Division, ORNL/TM-13470, 1998.
67. Blundo, C., De Santis, A., Herzberg, A., Kutten, S., Vaccaro, U., and Yung, M., Perfectly-secure key distribution for dynamic conferences, *Inform. Comput.*, 146(1), 1–23, 1998.
68. Poovendran, R., Ahmed, S., Corson, S., and Baras, J., A Scalable Extension of Group Key Management Protocol, Technical Report TR 98-14, Institute for Systems Research, 1998.
69. Chu, H., Qiao, L., and Nahrstedt, K., A Secure Multicast Protocol with Copyright Protection, in Proceedings of IS&T/SPIE Symposium on Electronic Imaging: Science and Technology, January 1999, pp. 460–471.
70. Scheikl, O., Lane, J., Boyer, R., and Eltoweissy, M., Multi-level Secure Multicast: The Rethinking of Secure Locks, in Proceedings of the 2002 ICPP Workshops on Trusted Computer Paradigms, Vancouver, BC, Canada, August 18–21, 2002, pp. 17–24.
71. Mittra, S., Iolus: A Framework for Scalable Secure Multicasting, in Proceedings of the ACM SIGCOMM '97, Cannes, France, September 1997, pp. 277–288.
72. Dondeti, L.R., Mukherjee, S. and Samal, A., A Dual Encryption Protocol for Scalable Secure Multicasting, presented at Fourth IEEE Symposium on Computers and Communications, Red Sea, Egypt, July 6–8, 1999.
73. Molva, R., and Pannetrat, A., Scalable Multicast Security in Dynamic Groups, in Proceedings of the 6th ACM Conference on Computer and Communications Security, Singapore, November 1999, pp. 101–112.
74. Setia, S., Koussih, S., and Jajodia, S., Kronos: A Scalable Group Re-keying Approach for Secure Multicast, presented at IEEE Symposium on Security and Privacy 2000, Oakland, CA, May 14–17, 2000.
75. Hardjono, T., Cain, B., and Doraswamy, N., A Framework for Group Key Management for Multicast Security, Internet Draft (work in progress), August 2000.
76. Ballardie, A., Scalable Multicast Key Distribution, RFC 1949, May 1996.
77. Briscoe, B., MARKS: Multicast Key Management Using Arbitrarily Revealed Key Sequences, presented at First International Workshop on Networked Group Communication, Pisa, Italy, November 17–20, 1999.
78. Dondeti, L.R., Mukherjee, S., and Samal, A., A Distributed Group Key Management Scheme for Secure Many-to-Many Communication, Technical Report, PINTL-TR-207-99, Department of Computer Science, University of Maryland, 1999.

79. Perrig, A., Efficient Collaborative Key Management Protocols for Secure Autonomous Group Communication, presented at International Workshop on Cryptographic Techniques and E-Commerce (CrypTEC '99), 1999, pp. 192–202.

80. Rodeh, O., Birman, K., and Dolev, D., Optimized Group Rekey for Group Communication Systems, presented at Network and Distributed System Security Symposium, San Diego, CA, February 3–4, 2000.

81. Kim, Y., Perrig, A., and Tsudik, G., Simple and Fault-Tolerant Key Agreement for Dynamic Collaborative Groups, in Proceedings of the 7th ACM Conference on Computer and Communications Security, November 2000, pp. 235–241.

82. Boyd, C., On Key Agreement and Conference Key Agreement, in Proceedings of Second Australasian Conference on Information Security and Privacy, in Second Australasian Conference, ACISP'97, Sydney, NSW, Australia, July 7–9, 1997, pp. 294–302.

83. Steiner, M., Tsudik, G., and Waidner, M., Cliques: A New Approach to Group Key Agreement, Technical Report RZ 2984, IBM Research, December 1997.

84. Becker, C., and Willie, U., Communication Complexity of Group Key Distribution, presented at 5th ACM Conference on Computer and Communications Security, San Francisco, CA, November 1998.

85. Dondeti, L.R., Mukherjee, S., and Samal, A., Comparison of Hierarchical Key Distribution Schemes, in Proceedings of IEEE Globecom Global Internet Symposium, Rio de Janerio, Brazil, 1999.

86. Bauer, M., Canetti, R., Dondeti, L., and Lindholm, F., MSEC Group Key Management Architecture, Internet Draft, IETF MSEC WG, September 8, 2003.

87. Yang, Y.R., Li, X.S., Zhang, X.B., and Lam, S.S., Reliable Group Rekeying: A Performance Analysis, in Proceedings of the ACM SIGCOMM '01, San Diego, CA, 2001, pp. 27–38.

88. Li, X.S., Yang, Y.R., Gouda, M.G., and Lam, S.S., Batch Rekeying for Secure Group Communications, in Proceedings of 10th International WWW Conference, Hong Kong, 2001, pp. 525–534.

89. Zhang, X.B., Lam, S.S., Lee, D.Y., and Yang, Y.R., Protocol Design for Scalable and Reliable Group Rekeying, in Proceedings of SPIE Conference on Scalability and Traffic Control in IP Networks, Vol. 4526, Denver, CO, 2001.

90. Moyer, M.J., Rao, J.R., and Rohatgi, P., Maintaining Balanced Key Trees for Secure Multicast, Internet Draft, June 1999.

91. Goshi, J. and Ladner, R.E., Algorithms for Dynamic Multicast Key Distribution Trees, presented at 22nd ACM Symposium on Principles of Distributed Computing (PODC '03), Boston, MA, 2003.

92. Dondeti, L.R., Mukherjee, S., and Samal, A., Survey and Comparison of Secure Group Communication Protocols, Technical Report, University of Nebraska–Lincoln, 1999.

93. Menezes, J., van Oorschot, P.C., and Vanstone, S.A., Handbook of Applied Cryptography, CRC Press, Boca Raton, Fl, 1997.

94. Moyer, M.J., Rao, J.R., and Rohatgi, P., A survey of security issues in multicast communications, IEEE Network, 12–23, November/December 1999.

95. Gennaro, R. and Rohatgi, P., How to Sign Digital Streams, in Advances in Cryptology — CRYPTO '97, 1997, pp. 180–197.

96. Wong, C.K. and Lam, S.S., Digital Signatures for Flows and Multicasts, in Proceedings of 6th IEEE International Conference on Network Protocols (ICNP '98), Austin, TX, 1998.

97. Rohatgi, P., A Compact and Fast Hybrid Signature Scheme for Multicast Packets, in Proceedings of 6th ACM Conference on Computer and Communications Security, Singapore, 1999.

98. Perrig, A., Canetti, R., Tygar, J.D., and Song, D., Efficient Authentication and Signing of Multicast Streams over Lossy Channels, in Proceedings of IEEE Symposium on Security and Privacy 2000, Oakland, CA, 2000.

99. Perrig, A., Canetti, R., Song, D., and Tygar, J.D., Efficient and Secure Source Authentication for Multicast, in Proceedings of Network and Distributed System Security Symposium, San Diego, CA, 2001.
100. Perrig, A., The BiBa One-Time Signature and Broadcast Authentication Protocol, in Proceedings of the 8th ACM Conference on Computer and Communications Security, Philadelphia, 2001.
101. Golle, P. and Modadugu, N., Authenticating Streamed Data in the Presence of Random Packet Loss, presented at Network and Distributed System Security Symposium Conference, 2001.
102. Miner, S. and Staddon, J., Graph-Based Authentication of Digital Streams, in Proceedings of the IEEE Symposium on Research in Security and Privacy, Oakland, CA, 2001, pp. 232–246.
103. Park, J.M., Chong, E.K.P., and Siegel, H.J., Efficient multicast stream authentication using erasure codes, *ACM Trans. Inform. Syst. Security*, 6(2), 258–285, 2003.
104. Pannetrat, A. and Molva, R., Efficient Multicast Packet Authentication, presented at 10th Annual Network and Distributed System Security Symposium, San Diego, CA, 2003.
105. Boneh, D., Durfee, G., and Franklin, M., Lower Bounds for Multicast Message Authentication, in *Proceedings of Eurocrypt 2001*, Lecture Notes in Computer Science Vol. 2045, Springer-Verlag, New York, 2001, pp. 437–452.
106. Podilchuk, C.I. and Delp, E.J., Digital watermarking: Algorithms and applications, *IEEE Signal Process. Mag.*, 33–46, July 2001.
107. Cox, I.J., Miller, M.L., and Bloom, J.A., *Digital Watermarking*, Morgan Kaufmann, 2002.
108. Lin, E.T., Eskicioglu, A.M., Lagendijk, R.L., and Delp, E.J., Advances in digital video content protection, *Proc. IEEE*, 2004.
109. Hartung, F. and Kutter, M., Multimedia watermarking techniques, *Proc. IEEE*, 87(7), 1079–1107, 1999.
110. Wolfgang, R.B., Podilchuk, C.I., and Delp, E.J., Perceptual watermarks for digital images and video, *Proc. IEEE*, 87(7), 1108–1126, 1999.
111. Brown, I., Perkins, C., and Crowcroft, J., Watercasting: Distributed Watermarking of Multicast Media, presented at First International Workshop on Networked Group Communication (NGC '99), Pisa, 1999.
112. Judge, P. and Ammar, M., WHIM: Watermarking Multicast Video with a Hierarchy of Intermediaries, presented at 10th International Workshop on Network and Operation System Support for Digital Audio and Video, Chapel Hill, NC, 2000.
113. Caronni, G. and Schuba, C., Enabling Hierarchical and Bulk-Distribution for Watermarked Content, presented at 17th Annual Computer Security Applications Conference, New Orleans, LA, 2001.
114. DeCleene, B.T., Dondeti, L.R., Griffin, S.P., Hardjono, T., Kiwior, D., Kurose, J., Towsley, D., Vasudevan, S., and Zhang, C., Secure Group Communications for Wireless Networks, in Proceedings of IEEE MILCOM 2001, McLean, VA, 2001.
115. Griffin, S.P., DeCleene, B.T., Dondeti, L.R., Flynn, R.M., Kiwior, D., and Olbert, A., Hierarchical Key Management for Mobile Multicast Members, Technical Report, Northrop Grumman Information Technology, 2002.
116. Zhang, C., DeCleene, B.T., Kurose, J., and Towsley, D., Comparison of Inter-Area Rekeying Algorithms for Secure Wireless Group Communications, presented at IFIP WG7.3 International Symposium on Computer Performance Modeling, Measurement and Evaluation, Rome, 2002.
117. Eskicioglu, A.M. and Delp, E.J., Overview of multimedia content protection in consumer electronics devices, *Signal Process.: Image Commn.*, 16(7), 681–699, 2001.
118. CPSA: A Comprehensive Framework for Content Protection, available at http://www.4Centity.com.
119. Content Scramble System, available at http://www.dvdcca.org.

120. Content Protection for Prerecorded Media, available at http://www.4Centity.com.
121. Content Protection for Recordable Media, available at http://www.4Centity.com.
122. 4C/Verance Watermark, available at http://www.verance.com.
123. High Definition Copy Protection, available at http://www.wired.com/news/tech nology/0,1282,41045,00.html.
124. Digital Transmission Content Protection, available at http://www.dtcp.com.
125. High-bandwidth Digital Content Protection, available at http://www.digital-CP.com.
126. http://www.vesa.org.
127. http://www.upnp.org.
128. Vevers, R. and Hibbert, C., Copy Protection and Content Management in the DVB, presented at IBC 2002 Conference, Amsterdam, 2002.
129. EIA-679B National Renewable Security Standard, September 1998.
130. OpenCable CableCARD Copy Protection System Interface Specification, available at http://www.opencable.com.
131. ATSC Standard A/70: Conditional Access System for Terrestrial Broadcast, available at http://www.atsc.org.
132. Proprietary conditional access system for DirecTV, http://www.directv.com.
133. Proprietary conditional access system for Dish Network, http://www.dishnetwork. com.
134. OpenCable System Security Specification, available at http://www.opencable.com.
135. Microsoft Windows Media DRM, available at http://www.microsoft.com/windows/ windowsmedia/drm.aspx.
136. Helix DRM, available at http://www.realnetworks.com/products/drm/index.html.
137. http://www.securemulticast.org.
138. http://www.wto.org.
139. Samuelson, P., Intellectual property and the digital economy: Why the anti-circumvention regulations need to be revised, *Berkeley Technol. Law J.*, 14, 1999.
140. Windows Media Rights Manager, available at http://msdn.microsoft.com.
141. Bell, A., The dynamic digital disk, *IEEE Spectrum*, 36(10), 28–35, 1999.
142. Bloom, J.A., Cox, I.J., Kalker, T., Linnartz, J.P.M.G., Miller, M.L., and Traw, C.B.S., Copy protection for DVD video, *Proc. IEEE*, 87(7), 1267–1276, 1999.
143. Eskicioglu, A.M., Town, J., and Delp, E.J., Security of digital entertainment content from creation to consumption, *Signal Process.: Image Commn.*, 18(4), 237–262, 2003.

2
Vulnerabilities of Multimedia Protection Schemes

Mohamed F. Mansour and Ahmed H. Tewfik

INTRODUCTION

The deployment of multimedia protection algorithms to practical systems has moved to the standardization and implementation phase. In the near future, it will be common to have audio and video players that employ a watermarking mechanism to check the integrity of the played media. The publicity of the detectors introduces new challenges for current multimedia security.

The detection of copyright watermarks is a binary hypothesis test with the decision boundary determined by the underlying test statistic. The amount of signal modification that can be tolerated defines a distortion hyperellipsoid around the representation of the signal in the appropriate multidimensional space. The decision boundary implemented by the watermark detector necessarily passes through that hyperellipsoid because the distortion between the original and watermarked signals is either undetectable or acceptable. Once the attacker knows the decision region, she can simply modify the signal within the acceptable limits and move it to the other side of the decision boundary.

The exact structure of the watermark is not important for our analysis because we focus on watermark detection rather than interpretation. What is important is the structure and operation of the detector. In many cases, especially for copyright protection purposes, the detection is based on thresholding the correlation coefficient between the test signal and the watermark. This detector is optimal if the host signal has a Gaussian distribution. The decision boundary in this case is a hyperplane, which can be estimated with a sufficient number of watermarked items.

The above scheme belongs to the class of *symmetric watermarking*, where the decoder uses the same parameters used in embedding. These parameters are usually generated using a *secret key* that is securely transmitted to the decoder. In contrast, *asymmetric watermarking* [1] employs different keys for encoding and decoding the watermark. In Reference 2, a unified decoding approach is introduced for different asymmetric watermarking schemes. The decoder uses a test statistic in quadratic form. Although the decision boundary is more complicated, it is still parametric and can be estimated using, for example, least square techniques.

Another important watermarking class is the quantization-based schemes. In this class, two or more codebooks are used to quantize the host signal. Each codebook represents an embedded symbol. The decision boundary in this case, although more complicated, is parameterized by the entries of each codebook. This boundary can be estimated using simple statistical analysis of the watermarked signal.

In this chapter, we propose a generic attack for removing the watermark *with minimum distortion* if the detector is publicly available. In this attack, the decision boundary is first estimated; then, the watermarked signal is projected onto the estimated boundary to remove the watermark with minimum distortion. We give implementations of the generic attack for the different watermarking schemes. Next, we propose a new structure for a watermark public detector that resists this attack. The structure is based on using a nonparametric decision boundary so that it cannot be estimated with unlimited access to the detector. The robustness degradation after this detector is minor and can be tolerated.

This chapter is organized as follows. In section "Vulnerability of Current Detection Algorithms", we describe the pitfalls of the current detectors when the detector is publicly available. In section "The Generic Attack", we introduce the generic attack, which aims at removing the watermark with minimum distortion. We propose possible implementations of the attack for the correlator and quantization-based detectors. In section "Secure Detector Structure", we describe the new detector structure with fractal decision boundary and provide a practical implementation of it. Also, we give an overview of the results that show that the distortion is essentially similar to the original detector.

VULNERABILITY OF CURRENT DETECTION ALGORITHMS

In this section, we analyze the security of the current watermark detection schemes. As mentioned earlier, all schemes share the common feature that the decision boundary is *parametric* (i.e., it can be fully specified by a finite set of parameters). The estimation of the parameters is possible in theory. For example, least square techniques can be applied efficiently if sufficient samples (points on the decision boundary) are available. In this section, we will review common detector schemes and describe their security gaps.

We will use the following notations in the remainder of the chapter. U is the original (nonwatermarked) signal, W is the watermark, X is the watermarked signal, and R is the signal under investigation (at the detector). The individual components will be written in lowercase letters and referenced by the discrete index n, where n is a two-element vector in the case of image (e.g., samples of the watermark will be denoted by $w[n]$), L will denote the signal length, and in the case of images, M and N will denote the numbers of rows and columns, respectively, and in this case $L = M \times N$. γ will denote the detection threshold, which is usually selected using the Neyman–Pearson theorem for optimal detection [3, chap. 3].

Correlation-Based Detectors

This detector is the most common and it is optimum for the class of additive watermark with white Gaussian probability density function (pdf) of the underlying signal. It is of fundamental importance; hence, it will be considered in detail. First, assume that the correlation is performed in the signal domain. The detector can be formulated as a binary hypothesis test,

$$H_0: \underline{X} = \underline{U}$$

$$H_1: \underline{X} = \underline{U} + \underline{W}$$

(2.1)

Without loss of generality, we assume that the detector removes the signal mean prior to detection. The log-likelihood test statistic is reduced after removing the common terms to

$$l(\underline{R}) = \underline{R}^* \cdot \underline{W} = (1/L) \sum_n r^*[n] \cdot w[n]$$

(2.2)

where the asterisk denotes the conjugate transpose of the matrix. H_1 is decided if $l(\underline{R}) > \gamma$, and H_0 is decided otherwise. For this detector, the probability distribution of $l(\underline{R})$ is approximately Gaussian if L is large by invoking the central limit theorem. If we assume that the watermark and the underlying signal are uncorrelated, then the mean and the variance of $l(\underline{R})$ are

$$E\{l(\underline{R})|H_0\} = 0$$

$$E\{l(\underline{R})|H_1\} = (1/L)\sum_n (w[n])^2$$

$$\text{Var}\{l(\underline{R})\} = (1/L^2)\sum_n\sum_m E(x^*[n]x[m])w^*[n] \cdot w[m], \qquad \text{under } H_0 \text{ and } H_1$$

Furthermore, if the signal samples are assumed independent then

$$\text{Var}\{l(\underline{R})\} = (1/L^2)\sum_n E\{|x[n]|^2\} \cdot (w[n])^2, \qquad \text{under } H_0 \text{ and } H_1 \qquad (2.3)$$

If the watermark $\underline{W}_c$ is embedded in the transform coefficients and if $\underline{R}_c$ is the transform of the test signal and T is the transformation matrix, then the correlation in the transform domain is

$$l_c(\underline{R}) = \underline{R}_c^* \cdot \underline{W}_c = (T \cdot \underline{R})^* \cdot TW = R^* \cdot T^*T \cdot \underline{W}.$$

If the transformation is orthogonal (which is usually the case), then $T^*T = I$ and

$$l_c(\underline{R}) = \underline{R}^* \cdot \underline{W}$$

where $\underline{W}$ is the inverse transform of the embedded watermark. Therefore, even if the watermark is embedded in the transform coefficients, the correlation in Equation 2.1 is still applicable. However, the watermark $\underline{W}$ will have arbitrary values in the time domain even if $\underline{W}_c$ has binary values.

For this detector, the decision boundary is a hyperplane in the multi-dimensional space R^L. It is completely parameterized by a single orthogonal vector (of length L), which can be estimated by L independent points on the hyperplane. The displacement of the hyperplane is equivalent to the threshold γ. Therefore, it is not important in estimating the watermark, as it will only lead to a uniform scaling of the amplitude of each component of the watermark. As we shall see below, the proposed attack sequentially removes an adaptively scaled version of the estimated watermark from the signal. Hence, the exact value of the scale factor is not important in the attack.

If more points are available, a least square minimization can be applied to estimate the watermark. In the next section, we will provide techniques for estimating this boundary using the simple LMS algorithm.

Some variations of the basic correlator detector were discussed in Reference 5. However, the decision boundary of the modified detectors becomes a composite of hyperplanes. This new decision boundary, although more complicated, is still parametric and inherits the same security problem.

Asymmetric Detector

Asymmetric detectors were suggested to make the problem of estimating the secret key for unauthorized parties more difficult. In Reference 2, four asymmetric techniques were reviewed and a unified form for the detector was introduced. The general form of the test statistic is

$$l(\underline{R}) = \underline{R}^T \cdot A \cdot \underline{R} = \sum_{n,m} a_{n,m} r[n] \cdot r[m] \qquad (2.4)$$

The decision boundary in this case is a multidimensional quadratic surface. This is more complicated than the hyperplane but still can be completely specified using a finite number of parameters, namely $\{a_{n,m}\}_{n,m=1:L}$. However, in this case, we will need at least L^2 samples to estimate the boundary rather than L.

Quantization-Based Detector

Techniques for data embedding using quantization are based on using two or more codebooks to represent the different symbols to be embedded. These codebooks can be simply scalar quantization with odd and even quantization indices. The most common schemes for quantization-based watermarking are quantized index modulation (QIM) and quantized projection, which will be discussed in the following subsections.

Quantized Index Modulation. In this class, multiple codebooks are used for embedding [6]. The data are embedded by mapping segments of the host signal to the cluster region of two or more codebooks. Each codebook represents a data symbol. The decoding is done by measuring the quantization errors between each segment and all codebooks. The extracted data symbol is the one that corresponds to the codebook with minimum quantization error. In general, the decision boundaries for each data segment can be completely specified by the centroids of the codebooks [6].

One implementation of particular importance in (QIM) embedding is dither modulation [7]. The embedding process is parameterized by three sets of parameters: $\{\Delta_k\}_{k=1:L}$, $\{\underline{d_1}(k)\}_{k=1:L}$, and $\{\underline{d_2}(k)\}_{k=1:L}$, where Δ_k is the quantization step for the kth component, $\underline{d_1}$ and $\underline{d_2}$ are pseudorandom sequences that correspond to 0 and 1, respectively, and $|\underline{d_1}(k)|$, $|\underline{d_2}(k)| < \Delta_k/2$ for $k=1:L$. The host signal is segmented into segments of length L. One bit is embedded per segment. The kth component of the ith segment is modified according to [6]

$$\underline{y_i}(k) = q(\underline{x_i}(k) + \underline{d_i}(k)) - \underline{d_i}(k), \qquad k = 1, 2, \dots, L \qquad (2.5)$$

where $q(\cdot)$ is a scalar quantization using quantization steps Δ_k and $\underline{d_i} \in \{\underline{d_1}, \underline{d_2}\}$. As discussed in Reference 6, the decision boundaries are determined by the values of $\{\Delta_k\}$. In general, the boundaries are symmetric cells around the reconstruction points. The quantization cells and the reconstruction points of any quantizer are shifted versions of the quantization cells and the reconstruction points of the other quantizer. The decoding is done by the following steps:

1. Adding $\underline{d_1}$ and $\underline{d_2}$ to the modified segments.
2. Quantizing the resulting segments using scalar quantization of each component in the segment with quantization steps $\{\Delta_k\}_{k=1:L}$.
3. Measuring the overall quantization error of each segment in both cases.
4. The bit that corresponds to the smaller quantization error (at each segment) is decided.

For the above scheme, the embedding process can be undone if $\{\Delta_k,$ $\underline{d_1}(k),\ \underline{d_2}(k)\}_{k=1:L}$ are estimated correctly. In Section "Attack on Quantization-Based Schemes", we provide an attack that unveils these parameters and removes the watermark with minimum distortion.

Note that, in Reference 6, 2 $\underline{d_2}(k)$ is expressed in terms of $\underline{d_1}(k)$ and Δ_k to maximize the robustness. In this case,

$$\underline{d_2}(k) = \underline{d_1}(k) + \Delta_k/2 \qquad \text{if } \underline{d_1}(k) < 0$$

$$= \underline{d_1}(k) - \Delta_k/2 \qquad \text{if } \underline{d_1}(k) \geq 0$$

This further reduces the number of unknowns and, hence, reduces the estimation complexity.

Quantized Projection Watermarking. This scheme is based on projecting the signal (or some features derived from it) on an orthogonal set of vectors. The projection value is then quantized using scalar quantization and the quantization index is forced to be odd or even to embed 1 or 0, respectively [8,9]. The embedding relation for modifying each vector v of the host signal is [8].

$$\tilde{v} = v - Pv + B_i \cdot \Delta_i \cdot Q_i\left(\frac{B_i^H v}{\Delta_i}\right) \tag{2.6}$$

where Δ_i is the quantization step of the projection value onto the ith base vector B_i (the ith column of the orthogonal matrix), P is the projection matrix to the subspace of B_i (i.e., $P = B_i B_i^H$), and Q_i is the quantization to the nearest even integer and odd integers (according to the corresponding data bit). Different orthogonal sets can be used in Equation 2.6. For example, in Reference 8, the columns of the Hadamard matrix are used as the orthogonal set.

Note that if a single vector is used in Equation 2.6 for embedding data (i.e., one bit per block) rather than the whole orthogonal set, then this is equivalent to the *hidden QIM* scheme [6], where the QIM occurs in the domain of the auxiliary vector B_i.

The decoding process is straightforward. Each vector v is projected onto the base vector B_i, and the projection value is rounded with respect to the quantization step Δ_i. If the result is even, then the corresponding bit is '0', and if it is odd, the corresponding bit is '1'. Note that if the vector length is fixed, then the transformation between different orthogonal sets is linear. The decision boundary in this case is parameterized by the set $\{\underline{B}_i, \Delta_i\}$.

Comments

In this section, we analyzed the structure of the decision boundary of common watermark detectors. They share the common feature that the decision boundary at the detector (between the two hypotheses of the presence and absence of the watermark) is parameterized by a finite set of parameters. These parameters change according to the underlying test statistics. However, all of them can be estimated when the detector is publicly available, as will be discussed in the next section.

THE GENERIC ATTACK

Introduction

The decision boundaries of common watermarking schemes, which were discussed in the previous section, can be estimated using, for example, least square techniques if a sufficient number of points on the boundary is available. The estimation problem may be nonlinear in some cases, but in theory, it can still be applied and efficient techniques can be proposed to optimize the estimation. This is easy when the attacker has unlimited access to the detector device even as a black box.

In this case, the attacker can make *slight* changes to the watermarked signal until reaching a point at which the detector is not able to detect the watermark. This point can always be reached, because a constant signal cannot contain a watermark. Therefore, by repeatedly replacing the samples of the watermarked signal by a constant value, the detector will respond negatively at some point. At this point, the attacker can go back and forth around the boundary until identifying a point on it with the required precision. Generating a large number of these points is sufficient for estimating the decision boundary for the cases discussed in the previous section.

Once the boundary is specified, any watermarked signal can be projected to the nearest point on the boundary to remove the watermark with the *smallest distortion*. For correlation-based detector, the estimation of the decision boundary is equivalent to estimating the watermark itself. Hence, the projection is equivalent to successively subtracting components of the watermark from the watermarked signal until the detector responds negatively. The attack idea is illustrated in Figure 2.1.

The boundary estimation for both correlator and asymmetric detectors is a linear estimation problem in terms of the unknown parameters, which are $\{w[n]\}_{n=1:L}$ for the correlation detector and $\{a_{n,m}\}_{m,n=1:L}$ for the asymmetric detector. In both cases, the problem can be solved using least square techniques. However, for large-sized data, this may be expensive for an average attacker. For example, for an image of size 512×512, each data vector for the correlator detector is of size $512^2 = 262{,}144$, and at least 512^2 independent vectors are needed for the least square problem to be well conditioned. The direct least square will require inversion of a matrix of size $512^2 \times 512^2$, which may be expensive. Even if recursive least square (RLS) is employed, it will require manipulation of a matrix of size $512^2 \times 512^2$ at each iteration step, which may be also impractical for the average attacker. However, the workhorse for adaptive filtering, namely the least mean square (LMS)

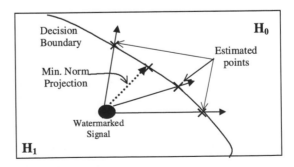

Figure 2.1. The generalized attack.

algorithm works extremely well in this case and gives satisfactory results. In the next subsection, we give an effective implementation of the generalized attack using the simple LMS algorithm for the class of correlation detectors.

For quantization-based embedding, the attack is simpler. Simple statistical analysis is sufficient to estimate the unknown parameters without need to recursion. The attack on quantization-based scheme is introduced in Section "Attack on Quantization-Based Schemes".

Attack on Linear Detectors

The problem of estimating the decision boundary for the correlator detector fits the system model of classical adaptive filtering, which is shown in Figure 2.2. In this model, the system tries to track the reference signal $d(t)$ by filtering the input $r(t)$ by the adaptive filter $w(t)$. The optimal solution for this problem for stationary input is the *Weiner* filter, which is impractical to implement if the input statistics are not known *a priori*. The LMS algorithm [10] approximates Weiner filtering and the basic steps of the algorithm are:

1. Error calculation: $e(t) = d(t) - r^*(t) \cdot w(t)$
2. Adaptation: $w(t+1) = w(t) + \mu \cdot r(t) \cdot e^*(t)$

For algorithm stability, the adaptation step μ should be less than $2/(\Sigma_i \lambda_i)$, where $\{\lambda_i\}_{i=1:L}$ are the eigenvalues of the data covariance matrix. A conservative value of μ that is directly derived from the data and guarantees stability is $\mu = 2/(L \cdot P)$, where P is the signal power.

First, we will describe the generalized attack if the attacker has knowledge of the correlation value in Equation 2.2 for every input. Later, this assumption will be relaxed to study the more practical case.

For this problem, $r(t)$ are the different samples of the watermarked signal after modification, $d(t)$ are the corresponding correlation values (output of the detector), and $w(t)$ is the estimated watermark at time t. It is important to have the input vectors as uncorrelated as possible. If we have a single watermarked signal, the different samples that are derived

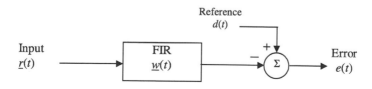

Figure 2.2. Adaptive filtering topology.

from it and are reasonably uncorrelated can be obtained by changing a random number of the signal values in a random order. For images, this is done by reordering the image blocks (or pixels) in a random fashion and replacing the first K blocks (where K is a random integer) by the image mean. If the watermark is binary, then the estimated coefficients should be approximated to binary values by taking the sign of each component. If the watermark is embedded in some transform coefficients, then the watermark values are not approximated. In Figure 2.3, we give an example of the convergence of the binary watermark components for the mandrill image of size 128×128. It represents the evolving of the matching between the estimated watermark and the actual watermark with the number of iterations. It should be noted that the shown convergence curve is obtained using a single watermarked signal. If more than one signal is employed in estimating the watermark, better convergence is obtained (see Figure 2.5).

After estimating the decision boundary, the watermark can be removed by successively subtracting the estimated watermark from the watermarked signal until the detector responds negatively. If the whole watermark components are subtracted without reaching the boundary, then they are subtracted again. This is typical if the strength of the embedded watermark is greater than unity. In Figure 2.4, an example of the projection is given for the watermarked *Pepper* image when the threshold is half-way between the maximum correlation (of the correct random sequence) and the second maximum correlation (of all other random sequences). We show the ideal case when the watermark is known exactly and the cases when only 90% and 80% of the watermark is recovered correctly. From Figure 2.4, it is noticed that the difference is very small among the three cases. However, with large errors in estimating the watermark (e.g., if the matching is less than 70%), the projection

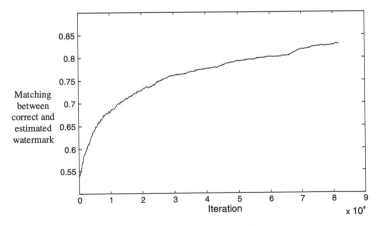

Figure 2.3. Convergence example of the LMS realization.

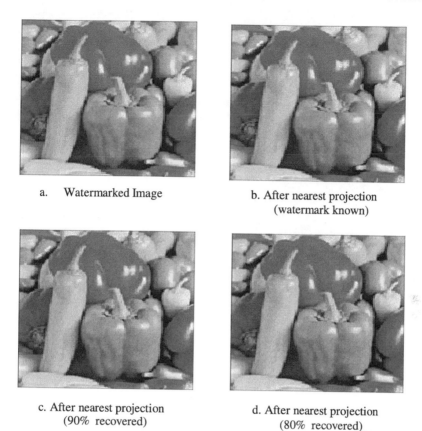

a. Watermarked Image

b. After nearest projection
(watermark known)

c. After nearest projection
(90% recovered)

d. After nearest projection
(80% recovered)

Figure 2.4. Example of watermark removal with minimum distortion.

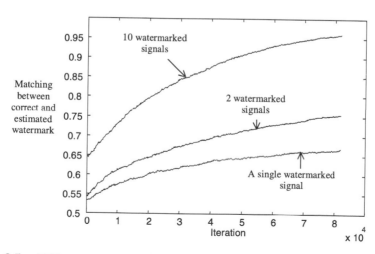

Figure 2.5. LMS convergence when only the detector decisions are observed.

becomes noticeable because subtracting the erroneous components boosts the correlation in Equation 2.2 and it needs more iterations to be canceled.

Now, we consider the practical case when the attacker does not have access to the internal correlation values but can only observe the detector decisions. In this case, the only points that can be determined accurately are the ones close to the boundary at which the detector changes its decision. As mentioned earlier, these points can be estimated by successively replacing the signal samples by a constant value until reaching a point at which the detector responds negatively. Starting from this point, small variations of the signal will cause the detector to go back and forth around the boundary until reaching a point on the boundary with the required precision.

The main problem in this case is the high correlation between the different modified samples of the same watermarked signal in the neighborhood of the decision boundary. This slows down the convergence significantly. However, convergence can be accelerated by employing multiple watermarked signals, and this is not a problem for public detectors. In Figure 2.5, we give a convergence example when a 1, 2, and 10 watermarked signals (all are images of size 128×128) are used for estimating the decision boundary.

The correlation threshold γ may be unknown to the attacker as well. However, if she uses any positive estimate of it, this will do the job because it will result in a scaled version of the watermark. The progressive projection that is described earlier will take care of the scaling factor.

A third possible scenario at the detector is when the detector decision is associated with a confidence measure about its correctness. This confidence measure is usually a monotonic function of the difference between the correlation value and the threshold. In principle, the attacker can exploit this extra information to accelerate the attack. Denote the confidence measure by η. In general we have

$$\eta = f(l(\underline{R}) - \gamma)$$

where $l(\underline{R})$ is the correlation value and γ is the detection threshold. Assume, without loss of generality, that η is negative when the watermark is not detected, that is, it is a confidence measure of negative detection as well. We consider the worst-case scenario when the attacker does not know the formula of the confidence measure. In this case, the attacker can view η as an estimate of the correlation value and proceed as if the correlation value is available. However, it was found experimentally that the threshold γ must be estimated along with the watermark to accelerate the convergence. Hence, the LMS iteration of the attack is modified as follows:

1. Calculate the confidence measure $\eta(t)$ of the image instance.
2. Error calculation: $e(t) = \eta(t) - [\underline{r}^*(t) \cdot \underline{w}(t) - \gamma(t)]$
3. Adaptation: $[\underline{w}(t+1), \gamma(t+1)] = [\underline{w}(t), \gamma(t)] + \mu \cdot e^*(t) \cdot [\underline{r}(t), -1]$

Note that, we do not need to find points exactly on the decision boundary because the confidence measure can be computed for any image, not only the instances on the decision boundary. This significantly accelerates the algorithm.

The above attack works well when the confidence measure is *linear* with the correlation. A nonlinear confidence measure does not fit the linear model of the LMS and results in substantial reduction in the algorithm speed. In Figure 2.6, we show the algorithm convergence when 10 watermarked images are used. The nonlinear function in the figure is quadratic in the difference $l(\underline{R}) - \gamma$. By comparing Figures 2.5 and 2.6, we note that the algorithm is accelerated only with the linear confidence measure. For the nonlinear confidence measure, it is better to ignore the confidence measure and run the attack with points only on the decision boundary, as discussed earlier.

If the watermark is embedded in the transform coefficients, the watermark will not be, in general, binary in the signal domain. In this case, the actual estimated values of the watermark components are used rather than their signs. However, if the attacker knows which transform

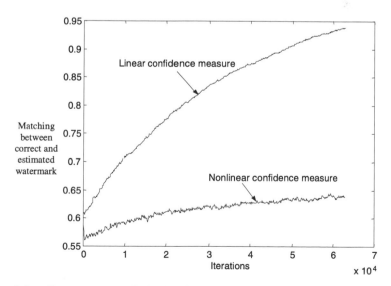

Figure 2.6. Convergence of the LMS attack when a confidence measure is available.

is used, then the above algorithm can be applied to the transformed signal rather than the signal.

Although we have used images to illustrate the attack, the algorithm discussed in this section works as well with audio and video watermarking. If the watermark is embedded in blocks, there is no need to get the blocks boundaries, as we assume that the watermark exists at each individual sample of the watermarked signal.

Attack on Quantization-Based Schemes

The basic idea of the attack on quantization-based schemes is to estimate the entries of the different codebooks and use them to remove the watermark with *minimum* distortion. It turns out that, from the attacker viewpoint, both quantization-based schemes are similar and do not need special consideration, as will be discussed in the following subsections.

Attack on QIM Scheme. The objective of the attack is to estimate Δ_k, $\underline{d}_1(k)$, and $\underline{d}_2(k)$ for $k = 1, 2, \ldots, L$, and then use the estimated parameters to remove the watermark with the minimum distortion.

First, we assume that the segment length L is known. Later, we will discuss how to estimate it. At the decoder, we have the modified segments in the form of Equation 2.5. If the correct quantization step Δ_k is used, then the quantization error for the kth component will be either $\underline{d}_1(k)$ or $\underline{d}_2(k)$. Hence, the histogram of the quantization error is concentrated around the correct value of $\underline{d}_1(k)$ and $\underline{d}_2(k)$. Based on this remark, the attack can be performed as follows:

1. Define a search range $[0, \Delta_{M(k)}]$ for the quantization step Δ_k. This range is determined by the maximum allowable distortion $\Delta_{M(k)}$ of each component.
2. Within this range, search is done for the best quantization step by performing the following steps for each candidate value of Δ_k:

    a. Each quantization step candidate is used to quantize the kth component of each segment. The quantization error is calculated. Positive quantization errors are accumulated together and negative quantization errors are accumulated together.

    b. For each quantization step candidate, the standard deviations of the positive and negative quantization errors are calculated.

    c. The best quantization step is the one that minimizes the sum of the standard deviations of the positive and negative quantization errors. A typical plot of the sum of the standard deviation for different values of Δ is shown in Figure 2.7.

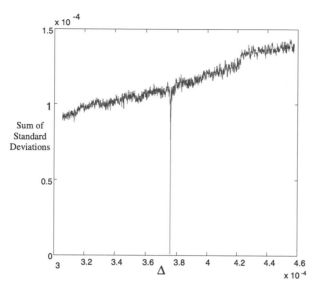

Figure 2.7. Example of Δ search.

3. After estimating Δ_k, $\underline{d}_1(k)$ and $\underline{d}_2(k)$ are simply the means of the positive and negative quantization errors, respectively. Note that, the signs of $\underline{d}_1(k)$ and $\underline{d}_2(k)$ are chosen such that the different components of $\underline{d}_1$ or $\underline{d}_2$ within each segment are consistent.

The estimation of L follows directly from the analysis of the quantization error. Note that for any quantization step Δ (even if it is incorrect), the quantization error pattern of similar components at different segments is quite similar. This is basically because the reconstruction points of these segments are discrete. Recall that the kth component of each segment is modified according to Equation 2.5. Assume that the quantization function with step Δ is $G(\cdot)$. The quantization error of the kth component after using Δ is

$$\underline{e}_i(k) = G(q(\underline{x}_i(k) + \underline{d}_i(k)) - \underline{d}_i(k)) - q(\underline{x}_i(k) + \underline{d}_i(k)) + \underline{d}_i(k) \qquad (2.7)$$

Let $z_k = G(q(\underline{x}_i(k) + \underline{d}_i(k)) - \underline{d}_i(k)) - q(\underline{x}_i(k) + \underline{d}_i(k))$. If $\underline{d}_i(k)$, $\Delta \ll \underline{x}_i(k)$, then z_k is approximately distributed as an uniform random variable (i.e., $z_k \sim U[-\Delta/2, \Delta/2]$). Hence, the autocorrelation function (ACF) of the quantization error at lags equal to a multiple of L will be approximately

$$\mathrm{ACF}_{e(k)}\ (nL) \approx (1/2) \sum_k (\underline{d}_1(k) + \underline{d}_2(k)) \qquad (2.8)$$

and the autocorrelation is almost zero elsewhere. This is illustrated in Figure 2.8 where we plot the ACF of the quantization error of a part of an audio signal when $\underline{d}_1$ has random amplitude and $|\underline{d}_1(k) - \underline{d}_2(k)| = \Delta_k/2$.

77

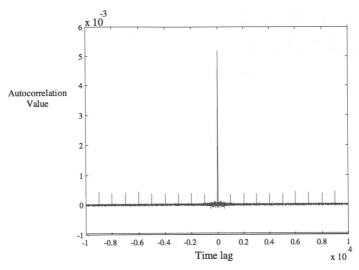

Figure 2.8. **Autocorrelation of the quantization error with incorrect Δ and $L = 1000$ [d_1 is arbitrary and $[\,|\,d_1(k) - d_2(k)\,| = \Delta_k/2]$.**

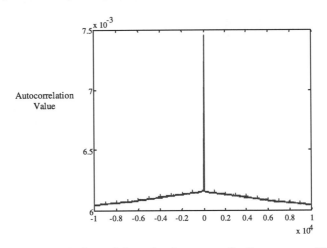

Figure 2.9. **Autocorrelation of the *absolute* quantization error with incorrect Δ and $L = 1000$ [$d_1(k), d_2(k)\,| = \pm \Delta_k/4]$.**

However, if $\underline{d}_1(k), \underline{d}_2(k) = \pm \Delta_k/4$, then the peaks at lags nL are much weaker. This can be treated by computing the ACF of the *absolute* quantization error rather than the quantization error itself. This is illustrated in Figure 2.9, where the peaks at multiple of nL are clear although weaker than the case with a dither vector with arbitrary amplitude.

Note that the procedure for estimating the key parameters of QIM embedding is not recursive. Hence, it is simpler than that of the correlation detector.

The removal of the watermark with minimum distortion after estimating the unknown parameters is straightforward. Note that $\underline{d}_1(k)$, $\underline{d}_2(k) \in [-\Delta_k/4, \Delta_k/4]$. For the specific implementation in reference 6, we have $|\underline{d}_1(k) - \underline{d}_2(k)| = \Delta_k/2$. The idea of the attack is to increase the quantization error for the correct dither vector (which corresponds to decreasing the quantization error for the incorrect dither vector). Let the kth component of the modified segment be y_k. The modified value after the attack is

$$z_k = y_k \pm \delta_k, \qquad k = 1, 2, \ldots, L \tag{2.9}$$

where δ_k is a positive quantity slightly larger than $\Delta_k/4$. Note that adding or subtracting δ_k is equally good, as it makes the quantization error of the incorrect dither vector smaller.

Rather than modifying the whole segment, it is sufficient to modify some components such that the quantization error of the incorrect dither vector is less than the quantization error of the correct dither vector. This can be achieved if a little more than half of the components of each segment is modified according to Equation 2.9. Moreover, not all of the segments are modified. The segments are modified progressively until the detector responds negatively.

In Figure 2.10, we give an example of the attack. In Figure 2.10, dither modulation is used to modify the 8×8 DCT coefficients of the image. The quantization step of each coefficient is proportional to the corresponding JPEG quantization step. The PSNR of the watermarked image is 40.7 dB. The attack as described above is applied with each component perturbed by an amount δ_k, whose amplitude is slightly larger than $\Delta_k/4$. If we assume that the detector fails when half of the extracted bits are incorrect, then only half of the blocks need to be modified. The

a. Watermarked Image   b. After removing watermark (PSNR = 44 dB)

Figure 2.10.   Example on attack on the QIM scheme.

resulting image is shown in Figure 2.10 and it has PSNR of 44 dB (compared to the watermarked image). In this attack, the watermark can be completely removed or it can be partially damaged to the degree that the detector cannot observe it.

Attack on the Quantized Projection Scheme. The projection values of a signal on a set of orthogonal vectors are equivalent to changing the basis for representing each vector from the standard set (which is the columns of an identity matrix) to the new orthogonal set. If the projection values are quantized according to Equation 2.6, then the reconstruction points will be discrete and uniformly spaced. Discrete reconstruction points of the projection values result in discrete reconstruction points in the signal domain. However, the quantization step in the signal domain is variable. It changes from one block to another according to the embedded data, as will be discussed.

Consider a projection matrix B. A vector of the projection values (of the ith block) $\underline{y}_i$ is related to the signal samples $\underline{x}_i$ by $\underline{y}_i = B \cdot \underline{x}_i$. Equivalently, we have $\underline{x}_i = B^T \underline{y}_i$. If the original quantization step vector is $\underline{q} = [\Delta_1 \cdots \Delta_L]$, then the discrete projection values will have the general form

$$\underline{y}_i = D_i \cdot \underline{q} \tag{2.10}$$

where D_i is a diagonal matrix with the quantization indices (that correspond to the embedded data) on its diagonal. In this case the reconstructed signal is $\underline{x}_i = B^T \cdot D_i \cdot \underline{q}$. If we assume that $\underline{x}_i$ is quantized by another quantization vector $\underline{q}' = [\Delta_1' \cdots \Delta_L']$, then we have $\underline{x}_i' = D_i' \cdot \underline{q}'$; that is,

$$D_i' \cdot \underline{q}' = B^T \cdot D_i \cdot \underline{q} \tag{2.11}$$

Now, the objective of the attack is to find $\underline{q}'$ and remove the watermark with the minimum distortion. The problem is that $\underline{q}'$ is data dependent, as noted from Equation 2.11. It has to be estimated within each block and this is difficult in general. However, if $\underline{q}'$ is averaged over all blocks, the average quantization vector $\underline{q}'_{av}$ will be a good approximation, as will be discussed. Note that the all the entries of $\underline{q}$ are positive. However, the entries of $\underline{q}'$, in general, may have negative signs. Estimating the signs of the quantization steps in the signal domain is necessary to be able to change the quantization values correctly in the transform coefficients.

The estimation of the block length is the same as described in the previous subsection because of the similarity in the distribution of the quantization noise at the corresponding samples of each block. This is illustrated in Figure 2.11, where we plot the ACF of the quantization error in the signal domain while the data are embedded by quantizing

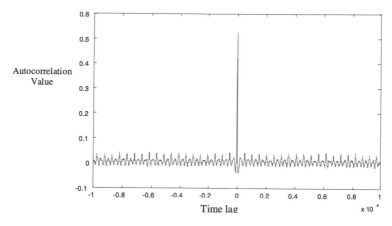

Figure 2.11. ACF with quantized projection embedding.

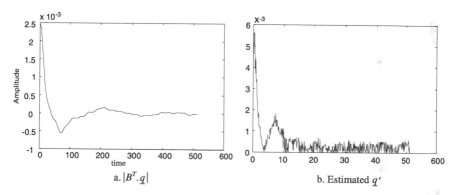

a. $|B^T \cdot \underline{q}|$ b. Estimated $\underline{q}'$

Figure 2.12. Optimum and estimated values of the average q'.

the DCT coefficients of blocks of size $L = 512$. The peaks at multiple of L are clear. This is analogous to Figure 2.8 for the QIM scheme.

The estimation of the average quantization step is the same as estimating Δ in the previous subsection. It is interesting to note that the relative amplitudes of the average quantization step is quite similar to $B^T \cdot q$, as illustrated in Figure 2.12, where we plot the average quantization step in the time domain and $B^T \cdot q$. In this example, the data are embedded in the first 40 DCT coefficients (with $L = 512$).

The estimation of the signs of $\underline{q}'$ is done by analyzing the signs of pairs of its entries. Note that we are interested in the *relative* signs of the coefficients rather than their absolute values. If the signs are estimated in reverse order (i.e., positive instead of negative, and the reverse), it will have the same effect of moving the quantization values in the transform domain to the adjacent slot. The idea of the estimation is to study the type of superposition between pairs of q' components. If the super-position is constructive, then the signs of the pair elements are the same and vice versa. First, consider the case with one embedded bit per segment (i.e., the data are projected onto one direction) and the projection value is quantized. The estimation of the signs proceeds as follows:

1. Select one coefficient in the signal domain and assume that its sign is positive.
2. Change the value of this coefficient by a sufficient amount δ such that the detector changes its decision.
3. Return the sample to its original value and then perturb it by $\delta/2$.
4. Perturb each other entry by $\delta/2$ (one at a time). If the perturbation results in changing the detector decision to its original value, then the sign of the corresponding entry is negative, otherwise it is positive.

The extension to multibit embedding is straightforward. The above procedure is applied for each embedded bit in the segment and the majority rule is applied to decide the sign of each component. In Figure 2.13, we give an example of the sign algorithm for the scheme in Figure 2.12.

After estimating the quantization step in the signal domain, the watermark is removed by perturbing each segment in the signal domain by $\pm q'_{av}/\alpha$. In Figure 2.14, we show typical results of the percentage damage of the watermark (i.e., the fraction of bits that are decoded incorrectly) vs the SNR of the output signal (after the attack). In this simulation, the SNR of the watermarked signal is 31.5 dB. Note that modifying the segments significantly results in moving to slots that match

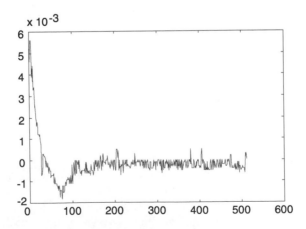

Figure 2.13. **Estimated q' after sign correction.**

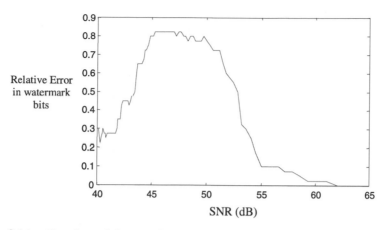

Figure 2.14.   Fraction of damaged watermark vs the output SNR.

the original slot. This results in decreasing the error rate, as noted in Figure 2.14.

Note that if multiple codebooks are used for data embedding (as described in Reference 6), with the entries of the codebooks replacing the original data segments, then the estimation is much simpler. In this case, each codebook can be constructed by just observing the segments of watermarked data and the detector decisions. Once all codebooks are estimated, the embedded data can be simply modified by moving each segment to another codebook.

Comments on Attacks on Quantized-Based Schemes. The proposed attack on quantization-based schemes is simpler than the LMS attack for linear detectors because it does not iterate. We assumed that, the data are embedded regularly in blocks of fixed size L. This is the typical scheme proposed in the original works that employ quantization for embedding (e.g., References 6 and 9). However, if the block boundaries are varying, then the attacker may use multiple watermarked items and perform the same statistical analysis on the different items. In this case, a unique quantization step is assumed for each sample and it is evaluated from the different watermarked items rather than blocks.

Discussion

The problem discussed in this subsection and the previous one motivated the search for decision boundaries that cannot be parameterized. For such boundaries, even if the attacker can change individual watermarked signals to remove the watermark, the modification will be *random* and the minimum distortion modification cannot be attained as earlier. The only choice for the attacker is to try to approximate the boundary numerically. This will require extra cost and the projection will be extremely difficult.

SECURE DETECTOR STRUCTURE

Introduction

In this section, we describe the structure of the decoder with nonparametric decision boundary. The new decision boundary is generated by changing the original decision boundary to a fractal curve (or any other nonparametric curve). Clearly, the resulting decoder is suboptimal. However, we will show that the degradation in the performance is slight.

In the next subsection, we give an overview of the suggested decoder structure. Then, we give a brief review of fractals generation and we provide a practical implementation procedure of the proposed decoder. Next, we show that the performance of the proposed detector is similar to the original detector. Finally, we establish the robustness of the proposed detector against the attack proposed in the previous section.

Algorithm

Rather than formulating the test statistic in a functional form, it will be described by a nonparameterized function. We select fractal curves to represent the new decision boundary. The fractal curves are selected because they are relatively simple to generate and store at the decoder. Random perturbation of the original decision boundary can also be used. However, in this case, the whole boundary needs to be stored at the decoder rather than the generation procedure in the fractal case.

Instead of the test statistic of Equations 2.2 and 2.4, we employ a fractal test statistic whose argument is R and has the general form

$$f(\underline{R}) > \text{threshold} \qquad (2.12)$$

where $f(\cdot)$ is a fractal function. The basic steps of the proposed algorithms are:

1. Start with a given watermarking algorithm and a given test statistic $f(x)$ (e.g., a correlation sum 2.2). The decision is positive if $f(x) > \gamma$.
2. Fractalize the boundary using a fractal generation technique, which will be discussed in the next subsection.
3. Use the same decision inequality but with a new test statistic using the new fractal boundary.
4. (Optional at the encoder) Modify the watermarked signal if necessary to assure that the *minimum distance* between the watermarked signal and the decision boundary is preserved.

Fractal Generation

This class of curves [11] has been studied for representing self-similar structures. Very complicated curves can be constructed using simple repeated structures. The generation of these curves is done by repetitively replacing each straight line by the generator shape. Hence, complicated shapes can be completely constructed by the initiator (Figure 2.15a), and the generator (Figure 2.15b). Different combinations of the generator and the initiator can be employed to generate different curves. This family of curves can be used to represent statistically self-similar random processes (e.g., the Brownian random walk [12, chap. 11]).

The family has the feature that it cannot be described using a finite set of parameters if the initiator and the generator are not known. Hence, they are good candidates for our particular purpose. Also, a Brownian motion process or a random walk can be used as well, but in this case, the whole curve should be stored at the detector.

Modifying the Decision Boundary

The most important step in the algorithm is modifying the decision boundary to have the desired fractal shape. In Figure 2.16, we give an example with a decision boundary of the correlator in R^2. The original boundary is a line, and the modified decision boundary is as shown in Figure 2.16.

There is a trade-off in designing the decision boundary. The maximum difference between the old decision and the modified one should be

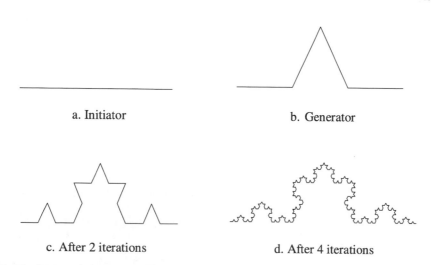

a. Initiator   b. Generator

c. After 2 iterations   d. After 4 iterations

Figure 2.15. Fraction example.

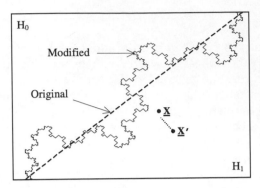

Figure 2.16. Example of modifying the decision boundary.

large enough so that the new boundary cannot be approximated by the old one. On the other hand, it should not be very large in order to avoid performance degradation.

After modifying the decision boundary, the watermarked signal X may need some modification to sustain the same shortest distance from X to the decision boundary. This is done by measuring the shortest distance between X and the new boundary and moving X along the direction of the shortest distance in the opposite direction (from X to X'). However, this modification is not critical for the performance, especially if the variance of the test statistic is small (which is usually the case) or if the distance between X and the original boundary is large compared to the maximum oscillation of the fractal curve.

Instead of evaluating the log-likelihood function of Equations 2.2 and 2.4, the fractal curve is evaluated for the received signal and is compared with the same threshold (as in Equation 2.10) to detect the existence of the watermark.

Practical Implementation

Instead of applying multidimensional fractalization, a simplified practical implementation that achieves the same purpose is discussed in this subsection. If the Gaussian assumption is adopted, then the test statistic in Equation 2.2 is optimal according to the Neyman–Pearson theorem [3, chap. 3]. Instead of this test statistic, two test statistics are used for the even and odd indexed random subsequences $r[2k]$ and $r[2k+1]$ (or any two randomly selected subsequences), respectively:

$$T_1(\underline{R}) = (2/L) \cdot \sum_k r[2k] \cdot w[2k], \qquad T_2(\underline{R}) = (2/L) \cdot \sum_k r[2k+1] \cdot w[2k+1]$$

$$\underline{T} = (T_1 \, T_2) \tag{2.13}$$

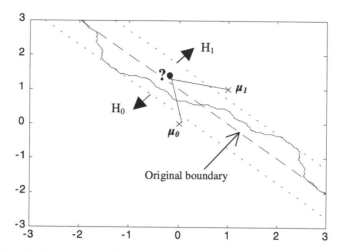

Figure 2.17. Detector operation.

Under H_0, $E(\underline{T}) = (0,0)$, and under H_1, $E(\underline{T}) = (1,1)$ if the watermark strength is assumed to be unity. In both cases, $\text{cov}(\underline{T}) = (2\sigma^2/L)I$, where σ^2 is the average of the variances of the signal coefficients. The Gaussian assumption of both T_1 and T_2 is reasonable if L is large by invoking the central limit theorem. Moreover, if the original samples are mutually independent, then T_1 and T_2 are also independent. The decision boundary in this case is a line (with slope -1 for the given means). If this line is fractalized, as discussed in the previous subsection, then the corresponding decision boundary in the multidimensional space will be also nonparametric.

The detection process is straightforward in principle but nontrivial. The vector $\underline{T}$ is classified to either hypothesis if it falls in its partition. However, due to the fractal nature of the boundary, this classification is not trivial. First, an unambiguous region is defined as shown in Figure 2.17 (i.e., outside the maximum oscillation of the fractal curve, which are the regions outside of the dotted lines). For points in the ambiguous area, we extend two lines between the point and means of both hypotheses. If one of the lines does not intersect with the boundary curve, then it is classified to the corresponding hypothesis. It should be emphasized that the boundary curve is stored at the detector and it should be kept secret.

Algorithm Performance

The technique proposed in the previous subsection is quite general. It can be applied to any watermarking scheme without changing the embedding algorithm. The algorithm performance is, in general, similar to

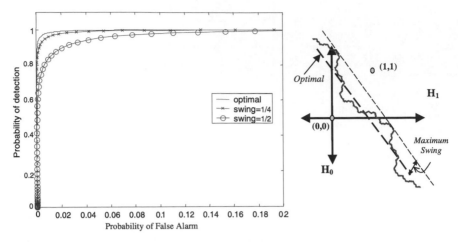

Figure 2.18. ROC of the proposed algorithm.

the performance of the underlying watermarking algorithm with its optimal detector. However, for some watermarked items, we may need to increase the watermark strength, as illustrated in Figure 2.16.

Note that, in practice, the variance of the two statistics is small, so that the performances under the old and the new decision boundaries are essentially the same. For example, for an image of size 512×512 and 8 bits/pixel, if we assume a uniform distribution of the pixel values, then the variance of both statistics will equal to 0.042 according to Equation 2.3. Hence, the joint pdf of the two statistics is very concentrated around the means of both hypotheses. Consequently, the degradation in the performance after introducing the new boundary is slight.

The receiver operating characteristic (ROC) of the proposed algorithm is close to the ROC of the optimal detector, and it depends on the maximum *oscillation* of the fractal boundary around the original one. In Figure 2.18, we illustrate the performance of the detector discussed in the previous subsection, when the mean is $(0, 0)$ under H_0 and is $(1, 1)$ under H_1, and the variance for both hypotheses is 0.1. As noticed from Figure 2.18, the performance of the system is close to the optimal performance, especially for a small curve oscillation. The degradation for a large perturbation is minor. This minor degradation can be even reduced by slightly increasing the strength of the embedded watermark.

Attacker Choices

In the discussion in Section "The Generic Attack", the estimation of the watermark is equivalent to the estimation of the decision boundary for correlator detector. After introducing the new algorithm, the two estimations are decoupled. The attacker has to *simultaneously* estimate the watermark and the nonparametric boundary. This makes the estimation process prohibitively complex.

The practical choice for the attacker is to assume that the fractal oscillation is so small that the new boundary can be approximated by a hyperplane and proceed as earlier. We tested the attack described in section "Attack on Linear Detectors" with the same encoder that is used in generating Figure 2.5 but using the new decision boundary. The points on the decision boundary were estimated in exactly the same way as earlier. We used the same 10 watermarked signals as the ones used in Figure 2.5. We used a fractal curve with a maximum swing of 0.25. The learning curve of the attack is *flat* around 67%, as shown in Figure 2.19. Comparing with Figure 2.5, it is clear that the attack is ineffective with the new algorithm. No information about the watermark is obtained. This even makes the estimation of the decision boundary more difficult.

To conclude the chapter, we compare the performance of the LMS attack when the correlator and fractal detectors are used. The comparison is in terms of the mean square error (MSE) between the watermarked image and the image after the attack. In Figure 2.20, we plotted the MSE of the modified image vs the number of iterations of the LMS attack. The same encoder parameters are used for both algorithms and the same number of images (10 images) is used for estimating the watermark. Note that the MSE for the fractal detector is higher because the new boundary is such that the *average* distance between the boundary and the centroid corresponding to H_1 increases. Note also that

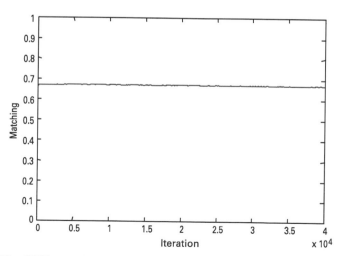

Figure 2.19. **LMS convergence for a fractal boundary with 10 watermarked images.**

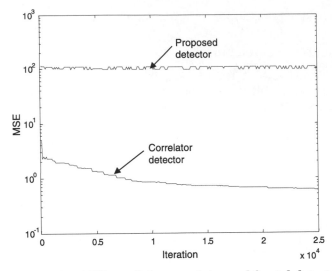

Figure 2.20. **MSE after LMS attack for correlator and fractal detectors.**

the MSE of the fractal detector is almost constant as expected from the learning curve in Figure 2.19.

DISCUSSION

In this work, we analyzed the security of watermarking detection when the detector is publicly available. We described a generalized attack for estimating the decision boundary and give different implementations of it. Next, we proposed a new class of watermark detectors based on a nonparametric decision boundary. This detector is more secure than traditional detectors especially when it is publicly available. The performance of the new algorithm is similar to the traditional ones because of the small variance of the test statistics.

The proposed detector can work with any watermarking schemes without changing the embedder structure. However, with some watermarking procedures, the strength of the watermark may need to be increased slightly to compensate for the new boundary.

REFERENCES

1. Furon, T. and Duhamel, P., Robustness of Asymmetric Watermarking Technique, in Proceedings of IEEE International Conference on Image Processing, 2000, Vol. 3, pp. 21–24.
2. Furon, T., Venturini, I., and Duhamel, P., An Unified Approach of Asymmetric Watermarking Schemes, in Proceedings of SPIE Conference on Security and Watermarking Multimedia Contents, 2001, pp. 269–279.

3. Kay, S., *Fundamentals of Statistical Signal Processing: Detection Theory*, Prentice-Hall, Englewood Cliffs, NJ, 1998.
4. Kolmogorov, A. and Fomin, S., *Introductory Real Analysis*, Dover, New York, 1970, chap. 4.
5. Mansour, M. and Tewfik, A. Secure Detection of Public Watermarks with Fractal Decision Boundaries, in Proceedings of XI European Signal Processing Conference, EUSIPCO-2002, 2002.
6. Chen, B. and Wornell, G., Quantization Index Modulation, a class of provably good methods for digital watermarking and information embedding, *IEEE Trans. Inform. Theory*, 47(4), 1423–1443, 2001.
7. Chen, B. and Wornell, G., Dither Modulation: A New Approach to Digital Watermarking and Information Embedding, in Proceedings of SPIE Conference on Security and Watermarking Multimedia Contents, 999, pp. 342–353.
8. Lan, T., Mansour, M., and Tewfik, A. Robust high capacity data embedding, *Proc. ICIP*, 2000.
9. Swanson, M., Zhu, B., and Tewfik, A., Data hiding for video-in-video, *Proc. ICIP*, 1996.
10. Haykin, S., *Adaptive Filter Theory*, Prentice-Hall, Englewood Cliffs, NJ, 1995.
11. Mandelbrot, B., *The Fractal Geometry of Nature*, W.H. Freeman, San Francisco, 1983.
12. Papoulis, A., Probability, Random Variables, and Stochastic Processes, 3rd ed., McGraw-Hill, New York, 1991.
13. Abramowitz, M. and Stegun, A., *Handbook of Mathematical Functions*, Dover, New York, 1972.
14. Cox, I. and Linnartz, J., Public Watermarks and Resistance to Tampering, in Proceedings of IEEE International Conference in Image Processing, ICIP, 1997, vol. 3, pp. 3–6.
15. Hernandez, J.R., Amado, M., and Perez-Gonzalez, F., DCT-domain watermarking techniques for still images: Detector performance analysis and a new structure, *IEEE Trans. on Image Process.*, 9(1), 55–68, 2000.
16. Kalker, T., Linnartz, J., and Dijk, M., Watermark Estimation Through Detector Analysis, in Proceedings of IEEE International Conference on Image Processing, ICIP, 1998, Vol. 1, pp. 425–429.
17. Linnartz, J. and Dijk, M. Analysis of the Sensitivity Attack Against Electronic Watermarks in Images, in Proceedings of 2nd International Workshop on Information Hiding, 1998, pp. 258–272.
18. Mansour, M. and Tewfik, A., Techniques for Data Embedding in Image Using Wavelet Extrema, in Proceedings of SPIE on Security and Watermarking of Multimedia Contents, 2001, pp. 329–335, 2001.
19. Press. W., et al., *Numerical Recipes in C*, Cambridge University Press, Cambridge, 1992.

Part II
Multimedia Encryption

3
Fundamentals of Multimedia Encryption Techniques

Borko Furht, Daniel Socek, and Ahmet M. Eskicioglu

INTRODUCTION

The recent advances in the technology, especially in the computer industry and communications, allowed a potentially enormous market for distributing digital multimedia content through the Internet. However, the proliferation of digital documents, multimedia processing tools, and the worldwide availability of Internet access have created an ideal medium for copyright fraud and uncontrollable distribution of multimedia content [1]. A major challenge now is the protection of the intellectual property of multimedia content in multimedia networks.

To deal with the technical challenges, two major multimedia security technologies are being developed:

1. Multimedia encryption technology to provide end-to-end security when distributing digital content over a variety of distribution systems

2. Multimedia watermarking technology as a tool to achieve copyright protection, ownership trace, and authentication

In this chapter, we present the current research efforts in multimedia encryption, whereas Chapter 7 gives an overview of multimedia watermarking techniques.

OVERVIEW OF MODERN CRYPTOGRAPHY

The three most important objectives of cryptography with respect to the multimedia information security include (1) confidentiality, (2) data integrity, and (3) authentication [2].

Confidentiality refers to the protection of information from unauthorized access. An undesired communicating party, called adversary, must not be able to access the communication material. *Data integrity* ensures that information has not been manipulated in an unauthorized way. Finally, *authentication* methods are studied in two groups: *entity* authentication and *message* authentication. Message authentication provides assurance of the identity of the sender of a message. This type of authentication also includes evidence of data integrity because if the data are modified during transmission, the sender cannot be the originator of the message. Entity authentication assures the receiver of a message of both the identity of the sender and his active participation.

Modern cryptographic techniques provide solutions for these three objectives. In general, there are two types of cryptosystem: (1) symmetric (private) key cryptosystems and (2) asymmetric (public) key cryptosystems.

Symmetric Key Cryptosystems

All classical cryptosystems (i.e., cryptosystems developed before the 1970s) are examples of symmetric key cryptosystems. In addition, most modern cryptosystems are symmetric as well. Some of the most popular examples of modern symmetric key cryptosystems include AES (Advanced Encryption Standard), DES (Data Encryption Standard), IDEA, FEAL, RC5, and many others. The AES algorithm will be discussed further in Section "Advanced Encryption Standard".

All symmetric key cryptosystems have a common property: They rely on a shared secret between communicating parties. This secret is used both as an encryption key and as a decryption key (thus the keyword "symmetric" in the name). This type of cryptography ensures only confidentiality and fails to provide the other objectives of cryptography.

Even more importantly, the disadvantage of symmetric key cryptography is that it cannot handle large communication networks. If a node in a communication network of n nodes needs to communicate confidentially with all other nodes in the network, it needs $n-1$ shared secrets. For a large value of n, this is highly impractical and inconvenient. On the other hand, an advantage over public key cryptosystems is that symmetric cryptosystems require much smaller key sizes for the same level of security. Hence, the computations are much faster and the memory requirements are smaller.

Public Key Cryptosystems

In public key cryptosystems, there are two keys: a public key, which is publicly known, and the private key, which is kept secret by the owner. The system is called "asymmetric" because the different keys are used for encryption and decryption — the public key and the private key. If data are encrypted with a public key, it can only be decrypted using the corresponding private key, and vice versa. Today, all public key cryptosystems rely on computationally intractable problems. For example, the cryptosystem RSA relies on the difficulty of factoring large integers, whereas the El-Gamal cryptosystem relies on the discrete logarithm problem (DLP), which is the problem of finding a logarithm of a group element with a generator base in a finite Abelian group.

Public key cryptosystems do not need to have a shared secret between communicating parties. This solves the problem of the large confidential communication network introduced earlier. In addition, public key cryptography opened the door for ways of implementing technologies to ensure all goals of cryptography. By means of combining public key cryptography, public key certification, and secure hash functions, there are protocols that enable digital signatures, authentication, and data integrity.

Due to the increase in processor speed and even more to smart modern cryptanalysis, the key size for public key cryptography grew very large. This created a disadvantage in comparison to symmetric key cryptosystems: Public key cryptography is significantly slower and requires a large memory capacity and large computational power. As an example, a 128-bit key used with the DES cryptosystem has approximately the same level of security as the 1024-bit key used with the RSA pubic key cryptosystem [3].

To solve these problems, researchers introduced different approaches. In order to decrease the key size so that public key cryptography can be used in smaller computational environments (such as smart cards or handheld wireless devices), Neil Koblitz introduced the idea of using a more exotic group in the public key underlying algebraic structure: the elliptic curve group. Elliptic curve cryptography (much of whose implementation is credited to CertiCom) enables smaller key sizes of public key cryptosystems that rely on DLP. The elliptic-curve-group algebra is much more complex, so the related cryptanalysis is much harder, resulting in smaller key requirements. Another solution came from public key cryptosystems that initially used more complex computational problem, such as the lattice reduction problem. The relatively new cryptosystem NTRU based on the algebra of a ring of truncated polynomials relies on the lattice reduction problem and it is the only public key cryptosystem that has the speed, memory, and computational complexity comparable to symmetric key cryptosystems. However, the security aspects of NTRU are yet to be investigated [4,5]. The most common implementation solution is to combine symmetric key cryptosystems with public key cryptography; namely to overcome the problems related to applying the symmetric key encryption only, the plaintext is encrypted using a fast symmetric key scheme, and only the secret key used for symmetric encryption is encrypted with the slow public key scheme such as RSA. In this way, all goals of cryptography can be achieved with a much better performance.

Cryptanalysis

Cryptanalysis is the art of deciphering an encrypted message, in whole or in part, when the decryption key is not known. Depending on the amount of known information and the amount of control over the system by the adversary (cryptanalyst), there are several basic types of cryptanalytic attack:

Ciphertext-only attack: The adversary only has access to one or more encrypted messages. The most important goal of a proposed cryptosystem is to withstand this type of attack.

Brute force attack: This is a type of ciphertext-only attack. It is based on exhaustive key search, and for well-designed cryptosystems, it should be computationally infeasible. By today's standards, 128-bit keys are considered secure against the brute force attack.

Known-plaintext attack: In this type of attack, an adversary has some knowledge about the plaintext corresponding to the given ciphertext. This may help determine the key or a part of the key.

Chosen-plaintext attack: Essentially, an adversary can feed the chosen plaintext into the black box that contains the encryption algorithm and the encryption key. The black box produces the corresponding ciphertext, and the adversary can use the accumulated knowledge about the plaintext-ciphertext pairs to obtain the secret key or at least a part of it.

Chosen-ciphertext attack: Here, an adversary can feed the chosen ciphertext into the black box that contains the decryption algorithm and the decryption key. The black box produces the corresponding plaintext, and the adversary tries to obtain the secret key or a part of the key by analyzing the accumulated ciphertext–plaintext pairs.

Advanced Encryption Standard

The AES algorithm is essentially a Rijndael [6] symmetric key cryptosystem that processes 128-bit data blocks using cipher keys with lengths of either 128, 192, or 256 bits. Rijndael is more scalable and can handle different key sizes and data blocksizes; however they are not included in the standard. Table 3.1 lists shows the basic AES parameters.

The AES algorithm's internal operations are performed on a two-dimensional array of bytes called the *State*. The State consists of 4 rows of bytes, each containing Nb bytes, where Nb is the blocksize divided by 32 (because the size of a word is 32 bits), which is, in the case of AES, equal to 4. At the beginning of the encryption or decryption procedure, the input array *in* is copied to the State array *s* according to the following rule: $s[r, c] = in[r + 4c]$, where $0 \leq r < 4$ and $0 \leq c < Nb$. Similarly, at the end of each procedure, the State is copied to the output array *out* as follows: $out[r + 4c] = s[r, c]$, where $0 \leq r < 4$ and $0 \leq c < Nb$. Clearly, in this notation, r stands for row and c for column.

AES Encryption. At the beginning of the AES encryption procedure, the input is copied to the State array using the above-described conventions. After an initial Round Key addition (see AES key schedule), the State array is transformed by implementing a round function 10, 12, or 14 times (depending on the key length), with the final round differing slightly from the first $Nr - 1$ rounds. The final State is then copied to the output as described above. Figure 3.1 shows the pseudocode for the AES encryption.

TABLE 3.1. The basic AES parameters

	Key length (Nk words)	Blocksize (Nb words)	Number of rounds (Nr)
AES-128	4	4	10
AES-192	6	4	12
AES-256	8	4	14

```
AesEncrypt (byte in[4*Nb], byte out [4*Nb], word w[Nb*(Nr+1)])
begin
        byte state[4,Nb]
        state = in
        AddRoundKey(state, w[0, Nb-1])

        for round = 1 step 1 to Nr-1
                SubBytes(state)
                ShiftRows(state)
                MixColumns(state)
                AddRoundKey(state, w[round*Nb, (round+1)*Nb-1])
        end for

        SubBytes(state)
        ShiftRows(state)
        AddRoundKey(state, w[Nr*Nb,(Nr+1)*Nb-1])

        out = state
end
```

Figure 3.1.   Pseudocode for the AES encryption algorithm.

TABLE 3.2.   AES encryption S-box: substitution values for the byte xy **(in hexadecimal representation)**

x								y								
	0	1	2	3	4	5	6	7	8	9	a	b	c	d	e	f
0	63	7c	77	7b	f2	6b	6f	c5	30	01	67	2b	fe	d7	ab	76
1	ca	82	c9	7d	fa	59	47	f0	ad	d4	a2	af	9c	a4	72	c0
2	b7	fd	93	26	36	3f	f7	cc	34	a5	e5	f1	71	d8	31	15
3	04	c7	23	c3	18	96	05	9a	07	12	80	e2	eb	27	b2	75
4	09	83	2c	1a	1b	6e	5a	a0	52	3b	d6	b3	29	e3	2f	84
5	53	d1	00	ed	20	fc	b1	5b	6a	cb	be	39	4a	4c	58	cf
6	d0	ef	aa	fb	43	4d	33	85	45	f9	02	7f	50	3c	9f	a8
7	51	a3	40	8f	92	9d	38	f5	bc	b6	da	21	10	ff	f3	d2
8	cd	0c	13	ec	5f	97	44	17	c4	a7	7e	3d	64	5d	19	73
9	60	81	4f	dc	22	2a	90	88	46	ee	b8	14	de	5e	0b	db
a	e0	32	3a	0a	49	06	24	5c	c2	d3	ac	62	91	95	e4	79
b	e7	c8	37	6d	8d	d5	4e	a9	6c	56	f4	ea	65	7a	ae	08
c	ba	78	25	2e	1c	a6	b4	c6	e8	dd	74	1f	4b	bd	8b	8a
d	70	3e	b5	66	48	03	f6	0e	61	35	57	b9	86	c1	1d	9e
e	e1	f8	98	11	69	d9	8e	94	9b	1e	87	e9	ce	55	28	df
f	8c	a1	89	0d	bf	e6	42	68	41	99	2d	0f	b0	54	bb	16

The `SubBytes()` transformation is essentially an S-box type of transform, which is a nonlinear byte substitution that operates independently on each byte of the State. The simplest representation of this S-box function is by the lookup table. The lookup table associated with `SubBytes()` is shown in Table 3.2.

In the `ShiftRows()` transformation, the bytes in the last three rows of the State are cyclically shifted over different numbers of bytes, whereas the first row is not shifted. The `ShiftRows()` transformation is defined as follows: $s'[r, c] = s[r, (c + shift(r, Nb))]$ mod Nb for $0 \leq r < 4$ and $0 \leq c < Nb$, where the shift value $shift(r, Nb)$ is defined in as follows: $shift(1, 4) = 1$, $shift(2, 4) = 2$, and $shift(3, 4) = 3$.

The `MixColumns()` transformation operates on the State column by column. The columns are considered as polynomials over the Galois Field GF(2^8) and multiplied modulo $x^4 + 1$ with a fixed polynomial $a(x) = \{03\}x^3 + \{01\}x^2 + \{01\}x + \{02\}$. Here, the notation $\{h_1 h_2\}$ denotes a byte in the hexadecimal format. This can be accomplished by applying the following four equations to the State columns:

1. $s'[0, c] = (\{02\} \bullet s[0, c]) \otimes (\{03\} \bullet s[1, c]) \otimes s[2, c] \otimes s[3, c]$
2. $s'[1, c] = s[0, c] \otimes (\{02\} \bullet s[1, c]) \otimes (\{03\} \bullet s[2, c]) \otimes s[3, c]$
3. $s'[2, c] = s[0, c] \otimes s[1, c] \otimes (\{02\} \bullet s[2, c]) \otimes (\{03\} \bullet s[3, c])$
4. $s'[3, c] = (\{03\} \bullet s[0, c]) \otimes s[1, c] \otimes s[2, c] \otimes (\{02\} \bullet s[3, c])$

Here, the $\bullet$ operation denotes multiplication in GF(2^8) modulo the polynomial $x^4 + 1$, where as the $\otimes$ denotes the usual bitwise XOR operation.

Finally, in the `AddRoundKey()` transformation, a Round Key is added to the State by a simple bitwise XOR operation. Each Round Key consists of Nb words from the key schedule (to be described in the AES Key Schedule section), each of which is each added into the columns of the State.

AES Key Schedule. In the AES algorithm, the initial input symmetric key is expanded to create a key schedule for each round. This procedure, given in Figure 3.2, generates a total of $Nb(Nr + 1)$ words that are used in

```
KeyExpansion (byte key[4*Nk], word w[Nb*(Nr+1)], Nk)
begin
        word temp

        i = 0
        while (i < Nk)
                w[i] = word(key[4*i], key[4*i+1], key[4*i+2], key[4*i+3])
                i = i+1
        end while
        i = Nk
        while (i < Nb * (Nr+1))
                temp = w[i-1]
                if (i mod Nk = 0)
                        temp = SubWord(RotWord(temp)) xor Rcon[i/Nk]
                else if (NK > 6 and i mod NK = 4)
                        temp = SubWord(temp)
                end if

                w[i] = w[i -NK] xor temp
                i = i+1
        end while
end
```

Figure 3.2. Pseudocode of the key schedule stage of the AES algorithm.

the `AddRoundKey()` procedure. In this algorithm, `SubWord()` is a function that takes a 4-byte input word and applies the S-box to each of the 4 bytes to produce an output word. The function `RotWord()` takes a word $[a_0, a_1, a_2, a_3]$ as input, performs a cyclic permutation, and returns the word $[a_1, a_2, a_3, a_0]$. The round constant word array, Rcon[i], contains the values given by $[x^{i-1}, \{00\}, \{00\}, \{00\}]$, with x^{i-1} being powers of x (x is denoted as $\{02\}$) in the field GF(2^8).

AES Decryption. AES decryption, the inverse of AES encryption, also consists of four stages in each round, which are simply the inverses of the transformations described earlier. Each AES decryption round performs the following four transformations: `InvShiftRows()`, `InvSubBytes()`, `InvMixColumns()`, and `AddRoundKey()`. The pseudocode for the AES decryption process is given in Figure 3.3

The `InvShiftRows()` is the inverse of the `ShiftRows()` transformation, which is defined as follows: $s[r, c] = s'[r, (c + shift(r, Nb)) \bmod Nb]$ for $0 \leq r < 4$ and $0 \leq c < Nb$. Similarly, `InvSubBytes()` is the inverse of the byte substitution transformation, in which the inverse S-box is applied to each byte of the State. The inverse S-box used in the `InvSubBytes()` transformation is presented in Table 3.3.

The `InvMixColumns()` is the inverse of the `MixColumns()` transformation. The transformation `InvMixColumns()` operates on the State column by column, treating each column as a polynomial over the field GF(2^8) modulo $x^4 + 1$ with a fixed polynomial $a^{-1}(x)$, given by $a^{-1}(x) = \{0b\}x^3 + \{0d\}x^2 + \{09\}x + \{0e\}$. This can be accomplished by applying the following four equations to the State columns:

1. $s'[0, c] = (\{0e\} \bullet s[0, c]) \otimes (\{0b\} \bullet s[1, c]) \otimes (\{0d\} \bullet s[2, c]) \otimes (\{09\} \bullet s[3, c])$

```
AesDecrypt (byte in[4*Nb], byte out [4*Nb], word w[Nb*(Nr+1)])
begin
        byte state[4,Nb]

        state = in

        AddRoundKey(state, w[Nr*Nb, (Nr+1) *Nb−1])

        for round = Nr−1 step −1 downto 1
            InvShiftRows(state)
            InvSubBytes(state)
            AddRoundKey(state, w[round*Nb, (round+1)*Nb−1])
            InvMixColumns (state)
        end for

        InvShiftRows(state)
        InvSubBytes(state)
        AddRoundKey(state, w[0, Nb−1])

        out = state
    end
```

Figure 3.3.   Pseudocode of AES decryption algorithm.

TABLE 3.3. **AES decryption S-box: substitution values for the byte** xy **(in hexadecimal representation)**

								y								
x	0	1	2	3	4	5	6	7	8	9	a	b	c	d	e	f
0	52	09	6a	d5	30	36	a5	38	bf	40	a3	9e	81	f3	d7	fb
1	7c	e3	39	82	9b	2f	ff	87	34	8c	43	44	c4	de	e9	cb
2	54	7b	94	32	a6	c2	23	3d	ee	4c	95	0b	42	fa	c3	4e
3	08	2e	a1	66	28	d9	24	b2	76	5b	a2	49	6d	8b	d1	25
4	72	f8	f6	64	86	68	98	16	d4	a4	5c	cc	5d	65	b6	92
5	6c	70	48	50	fd	ed	b9	da	5e	15	46	57	a7	8d	9d	84
6	90	d8	ab	00	8c	bc	d3	0a	f7	e4	58	05	b8	b3	45	06
7	d0	2c	1e	8f	ca	3f	0f	02	c1	af	bd	03	01	13	8a	6b
8	3a	91	11	41	4f	67	dc	ea	97	f2	cf	ce	f0	b4	e6	73
9	96	ac	74	22	e7	ad	35	85	e2	f9	37	e8	1c	75	df	6e
a	47	f1	1a	71	1d	29	c5	89	6f	b7	62	0e	aa	18	be	1b
b	fc	56	3e	4b	c6	d2	79	20	9a	db	c0	fe	78	cd	5a	f4
c	1f	dd	a8	33	88	07	c7	31	b1	12	10	59	27	80	ec	5f
d	60	51	7f	a9	19	b5	4a	0d	2d	e5	7a	9f	93	c9	9c	ef
e	a0	e0	3b	4d	ae	2a	f5	b0	c8	eb	bb	3c	83	53	99	61
f	17	2b	04	7e	ba	77	d6	26	e1	69	14	63	55	21	0c	7d

2. $s'[1, c] = (\{09\} \bullet s[0, c]) \otimes (\{0e\} \bullet s[1, c]) \otimes (\{0b\} \bullet s[2, c]) \otimes (\{0d\} \bullet s[3, c])$

3. $s'[2, c] = (\{0d\} \bullet s[0, c]) \otimes (\{09\} \bullet s[1, c]) \otimes (\{0e\} \bullet s[2, c]) \otimes (\{0b\} \bullet s[3, c])$

4. $s'[3, c] = (\{0b\} \bullet s[0, c]) \otimes (\{0d\} \bullet s[1, c]) \otimes (\{09\} \bullet s[2, c]) \otimes (\{0e\} \bullet s[3, c])$

Finally, the `AddRoundKey()` is identical to the one from the AES encryption procedure because this transformation is its own inverse.

Today, the security level of AES is substantial because there are no known effective attacks discovered thus far. For more information about AES, see Reference 6.

INTRODUCTION TO MULTIMEDIA SECURITY

Multimedia security in general is provided by a method or a set of methods used to protect multimedia content. These methods are heavily based on cryptography and they enable either communication security or security against piracy (Digital Rights Management [DRM] and watermarking), or both. Communication security of digital images and textual digital media can be accomplished by means of standard symmetric key cryptography. Such media can be treated as a binary sequence and the complete data can be encrypted using a cryptosystem such as AES or DES [7]. In general, when the multimedia data are static (not a real-time streaming), we can treat it as regular binary data and use the conventional encryption techniques. Encrypting the entire multimedia stream using standard encryption methods is often referred to as the *naive approach*.

However, due to variety of constraints (such as the near real-time speed, etc.), communication security for streaming audio and video media is harder to accomplish. Communication encryption of video and audio multimedia content is not simply the application of established encryption algorithms, such as DES or AES, to its binary sequence. It involves careful analysis to determine and identify the optimal encryption method when dealing with audio and video data. Current research is focused on modifying and optimizing the existing cryptosystems for real-time audio and video. It is also oriented toward exploiting the format-specific properties of many standard video and audio formats in order to achieve the desired speed and enable real-time streaming. This is referred to as *selective encryption*.

For textual data, most applications of still images and some low-quality audio and video streaming multimedia, we can still apply real-time packet encryption by means of SRTP (Secure Real-time Transport Protocol) [8], which is based on AES and encrypts the entire multimedia bit stream. This is essentially an application of the naive approach.

Deciding upon what level of security is needed is harder than it looks. To identify an optimal security level, we have to carefully compare the cost of the multimedia information to be protected and the cost of the protection itself. If the multimedia to be protected is not very valuable in the first place, it is sufficient to choose a relatively light level of encryption. On the other hand, if the multimedia content is highly valuable or represents government or military secrets, the cryptographic security level must be the highest possible.

For many real-world applications such as pay-per-view, the content data rate is typically very high, but the monetary value of the content may not be high at all. Thus, very expensive attacks are not attractive to adversaries, and light encryption (often called *degradation*) may be sufficient for distributing MPEG video. For these applications, DRM is of much more interest. In contrast, applications such as videoconferencing or videophone (or even the Internet phone) may require a much higher level of confidentiality. If the videoconference is discussing important industrial, governmental, or military secrets, then the cryptographic strength must be substantial. Maintaining such a high level of security and still keeping the real-time and limited-bandwidth constraints is not easy to accomplish.

Comprehensive survey studies of multimedia encryption techniques are given in References 9–11. In Reference 10, Liu and Eskicioglu present a comprehensive classification that summarizes most of the presented algorithms. An updated version of that classification is shown in Table 3.4.

VIDEO ENCRYPTION TECHNIQUES

In this section, we discuss different research directions, which were taken in the area of video communication encryption. As discussed earlier, a naive approach can be used to encrypt multimedia content. The whole multimedia stream is encrypted using a symmetric key cryptosystem. However, even the fastest modern symmetric schemes, such as DES or AES, are computationally very expensive for many real-time video and audio data.

Video Scrambling

Scrambling is a popular method of applying fast, yet very insecure distortion of the video signal. In general, scrambling was the product of an immediate industrial need by cable companies for a fast solution to make the free viewing of paid cable channels more difficult than doing nothing to the video signal.

Even though there is no distinct definition of scrambling, it refers to the simplest possible encryption attempts such as simple substitution or simple transposition ciphers whose security in the modern world is the lowest possible. Early work on signal scrambling was based on using an analog device to permute the signal in the time domain or distort the signal in the frequency domain by applying filter banks or frequency converters [29]. However, these schemes seriously lack security and they are extremely easy to crack using modern computers. As far as the attacker is concerned, these methods represent a temporary inconvenience in getting the original video or audio signal.

In many instances, this is still a solution that serves cable companies around the world quite effectively because most people lack the knowledge and engineering skills to crack the scrambled video signal, but the modified cable boxes that can unscramble all scrambled material are quite easy to manufacture. However, scrambling is out of the question for applications that require more serious level of security.

Selective Video Encryption

In order to meet real-time constraint for audio and video multimedia playback, *selective encryption* techniques have been proposed. The basic idea of selective encryption is to encrypt only a portion of the

TABLE 3.4. Classification of selective encryption schemes

Type of data	Domain	Proposal	Encryption algorithm	What is encrypted?
Image	Frequency domain	Cheng and Li, 2000 [12]	No algorithm is specified	Pixel-and set-related significance information in the two highest pyramid levels of SPIHT
		van Droogenbroeck and Benedett, 2002 [13]	DES, triple DES, and IDEA	Bits that indicate the sign and magnitude of the non-zero DCT coefficients
		Pommer and Uhl, 2003 [14]	AES	Subband decomposition structure
	Spatial domain	Cheng and Li, 2000 [12]	No algorithm specified	Quadtree structure
		Droogenbroeck and Benedett, 2002 [13]	Xor	Least significant bitplanes
		Podesser, Schmidt, and Uhl, 2002 [15]	AES	Most significant bitplanes
Video	Frequency domain	Meyer and Gadegast, 1995 [16]	DES, RSA	Headers, parts of I-blocks, all I-blocks, I-frames of the MPEG stream
		Spanos and Maples, 1995 [17,18]	DES	I-frames, sequence headers and ISO end code of the MPEG stream
		Tang, 1996 [19]	Permutation, DES	DCT coefficients
		Qiao, and Nahrstedt, 1997 [20]	xor, permutation, IDEA	Every other bit of the MPEG bit stream
		Shi and Bhargava, 1998 [21]	xor	Sign bit of DCT coefficients

	Reference	Algorithm	Description
	Shi, Wang, and Bhargava, 1999 [22]	IDEA	Sign of motion vectors
	Alattar, A-Regib, and Al-Semari, 1999 [23]	DES	Every nth I-macroblock, headers of all the predicted macroblocks, header of every nth predicted macroblock
	Shin, Sim, and Rhee, 1999 [24]	Permutation, RC4	Sign bits of DC coefficients of I pictures; random permutation of the DCT coefficients
	Cheng and Li, 2000 [12]	No algorithm specified	Pixel- and set-related significance information in the two highest pyramid levels of SPIHT in the residual error
	Wen, Severa, Zeng, Luttrell, and Jin, 2002 [25]	DES, AES	The information-carrying fields, either fixed length code (FLC) codewords or variable length code (VLC) codewords
	Zeng and Lei, 2002 [26]	Permutation, xor	Selective bit scrambling, block shuffling, block rotation of the transform coefficients (wavelet and JPEG) and JPEG motion vectors
	Wu and Mao, 2002 [27]	Any modern cipher, random shuffling on bit-planes in MPEG-4 FGS	Bit stream after entropy coding, quantized values before run-length coding (RLC) or RLC symbols, intra-bit-plane shuffling in MPEG-4 FGS
Spatial domain Entropy codec	Cheng and Li, 2000 [12]	No algorithm specified	Quadtree structure of motion vectors and quadtree structure of residual errors
	Wu and Kuo, 2000 [28]; Wu and Kuo, 2001 [29]	Multiple Huffman tables, multiple state indices in the QM coder	Encryption of data by multiple Huffman coding tables and multiple state indices in the QM coder
Speech Compressed domain	Wu and Kuo, 2000 [28]	DES, AES	Most significant bits of the LSP codebook indices, the lag of pitch predictor, the pitch gain vectors, the fixed codebook gains, and the VAD mode flag coefficients of the G.723.1 speech codec

(Continued)

TABLE 3.4. Continued

Type of data	Domain	Proposal	Encryption algorithm	What is encrypted?
		Servetti and De Martin, 2002 [30]	Not specified	1st stage vector of LSP quantizer, 2nd stage lower vector of LSP quantizer, pitch delay 1st subframe, pitch delay 2nd subframe, gain codebook (stage 1) 1st subframe, gain codebook (stage 2) 1st subframe, gain codebook (stage 1) 2nd subframe, gain codebook (stage 2) 2nd subframe
Audio	Compressed domain	Servetti, Testa, and De Martin, 2003 [31]	Not specified	File header (that contains the original *table_select*, and *big_value* values) at the beginning of the audio track; a subset of region1 bits and a subset of region2 bits
		Thorwirth, Horvatic, Weis, and Zhao, 2000 [32]	Not specified	Audio quality layers associated with the compressed audio data

Source: Adopted from Liu, X. and Eskicioglu, A.M., IASTED International Conference on Communication, International and Information Technology (CIIT 2003) 2003. With permission.

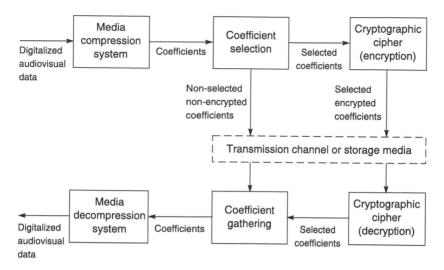

Figure 3.4.   A common selective encryption scheme.

compressed bit stream. A common approach in selective encryption is to integrate compression with encryption, as shown in Figure 3.4.

For example, we can select only the most important coefficients from either final or intermediate steps of a compression process and encrypt those coefficients. Less important coefficients are left unencrypted. Still, some schemes prefer to lightly encrypt even the less important coefficients. By "light" encryption we mean applying a simple encryption scheme whose speed is high and whose security level is low (there is always a trade-off between the speed and security). In this manner, we define *variable encryption* as an approach that applies different encryption schemes with different security strengths. Selective encryption can be treated as a special case of variable encryption [11].

SECMPEG by Meyer and Gadegast, 1995. In 1995, Meyer and Gadegast introduced the selective encryption method called *Secure MPEG (SECMPEG)*, designed for the MPEG-1 video standard [16]. The SECMPEG contains four levels of security. At the first level, SECMPEG encrypts the headers from the sequence layer to the slice layer, and the motion vectors and DCT blocks are unencrypted. At the second level, the most relevant parts of the I-blocks are additionally encrypted (upper left corner of the block). At the third level, SECMPEG encrypts all I-frames and all I-blocks. Finally, at the fourth level, SECMPEG encrypts the whole MPEG-1 sequence (the naive approach). The authors chose DES symmetric key cryptosystem, which was the natural choice, given that this cryptosystem had been around since 1976 and was the official symmetric encryption algorithm standardized by NIST and adopted by

109

the U.S. government. Because DES is a symmetric key cryptosystem, it could only be used to achieve confidentiality. Meyer and Gadegast targeted solving the problem of data integrity as well. For that reason, the Cyclic Redundancy Check (CRC) was incorporated as a low-level solution to integrity. The real data integrity mechanisms that included public key cryptography and cryptographically good hash functions such as MD4, MD5, or SHA were left for further research.

The selective encryption in SECMPEG (levels 1, 2, and 3) has some weaknesses [33,34]. It is shown that even though a single P- or B-frame carries almost no information without the corresponding I-frame, a series of P- or B-frames can tell much if their base I-frames are correlated. The experiments by Agi and Gong showed that encrypting I-frames only might leave visible I-blocks in other frames [33]. The same authors then proposed a few trade-off improvements: increasing the frequency of the I-frames or encrypting all P- and B-frames. These improvements decrease speed and further disrupt the compression ratio. Because SECMPEG introduces changes to the MPEG-1 format, a special encoder and decoder is needed to handle SECMPEG streams.

Nevertheless, the SECMPEG paper [16] and implementation by Meyer and Gadegast was one of the first important research initiatives for selective encryption of multimedia streams. These authors were among the first to realize the benefits of encrypting only selected bits in a video bit stream. Their experiments confirmed that the visual disruption was substantial only by encrypting the DC coefficients and the first three to eight AC coefficients in the 8×8 JPEG block of the MPEG-1 I-frames.

Aegis by Maples and Spanos, 1995. Another important research achievement on selective encryption of a MPEG video stream appeared in 1995 by Maples and Spanos [17,18], and it introduced the secure MPEG video mechanism called *Aegis*. Aegis was initially designed for MPEG-1 and MPEG-2 video standards. It encrypts I-frames of all MPEG groups of frames in an MPEG video stream, and B- and P-frames are left unencrypted. In addition, Aegis also encrypts the MPEG video sequence header, which contains all of the decoding initialization parameters that include the picture width, height, frame rate, bit rate, buffer size, and so forth. Finally, this method also encrypts the ISO end code (i.e., the last 32 bits of the stream). As a result, the bit stream's MPEG identity is further concealed. The Aegis' behind-the-scenes encryption engine chosen by Maples and Spanos was DES. In a way, Aegis is very similar to the level 3 of SECMPEG by Meyer and Gadegast, and the security weaknesses that are applicable to SECMPEG are carried over to Aegis.

As discussed earlier, Agi and Gong criticized encrypting only I-frames by Aegis and SECMPEG, in view of the fact that their experiments showed that it was possible to discern some aspects from the scene from the decoded P- and B-frames [33], For example, they were able to conclude that the talking head was present at the scene. Alattar and Al-Regib presented a similar argument regarding the security of SECMPEG in Reference 35. It is important to notice that this type of security is not particularly good for applications where the security is one of the top priorities (such as important military or business teleconferencing), but it might be sufficient for the quality degradation of the entertainment videos, such as the pay TV broadcast.

Aegis is also considered as one of the pioneering research accomplishments in the area of selective video encryption. The experimental results for selective encryption published by Maples and Spanos in References 17 and 18 confirmed the enormous gain in speed over the naive approach.

Zigzag Permutation Algorithm by Tang, 1996. Tang's *Zigzag permutation algorithm* [19] is based on embedding the encryption into the MPEG compression process. The JPEG images and the I-frames of MPEG video undergo a zigzag reordering of the 8×8 blocks. The zigzag pattern forms a sequence of 64 entries that are ready to enter entropy-encoding stage. The main idea of this approach is to use a random permutation list to map the individual 8×8 blocks to a 1×64 vector. The algorithm consists of three stages:

- *Stage 1:* A list of 64 permutations is generated.
- *Stage 2:* Splitting procedure on an 8×8 block is performed as follows: We denote the DC coefficient by an eight-digit binary number with digits $b_7b_6b_5b_4b_3b_2b_1b_0$. This binary sequence is then split into two equal halves: $b_7b_6b_5b_4$ and $b_3b_2b_1b_0$. Finally, we place the number $b_7b_6b_5b_4$ into the DC coefficient and then assign the number $b_3b_2b_1b_0$ to AC_{63} (the last AC coefficient), which is the least important value in the block. This will result in no visible degradation of the quality.
- *Stage 3:* The random permutation is applied to the split block.

However, the zigzag permutation algorithm is not particularly secure [34,36]. Qiao and Nahrstedt introduced two types of attack on zigzag permutation: a known-plaintext attack and a ciphertext-only attack. A known-plaintext attack is particularly applicable to the videos with known clips such as blank screens, the movie rating screens, the MGM roaring lion, and so forth. If these known frames are used as a comparison, the adversary can easily generate the secret permutation list.

Because Tang realized the significant vulnerability to known-plaintext attack, he introduced an improvement of the zigzag permutation technique by using the binary coin-flipping sequence method together with two different permutation lists. Two different permutation lists are generated, and for each 8×8 block, a coin is randomly flipped. The outcome event determines which list to apply (i.e., if the head is flipped we apply one list and if the tail is flipped we apply the other list). As shown in References 34 and 36, this addition turns out to be useless. If we know some of the original frames in advance (known plaintext), then, by simple comparison, both permutation lists can be found. Because of certain statistical properties of the blocks (upper left corner gathering of the AC coefficients within a block), we can easily determine which list of permutations to use.

In addition, Qiao and Nahrstedt showed that Tang's splitting method is just a subset of SECMPEG with a second level; thus, its security is not good. Finally, Qiao and Nahrstedt showed that Tang's zigzag permutation algorithm is susceptible to the ciphertext-only attack. The attack relies on the statistical properties of the discrete cosine transform (DCT) coefficients, for which most nonzero terms are gathered in the top left corner of the I-block. Statistical analysis shows the following observation [34]: The DC coefficient always has the highest frequency value, the frequencies of AC_1 and AC_2 coefficients are among the top six, and frequencies of AC_3 and AC_4 are among the top ten. Applying these observations, one can reconstruct the original video to a relatively good accuracy using only five DCT coefficients. Therefore, zigzag permutation cipher seriously lacks the desired level of security.

Video Encryption Algorithm by Qiao and Nahrstedt, 1997. The *Video Encryption Algorithm* (VEA) by Qiao and Nahrstedt [20] is constructed with the goal of exploiting the statistical properties of the MPEG video standard. The algorithm consists of the following four steps:

- *Step 1:* Let the $2n$-byte sequence, denoted by $a_1a_2\ldots a_{2n}$, represent the chunk of an I-frame
- *Step 2:* Create two lists, one with odd indexed bytes $a_1a_3\ldots a_{2n-1}$, and the other with even indexed bytes $a_2a_4\ldots a_{2n}$
- *Step 3:* xor the two lists into an n-byte sequence denoted by $c_1c_2\ldots c_n$
- *Step 4:* Apply the chosen symmetric cryptosystem E (e.g., DES or AES) with the secret key *KeyE* on either the odd list or even list and, thus, create the ciphertext sequence $c_1c_2\ldots c_n E_{KeyE}(a_1a_3\ldots a_{2n-1})$ or $c_1c_2\ldots c_n E_{KeyE}(a_2a_4\ldots a_{2n})$, respectively.[1]

---

[1] The authors' choice was to encrypt the even sequence, creating the ciphertext $c_1c_2\ldots c_n E_{KeyE}(a_2a_4\ldots a_{2n})$.

Clearly, the decryption mechanism at the other end of the communication channel consists of two easy steps: Apply the symmetric cryptosystem E with the appropriate key to the second half of the ciphertext to obtain the first half of the original sequence, and xor this result with the first half of the ciphertext to obtain the other half of the original sequence.

Even though the security of the system is based on the security of the chosen underlying cryptosystem, if an adversary obtains either even or odd list, the system is completely broken. Thus, the authors suggest an even more secure approach. Instead of dividing the list into the even and odd lists, the random $2n$-bit key, called *KeyM*, is generated and used to split the $2n$-byte chunk of MPEG stream into two lists. The method of partitioning the $2n$-byte MPEG sequence into two sub-sequences is as follows: if the ith bit in *KeyM* is 1, then the ith byte of the MPEG sequence goes into the first subsequence, and, similarly, if the ith bit in *KeyM* is 0, then the ith byte of the MPEG sequence goes into the second subsequence. It is required that *KeyM* be a binary sequence with exactly n ones and n zeros, in order to ensure that the two lists are of equal length. The suggested value for n are 64 or 128. By applying this approach, the attacker will have a very hard time ordering the resulting sequence of bytes in the case that the encrypted subsequence is somehow cracked, assuming that the binary sequence *KeyM* remained secret. This means that *KeyM* must be securely transmitted to the receiver before the decryption can take place.

Another consideration is that a possible pattern repetition in the $2n$ stream chunk represents a significant security hole in the method proposed be Qiao and Nahrstedt. According to the statistical analysis of MPEG stream sequences, it was observed that the nonrepeating patterns have a lifetime over only one 1/16 chunk [20]. Therefore, to obtain the nonrepeating pattern with a length of 1/2 frame, which consists of eight such chunks, it is sufficient to shuffle only these eight chunks. However, each of these chunks must be shuffled by using different keys, which are here denoted by Key_1, Key_2, ..., Key_8. The authors suggest using the permutations of degree 32 (even though degree 24 was shown to be sufficient), so that each Key_i is stored as a random permutation of degree 32. Hence, the length of a Key_i is 20 bytes, resulting in the total length of 160 bytes for Key_1, Key_2, ..., Key_8. Once the keys are generated, the Key_1 permutation is applied to the first 32 bytes of the even list, the Key_2 permutation is applied to the next 32 bytes, and so on, repeating this process after the eighth key. This procedure will ensure nonrepetition in the MPEG stream, thus significantly decreasing the vulnerability of the system to the repetitive pattern attacks. Furthermore, the increase in the bit-stream size caused by including these extra keys is at a negligible level.

Finally, Qiao and Nahrstedt proposed the use of one more additional key, denoted by *KeyF*. The experiments have shown that the pattern of selecting the even list and the odd list should be changed for every frame in order to ensure the anticipated security level [20]. For this reason, a 64-bit key *KeyF* is randomly generated for every single frame in the MPEG sequence. However, this key is subject to the constraint that every four consecutive bits represent a unique number from 0 to 15 (i.e., so that any *KeyF* represents a permutation of degree 16). To get the new keys for each frame, we permute *KeyM* and the *Key$_i$*'s repeatedly using the permutation defined by *KeyF* assigned to that frame.

In order to securely transmit these additional keys (*KeyM*, *Keyi*'s and *KeyF*'s), they must be encrypted using the encryption function E with *KeyE*. As a final point, the secret key *KeyE* has to be securely transmitted to the decoder's side (the receiver), possibly by using public key cryptography or by using a separate secure channel. Alternatively, the separate secure channel can be used for the transmission of keys

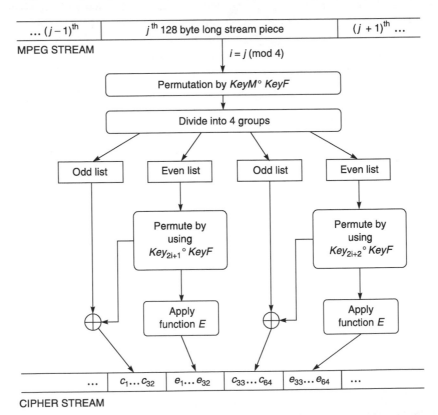

Figure 3.5. Selective video encryption algorithm by Qiao and Nahrstedt. (From Qiao, L. and Nahrstedt, K., Proceedings of the 1st International Conference on Imaging Science, Systems and Technology (CISST '97), 1997. With permission.)

KeyM and the *Key_i*'s. The diagram of the proposed algorithm is shown in Figure 3.5.

The experiments that were originally performed in 1997 were mainly using DES and IDEA cryptosystems; however, the newer choice for the underlying cryptosystem *E* would probably be the recently standardized AES cryptosystem (Rijndael). The security of the method proposed by Qiao and Nahrstedt is very close to the security of the encryption scheme *E* that is internally used. The speed of this algorithm is roughly a half of the speed of the naive algorithm, but that is arguably still a large amount of computation for high-quality real-time video applications that have high bit rates [29].

Video Encryption Algorithms by Shi, Wang, and Bhargava, 1998 and 1999. In Reference 37, Shi, Wang, and Bhargava classified their previous work into four different video encryption algorithms: *Algorithm I*, *Algorithm II* (VEA), *Algorithm III* (MVEA), and *Algorithm IV* (RVEA). The original work was published by the same authors years earlier [21,22]. The basic idea of these algorithms is to apply selective encryption on selective coefficients in the JPEG/MPEG schemes, as illustrated in Figure 3.6.

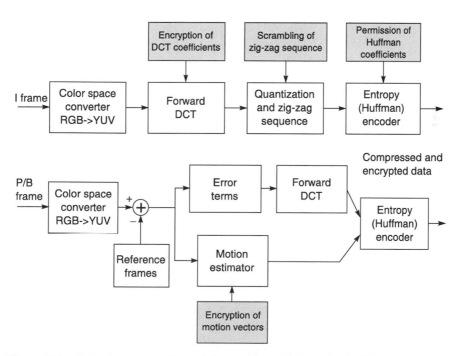

Figure 3.6. Selective encryption of selective coefficients in MPEG/JPEG coding schemes: DCT and its zigzag sequence, coefficients in the Huffman table, and motion vectors.

Algorithm I. The first algorithm, denoted simply by Algorithm I, uses the permutation of Huffman codewords in the I-frames. This method incorporates encryption and compression in one step. The secret part of the algorithm is a permutation π, which is used to permute the standard JPEG/MPEG Huffman codeword list. In order to save the compression ratio, the permutation π must be such that it only permutes the codewords with the same number of bits. In addition, the distance between the original and the permuted codeword list must be greater than the encryption quality parameter β, which is chosen by the user.

The security of Algorithm I is not particularly good. In Reference 38, it is shown that Algorithm I is highly vulnerable to both known-plaintext attack and ciphertext-only attack. If some of the video frames are known in advance (such as standard introductory jingles and similar), one can reconstruct the secret permutation π by comparing the original and encrypted frames. Vulnerability to this type of attack was also discussed in Reference 37. However, Algorithm I is also subject to ciphertext-only attack. The low-frequency error attack can be applied on ciphertext produced by Algorithm I [38]. Basically, because permutation π has a special form (i.e., it only shuffles codewords with the same length), most of the security comes from shuffling the 16-bit codewords in the AC coefficient entropy table. However, because there is a very limited number of codewords with length of less than 16 bits, it is very easy to reconstruct all of the DC coefficients and most frequent AC coefficients (as these will be encoded with less than 16-bit codewords). In other words, the only hard part would be to figure out how the permutation π shuffles the 16-bit codewords. However, these appear to be extremely rare, and the reconstructed video may be of almost the same quality as the original. Furthermore, the rare pixels that do contain 16-bit codewords can easily be interpolated to get rid of the noise effect and to create the reconstruction image that, at least to the human visual system, appears as flawless as the original.

Algorithm II (VEA). The Algorithm II (VEA) uses the following selective encryption observation: it is sufficient to encrypt only the sign bits of the DCT coefficients in an MPEG video. Algorithm II simply xors the sign bits of the DCT coefficients with a secret m-bit binary key $k = k_1 k_2 \ldots k_m$. The effect of this scheme is that the encryption function randomly changes the sign bits of the DCT coefficients depending on the corresponding bit value of k. Let us denote the sign bits of the DCT coefficients in an MPEG stream that belong to the same GOP (Group of Pictures) by $s_1 s_2 \ldots s_n$. Then, a bit s_i will not be changed if the value of the key bit $k_{i \pmod m}$ is 0, and it will be flipped if the value of the key bit $k_{i \pmod m}$ is 1. The next GOP would simply reuse the secret key, which serves for resynchronization purposes. The ability to resynchronize the video stream is important in the case of unreliable networks, transmission errors and VCR-like functionality such as fast forwarding or rewinding.

The security of Algorithm II depends on the length of the key. The authors encourage the use of a long binary key; however, too long of a key may be infeasible and impractical. On the other hand, if the key is short, the system can be easily broken. If the key is as long as the video stream and it is unique and used only once, it would correspond to the Vernam cipher (also referred to the one-time pad), which is known to be absolutely secure. However, this is highly impractical for mass applications such as VOD (Video on Demand). On the other hand, if the key is too short, the whole method simply becomes the Vigenére-like cipher, a well-studied classical cipher for which there are many attacks developed (e.g., the Kasiski analysis) [38]. In addition, Vigenére is highly sustainable to the known-plaintext attack (one can literally read off the secret key). The authors of Reference 17 suggest the use of the pseudorandom generator that generates a stream of pseudorandom bits to be the key k of arbitrary length. The trouble with this approach is the security, randomness, and speed of this P-RNG (Pseudo-Random Number Generator). The additional issues include the synchronization of the P-RNG and the secure transmission of the seed (the secret key) for the P-RNG.

Algorithm III (MVEA). Algorithm III (MVEA) is an improvement to Algorithm II (VEA). It includes the following addition: The sign bits of differential values of motion vectors in P- and B-frames can also be randomly changed. This type of improvement makes the video playback more random and more nonviewable. When the sign bits of differential values of motion vectors are changed, the directions of motion vectors change as well. In addition, the magnitudes of motion vectors change, making the whole video very chaotic. The authors found that the encryption of sign bits of motion vectors makes the encryption of sign bits of DCT coefficients in B- and P-frames unnecessary.

Furthermore, the original Algorithm III (MVEA) was designed to encrypt only the sign bits of DC coefficients in the I-frames of the MPEG video sequence, leaving the AC coefficients unencrypted. This significantly reduces the computational overhead, but it opens up new security risks; namely because the DC coefficient and the sum of all AC coefficients within the block are related, an adversary may use the unencrypted AC coefficients to derive the unknown (encrypted) DC coefficients [22,37]. For that reason, the authors recommend encrypting all DCT coefficients in the I-frames for applications that need higher level of security.

Just like Algorithm II (VEA), Algorithm III (MVEA) relies on the secret m-bit key k. Also, the resynchronization is done at the beginning of a GOP.

Unfortunately, the fundamental security issues and problems that are applicable to VEA are also applicable to MVEA.

Algorithm IV (RVEA). Finally, Algorithm IV (RVEA) is a significantly more secure approach than the previous three algorithms. This approach is considered to be robust under both ciphertext-only and known-plaintext. The difference between the RVEA and MVEA/VEA algorithms is that RVEA uses conventional symmetric key cryptography to encrypt the sign bits of DCT coefficients and the sign bits of motion vectors. The conventional cryptosystems are mathematically well understood and thoroughly tested by the experts in the field, which definitely adds to the security aspect of RVEA. The selective approach significantly speeds up the process of conventional encryption by only encrypting certain sign bits in the MPEG stream. The experiments show that the encryption quality is quite good considering the amount of information changed.

In Algorithm IV (RVEA), the sign bits of DCT coefficients and motion vectors are simply extracted from the MPEG video sequence, encrypted using a fast, conventional cryptosystem such as AES, and then restored back to their original position in the encrypted form. The effect of this is similar to VEA/MVEA, for which the sign bits are either flipped or left unchanged. The authors limited the number of bits for encryption to at most 64 for each MPEG stream macroblock, for the purposes of reducing and bounding the computation time. Next, we describe exactly how these sign bits are selected for encryption.

Each MPEG video slice consists of one or more *macroblocks*. The macroblock corresponds to a 16×16 pixel square, which is composed of four 8×8 pixel luminance blocks denoted by Y_1, Y_2, Y_3 and Y_4, and two 8×8 chrominance blocks Cb and Cr. Each of these six 8×8 blocks can be expressed as a sequence $ba_1a_2 \ldots a_n$, where n is at most 64 since b is the code for DC coefficient and a_i is the code for a non-zero AC coefficient. Then, for each such 8×8 block, we can define the sequence $\beta \alpha_1 \alpha_2 \ldots \alpha_n$, which corresponds to the sign bits of the matching DCT coefficients b, a_1, $a_2, \ldots$, and a_n. Figure 3.7 shows the order of selection of the sign bits for each macroblock. The obvious reason for this selection order is that the DC and higher-order AC coefficients are carrying much more information that the low-order AC coefficients.

In conclusion, RVEA only encrypts the fraction (typically about 10%) of the whole MPEG video by using the conventional secure cryptographic schemes such as DES, IDEA, AES, and so forth. Therefore, Algorithm IV (RVEA) is a much better method than the previous three algorithms in terms of security. Furthermore, it saves up to 90% of the computation time compared to the naive approach.

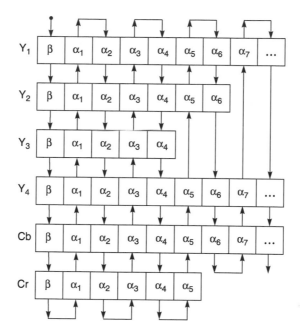

Figure 3.7.   The bit selection order in RVEA. (From Bhargava, B., Shi, C., and Wang, Y. (http://raidlab.cs.purdue.edu/papers/mm.ps)

Video Encryption Methods by Alattar, Al-Regib, and Al-Semari, 1999. In 1999, Alattar, Al-Regib, and Al-Semari presented three methods for selective video encryption based on DES cryptosystem [23]. These methods, called simply *Method I*, *Method II*, and *Method III*, were computationally improved versions of the previous work from two of the co-authors [35], which is referred to as *Method 0*.

The first algorithm (Method 0), proposed by Alattar and Al-Regib in [35], essentially encrypts all macroblocks from I-frames and the headers of all prediction macroblocks using the DES cryptosystem. This method performs relatively poorly because encryption is carried out on 40–79% of the MPEG video stream [11].

In Method I, the data of every nth macroblock from the I-frame of MPEG video stream is encrypted using the DES cryptosystem, and the information from all of the other I-frame macroblocks is left unencrypted. The value of n was not specified and it can be chosen depending on the application needs. If the value of n is 2, then the encryption is performed on approximately one-half of all I-frame macroblocks, but the security level is higher. On the other hand, if the value of n is higher,

119

the computational savings are greater, yet the security level is lower. An important observation is that even though a certain number of I-macroblocks are left unencrypted, they are not expected to reveal any information about the encrypted ones [23]. However, this method does not hide the motion information and the authors suggest an improved algorithm, which will be discussed next.

To improve the security of Method I, Alattar, Al-Regib, and Al-Semari suggested Method II, which additionally encrypts the headers of all predicted macroblocks using DES. Because DES is a block cipher that operates on 64-bit blocks, a 64-bit segment starting from the header of a predicted macroblock is processed in the beginning. This segment may include exactly the whole header (which is the rare case when the header size is equal to 64 bits), a part of the header (when header size is larger than 64 bits), or the whole header along with the part of the macroblock data (when the header size is smaller than 64 bits). In the case when the encrypted segment contains a part of the header, an adversary would have serious problems with synchronization, which adds to the security regarding motion vectors [23]. The security is further increased if the encrypted segment also contains a part of the macroblock data. The computation performed using Method II is clearly faster than that of Method 0, but slower than that of Method I.

Finally, Alattar, Al-Regib, and Al-Semari proposed Method III to reduce the amount of computation of Method II; namely instead of encrypting all predicted macroblocks, the encryption in Method III is performed on every nth predicted macroblock, along with encrypting every nth I-macroblock.

Partial Encryption Algorithms for Videos by Cheng and Li, 2000. The partial encryption schemes for still images introduced by Cheng and Li are also extended to video [12]. The approaches proposed by Cheng and Li are not suitable for JPEG image compression and, thus, also not suitable for the MPEG video compression standard. Instead, the partial encryption algorithms are designed for video compression methods, which use either quadtree compression or wavelet compression based on zerotrees for video sequence intraframes, motion compensation, and residual error coding. For example, the proposed approach is applicable to video that is based on the Set Partitioning in Hierarchical Trees (SPIHT) image compression algorithm, which is an implementation of zerotree wavelet compression.

Cheng and Li's partial encryption algorithms are designed to disguise the intraframes (I-frames), the motion vectors, and the residual error code of the given video sequences. In both quadtree-compression- and wavelet-compression-based videos, all I-frames are encrypted using the previously discussed methods for partial encryption of still images by Cheng and Li [12].

In addition, it is also important to encrypt the motion vectors. If the motion vector information is unencrypted, the adversary may be able to use an image frame to obtain approximations to the successive frames. Almost all motion estimation algorithms divide the frame into blocks and try to predict their movement (the position in the next frame) by constructing the estimated motion vector for each block. The blocks that belong to the same large object often have identical motion vectors and it is efficient to encode these vectors together. The authors employ video encryption algorithms that use a quadtree for merging these blocks [12]. Then, quadtree partial encryption is used to encrypt the motion vectors.

Finally, for security purposes, it is important to encrypt the residual error as well. Unencrypted residual error may reveal the outline of a moving object. The residual error is often treated as an image frame and then compressed using some standard image compression algorithm. Again, we restrict ourselves to video compression algorithms that use either a quadtree- or wavelet-based image compression algorithm to compress the residual error frames. Thus, partial encryption schemes for both quadtree and wavelet compression can be applied to the residual error encryption.

MHT-Encryption Scheme and MSI-Coder by Wu and Kuo, 2000 and 2001. In their recent work [28,29], Wu and Kuo proposed two selective encryption algorithms for MPEG video sequences: the *MHT-encryption scheme* (where MHT stands for Multiple Huffman Tables) and the *MSI-coder* (where MSI stands for Multiple State Indices).

The MHT-encryption scheme is based on changing the standard Huffman table to a different one based on a secret key. Shi, Wang, and Bhargava's Algorithm I, which was discussed earlier, proposed a similar idea. The basic procedure for generating the entropy code based on different Huffman trees (the "encryption") is summarized in the following three steps:

1. Generate 2^k different Huffman tables, numbered from 0 to $2^k - 1$.
2. Generate a random vector $P = (p_1, p_2, \ldots, p_n)$, where each p_i is a k-bit integer ranging from 0 to $2^k - 1$.
3. For the ith symbol in the Huffman entropy encoding stage, use the Huffman table indexed by $p_{(i-1(\text{mod } n))+1}$ to encode it.

As the compression ratio should not be affected, it is important to select optimal Huffman tables that perform similarly to the standard one. This, of course, creates an additional constraint that may be a security risk [38]. The authors propose creating a number of so-called train-ing multimedia elements (images or audio data) that can be used to represent the majority of the multimedia data in the world, so that the associated Huffman code that can be easily generated is optimal for general use. The authors' experimental results show a relatively small increase in the size of output images on average, which means that in this approach, the compression performance will be satisfactory, even though the symbols are chosen from multiple Huffman tables.

Another method for obtaining optimal Huffman trees is proposed in Reference 28 and 29; namely, only four base optimal Huffman trees are generated, and the other optimal tables are produced by what is referred to as a *Huffman tree mutation process*. The label pairs are permuted using the secret permutation, as illustrated by the example in Figure 3.8. Clearly, there is no change in the optimality. In addition, the authors propose two slight security enhancements for the MHT-encryption scheme to increase the resistance against known-plaintext and chosen-plaintext attacks (see Reference 29).

The second method (MSI-coder) is designed for video codecs that use QM coding instead of Huffman coding. The QM coder is an adaptive coder based on low-cost probability estimation. It dynamically adjusts the statistical model to a received binary sequence symbols, whose probability is updated at a low computational cost via table lookup. In the standard QM coder, there is a total of 113 possible values for the state index, whose initial value is 0, indicating the equal probabilities of 0's and 1's. Because there are only 113 possibilities to initialize the state index, initializing it to a different "secret" value is pretty useless. Therefore, the MSI-coder uses four indices, which are initially set to different secret values and used alternatively in a secret order. The MSI-coder consists of the following four steps:

1. Generate a random key $K = \{(s_0, s_1, s_2, s_3), (p_0, p_1, \ldots, p_{n-1}), (o_0, o_1, \ldots, o_{n-1})\}$, where each s_i is a 4-bit integer, whereas each p_i and o_i is a 2-bit integer.

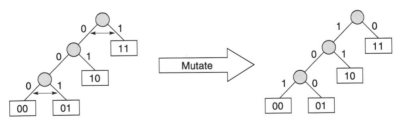

Figure 3.8. The Huffman tree mutation process. (From Wu, C.-P. and Kuo, C.-C.J. in SPIE International Symposia on Information Technologies 2000, pp. 285–295; Wu, C.-P. and Kuo, C.-C.J., in Proceedings of SPIE Security and Watermarking of Media Content III, 2001, vol. 4314.)

2. Initialize four state indices (I_0, I_1, I_2, I_3) to (s_0, s_1, s_2, s_3).
3. To encode the ith bit from the input, we use the index $Ip_{i \pmod n}$ to determine the probability estimation value Q_e.
4. If the state update is required after encoding the ith bit from the input, all state indices are updated except for $Io_{i \pmod n}$.

The not-so-obvious step 4 is included to ensure that the state indices will become different even if they become the same at some point. By experiments, Wu and Kuo showed that the MSI-encoded data is increased in size by only 0.82–4.16% when compared to the data encoded with the original QM coder. Like the previous scheme, the MSI-coder can similarly be slightly enhanced against the known-plaintext and chosen-plaintext attacks [29].

Format-Compliant Configurable Encryption by Wen et al., 2002. This approach, introduced in Reference 25, is a generalization of the ideas of selective encryption into a format-compliant method. It stresses the importance of maintaining the compliance to a standard (such as MPEG-4 codec). In this model, the data are divided into information-carrying and not- information-carrying portions. We only need to encrypt the information-carrying fields. These fields are either fixed-length code (FLC) codewords or variable-length code (VLC) codewords. To achieve the format compliance, we extract the bits that are chosen to be encrypted, concatenate them, and then encrypt with a secure scheme such as DES. Finally, we put the encrypted bits back to their original positions. For FLC coded fields, this operation most likely creates a compliant bit stream. If it does not, we apply the same approach for VLC coded fields. For VLC coded fields, encrypting concatenated codewords may not result in another valid codeword. In order to still maintain compliance, we assign a fixed-length index to each codeword in the VLC codeword list, encrypt the index concatenation, and map it back to the codewords. This, however, introduces the overhead, but the percentage of overhead is content dependent. In their proposal, Wen et al. stress the importance of preserving the format of standard multimedia codecs when including the additional filters such as encryption and decryption.

Selective Scrambling Algorithm by Zeng and Lei, 2002. Zeng and Lei proposed the frequency domain scrambling algorithm that groups the transform coefficients into blocks or segments that undergo some or all of the following three operations: (1) selective bit scrambling, (2) block shuffling, and (3) block rotation [27].

Most of the video/image compression algorithms are based on either the wavelet transform or the DCT. Therefore, the selective scrambling algorithm proposed by Zeng and Lei was designed for two types of compression: wavelet transform based and 8×8 DCT based.

In the case of wavelet-transform-based codecs, selective bit scrambling is performed on the sign bits because they are not highly compressible. Other kinds of bits are refinement bits and significance bits. Significance bits are the worst candidates for selective scrambling because they have the lowest entropy and, therefore, they are highly compressible. On the other hand, refinement bits are somewhere in the middle as far as compressibility is concerned and, thus, they can be scrambled too, but the degradation level is not as high as in sign bit scrambling. In the block shuffling procedure, each subband is divided into a number of blocks of equal size, but for different subbands, the blocksize can vary. These blocks can be shuffled using the table that is generated by some secret key. Visual effects of this procedure are dramatic, whereas the output data increases by only about 5% [26]. Finally, the set of blocks can further be rotated by a cyclic shift controlled by the secret key.

For 8×8 DCT-based algorithms, each I-frame is divided into segments, each containing several macroblocks. Within each segment, the DCT coefficients with the same frequency location are shuffled. In addition, the sign bits of each coefficient can be encrypted. They are flipped "randomly" according to some secret key. The authors suggest that scrambling the P- and B-frames, as well as the motion vectors, is a good idea for security purposes, because they give some information about the original I-frames. If that is computationally expensive, at least the I-blocks in P- and B-frames should be scrambled.

IMAGE ENCRYPTION TECHNIQUES

When dealing with still images, the security is often achieved by using the naive approach to completely encrypt the entire image. However, there are a number of applications for which the naive-based encryption and decryption represent a major bottleneck in communication and processing. For example, a limited bandwidth and processing power in small mobile devices calls for a different approach. Additionally, different image encryption techniques are used as a basis for selective video encryption. In this section, we introduce a few newly proposed techniques for image encryption, based on encrypting only certain parts of the image in order to reduce the amount of computation. This is referred to as *selective image encryption*. A similar idea was already explored in the previous section.

Partial Encryption Algorithms by Cheng and Li, 2000

In Reference 12, Cheng and Li proposed partial encryption methods that are suitable for images compressed with two specific classes of compression algorithms: quadtree-compression algorithms and wavelet-compression algorithms based on zerotrees. The encryption/decryption function is not specified, but, in general, the user can select the conventional cryptosystems such as IDEA, AES, and so forth.

The quadtree image compression produces the quadtree structure and the parameters describing each block in the tree. For simplicity of argument, it is assumed that the only parameter that describes each block is the average intensity. The intensity for each block does not provide enough information about the original image, but the quadtree decomposition allows the reconstruction of outlines of objects in the original frame [12]. Therefore, the quadtree encryption method proposed by Cheng and Li encrypts only the quadtree structure, and the block intensities, located at leaf nodes of a quadtree, are left unencrypted. The quadtree partial encryption can be applied to both lossless and lossy image compression. Another consideration is the ordering of the leaf nodes for the transmission. Figure 3.9 illustrates the different leaf orderings to be discussed. The four branches of each tree node in Figure 3.9 correspond to NE (northeast), SW (southwest), SE (southeast), and NE (northeast) quadrants in that order; the black leaf node has value 0 and the white leaf node has value 1. Vulnerability to certain cryptanalytical attacks exists if the ordering of leaf nodes is done using Leaf Ordering I, where leafs are encoded via the inorder traversal of the quadtree. If we would apply Leaf Ordering I to the tree in Figure 3.9, the result would be binary sequence 0010011011110010. Therefore, the authors suggest using Leaf Ordering II, where leafs are encoded via the opposite of breadth-first traversal of the tree. This increases the security level [12]. For illustration, applying Leaf Ordering II to the quadtree from Figure 3.9 would result in the binary sequence 1111000001100101.

The wavelet-compression algorithm based on zerotrees generally transmits the structure of the zerotree, along with the significant coefficients. A typical example of this kind of compression algorithm is the SPIHT compression algorithm, which transmits the significance of the

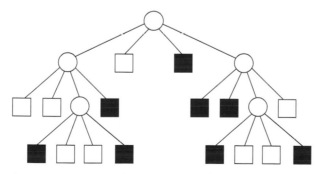

Figure 3.9. **An example of a quadtree for a black-and-white (binary) image.** (From Cheng, H. and Li, X., *IEEE Trans. Signal Process.*, 48(8), 2439–2457, 2000. With permission.)

coefficient sets that correspond to the trees of coefficients. The SPIHT algorithm uses the significance of the coefficient sets to reconstruct the zerotree structure. In addition, the structure of the tree strongly affects the execution of the SPIHT algorithm. The SPIHT-encoded image contains different types of data such as the sign bits, the refinement bits, and the significance bits (significance of pixels and significance of sets). Even the small amount of incorrect information at the beginning of the encoded block containing the significance bits causes failure in the decoding process. Incorrect significance bits often cause the misinterpretation of future bits, unlike the incorrect sign bits or refinement bits, which do not. Therefore, Cheng and Li propose encrypting only the significance information of pixels and sets, as well as the initial threshold parameter n that determines which coefficients are significant [12]. Furthermore, the wavelet transform creates a hierarchy of coefficient bands called *pyramid decomposition*, and the number in the band label is referred to as the *pyramid level*. The authors restrict themselves to encrypting only those significance bits that are in the two highest pyramid levels, because all other pixels and sets are derived by decomposing the sets from the two highest pyramid levels. The information about the significant bits from the first two pyramid levels cannot be deduced just by observing the lower levels.

Selective Encryption Methods for Raster and JPEG Images by Droogenbroeck and Benedett, 2002

In 2002, Droogenbroeck and Benedett proposed selective encryption methods for uncompressed (raster) images and compressed (JPEG) images [13].

An uncompressed (raster) gray-level image defines eight bitplanes. The highest (most significant) bitplanes are highly correlated to the original gray-level image. On the other hand, the least significant bitplanes appear randomlike. Encrypting only the bitplanes that contain nearly uncorrelated values would decrease the vulnerability to the known-plaintext attacks [13]. The authors propose the encryption scheme that consists of xoring the selected bitplanes with a key that has the same number of bits as the bits that are to be encrypted. According to Droogenbroeck and Benedett, at least four to five least significant bitplanes should be encrypted to achieve the satisfactory visual degradation of the image. It is important to note that the partial degradation of the image is not sufficient for applications where high security is a must.

The second method is designed to selectively encrypt the JPEG compressed images. The Huffman entropy encoding creates the symbols based on the run-length coding. In particular, zero coefficients are encoded into the symbols that contain the run of zeros and the magnitude categories for the nonzero coefficients that terminate the runs. These symbols are encoded with the 8-bit Huffman codewords, followed by the appended bits that specify the sign and magnitude of the nonzero coefficients. In the proposed method, only the appended bits that correspond to the selected AC coefficients are encrypted. The DC coefficients are left unencrypted, because their values are highly predictable [13]. On the other hand, the codewords are left unencrypted for synchronization purposes.

Selective Bitplane Encryption Algorithm by Podesser, Schmidt, and Uhl, 2002

In Reference 15, Podesser, Schmidt, and Uhl proposed a selective encryption algorithm for the uncompressed (raster) images, which is in contrast with the first method by Droogenbroeck and Benedett [13]. In the raster image that consists of eight bitplanes, Schmidt and Uhl's algorithm encrypts only the most significant bitplanes. The proposed underlying cryptosystem for this method was AES. However, without loss of generality, any fast conventional cryptosystem can be chosen.

After performing the experiments, Podesser, Schmidt, and Uhl came to the same conclusion in Reference 15 as Droogenbroeck and Benedett did in Reference 13; that is, encrypting only the most significant bitplane (MSB) is not secure. Podesser, Schmidt, and Uhl argue that the MSB bitplane can be reconstructed with the aid of the unencrypted remaining bitplanes. Therefore, they suggest encrypting at least two or four bitplanes. Encrypting only two bitplanes is sufficient if severe alienation of the image data is acceptable, whereas encrypting four bitplanes provides higher security. It is up to the user to determine if the degradation method by encrypting only two bitplanes is sufficient for the given application or if the more secure approach of encrypting four bitplanes is needed.

AUDIO AND SPEECH ENCRYPTION TECHNIQUES

In many applications, the audio sequences need confidentiality during transmission. Sometimes, it is enough to apply the naive approach, but, in many instances, this is too computationally expensive (e.g., in small mobile devices). As far as security is concerned, perhaps the most important type of audio data is speech. Unlike in the case of music files and similar entertainment audio sequences, in many applications speech requires a substantial level of security. This section discusses the methods for speech protection, as well as the protection of general audio compressed sequences such as MP3.

Encryption of Compressed Speech

Traditional speech scrambling has a long history. As in the case of analog video signal, the early work was based on analog devices that simply

127

permute speech segments in the time domain or distort the signal in the frequency domain by applying inverters and filter banks. These schemes are very insecure by today's computing standards [28]. Today's research is shifted toward securing the speech in the digital form, by applying partial or selective encryption [28,30,31].

Selective Encryption Algorithm for G.723.1 Speech Codec by Wu and Kuo, 2000. One of the most popular digital speech codecs is the ITU recommendation G.723.1 compression standard. It has a very low bit rate and it is extremely suitable for voice communications over the packet-switching-based networks. It is also a part of the ITU H.324 standard for videoconferencing/telephony over the regular public telephone lines.

The compression is based on the analysis-by-synthesis method. The encoder incorporates the entire decoder unit, which synthesizes a segment of speech according to given input coefficients. The encoder then changes these coefficients until the difference between the original speech and the synthesized speech is within the acceptable range. The decoding is performed with three decoders: the LSP decoder, the pitch decoder, and the excitation decoder; the G.723.1 coefficients can be categorized depending on the decoder to which they are fed. In addition, this codec can work in either 6.3-Kbps or 5.3-Kbps modes.

In Reference 28, Wu and Kuo suggest applying selective encryption to the most significant bits of all important G.723.1 coefficients. They identified the following coefficients as the important ones: the LSP codebook indices, the lag of pitch predictor, the pitch gain vectors, the fixed codebook gains, and the VAD mode flag. The total number of selected bits for encryption is 37 in each frame, which is less than one-fifth of the entire speech stream at the 6.3-Kbps rate and less than one-fourth of the entire speech stream at the 5.3-Kbps rate.

Perception-Based Partial Encryption Algorithm by Servetti and De Martin, 2002. In 2002, Servetti and De Martin published a perception-based algorithm for partial encryption of telephone bandwidth speech [30]. The algorithm was implemented for the ITU-T G.729 codec for a rate of 8 Kbps.

Servetti and De Martin suggested two methods for partial telephony speech encryption. The goal of the method was to have low security but high bit rate (similar to the degradation schemes for videos and images). The purpose was to degrade the speech enough to prevent immediate eavesdropping. However, more substantial cryptanalysis could reveal the encrypted speech. The second algorithm encrypts more of the bit stream and its goal is to provide more security. Servetti and De Martin argue that even though it encrypts only about a half of the bit stream, the algorithm's security level is comparable to that of the naive algorithm.

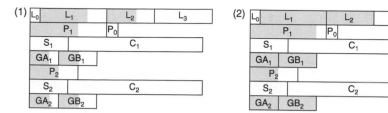

Figure 3.10. Partial encryption for G.729 speech codec (grayed bits are selected for encryption): (1) low-security algorithm and (2) high-security algorithm. (From Servetti, A. and De Martin, J.C., *IEEE Trans. Speech Audio Process.*, 637–643, 10(8), 2002. With permission.)

There is a total of 15 segments from the ITU-T G.729 codec output: L_0 (1 bit), L_1 (7 bits), L_2 (5 bits), L_3 (5 bits), P_1 (8 bits), P_0 (1 bit), S_1 (4 bits), C_1 (13 bits), GA_1 (3 bits), GB_1 (4 bits), P_2 (5 bits), S_2 (4 bits), C_2 (13 bits), GA_2 (3 bits), and GB_2 (4 bits). Rather than describing it in words, the selected bits for encryption in Servetti and De Martin's low- and high-security algorithms are shown in Figure 3.10. The experiments showed that by applying the high-security selective encryption algorithm, the speech is highly scrambled, it cannot be reconstructed from the known bits, and the total computational savings is more than 50%.

Encryption of Compressed Audio

There are numerous applications for which the general audio data needs to be protected and the expensive naive approach might be infeasible. This demand created some research initiatives toward developing selective encryption approaches for compressed audio streams.

MP3 Security Methods by Thorwirth, Horvatic, Weis, and Zhao, 2000. Thorwirth et al. proposed a selective encryption algorithm for perceptual-audio-coding (PAC)-based compression standards, such as MPEG-I Layer-3 (MP3) [32]. In their proposal, the authors concentrated on analyzing encryption of MP3-encoded audio files.

In the late 1980s, the researchers at Fraunhofer Institute developed the audio coding technique called MPEG-I Layer-3, or simply MP3. This codec achieved the stunning compression ratio of $1:10$ to $1:12$ while still maintaining the CD sound quality. It is a PAC that uses the perceptual masking algorithms to discard the redundancy from the raw audio digital signal. All components of the raw audio signal that are inaudible to the human hearing system are identified as redundant data by the perceptual models. The compression in MP3 audio standard is based on the Huffman entropy code.

The hearable frequency spectrum is divided into the segments called the *audio quality layers*. The 20 Hz to 4 kHz range represents the lowest audio quality layer, whereas the CD quality range is 20 Hz to 22 kHz. The authors propose encrypting the audio quality layers associated with the compressed audio data, in order to enforce audio quality control by means of encryption. The algorithm identifies the frequency spectrum boundaries that are used to determine the audio quality layers. Each known layer is then grouped into the blocks of equal size and a block cipher is applied. After the encryption stage, the encrypted data can be easily plugged back to the MP3 bit stream, thus preserving the format. Such encrypted MP3 sequences are decodable and playable by any valid MP3 player [32].

SUMMARY

In this chapter, many of the current important multimedia encryption techniques were presented and analyzed. In brief, the best way of protecting multimedia data is by means of the naive algorithm (i.e., by encrypting the entire multimedia bit sequence using a fast conventional cryptosystem). Yet, in many instances, this is not possible due to the expensive computational overhead introduced by the encryption and decryption stages. Much of the past and current research targets encrypting only a carefully selected part of the multimedia bit stream in order to reduce the computational load and yet keep the security level high. Many of the proposed schemes only achieved moderate to low security, which may find applications in which quality degradation is preferred over absolute security. However, only few of the proposed methods promise to achieve substantial security, which is the number one requirement in many multimedia applications. It is also worthwhile mentioning that many of the above-proposed audiovisual encryption techniques are flexible in terms of selecting the underlying cryptosystem, regardless of the choice made by the authors. In general, a well-studied, fast, and secure conventional cryptosystem should be chosen. Surely, AES is a good candidate, but there are numerous other successful modern cryptosystems with similar and, in some aspects, even better performances.

Further research should be directed toward developing substantially secure encryption schemes for high bit rate and high-quality audiovisual data. Furthermore, the experts should investigate the security aspects of the newly proposed multimedia encryption techniques.

REFERENCES

1. *IEEE Transactions on Circuits and Systems for Video Technology* (Special issue on authentication, copyright protection, and information hiding), 13(8), 2003.
2. Eskicioglu, A.M., Protecting intellectual property in digital multimedia networks, *IEEE Computer*, (Vol. 36, No. 7), 39–45, 2003.

3. van Oorschot, P.C., Menezes, A.J., and Vanstone, S.A., *Handbook of Applied Cryptography*, CRC Press, Boca Raton, FL, 1997.
4. Seidel, T., Socek, D., and Sramka, M., Parallel Symmetric Attack on NTRU using Non-Deterministic Lattice Reduction, Designs, Codes and Cryptography, Kluwer Academic Publishers, May 2004, vol. 32, iss. 1–3, pp. 369–379(11).
5. Howgrave-Graham, N., Nguyen, P., Pointcheval, D., Proos, J., Silverman, J.H., Singer, A., and Whyte, W., The Impact of Decryption Failures on the Security of NTRU Encryption, in Proceedings of Crypto 2003, Santa Barbara, CA, 2003.
6. Daemen, J. and Rijmen, V., AES Proposal: Rijndael, AES Algorithm Submission, September 3, 1999, available at http://www.nist.gov/CryptoToolkit.
7. Stinson, D.R., *Cryptography Theory and Practice*, CRC Press, Boca Raton, FL, 2002.
8. McGrew, D., Naslund, M., Norman, K., Blom, R., Carrara, E., and Oran, D., The Secure Real time Transport Protocol (SRTP), RFC 3711, March 2004.
9. Furht, B. and Socek, D., Multimedia Security: Encryption Techniques, IEC Comprehensive Report on Information Security, International Engineering Consortium, Chicago, IL, 2003.
10. Liu, X. and Eskicioglu, A.M., Selective Encryption of Multimedia Content in Distribution Networks: Challenges and New Directions, Presented at IASTED International Conference on Communications, Internet and Information Technology (CIIT 2003), Scottsdale, AZ, 2003.
11. Lookabaugh, T., Sicker, D.C., Keaton, D.M., Guo, W.Y., and Vedula, I., Security Analysis of Selectively Encrypted MPEG-2 Streams, Presented at Multimedia Systems and Applications VI Conference, Orlando, FL, 2003.
12. Cheng, H. and Li, X., Partial encryption of compressed images and video, *IEEE Trans. Signal Process.*, 48(8), 2439–2451, 2000.
13. Van Droogenbroeck, M. and Benedett, R., Techniques for a Selective Encryption of Uncompressed and Compressed Images, in Proceedings of Advanced Concepts for Intelligent Vision Systems (ACIVS) 2002, Ghent, 2002.
14. Pommer, A. and Uhl, A., Selective encryption of wavelet-packet encoded image data, *ACM Multimedia Syst. J.*, (Vol. 9, No. 3), pp. 279–287, 2003.
15. Podesser, M., Schmidt, H.-P., and Uhl, A., Selective Bitplane Encryption for Secure Transmission of Image Data in Mobile Environments, Presented at 5th Nordic Signal Processing Symposium, Hurtigruten, Norway, 2002.
16. Meyer, J. and Gadegast, F., Security Mechanisms for Multimedia Data with the Example MPEG-1 Video, Project Description of SECMPEG, Technical University of Berlin, 1995.
17. Maples, T.B. and Spanos, G.A., Performance Study Of Selective Encryption Scheme For The Security Of Networked Real-Time Video, in Proceedings of the 4th International Conference on Computer and Communications, Las Vegas, NV, 1995.
18. Spanos, G.A. and Maples, T.B., Security for Real-Time MPEG Compressed Video in Distributed Multimedia Applications, Presented at Conference on Computers and Communications, 1996, pp. 72–78.
19. Tang, L., Methods for Encrypting and Decrypting MPEG Video Data Efficiently, in Proceedings of the 4th ACM International Multimedia Conference, Boston, 1996, pp. 219–230.
20. Qiao, L. and Nahrstedt, K., A New Algorithm for MPEG Video Encryption, in Proceedings of the 1st International Conference on Imaging Science, Systems and Technology (CISST '97), 1997, pp. 21–29.
21. Shi, C. and Bhargava, B., A Fast MPEG Video Encryption Algorithm, in Proceedings of the 6th International Multimedia Conference, Bristol, U.K., 1998.
22. Shi, C., Wang, S.-Y., and Bhargava, B., MPEG Video Encryption in Real-Time Using Secret key Cryptography, in 1999 International Conference on Parallel and Distributed Processing Techniques and Applications (PDPTA'99), Las Vegas, NV, 1999.

23. Alattar, A.M., Al-Regib, G.I., and Al-Semari, S.A., Improved Selective Encryption Techniques for Secure Transmission of MPEG Video Bit-Streams, in Proceedings of the 1999 International Conference on Image Processing (ICIP '99), Kobe, 1999, Vol. 4, pp. 256–260.

24. Shin, S.U., Sim, K.S., and Rhee, K.H., A Secrecy Scheme for MPEG Video Data Using the Joint of Compression and Encryption, Lecture Notes in Computer Science Vol. 1729, Springer-Verlag, Berlin, 1999, pp. 191–201.

25. Wen, J., Severa, M., Zeng, W., Luttrell, M., and Jin, W., A format-compliant configurable encryption framework for access control of video, *IEEE Trans. Circuits Syst. Video Technol.*, 12(6), 545–557, 2002.

26. Zeng, W. and Lei, S., Efficient frequency domain selective scrambling of digital video, *IEEE Trans. Multimedia*, 5(1), 118–129, 2002.

27. Wu, M. and Mao, Y., Communication-Friendly Encryption of Multimedia, Presented at 2002 International Workshop on Multimedia Signal Processing, St. Thomas, 2002.

28. Wu, C.-P. and Kuo, C.-C.J., Fast Encryption Methods for Audiovisual Data Confidentiality, Presented at SPIE International Symposia on Information Technologies 2000, Boston, MA, 2000, pp. 284–295.

29. Wu, C.-P. and Kuo, C.-C.J., Efficient Multimedia Encryption via Entropy Codec Design, in Proceedings of SPIE Security and Watermarking of Multimedia Content III, San Jose, CA, 2001,Vol. 4314.

30. Servetti, A. and De Martin, J.C., Perception based partial encryption of compressed speech, *IEEE Trans. Speech Audio Process.*, 10(8), 637–643, 2002.

31. Servetti, A., Testa, C., and De Martin, J.C., Frequency-Selective Partial Encryption of Compressed Audio, Presented at IEEE International Conference on Aucoustics, Sppech and Signal Processing, Hong-Kong, 2003.

32. Thorwirth, N.J., Horvatic, P., Weis, R., and Zhao, J., Security Methods for MP3 Music Delivery, Presented at Conference Record of the Thirty-Fourth Asilomar Conference on Signals, Systems and Computers, 2000, Vol. 2, pp. 1831–1835.

33. Agi, I. and Gong, L., An Empirical Study of Secure MPEG Video Transmission, in Proceedings of the Symposium on Network and Distributed Systems Security, 1996.

34. Qiao, L. and Nahrstedt, K., Comparison of MPEG encryption algorithms, *Int. J. Computer Graphics*, 22(3), 1998.

35. Alattar, A. and Al-Regib, G., Evaluation of Selective Encryption Techniques for Secure Transmission of MPEG Video Bit-streams, in *Proceedings of the IEEE International Symposium on Circuits and Systems*, 1999, Vol. 4, pp. IV-340–IV-343.

36. Qiao, L., Nahrstedt, K., and Tam, I., Is MPEG Encryption by Using Random List Instead of Zigzag Order Secure? in IEEE International Symposium on Consumer Electronics, 1997.

37. Bhargava, B., Shi, C., and Wang, Y., MEPG Video Encryption Algorithms, available at http://raidlab.cs.purdue.edu/papers/mm.ps, 2002.

38. Seidel, T., Socek, D., and Sramka, M., Cryptanalysis of Video Encryption Algorithms, in Proceedings of the 3rd Central European Conference on Cryptology TATRACRYPT 2003, Bratislava, Slovak Republic, 2003.

4
Chaos-Based Encryption for Digital Images and Videos

Shujun Li, Guanrong Chen, and Xuan Zheng

INTRODUCTION

Many digital services, such as pay TV, confidential videoconferencing, medical and military imaging systems, require reliable security in storage and transmission of digital images and videos. Because of the rapid progress of the Internet in the digital world today, the security of digital images and videos has become more and more important. In recent years, more and more consumer electronic services and devices, such as mobile phones and personal digital assistant (PDA), have also started to provide additional functions of saving and exchanging multimedia messages. The prevalence of multimedia technology in our society has promoted digital images and videos to play a more significant role than the traditional dull texts, which demands a serious protection of users' privacy. To fulfill such security and privacy needs in various applications, encryption of images and videos is very important to frustrate malicious attacks from unauthorized parties.

From the early 1990s, many efforts have been made to investigate specific solutions to image and video encryption. Due to the tight relationship between chaos theory and cryptography, chaotic cryptography has also been extended to design image and video encryption schemes. This chapter focuses on different image and video encryption algorithms based on chaos and clarifies some experiences, lessons, and principles of using chaos to design such encryption schemes. To start, a comprehensive discussion on the state of the art of image and video encryption will first be given as background and motivation for the discussion of the chapter.

The organization of this chapter is as follows. In section "Image and Video Encryption: Preliminaries", some preliminaries on image and video encryption are given. Section "Image and Video Encryption: A Comprehensive Survey" is a comprehensive review on today's image and video encryption technology without using chaos theory. Chaos-based image and video encryption schemes are then surveyed in section "Chaos-Based Image and Video Encryption". Some experiences, lessons, and principles drawn from existing chaos-based image and video encryption algorithms are commented on and discussed in section "Experiences and Lessons". The last section concludes the chapter.

IMAGE AND VIDEO ENCRYPTION: PRELIMINARIES
A Brief Introduction to Modern Cryptography

To facilitate the following discussion, in this subsection, we give a brief introduction to the basic theory of modern cryptography [1].

An encryption system is also called a *cipher* or a *cryptosystem*. The message for encryption is called *plaintext*, and the encrypted message is called *ciphertext*. Denote the plaintext and the ciphertext by P and C, respectively. The encryption procedure of a cipher can be described as $C = E_{K_e}(P)$, where K_e is the encryption key and $E(\cdot)$ is the encryption function. Similarly, the decryption procedure is $P = D_{K_d}(C)$, where K_d is the decryption key and $D(\cdot)$ is the decryption function. Following Kerckhoffs' principle [1], the security of a cipher should only rely on the decryption key K_d, because adversaries can recover the plaintext from the observed ciphertext once they get K_d. See Figure 4.1 for a diagrammatic view of a cipher.

There are two kinds of cipher following the relationship of K_e and K_d. When $K_e = K_d$, the cipher is called a *private key cipher* or a *symmetric cipher*. For private key ciphers, the encryption–decryption key must be transmitted from the sender to the receiver via a separate secret channel. When $K_e \neq K_d$, the cipher is called a *public key* cipher or an *asymmetric*

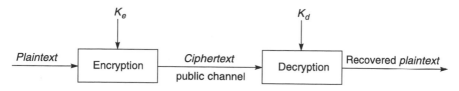

Figure 4.1. Encryption and decryption of a cipher.

cipher. For *public key* ciphers, the encryption key K_e is published and the decryption key K_d is kept private, for which no additional secret channel is needed for key transfer.

According to the encryption structure, ciphers can be divided into two classes: *block ciphers* and *stream ciphers*. Block ciphers encrypt the plaintext block by block and each block is mapped into another block with the same size. Stream ciphers encrypt the plaintext with a pseudorandom sequence (called *keystream*) controlled by the encryption key.

A cryptographically secure cipher should be strong enough against all kinds of attack. For most ciphers, the following four attacks should be tested: (1) *ciphertext-only attack* — attackers can get the ciphertexts only; (2) *known-plaintext attack* — attackers can get some plaintexts and the corresponding ciphertexts; (3) *chosen-plaintext attack* — attackers can choose some plaintexts and get the corresponding ciphertexts; (4) *chosen-ciphertext attack* — attackers can choose some ciphertexts and get the corresponding plaintexts. It is known that many image and video encryption schemes are not secure enough against known- or chosen-plaintext attack, as shown in the following subsections.

The Need for Image and Video Encryption Schemes

The simplest way to encrypt an image or a video is perhaps to consider the two-dimensional (2-D) or 3-D stream as a 1-D data stream and then encrypt this 1-D stream with any available cipher [2–5]. Following References 6–8, such a simple idea of encryption is called *naive encryption*. Although naive encryption is sufficient to protect digital images and videos in some civil applications, the following issues have to be taken into consideration when advanced encryption algorithms are specially designed for sensitive digital images and videos, because their special features are very different from texts:

1. *Trade-off between bulky data and slow speed*: Digital images and videos are generally bulky data of large sizes, even if they

135

are efficiently compressed (see discussion below). Because the encryption speed of some traditional ciphers is not sufficiently fast, especially for software implementations, it is difficult to achieve fast and secure real-time encryption simultaneously for large-sized bulky data.

2. *Trade-off between* encryption and compression: If encryption is applied before compression, the randomness of ciphertexts will dramatically reduce the compression efficiency. Thus, one has to apply encryption after compression, but the special and various image and video structures make it difficult to embed an encryption algorithm into the integrated system. For example, some popular compression standards (such as MPEG-x) are antagonistic toward selective encryption [9]; that is, there exist notable trade-offs between the compression and the encryption [10].

3. *Dependence of encryption on compression*: The lossy compression technique is widely used for images and videos to dramatically reduce the size of the data for encryption and it is natural to expect that the design and implementation of fast encryption schemes will be easier. However, it was pointed out [11] that the encryption cannot benefit much from such a data reduction, because generally the time consumed by the compressor is much longer than that by the encryption algorithm. This implies that the encryption efficiency depends heavily on the involved compression algorithm.

4. *Incapability of compression to reduce data size*: In general, lossy compression of images and videos is not acceptable in some applications due to legal considerations. Medical imaging systems are well-known examples, for which the diagnostic images and videos of all patients are required to be stored in lossless forms. In this case, the only choice is to use lossless compression or to leave the images and videos uncompressed, which emphasizes, once again, the trade-off between bulky data and slow encryption speed.

5. *Intractable high redundancy*: There exists high (short distance and long distance) redundancy in uncompressed images and videos, which may make block ciphers running in the ECB (electronic code book) mode fail to conceal all visible information in some plain-images and plain-videos. As a result, block cipher running in the CBC (cipher block chaining) mode (or other modes [1]) or stream ciphers should be used to encrypt uncompressed images and videos with high redundancy.

- If an image contains an area with a fixed color, which means large redundancy, the edge of this area will be approximately

(a)

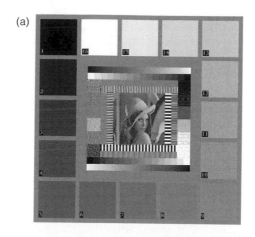

(b)

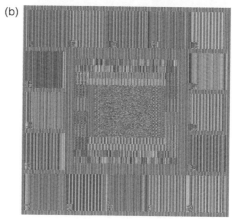

Figure 4.2. An uncompressed plain image (a) containing many areas with fixed gray levels and its corresponding cipher image (b) encrypted by 128-bit Advanced Encrytion Standard (AES) running in the ECB mode.

preserved after encryption. This is because those consecutive identical pixels lead to the same repeated patterns when a block cipher is used in the ECB mode. See Figure 4.2 for a real example of this phenomenon.

- Interestingly, a recent work [12] reported that encrypted BMP images (uncompressed) are less random than encrypted JPEG images (compressed), which implies that the high redundancy cannot be removed efficiently after encryption.

6. *Loss of the avalanche property*: Apparently, a direct use of traditional ciphers in image and video encryption cannot provide the avalanche property at the level of the image and video frame. For example, for two images with only 1-bit difference at the

position (i,j), all cipher-pixels except for the ones at (i,j) will be identical in the ECB mode, and all cipher-pixels before (i,j) will be identical in the CBC mode. To maintain the avalanche property, special algorithms should be developed.

7. *New concepts of security and usability*: The diverse multimedia services need different security levels and usability requirements, some of which should be defined and evaluated with human vision capabilities. A typical example is the so-called perceptual encryption, with which only partial visible information is encrypted and the cipher-image and cipher-video gives a rough view of the high-quality services.

Some Special Features of Image and Video Encryption Schemes

In image and video encryption systems, some features are required to support special functions of diverse multimedia services in different environments. These features are generally realized with a combination of compression and encryption and impose some limits on the design of the image and video encryption algorithms.

- *Format compliance* (also called *transcodability* [13], *transparency*, or *syntax-awareness* [14]): This feature means that the encrypted image and video are still decodable at the receiver end without knowledge about the decryption key. For online stream multimedia services, especially those running in wireless environments, the transparency property is desirable to eliminate the problems caused by loss or uncorrected data. The transparency property is also useful for the ease of concatenating other postprocessing devices (such as digital watermarking) to the whole compression or encryption system errors [14,15]. To achieve transparency, the encryption procedure should not destroy the syntax structure of the encrypted file or stream; that is, the descriptive information of the file or stream should be left unencrypted.

- *Scalability*: Scalability means multilevel security for different applications with flexible parameter settings [16–25]. The embedded multilayer structure (i.e., the fine granularity scalability [FGS]), of JPEG2000 images [26] and MPEG-2/4 videos [27] makes scalable encryption easy and natural. The basic idea in realizing scalability is to encrypt partial layers and partial data in selected layers. Scalability can be considered as a control mechanism for the visual quality of the encrypted image and video.

- *Perceptibility*: Perceptibility means partial encryption of visible information of the plain image or plain video, which is useful for pay-after-trial services of digital multimedia, such as pay TV and video-on-demand (VoD) services [13]. It is a generalization of scalable (multilayered) encryption and does not depend on the embedded multilayered structure of the encrypted image and video. Two typical perceptual encryption schemes for JPEG images and MP3 music were proposed in References 28 and 29, respectively. The selective image encryption schemes proposed in References 30–35 and video scrambling schemes in References 36 and 37 are also examples of perceptibility, although this term was not explicitly used.

- *Error tolerability*: The concept of *error-tolerating* (or *error-resilient*) encryption was investigated in References 22 and 38 and also mentioned in References 14 and 15. It is undesirable if an encrypted stream cannot be decoded when some bit errors are introduced, which frequently occurs in multimedia applications, particularly in wireless environments (error-correction mechanism may fail in some situations). However, the avalanche property of good ciphers means high sensitivity to errors, which may lead decryption to fail in some cases. To solve this problem, the idea of selective encryption was used in Reference 22 to a provide better error-tolerating property; in Reference 38, the possibility of designing an *error-preserving* encryption algorithm was studied.

Generally speaking, there exist tight relationships among the above-discussed features: (1) Transparency and the idea of *selective encryption* (see section "Selective Encryption" for more details) are requirements of scalability and perceptibility and (2) multilevel security is achieved by providing perceptibility in some scalable encryption schemes.

IMAGE AND VIDEO ENCRYPTION: A COMPREHENSIVE SURVEY

Generally speaking, there are two basic ways to encrypt digital image: in the spatial domain or in the transform domain. Because digital videos are generally compressed in the DCT (discrete cosine transform) domain, almost all video encryption algorithms work in DCT domain. Due to the recent prevalence of the wavelet compression technique and the adoption of the wavelet transform in JPEG2000 standard [26], in recent years image and video encryption algorithms working in the wavelet domain also attracted some attention [35,39–43]. In addition, some novel image and video compression algorithms have also been proposed to realize joint compression–encryption schemes.

Although many efforts have been devoted to better solutions for image and video encryption, the current security analyses of many schemes are not sufficient, especially for the security against known-plaintext and chosen-plaintext attacks. What is worse, many selective encryption schemes are, indeed, insecure against ciphertext-only attack, due to the visible information leaking from unencrypted data.

In this section, we provide a comprehensive survey on image and video encryption schemes without using chaos theory, as a background of chaos-based encryption schemes. Before that, a widely used idea in the image and video encryption community, called selective (or partial) encryption, was introduced in order to facilitate the security evaluations of selective image and video encryption schemes to be carried out later.

Selective Encryption

Since the 1990s, selective encryption has been widely suggested and adopted as a basic idea for the encryption of digital images and videos, aiming to achieve a better trade-off between the encryption load and the security level. In Reference 22, it was pointed out that selective encryption is also useful for realizing the error-tolerating property in wireless video transmission. The MPEG-4 IPMP (Intellectual Property Management and Protection) extensions standard also supported selective encryption [44]. In addition, as mentioned earlier, scalability and perceptibility are generally realized via selective encryption.

Some reviews of selective encryption methods and their performances can be found in References 9 and 45–47, among which Reference 47 gave a partial list of some representative selective encryption methods and Reference 9 analyzed the potential insecurity of different selective syntactic elements in the MPEG-2 video stream against possible attacks. Although it seems that selective encryption is a good idea for the design of high-efficiency image and video encryption schemes, some essential defects of selective encryption have been pointed out, showing its incapability to provide a satisfactory balance between security and usability. For selective encryption methods of MPEG videos, for instance, the following defects have been clarified [9–11,14,47–50]:

1. Although the energy of a compressed image concentrates on lower DCT coefficients, the visible information does not concentrate on partial coefficients, but scatters over all DCT coefficients. Actually, only one DCT coefficient is enough to recover some visible information of the concerned plain-frame, as shown in Figure 2d of Reference 10 and Figure 5.2 of Reference 51 on some experimental images. By setting all encrypted DCT coefficients to fixed values, it is possible for an attacker to recover a rough view of the plain-frame. A more detailed discussion on this defect can be found in Section IV of Reference 14, where this defect is used to realize the so-called *error-concealment attack* (also called *perceptual attack* [9]).

2. The scattering effect of visible information also exists in different bits of DCT coefficients, which makes it insecure to partially

encrypt significant bits of DCT coefficients. In Figure 3 of Reference 10, it was shown that neither encrypting the sign bits nor encrypting multiple significant bits are secure enough against ciphertext-only attack utilizing the unencrypted bits.

3. Only encrypting I-frames of a video cannot provide sufficient security against ciphertext-only attack. The unencrypted I-blocks and motion vectors in B- and P-frames can be still used to uncover partial visible information of the original videos.

4. Because all I-frames occupy about 30–60% or even more of an MPEG video [8,25], the reduction of computation load of encrypting I-frames only is not significant. If all I-macroblocks in B- and P-frames are also encrypted, the encryption load will be close to that of full encryption. In Reference 11, it was further pointed out that selective encryption of compressed images and videos cannot significantly save the overall processing time, because the compression procedure generally consumes more time than the encryption of the compressed images and videos.

5. If the selective encryption is exerted before the entropy-coding stage, the compression performance will decrease, whereas if the selective encryption is exerted after the entropy-coding stage, the format-compliance may be lost. If the entropy-coding algorithm is kept secret, the secret entropy codec is generally insecure against plaintext attack, because the sizes of the Huffman tables are too small from the cryptographical point of view.

Apparently, the first two defects of selective encryption of MPEG videos are due to the orthogonality of DCT. Because all transforms used in image and video compression algorithms have the orthogonal property, the two defects exist for all transform-based image and video compression algorithms. The first two defects on uncompressed images were also pointed out in Reference 50, and the second defect was used in Reference 33 to achieve perceptual encryption with controllable quality of cipher-images.

In Reference 9, a comparison of the performances of selectively encrypting different syntactic elements of MPEG-2 was given, and three kinds of relationship between the compressor and the selective encryption system were clarified: cooperative, neutral, and antagonistic. Although it was thought in Reference 9 that the relationship between the MPEG compressor and almost selective MPEG encryption schemes is neutral, the aforementioned defects of selective MPEG encryption imply that the MPEG compressor plays an antagonistic role with respect to the selective encryption.

In summary, considering the security defects of selective image and video encryption, it is difficult to simply use the idea of selective encryption alone to achieve a high security level. From a conservative point of view, the meaningful use of selective encryption is to realize perceptual encryption (i.e., to degrade the visible quality of the encrypted plain-image and plain-video), rather than to provide cryptographically strong security. It has been argued that selective encryption can work well with model-based compression algorithms, such as context-based arithmetic coding [53] and model-based speech coding [10], but further studies are still needed to confirm the security level of the selective encryption in such compression models.

Joint Image and Video Encryption Schemes

Due to the similarity between MPEG videos and JPEG images, most MPEG encryption methods can be used to encrypt JPEG images directly [33,54,55]. On the other hand, some encryption techniques proposed for images can also be extended to encrypt videos with similar structures. Therefore, the distinction between image encryption and video encryption is not prominent. In this subsection, we discuss those schemes lying on their boundary: the joint image and video encryption schemes.

- Almost all *DCT-based encryption schemes* can be used to encrypt both JPEG images and MPEG videos, although most DCT-based encryption schemes were originally proposed for MPEG videos, with a few for JPEG images [33,55]. Video encryption schemes for MPEG will be further discussed in section "Video Encryption Schemes".

- Because entropy coding is widely used in image and video compression algorithms, the idea of *making entropy codec secret* can work for both image and video encryptions.

  In References 56 and 57, the secretly permuted Huffman table is used to encrypt the input image or video stream. This scheme is not secure enough against known-plaintext or chosen-plaintext attacks and its key space is too small.

  In Reference 10, multiple Huffman tables (MHTs) are employed and a random index is used as the secret key to select a Huffman table for each codeword. This scheme cannot resist the chosen-plaintext attack. Later, two methods were proposed to enhance its security: secretly inserting pseudorandom bits [58] and changing the random index frequently [58,59].

  In Reference 60, another encryption scheme based on secret Huffman tables was proposed. Random flipping of the last bit of each codeword is introduced to further adaptively change the Huffman table.

In References 14 and 15, the Huffman tables are left untouched and the secret entropy codec is achieved as follows: Map each plain-codeword to an index representing its position in the Huffman table, encrypt the index, and then remap the encrypted index back to generate the cipher-codeword.

- All *wavelet-based* image *encryption schemes* can be extended to encrypt videos based on wavelet compression algorithms, such as motion JPEG2000 [26]. Wavelet-based image encryption schemes will be surveyed in the next subsection.

    In References 61 and 62, it was suggested to use the following three operations for encryption of wavelet-compressed videos: selective bit scrambling, block shuffling, and block rotation. It appears that this scheme cannot resist the chosen-plaintext attack.

    In References 51, 63, and 64, the use of selective encryption in SPIHT-encoded images was studied, in which SPIHT coding is cooperative with the selective encryption.

- In References 51 and 63–65, the use of selective encryption in *quadtree decomposition* of digital images was discussed in detail. It was pointed out that Leaf Ordering I (from top to bottom) is not sufficiently secure in the selective encryption framework. This scheme was also extended to selective video encryption.

- In Reference 53, a selective encryption scheme based on *context-based arithmetic coding* of DWT (discrete wavelet transform) coefficients was proposed. This scheme uses a traditional cipher to selectively encrypt partial leading bits of the compressed bit stream. Without knowledge of the leading bits, it is cryptographically difficult to decompress the bit stream due to the sensitivity of the decompression to the leading bits (i.e., to the modeling context). This scheme is a good example that exhibits the cooperative relationship between compression and selective encryption.

Image Encryption Schemes

- One of the simplest idea to selectively encrypt images in the spatial domain is to *encrypt one or more bitplanes* [33,66,67]. Because the MSB (most significant bit) contains the most defining information on the image, it serves as the first bitplane for encryption. As a result, the selection order is generally from MSB to LSB least significant bit [66,67], but the order should be reversed for perceptual encryption schemes so as to reserve some visible information [33].

- In Reference 68, a *VQ-based image encryption* algorithm was proposed. In this scheme, the codebook is first extended, then transformed to another coordinate system, and, finally, permuted to make a shuffled codebook in a larger size. The parameters used in the above three processes are encrypted with a traditional cipher. It was claimed [68] that this scheme can resist all kinds of known attack, and the encryption load is very light.

- In Reference 39, *secret permutations* were suggested to shuffle the *DWT coefficients* in each subband of wavelet compressed images. An enhanced version of this scheme suggested encrypting the lowest subband with a traditional cipher. In Reference 69, some permutation methods of DWT coefficients were compared, and a novel method was proposed to realize a stronger confusion of shuffled coefficients. Because of the scattering effect of visible information in all DWT coefficients (see Figures 1 and 3 of Reference 39), such permutations alone are not secure enough against the ciphertext-only attack, if not all subbands are permuted. Also, the secret permutations can be reconstructed under known-plaintext and chosen-plaintext attacks.

- In References 32 and 35, several perceptual encryption schemes were proposed by *selectively encrypting partial DWT coefficients*. The bit-shift encryption function used in Reference 35 is not secure enough against known-plaintext and chosen-plaintext attacks. In References 67 and 76 the use of selective encryption in JPEG2000 image encryption was discussed. It was pointed out that at least 20% of encrypted data is needed to provide sufficient security.

- In Reference 43, *encryption of sign bits of partial DWT coefficients* was suggested. This scheme is not secure due to the aforementioned second defect of selective encryption.

- In Reference 71, the idea of *encrypting wavelet filters* was proposed to develop a new encryption system. Later, in References 40, 41, and 72, this algorithm was studied with wavelet packet decomposition and it was shown that such an encryption scheme is not secure enough against the ciphertext-only attack based on a heuristic cryptanalytic method.

- In References 41, 42, 73, and 74, the possibility of *encrypting the quadtree decomposition structure* was studied for images compressed with *wavelet packet algorithms*. Uniform scalar quantization of wavelet coefficients was found insecure against the ciphertext-only attack, and the zerotree coder was suggested to provide higher security.

- *Encryption schemes based on fractal compression* were proposed in References 30 and 31. Selected parameters are encrypted to achieve perceptibility.

- In Reference 75, an image encryption scheme was proposed based on the so-called *base-switching lossless compression algorithm*. The base-switching compression is a simple lossless algorithm, which compresses each 3×3 block of the image into a 3-tuple data (m, b) and an integer with nine b-base bits. The base b of each block is encrypted with a secret polynomial function to generate the cipher-image. Apparently, this scheme cannot resist known-plaintext and chosen-plaintext attacks, because using a polynomial function is not sufficiently secure from the cryptographical point of view.

Video Encryption Schemes

MPEG Encryption Schemes

- The most frequently used idea of MPEG encryption is to *encrypt selective frames, macroblocks, DCT coefficients*, and *motion vectors*. The following is a list of selective data for encryption in different schemes (from light to heavy encryptions):

  All header information (and partial blocks) [76]

  Selective AC coefficients of Y/V blocks in all I-frames [77]

  Selective leading DCT coefficients of each block [78]

  Selective DCT coefficients of each block and motion vectors[61,62]

  All or selective motion vectors [36]

  All I-frames[34,79] and the header of the MPEG video sequence [48,80]

  All I- and P-frames [79], or all I-frames and I-macroblocks in B- and P-frames [76]

  All or selective I-macroblocks and the headers of all predicted macroblocks [81].

Note that almost all selective MPEG encryption schemes have the defects mentioned in section "Selective Encryption" and the encryption of the headers will cause the loss of format-compliance.

- Another frequently used idea is to *secretly permute* all *or selective macroblocks, blocks, DCT coefficients* and *motion vectors*.

  In Reference 54, DCT coefficients are secretly permuted within each block, and the DC coefficient is split into two halves and the 63rd AC coefficient is set to be the higher half of the DC coefficient.

  In References 61 and 62, the following three operations are combined to encrypt images: secret permutations of DCT coefficients, selective encryption of DCT coefficients, and motion vector scrambling. In this scheme, several blocks

145

compose a segment, and DCT coefficients in different blocks at the same frequency are secretly permuted.

In Reference 82, the secret permutations of DCT coefficients and the encryption of sign bits of DC coefficients are combined to realize a selective encryption scheme.

In References 14 and 83, different basic units are secretly permuted: macroblocks, 8 × 8 blocks, and run-level codewords. The secret permutation tables are pseudorandomly generated from some encrypted local-content-specific bits.

The first three of the above schemes are all insecure against ciphertext-only attack [6,7,49,84,85] and known-plaintext and chosen-plaintext attacks [6,7,49,84]. Another disadvantage of the secret permutation of DCT coefficients is the expansion of the video size [61,62].

- In References 57 and 86–88, three encryption schemes were proposed that *encrypt selective sign bits of DCT coefficients and motion vectors*, which are respectively called VEA (video encryption algorithm) [87], MVEA (modified VEA) [86], and RVEA (real-time VEA) [88]. Note that the encryption of sign bits was also involved in other designs [14,82].

  VEA is a simple cipher XORing all sign bits with a repeated m-bit key. It is too simple to resist ciphertext-only attack and known-plaintext and chose-plaintext attacks.

  MVEA is a simple modification of VEA, in which only sign bits of DC coefficients in I-frames and sign bits of motion vectors in B- and P-frames are XORed with the secret key. In fact, MVEA is weaker than VEA [57, appendix].

  RVEA is a combined version of VEA and MVEA, where the XOR operations are replaced by a traditional cipher. For each macroblock, at most 64 sign bits are encrypted with the order from low frequency to high frequency (see Figure 5 of Reference 57). Although RVEA is the strongest VEA cipher, the attempt of using unencrypted AC coefficients to reconstruct some visible information is still possible.

Generic Video Encryption Schemes

- In References 6,7, and 52, another VEA was proposed to encrypt video stream: divide each 128-byte piece of the plain-video into two 64-byte lists: an Odd List and an Even List, and the Even List is XORed by the Odd List and then encrypted. This VEA can reduce the encryption load by 50%. An extended VEA with a lower encryption cost was proposed in Reference 22 to support the error-tolerability property.

- In Reference 89, a simple video encryption scheme was proposed that applies a *secret linear transformation* on each pixel value. All pixels in the same macroblock are encrypted with the same secret parameters set. In Reference 37, a similar encryption scheme was proposed. Both schemes are insecure against known-plaintext and chosen-plaintext attacks.

- In Reference 90, *combining a fast (stream) cipher and a secure (block) cipher* was suggested to achieve overall fast encryption of digital videos. The encryption procedure can be described as follows. Assume that C and M respectively mean the ciphertext and the plaintext, and that **SE** and **FE** respectively mean the secure cipher and the fast cipher,

$$C = \{\mathbf{SE}(K, M_{ik+1}), \mathbf{FE}(K_{i+1}, M_{ik+2}M_{ik+2} \cdots M_{(i+1)k})\}_{i=0}^{t}$$

where $K_{i+1} = \mathbf{SE}(K, \mathbf{SE}(K, M_{ik+1}))$. Two ciphers with high encryption speeds were designed. Because partial plaintexts are selected to be encrypted with a securer cipher but others are selected to be encrypted by a faster cipher, such a mechanism can be regarded as a generalized version of selective encryption.

CHAOS-BASED IMAGE AND VIDEO ENCRYPTION

Due to the tight relationship between chaos and cryptography, in the past two decades it was widely investigated how to use chaotic maps to construct cryptosystems. For a comprehensive survey of digital chaos ciphers, see Chapter 2 of Reference 91. Basically, there are two typical ways to use chaos in image and video encryption schemes: (1) use chaos as a source to generate pseudorandom bits with desired statistical properties to realize secret encryption operations and (2) use 2-D chaotic maps (or fractallike curves) to realize secret permutations of digital images and frames. The first way has been widely used to design chaotic stream ciphers, whereas the second has been specially employed by chaos-based image encryption schemes.

Image Encryption Schemes Based on 2-D Chaotic Maps

The idea of using 2-D chaotic maps to design permutation-based image encryption schemes was initially proposed in References 92–95 and later systematized in References 96–98. Assuming that the size of the plain-image is $M \times N$, the encryption procedure can be described as follows (see Figure 4.3):

- Define a discretized and invertible 2-D chaotic map on an $M \times N$ lattice, where the discretized parameters serve as the secret key.

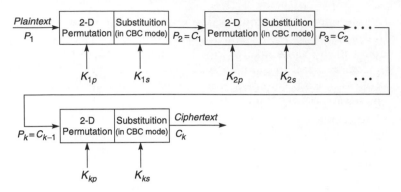

Figure 4.3. **A general framework of image encryption systems based on 2-D chaotic permutations (substitution part is run in CBC mode to resist chosen-plaintext attack).**

- Iterate the discretized 2-D chaotic maps on the plain-image to permute all pixels.
- Use a substitution algorithm (cipher) so as to modify the values of all pixels to flatten the histogram of the image (i.e., to enable the confusion property).
- Repeat the permutation and the substitution for k rounds to generate the cipher-image.

To resist the chosen-plaintext attack, the substitution part should be used in context-sensitive modes (generally in CBC mode). This provides the needed diffusion property for the cipher-image with respect to small changes in the plain-image. The most favorable 2-D chaotic map for discretization is the Baker map whose continuous form is defined as follows:

$$B(x, y) = \begin{cases} (2x, y/2), & 0 \leq x < 1/2 \\ (2x - 1, y/2 + 1/2), & 1/2 \leq x < 1 \end{cases}$$

It is also possible to use the Cat map and the Standard map as discussed in References 97 and 98, but it was found that these two maps are not as good as the Baker map.

In Reference 99, it was pointed out that there exist some weak keys in Fridrich's image encryption scheme [96–98]. This problem is caused by the short recurrent-period effect of the discretized Baker map: for some keys, the recurrent period of the chaotic permutation is only 4. Although this issue was discussed from a statistical point of view in Section 5 of Reference 98 and Section 4.2 of Reference 97, the existence of such a short period was not noticed. To overcome this defect, in Reference 100, a modified Fridrich's image encryption scheme was proposed. An extra permutation (shifting pixels in each row) is introduced to avoid a short recurrent period of the discretized chaotic permutation. The existence of weak keys signifies the importance of the dynamical properties of the discretized chaotic permutations. However, until now, only limited results on a subset of all chaotic possible permutations have been reported [101]; therefore, further theoretical research is needed to clarify this issue and its negative effect on the security of the image encryption schemes.

Recently, 2-D discretized chaotic maps were generalized to 3-D counterparts: the 3-D Baker map in References 102–104 and the 3-D Cat map in Reference 105. Based on the proposed 3-D discretized chaotic maps, after rearranging the 2-D plain-image into a 3-D lattice (over which the 3-D chaotic map is iterated), the plain-image can be encrypted in a procedure similar to the original Fridrich's scheme. Another difference of the two schemes from the Fridrich's scheme is that a chaotic pseudo-random number generator (PRNG) based on the 1-D Logistic map is used to realize the substitution algorithm.

In References 106 and 107, a different discretized rule for the Baker map was proposed and a generalized version of the original Scharinger–Fridrich image encryption schemes [92–98] was developed. For a $2^l \times 2^l$ image, the new discretized Baker map was controlled by a 2^{2l-1}-size sequence $\{b_k^{(1)}, \ldots, b_k^{(2^{2l-1})}\}$, which determines 2^{2l-1} rotation directions of local pixels. Let the encryption key be of 2^{2l-1} addresses $\{a_0^{(1)}, \ldots, a_0^{(2^{2l-1})}\}$. The encryption procedure is realized by running the following two combined operations n times:

1. Use selected bits from the pixel values indexed by the 2^{2l-1} addresses to generate $\{b_k^{(1)}, \ldots, b_k^{(2^{2l-1})}\}$, so as to control the secret permutations

2. Use selected bits from the Y coordinates of the permuted pixels in the last round to substitute the pixel values in the current round.

The above two operations are similar to those used in the original Scharinger–Fridrich schemes. This generalized encryption scheme has two special features: (1) The decryption key is different from the encryption key and (2) the decryption key depends on both the encryption key and the plain-image. In fact, the decryption key is a permutation of the addresses containing in the encryption key. A defect of this scheme is that its key size is too long.

Image Encryption Schemes Based on Fractallike Curves

The permutations defined by discretized 2-D chaotic maps can also be generated from a large group of fractallike curves, such as the Peano–Hilbert curves [108]. Due to the tight relationship between chaos and fractals, permutation encryption schemes based on fractallike curves can be designed, which are further discussed in this subsection.

This basic idea of using fractallike curves to generate noiselike images can be retrospected to early work in image-understanding on noiselike coding of associate memories [109]. In such noiselike associative memories, associative keys are used to store and retrieve data, and the keys can be generated in a manner similar to image encryption. For example, in Reference 108, the Peano–Hilbert curve is used to permute all pixels and a pseudorandom mechanism is used to substitute the gray values. In Reference 110, the idea of using chaos to generate keys for noiselike coding memories was also suggested.

In Reference 111, a simple video scrambling scheme was proposed, which encrypts images or uncompressed video frames by scanning all pixels with a secret space-filling curve (SFC). Although it is called a video scrambling method, it actually is an image scrambling algorithm. This scheme can be broken under ciphertext-only attack, as reported in Reference 112. In addition, it is insecure against known-plaintext and chosen-plaintext attacks, because the secret permutation can be reconstructed easily with some pairs of plain-images and cipher-images.

Since the 1990s, a number of image and video encryption schemes based on SCAN, which is a fractal-based image processing language [113], have been proposed [114–121] and realized in hardware [122–124]. All SCAN-based image and video encryption schemes can be divided into two generations: The first generation uses secret permutations to encrypt plain-images and the second combines secret permutations and the substitution algorithm. It has been pointed out that the first generation is not secure enough against known-plaintext and chosen-plaintext attacks [51,64,125,126], and the second can enhance the security of the first generation. In all SCAN-based encryption schemes, the secret permutations are generated from 13 different SCAN patterns and 24 different partition patterns, where the SCAN words serve as the secret key. Figure 1 to Figure 3 of Reference 121 illustrate details about how these SCAN patterns are combined to generate a secret permutation (or a scan order) of the image. In the following, a brief introduction is given to all SCAN-based image and video encryption schemes.

- *The first generation* [114–119]:

  In References 114–116, SCAN language was used to design a SCAN transposition cipher (STC), which is then combined with a stream cipher to compose a product cipher. Following the analyses given in References 51, 64 and 215, STC is insecure against known-plaintext and chosen-plaintext attacks and its security is ensured by another stream cipher.

In Reference 117, the SCAN language is combined with quadtree decomposition to encrypt images, where the SCAN language is used to permute the order of four nodes in each level of the quadtree structure. This scheme was insecure due to many security weaknesses, which were discussed in detail in Reference 51.

In Reference 118, the SCAN language is slightly modified and then combined with the 2DRE (two-dimensional run-encoding) technique to encrypt binary images. It was pointed out that this cryptosystem cannot resist the known-plaintext attack [126].

In Reference 119, the SCAN language is used to compress the image into a shorter binary string, then rearrange the compressed string into multiple $2^n \times 2^n$ blocks, and, finally permute each $2^n \times 2^n$ block with two different SCAN patterns m times. Apparently, this scheme corresponds to a single STC, which is insecure against known-plaintext and chosen-plaintext attacks [51,64,126].

- *The second generation* [120,121]: The following substitution mechanism is exerted after permutation, so as to resist known-plaintext and chosen-plaintext attacks (just like those image encryption schemes based on 2-D chaotic maps):

$$C[j] = (B[j] + (C[j-1]+1) \cdot R[j]) \quad \mathrm{mod}\, 256$$

where $R[j]$ is a pseudorandom integer between 0 and 255. Apparently, the substitution algorithm corresponds to a stream cipher with ciphertext feedback. The mixing of different SCAN patterns (two secret SCAN patterns and two public SCAN patterns — the sprial pattern $s0$ and the diagonal pattern $d0$) and the substitution part makes the new SCAN-based encryption scheme much securer against known-plaintext and chosen-plaintext attacks.

Image Encryption Schemes Based on 1-D Chaotic Maps

All chaos-based cryptosystems using 1-D chaotic maps can be used to encrypt digital images. It is common to use images to show visible performances of the designed cryptosystems [127–130]. In this subsection, discussion is focused on the chaos-based encryption algorithms specially designed for digital images.

Yen et al.'s Image Encryption Schemes. Yen et al. [131–147] proposed a series of chaos-based image and video encryption schemes. In these proposed algorithms, the employed chaotic map is the Logistic map $f(x) = \mu x(1-x)$ realized in finite computing precision, and the secret key always includes the initial condition $x(0)$ and the control parameter μ (for some schemes, the secret key includes some additional parameters).

All of these encryption schemes follow a similar basic idea of using chaos to realize encryption:

- Run the Logistic map to generate a pseudorandom binary sequence $\{b(i)\}$ from the n-bit representation of each chaotic state $x(k) = 0.b(n \cdot k)b(n \cdot k + 1) \cdots b(n \cdot k + n - 1)$, where n may be less than the finite computing precision.
- Use the chaotic binary sequence $\{b(i)\}$ to pseudorandomly control permutations and value substitutions of all pixels.

Note that the generated sequence $\{b(i)\}$ is generally not balanced; that is, the number of 0's is different from that of 1's, because the variant density function of the Logistic map is not uniform[148]; that is, the Logistic map is not a good choice for encryption, so it is better to use other 1-D chaotic maps with uniform variant density. However, in the following, one can see that Yen et al.'s encryption schemes are not secure even when $\{b(i)\}$ satisfies the balance property.

- BRIE (*Bit Recirculation Image Encryption*) [137]: $\{b(i)\}$ is used to pseudorandomly control shift operations exerted on each pixel. This scheme has been cryptanalyzed in Reference 149. Due to the essential defects of rotation shift operations, the encryption performance of this scheme is rather poor when some specific secret keys are used. Also, it cannot resist known-plaintext and chosen-plaintext attacks, because a mask image that is equivalent to the secret key can be derived from a pair of plain-image and cipher-image
- CKBA (*Chaotic Key-Based Algorithm*) [138]: $\{b(i)\}$ is used to pseudorandomly XOR (or NXOR) each pixel with *key1* or *key2*, where *key1* and *key2* are also included in the secret key. According to Reference 150, CKBA cannot resist known-plaintext and chosen-plaintext attacks. Only one pair of known-plain-image or chosen plain-image and cipher-image is enough to reconstruct *key1* and *key2* and then derive $\{b(i)\}$. From $\{b(i)\}$, $x(0)$ and an approximate value of μ can also be found.
- HCIE (*Hierarchical Chaotic Image Encryption*) [131,133,139]: In this scheme, an $M \times N$ plain-image is divided into $S_M \times S_N$ blocks for encryption, where $\sqrt{M} \leq S_M \leq M$ and $\sqrt{N} \leq S_N \leq N$. $\{b(i)\}$ is used to pseudorandomly control *no*-round $4(S_M + S_N) - 2$ shift operations with four different directions to permute all blocks and all pixels in each block. The CIE (Chaotic Image Encryption) algorithm proposed in Reference 132 is a simplified version of HCIE. Because HCIE is a permutation-only image encryption scheme, it cannot resist the chosen-plaintext attack, and a mask image that is equivalent to the key can still be reconstructed from a small number of pairs of plain-images and cipher-images.

- MLIE (*Mirror-Like Image Encryption*) [134] and CMLIE (*Chaotic MLIE*) [141,142]: $\{b(i)\}$ is used to pseudorandomly control four different mirrorlike swapping operations within the images. Because the two schemes are both permutation-only image encryption schemes, they cannot resist known-plaintext and chosen-plaintext attacks.

- CNNSE (*Chaotic Neural Network for Signal Encryption*) [135,140,143]: $\{b(i)\}$ is used to control the weights of a neutral network, which is then used to encrypt each pixel bit by bit. The final function of the chaotic neutral network is very simple: $d_i'(n) = d_i(n) \oplus b(8 \times n + i)$, where $d_i(n)$ and $d_i'(n)$ respectively represent the ith plain-bit of the nth plain-pixel and the ith cipher-bit of the nth cipher-pixel. Apparently, this scheme works like a stream cipher, and so CNNSE is vulnerable to known-plaintext and chosen-plaintext attack: only one known-plaintext or chosen plaintext is enough to reveal the sequence b and then to derive the secret key. It is the least secure image encryption scheme among all those proposed by Yen et al.

- TDCEA (*The 2-D Circulation Encryption Algorithm*) [144]: This scheme is an enhanced version of BRIE [136,137]. For a bit-matrix M composed of eight consecutive pixel values, $\{b(i)\}$ is used to make bit-shift operations in four different directions. TDCEA has better capability against the known-plaintext attack than BRIE, but it is still not secure enough against the chosen-plaintext attack.

- DSEA (*Domino Signal Encryption Algorithm*) [145]: For the nth plain-byte $f(n)$, `true_key=initial_key` if $i \bmod L = 0$, else `true_key=` $f'(n-1)$, where $f'(n)$ denotes the nth cipher-byte. $\{b(n)\}$ is used to pseudo-randomly select an encryption function from XOR and NXOR: If $b(n)=1$, then $g'(n)=g(n)$ XOR `true_key`, else $g'(n)=g(n)$ NXOR `true_key`. Because a NXOR $b = a$ XOR $\bar{b}$, DSEA cannot resist known-plaintext and chosen-plaintext attacks.

- RSES (*Random Seed Encryption Subsystem*) [147] and RCES (*Random Control Encryption Subsystem*) [146]: RSES and RCES are actually the same encryption scheme, which is an enhanced version of CKBA [138]. The two masking keys, *key*1 and *key*2, used in CKBA become time-variant and are pseudorandomly controlled by $\{b(i)\}$. $\{b(i)\}$ are also used to control swap operations exerted on neighboring pixels. Although this scheme is much more complex than CKBA, it is still insecure against known-plaintext and chosen-plaintext attacks [151].

Other Image Encryption Schemes on 1-D Chaotic Maps.

- In Reference 152, another permutation-only image encryption scheme was proposed for binary images. The 1-D chaotic map $x(k+1) = \sin(a/x(k))$ is used to pseudorandomly permutate all pixels. Apparently, it cannot resist known-plaintext and chosen-plaintext attacks, but it is generally difficult to derive the secret key by breaking the permutation matrix.

- In References 103 and 104, the Logistic map is used as a chaotic stream cipher (more than a chaotic PRNG) to mask the SPIHT-encoding bit stream of a wavelet-compressed image. To resist known-plaintext and chosen-plaintext attacks, three different masking operations are used and the selected operation at each position is dependent on previous cipher-bits.

A Related Topic: Chaos-Based Watermarking

Due to the similarity between image encryption and image watermarking, some ideas of chaos-based image encryption have also been used in the area of digital watermarking [153–157]. Generally speaking, there are two ways to use chaos in a watermarking system: directly generating watermark signals (pattern) from chaos and mixing (2-D) watermark patterns via chaos. The first way corresponds to chaotic PRNG, and the second corresponds to chaotic permutations of 2-D images (i.e., the secret permutation used in image encryption schemes based on 2-D chaotic maps). At present, almost all chaos-related techniques used in watermarking systems are borrowed from chaotic cryptography (especially chaos-based image encryption), but it can be expected that some new ideas in chaos-based digital watermarking will also benefit image encryption community in the future.

Chaos-Based Video Encryption

- *SCAN-based encryption for losslessly compressed videos*: The new generation of SCAN-based image encryption schemes were also extended to encrypt videos [120,121,123,124]. The plain-video is compressed with a frame-difference-based light lossless compression algorithm and then encrypted with the new SCAN-based encryption algorithm.
- *Yen et al.'s encryption schemes for MPEG videos*: Some of Yen et al.'s image encryption schemes [134–136,138,140–143,145,146] were recommended for encrypting MPEG videos, thanks to their high encryption speed. The idea of selective encryption was also used in References 136 and 138: Motion vectors are selected as data for encryption using BRIE or CKBA. However, the insecurity of Yen et al.'s encryption schemes makes their use in video encryption unacceptable.
- *Chiaraluce et al.'s chaos-based H.263+ encryption scheme*: In Reference 158, a chaotic video encryption scheme was proposed for encrypting the following selective data of an H.263+ encoded

video: most sign bits of DC coefficients, the AC coefficients of I-macroblocks, sign bits of AC coefficients of P-macroblocks, and sign bits of motion vectors. The proposed scheme is a stream cipher based on three chaotic maps: the skew tent map, the skew sawtooth map, and the discretized Logistic map. The outputs of the first two chaotic maps are added and then the addition is scaled to be an integer between 0 and 255. Each scaled integer is used as the initial condition of the third map to generate a 64-size key stream to mask the plaintext with XOR operation. To further enhance the security against known-plaintext chosen-plaintext attacks, it was suggested to change the key every 30 frames.

- *CVES (Chaotic Video Encryption Scheme)*: In Reference 159, a chaos-based encryption scheme called CVES using multiple chaotic systems was proposed. CVES is actually a fast chaotic cipher that encrypts bulky plaintext frame by frame. The basic idea is to combine a simple chaotic stream cipher and a simple chaotic block cipher to construct a fast and secure product cipher. In Chapter 9 of Reference 91, it was pointed out that the original CVES is not sufficiently secure against the chosen-plaintext attack, and an enhanced version of CVES was proposed by adding ciphertext feedback. Figure 4.4 gives a diagrammatic view of the enhanced CVES, where CCS and ECS(1) to ECS(2^n) are all piecewise linear chaotic maps, and m-LFSR$_1$ and m-LFSR$_2$ are the perturbing PRNG of CCS and ECS respectively. The encryption procedure of CVES can

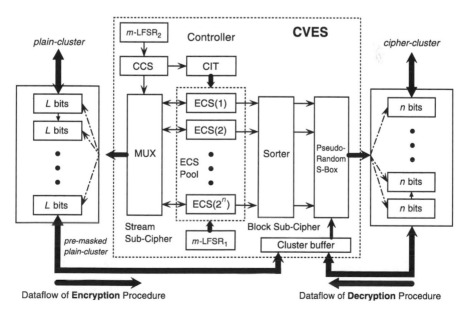

Figure 4.4. The enhanced CVES proposed in Chapter 9 of Reference 91.

be described as follows. Each plain-block is first XORed by a chaotic signal pseudorandomly selected from chaotic orbits of the 2^n ECSs (Encryption Chaotic Systems) and then substituted by a pseudorandom S-box generated from all chaotic orbits of the 2^n ECSs. Initial tests have shown that the encryption speed of CVES is competitive with most traditional ciphers and is only a little slower than the AES reference code.

EXPERIENCES AND LESSONS

From the above survey on chaos-based image and video encryption schemes, one can learn about some experiences and also lessons on how to design fast and secure image and video encryption schemes with or without using chaos. Although some experiences and lessons are merely common rules in modern cryptography, for some reasons they have not been widely accepted by the image and video encryption community. In this section, we summarize these learned experiences and lessons in the following list of remarks.

Remark 1. *Permutation-only image/video encryption schemes are generally insecure against known-plaintext and chosen-plaintext attacks.*

Under known-plaintext and chosen-plaintext attacks, a mask image that is equivalent to the secret key can be obtained to decrypt cipher-images encrypted with the same key. In addition, permutation-only image ciphers cannot change plain-histograms, which contain some useful information about the plain-images; for example, cartoon pictures and real photos have different histograms and photos of human faces usually have narrower histograms than photos of natural scenes.

Remark 2. *Secret permutation is NOT a prerequisite of secure image and video encryption schemes.*

This is because the substitution part running in CBC mode is sufficient for achieving confusion and diffusion properties and for ensuring security against the chosen-ciphertext attack. In fact, comparing Figure 8 and Figure 13 of Reference 98, one can see that the substitution algorithm can make the cipher-image look "chaotic" much faster than a chaotic permutation. This implies the incapability of chaotic permutations. It is believed that the main reason for using secret chaotic permutation in a image and video encryption schemes is that secret permutation may be useful for increasing the computational complexity of a potential chosen-plaintext attack. With the use of secret permutation, the statistical cryptanalysis of the concerned encryption scheme will become much more complicated or impractical.

Remark 3. *Ciphertext feedback is very useful for enhancing the security against known-plaintext and chosen-plaintext attacks.*

Most image encryption schemes based on 2-D chaotic maps, the new generation of SCAN-based image encryption schemes, and the enhanced CVES all utilize the ciphertext feedback mechanism to resist the chosen-plaintext attack.

Remark 4. *Combining the ciphertext feedback and a (secret) permutation can achieve the diffusion (i.e., avalanche) property at the level of image and video frame.*

Although secret permutations are commonly used together with the ciphertext feedback to provide the diffusion property, the permutation can actually be publicized. Apparently, using a public permutation only means that the secret key of the secret permutation is publicized so that the security relies on other parts of the whole cryptosystem.

Remark 5. *Combining a simple stream cipher and a simple block cipher running in the CBC mode can yield a secure and fast encryption scheme, where the stream cipher is used to create the confusion property, the block cipher is used to realize the diffusion property, and the ciphertext feedback in the CBC mode is used to resist the chosen-plaintext attack.*

The modified CVES proposed in Chapter 9 of Reference 91 is a one-round version of such a simple but fast encryption scheme. In CVES, the avalanche property is sacrificed in order to increase the encryption speed. The loss of the avalanche feature may not influence the security of the schemes, however.

Remark 6. *The diffusion methods used in most chaos-based encryption schemes are too slow due to the multiple time-consuming iterations of chaotic permutations.*

In fact, only two rounds are enough to achieve diffusion: Encrypt the data in CBC mode from top to bottom and then encrypt the partially encrypted data again in the CBC mode from bottom to top.

Remark 7. *Selective encryption may provide enough security if the unencrypted data are dependent on the encrypt data.*

An example of such selective encryption schemes is the one proposed in Reference 53, which is based on the context-sensitive compression algorithm. More studies are needed to clarify the relationship between the selective encryption and the model-based compression algorithms.

Remark 8. *Although the selective encryption is generally insecure from the cryptographical point of view, it can be generalized to provide a more general way for designing fast and secure video encryption* *schemes: Use a slow (but stronger) cipher to encrypt selective data and use a fast (but weaker) cipher to encrypt the rest data.*

Two video ciphers were constructed in Reference 90 to show the potential of this idea. For the case that the compression algorithm is antagonistic to the simple selective encryption, this generalized version of selective encryption will be very useful as a replacement for the simple one.

CONCLUSIONS

Image and video encryption plays a more and more important role in today's multimedia world. Although many encryption schemes have been proposed to provide security for digital images and videos, some of them are too weak to resist various attacks designed by cryptanalysts. Basically, many efforts have been devoted to study the security issue, but for multimedia, the security is still not strong from a serious cryptographic point of view. To design truly secure image and video encryption schemes, the classical cryptology must be employed.

As an emerging tool for the design of digital ciphers, chaos theory has been widely investigated, especially to develop image and video encryption algorithms. The simplicity of many discrete chaotic maps and the well-established chaos theory make it possible to approach practically good solutions to image and video encryption. The success and failure of chaos-based encryption schemes have led to some valuable experiences and lessons, which can be used as the fundamentals of future research on the chaos-based multimedia encryption technology. At this point, chaos theory for image and video encryption appears to be promising but not yet mature. More efforts are needed for its further development toward a marketable future.

ACKNOWLEDGMENTS

The authors would like to thank Dr. Gonzalo Álvarez at Instituto de Física Aplicada of Consejo Superior de Investigaciones Científicas (Spain), Dr. Ennio Gambi at Universitá di Ancona (Italy), and Mr. Chengqing Li at Zhejiang University (China) for their valuable comments on the manuscript of this chapter.

REFERENCES

1. Schneier, B., *Applied Cryptography: Protocols, Algorithms, and Souce Code in C*, 2nd ed., John Wiley & Sons, New York, 1996.
2. Dang, P.P. and Chau, P.M., Image encryption for secure internet multimedia applications, *IEEE Trans. Consumer Eletron.*, 46(3), 395–403, 2000.
3. Dang, P.P. and Chau, P.M., Implementation IDEA algorithm for image encryption, in *Mathematics and Applications of Data/Image Coding, Compression, and Encryption III*, Proc. SPIE Vol. 4122, Mark S. Schmalz, Ed., 2000, pp. 1–9.

4. Dang, P.P. and Chau, P.M., Hardware/software implementation 3-way algorithm for image encryption, in *Security and Watermarking of Multimedia Contents II*, Proc. SPIE Vol. 3971, Ping W. Wong and Edward J. Delp III, Eds., 2000, pp. 274–283.

5. McCanne, S. and Van Jacobson. vic: A flexible framework for packet video, in *Proceedings of 3rd ACM International Conference on Multimedia*, 1995, pp. 511–522.

6. Qiao, L., *Multimedia Security and Copyright Protection*. Ph.D. thesis, University of Illinois at Urbana–Champaign, Urbana, IL, 1998.

7. Qiao, L. and Nahrsted, K., Comparison of MPEG encryption algorithms, *Computers Graphics*, 22(4), 437–448, 1998.

8. Agi, I. and Gong, L., An empirical study of secure MPEG video transmission, in Proceedings of ISOC *Symposium on Network and Distributed Systems Security (SNDSS'96)*, 1996, pp. 137–144.

9. Lookabaugh, T.D., Sicker, D.C., Keaton, D.M., Guo, W.Y., and Vedula, I., Security analysis of selectively encrypted MPEG-2 streams, in *Multimedia Systems and Applications VI*, Proc. SPIE Vol. 5241, Andrew G. Tescher, Bhaskaran B. Vasudev, V. Michael Bove Jr., and Ajay Divakaran, Eds., 2003, pp. 10–21.

10. Wu, C.P. and Kuo, C.-C.-J., Fast encryption methods for audiovisual data confidentiality, in *Multimedia Systems and Applications III*, SPIE Vol. 4209, 2001, pp. 284–285.

11. Pommer, A. and Uhl, A., Application scenarios for selective encryption of visual data, in *Proceedings of Multimedia and Security Workshop of 10th ACM International Conference on Multimedia*, 2002, pp. 71–74.

12. Ziedan, I.E., Fouad, M.M., and Salem, D.H., Application of data encryption standard to bitmap and JPEG images, in *Proceedings of 12th National Radio Science Conference (NRSC'2003)*, 2003, pp. C16/1–C16/8.

13. Macq, P.M., and Quisquater, J.J., Cryptology for digital TV broadcasting, *Proc. IEEE*, 83(6), 944–957, 1995.

14. Wen, J., Severa, M., Zeng, W., Luttrell, M.H., and Jin, W., A format-compliant configurable encryption framework for access control of video, *IEEE Trans. Circuits Syst. Video Technol.*, 12(6), 545–557, 2002.

15. Wen, J., Severa, M., Zeng, W., Luttrell, M.H., and Jin, W., A format-compliant configurable encryption framework for access control of multimedia, *in Proceedings of IEEE 4th Workshop on Multimedia Signal Processing (MMSP'2001)*, 2001, pp. 435–440.

16. Wee, S.J. and Apostolopoulos, J., Secure scablable streaming and secure transcoding with JPEG2000, in *Proceedings of IEEE International Conference on Image Processing (ICIP'2003)*, Vol. I, 2003, pp. 205–208.

17. Yu, H.H. and Yu, X., Progressive and scalable encryption for multimedia content access control, in *Proceedings of IEEE International Conference on Communications (ICC'2003)*, Vol. 1, 2003, pp. 547–551.

18. Yuan, C., Zhu, B.B., Wang,Y., Li, S. and Zhong, Y., Efficient and fully scalable encryption for MPEG-4 FGS, in *Proceedings of IEEE International Symposium on Circuits and Systems (ISCAS'2003)*, Vol. II, 2003, pp. 620–623.

19. Eskicioglu, A.M. and Delp, E.J., An integrated approach to encrypting scabable video, in *Proceedings of IEEE International Conference on Multimedia and Expo (ICME'2002)*, 2002, pp. 573–576.

20. Wee, S.J. and Apostolopoulos, J.G., Secure scablable streaming enabling transcoding without decryption, in *Proceedings of IEEE International Conference on Image Processing (ICIP'2001)*, Vol. 1, 2001, pp. 437–440.

21. Wee, S.J. and Apostolopoulos, J.G., Secure scablable video streaming for wireless networks, in *Proceedings of IEEE International Conference on Acoustics, Speech, and Signal Processing (ICASSP'2001)*, Vol. 4, 2001, pp. 2049–2052.

22. Tosun, A.S. and Feng, W., Lightweight security mechanisms for wireless video transmission, in *Proceedings of IEEE International Conference on Information Technology: Coding and Computing*, 2001, pp. 157–161.

23. Tosun, A.S. and Feng, W., Efficient multi-layer coding and encryption of MPEG video streams, in *Proceedings of IEEE International Conference on Multimedia and Expo (ICME'2000)*, 2000, pp. 119–122.

24. Kunkelmann, T. and Horn, U., Partial video encryption based on scalable coding, in *Proceedings of 5th International Workshop on Systems, Signals and Image Processing (IWSSIP'98)*, 1998.

25. Dittmann, J. and Steinmetz, A., Enabling technology for the trading of MPEG-encoded video, in *Information Security and Privacy: Second Australasian Conference (ACISP'97) Proceedings*, Lecture Notes in Computer Science, Vijay Varadharajan, Josef Pieprzyk, and Yi Mu, Eds., Vol. 1270, Springer-Verlag, New York, 1997, pp. 314–324.

26. Taubman, D.S. and Marcellin, M.W., JPEG2000: Standard for interactive imaging, *Proc. IEEE*, 90(8), 1336–1357, 2002.

27. Li, W., Overview of fine granularity scalability in MPEG-4 video standard, *IEEE Trans. Circuits Syst. Video Technol.*, 11(3), 301–317, 2001.

28. Torrubia, A. and Mora, F. Perceptual cryptography of JPEG compressed images on the JFIF bit-stream domain, in *Digest of Technical Papers of IEEE International Conference on Consumer Electronics (ICCE'2003)*, 2003, pp. 58–59.

29. Torrubia, A. and Mora, F., Perceptual cryptography on MPEG layer III bit-streams, *IEEE Trans. Consum. Eletron.*, 48(4), 1046–1050, 2002.

30. Roche, P., Dugelay, J.-L., and Molva, R., Multi-resolution access control algorithm based on fractal coding, in *Proceedings of IEEE International Conference on Image Processing (ICIP'96)*, Vol. 3, 1996, pp. 235–238.

31. El-Khamy, S.E. and Abdou, H.E.M., A novel secure image coding scheme using fractal transformation, in *Proceedings of 15th (Egypt) National Radio Science Conference (NRSC'98)*, 1998, pp. C23/1–C23/9.

32. Grosbois, R., Gerbelot, P., and Ebrahimi, T., Authentication and access control in the JPEG 2000 compressed domain, in *Applications of Digital Image Processing XXIV*, Proc. SPIE, Vol. 4472, Andrew G. Tescher, Ed., 2001, pp. 95–104.

33. Van Droogenbroeck, M. and Benedett, R., Techniques for a selective encryption of uncompressed and compressed images, in *Proceedings of Advanced Concepts for Intelligent Vision Systems (ACIVS'2002)*, 2002, pp. 90–97.

34. Simitopoulos, D., Zissis, N., Georgiadis, P., Emmanouilidis, V. and Strintzis, M.G., Encryption and watermarking for the secure distribution of copyrighted MPEG video on DVD. *Multimedia Syst.*, 2003, 9(3), 217–227.

35. Kiya, H., Imaizumi, D. and Watanabe, O., Partial-scrambling of images encoded using JPEG2000 without generating marker codes, in *Proceedings of IEEE International Conference on Image Processing (ICIP'2003)*, Vol. III, 2003, pp. 205–208.

36. Bodo, Y., Laurent, N. and Dugelay, J.-L., A scrambling method based on disturbance of motion vector, in *Proceddings of 10th ACM International Conference on Multimedia*, 2002, pp. 89–90.

37. Pazarci, M. and Dipçin, V., A MPEG2-transparent scrambling technique, *IEEE Trans. Consumer Eletron.*, 48(2), 345–355, 2002.

38. Tosun, A.S. and Feng, W., On error preserving encryption algorithms for wireless video transmission,in *Proceedings of 9th ACM International Conference on Multimedia*, 2001, pp. 302–308.

39. Uehara, T., Safavi-Naini, R., and Ogunbona, P., Securing wavelet compression with random permutations, in *Proceedings of IEEE Pacific-Rim Conference on Multimedia (IEEE-PCM'2000)*, 2000, pp. 332–335.

40. Pommer, A. and Uhl, A., Wavelet packet methods for multimedia compression and encryption, in *Proceedings of IEEE Pacific Rim Conference on Communications, Computers and Signal Processing (PACRIM'2001)*, Vol. 1, 2001, pp. 1–4.

41. Pommer, A., *Selective Encryption of Wavelet-Compressed Visual Data*, Ph.D. thesis, University of Salzburg, Austria, 2003.

42. Pommer, A. and Uhl, A., Selective encryption of wavelet-packet encoded image data: Efficiency and security, *Multimedia Syst.*, 9(3), 279–287, 2003.

43. Seo, Y.-H., Kim, D.-W., Yoo, J.-S., Dey, S., and Agrawal, A., Wavelet domain image encryption by subband selection and data bit selection, in *Proceedings of World Wireless Congress (WCC'03/3GWireless'2003)*, 2003.

44. ISO, MPEG4 IPMP (Intellectual Property Management and Protection) Final Proposed Draft Amendment (FPDAM), ISO/IEC 14496-1:2001/AMD3, ISO/IEC JTC 1/SC 29/WG11 N4701, 2002.

45. Kunkelmann, T., Reinema, R., Steinmetz, R., and Blecher, T., Evaluation of different video encryption methods for a secure multimedia conferencing gateway, in *From Multimedia Services to Network Services: 4th International COST 237 Workshop Proceedings*, Lecture Notes in Computer Science Vol. 1356, André V. Dantine and Christophe Diot, Eds., Springer-Verlag, Berlin, 1997, pp. 75–89.

46. Kunkelmann, T., and Horn, U., Video encryption based on data partitioning and scalable coding—a comparison, in *Interactive Distributed Multimedia Systems and Telecommunication Services: 5th International Workshop (IDMS'98) Proceedings*, Lecture Notes in Computer Science Vol. 1483, Thomas Plagemann and Vera Goebel, Eds., Springer-Verlag, Berlin, 1998, pp. 95–106.

47. Liu, X., and Eskicioglu, A.M., Selective encryption of multimedia content in distribution networks: Challenges and new directions, in *Proceedings of IASTED International Conference on Communications, Internet and Information Technology (CIIT'2003)*, 2003.

48. Spanos, G.A. and Maples, T.B., Security for real-time MPEG compressed video in distributed multimedia applications, in *Proceedings of IEEE 15th Annual International Phoenix Conference on Computers and Communications*, 1996, pp. 72–78.

49. Dolske, J.H. Secure MPEG video: Techniques and pitfalls. available online at http://www.dolske.net/old/gradwork/cis788r08/, 1997.

50. Skrepth, C.J. and Uhl, A., Selective encryption of visual data, in *Proceedings of IFIP TC6/TC11 6th Joint Working Conference on Communications and Multimedia Security (CMS'2002)*, 2002, pp. 213–226.

51. Cheng, H.C.H., Partial Encryption for Image and Video Communication, Master's thesis, University of Alberta, Edmonton, Alberta, Canada, 1998.

52. Qiao, L. and Nahrstedt, K., A new algorithm for MPEG video encryption, in *Proceedings of 1st International Conference on Imaging Science, Systems, and Technology (CISST'97)*, 1997, pp. 21–29.

53. Wu, X. and Moo, P.W., Joint image/video compression and encryption via high-order conditional entropy coding of wavelet coefficients, in *Proceeding of IEEE Conference on Multimedia Computing and Systems (CMS'99)*, 1999, pp. 908–912.

54. Tang, L. Methods for encrypting and decrypting MPEG video data efficiently, in *Proceedings of 4th ACM Internationa of Conference on Multimedia*, 1996, pp. 219–230.

55. Kailasanathan, C., Safavi-Naini, R., and Ogunbona, P., Compression performance of JPEG encryption scheme, in *Proceedings of 14th International Conference on Digital Signal Processing (DSP'2002)*, Vol. 2, 2002, pp. 1329–1332.

56. Shi, C. and Bhargava, B., Light-weight MPEG video encryption algorithm, in *Proceedings of International Conference on Multimedia (Multimedia'98, Shaping the Future)*, 1998, pp. 55–61.

57. Bhargava, B., Shi, C., and Wang, S.-Y., MPEG video encryption algorithm, *Multimedia Tools Applic.*, 24(1), 57–79, 2004.

58. Wu, C.-P. and Kuo, C.-C.J., Efficient multimedia encryption via entropy codec design, in *Security and Watermarking of Multimedia Contents III*, Proc. SPIE, Vol. 4314, Sethuraman Panchanathan, V. Michael Bove, Jr., and Subramania, I. Sudharsanan, Eds., 2001, pp. 128–138.

59. Xie, D. and Kuo, C.-C.J., An enhanced MHT encryption scheme for chosen plaintext attack, in *Internet Multimedia Management Systems IV*, Proc. SPIE, Vol. 5242, John R. Smith, Sethuraman Panchanathan and Tong Zhang, Eds., 2003, pp. 175–183.

60. Kankanhalli, M.S., and Guan, T.T., Compressed-domain scrambler/descrambler for digital video, *IEEE Trans. Consumer Eletron.*, 48(2), 356–365, 2002.

61. Zeng, W. and Lei, S., Efficient frequency domain video scrambling for content access control, in *Proceedings of 7th ACM International Conference on Multimedia*, 1999, pp. 285–294.

62. Zeng, W. and Lei, S., Efficient frequency domain selective scrambling of digital video, *IEEE Trans. Multimedia*, 5(1), 118–129, 2003.

63. Li, X., Knipe, J., and Cheng, H., Image compression and encryption using tree structures, *Pattern Recogn. Lett.*, 18(8), 2439–2451, 1997.

64. Cheng, H. and Li, X., Partial encryption of compressed images and videos, *IEEE Trans. Signal Process.* 48(8), 2439–2451, 2000.

65. Cheng, H. and Li, X., On the application of image decomposition to image compression and encryption, in *Communications and Multimedia Security II: Proceedings IFIP TC6/TC11 International Conference (CMS'96)*, 1996, pp. 116–127.

66. Podesser, M., Schmidt, H.-P., and Uhl, A., Selective bitplane encryption for secure transmission of image data in mobile environments, in *Proceedings of 5th IEEE Nordic Signal Processing Symposium (NORSIG'2002)*, 2002.

67. Norcen, R., Podesser, M., Pommer, A., Schmidt, H.-P., and Uhl, A., Confidential storage and transmission of medical image data, *Computers in Biolo. Med.*, 33(3), 277–292, 2003.

68. Chang, C.-C., Hwang, M.-S., and Chen, T.-S., A new encryption algorithm for image cryptosystems, *J. Syst. Software*, 58(2), 83–91, 2001.

69. Lian, S. and Wang, Z., Comparison of several wavelet coefficients confusion methods applied in multimedia encryption, in *Proceedings of International Conference on Computer Networks and Mobile Computing (ICCNMC'2003)*, 2003, pp. 372–376.

70. Norcen, R. and Uhl, A., Selective encryption of the JPEG2000 bitstream, in *Proceedings IFIP TC6/TC11 7th Joint Working Conference on Communications and Multimedia Security (CMS'2003)*, Lecture Notes in Computer Science Vol. 2828, Antonio Lioy and Daniele Mazzocchi, Eds., Springer-Verlag, Berlin, 2003, pp. 194–204.

71. Vorwerk, L., Engel, T., and Meinel, C., A proposal for a combination of compression and encryption, in *Visual Communications and Image Processing 2000*, Proc. SPIE, Vol. 4067, King N. Ngan, Thomas Sikora and Ming-Ting Sun, Eds., 2000, pp. 694–702.

72. Pommer, A. and Uhl, A., Multimedia soft encryption using NSMRA wavelet packet methods: Parallel attacks, in *Proceedings of International Workshop on Parallel Numerics (Par-Num'2000)*, 2000, pp. 179–190.

73. Pommer, A. and Uhl, A., Selective encryption of wavelet packet subband structures for obscured transmission of visual data, in *Proceedings of 3rd IEEE Benelux Signal Processing Symposium (SPS'2002)*, 2002, pp. 25–28.

74. Pommer, A. and Uhl, A., Selective encryption of wavelet packet subband structures for secure transmission of visual data, in *Proceedings of Multimedia and Security Workshop of 10th ACM International Conference on Multimedia*, 2002, pp. 67–70.

75. Chuang T.-J. and Lin, J.-C., New approach to image encryption, *J. Electron. Imaging*, 7(2), 350–356, 1998.

76. Meyer, J. and Gadegast, F., *Security Mechanisms for Multimedia-Data with the Example MPEG-1 Video*. Berlin, 1995; available online at http://www.gadegast.de/frank/doc/secmeng.pdf.

77. Taesombut, N., Huang, R. and Rangan, V.P., A secure multimedia system in emerging wireless home networks, in *Communications and Multimedia Security (CMS'2003)*, Lecture Notes in Computer Science, Vol. 2828, Antonio Lioy and Daniele Mazzocchi, Eds., 2003, pp. 76–88.

78. Kunkelmann, T. and Reinema, R., A scalable security architecture for multimedia communication standards, in *Proceedings of IEEE International Conference on Multimedia Computing and Systems (ICMCS'97)*, 1997, pp. 660–661.

79. Li, Y., Chen, Z., Tan, S.-M. and Campbell, R.H., Security enhanced MPEG player, in *Proceedings of International Workshop on Multimedia Software Development (MMSD'96)*, 1996, pp. 169–175.

80. Spanos, G.A. and Maples, T.B., Performance study of a selective encryption scheme for the security of networked, real-time video, in *Proceedings of 4th International Conference on Computer Communications and Networks (IC3N'95)*, 1995, pp. 2–10.

81. Alattar, A.M. and Al-Regib, G.I., Evaluation of selective encryption techniques for secure transmission of MPEG video bit-streams, in *Proceedings of IEEE International Symposium on Circuits and Systems (ISCAS'99)*, Vol. IV, 1999, pp. 340–343.

82. Shin, S.U., Sim, K.S. and Rhee, K.H., A secrecy scheme for MPEG video data using the joint of compression and encryption, in *Information Security: Second International Workshop (ISW'99) Proceedings*, Lecture Notes in Computer Science, Vol. 1729, Masahiro Mambo and Yuliang Zheng, Eds., Springer-Verlag, Berlin, 1999, pp. 191–201.

83. Zeng, W., Wen, J. and Severa, M., Fast self-synchronous content scrambling by spatially shuffling codewords of compressed bitstream, in *Proceedings of IEEE International Conference on Image Processing (ICIP'2002)*, Vol. III, 2002, pp. 169–172.

84. Qiao. L. and Nahrstedt, K., Is MPEG encryption by using random list instead of ZigZag order secure? in *Proceedings of IEEE International Symposium on Consumer Electronics (ISCE'97)*, 1997, pp. 332-335.

85. Uehara, T. and Safavi-Naini, R., Chosen DCT coefficients attack on MPEG encryption schemes, in *Proceedings of IEEE Pacific-Rim Conference on Multimedia (IEEE-PCM'2000)*, 2000, pp. 316-319.

86. Shi, C. and Bhargava, B., An efficient MPEG video encryption algorithm, in *Proceedings of IEEE Symposium on Reliable Distributed Systems*, 1998, pp. 381–386.

87. Shi, C. and Bhargava, B., A fast MPEG video encryption algorithm, in *Proceedings of 6th ACM International Conference on Multimedia*, 1998, pp. 81–88.

88. Shi, C., Wang, S.-Y., and Bhargava, B., MPEG video encryption in real-time using secret key cryptography, in *Proceedings of International Conference on Parallel and Distributed Processing Techniques and Applications (PDPTA'99)*, 1999, pp. 191–201.

89. Pazarci, M. and Dipçin, V., A MPEG-transparent scrambling technique, *IEEE Trans. Consumer Eletron.*, 48(2), 2002, 345–355.

90. Bao, F., Deng, R., Feng, P., Guo, Y., and Wu, H., Secure and private distribution of online video and some related cryptographic issues, in *Information Security and Privacy: The 6th Australasian Conference (ACISP 2001) Proceedings*, Lecture Notes in Computer Science, Vol. 2119, Vijay Varadharajan and Yi Mu, Eds., 2001, pp. 190–205.

91. Li, S., *Analyses and New Designs of Digital Chaotic Ciphers*. Ph.D. thesis, School of Electronics & Information Engineering, Xi'an Jiaotong University, Xi'an, China, 2003; available http://www.hooklee.com/pub.html (in both Chinese and English).

92. Pichler, F. and Scharinger, J., Finite dimensional generalized Baker dynamical systems for cryptographic applications,in *Proceedings of 5th International Workshop on Computer Aided Systems Theory (EuroCAST'95)*, Lecture Notes in Computer Science, Vol. 1030, Franz Pichler, Robert Moreno-Díaz and Rudolf F. Albrecht, Eds., 1996, pp. 465–476.

93. Scharinger, J., Fast encryption of image data using chaotic Kolmogorov flows, in *Storage and Retrieval for Image and Video Databases V*, Proc. SPIE, Vol. 3022, Ishwar K. Sethi and Ramesh C. Jain, Eds., 1997, pp. 278–289

94. Scharinger, J., Secure and fast encryption using chaotic Kolmogorov flows, in *Proceedings of IEEE Information Theory Workshop (ITW'98)*, 1998, pp. 124–125.

95. Scharinger, J., Fast encryption of image data using chaotic Kolmogorov flows, *J. Electronic Imaging*, 7(2), 318–325, 1998.

96. Fridrich, J., Image encryption based on chaotic maps, in *Proceedings of IEEE International Conference on Sysytems, Man and Cybernetics (ICSMC'97)*, Vol. 2, 1997, pp. 1105–1110.

97. Fridrich, J., Secure image ciphering based on chaos, Technical Report RL-TR-97-155, the Information Directorate (former Rome Laboratory) of the Air Force Research Laboratory, New York, 1997.

98. Fridrich, J., Symmetric ciphers based on two-dimensional chaotic maps, *Int. J. Bifurcat. Chaos*, 8(6), 1259–1284, 1998.

99. Salleh, M., Ibrahim, S. and Isinn, I.F., Ciphering key of chaos image encryption, in *Proceedings of International Conference on Artificial Intelligence in Engineering and Technology (ICAIET'2002)*, 2002, pp. 58–62.

100. Salleh, M., Ibrahim, S., and Isinn, I.F., Enhanced chaotic image encryption algorithm based on Baker's map, in *Proceedings of IEEE Int. Symposium on Circuits and Systems (ISCAS'2003)*, Vol. II, 2003, pp. 508–511.

101. Qi, D., Zou, J. and Han X., A new class of scrambling transformation and its application in the image information covering. *Sci. in China E (Engl. Ed.)*, 43(3), 304–312, 2000.

102. Mao, Y., Chen, G. and Lian, S., A novel fast image encryption scheme based on 3D chaotic Baker maps, *Int. J. Bifurcat. Chaos*, in press.

103. Mao, Y. and Chen, G., Chaos-based image encryption, in *Handbook of Computational Geometry for Pattern Recognition, Computer Vision, Neural Computing and Robotics*, Bayco-Corrochano, E., Ed., Springer-Verlag, Heidelberg, 2004.

104. Mao, Y., *Research on Chaos-based Image Encryption and Watermarking Technology*. Ph.D. thesis, Nanjing University of Science & Technology, Nanjing, China, 2003 (in Chinese).

105. Mao, Y., Chen, G. and Chui, C.K., A symmetric image encryption scheme based on 3D chaotic Cat maps, *Chaos, Solitons Fractals*, 21(1), 749–761, 2004.

106. Miyamoto, M., Tanaka, K. and Sugimura, T., Truncated Baker transformation and its extension to image encryption, in *Mathematics of Data/Image Coding, Compression, and Encryption II*, Proc. SPIE, Vol. 3814, Mark S. Schmalz, Ed., 1999, pp. 13–25.

107. Yano, K. and Tanaka, K., Image encryption scheme based on a truncated Baker transformation, *IEICE Trans. Fundamentals*, E85-A(9), 2025–2035, 2002.

108. Zunino, R., Fractal circuit layout for spatial decorrelation of images, *Electron. Lett.*, 34(20), 1929–1930, 1998.

109. Bottini, S., An algebraic model of an associative noise-like coding memory, *Biol. Cybern.*, 36(4), 221–228, 1980.

110. Parodi, G., Ridella, S. and Zunino, R., Using chaos to generate keys for associative noise-like coding memories, *Neural Networks*, 6(4), 559–572, 1993.

111. Matias, Y. and Shamir, A., A video scrambing technique based on space filling curve (extended abstract), in *Advances in Cryptology–Crypto'87*, Lecture Notes in Computer Science, Vol. 293, Carl Pomerance, Ed., Springer-Verlag, Berlin, 1987, pp. 398–417.

112. Bertilsson, M., Brickell, E.F., and Ingemarson, I., Cryptanalysis of video encryption based on space-filling curves, in *Advances in Cryptology–EuroCrypt'88*, Lecture Notes in Computer Science, Vol. 434, Jean-Jacques Quisquater and Joos Vandewalle, Eds., Springer-Verlag, Berlin, 1989, pp. 403–411.

113. Bourbakis, N.G. and Alexopoulos, C., A fractal-based image processing language: Formal modeling, *Pattern Recogn.*, 32(2), 317–338, 1999.

114. Bourbakis, N.G. and Alexopoulos, C., Picture data encryption using SCAN patterns, *Pattern Recogn.*, 25(6), 567–581, 1992.

115. Alexopoulos, C., Bourbakis, N.G., and Ioannou, N., Image encryption method using a class of fratcals, *J. Electron. Imaging*, 4(3), 251–259, 1995.

116. Bourbakis, N.G., Image data compression-encryption using G-SCAN pattern, in *Proceedings of IEEE International Conference on Systems, Man and Cybernetics*, Vol. 2, 1997, pp. 1117–1120.
117. Chang, H.K.-C. and Liu, J.-L., A linear quadtree compression scheme for image encryption, *Signal Process.: Image Commn.*, 10(4), 279–290, 1997.
118. Chung, K.-L. and Chang, L.-C., Large encryption binary images with higher security, *Pattern Recogn. Lett.*, 19(5–6), 461–468, 1998.
119. Maniccam, S.S. and Bourbakis, N.G., SCAN based lossless image compression and encryption, in *Proceedings of IEEE International Conference on Information Intelligence and Systems (ICIIS'99)*, pp. 490–499, 1999.
120. Bourbakis, N.G. and Dollas, A., SCAN-based compression–encryption-hiding for video on demand, *IEEE Multimedia*, 10(3), 79–87, 2003.
121. Maniccam, S.S. and Bourbakis, N.G., Image and video encryption using SCAN pattern, *Pattern Recogn.*, 37(4), 725–737, 2004, http://www.sciencedirect.com, 2003.
122. Bourbakis, N.G., Brause, R., and Alexopoulos, C., SCAN image compression/ encryption hardware system, in *Digital Video Compression: Algorithms and Technologies 1995*, Proc. SPIE Vol. 2419, Arturo A. Rodriguez, Robert J. Safranek and Edward J. Delp III, Eds., 1995, pp. 419–428.
123. Dollas, A., Kachris, C., and Bourbakis, N.G., Performance analysis of fixed, reconfigurable, and custom architectures for the SCAN image and video encryption algorithm, in *Proceedings of 11th Annual IEEE Symposium on Field-Programmable Custom Computing Machines (FCCM'2003)*, 2003, pp. 19–28.
124. Kachris, C., Bourbakis, N.G., and Dollas, A., A reconfigurable logic-based processor for the SCAN image and video encryption algorithm, *Int. J. Parallel Program.*, 31(6), 489–506, 2003.
125. Jan, J.-K. and Tseng, Y.-M., On the security of image encryption method, *Inf. Process. Lett.*, 60(5), 261–265, 1996.
126. Chang, C.-C., and Yu, T.-X., Cryptanalysis of an encryption scheme for binary images, *Pattern Recogn. Lett.*, 23(14), 1847–1852, 2002.
127. Roskin, K.M. and Casper, J.B., From chaos to cryptography, http://xcrypt.theory. org/paper, 1999.
128. Murali, K., Yu, H., Varadan, V., and Leung, H., Secure communication using a chaos based signal encryption scheme, *IEEE Trans. Consumer Eletron.*, 47(4), 709–714, 2001.
129. Peng, J., Liao, X., and Wu, Z., Digital image secure communication using Chebyshev map sequences, in *Proceedings of IEEE International Conference on Communications, Circuits and Systems and West Sino Expositions (ICCCS'2002)*, 2002, pp. 492–496.
130. Murali, K., Leung, H., Preethi, K.S., and Mohamed, I.R., Spread spectrum image encoding and decoding using ergodic chaos, *IEEE Trans. Consuming Eletron.*, 49(1), 59–63, 2003.
131. Yen, J.-C. and Guo, J.-I., A new hierarchical chaotic image encryption algorithm and its hardware architecture, in *Proceedings of 9th (Taiwan) VLSI Design/CAD Symposium*, 1998.
132. Yen, J.-C. and Guo, J.-I., A new chaotic image encryption algorithm, in *Proceedings (Taiwan) National Symposium on Telecommunications*, 1998, pp. 358–362.
133. Guo, J.-I., Yen, J.-C., and Jen-Chieh Yeh., The design and realization of a new hierarchical chaotic image encryption algorithm, in *Proceedings of international Symposium on Communications (ISCOM'99)*, 1999, pp. 210–214.
134. Guo, J.-I. and Yen, J.-C. A new mirror-like image encryption algorithm and its VLSI architecture, in *Proceedings of 10th (Taiwan) VLSI Design/CAD Symposium*, 1999, pp. 327–330.

135. Yen, J.-C. and Guo, J.-I., A chaotic neural network for signal encryption/decryption and its VLSI architecture, in *Proceedings of 10th (Taiwan) VLSI Design/CAD Symposium*, 1999, pp. 319–322.
136. Yen, J.-C. and Guo, J.-I., A new MPEG/encryption system and its VLSI architecture, in *Proceedings of international Symposium on Communications (ISCOM'99)*, 1999, pp. 215–219.
137. Yen, J.-C. and Guo, J.-I., A new image encryption algorithm and its VLSI architecture, in *Proceedings of IEEE Workshop on Signal Processing Systems (SiPS'99)*, 1999, pp. 430–437.
138. Yen, J.-C. and Guo, J.-I., A new chaotic key-based design for image encryption and decryption, in *Proceedings of IEEE Int. Symposium on Circuits and Systems 2000*, Vol. 4, 2000, pp. 49–52.
139. Yen, J.-C. and Guo, J.-I., Efficient hierarchical chaotic image encryption algorithm and its VLSI realisation, *IEEE Proc. Vis. Image Signal Process.*, 147(2), 167–175, 2000.
140. Su, S., Lin, A., and Yen, J.-C., Design and realization of a new chaotic neural encryption/decryption network, in *Proceedings of IEEE Asia–Pacific Conference on Circuits and Systems (APCCAS'2000)*, 2000, pp. 335–338.
141. Yen, J.-C. and Guo, J.-I., A new chaotic mirror-like image encryption algorithm and its VLSI architechture, *Pattern Recogn. Image Anal. (Adv. Math. Theory Applic.)*, 10(2), 236–247, 2000.
142. Guo, J.-I., Yen, J.-C., and Lin, J.-I., The FPGA realization of a new image encryption/decryption design, in *Proceedings of 12th (Taiwan) VLSI Design/CAD Symposium*, 2001.
143. Yen, J.-C. and Guo, J.-I.,The design and realization of a chaotic neural signal security system. *Pattern Recogn. and Image Anal. (Adv. in Math. Theory, Applic)*, 12(1), 70–79, 2002.
144. Yen, J.-C. and Guo, J.-I., Design of a new signal security system, in *Proceedings of IEEE International Symposium on Circuits and Systems (ISCAS'2002)*, Vol. IV, 2002, pp. 121–124.
145. Yen, J.-C. and Guo, J.-I., The design and realization of a new domino signal security system, *J. Chin. Inst. Electr. Engi. (Trans. Chin. Inst. Engi. E)*, 10(1), 69–76, 2003.
146. Chen, H.-C., and Yen, J.-C., A new cryptography system and its VLSI realization, *J. Systems Architechture*, 49(7–9), 355–367, 2003.
147. Chen, H.-C., Yen, J.-C., and Guo, J.-I., Design of a new cryptography system, in *Advances in Multimedia Information Processing — PCM 2002: Third IEEE Pacific Rim Conference on Multimedia Proceedings*, Lecture Notes in Computer Science, Vol. 2532, Yung-Chang Chen, Long-Wen Chang and Chiou-Ting, Hsu, Eds., Springer-Verlag, Berlin, 2002, pp. 1041–1048.
148. Lasota, A. and Mackey, M.C., *Chaos, Fractals, and Noise: Stochastic Aspects of Dynamics*, 2nd ed., Springer-Verlag, New York, 1997.
149. Li, S. and Zheng, X., On the security of an image encryption method, in *Proceedings of IEEE International Conference on Image Processing (ICIP'2002)*, Vol. 2, 2002, pp. 925–928.
150. Li, S. and Zheng, X., Cryptanalysis of a chaotic image encryption method, in *Proceedings of IEEE International Symposium on Circuits and Systems (ISCAS'2002)*, Vol. II, 2002, pp. 708–711.
151. Li, S., Li, C., Chen, G., and Mou, X., Cryptanalysis of the RCES/RSES image encryption scheme, Submitted to *IEEE Trans. Image Processing*, 2004.
152. Belkhouche, F. and Qidwai, U., Binary image encoding using 1D chaotic maps, in *Proceedings of Annual Conference of IEEE Region 5*, 2003, pp. 39–43.
153. Voyatzis, G. and Pitas, I., Digital image watermarking using mixing systems, *Computers Graphics*, 22(4), 405–416, 1998.

154. Scharinger, J., Secure digital watermark generation based on chaotic Kolmogorov flows, in *Security and Watermarking of Multimedia Contents II*, Proc. SPIE, Vol. 3971, Ping W. Wong and Edward J. Delp III, Eds., 2000, pp. 306–313.
155. Nikolaidis, N., Tsekeridou, S., Nikolaidis, A., Tefas, A., Solachidis, V., and Pitas, I., Applications of chaotic signal processing techniques to multimedia watermarking, in *Proceedings of IEEE workshop on Nonlinear Dynamics in Electronic Systems*, 2000, pp. 1–7.
156. Lou, D.-C., Yin, T.-L., and Chang, M.-C., An efficient steganographic approach, *Computer Syst. Sci. Eng.*, 17(4/5), 263–273, 2002.
157. Tefas, A., Nikolaidis, A., Nikolaidis, N., Solachidis, V., Tsekeridou, S., and Pitas, I., Performance analysis of correlation-based watermarking schemes employing Markov chaotic sequences, *IEEE Trans. Signal Process.*, 51(7), 1979–1994, 2003.
158. Chiaraluce, F., Ciccarelli, L., Gambi, E., Pierleoni, P., and Reginelli, M., A new chaotic algorithm for video encryption, *IEEE Trans. Consumer Eletron.*, 48(4), 838–844, 2002.
159. Li, S., Zheng, X., Mou, X., and Cai, Y., Chaotic encryption scheme for real-time digital video, in *Real-Time Imaging VI*, Proc. SPIE, Vol. 4666, Nasser Kehtarnavaz, Ed., 2002, pp. 149–160.

5

Key Management and Protection for IP Multimedia

Rolf Blom, Elisabetta Carrara, Fredrik Lindholm, Karl Norrman, and Mats Näslund

INTRODUCTION

IP multimedia services have, as the name indicates, the Internet Protocol (IP) Suite as a common communication platform. However, the open Internet has achieved a poor reputation with respect to security and, today, security problems end up in media headlines and users have become much more security-aware. Thus, strong security of multimedia applications will be an absolute requirement for their general and widespread adoption. This chapter describes a general framework for how to secure multimedia services.

The popularity of multimedia applications is not only a growing market for the fixed-access Internet users, but it is also an increasing and promising market in the emerging third generation (3G) of mobile communications. This calls for more adaptive multimedia applications, which can work both in a more constrained environment (such as cellular or modem access) and broadband networks. Environments with a mixture of different types of network are usually referred to as heterogeneous networks.

0-8493-2773-3/05/$0.00+1.50
© 2005 by CRC Press

The main topic of interest in this chapter is restricted to the protection of streaming media (e.g., audio and video sessions) in heterogeneous environments. Protection of downloadable content and session establishment is also handled, but only to give the needed context. The focus is on a few specific IP multimedia applications and their security, the key management protocol Multimedia Internet KEYing (MIKEY) [1], and the media security protocol Secure Real-time Transport Protocol (SRTP) [2]. Both of these protocols have been developed in the Internet Engineering Task Force (IETF).

A basic overview of the main threats for IP multimedia applications is provided together with a brief summary of common countermeasures. It is often critical to provide security from a service acceptance point of view, but it may also have major impact on service performance. There are, in particular, some constrained environments where security solutions may encounter problems (e.g., in the emerging heterogeneous networks as documented later). SRTP and MIKEY together can implement an efficient security solution for IP multimedia applications in heterogeneous environments. Basic scenarios and a walk-through of typical multimedia session instances are described to show how the two security protocols are used.

Some parts of the chapter assume that the reader is somewhat familiar with cryptography, and we refer to Reference 3 for extended background on this area.

Threats and Countermeasures

There are several security threats that can undermine the deployment of multimedia applications. Some of the more urgent threats to prevent are the following:

Unauthorized access to services

Unauthorized access to sensitive data (violation of confidentiality)

Message manipulation

Rogue or replayed messages

Disturbance or misuse of network services (often leading to reduced availability)

Common security countermeasures against such threats are the following:

User authentication, which can be used to mitigate unauthorized access to services and which is commonly achieved by using digital signatures

Confidentiality protection, which can be achieved by encrypting data

Integrity protection, which can be achieved by applying a so-called Message Authentication Code (MAC) on the message or by applying a digital signature

Replay protection, which is usually achieved by keeping track of previously received messages in combination with integrity protection of the messages

Assertion of data origin authentication is an important security function in certain group-communication scenarios. Integrity protection alone does not always guarantee that the message actually comes from a given source. It does so in peer-to-peer communication, where integrity protection is also a form of data origin authentication (though repudiable [i.e., the sender can deny the transmission of the message]), as only the two peers know the secret key. However, in a group scenario, where the key is shared among the members, integrity protection can only guarantee that the message comes (unmodified) from a member of the group. A malicious member could alter the message sent from another member or could send the message impersonating another member and would not be discovered. Data origin authentication is commonly provided by use of certificates and digital signatures. This also guarantees nonrepudiation, a security function that, however, is seldom implemented for media or signaling traffic, as it is computationally expensive.

Finally, the threat of misusing content in the form of duplicating it, copying it, and so forth may be an important aspect in applications requiring Digital Rights Management (DRM), but it is largely out of scope of this chapter. However, it should be noted that a security protocol, coupled with proper key management, can serve as a central ingredient in providing DRM.

Basic Mechanisms

The security of data in transmission is implemented by a *security protocol*. A security protocol provides rules and formatting of the packet so that security functions such as encryption and integrity protection can be applied to the transmitted message.

There exist a few different mechanisms to secure general IP communication. The IP Security Protocol (IPsec) [4] is a well-known security protocol for protecting IP traffic, operating at the network layer. Virtual Private Networks (VPNs) based on IPsec are a common way to secure traffic, especially network-to-network traffic. The International Telecommunication Union (ITU) H.323 [5] forms a set of protocols for voice, video, and data conferencing over packet-based networks and also defines its security solution in the H.235 specification [6]. H.235 uses an application security protocol for protecting the media traffic. A later version of H.235 (version 3) [7] adopts, as well, the IETF protocol SRTP.

The parameters needed by the security protocols, such as keys and which cryptographic algorithms to use, are exchanged via a *key management protocol* executed prior to the data transmission, typically at session setup time.

The Transport Layer Security (TLS) protocol [8] is functioning both as a key management protocol and as a security protocol mostly for applications using the Transport Control Protocol (TCP) and may, for example, be used to secure signaling protocols like H.323 and the Session Initiation Protocol (SIP) [9]. The Internet Key Exchange (IKE) [10] is the key management protocol used for exchanging parameters for IPsec. H.323 exchanges the security parameters for securing the media session within the signaling protocol (already protected by TLS or IPsec). IETF has developed a few key management protocols for multicast and groups, among them is MIKEY, which is also adopted in the new version of H.235.

In this chapter, the main focus is on IP multimedia applications based on IETF-based protocols, such as SIP for signaling and the Real-Time Transport Protocol (RTP) [11] together with the Real-Time Transport Control Protocol (RTCP) [11] for the media transport. Figure 5.1 gives an overview of the protocols and their placement in the stack in relation to other protocols. As already discussed, security may be applied on different levels, either at the network layer by applying IPsec or higher

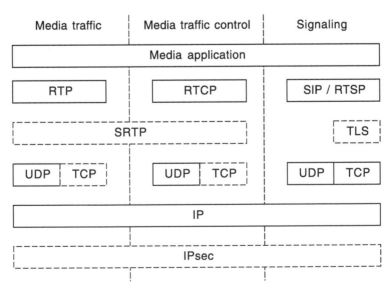

Figure 5.1. Overview of the IP multimedia protocols and their relations to other protocols.

up in the stack by, for example, using SRTP to protect the RTP traffic and TLS to protect the signaling traffic.

Security in Heterogeneous Networks

The design of protocols suitable for heterogeneous environments forces the designers to consider worst-case scenarios, in terms of bandwidth limitation, bit errors, round-trip delays, and so forth. Hence, the heterogeneous environments pose constraints, both on the multimedia applications as such and on the security solutions. In other words, when a communication path consists of several links with very different characteristics, it is the most restrictive link (with respect to bandwidth, bit-error rate, etc.) that dictates the requirements when end-to-end protection is desired. Indeed, true end-to-end protection is the only method that will guarantee security in this case, as one usually cannot be sure that all "hops" along the path are sufficiently secured. Although it is, in general, possible to apply security at lower layers (e.g., secure VPNs based on IPsec), applying security at the application layer allows one to better tune the security protocols to a given scenario and application.

A first concern when designing protocols for heterogeneous environments is that the addition of security should not affect the service in any drastic way (e.g., in terms of convenience and resource consumption). In cellular networks, spectrum efficiency is clearly a major factor for the though-break of any service. To a great extent, this has been the motivation for the work in (robust) header compression for IP protocols, particularly important for applications with short packets such as Voice over IP over Wireless. Encryption may obstruct header compression when applied before it, as it introduces randomness, hence effectively causing a major increase in bandwidth cost. Other common properties to strive for when designing a security protocol are low bandwidth overhead, low number of extra round-trips, and low computational cost.

SCENARIOS

There are many different types of multimedia application, all with different characteristics and requirements. The targeted environments, with all the assumptions that can be made of the underlying network and end devices, also affect these applications. When considering security, it may not always be possible to use one single solution for all types of multimedia application; instead, the scenario and environment may force different types of solution. As an example, an application may use TLS to secure downloadable content. However, the application would not be able to reuse the TLS session to also secure the streaming session that uses RTP over UDP (instead it could use SRTP to secure the RTP session).

This section gives an overview of the scenarios considered in this chapter and also specific security requirements that these scenarios may put on a security solution. For all these scenarios, it is assumed that the network environment is of a heterogeneous nature.

Conversational Multimedia Scenario

In a conversational multimedia scenario, two (or more) users interact with each other in real time, usually using, for example, a voice/video-over-IP application between the involved parties. To be able to send the media between the involved parties, the users must first locate and contact each other using some type of session establishment protocol (see section "Session Establishment"). The session establishment protocols are specially designed to handle the media sessions between the parties, which include operations such as creating new media sessions, tearing down media sessions, and negotiating codecs and other media-specific parameters. As will be shown in section "Protecting the IP Multimedia Sessions", the session establishment can also play an important role for the media security establishment.

In the example scenario of Figure 5.2, Alice uses SIP to establish a call to Bob (see also section "Session Establishment" for more discussions about SIP). This is done by first contacting Alice's SIP proxy, which, in turn, locates Bob's SIP proxy. As Bob's SIP proxy knows the whereabouts of Bob, it can, in turn, contact Bob. When Bob answers and accepts the incoming call, this is also signaled with SIP (along the same path). In general, SIP signaling is sent in a hop-by-hop fashion with the SIP proxies as intermediate nodes.

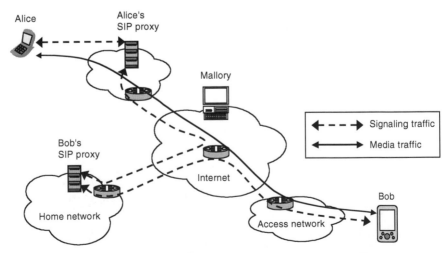

Figure 5.2. Example of an establishment of a two-party SIP call.

The media traffic is, in contrast to the signaling traffic, sent directly between the end nodes. The reason why this can be done is simply because when the session establishment has been executed, the parties will have the direct contact information of the others.

In an unprotected environment, Mallory (the malicious attacker) can launch many different types of attack. The unprotected session establishment is vulnerable to attacks such as media session redirection or hijacking (i.e., directing the media streams to a destination other than the intended one), fake calls, and different types of Denial of Service attack (e.g., by tearing down established media sessions). The unprotected media itself is vulnerable to attacks such as eavesdropping and message manipulation.

Small-Size Group Scenario

In a two-party conference, it is sometimes desired to have the possibility of inviting further participants in the call in an ad hoc manner. The most straightforward approach is, as in the example of Figure 5.3, to have one of the current participants invite the new member and also handle the media flows between the new and existing members. This scenario has its limitations, as it will put high resource demands (both on the processing power and the available bandwidth) on the members as the number of members in the group increases. From a security point of view, this approach will not need much extra security functionality compared to the normal conversational multimedia scenario. However,

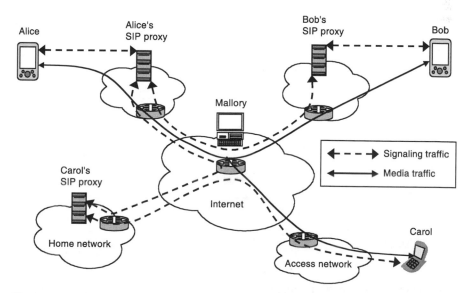

Figure 5.3. Example of a two-party SIP call with the invitation of a third party.

the member controlling the group will need to have one security relationship with each participant to protect the direct communication with the other group members. An additional requirement may be that of backward (forward) security: A party that joins (leaves) the group must not have access to previous (future) communication. This is a key management issue.

One way to circumvent the scalability problem is to use a multicast conference. In that case, the media traffic will be sent on the multicast channel to all available members, and only the signaling traffic will need to be sent on a peer-to-peer basis between the member handling the group and all other members. However, it should be noted that multicast groups are usually prearranged and are not generally setup on an ad hoc basis. From a security perspective, a multicast conference will imply demands, both on the security protocol for the media (e.g., it must be possible to use for multicast delivery) and on the key management protocol (i.e., it must be possible to deliver a group key that can be used by all members).

In the scenario of Figure 5.4, a centralized group controller is used, commonly referred to Multipoint Control Unit (MCU). The MCU is the contact point for all members in the group. A benefit of this approach is, of course, that the media mixing is done on a dedicated server and that the conference is not dependent on the equipment and availability of one member only.

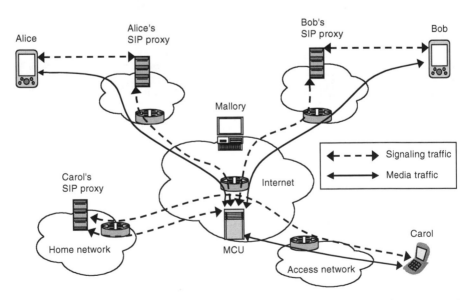

Figure 5.4. Example of a multiparty SIP conference with a centralized control unit.

Securing the media traffic and signaling will not be much different from the conversational multimedia scenario, as individual security relationships between the MCU and the members are needed. However, this scenario assumes that the members can trust the MCU to act fairly and not leak information (so any server could not function as a secure MCU).

One issue that has not been treated so far for these scenarios is authorization for the groups (i.e., who is allowed to join the group and what are they privileged to do). In a peer-to-peer environment, it is common practice to base the authorization on preshared keys, certificates, and so forth. In a group, this can sometimes be more complicated, as, for example, two communicating parties may have different authorization policies. One simple approach for handling group authorization is to allow the group controller to be the only one handling these policies (assuming the existence of a group controller). In the case where no group controller exists or the controller delegates the possibility of inviting new members to other controllers or members, the group authorization policies are negotiated and distributed when the group is set up.

Streaming Scenario

Figure 5.5 shows a streaming scenario where one sender distributes media to multiple receivers (using a normal client–server model). The media distribution can be done either by using unicast distribution (e.g., by a video-on-demand application) or by using multicast distribution (e.g., Web TV broadcast application). Signaling is often done on

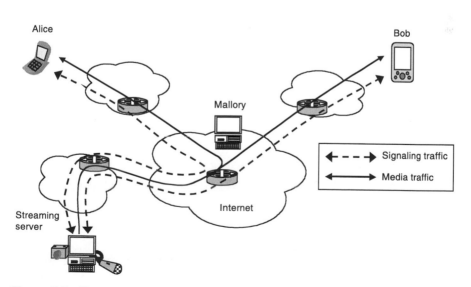

Figure 5.5 Example of a streaming session with one server and two clients.

an individual basis between the client and the server. From a security perspective, this scenario is quite similar to the multicast group scenario (without a centralized controller).

SESSION ESTABLISHMENT

The scenarios described above show the use of signaling protocols for the setup of the communication. Two major standard signaling protocols are the ITU H.323 and the IETF SIP. They are both intended for setup, control, and tear down of sessions, although their designs are based on very different network models. The IETF RTSP [12] is another example of signaling protocol, aimed at client-to-server streaming communication. These signaling protocols often carry a protocol describing the session parameters, such as codecs and ports. One such protocol is the Session Description Protocol (SDP) [13]. There are then cases where the session is just announced (e.g., for multicast communications), where announcement protocols like SAP [14] are used.

The threats against signaling protocols are clearly many, as they control the entire session; it is, therefore, particularly important to provide them with protection. The security applicable to signaling protocols is strongly dependent on the protocol architecture and is defined in the correspondent standards. H.323 (with its security solution, H.235) and SIP make use of hop-by-hop IPsec or TLS, and SIP defines the use of the Secure Multipurpose Internet Mail Extensions (SMIME) protocol [15] for end-to-end protection. The Real Time Streaming Protocol (RTSP) currently uses HTTP security, whereas SAP has no real security solution defined.

A trust model identifies the relation between different actors in a system. The trust model is very dependent on the setting in which the application is used. However, some generalization can be made. Figure 5.6 depicts the trust model for a general SIP call. As shown, there may be a need to have one trust relationship with and between all the proxies to secure the signaling traffic and another trust relationship to secure the media traffic on an end-to-end basis. The reason for this is that the proxies need to be able to read parts of the signaling messages in order to route the message correctly (in some situations, the proxies may also need to modify parts of the message). Thus, to "feel secure," one needs to trust each pair of proxies to sufficiently protect the signaling between them against "third parties" and also that they do not try to read or manipulate the signaling in a malicious way. However, as the media traffic is sent directly between the two communicating parties, the proxies will not need to access the media traffic, which, therefore, can be sent protected end-to-end. Sometimes, mixers or translators

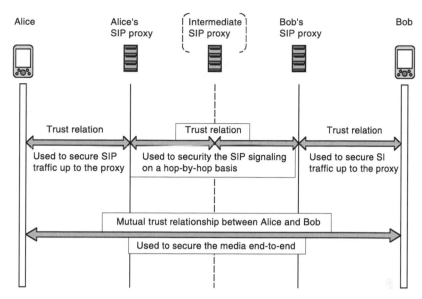

Figure 5.6. A typical trust model for a SIP call.

are needed to harmonize the media communications; in this case, the middlebox typically operates as security endpoints as it needs to have full access to the media.

MULTIMEDIA INTERNET KEYING

Any security protocol providing security functions such as confidentiality and integrity protection requires that the sender and receivers have established a so-called Security Association (SA), which describes how to enable and use the security protocol in a trusted manner. The SA is generally represented by a shared set of data and security parameters (e.g., the cryptographic key used and the encryption/ authentication algorithm used). Typically, these parameters are agreed upon with the aid of a key management protocol. The term "key management protocol" is a slightly misleading but commonly used term for protocols that establish security associations between two or more entities. This section describes the key management protocol Multimedia Internet KEYing (MIKEY).

Overview

A multimedia application in general includes more than one media session that needs protection; for example, a conversational multimedia application can be expected to have two audio streams (one in each

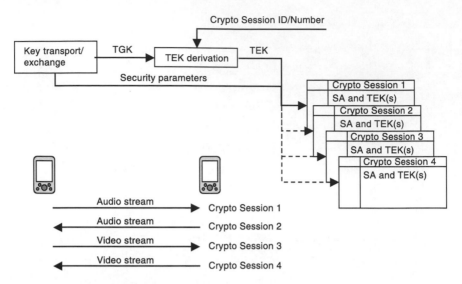

Figure 5.7.   The tasks and main objects of MIKEY.

direction) and two video streams (also one in each direction) between the involved parties. To protect these media sessions, a security protocol needs to be applied for each separate session (e.g., SRTP in the case of RTP traffic; see also section "The Secure Real-Time Transport Protocol"). It would be inconvenient to separately set up a security association for each protected media session, as that would increase both computational cost and the required number of message exchanges between the parties. MIKEY allows setting up security associations for multiple protected media sessions in one run of the protocol. In MIKEY, a protected media session is generally referred to as a Crypto Session (CS). A collection of CSs is called a Crypto Session Bundle (CSB) (e.g., corresponding to a multimedia session).

The main target of MIKEY (see Figure 5.7) is to establish a SA and a corresponding set of Traffic Encryption Keys (TEK) for each CS. This is accomplished by first running the key transport and exchange mechanism, which establishes the SA and so-called TEK Generation Keys (TGK) at all entities involved in the communication. When the TGK is in place, individual TEKs for each CS can be derived from these TGKs using a predetermined key derivation function.

The SA includes parameters required to protect the traffic for the security protocol such as algorithms used and key lifetimes. What parameters actually go into the SA is dependent on the security protocol used. Note that different media streams within the same multimedia session could for example use completely different protection algorithms or different key sizes.

Key Establishment Methods

Because the scope of MIKEY includes thin devices as well as computationally stronger ones, the main goal is to provide strong security but inducing as small a computational load as possible. Furthermore, the number of round-trips is a crucial factor for wireless networks, in particular for cellular networks, where the round-trip times can be quite long. To address both the most extreme scenario and the more general one, MIKEY provides three different key establishment methods. These methods are a transport method based on preshared symmetric keys, an asymmetric-key-based transport method, and a Diffie–Hellman (DH)-based key agreement method.

The messages used by MIKEY to establish a CSB is constructed by several building blocks, called payloads. For instance, each message does first include a header payload, which includes information on the crypto session identifier. After the header, additional payloads are added containing information such as timestamp, encrypted TGKs and signature. This subsection will not go into depth on the payload and encoding of the messages, but will concentrate on the cryptographic characteristic of the different key establishment methods.

Preshared Key Method. The preshared key method is based on the assumption that the involved parties have established a trust relationship by agreeing on a preshared key. The agreement of the preshared key is outside the scope for the protocol but can, for example, be done by using another cryptographic protocol or by physical distribution of the key (e.g., using smart cards or CDs), other types of out-of-band signaling, or a previous run of MIKEY. The preshared key, K, is used to derive one encryption key, K_E, and one authentication key, K_A, in a cryptographically independent manner (see Section "Deriving Keys"). These keys are later used to protect the MIKEY message.

The protocol is executed, as shown in Figure 5.8, by the Initiator creating a message P including the CSB identifier (ID_{CSB}) to which the

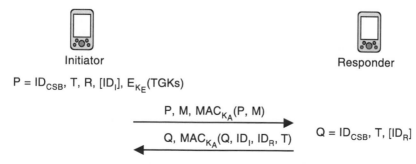

Initiator   Responder

$P = ID_{CSB}, T, R, [ID_I], E_{K_E}(TGKs)$

$P, M, MAC_{K_A}(P, M)$

$Q, MAC_{K_A}(Q, ID_I, ID_R, T)$ $Q = ID_{CSB}, T, [ID_R]$

Figure 5.8. Message sequence chart for the preshared key method.

message corresponds, a timestamp (T), optionally its own identity (ID_I), a random number (R), and the TGKs protected using the encryption key K_E (E_{K_E} (TGKs)). The message is then sent together with other required information (M), such as the security protocol parameters, to the Responder. The entire communication is integrity protected using a MAC, which is keyed using the authentication key, K_A.

The Responder verifies the MAC and, if correctly verified, processes the MIKEY message. (This order, verify-then-process, improves resistance to denial-of-service attacks.) The Responder can then extract the TGKs and other necessary information to create the SA out of the MIKEY message. If requested by the Initiator, the Responder creates a verification message, including a MAC, which is sent back to the Initiator as a proof that the Responder correctly received the keys and parameters.

This method uses no computationally demanding operations and can be performed in less than or equal to one round-trip. Furthermore, the messages are short, because only symmetric key cryptography is involved, reducing the delay over delay-sensitive links. Hence, it is highly suitable for thin devices connected over cellular networks. A drawback is, of course, the fact that a preshared key must be in place, but there are several scenarios where this is likely to be the case, such as the streaming scenario described earlier.

Public-Key Method. The public-key method, outlined in Figure 5.9, is very similar to the preshared-key method, with the main difference being that there is no longer any dependency on prior establishment of a shared key. The Initiator, instead, uses public keys and a so-called envelope approach; that is, the Initiator chooses a (pseudo-)randomly generated symmetric envelope key K_{ENV}, which is used to derive the keys K_E and K_A. The encryption key, K_E, is used as in the preshared-key

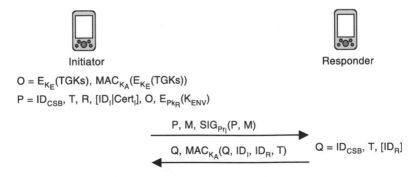

Figure 5.9. Message sequence chart for the public-key method.

case, to encrypt the TGKs, and the authentication key is similarly used to compute a MAC over the encrypted TGKs (compare this with the preshared-key case, where the MAC is computed over the entire message). The main difference to the preshared case is that the envelope key K_{ENV} is encrypted using the Responder's public key Pk_R, which results in the public-key envelope $E_{Pkr}(K_{ENV})$.

In other words, using the public key of the Responder, the Initiator can securely send a key (K_{ENV}), which then can serve the same purpose as the preshared key did in the previous case. It might seem simpler to protect the TGK directly using the public key Pk_R. The rationale for the two-layer approach, introducing the intermediate envelope key, is that no matter how much "TGK material" is transported, one single public-key operation suffices to protect all of the TGK material. Put differently, it is a trade of possible multiple public key operations for symmetric key operations, a significant saving.

When the Initiator sends the message to the Responder, the message is also signed using the Initiators signing key (Pr_I). The Responder can authenticate the Initiator by verifying the signature. As for the preshared-key method, the Responder can obtain all TGKs and other necessary information to create the SAs from the received message. If requested by the Initiator, the Responder creates a verification message, equivalent to the one created in the preshared-key method, which is then sent back to the Initiator.

The public-key-based method includes both an asymmetric key encryption and a signing, in addition to the symmetric key operations required to encrypt the TGKs. The Responder has to verify the signature and possibly also the corresponding certificate/certificate chain, the latter to verify the authenticity of the Responder's public key. This makes this method far heavier computationalwise, compared to the preshared-key method. In a typical case (using RSA), the public-key operations of decryption and signature generation imply modular arithmetic (exponentiation) on very large numbers. This is the price that is paid for the greater flexibility of not having the need of preshared keys in place prior to running MIKEY. Note, though that once the envelope key is in place, this key can be used as a preshared key for future sessions.

Diffie–Hellman Method. Multimedia Internet KEYing protocol also includes a DH method [3], which in contrast to the other two methods, is not based on key transport. Instead, the DH method requires that both parts participate in the creation of the TGK. The method relies on a "cryptographically suitable" finite cyclic group, G, generated by an element, g, in this group, and where g^x denotes g multiplied by itself x times (e.g., $g^3 = g \cdot g \cdot g$).

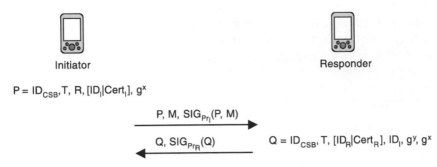

Figure 5.10. **Message sequence chart for the Diffie–Hellman method.**

The DH method is depicted in Figure 5.10. The Initiator starts with creating a message P including the CSB identifier (ID_{CSB}), a timestamp (T), a random number (R), optionally its own identifier (ID_I) or certificate ($Cert_I$), and its public DH value (g^x). The DH value is computed using the generator g of a chosen DH group and a (pseudo-)randomly chosen and secret value x. The message is then signed by the Initiator and sent to the Responder.

The Responder authenticates the Initiator by verifying the signature in the received message. If correctly verified, the Responder creates a response message, similar to the Initiator's message, but also including its own public DH value (g^y).

The Initiator authenticates the Responder by verifying the signature included with the response message. Once each part has both g^x and g^y, they can independently compute the TGK as g^{xy}. Note that anyone listening in on the communication channel will be able to retrieve both g^x and g^y, but under the assumption on the hardness of the DH problem, this information is not sufficient to efficiently compute g^{xy}.

The DH method is clearly limited compared to the transport methods, in the sense that it can only be used in peer-to-peer communications (it is not possible to establish group keys) and that it can only provide one single TGK. Furthermore, it is usually even more computation-intensive than the public-key-based method, because it (compared to, for example, RSA public-key methods) requires one additional modular exponentiation. Also, note that the second message may introduce extra delay compared to the other methods, as the message is not optional to send and, most importantly, it is larger and more demanding to generate than for the other methods.

Deriving Keys

Many times, it is desired to have the possibility to derive one or more keys from a "Master key." This is usually to avoid an "avalanche" of keys need to be created and communicated for different media sessions and purposes. However, to avoid decreasing the security of the system, such a derivation cannot be done using any method, but must be done in a way so that one derived key cannot be used to obtain another derived key, nor the Master key. The first property is usually referred to as an "independence" property and the latter one is usually referred to as Forward Secrecy.

Multimedia Internet KEYing utilizes a key derivation function in two different contexts. In the first context, it is used to derive independent TEKs for multiple protection protocols from one single TGK (without compromising the security); that is, the TGK acts as a Master key for a number of TEKs. In the second context, it is used to compute the encryption key (K_E) and the authentication key (K_A), which are used to protect the messages sent during the key establishment phase in the preshared-key and public-key methods. These keys are the result of applying the key derivation function to either the preshared key (preshared method) or to the envelope key (public-key method). In the DH method of the protocol, the derivation function is used with the DH key, g^{xy}, to directly derive TEKs.

One of the common ways to fulfill the basic security requirements for the key derivation functions is to use a so-called pseudorandom function (PRF). Intuitively, this is a keyed function, $f(k, x)$, such that even after observing a large number of input–output pairs, $(x, f(k, x))$, an adversary gains no significant advantage in predicting the output corresponding to a new input, x' (nor does he learn k). This means that even if the adversary learns one key generated by the PRF, other keys generated are still secure, and these are precisely the desired properties we described earlier.

The PRF in MIKEY is quite similar to the one used in TLS. The differences lie in how the input key data is split and the MAC used. The outline of the PRF is depicted in Figure 5.11. The PRF takes a key (inkey) of specified length and a specific label (a "string") as inputs and then returns a derived key of specified length. The PRF splits the inkey into blocks of 256 bits, and each block is fed into the so-called P-function together with the label. All outputs from each of the computations are then combined using bitwise Exclusive OR (XOR), denoted $\otimes$ in Figure 5.11, which results in the final key (outkey).

The P-function takes a so-called p-inkey and a seed as input and produces a bit string of required length. The P-function is built upon the HMAC function [16], where at a first layer, HMAC is computed using a so-called Output Feedback Mode with the seed as initial input value and the p-key as key. Each output of the HMAC in the first layer is then concatenated (denoted $\oplus$ in Figure 5.11) with the seed and fed into

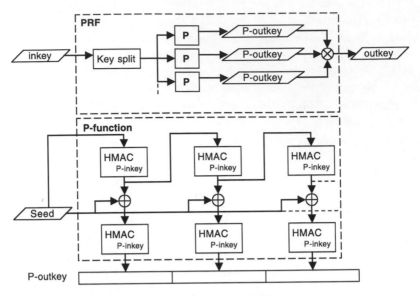

Figure 5.11. Overview of the PRF used in MIKEY.

another "masking" layer of HMACs, which produces the final bit string of the P-function.

THE SECURE REAL-TIME TRANSPORT PROTOCOL

The Secure Real-Time Transport Protocol (SRTP) [2] is an application layer security protocol designed by the IETF community for RTP. SRTP secures RTP as well as the control protocol for RTP (RTCP). This section goes through the building blocks of SRTP and its applications.

The SRTP is not the only mechanism that can be used to secure RTP. For example, IPsec may be used as network security protocol. However, as mentioned, applying security at the application level allows tuning of the security protocol to the need of the application within a particular scenario (e.g., to reduce bandwidth consumption).

Protocol Overview

The SRTP is a profile of the Audio-Video Profile (AVP) of RTP (called Secure AVP [SAVP]), as it defines extensions to RTP and RTCP specifically to secure the applications. SRTP is built as a framework, following general practice to allow for future extensions of the protocol. This makes it possible to add new, possibly even more efficient algorithms in the future and it ensures that it is possible to replace the current cryptographic algorithms if they reveal weaknesses in the future.

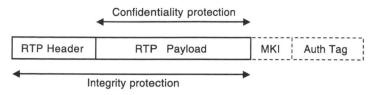

Figure 5.12. Overview of the SRTP packet.

The security parameters needed by SRTP are collected in the so-called cryptographic context. Parameters in the cryptographic context are, for example, the indication of the cryptographic algorithms used and the Master key. Some of the parameters are fixed during the session; others are updated per packets (e.g., replay protection information).

Protecting RTP. One of the main objectives of SRTP is to provide confidentiality, integrity, and replay protection for the RTP media stream. This is done by using a per-packet protection method, which is outlined in Figure 5.12. The confidentiality mechanism covers the original RTP payload, whereas the integrity protection covers both the RTP payload and the RTP header.

An authentication tag is added after the RTP packet when integrity protection is desired. The authentication tag is created by computing a MAC over the RTP packet. The length of the authentication tag is variable and may, for example, be 32 or 80 bits (depending on the level of security needed vs the desired bandwidth usage).

As an additional feature, a so-called Master Key Identifier (MKI) may be added to point to the current Master key used for the confidentiality and integrity protection. The length of the MKI is dependent on the number of expected active Master keys but it would normally be in the order of 1 to 2 bytes. The MKI provides a simple way for the receiver to retrieve the correct key to process a received packet — in particular, in scenarios where the security policy is such that the sender changes the key frequently. In other, more simple configurations, such as with one stream per key and low-rate rekeying, the MKI can be omitted. Readers familiar with protocols such as IPsec can think of the MKI as a Security Parameter Index (SPI), with the difference that the MKI only identifies the key, not the other parameters.

Each SRTP packet is associated with an Index number. This Index number is used for the encryption algorithm and key synchronization and in conjunction with the integrity protection to obtain replay protection. The Index number is a 48-bit number, where the lower 16 bits are the RTP sequence number (sent in the RTP packet header) and the remaining high-order 32 bits are a "rollover counter" (ROC), internal to SRTP.

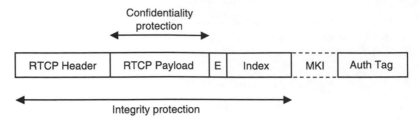

Figure 5.13. Overview of the SRTCP packet.

ROC always starts at zero and is, therefore, only communicated explicitly when a receiver joins an ongoing session. The ROC is updated locally by observing wrap-arounds in the RTP sequence number space.

Protecting RTCP. The SRTP also provides confidentiality, integrity, and replay protection of the RTCP traffic (see Figure 5.13). Although optional for RTP, integrity protection is mandatory for RTCP. This is because RTCP is a control protocol, and, as such, it can inflict actions like termination of the session. The integrity of certain RTCP messages is therefore critical.

The definition of the Secure RTCP (SRTCP) packet is different from SRTP because the RTCP header does not carry a sequence number, which otherwise could have been reused. Therefore, SRTCP adds a 31-bit field at the end of the RTCP packet to synchronize the security processing, called the SRTCP Index. One extra bit preceding the SRTCP Index is the E-bit, a flag indicating if the packet is encrypted or not. This indication is required, as RTP allows the split of a compound RTCP packet into two lower-layer packets, of which one is encrypted and one is sent in the clear to facilitate network monitoring. Similarly to the SRTP packet construction, a MKI may be appended; the authentication tag is, as mentioned, always appended.

Efficiency. Because SRTP targets heterogeneous environments, one major design principle is bandwidth efficiency, which motivates several of the design choices. Limited message expansion is achieved by reusing existing fields for synchronization and by optionally adding variable-length fields, whose length can be policed according to the use case.

The encryption algorithms currently defined in SRTP are used in a way so that no additional field is needed (unless the MKI is used), nor is there any expansion of the payload due to padding. However, this may not hold with other encryption algorithms, possible to add in the future, for example, block ciphers, which require padding and explicit synchronization fields.

To allow further improvement in bandwidth saving, SRTP does not obstruct header compression of the RTP packet; that is, it is possible to apply header compression of the whole IP/UDP/RTP header while using SRTP. In order not to obstruct header compression, SRTP does not encrypt the headers; however, the RTP header may be integrity protected by SRTP (without conflicting with header compression). At the same time, this means that information such as payload type is not kept confidential, which, however, seems a reasonable compromise in most cases.

Note that the bandwidth factor is less critical with RTCP; therefore, adding some bytes and an authentication tag per packet is affordable. In fact, RTCP packets may be quite long and are sent less frequently than RTP packets, and RTP has rules for the bandwidth allocation to RTCP in a given session. Keeping in mind how many bytes SRTP adds to the RTCP packet allows maintaining the same bandwidth sharing ratio (SRTCP packets are simply transmitted with slightly longer packet intervals).

Key Usage

Several cryptographic keys are needed by SRTP for different purposes. To make the key handling more manageable both for SRTP and the key management protocol, only one key (a so-called Master key) is required, from which SRTP can dynamically derive all other keys needed for the current session ("key splitting" procedure).

The SRTP may require as many as six session keys per instance of the protocol, according to which security functions are provided (a triplet for securing RTP and the equivalent but distinct triplet for RTCP):

A session encryption key, to encrypt the data

A so-called session salt key, when required by the underlying encryption transform

A session authentication key, to integrity protect the data

Similar to the MIKEY key derivation function (see section "Deriving Keys"), SRTP derives session keys using a PRF with the Master key and a specific Label as input. The selected unique value of the Label guarantees the uniqueness and independence of each generated session key. The PRF is different from the one in MIKEY and is currently implemented as the Advanced Encryption Standard (AES) [17] in counter mode, as that is also the default encryption algorithm (see section "SRTP Transforms") and thus reduces code size. Although it might seem desirable to reuse code also between MIKEY and SRTP, MIKEY needs to have a slightly more flexible PRF to cater for more general use. Conversely, using the MIKEY PRF in SRTP would be a general overkill, because SRTP could be running together with another key management protocol. Still, note that many of the transforms in MIKEY and SRTP indeed are the same (e.g., AES and HMAC).

SRTP Processing

The SRTP works logically as a so-called "bump in the stack" implementation: at the sender side, it intercepts the RTP packet, forms the SRTP packet from it, and forwards it to the lower layers of the stack for transmission. Conversely, at the receiver side, SRTP intercepts the packet, performs the needed security processing, and (if successful) forwards the RTP packet to the upper layer of the stack.

When an SRTP packet is to be transmitted or received, at first the cryptographic context under which the packet is secured has to be retrieved by looking at the triplet SSRC (which is a field in the RTP packet header to identify the source), destination IP address, and destination transport port number (from the UDP packet header); that is, the sender and receiver maintain a database of cryptographic contexts, indexed by this triplet.

On the sender side (see Figure 5.14), the correct session keys are derived from the Master key in the cryptographic context, by running the pseudorandom function with appropriate input. The packet then may be encrypted and an authentication tag may be added for integrity protection. Possibly a MKI field is also appended, to point to the correct Master key.

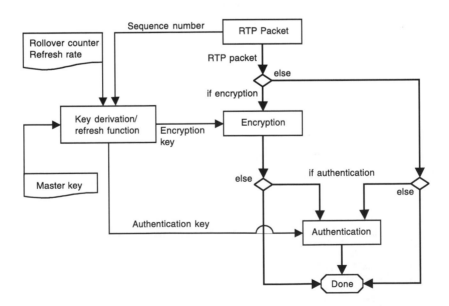

Figure 5.14.   A simplified state chart for SRTP sender processing.

On the receiver side, after the cryptographic context retrieval (possibly using the MKI to identify the Master key), the correct session keys are derived from the Master key. The receiver first checks if the packet has been replayed, by comparing the Index of the packet that has just arrived against the replay list, and accepting it if not seen earlier. Replay protection is achieved by implicitly authenticating the RTP sequence number; hence, replay protection can only be securely implemented when integrity protection is provided. Then, the integrity of the packet is verified (when integrity protection is provided). If and only if the replay and integrity check are successful, the payload is finally decrypted (when confidentiality is enabled), replay information is updated, and the packet is passed to the upper layer application.

SRTP Transforms

An important aspect is the choice of the cryptographic algorithms, also called transforms. SRTP defines some default transforms to promote interoperability. Efficient algorithms are chosen to maintain low computational cost and a small footprint in code size, to accommodate the heterogeneous scenarios. As mentioned, transform reuse is also important.

The algorithm defined for integrity protection is HMAC, based on the Secure Hash Algorithm (SHA1) [18], a common and rather efficient transform. Coverage of the HMAC is the RTP header and RTP payload, together with the local rollover counter. The HMAC is keyed using the session authentication key. The output is an authentication tag of variable length that is appended to the SRTP packet.

There are currently two different ciphers defined in SRTP. Both are stream ciphers, a class of ciphers that is tolerant to bit errors, which may occur in wireless environments. They also conserve payload size without adding any padding. Both ciphers operate on each packet independently: Encryption/decryption of one packet does not depend on preceding packets, to allow for possible packet loss and re-ordering of packets. The two stream ciphers are based on the AES block cipher operating in special modes.

The inputs to the stream ciphers are the session encryption key and some synchronization data, called the Initialization Vector (IV). The output of the stream cipher is a keystream that is then XORed bit by bit with the message to encrypt. This generates the ciphertext (encrypted data) that is sent in the SRTP payload.

The SRTP default encryption transform is AES in counter mode, with the IV derived as a function of the SRTP Index, the SSRC field carried in the RTP header (to allow key sharing between streams), and a Salt key (i.e., a random value agreed upon by the key management and needed by stream ciphers to defeat certain attacks).

An optional cipher is defined: AES in f8 mode. AES in f8 mode operates similarly to AES in counter mode. The IV is a function of the SRTP Index, Salt key, and SSRC, as with AES in counter mode, but also of other fields from the RTP packet header. This provides an implicit authentication of those fields when "real" integrity protection is not provided (modifications of those fields would result in erroneous decryption).

There are, in general, two security aspects that need to be considered when applying a stream cipher in SRTP. First, rekeying must be triggered as soon as one of the streams protected by a given key reaches 2^{48} SRTP packets or 2^{31} SRTCP packets (whichever occurs first). This is because stream ciphers can, if not used correctly, generate a so-called two-time pad: The same output keystream is used more than once to encrypt different content. In this case, an attacker could get information by applying bitwise XOR between the two ciphertexts, which will result in the XOR of the two plaintexts. Note that even at very high packet rates, the above-mentioned limits possess no practical problems, as they correspond to a quite long session time period. Furthermore, some parameter of cryptographic significance should be made unique per stream and packet. One easy way is to use different Master keys per each stream. Two-time pads could also happen by key management mistakes. To avoid two-time pads, automatic key management is needed, as manual key management can more easily lead to mistakes.

SRTP and Key Management

The SRTP is a security protocol independent of any specific key management protocol. This means that any general key management protocol may be used to negotiate the parameters needed for SRTP to work. This desire of modularity is also one of the reasons why SRTP only needs one Master key (which is then internally split into the necessary session keys).

Rekeying. Rekeying, which means changing the Master key during an ongoing session, is done by retrieving a new Master key by the key management protocol. Rekeying during the session lifetime may be triggered for different reasons. For example, in group-communication scenarios, a member leaving the group may trigger such an event, to ensure that the leaving member cannot listen into future group communications (forward security). Similarly, the group policy may require rekeying when a new member joins a group (backward security).

There are other practical reason why rekeying may be needed. For the currently defined transforms in SRTP, rekeying has to be triggered after at most 2^{48} RTP (or 2^{31} RTCP) packets have been secured. Finally, rekeying may simply be performed for security reasons, such as an unfortunate compromise of the Master key, or simply a conservative security policy that wants to strongly limit the amount of ciphertext under the same Master key. In the end, it is always a matter of policy of the key management protocol when the rekeying is actually triggered.

When a rekey event occurs, it is important that all of the parties involved in the secure communication knows when to change the Master key. In SRTP, two different methods can be used. The most straightforward one, suitable in any scenario, is by adding the MKI field to each packet, as a pointer to the correct Master key. A new MKI value in a SRTP packet indicates that the sender has switched the Master key and which key he used. The mapping between each MKI value and Master key is done, for example, by the key management protocol.

The alternative method for supporting rekeying is by SRTP Indices, called the <From, To> pair, indicating the range of packets (48-bit Indices) where the key is active. The new Master key is changed after the packet corresponding to the Index To. The advantage of this method is that no field is added to the packet. Such a method, however, is smoothly applicable only to simple cases — in particular when each Master key is used for a single stream. As a matter of fact, with multiple streams there is no defined way to indicate when to change the Master key simultaneously on all the streams. Indeed, also in case of a single stream, the switching of the key on SRTCP is done as soon as the key is changed on the correspondent RTP stream, already quite an approximation.

Key Refresh. The key derivation provides an additional feature: key refresh. After the mandatory Master key "splitting" into session keys, it is possible to periodically further run the key derivation function, generating fresh session keys (independent from previous session keys). This is clearly different from rekeying, as the latter requires running the key management protocol (often an expensive operation). Use of fresh session keys prevents producing large amount of ciphertext with a single fixed session key, from which an attacker potentially could take advantage.

Furthermore, a compromised session key discloses only the data protected by that key, as other session keys derived from the same Master key are not compromised, due to the use of the pseudorandom function (although compromising a Master key is catastrophic for all of the sessions under it). Therefore, the key refresh is a welcome security enhancement that can be implemented by already existing code (the key derivation function) and does not require running the key management protocol.

PROTECTING THE IP MULTIMEDIA SESSIONS

Multimedia Internet KEYing and SRTP are two components that can be used to create an overall secure multimedia solution. When used together, they provide an end-to-end security solution of the media sessions. In a practical setting, instead of running it separately, MIKEY is generally transported within the session establishment protocol. This is done to reduce the number of round-trips and the overall delay.

Figure 5.15 describes how MIKEY can be used together with SIP to establish a secure call. The MIKEY message as such is carried within the SDP part in a SIP message. The SIP INVITE and the answer to this (the 200 OK message) carry the two-message MIKEY exchange [19]. Because MIKEY provides its own security in an end-to-end fashion, it is independent of the security of the underlying transport protocol. SRTP is used to efficiently protect the RTP sessions within the multimedia session, and its security parameters (cryptographic context) are therefore exchanged by MIKEY. The MIKEY Crypto Session is typically formed by an RTP stream and the corresponding RTCP stream. In particular, MIKEY distributes the SRTP Master key as the MIKEY Traffic Encryption Key (TEK). It can be noted that the approach of Figure 5.15 is inline

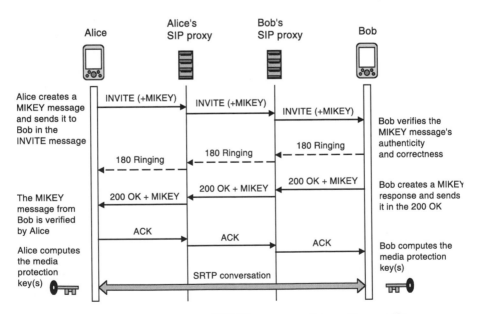

Figure 5.15. SIP signaling, using MIKEY to set up the media security.

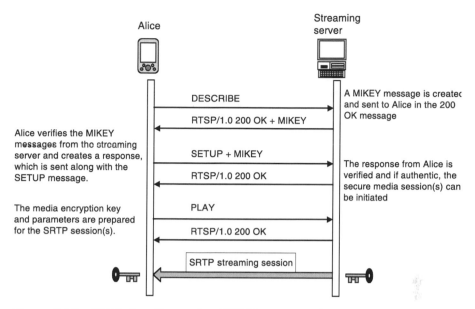

Figure 5.16.   RTSP signaling, using MIKEY to set up the media security.

with the scenario of section "Conversational Multimedia Scenarios" and can easily be extended to work in scenarios such as the group scenarios of section "Small-Size Group Scenario".

An alternative to MIKEY is to transport the security parameters needed for the security protocol instantiation directly within SDP, completely relying on the security offered by it [20] (i.e., if SIP/SDP is only protected in a hop-by-hop fashion between the SIP proxies, with this approach it is not possible to offer end-to-end security for the media as the keys will be exposed in each proxy). Hence, the security of this approach is conditioned by the trust model of SIP itself.

Figure 5.16 shows similarly how MIKEY can be integrated within RTSP. The streaming server sends the first MIKEY message within the answer to the RTSP DESCRIBE message. The client completes the MIKEY exchange using the RTSP SETUP message. When the exchange has completed successfully, both client and server will have the security parameters and keys needed to set up the SRTP session in a correct way. It can be noted that the approach of Figure 5.16 is inline with the scenario previously described.

In general, the use of MIKEY with SRTP creates a security solution that can be integrated with many different session establishment protocols and in a multitude of different scenarios. As noted, MIKEY and SRTP can be used together for efficient end-to-end protection of the media traffic.

The overall delay introduced by adding security is a crucial aspect to consider — in particular in the setup phase when the user is waiting for the communication to begin. The integration of MIKEY in the session establishment protocol together with its ability to set up the security for multiple security associations in one run are some of the important factors that keeps the additional delay for the security setup low.

REFERENCES

1. Arkko, J., et al., MIKEY: Multimedia Internet KEYing, RFC 3830, Internet Engineering Task Force, 2004.
2. Baugher, M., et al., The Secure Real-time Transport Protocol, RFC 3711, Internet Engineering Task Force, 2004.
3. Menezes, A.J., van Oorschot, P.C., and Vanstone, S.A., *Handbook of Applied Cryptography*, CRC Press, Boca Raton, FL, 1997.
4. Kent, S. and Atkinson, R., Security Architecture for the Internet Protocol, RFC 2401, Internet Engineering Task Force, 1998.
5. Packet-Based Multimedia Communications Systems, ITU-T H.323, Telecommunication Standardization Sector of ITU, 2003.
6. Security and Encryption for H-Series (H.323 and other H.245-Based) Multimedia Terminals, ITU-T H.235, Telecommunication Standardization Sector of ITU, 2003.
7. Usage of the MIKEY Key Management Protocol for the Secure Real Time Transport Protocol (SRTP) within H.235, Draft ITU-T Recommendation H.235, Annex G, Telecommunication Standardization Sector of ITU, 2003.
8. Dierks, T. and Allen, C., The TLS Protocol, Version 1.0, RFC 2246, Internet Engineering Task Force, 1999.
9. Rosenberg, J., et al., SIP: Session Initiation Protocol, RFC 3261, Internet Engineering Task Force, 2002.
10. Harkins, D. and Carrel, D., The Internet Key Exchange (IKE), RFC 2409, Internet Engineering Task Force, 1998.
11. Schulzrinne, H., et al., RTP: A Transport Protocol for Real-Time Applications, RFC 3550, Internet Engineering Task Force, 2003.
12. Schulzrinne, H., Rao, A., and Lanphier, R., Real Time Streaming Protocol (RTSP), RFC 2326, Internet Engineering Task Force, 1998.
13. Handley, M. and Jacobson, V., SDP: Session Description Protocol, RFC 2327, Internet Engineering Task Force, 1998.
14. Handley, M., Perkins, C., Whelan, E., Session Announcement Protocol, RFC 2974, Internet Engineering Task Force, 2000.
15. Ramsdell, B., S/MIME Version 3 Message Specification, RFC 2633, Internet Engineering Task Force, 1999.
16. Krawczyk, H., Bellare, M., and Canetti, R., HMAC: Keyed-Hashing for Message Authentication, RFC 2104, Internet Engineering Task Force, 1997.
17. NIST, Advanced Encryption Standard (AES), Federal Information Processing Standard Publications (FIPS PUBS) 197, 2001.
18. NIST, Secure Hash Standard, Federal Information Processing Standard Publications (FIPS PUBS) 180–1, 1995.
19. Arkko, J., et al., Key Management Extensions for Session Description Protocol (SDP) and Real Time Streaming Protocol (RTSP), Internet Engineering Task Force, 2003.
20. Andreasen, F., Baugher, M., and Wing, D., Session Description Protocol Security Descriptions for Media Streams, Internet Engineering Task Force, 2003.

6

Streaming Media Encryption

Heather Yu

Imagine being able to use your home PC as a control center from which you can direct audio or video content (music, movies, and so on) from the Internet or your hard drive to play on your stereo or TV. Further imagine sitting on your couch with friends and family viewing your latest vacation pictures on your TV — a slide show streamed directly from your PC. Digital content, broadband access, and wired and wireless home networks are ushering in a new digital media age that will make such things possible.

— M. Jeronimo and J. Weast, *UPnP Design by Example*, Intel Press, U.S. 2003

Advances in multimedia computing, compression, and communication are providing us with growing business opportunities and demands for multimedia services. For many multimedia services, security is an important component to restrict unauthorized content access and distribution. Today, high-quality audio and video are a reality on the Web. Digital audio and video streaming services are increasingly available to average consumers. Considering the facts that (1) a typical consumer-use personal computer (PC) has enough CPU power to decode a compressed video to a full screen in real time, (2) the number of broadband (Digital Subscriber Line [DSL] and cable modem) Internet users is in the millions, and (3) the audio and video quality at low and medium resolutions over such broadband (> 400 kbps) connections is good enough to find a wide acceptance among consumers, streaming media application is becoming a serious challenger of broadcast and recorded audio and video applications. Consequently, various content providers face serious new challenges protecting the rights of their digital contents, which is an important part in building up a successful streaming media business model. In this chapter, we will discuss an important aspect of streaming media content protection: streaming media encryption to ensure confidentiality and assist content access control.

STREAMING MEDIA, AN EXCITING APPLICATION ON THE INTERNET

What Is Streaming Media?

Streaming media [1] (more precisely, streaming digital media) refers to digital media being transmitted through a network, from a server to a client, in a streaming or continuous fashion. The streamed data are transmitted by a server application and received and displayed in real time by client applications. These applications can start displaying video or playing back audio as soon as enough data have been received. In a packet-switch network, the server breaks the media into packets that are routed and delivered over the network. At the receiving end, a series of time-stamped packets, called stream, is reassembled by the client and the media is played as it comes in. From the end user's perspective, streaming differs from conventional file download in that the client plays the media as it arrives, rather than waiting for the entire media to be downloaded before it can be played. In fact, a client may never actually download a streaming media; it may simply reassemble the media's packets, buffer, play, and then discard them. To an end user, the media stream should come in continuously without interruption. This is, indeed, one of the most important goals for which a media streaming system is designed.

In this chapter, we discuss digital media transmission through a packet-switch network, such as the Internet only. For simplicity, we use the term *streaming media* (or SM) instead of streaming digital media throughout this chapter, unless otherwise specified.

What Types of Media Can be Streamed?

Conceptually, all types of digital media should be able to be streamed. For example, text, image, audio, video, software, and three-dimensional (3-D) data stream all can be streamed. In this chapter, though, we focus on time-varying media types, such as audio and video.

Streaming Media System

A sample generic streaming media system is illustrated in Figure 6.1, which represents the model for both on-demand and live streaming media services. In on-demand mode, a precoded medium is archived on a storage system and served to the client on demand, whereas in live streaming, the media stream is coded, packetized, and streamed to the client in real time.

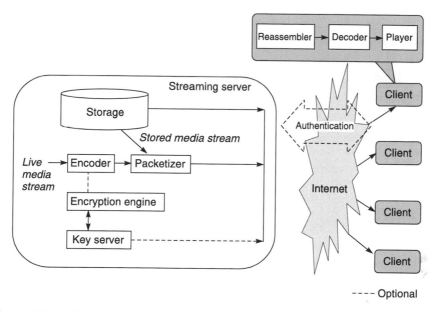

Figure 6.1.   General architecture of a streaming media system.

Notice that the Internet is a best-effort network that is heterogeneous in nature. Packet loss and reordering caused by a constantly changing bandwidth and congestion condition may result in undesirable playback experience at the client due to packet loss. Rate adaptation and error control schemes, along with a suitable protocol, are thus used to offer better user streaming media experiences [1].

A generic streaming media service may include the following basic components:

- Streaming media request
- Coding (compression)
- Packetization
- Packets transmission
- Rate adaptation
- Error control
- Reassembling
- Decoding (decompression)
- Client device playback

When security is required, the system should comprise additional components such as authentication, encryption, and key management. A good secure streaming media system should secure the end-to-end system at the server, during transmission, and within the client player.

199

A generic secure streaming media service requires some or all of the following additional elements:

- User and server authentication
- Media encryption
- Rights management
- Key distribution

Protocol

To transport streaming media over the Internet, appropriate protocol aiming at delivery of delay sensitive media streams is needed. Protocols for real-time and streaming media, such as the Real-time Transport Protocol (RTP) [2], are thus created.

The RTP is a standard protocol for the end-to-end network transport of real-time data, including audio and video. It can be used for both types of streaming media: media-on-demand and live media. RTP consists of a data part and a control part. The latter is called RTCP. In RTP, timing information and a sequence number contained in the RTP header are used to cope with packet loss and reordering. This allows the receivers to reconstruct the timing produced, so that chunks of streaming media are continuously played out. This timing reconstruction is performed separately for each source of RTP packets in the streaming media. The sequence number can also be used by the receiver to estimate how many packets are being lost. Furthermore, reception reports using RTCP are used to check how well the current streaming media is being received and may be used to control adaptive encodings.

Now, let us look at how streaming media encryption can be done in various scenarios.

STREAMING MEDIA ENCRYPTION
Challenges of Streaming Media Encryption

Due to the large size of multimedia data streams, the time constraint for continuous playback of streaming media, and the potential cost overhead for decryption at the client device, streaming media encryption imposes challenges in many applications.

Real-Time Constraint. To transfer huge amounts of time-critical data (e.g., video), it is essential that all security features be implemented with minimal delay and jitter. Otherwise, it may easily amount to non-negligible loss of bandwidth. Thus, new encryption implementations that are fast to execute are needed.

To resolve the real-time constraint, selective encryption was proposed [3–8]; that is, only some parts of the entire bit stream are encrypted while the rest are left in the clear. For instance, only I-frames or I-frames plus the I-blocks in P- and B-frames of a MPEG video are encrypted. Another algorithm is to encrypt only the sign bits and leave the rest in the clear.

As we introduced earlier, at the protocol level, RTP uses timing information and sequence number to ensure correct continuous playback of the media stream at the client.

Potential Cost Increase. The traditional online encryption model was designed to serve small amounts of data, such as credit card information (CCI), typically exchanged in a secure transactional manner. However, audio and video files are often multiple megabytes or more, larger than those data files like CCI by many orders of magnitude. Obviously, the larger the file is, the more data need to be encrypted online, the more processing power is required, and, hence, the higher the cost is. Traditional small data stream based online encryption model does not address the computation intensity problem caused by large data set. Because content is being viewed while streaming, the client player must be able to perform reassembling, decoding, decryption, error control, and reencryption if specified by usage. Computational overhead and additional hardware requirement at the client device is entailed. This processor-intensive routine can significantly deter the end-user experience if the computational overhead caused by the decryption and reencryption processes is more than the client device can handle. Yet, for average consumer streaming media applications, low decryption and reencryption cost overhead is critical for a successful business model. The essence of complexity trades security makes it harder to design a secure streaming media system with minimum additional client device cost. These imply the need for new ways to handle video and audio encryption.

To lower the cost, selective and light-weight encryption schemes have both been proposed. For instance, RTP allows partial packet encryption using a default algorithm — Data Encryption Standard (DES) — which is very fast. Details on RTP encryption will be discussed in section "Encryption in RTP".

Potential Bit-Rate Increase. When a bit stream is scrambled, the original format of the bit stream may be compromised if care is not taken. This is especially serious for compressed multimedia data streams. If scrambling destroys certain inherent structure, compression efficiency can be compromised. With a fixed channel bandwidth, user experience could noticeably worsen due to quality reduction or playback delay at the client machine caused to bit-rate increase. Let us look at a simple example. Assume we encrypt only the I-frames of MPEG video using intrablock discrete cosine transform (DCT) coefficient shuffling; that is we shuffle the DCT coefficients within each DCT block. Assume a low-bit-rate video transmission over a wireless network. As a result of shuffling, some

201

clustered zero coefficients may be shuffled apart that results in a considerable bit-rate increase.

To guarantee full compatibility with any decoder, the bit stream should only be altered (encrypted) in ways that do not compromise the compliance to the original format. This principle is referred to as *format compliance* [8]. Furthermore, a careful design of the RTP packetization may avoid bit-rate increase.

The Rate Variation Challenge. Consider the following consumer application scenario: A streaming video server streams video to clients with various playback devices, such as DTV, desktop PC, personal digital assistant (PDA), and cellular phone. To ensure access control to premium content, the server stores an encrypted copy on the server. Upon receiving access requests, the server needs to send the encrypted medium stream, at the rate suitable for the network channel condition and receiver device capability, to the client (see Figure 6.2). To meet the scalability requirement for plaintext audio and video streaming, simulcast as well as multiple description coding (MDC), scalable and fine-grained scalable (FGS) [9,10] compression algorithms were proposed to provide scalable access of multimedia with interoperability between different services and flexible support to receivers with different device capabilities. With simulcast, multiple bit streams of multiple bit rates for the same content are generated. Unlike simulcast, scalable coding and MDC-based schemes do not require the server to compute and save N copies of the source medium data stream. With MDC, multiple

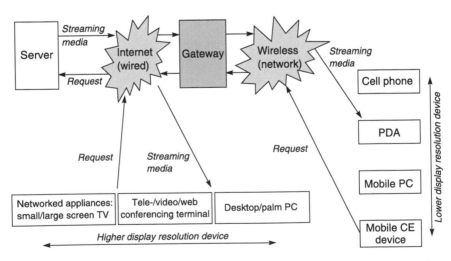

Figure 6.2. Streaming media over heterogeneous networks to various devices, a general layout.

descriptions of the content are generated and sent over different channels to the receiver. If all descriptions are received, the decoder can reconstruct the source medium data stream using all of them to achieve minimum distortion. If only some of the descriptions are received, the decoder can still reconstruct the source medium data stream, but with a higher level of distortion. The reconstruction quality is proportional to the number of descriptions received. The more descriptions the decoder can receive, the less distortion it achieves. However, if the decoder receives a partial description or it cannot receive all bits of the description correctly in time, that description may be rendered useless. With scalable coding, a bit stream is partially decodable at a given bit rate. With FGS coding, it is partially decodable at any bit rate within a bit-rate range to reconstruct the medium signal with the optimized quality at that bit rate. In most applications, FGS is more efficient compared with simulcast [9] and it is one of the most efficient schemes with continuous scalability.

The format-compliant requirement indicates that encryption has to be compliant to the scalable coding schemes to provide adequate quality of service. For example, if a streaming medium compressed using scalable coding needs to be protected and nonscalable cryptography algorithms are used, the advantages of scalable coding may be lost. Previously proposed selective encryption algorithms can often be easily modified to offer two-level scalability. However, they are not specifically designed to be compatible with MDC or FGS coding. If a medium compressed using FGS coding needs to be protected and previous non-FGS-compatible cryptography algorithms are used, the advantages of FGS coding are lost. Also, observe that if an uncompressed medium is encrypted with non-FGS-compatible schemes and transmitted through a dynamic channel to various unknown receivers, reconstruction with optimized quality will be difficult to achieve.

Scalable cryptography schemes can also offer means for multilevel access control of multimedia content. A traditional security system commonly offers two states: access authorization and access denial. With scalable encryption algorithms, a single encrypted bit stream can suggest multiple levels of access authorization — for example, access denial, general preview access authorization that allows a preview of sample or part of the movie in low resolution on mobile handheld devices, high-quality preview access authorization that grants access to club members a sneak preview of the content in high resolution, and full content with quality adapted to device capability.

In section "Scalable Streaming Media Encryption", we argue, in detail, how scalable encryption can be achieved. Several algorithms in the literature will be addressed.

Further, recall that RTP uses reception report to control adaptive encoding. This is helpful in rate adaptation because of network congestion to provide optimal Quality of Service (QoS). For instance, in an audio conferencing application, each conference participant sends audio data in small chunks of, say, 20 msec duration. Each chunk of audio data is preceded by an RTP header; the RTP header and data are, in turn, contained in a User Datagram Protocol (UDP) packet. The RTP header indicates what type of audio encoding (such as Pulse Code Modulation (PCM), Adaptive Differential PCM (ADPCM) or Linear Predictive coding (LPC)) is contained in each packet so that senders can change the encoding during a conference, for example, to accommodate a new participant that is connected through a low-bandwidth link or react to indications of network congestion.

The Dynamic Network Challenge. In another scenario, a song is being streamed to a client through the Internet. As we already know, the Internet, like other packet switch networks, may lose and reorder packets and delay them by variable amounts of time. These hinder the correct reception, decoding, playback of some portions of the audio and may affect the playback quality significantly. If privacy/confidentiality is desired, the data stream and the control packets (in RTP) encryption and streaming pose challenges. If the server encrypts each and every bit of the audio using a long block cipher and sends it to the receiver where it will be decrypted at the same rate as encrypted, it requires correct reception of each and every bit of the encrypted message to avoid subsequent plaintext being improperly decrypted or rendered useless. To do that three or more assumptions are made: the channel capacity is known; the receiver can receive all the bits correctly in time for decryption; and the receiver is capable of reconstructing the media in time for streaming media playback. However, due to the dynamic characteristics of the Internet and the error prone property of some networks, those assumptions are challenging. From consumer's perspective, it is desirable that the changing network condition will not result in streaming interruption but only quality adaptation to the changing network conditions. That is, encryption should preserve the loss resilient capability of the streaming media system.

At the end of section "Scalable Streaming Media Encryption", how to design a loss resilient encryption scheme with the scalability of a streaming media system preserved is addressed and sample algorithm in given.

The Transcoding Challenge. In a heterogeneous environment, where streaming media are transmitted through time-varying communication channels, transcoders are often used at intermediate network nodes to perform transcoding operations on compressed bit streams to provide proper QoS for different clients. If a transcoder requires decryption and reencryption of the encrypted streaming media for transcoding operations, an additional security threat at the intermediate network node is imposed. This further complicates the problem and signifies the need for a new set of cryptography implementation that are suitable for video and audio streaming applications.

Transcoding safe streaming media encryption has been reported in the literature. We will discuss one such algorithm in section "Scalable Streaming Media Encryption" as well.

Other Challenges. Video cassette recorder(VCR)-like function, end-to-end system security, interoperability, upgradeability, and renewability are some other challenges of streaming media encryption.

User requirement for VCR-like function, such as fast forwarding or rewinding, makes the problem even more complex. With traditional encryption methods, the entire file must be processed through the encryption cycle from the beginning to move to other locations within the file. The result can be a slow, linear process that diminishes the consumer experience.

An end-to-end content protection system ensures safe delivery and playback of encrypted streaming media in the entire process. It reduces the risk of security breach and, hence, encourages the adaptation of new streaming media applications by various content providers.

The availability of interoperability, upgradability, and renewability in a streaming media system potentially reduces both manufacture and customer costs in the long run. It is therefore desirable. RTP specifies a default encryption algorithm for maximum interoperability. In the meantime, it allows the use or upgrade of other encryption algorithms. In addition, it makes use of profiles for additional payload encryption, which allows encrypting only the data while leaving the headers in the clear for applications where that is desired.

A Brief Summary of Streaming Media Encryption System Requirements

There are a number of reasonable requirements that are necessary to maintain a user-friendly, yet secure, end-user experience. A secure streaming media system:

- Should be secure but low cost in implementation to appeal to more content creators and providers to present their media online via streaming means
- Should not reduce the playback quality of the streaming media (i.e., should not impact continuous playback, loss resilient capability, and scalability of the system)

- Should sustain current and new heterogeneous environment to attract more applications and more customers
- Should be extendable from PCs to mobile devices and still remain secure, for flexible new business models
- Should be easily renewable
- Should be able to preserve entertainment like experience; users should be able to fast forward or rewind content without any degradation on the viewing or playback experience

The rest of the chapter is dedicated to the discussion on how to design streaming media schemes, on both system and algorithm levels, to meet the requirements. First, we look at RTP encryption in the next subsection. Then, a sample streaming media encryption system that is RTP compliant is discussed. Several example algorithms with scalability and further with loss resilient scalability to meet the secure streaming media transmission requirements are addressed last.

Encryption in RTP

Today, RTP is often used through Internet that cannot be considered secure. To ensure the confidentiality of the media content, encryption has to be used. To support on-demand and live streaming media services, a streaming media system needs to support preencryption and live-encryption, respectively. Further, the system must be capable of delivering the decryption key to the authorized clients securely.

In RTP, when encryption is desired, all of the octets that will be encapsulated for transmission in a single lower-layer packet are encrypted as a unit. The presence of encryption and the use of the correct key are confirmed by the receiver through header or payload validity checks. The default encryption algorithm in RTP is specified to be the DES algorithm in cipher block chaining (CBC) mode. We will discuss CBC in Section "Scalable Streaming Media Encryption". Strong encryption algorithms, such as Triple-DES, can be used in place of the default algorithm for better security. In addition, profiles may define additional payload types for encrypted encodings. A framework for encryption of RTP and RTCP streams is provided in SRTP, the Secure Real-time Transport Protocol [11]. SRTP is a profile of RTP that defines a set of default cryptographic transforms. It allows new transforms to be introduced in the future. With appropriate key management, SRTP is secure for unicast and multicast RTP applications.

In RTP, a mixer is used to perform remixing of RTP streams, including encrypted streams. They are able to decrypt and reencrypt streams. Translators are another type of application-level device in RTP. They perform payload format conversions, tunnel the packets through

206

firewalls, and add or remove encryption to enable the coexistence of the different networking technologies. Because in many applications not all information should be confidential, the use of mixer and translator can help to optimize network usage. As RTP is often used for transferring huge amounts of time-critical data (e.g., video), it is essential that all security features be implemented with minimal delay and jitter. It should be evident that with huge transmission rates, even a small timing overhead easily amounts to huge loss of bandwidth [12]. RTP, implemented on an application bases, provides the flexibility to allow splitting of packets into encrypted and unencrypted parts.

Interested readers are directed to IETF's Real-time Transport Protocol [2] and the Secure Real-time Transport Protocol Internet draft [11] documents for more details about the protocols.

A Sample Streaming Media Encryption System

Encryptionite [12] by SecureMedia uses the Indexed Encryption technique for secure streaming media distribution. Based on the Diffie–Hellman cryptographic algorithm, it is fast (to provide good continuity at playback and low-cost overhead at client device) and can accommodate lost, garbled, or out-of-order packets. It works as follows:

Both the server and the client contain the same set of unique packet keys. The packet keys are contained in a virtual key chain called the *shared index*. The first step in the initialization process occurs when the client generates a private encryption key based on a true random number generation process that uses the client's mouse movement patterns. In the next step, a one-way function is applied to the client's private key to derive the public key. Unique key sets are generated for each client session, assuring that software keys are not compromised by remote hackers. The client's public key is passed to the server, where it is used to create a synchronization token. The synchronization token determines which key in the index of keys will be used to begin the media encryption process. After the synchronization token is securely passed back to the client, the client index is synchronized with the server index. The client and the server are now ready to transmit secure packets of data. Each media packet is then encrypted based on the key indexing system. Using the unique packet ID associated with every packet, the encryption index will select a strong symmetric key that is derived from a mathematical relationship between the "home base," a randomly selected key determines where in the universe of Encryptonite keys the encryption process should begin, and the packet ID (see Figure 6.3). The encryption index is claimed to be secure because the "home base" is determined from a true random process on the client's system and is securely transmitted to the server. Only the client and the server know

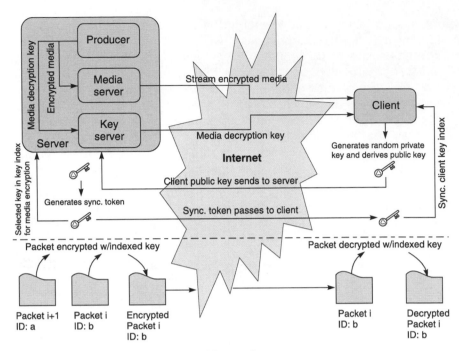

Figure 6.3. **Encryptionite system illustration.**

which key is the starting key. Packet IDs by themselves cannot determine the "home base" because they contain no information that points to the starting key.

Figure 6.3 illustrates both the initialization and the media transmission processes of Encryptionite. Since the packets do not contain any keys, they can be streamed at higher rates (lower bit-rate increase) compared to those systems that place the decryption key inside the packets. In addition, in Encryptionite, packets do not need to be received in any particular order (partially loss resilient) because packet encryption is not based on a packet's relationship to other packets.

When the client receives an encrypted packet, the packet ID is used to determine which key should be used to decrypt the packet. All keys located within the encryption index are unique, random, and not mathematically related. Each packet in a stream has its own unique encryption key. In the rare case that one packet is decrypted by a hacker, that person would only have access to a single packet. Hacking additional packets would be a time-consuming and difficult process. The likelihood of a successful hack is further complicated by extrastrong encryption keys from 612 bits to over 2200 bits (application-adequate security level).

The system process is as follows:

1. Client and server key indexes are synchronized (see Figure 6.3).
2. Each packet of data is encrypted by keys that are determined based on a relationship between the starting key and the packet ID.
3. Each encrypted packet is sent to the client without key information.
4. The client receives the encrypted packet and decrypts it using the appropriate key as determined by the relationship between the packet ID and the starting key.
5. The packet of media is ready to be displayed.

Scalable Streaming Media Encryption

In section "Challenges of Streaming Media Encryption", we discussed the demand for scalable streaming media encryption schemes to cope with the heterogeneous environment, which can provide scalability to various multimedia content, different network topology, changing bandwidth, and diverse device capabilities. In the subsequent sections we shall discuss several scalable encryption schemes that are suitable for streaming media applications, starting with the basic algorithms for stream message encryption.

Digital Data Stream Encryption. Block cipher and stream cipher are two basic encryption algorithms.

> Block ciphers operate on data with a fixed transformation on large blocks of plaintext data; stream ciphers operate with a time-varying transformation on individual plaintext digits [13].

Stream ciphers (SCs), which convert plaintext to ciphertext 1 bit at a time [14], offer means to decrypt an earlier portion of the ciphertext without the availability of the later portion. So does cipher block chaining (CBC) [14]. Therefore, the most intuitive way to provide scalable distribution of uncompressed and some types of compressed media using a single encrypted bit stream is to prioritize data, bit-by-bit or block-by-block, and encrypt the bit stream using SCs or CBC.

Cipher Block Chaining. In the CBC mode, the plaintext bit stream is first partitioned into blocks of plaintext bitstreams. The ciphertext of a current block is a function of the current plaintext block and the previous ciphertext block. In other words, the ciphertext of a current block is a function of all the previous blocks of plaintext bit streams. For instance, the current plaintext block bit stream P^i may be XORed with the previous ciphertext block bit stream C^{i-1} before it is encrypted. Figure 6.4 illustrates the idea. Noticeably, the feedback in CBC causes error propagation; that is, a single error in the ciphertext bit stream will result in incorrect decryption of all the subsequent ciphertext blocks.

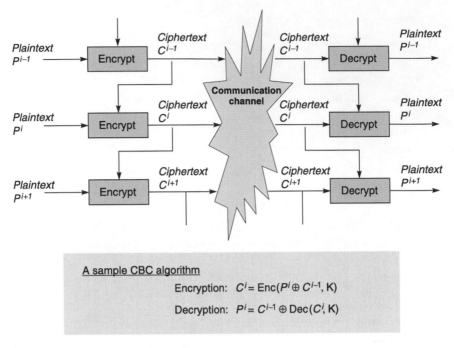

Figure 6.4.   Cipher block chaining.

Stream Cipher.   Stream cipher is to encrypt the plaintext 1 bit at a time. Figure 6.5a illustrates a basic SC architecture. Assume $P = P_1, P_2, \ldots, P_M$ is the plaintext bit stream, $C = C_1, C_2, \ldots, C_M$ is the ciphertext bit stream, and a key stream generator outputs a stream of key bits $K = K_1, K_2, \ldots, K_M$; a simple SC example is given in Figure 6.5b. The ciphertext bit stream C_i is produced by XORing the plaintext bit stream P_i and the key stream K_i bit by bit: $C_i = P_i \oplus K_i$, $i \in [1, M]$. If the medium stream is encrypted Most Significant Bit (MSB) first and Least Significant Bit (LSB) last, it can be truncated $(C_1, C_2, \ldots, C_M, \quad M < N)$, transmitted, and reconstructed $(P_i = C_i \oplus K_i = (P_i \oplus K_i) \oplus K_i = P_i \oplus (K_i \oplus K_i) = P_i \oplus \quad 0, \quad i \in [1, M])$ at corresponding scale with corresponding quality. For an 8-bit gray-scale image, a maximum of eight scales can be achieved using the above simple algorithm. To provide finer-grained scalability, one can take advantage of various transformations, such as DCT and wavelet transformation, and a fine-tuned algorithm that can prioritize the DCT and wavelet coefficients to provide optimized quality at any given bit-rate.

Clearly, in a SC system, the system security very much depends on the key stream, which is often generated using a key stream generator. The closer the key stream is to random, the harder it is to break it.

In a synchronous SC, the encryption and the decryption key streams are identical but independent of the plaintext data stream. When there is

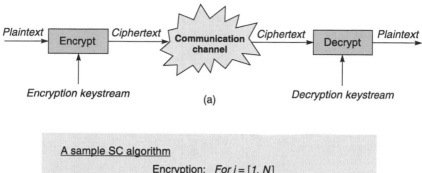

(a)

A sample SC algorithm

Encryption: $For\ i = [1, N]$
$\qquad\qquad C_i = P_i \oplus K_i$

Decryption: $For\ i = [1, N]$
$\qquad\qquad P_i = C_i \oplus K_i$
$\qquad\qquad = P_i \oplus K_i \oplus K_i$
$\qquad\qquad = P_i$

(b)

Figure 6.5.   Stream cipher.

any error, even if it is just 1 bit, in either the ciphertext bit stream or the key stream, a corresponding error will occur in the deciphered plaintext bit stream without error propagation. However, when there is a ciphertext bit loss, the subsequent ciphertext bit stream will decrypt incorrectly.

In a self-synchronizing SC, the key stream is a function of a fixed number of previous ciphertext bits; that is, the output ciphertext at encryption is fed back to the encryption engine to encrypt a subsequent bit.

Note that it is important to choose a SC (e.g., choose the use of synchronization mode, feedback or forward mode, and the key length) based on the security requirement and time constraint of each application.

Next, let us look at how CBC and SC can be used to facilitate secure transcoding for encrypted streaming media through a sample scheme.

A Scalable Streaming Media Encryption Scheme That Enables Transcoding Without Decryption. Wee and Apostolopoulos [15] proposed a scalable streaming media encryption scheme, SSS, to enable transcoding without decryption. It utilizes CBC or SC to achieve progressive decryption ability. First, the input video frame is segmented into regions. Then, each region is coded into scalable video data and header data. Next, the scalable video data are encrypted with CBC- or SC-based progressive encryption, which allows truncation of the scalable video data stream and quality adaptation in accordance. Finally, secure scalable packets are created by combining the unencrypted header data with the progressively encrypted scalable video data. The unencrypted header contains information, such as recommended truncation points or hints, for subsequent transcoding and decoding operations.

The SSS transcoders simply read the unencrypted header data at the beginning of each packet, then use that information to discard packets or truncate packets at the appropriate locations. The SSS decoders then decrypt and decode the received packets; the resolution and quality of the reconstructed video will depend on the transcoding operation.

A FGS Streaming Media Encryption Algorithm. In Reference 16, a scalable encryption scheme for MPEG-4 FGS [9] is proposed. It can be used for streaming media to provide suitable fine-grained scalability. The algorithm provides a lightweight encryption that preserves the full fine-grained scalability and the error resilience performance of MPEG-4 FGS after encryption. It applies different encryption schemes to the base layer and the enhancement layer. The base layer is encrypted in either a selective or a full-encryption mode. Different encryption algorithms can be selected based on the application requirement. In Reference 16, the Chain and Sum encryption algorithm is chosen because it is fast with the least compression overhead. In the selective encryption mode, the DC values of known number of bits, the sign bits of DCT coefficients, and the motion vector sign bits, as well as the motion vector residues, are extracted from each base Video Object Plane (VOP) to form an encryption unit that they call the encryption cell. After encryption, the corresponding bits in the ciphertext are placed back to write over the original bits. In the full-encryption mode, the entropy-coded video data, except the VOP header, form an encryption cell for each base VOP. The VOP start code serves as the separator header for each encryption cell.

A lightweight selective encryption is applied to the enhancement layer to make the encryption transparent to intermediate processing stages. In this way, the FGS fine-granularity scalability is fully preserved in an encrypted FGS stream. A content-based key K_c of a VOP is first generated. K_c combined with the video encryption key K_v and a fixed text key K_T is used as the input to a stream cipher to generate a random binary matrix of the same size as the VOP. The DCT sign bits, the motion vector sign bits, and the motion vector residues are encrypted by XORing the random bit at the same position in the random binary matrix. These ensure correct alignment of received packets with the random bits when packet loss presents. Because the proposed encryption is applied after entropy coding, negligible compression overhead is imposed. Because the enhancement layer is encrypted using a SC, error propagation is prevented. These ensure the same fine-granularity scalability and error resilience performance as an unencrypted FGS video stream. Because both the base-layer and the enhancement-layer encryption algorithms selected in Reference 16 are fast, no significant computational overhead is introduced. Hence, real-time continuous media streaming and playback at the client is enabled.

Achieving Loss Resilient Scalability. The advantage of the above-discussed FGS SM encryption algorithm lies in its means to preserve the fine-grained scalability for MPEG-4 video stream transmission over heterogeneous networks, without departing from the streaming media encryption basic requirements. Introducing encryption service does not drop or reduce the scale of quality of service offered by FGS coding, which is optimally adapted to the network condition and device capability. The disadvantage it suffers from is that it is only partially loss resilient. If a base-layer packet is lost, no reconstruction is possible at the receiver end. Recall that Multiple Description Coding (MDC) [10] provides loss resilient scalability by generating multiple descriptions of the content and transmitting over different channels to the receivers. If all descriptions are received, the decoder can reconstruct the source medium data stream using all of them to achieve minimum distortion. If only some of the descriptions are received, the decoder can still reconstruct the source medium data stream, but with a higher level of distortion. The reconstruction quality is proportional to the number of descriptions received. The more descriptions the decoder can receive, the less distortion it achieves. This implies the possibility of combining the MDC-based encryption algorithm with the FGS-based encryption algorithm for a hybrid encryption algorithm to achieve better loss resilience with application suitable scalability [17]. Figure 6.6 demonstrates one possible configuration. For ease of presentation, a sample algorithm for noncompressed medium data stream is presented:

Algorithm I:

- Generate M descriptions: $D_1, D_2, \ldots, D_M$.
- For $m = [1, M]$, decompose description D_m into base layer D_{bm} and enhancement layer D_{em}: $D_m = D_{bm} + D_{em}$.
- Partition $D_m = D_{bm} + D_{em}$ into $Q_m = Q_{bm} + Q_{em}$ units.
- Encrypt each unit of D_{bm}, $D^q{}_{bm} = \text{Enc}(K_{enc}, D^q{}_{bm})$ using any encryption algorithm with suitable security and computational complexity for the application. In the RTP default mode, DES in the CBC mode should be used. When SRTP is used, AES can be used.
- Encrypt each unit of D_{em}, $D^q{}_{em} = \text{Enc}(K_{enc}, D^q_{em})$ using the FGS SM encryption algorithm such as the one described in the last section.
- Packetize each unit of the stream D^q_{bm} and D^q_{em}.

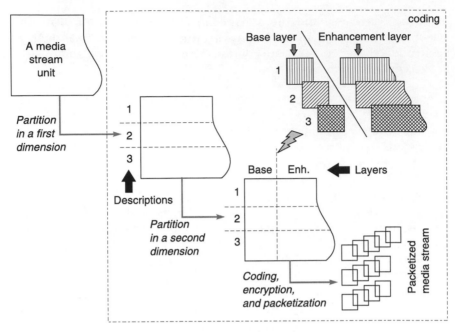

Figure 6.6. **A three-description, two-layer hybrid configuration.**

At the decoder, assume descriptions D'_j, $D'_{j+1}, \ldots, D'_J$ $(J-j) < M$, $j > 0$, $J > 0$, $Q'_j \geq Q_{bm}$ are received, where Q'_j denotes the number of errorless packets, which include all base layer packets of D'_j and $Q'_j - Q_{bm}$ enhancement layer packets, received for description D'_j:

- Decrypt D'_{bj} using all the Q_{bm} packets of D'_j.
- Decrypt D'_{ej} using the $Q'_j - Q_{bm}$ packets of D'_j.
- For $m = [j, J]$, construct each description.
- Reconstruct the media stream using descriptions $D_j, D_{j+1}, \ldots, D_J$.

Assume a base-layer packet $D^{q'}{}_{bm}$, $j < m < J$, is lost during transmission, reconstruction is done using packets $D'_j, \ldots, D'_{m-1}, D'_{m+1}, \ldots, D'_J$. The reconstruction quality is proportional to the total number of descriptions $J - j - 1$ received.

Alternatively, a media stream can be partitioned into base layer and enhancement layer first. The base layer can be further partitioned into multiple descriptions. In this way, base-layer error resilient capability is achieved through multiple description encoding and encryption whereas the enhancement layer error resilient capability may be achieved similar to that was proposed in Reference 16.

Assume a base-layer packet D_m, $1 \leq m \leq M$, is lost during transmission and a base layer is reconstructed using packets $D'_1, D'_2, \ldots, D'_{m-1}$,

$D'_{m+1}, \ldots, D'_M$. Also assume that the media stream can be reconstructed using the received base- and enhancement-layer packets. Subsequently, quality adaptation to the channel condition and device capability with minimum distortion is achieved.

Notice that MDC-based encryption algorithm by itself may be developed for a scalable streaming media encryption system. It will offer a certain level of loss resilient scalability. However, if the decoder receives a partial description or it cannot receive all bits of the description correctly in time, that description may be rendered useless. The disadvantage is that many descriptions have to be generated to achieve fine-granularity scalability.

SUMMARY

The availability of streaming media services is providing us with an exciting new set of applications. The market for Internet audio and video streaming has been expending quickly since the mid-1990s. Today, it is already a billion dollar market and it continues to grow. Because the Internet is heterogeneous in nature with no QoS guarantee, especially for continuous media, it poses substantial challenges for streaming media applications and businesses. Meanwhile, traditional cryptography systems are not intended for large continuous media, such as video and audio, and are not designed for streaming media services in a heterogeneous environment with non-interrupted continuous playback, random access capability, and loss resilient user experiences. These make the design of a secure streaming media system non-trivial. In this chapter, we presented an overview of streaming media encryption and addressed several critical issues for a successful system implementation. Sample system architecture and algorithms are given as references for readers.

The coverage of this chapter is by no means complete. Nevertheless, we hope it can provide application developers with the knowledge of existing and proposed solutions for streaming media encryption and can offer researchers some better understanding toward secure steaming media architecture, algorithm, the associated requirements, and remaining challenges for different secure streaming media applications.

Interested readers can also refer to the following literature for additional information about streaming media technologies, including streaming media encryption. In Reference18, an encryption algorithm for multiple access control of MPEG-4 FGS coded stream is proposed. SMIL, Synchronized Multimedia Integration Language, is the standard markup language for streaming multimedia presentations. It was approved by the World Wide Web Consortium in 1998. SMIL is the language used to lay out audio and video presentations like those found using RealPlayer G2 and, now, QuickTime 4.1. Examples of what SMIL can do are applications like interactive video, video-on-demand, online training, and any other conceivable opportunity where streaming rich media could captivate given audience. References 19 and 20 provide a good introduction to SMIL, and Reference 21 is the standard SMIL specification by W3C. In the meantime, information about Secure XML, which includes encryption, authentication, and key management for enabling the secure transmission of data across the World Wide Web can be found in Reference 22. In addition, the Web sites listed in References 23–26 offer a comprehensive set of information on streaming media, which includes streaming media security and other technologies related to streaming media.

REFERENCES

1. Lu, J., Signal processing for Internet video streaming: A review, *Image and Video Communications and Processing*, Proc. SPIE, 2000.
2. Schulzrinne, H., Casner, S., Frederick, R., and Jacobson, V., RTP: A Transport Protocol for Real-Time Applications, IETF Internet Draft, RFC-1889, 1996.
3. Agi, I. and Gong, L., An Empirical Study of MPEG Video Transmissions, in Proceedings of Internet Society Symposium on Network and Distributed System Security, 1996.
4. Li Y., Chen, Z., Tan, S., and Campbell, R., Security Enhanced MPEG Player, in Proceedings of IEEE First International Workshop on Multimedia Software Development, 1996.
5. Qiao, L. and Nahrstedt, K., A New Algorithm for MPEG Video Encryption, in Proceedings of the First International Conference on Imaging Science, Systems and Techonology, 1997.
6. Wu, C.-P. and Kuo, C.-C.J., Fast encryption methods for audiovisual data confidentiality, in Proceedings of SPIE Voice, Video and Data Communications, Boston, MA, November, 2000.
7. Servetti, A. and De Martin, J.C., Perception-Based Selective Encryption of G.729 Speech, in Proceedings of IEEE International Conference on Acoustics, Speech, and Signal Processing 2002, 2002.
8. Zeng, W., Format-Compliant Selective Scrambling for Multimedia Access Control, in Proceedings of IEEE International Conference on Information Technologies: Coding and Computing 2002, 2002.
9. Li, W., Overview of fine granularity scalability in MPEG-4 video standard, *IEEE Trans. Circuits Syst. Video Technol.* 11(3), 301–317, 2001.
10. Goyal, V. K., Multiple description coding: Compression meets the network, *IEEE Signal Process. Mag.* 18(5), 74–93, 2001.
11. Baugher, M., McGrew, D., Carrara, E., Naslund, M., and Norrman, K, The Secure Real-time Transport Protocol, IETF Internet Draft, 2003.
12. SecureMedia—Encryptonite product brief, http://www.stelzner.com/PDF/Secure Media-PB.pdf, SecureMedia, 2000.
13. Hallivuori, V., Real-time Transport Protocol (RTP) security, Helsinki University of Technology, Telecommunications Software and Multimedia Laboratory, http://kotiweb.kotiportti.fi/vhallivu/files/rtp_security.pdf. 2000.
14. Schneier, B., *Applied Cryptography: Protocols, Algorithms, and Source Code in C*, 2nd Ed., John Wiley & Sons, New York, 1996.
15. Wee, S. J. and Apostolopoulos, J.G., Secure Scalable Streaming Enabling Transcoding Without Decryption, Proceedings of IEEE International Conference on Image Processing, 2001.

16. Yuan, B., Zhu, B., Wang, Y., Li, S., and Zhong, Y., Efficient and Fully Scalable Encryption for MPEG-4 FGS, IEEE International Symposium on Circuits and Systems, Bangkok, 2003.
17. Yu, H., On scalable encryption for mobile consumer multimedia applications, in Proceedings of IEEE International Conference on Communications, 2004.
18. Yuan, C., Zhu, B. B., Su, M., Wang, X., Li, S., and Zhong, Y., Scalable Access Control Enabling Rate Shaping without Decryption for MPEG-4 FGS Video, in Proceedings of IEEE International Conference on Image Processing, Barcelona, 2003.
19. SMILe with Rich Media, white paper by Confluent Technologies, http://www.smilsoftware.com/docs/SMILwithRichMedia.pdf.
20. Creat Rich Media and SMILe, white paper by Confluent Technologies, http://www.smilsoftware.com/docs/RichMediaandSMIL.pdf.
21. W3C Synchronized Multimedia Working Group, Synchronized Multimedia Integration Language (SMIL) 1.0 Specification, W3C REC-smil-19980615, W3C Recommendation, 1998.
22. Eastlake, D. E., and Niles, K., *Secure XML: The New Syntax for Signatures and Encryption*, Addison-Wesley, Reading, MA, 2002.
23. Streaming Media Land, http://streamingmedialand.com/technology_frameset.html.
24. Streaming Media Inc., http://www.streamingmedia.com/.
25. Streaming Media World, http://smw.internet.com/.
26. Streaming Magazine, http://www.streamingmagazine.com/.

Part III
Multimedia Watermarking

7

Survey of Watermarking Techniques and Applications

Edin Muharemagic and Borko Furht

INTRODUCTION

A recent proliferation and success of the Internet together with the availability of relatively inexpensive digital recording and storage devices have created an environment in which it became very easy to obtain, replicate, and distribute digital content without any loss in quality. This has become a great concern to the multimedia content (music, video, and image) publishing industries, because technologies or techniques that could be used to protect intellectual property rights for digital media and prevent unauthorized copying did not exist.

Although encryption technologies can be used to prevent unauthorized access to digital content, it is clear that encryption has its limitations in protecting intellectual property rights: Once a content is decrypted, there is nothing to prevent an authorized user from illegally replicating digital content. Some other technology was obviously needed to help establish and prove ownership rights, track content usage, ensure authorized access, facilitate content authentication, and prevent illegal replication.

0-8493-2773-3/05/$0.00+1.50
© 2005 by CRC Press

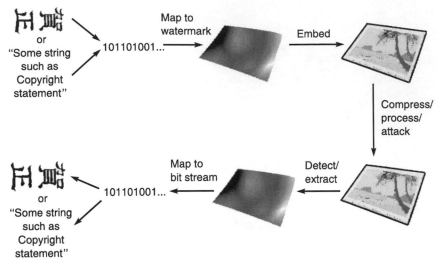

Figure 7.1. General framework of digital watermarking systems.

This need attracted attention from the research community and industry leading to a creation of a new information-hiding form, called *digital watermarking*. The basis idea is to create a metadata containing information about the digital content to be protected and hide it within that content. The information to hide, the metadata, can have different formats. For example, it may be formatted as a character string or a binary image pattern, as illustrated in Figure 7.1. The metadata is first mapped into its bit stream representation and then into a *watermark*, a pattern of the same type and dimension as the *cover work* (the digital content to be protected). The watermark is then embedded into the cover work. The embedded watermark should be imperceptible and it should be robust enough to survive not only most common signal distortions but also distortions caused by malicious attacks.

It is clear that digital watermarking and encryption technologies are complementing each other and that a complete multimedia security solution depends on both. This chapter provides and overview of the image watermarking techniques and it describes various watermarking application scenarios.

DIGITAL WATERMARKING TECHNIQUES

Digital Watermarking Systems

A digital watermarking system consists of two main components: *watermark embedder* and *watermark detector*, as illustrated in Figure 7.2. The embedder combines the *cover work* C_O, an original copy of digital media (image, audio, video), and the *payload P*, a collection of bits

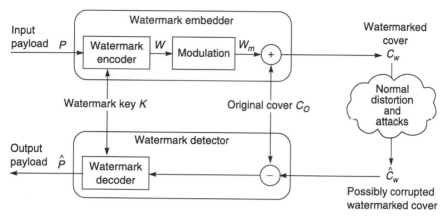

Figure 7.2. Digital watermarking systems with informed detection.

representing metadata to be added to the cover work, and it creates the *watermarked cover* C_W. The watermarked cover C_W is perceptually identical to the original C_O but with the payload embedded. The difference between C_W and C_O is referred to as *embedding distortion*. The payload P is not directly added to the original cover C_O. Instead, it is first encoded as a *watermark* W, possibly using a secret key K. The watermark is then modulated and/or scaled, yielding a *modulated watermark* W_m, to ensure that embedding distortion will be small enough to be imperceptible.

Before it gets to a detector, the watermarked cover C_W may be subjected to different types of processing, yielding *corrupted watermarked cover* $\hat{C}_w$. This corruption could be caused either by various distortions created by normal signal transformations, such as compression, decompression, D/A and A/D conversions, or by distortions introduced by various malicious attacks. The difference between $\hat{C}_W$ and C_W is referred to as *noise N*.

The watermark detector either extracts the payload $\hat{P}$ from the corrupted watermarked cover $\hat{C}_W$ or it produces some kind confidence measure indicating how likely it is for a given payload P to be present in $\hat{C}_w$. The extraction of the payload is done with help of a watermark key K.

Watermark detectors can be classified into two categories, *informed* and *blind*, depending on whether the original cover work C_O needs to be available to the watermark detection process or not. *Informed detector*, also known as a nonblind detector, uses the original cover work C_O in a detection process. The *blind detector*, also known as an oblivious detector, does not need the knowledge of the original cover C_O to detect a payload.

Watermarking as Communication

The watermarking system, as presented in the previous section, can be viewed as some form of communication. The payload message P, encoded as a watermark W, is modulated and transmitted across a communication channel to the watermark detector. In this model, the cover work C_O represents a communication channel and, therefore, it can be viewed as one source of noise. The other source of noise is a distortion caused by normal signal processing and attacks.

Modeling watermarking as communication is important because it makes it possible to apply various communication system techniques, such as modulation, error correction coding, spread spectrum communication, matched filtering, and communication with side information, to watermarking.

Those techniques could be used to help design key building blocks of a watermarking system which deal with the following:

- How to embed and detect one bit
- What processing/embedding domain to use
- How to use side information to ensure imperceptibility
- How to use modulation and multiplexing techniques to embed multiple bits
- How to enhance robustness and security, where robustness can be defined as a watermark resistance to normal signal processing, and security can be defined as a watermark resistance to intentional attacks

Embedding One Bit in the Spatial Domain

It is a common practice in communication to model channel noise as a random variable whose values are drawn independently from a Normal distribution with zero mean and some variance, σ_n^2 This type of noise is referred to as Additive White Gaussian Noise (AWGN). It is also known from communication theory that the optimal method for detecting signals in the presence of AWGN is matched filtering, which is based on computing a linear correlation between transmitted and received signals and comparing it to a threshold.

Applying those two ideas to watermarking, yields a simple, spatial domain image watermarking technique with blind detection, which is illustrated in Figure 7.3. The watermark is created as an image having the same dimensions as the original cover image C_O with the luminance values of its pixels generated as a key-based pseudorandom noise pattern drawn from a zero-mean, unit-variance Normal distribution, $N(0, 1)$. The watermark is then multiplied by the embedding strength factor s and

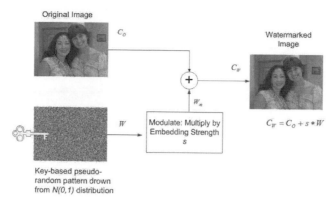

Figure 7.3. Watermark embedding procedure.

added to the luminance values of the cover image pixels. The embedding strength factor is used to impose a power constraint in order to ensure that once embedded, the watermark will not be perceptible. Note that once the embedding strength factor s is selected, it is applied globally to all cover images that need to be watermarked. Also note that this embedding procedure creates the watermark W independently of the cover image C_O.

According to the model of Digital Watermarking System depicted in Figure 7.2, the watermark detector will work on a received image C, which could be represented either as $C = \hat{C}_W = C_O + W_m + N$, if the image was watermarked, or as $C = C_O + N$ otherwise, where N is a noise caused by normal signal processing and attacks.

To detect the watermark, a detector has to detect the presence of the signal W in the received, possibly watermarked, image C. In other words, the detector has to detect the signal W in the presence of noise caused by C_O and N. Assuming that both C_O and N are AWGN, the optimal method of detecting watermark W in the received image C is based on computing the linear correlation between W and C:

$$\mathrm{LC}(W, C) = \frac{1}{I \cdot J} W \cdot C = \frac{1}{I \cdot J} \sum_{i,j} w_{ij} c_{ij} \tag{7.1}$$

where w_{ij} and c_{ij} represent pixel values at location i,j in W and C and I and J represent the image dimensions.

If the received image C was watermarked (i.e., if $C = C_O + W_m + N$), then

$$\mathrm{LC}(W, C) = \frac{1}{I \cdot J} (W \cdot C_O + W \cdot W_m + W \cdot N) \tag{7.2}$$

225

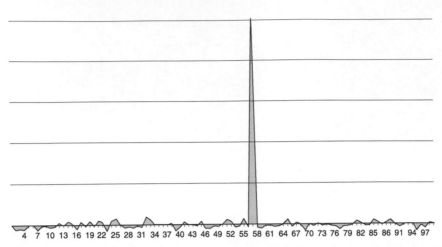

Figure 7.4. **Correlation values for a pseudorandom pattern generated with seed = 57 correlated with pseudorandom patterns generated with other seeds.**

Because we assumed that C_O and N were AWGN and we have created the watermark W as AWGN, the additive components of linear correlation, $W \cdot C_O$ and $W \cdot N$, are expected to have small magnitudes and the component $W \cdot W_m = sW \cdot W$ is expected to have a much larger magnitude. This is illustrated in Figure 7.4, where it is shown that AWGNs generated as pseudorandom patterns using different keys (i.e., seeds) have a very low correlation with each other, but a high correlation with itself. Therefore, if a calculated linear correlation $LC(W, C)$ between the received image C and watermark W is small, then a conclusion can be made that the image C was not watermarked. Otherwise, the image C was watermarked. This decision is usually made based on a threshold T, so that if $LC(W, C) < T$, the watermark W is not detected in C, and if $LC(W, C) > T$, the watermark W is detected in C.

A watermark detection procedure based on threshold is illustrated in Figure 7.5. Two curves represent distribution of linear correlation (LC) values calculated for the set of unmarked images (the curve that peaks for the detection value 0), and for the set of watermarked images (the curve that peaks for the detection value 1). For a selected threshold value T, the portion of the curve for the unmarked images to the right of the threshold line T, represents all tested unmarked images, which will be erroneously detected as marked images, and the portion of the curve for the marked images to the left of the threshold line T represents watermarked images which will erroneously be declared as unmarked. The former error is called a *false-positive error* and the latter is called a *false-negative error*. The false-negative error rate can also be seen as a measure of efficiency of the watermarking system because it

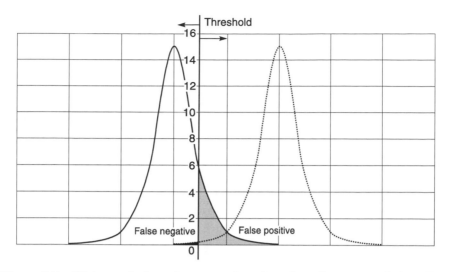

Figure 7.5. **Watermark detection procedure based on linear correlation: The left-hand curve represents distribution of LC values when no watermark has been embedded; the right-hand curve represents distribution of LC values when a watermark was embedded.**

can be seen as a failure rate of the embedder to embed a detectable watermark.

False-positive and false-negative errors occur because original cover images C_O are not accurately modeled as AWGN and, consequently, they can have a high correlation with the watermark signal W. Several proposed watermarking systems are based on this technique [1–5].

Patchwork: Another Spatial Domain Watermarking Technique

Patchwork is another spatial domain watermarking technique designed to imperceptibly embed a single bit of information in a cover image [1]. It is a statistical method that embeds a watermark by changing the statistical distribution of the luminance values in the set of pseudo-randomly selected pairs of image pixels. This technique is based on an assumption that luminance values of an image have the following statistical property: $\sum_n (a_i - b_i) \approx 0$ where $A = \{a_i\}_1^n$ and $B = \{b_i\}_1^n$ are two patches of pseudorandomly selected image pixels. A watermark is embedded by increasing the brightness of pixels that belong to the patch $A = \{a_i\}_1^n$ and, accordingly, decreasing the brightness of pixels that belong to the patch $B = \{b_i\}_1^n$ In other words, after pseudo-randomly selecting patches $A = \{a_i\}_1^n$ and $B = \{b_i\}_1^n$ luminance values of the selected pixels are modified according to the following formula: $\tilde{a}_i = a_i + \delta \wedge \tilde{b}_i = b_i - \delta$. This modification creates a unique

227

statistic that indicates the presence or absence of a watermark. The watermark detector will select the same n pairs of pixels belonging to two patches and it will compute $\Delta = \sum_n (\tilde{a}_i - \tilde{b}_i)$. If $\Delta = 2n\delta$, the image is watermarked; otherwise, it is not.

This watermarking technique creates a watermark independently of the cover image and it uses blind detection (i.e., the detector does not require the original cover image in order to be able to determine whether the image has been watermarked or not). The watermark detection is based on linear correlation because the detection process described is equivalent to correlating the image with a pattern consisting of 1's and -1's, where the pattern contains a 1 for each pixel from the patch $A = \{a_i\}_1^n$ and a -1 for each pixel from the patch $B = \{b_i\}_1^n$.

Watermarking in Transform Domains

A watermark can be embedded into the cover image in a spatial domain, and we have described two such techniques earlier. Alternatively, a watermark embedding operation can be carried out in a transform domain, such as discrete Fourier transform (DFT) domain, the full-image (global) discrete cosine transform (DCT) domain, the block-based DCT domain, the Fourier–Mellin transform domain, or the wavelet transform domain.

Transform domains have been extensively studied in the context of image coding and compression, and many research results seem to be very applicable to digital watermarking. From the theory of image coding we know that in most images the colors of neighboring pixels are highly correlated. Mapping into a specific transform domain, such as DCT or discrete wavelet transform (DWT), serves two purposes. It should decorrelate the original sample values and it should concentrate the energy of the original signal into just a few coefficients. For example, when a typical image is mapped into the spatial-frequency domain, the energy is concentrated in the low-index terms, which are very large compared to the high-index terms. This means that a typical image is dominated by the low-frequency components. Those low frequencies represent the overall shapes and outlines of features in the image and its luminance and contrast characteristics. High frequencies represent sharp edges and crispness in the image but contribute little spatial-frequency energy. As an example, a typical image might contain 95% of the energy in the lowest 5% of the spatial frequencies of the two dimensional DCT domain. Retention of these DCT components, together with sufficiently many higher-frequency components to yield an image with enough sharpness to be acceptable to the human eye, was the objective for creation of an appropriate quantization table to be used for JPEG compression.

As we will see in some of the following sections, the selection of a specific transform domain to use for watermarking has its own reasons and advantages.

Watermarking in DCT Domain and Spread Spectrum Technique. The DCT domain has been used extensively for embedding a watermark for a number of reasons. Using the DCT, an image is divided into frequency bands, and the watermark can be conveniently embedded in the visually important low- to middle-frequency bands. Sensitivities of the human visual system to changes in those bands have been extensively studied in the context of JPEG compression, and the results of those studies can be used to minimize the visual impact of the watermark embedding distortion. Additionally, requirements for robustness to JPEG compression can be easily addressed because it is possible to anticipate which DCT coefficients will be discarded by the JPEG compression scheme. Finally, because JPEG/MPEG coding is based on a DCT decomposition, embedding a watermark in the DCT domain makes it possible to integrate watermarking with image and video coding and perform real-time watermarking applications.

An efficient solution for watermarking in the global DCT domain was introduced by Cox et al. [6] and it is based on spread spectrum technology. The general spread spectrum system spreads a narrow-band signal over a much wider frequency band so that the signal-to-noise ratio (SNR) in a single frequency band is low and appears like noise to an outsider. However, a legitimate receiver with precise knowledge of the spreading function should be able to extract and sum up the transmitted signals so that the SNR of the received signal is strong.

Because, as we pointed out earlier, a watermarking system can be modeled as communication where the cover image is treated as noise and the watermark is viewed as a signal that is transmitted through it, it was only natural to try to apply techniques that worked in communications to watermarking. The basic idea is to spread the watermark energy over visually important frequency bands, so that the energy in any one band is small and undetectable, making the watermark imperceptible. Knowing the location and content of the watermark makes it possible to concentrate those many weak watermark signals into a single output with a high watermark-to-noise ratio (WNR). The following is a high-level overview of this watermarking technique.

The watermark is embedded in the first n lowest frequency components $C = \{c_i\}_1^n$ of a full-image DCT in order to provide high level of robustness to JPEG compression. The watermark consists of a sequence of real numbers $W = \{w_i\}_1^n$ drawn from a Normal distribution $N(0, 1)$, and it is embedded into the image using the formula

$\tilde{c}_i = c_i(1 + sw_i)$, where s is the watermark embedding strength factor. Watermark detection is performed using the following similarity measure:

$$\text{sim}(W, W') = \frac{W \cdot W'}{\sqrt{W' \cdot W'}} \tag{7.3}$$

where W' is the extracted watermark, calculated as

$$\{w'_i\}_1^n = \left\{\left(\frac{\tilde{c}_i}{c_i} - 1\right)\Big/s\right\}_1^n \tag{7.4}$$

where the $\tilde{c}_i$ components are extracted from the received, possibly watermarked, image and the c_i components are extracted from the original cover image. The watermark is said to be present in the received image if $\text{sim}(W, W')$ is greater than the given threshold.

Because the original image is needed for calculation of the extracted watermark W', which is used as part of the watermark presence test, this watermarking system falls into the category of systems with informed detectors.

The authors used an empirically determined value of 0.1 for the embedding strength factor s and chose to spread the watermark across 1000 lowest-frequency non-DC DCT coefficients ($n = 1000$). Robustness tests showed that this scheme is robust to JPEG compression to the quality factor of 5%, dithering, fax transmission, printing–photocopying–scanning, multiple watermarking, and collusion attacks.

Watermarking in the Wavelet Domain. With the standardization of JPEG-2000 and a decision to use wavelet-based image compression instead of DCT-based compression, watermarking techniques operating in the wavelet transform domain have become more attractive to the watermarking research community. The advantages of using the wavelet transform domain are an inherent robustness of the scheme to the JPEG-2000 lossy compression and the possibility of minimizing computation time by embedding watermarks inside of a JPEG-2000 encoder. Additionally, the wavelet transform has some properties that could be exploited by watermarking solutions. For example, wavelet transform provides a multiresolution representation of images, and this could be exploited to build more efficient watermark detection schemes, where watermark detection starts from the low-resolution subbands first, and only if detection fails in those subbands, it explores the higher-resolution subbands and the additional coefficients it provides.

Zhu et al. [7] proposed a unified approach to digital watermarking of images and video based on the two-dimensional (2-D) and three-dimensional (3-D) discrete wavelet transforms. This approach is very similar that of Cox et al. [6] presented earlier. The only difference is that Zhu generates a random vector with $N(0, 1)$ distribution and spreads it across coefficients of all high-pass bands in the wavelet domain as a multiresolution digital watermark, whereas Cox et al. do it only across a small number of perceptually most important DCT coefficients. The watermark added to a lower resolution represents a nested version of the one corresponding to a higher resolution and the hierarchical organization of the wavelet representation allows detection of watermarks at all resolutions except the lowest one. The ability to detect lower-resolution watermarks reduces computational complexity of watermarking algorithms because fewer frequency bands are involved in the computation. It also makes this watermarking scheme robust to image and video down sampling operation by a power of 2 in either space or time.

Watermarking in the DFT Domain. The discrete Fourier transform of an image is generally complex valued and this leads to a magnitude and phase representation for the image. Most of the information about any typical image is contained in the phase and the DFT magnitude coefficients convey very little information about the image [8].

Adding a watermark to the phase of the DFT, as it was proposed in Reference 9, improves the robustness of the watermark because any modification of those visually important image components in an attempt to remove the watermark will significantly degrade the quality of the image. Another reason to modify or modulate the phase coefficients to add a watermark is based on communications theory, which established that modulating the phase is more immune to noise than modulating the amplitude. Finally, the phase-based watermarking was also shown to be relatively robust to changes in image contrast [9].

Adding a watermark to the DFT magnitude coefficients and ignoring the phase was proposed in Reference 8. Embedding a watermark in the DFT magnitude coefficients, which convey very little information about an image, should not introduce a perceptible distortion. However, because modifications of the DFT magnitude coefficients are much less perceptible than phase modifications, one would expect that good image compressors would give much higher importance to preserving the DFT phase than the DFT magnitude, rendering the DFT magnitude-based watermarking system vulnerable to image compression. The authors reported the surprising result that all major compression schemes (JPEG, set partitioning in heirarchial trees (SPIHT), and MPEG) preserved the DFT magnitude coefficients as well as preserving the DFT phase.

Another reason for using the DFT magnitude domain for watermarking is its translation- or shift-invariant property. A cyclic translation of an image in the spatial domain does not affect the DFT magnitude, and because of that, the watermark embedded in the DFT magnitude domain will be translation-invariant.

Image translation, as well as image scaling and rotation, generally do not affect a perceived image quality. However, translation or any other geometrical transformation desynchronizes the image and thus makes the watermarks embedded using techniques described in the previous subsections undetectable. To make the watermark detectable after a geometrical transformation has been applied to the watermarked image, the watermark needs to be synchronized, and the synchronization process consists of an extensive search over a large space that covers all possible x-axis and y-axis translation offsets, all possible angles of rotation, and all possible scaling factors. An alternative to searching for synchronization during the watermark detection process was proposed in Reference 10. The basic idea is to avoid a need for synchronization search by transforming the image into a new workspace which is invariant to specific geometrical transformations, and embedding the watermark in that workspace. This is shown in Figure 7.6. The authors in Reference 10 proposed embedding the watermark in the Fourier–Mellin transform domain, which is invariant to translation, scaling, and rotation. The Fourier–Mellin transform is computed by taking the Fourier transform of a log-polar map. A log-polar mapping is defined as

$$u = e^{\mu} \cos(\theta)$$
$$v = e^{\mu} \sin(\theta)$$

$$(7.5)$$

It provides one-on-one mapping between $(u,v) \in \Re^2$, and (μ,θ), $\mu \in \Re$, $\theta \in (0, 2\pi)$, spaces and scaling and rotation in the (u, v) space convert into a translation in the (μ, θ) space. The (μ, θ) space is converted into the DFT magnitude domain to achieve translation invariance, and the

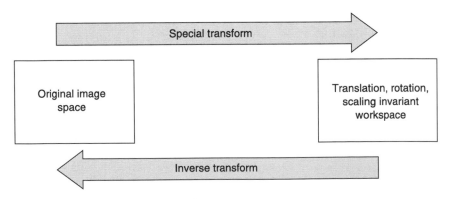

Figure 7.6. **Robustness to geometric transformations: Embed the watermark inside a new workspace that is invariant to translation, rotation, and scaling.**

watermark can be embedded in that domain, perhaps using one of techniques we described so far.

Watermarking with Side Information: Informed Embedding

The embedding components of the watermarking systems described so far create a watermark W independently of the cover C_O. The embedder depicted in Figure 7.3, for example, pseudorandomly generates the watermark pattern first and then multiplies each watermark pixel value with a global embedding strength factor s and adds it to the cover C_O. The global embedding strength factor s is also selected independently of the cover C_O and it is used to control a trade-off between watermark robustness and its transparency or imperceptibility. Increasing the embedding strength factor s will increase the energy of the embedded watermark, resulting in higher robustness. However, it will also increase embedding distortion, resulting in a less transparent watermark and causing a loss of fidelity of the watermarked image C_W compared to the original cover C_O.

The embedder obviously has access to the original cover image C_O in order to be able to embed the watermark into it. However, even though the embedder has access to the original cover image, the watermarking systems described so far did not take advantage of that information. In this subsection, we will see how watermarking systems can use the information about the original cover image to improve watermark embedding performance. This kind of watermarking system is called, a watermarking system with *informed embedding*, and the model is presented in Figure 7.7.

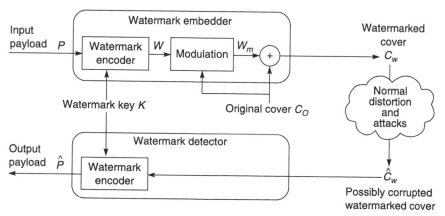

Figure 7.7. Digital watermarking system with informed embedding and blind detection.

We will first look into how we could improve the effectiveness of the watermarking embedder depicted in Figure 7.3 by having the embedder take into consideration the original cover image and calculate the embedding strength factor s according to this information.

We have seen that the original watermarking system was not 100% effective because it had a nonzero false-negative rate. Because a watermark was created independently of the cover image, it was to be expected that some cover images would interfere with the watermark in such a way that the embedded watermark would not be detectable. The informed embedding can be used to create a watermarking system that yields 100% effectiveness. This can be achieved by adjusting the embedding strength s for each individual cover image, so that every watermarked cover C_W will have a fixed-magnitude linear correlation with the watermark W. In other words, the embedder will select the embedding strength factor s to ensure that $LC(W, C_W) > T$ is always correct.

The design of a watermarking system with informed embedding can be cast as an optimization problem that could be stated as follows: Given an original cover C_O, select the embedding strength s to maximize a specific important property, such as fidelity, robustness, or embedding effectiveness, while keeping the other property or properties fixed. For example, informed embedding can be used to improve the robustness of the watermark while maintaining a fixed fidelity. The objective is to maximize the energy of the watermark signal, without increasing *perceptible* distortion of the watermarked signal. This can be done by taking advantage of imperfections of human visual system (HVS) and its inability to recognize all the changes equally. The characteristics of HVS and its sensitivities to frequency and luminance changes, as well as its masking capabilities have been captured into various *perceptual models* (i.e., models of HVS). Those models are then used as part of watermark embedding algorithms to help identify areas in the original cover image where the watermark embedding strength factor can be locally increased without introducing a perceptible change.

Much of research has been done over the years to understand how HVS responds to frequency and luminance changes. The frequency sensitivity refers to the eye's response to spatial-, spectral- or time-frequency changes. Spatial frequencies are perceived as patterns or textures, and spatial-frequency sensitivity is usually described as the eye's sensitivity to luminance changes [11]. It has been shown that an eye is the most sensitive to luminance changes in the midrange spatial frequencies and that sensitivity decreases at lower and higher spatial frequencies. The pattern orientation affects sensitivity as well and the eye is the most sensitive to vertical and horizontal lines and edges and least sensitive to lines and edges with a 45° orientation. Spectral frequencies are perceived as colors, and the human eye is least sensitive to changes in blue color. Hurtung and Kutter [12] took into consideration the color sensitivity of the HVS and proposed a solution where the watermark is added to the blue channel of an RGB image. Temporal frequencies are perceived as motion or flicker, and it has been demonstrated that eye sensitivity decreases very quickly as temporal frequencies exceed 30 Hz.

A number of solutions have been proposed in which the frequency sensitivity of the HVS is exploited to ensure that the watermark is imperceptible. Those solutions use transform domain (e.g., DCT, DFT, wavelet), and the watermark is added directly into the transform coefficients of the image. Even when a global embedding strength factor s is used by the embedder, the watermark embedding algorithm can be changed to take into account local characteristics of the cover C_o as follows. If $W = \{w_i\}$ is the watermark, $C_O = \{c_i\}$ is the cover image, $C_W = \{\tilde{c}_i\}$ is the watermarked image, and s represents the global embedding strength, then the embedder can embed the watermark using the following formula: $\tilde{c}_i = c_i(1 + sw_i)$. Here, the amount of change is clearly dependent on characteristics of the cover image C_O. Contrast that with the embedding formula $\tilde{c}_i = c_i + sw_i$ that we used earlier, where the amount of change was the same irrespective of the magnitude of the c_i coefficients.

More advanced embedding algorithms have been created by taking full advantage of characteristics of the HVS. For example, it is known that different spectral components may have different levels of tolerance to modification and it is also known that the presence of one signal can hide or mask the presence of another signal. Those characteristics of the HVS can be exploited as well to create an efficient image-adaptive solution. A single embedding strength factor, s, will not be appropriate in that case. Instead, the more general watermark embedding formula $\tilde{c}_i = c_i(1 + s_iw_i)$ should be used. Different image-adaptive solutions select multiple scaling parameters s_i different ways. Wolfgang et al. [13] present a couple of image-adaptive watermarking solutions.

Watermarking with Side Information: Informed Coding

The watermarking systems we described earlier use the cover information as part of the watermark embedding operation. The watermark W is created independently of the cover C_O and then it is locally amplified or attenuated depending on the local characteristics of the cover C_O and based on perceptual models of sensitivities and masking capabilities of the HVS. Because the watermark W is created independently of the cover C_O, it is clear that those algorithms do not take full advantage of all the side information about the cover C_O available to the watermark embedder.

Instead of creating the watermark independently of the cover C_O and then modifying it based on local characteristics of the cover C_O in an attempt to minimize interference and distortion, the embedder can use the side information about the cover C_O during the watermark encoding and creation process to choose between several available alternative watermarks and select the one that will cause the least distortion of the cover C_O. This technique is referred to as Watermarking with Informed Coding.

Watermarking with Informed Coding was inspired by theoretical results published by Max Costa in his "Writing on Dirty Paper" report [14] on the capacity of a Gaussian channel having interference that is known to the transmitter. Costa described the problem using a dirty paper analogy, which could be stated as follows [11]: Given a sheet of paper covered with independent dirt spots having normally distributed intensity, write a message on it using a limited amount of ink and then send the paper on its way, to a recipient. Along the way, the paper acquires more normally distributed dirt. How much information can be reliably sent, assuming that the recipient cannot distinguish between ink and dirt? This problem is illustrated in Figure 7.8.

Costa showed that the capacity of his dirty paper channel is given by

$$C = \frac{1}{2}\log_2\left(1 + \frac{P}{\sigma_N^2}\right) \qquad (7.6)$$

where P represents the power constraint imposed on the transmitter (i.e., there was a limited amount of ink available to write a message) and σ_N^2 represents a variance of the second source of noise. Surprisingly enough, the first source of noise, the original dirt on the paper, does not have any effect on the channel capacity.

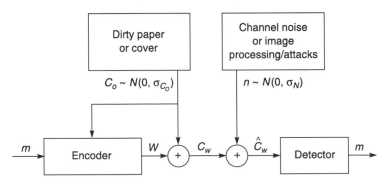

Figure 7.8. Dirty paper channel studied by Costa. There are two noise sources, both AWGN. The encoder knows the characteristics of the first noise source (dirty paper or the original cover) before it selects (watermark) W.

Costa's dirty paper problem can also be viewed as watermarking system with blind detection. The message to be written is a watermark W, the message is written on (embedded in) the first coat of dirt, the cover C_O, the limited amount of ink can be interpreted as a power constraint that ensures fidelity, and the second noise source, n, represents distortions caused by normal signal processing and attacks. Because the dirty paper channel can be cast as a watermarking system, Costa's results attracted much attention in the watermarking research community because they established an upper bound for the watermark capacity and demonstrated that capacity does not depend on the interference caused by the cover C_O.

The watermark systems inspired by Costa's work are based on the following principle: Instead of having one watermark for each message, have several alternatives available (the more, the better) and select the one with minimum interference with the cover C_o. Unfortunately, straightforward implementation of this principle is not practical. The problem is that both a watermark embedder and a watermark detector are required to find the closest watermark to a given vector representing the possibly distorted cover C_O for every message or payload. For a large number of messages and a large number of watermarks for each message, computational time and storage requirements are simply too high.

As a solution to this problem, we need to use watermarks that are structured in such a way as to allow efficient search for closest watermark to a given cover C_O [15]. Quantizing to a lattice has been identified as an appropriate tool, and most of the work related to the watermarking with informed coding is based on using lattice codes, where watermarks represent points in a regular lattice.

Chen and Wornell [16] have proposed watermark embedding based on that principle. Their method called quantization index modulation (QIM) is based on the set of N-dimensional quantizers, one quantizer for each possible message m that needs to be transmitted. The message to be transmitted determines the quantizer to use. The selected quantizer is then used to embed the information by quantizing the cover C_O. The quantization of C_O can be done in any domain (i.e., spatial, DCT, etc.). A distortion can be controlled by selecting an N-dimensional quantization point closest to the cover C_O. In the decoding process, a distance metric is evaluated for all quantizers and the one with the smallest distance from the received image $\widetilde{C}_W$ identifies the embedded information. This is illustrated in Figure 7.9, for a one-dimensional case and two uniform, scalar quantizers representing two different messages, *m1* and *m2*. Those two messages could be used to represent two distinct values of one bit: 0 and 1. The watermarking system based on QIM was shown to have better performance than other watermarking

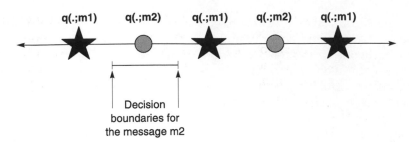

Figure 7.9. **Quantization index modulation information embedding and detection.**

systems based on the standard spread spectrum modulation, which are not image adaptive.

There are other possible ways to partition the space and there are other possible ways to embed a watermark. For example, it is possible to embed a watermark by enforcing a desired relationship, as proposed by Koch et al. [17]. This solution uses the relationship between DCT coefficients to embed a watermark as follows: An image is partitioned into 8×8 blocks, and a pair of DCT coefficients is selected to represent a single bit in each block. Let us say out of the set of DCT coefficients $(a_{11}, \ldots, a_{88})$, the selected pair is (a_{ij}, a_{mn}). The mutual relationship between two coefficients can then be used to represent the value of 1 bit. For example, $a_{ij} < a_{mn}$ can be interpreted as bit 1. The bit embedding process then consists of making appropriate changes to the pair of coefficients, if needed, to ensure that the relationship between coefficients is correct for the bit we want to embed. For example, assuming that the relationship $a_{ij} < a_{mn}$ represents bit value 1, the embedding algorithm can be described as follows: To embed a bit value 1 into the 8×8 block, check the relationship between coefficients in the pair. If $a_{ij} < a_{mn}$, nothing needs to be done because the relationship already indicates a correct bit value. Otherwise, modify coefficients appropriately to enforce the desired relationship $a_{ij} < a_{mn}$. In order to strike the balance between robustness and possible image degradation caused by modifications of the coefficients, the pair is selected from midrange frequencies. This approach shows good robustness to JPEG compression down to a quality factor of 50%.

Watermarking and Multibit Payload

A watermark designed to carry only 1 bit of information is typically created as a pseudorandom noise drown from Gaussian, as we described earlier. The detector extracts the embedded bit by verifying whether the watermark is present. Most watermarking applications, however, require more than 1 bit of information to be embedded.

The information rate of the watermarking system can be increased by introducing additional watermarks and mapping each individual watermark to a different bit string (multibit message). For example, to support 4-bit messages, one would need $2 = 16$ different watermarks, each one mapping to a different 4-bit message. The message is detected by computing a detection value for each of 16 watermarks and selecting the one with the highest detection value. This technique is known as *direct message coding*. It works well for short messages, but it is not practical for longer bit strings. For example, in order to embed 16 bits of information, the watermarking system would need $2^{16} = 65,536$ different watermarks. Those watermarks would have to be created with the maximum possible separation to avoid a situation where a small corruption of the watermarked image would lead to erroneous watermark detection. Ensuring that 65,536 watermarks are far apart from one another is not easy. Additionally, the detector would have to compare a test image against 65,536 different watermarks even if it only had to check for the watermark absence.

An alternative to direct message coding is a technique in which a different watermark represents each individual bit of the multibit message. A multibit message can be embedded into a cover image by adding watermarks representing individual bits of the multibit message to the cover, one by one. This is the approach used in Reference 17, as we described earlier. More generally, this technique can be presented as the one where watermarks representing individual bits of a multibit message are first combined together into a single watermark representing the whole message and then added into the cover image.

Watermarks can be combined together in a couple of different ways. For example, they could be tiled together in such a way that any individual tile is a watermark representing an individual message bit. This is equivalent to space division multiplexing. Alternatively, an approach equivalent to frequency division multiplexing could be used, where watermarks representing individual message bits would be placed into disjoint frequency bands. Or, most generally, an approach analogous to code division multiplexing in spread spectrum communications could be used. In this approach, each bit is spread across the whole image. The watermarks representing individual bits can be combined together without interfering with each other because they are selected to be mutually orthogonal.

Classification of Watermarking Systems

Watermarking systems can be classified according to many different criteria. For example, depending on whether a watermark embedder uses side information or not, watermarking systems can be categorized into the systems with blind and informed embedders. The informed embedder category can further be divided into embedders with informed embedding and with informed coding depending on whether side information is used to optimize the watermark embedding operation of an independently generated watermark or it is used to help select the most appropriate watermark for the given cover work. Watermarking systems can also be categorized into systems with informed and blind detectors, depending on whether the original cover work is needed to be able to detect the watermark. Other classifications are possible as well. Watermarking systems can be classified based on how a watermark gets merged with the cover work to create the watermarked cover, what technology is used to minimize perceptible distortion of the watermarked cover, whether watermarks are manipulated in spatial or transform domains, or based on how they implement support for multibit messages. The classification summary is shown in the Table 7.1.

Evaluation of Watermarking Systems

Once a watermarking system has been designed and implemented, it is important to be able to objectively evaluate its performance. This evaluation should be done in such a way to be able to compare results against other watermarking systems designed for the same or similar purpose [18–20].

By definition, watermarking is a technique for embedding a watermark into a cover work imperceptibly and robustly. Therefore, the quality of a new watermarking system can be measured by, for example, evaluating those two properties and comparing results against an equivalent set of measures obtained by evaluating other watermarking systems. However, how does one objectively measure whether a distortion introduced by embedding a watermark is perceptible or not? This is not easy. As we will see, watermark imperceptibility can be evaluated either using subjective evaluation techniques involving human observers or using some kind of distortion or distance metrics. The former cannot be automated and the later is not always dependable. Watermark robustness is easier to evaluate thanks to the existence of standardized benchmark tests. Those tests are designed to create various distortions to the watermarked cover under tests, so that it is possible to measure watermark detection rate under those conditions.

In addition to imperceptibility and robustness, watermarks have other properties that may need to be evaluated as well. We will address those later, after we look into techniques one can use to evaluate watermark imperceptibility.

Table 7.1. Classification of watermaking systems

Criteria	Categories	Characteristics
Watermark embedding	Blind	Watermark is selected independently of the of the cover work and it is embedded independently of the cover work.
	Informed embedding	Watermark is selected independently of the cover work, but the cover work information is used to optimize the watermark embedding operation.
	Informed coding	Watermark is selected based on the cover information. This is done by having a number of different watermarks available for embedding, and selecting the closest one to the given cover work. The closest watermark is the one with the minimum interference with the given cover work.
Watermark and cover merging	Addition	Watermark signal is simply added to the cover work. This addition could either be blind or informed embedding. Watermark signal can be added to the luminance channel or to the color channels, and this addition can take place in different workspace domains.
	Quantization	Informed coding that uses lattice codes to allow efficient search for the closest watermark to a given cover work. The watermark candidates represent points in a regular lattice, and the cover work is quantized to that lattice.

(Continued)

Table 7.1. Continued

Criteria	Categories	Characteristics
	Masking	Informed embedding that takes advantage of the properties of the HVS to optimize the watermark embedding operation. Optimization could maximize watermark energy while keeping a visual distortion of the watermarked cover constant, or alternatively, it could minimize visual distortion while keeping the watermark energy constant.
Main technologies	Spread spectrum	Addition-based watermarking method that uses spread spectrum technology to maximize security and robustness of the embedded watermark and minimize distortion of the watermarked cover. The watermark energy is spread across visually important frequency bands, so that the energy in any one band is small and undetectable, making the embedded watermark imperceptible. However, knowing the location and the content of the watermark makes it possible to concentrate those many weak watermark signals into a single signal with high watermark to noise ration.
	Quantization index modulation	Quantization-based watermarking method which uses a set of N-dimensional quantizers, one quatizer for each possible message m (i.e., watermark W) that needs to be transmitted.
Watermark detection	Blind or oblivious	Watermarking system that does not require the original cover work to be able to detect the embedded watermark.
	Informed	Watermarking system that uses the original cover work in the watermark detection process.

Workspace domain	Spatial		Watermark embedding takes place in the spatial domain.
	Transform	DCT	Watermark embedding takes place either in a global or block DCT domain. In a DCT domain, the Cover Work is divided into frequency bands, and the watermark can be embedded either into low, medium, or high frequencies depending on how robust and perceptible the embedded watermark is required to be. Additionally, watermarks embedded in the DCT domain are inherently robust to the JPEG lossy compression.
		Wavelet	Watermark embedding takes place in the wavelet transform domain, which provides multiresolution representation of the cover work. Embedding watermarks hierarchically, starting from the low-resolution subbands first, makes it possible to implement successive decoding of the watermark, where higher-resolution subbands are consulted only if watermark was not detected in the lower-resolution subbands. Additionally, watermarks embedded in the wavelet transform domain are inherently robust to the JPEG-2000 lossy compression.
		DFT	Watermark embedding takes place in the DFT domain, where the watermark signal is added either to the phase coefficients or magnitude coefficients of the DFT. The phase coefficients have been used for robustness, because the phase contains most of the energy of the typical image. The magnitude coefficients have been used because they are not affected by image cyclic translation.

(Continued)

Table 7.1. Continued

Criteria	Categories	Characteristics
Multibit watermarking	Direct message coding	Each multibit message is mapped to an individual, uniquely detectable watermark.
	Bit coding	
	Space division	Individual message bits are mapped into watermarks. The cover work is divided in space into equal-sized blocks, and watermarks representing individual message bits are embedded into different blocks, one watermark per block.
	Frequency division	Individual message bits are mapped into watermarks, and watermarks representing those individual message bits are placed into disjoint frequency bands.
	Code division	Individual message bits are mapped into watermarks, and watermarks representing those individual message bits are spread across the whole cover work. The embedded watermarks will not interfere with each other because they have been selected to be mutually orthogonal.

Evaluation of Imperceptibility. Imperceptibility of an embedded watermark can be expressed either as a measure of *fidelity* or *quality*. Fidelity represents a measure of similarity between the original and watermarked cover, whereas quality represents an independent measure of acceptability of the watermarked cover. The most accurate tests of fidelity and quality are subjective tests that involve human observers. Those tests have been developed by psychophysics, a scientific discipline whose goal is to determine the relationship between the physical world and people's subjective experience of that world. An accepted measure for evaluation of the level of distortion is a *Just Noticeable Difference* (JND) and it represents a level of distortion that can be perceived in 50% of experimental trials. Thus, one JND represents a minimum distortion that is generally perceptible.

Watermark perceptibility can be measured using different experiments developed as a result of various psychophysics studies. One example is the so-called two-alternative, forced-choice test. In this procedure, human observers are presented with a pair of images, one original and one watermarked, and they must decide which one has the higher quality. Statistical analysis of responses provides some information about whether the watermark is perceptible. For example, if the fidelity of the watermarked image is high, meaning that it is very similar to the original, the random responses will be received and we will see approximately 50% of the observers selecting the original image as the higher-quality one and 50% of the observers selecting the watermarked image as the higher-quality image. This result can be interpreted as zero JND. As we increase the watermark strength, the perceptible distortion will increase, and with that, the ratio of observers identifying the original image as the higher quality one will increase as well. Once this ratio gets to 75%, we have reached a distortion equivalent to one JND. Variations of that test are possible, and more information about it can be found in Reference 11.

Another, more general approach allows observers more options in their choice of answers. Instead of selecting the higher-quality image, observers are asked to rate the quality of the watermarked image under test. One example of quality scale that can be used to evaluate perceptibility of an embedded watermark is the one recommended by the ITU-R Rec. 500, in which a quality rating depends on the level of impairment a distortion creates. The recommended scale has five quality levels that go from excellent to bad, and those quality levels correspond to impairment descriptions that go from imperceptible distortion to very annoying distortion.

These subjective tests can provide very accurate measure of perceptibility of an embedded watermark. However, they can be very expensive to administer, they are not easily repeatable, and they cannot be automated.

An alternative approach is an automated technique for quality measure based on a model of the HVS. One such model was proposed by Watson [21]. Watson's model estimates the perceptibility of changes in terms of changes of individual DCT blocks and then it pools those estimates into a single estimate of perceptual distance $D(C_O, C_W)$, where C_O is the original image and C_W is a distorted version of C_O.

The model has three components: sensitivity table, luminance masking, and contrast masking. The sensitivity table, derived in Reference 22, specifies the amount of change for each individual DCT coefficient that produces one JND. However, it is known that sensitivity to coefficient change depends on the luminance value, so that in bright background, DCT coefficients can be changed by a larger amount before producing one JND. In other words, the bright background can mask more noise than the dark background. To account for this, Watson's model adjusts the sensitivity table S_{ij} for each block k, according to the block's DC term, as follows:

$$SL(i,\ j,\ k) = S(i,\ j)\left[\frac{C_O(0,0,k)}{\overline{C}_O}\right]^{\alpha} \qquad (7.7)$$

where $C_O(0,0,k)$ is the DC values of the kth block, α is a constant with a suggested value of 0.649, and $\overline{C}_O$ is the average of the DC coefficients in the image.

The third component of the model, the contrast masking, represents the reduction in visibility of change in one frequency due to the energy present in that frequency. The contrast masking is accounted for as follows:

$$SLC(i,\ j,\ k) = \max\{SL(i,\ j,\ k) | C_O(i,\ j,\ k)|^{v(i,\ j)} SL(i,\ j,\ k)^{1-v(i,\ j)}\} \qquad (7.8)$$

where $v(i,j)$ is a constant between 0 and 1 and may be different for each frequency coefficient. Watson uses a value of 0.7 for all i and j. The $SLC(i,j,k)$ represents the amounts by which individual terms of the block DCT may be changed before resulting in one JND.

To compare the original image C_O and a distorted image C_W, the model first computes the difference between corresponding DCT coefficients,

$$e(i,\ j,\ k) = C_W(i,\ j,\ k) - C_O(i,\ j,\ k) \qquad (7.9)$$

246

and then uses it to calculate the error in the i, jth frequency of the block k as a fraction of one JND given by

$$d(i, j, k) = \frac{e(i, j, k)}{\text{SLC}(i, j, k)} \qquad (7.10)$$

Those individual errors are then combined, or pooled together, into a single perceptual distance measure:

$$D(C_O, C_W) = \left(\sum_{i, j, k} |d(i, j, k)|^p \right)^{1/p} \qquad (7.11)$$

where Watson recommends a value of $p = 4$.

In general, modeling of HVS is very complex and the resulting quality metrics did not show clear advantage over simple distortion metrics so far [23].

The distortion metrics is yet another alternative. It is based on measuring distortion caused by embedding a watermark, and it is very easy to apply. The distortion can be represented as a measure of difference or distance between the original and the watermarked signal. One of the simplest distortion measures is the mean squared error (MSE) function defined as

$$\text{MSE}(C_W, C_O) = \frac{1}{N} \sum_N (c_w[i] - c_o[i])^2 \qquad (7.12)$$

The most popular distortion measures are the signal-to-noise ratio defined as

$$\text{SNR}(C_W, C_O) = \sum_N c_o^2[i] \Big/ \sum_N (c_o[i] - c_w[i])^2 \qquad (7.13)$$

and the peak SNR defined as

$$\text{PSNR}(C_O, C_W) = \max_N c_o^2[i] \Big/ \sum_N (c_o[i] - c_w[i])^2 \qquad (7.14)$$

For a more detailed list of distortion measures, see Reference 18.

Distortion metric tests are simple and popular. Their advantage is that they do not depend on subjective evaluations. Their disadvantage is that they are not correlated with human vision. In other words, a small distance between the original and the watermarked signal does not always guaranty high fidelity of the watermarked signal.

247

Wang and Bovik [23] have proposed a new quality metric called the *Universal Image Quality Index*. The index is calculated by modeling any image distortion as a combination of the following three factors: loss of correlation, luminance distortion, and contrast distortion. The new index is mathematically defined and it is not explicitly based on the HVS model. The authors claim that it performs significantly better than the widely used MSE distortion metric. It is defined as

$$Q = \frac{4\sigma_{xy}\bar{x}\bar{y}}{(\sigma_x^2 + \sigma_y^2)(\bar{x}^2 + \bar{y}^2)}$$

(7.15)

where x is the original image, y is a distorted version of x, and

$$\bar{x} = \frac{1}{N}\sum x_i, \quad \bar{y} = \frac{1}{N}\sum y_i,$$

$$\sigma_x^2 = \frac{1}{N-1}\sum (x_i - \bar{x})^2, \quad \sigma_y^2 = \frac{1}{N-1}\sum (y_i - \bar{y})^2$$

Evaluation of Other Properties. The robustness property can be evaluated by applying various kinds of "normal" signal distortions and attacks that are relevant for the target application. The robustness can be assessed by measuring detection probability of the watermark after signal distortion. This is usually done using standardized benchmarking tests, and we will provide more information about it in the next subsection.

Reliability can be evaluated by assessing the watermark detection error rate. This can be done either analytically, by creating models of watermarking systems under test, or empirically, by running a number of tests and counting the number of errors. As we stated earlier, false-positive and false-negative errors are interrelated and it is not possible to minimize both probabilities (or error rates) simultaneously. Because of that, those two errors should always be measured and presented together — for example, using a receiver operating characteristics (ROC) curve.

Capacity is an important property because it has a direct negative impact on watermark robustness. A higher capacity (the amount of information being embedded) causes a lower watermark robustness. Capacity can be assessed by calculating the ratio of capacity to reliability. This can be done empirically by fixing one parameter (e.g., payload size) and determining the other parameter (e.g., error rate). Those results can then be used to estimate the theoretical maximum capacity of the watermarking system under consideration. Because a requirement for capacity depends on the application, the question is "How important it is to estimate the excess capacity capability of the watermarking system under consideration?". It may be important because the excess capacity can be traded for improvements in reliability. This can be done by using the excess payload bits for error detection and correction.

Another property that may need to be taken into consideration is the watermark access unit or granularity. It represents the smallest part of an audiovisual signal needed for reliable detection of a watermark and extraction of its payload. In the case of image watermarking for example, this property can be evaluated by using test images of different sizes.

In general, in order to obtain statistically valid results, it is important of ensure that (1) the watermarking system under consideration is tested using a large number of test inputs, (2) the set of test inputs is representative of what is expected in the operating environment (application), and (3) the tests are executed multiple times using different watermarking keys.

Benchmarking. There are a number of benchmarking tools that have been created to standardize watermarking system evaluating processes.

Stirmark is a benchmarking tool for digital watermarking designed to test robustness. For a given watermarked input image, Stirmark generates a number of modified images that can then be used to verify if the embedded watermark can still be detected. The following image alterations have been implemented in Stirmark Version 3.1: cropping, flip, rotation, rotation scale, sharpening, Gaussian filtering, random bending, linear transformations, aspect ratio, scale changes, line removal, color reduction, and JPEG compression. More information about is can be found at www.watermarkingworld.org.

Checkmark is a benchmarking suite for digital watermarking developed on Matlab under UNIX and Windows. It has been recognized as an effective tool for the evaluation and rating of watermarking systems. Checkmark offers some additional attacks not present in Stirmark. Also, it takes the watermark application into account, which means that the scores from individual attacks are weighted according to their importance for a given watermark purpose. The following image alterations have been implemented: wavelet compression (JPEG 2000 based on Jasper), projective transformations, modeling of video distortions based on projective transformations, warping, copy, template removal, denoising (midpoint, trimmed mean, soft and hard thresholding, Wiener filtering), denoising followed by perceptual remodulation, nonlinear line removal, and collage. More information about is can be found at http://watermarking.unige.ch/Checkmark/.

Optimark is a benchmarking tool developed to address some deficiencies recognized in Stirmark 3.1. Some of its features are graphical user interface, detection performance evaluation using multiple trials utilizing different watermarking keys and messages, ROC curve, detection and embedding time evaluation, payload size evaluation, and so on. More information about this can be found at http://poseidon.csd.auth.gr/optimark.

Certimark is a benchmarking suite developed for watermarking of visual content and a certification process for watermarking algorithms. It has been created as a result of a large research project funded by the European Union. More information about this can be found at www.certimark.org.

APPLICATIONS OF DIGITAL WATERMARKING

Very frequently there is a need to associate some additional information with a digital content, such as music, image or video. For example, a copyright notice may need to be associated with an image to identify a legal owner of that image, or a serial number may need to be associated with a video to identify a legitimate user of that video, or some kind of identifier may need to be associated with a song to help find a database from which more information about it can be obtained. This additional information can be associated with a digital content by placing it in the header of a digital file, or for images, it can be encoded as a visible notice. Storing information in the header of a digital file has a couple of disadvantages. First, it may not survive a file format conversion and, second, once an image is displayed or printed, its association with the header file and information stored in it is lost. Adding a visible notice to an image may not be acceptable if it negatively affects the esthetics of the image. This could be corrected to some extent by making the notice as small as possible or moving it to a visually insignificant portion of the image, such as the edge. However, once on the edge, this additional information can easily be cropped off, either intentionally or unintentionally.

This is exactly what happened with an image of Lena Soderberg after its copyright notice was cropped off (see Figure 7.10). The image was originally published as a Playboy centerfold in November 1972. After the image has been scanned for use as the test image, most of it has been cropped, including the copyright notice, which was printed on the edge of the image. The "Lena" image became probably the most frequently used test image in image processing research and appeared in a number of journal articles without any reference to its rightful owner, Playboy Enterprises, Inc.

Figure 7.10. The Lena image used as a test image on the left and the cropped part of the original image, which identifies the copyright owner, Playboy Enterprises, Inc. on the right.

Digital watermarking seems to be the suitable method for associating this additional information, the metadata, with a digital work. The metadata is imperceptibly embedded as a watermark in a digital content, the cover work, and it becomes inseparable from it. Furthermore, because watermarks will go through the same transformations as the cover work they are embedded in, it is sometimes possible to learn whether and how the content has been tampered with by looking into the resulting watermarks.

Classification of Digital Watermarking Applications

There are a number of different watermarking application scenarios and they can be classified in a number of different ways. The classification presented in Table 7.2 is based on the type of information conveyed by the watermark [24]. In the following subsection, we will provide a more detailed explanation of possible application scenarios involving watermarking.

Digital Watermarking for Copyright Protection

Copyright protection appears to be one of the first applications for which digital watermarking was targeted. The metadata in this case contains information about the copyright owner. It is imperceptibly embedded as a watermark in the cover work to be protected. If users of digital content (music, images, and video) have an easy access to watermark detectors, they should be able to recognize and interpret the embedded watermark and identify the copyright owner of the watermarked content.

An example of one commercial application created for that purpose is Digimarc Corporation's ImageBridge Solution. The ImageBridge

Table 7.2.   Classes of watermarking applications

Application Class	Watermark Purpose	Application Scenarios
Protection of intellectual property rights	Conveys information about content ownership and intellectual property rights	Copyright protection Copy protection Fingerprinting
Content verification	Ensures that the original multimedia content has not been altered and helps determine the type and location of alteration	Authentication Integrity checking
Side-channel information	Represents the side channel used to carry additional information	Broadcast monitoring System enhancement

watermark detector is made available in a form of plug-ins for many popular image processing solutions, such as Adobe PhotoShop or Corel PhotoPaint. When a user opens an image using an Digimark-enabled application, Digimarc's watermark detector will recognize a watermark. It will then contact a remote database using the watermark as a key to find a copyright owner and his contact information. An honest user can use that information to contact the copyright owner to request permission to use the image.

We have shown earlier how an invisibly embedded watermark can be used to identify copyright ownership. It would be nice if an embedded watermark could be used to prove the ownership as well, perhaps even in a court of law. We can envision the following scenario: A copyright owner distributes his digital content with his invisible watermark embedded in it. In the case of a copyright ownership dispute, a legal owner should be able to prove his ownership by demonstrating that he owns the original work and that the disputed work has been derived from the original by embedding a watermark into it. This could be done by producing the original work together with the watermark detector and having the detector detect the owner's watermark in a disputed work. Unfortunately, it appears that the above scenario can be defeated under certain assumptions, and because of that, watermarking has not been accepted yet as a technology dependable enough to be used to prove the ownership. One potential problem is related to the availability of the watermark detector. It has been demonstrated that if a detector is widely available, then it is not possible to protect watermark security. In other words, if a detector is available, it is always possible to remove an embedded watermark. This can be achieved by repeatedly making imperceptible changes to the watermarked work, until a watermark detector fails to detect the watermark. Once the watermark is removed, the original owner will no longer be able to prove his ownership. Even if the original watermark cannot be removed, Craver et al. [25–27] demonstrated that, under certain conditions, it is possible to add another watermark to an already watermarked image in such a way as to make it appear that this second watermark is present in all copies of disputed image, including the original image. This is known as an ambiguity attack and it could be used not only to dispute the ownership claims of the rightful copyright owner but also to make new ownership claim to the original digital content.

Digital Watermarking for Copy Protection

The objective of a copy protection application is to control access to and prevent illegal copying of copyrighted content. It is an important application, especially for digital content, because digital copies can be easily made, they are perfect reproductions of the original, and they can easily and inexpensively be distributed over the Internet with no quality degradation.

There are a number of technical and legal issues that need to be addressed and resolved in order to create a working copy protection solution. Those issues are difficult to resolve in open systems, and we are not aware of the existence of an open-system copy protection solution. Copy protection is feasible in closed, proprietary systems, and we will describe one proprietary solution, the digital versatile disk (DVD) copy protection solution [28].

The DVD copy protection system has a number of components designed to provide copy protection at several levels. The Content Scrambling System (CSS) encrypts MPEG-2 video and makes it unusable to anyone who does not have a decoder and a pair of keys required to decrypt it. However, once the video has been decrypted, CSS does not provide any additional protection for the content.

Additional mechanisms have been put in place to provide extra protection for the decrypted (or unscrambled) video. For example, the analog protection system (APS) prevents an unscrambled video displayed on television from being recorded on an analog device, such as VCR. APS does it by modifying national television system committee (NTSC)/ phase alternating line (PAL) signals in such a way that video can still be displayed on television but it cannot be recorded on a VCR.

There was also a need to support limited copying of video content. For example, a customer should be able to make a single copy of the broadcast video for later viewing (a.k.a. time-shifting recording), but he should not be able to make additional copies. The Copy Control

Management System (CCMS) has been designed to provide that level of copy control by introducing and supporting three rules for copying: Copy_Free, Copy_Never, and Copy_Once. Two bits are needed to encode those rules and the bits are embedded into the video frames in the form of watermarks.

It is obvious that this copy control mechanism will work only if every DVD recorder contains a watermark detector. The problem is how to ensure that every DVD recorder will have the watermark detector, because there does not seem to exist a natural economic incentive for DVD manufacturers to increase a production cost of their product by incorporating watermark detectors in DVD recorders. After all, a perceived market value of a DVD recorder with a watermark detector may be lower compared with a recorder without it, because a customer would rather have a device that can make illegal copies.

One solution to this problem could be to force DVD manufacturers to add watermark detectors in their devices by law. Because such a law does not exist, and even if it did, it would be very difficult to enforce it across every country in the world, an alternative solution was needed. The solution that has been adopted for DVD systems is based on the patent license. Basically, the DVD encryption patent license makes it mandatory to use watermark detectors in the patent-compliant devices.

The patent-license approach ensures that compliant devices will use watermark detectors and prevent illegal copying, but it also makes it legal to manufacture noncompliant devices, the devices which do not implement the patented decryption, and, therefore, do not have to implement a watermark detector. Consequently, the DVD copy control mechanism does not prevent all possible illegal copying.

Interaction of encryption and copy control, combined with the playback control, a mechanism that allows a DVD compliant device to detect illegal copies, is used to create a solution where only illegal copies can be played on noncompliant devices and only legal copies can be played on compliant devices, as illustrated by Figure 7.11. The objective of this scheme is to ensure that one device will not be able to play both legal and illegal content. If a customer wants to play both legal and illegal copies, he will have to purchase both compliant and noncompliant devices. However, if one of the two has to be selected, the hope is that most customers will choose a compliant one.

Digital Watermarking for Fingerprinting

There are some applications where the additional information associated with a digital content should contain information about the end user, rather than about the owner of a digital content. For example, consider

254

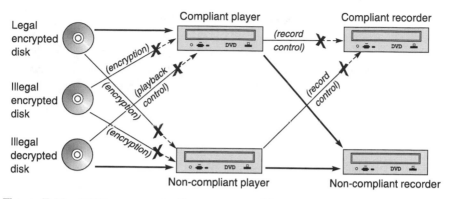

Figure 7.11. DVD copy protection systems with watermarking.

what happens in a film-making environment. During the course of film production, the incremental results of work are usually distributed each day to a number of people involved in a movie-making activity. Those distributions are known as film dailies and they are confidential. If a version is leaked out, the studio would like to be able to identify the source of the leak. The problem of identifying the source of a leak can be solved by distributing slightly different copies to each recipient, thus uniquely associating each copy with a person receiving it.

As another example, consider a digital cinema environment, an environment in which films are distributed to cinemas in digital format instead of via express mail in the form of celluloid prints. Even though digital distribution of films could be more flexible and efficient and less expensive, film producers and distributors are slow to adopt it because they are concerned about potential loss of revenue caused by illegal copying and redistribution of films. Now, if each cinema receives a uniquely identifiable copy of a film, then if illegal copies have been made, it should be possible to associate those copies with the cinema where they have been made and initiate an appropriate legal action against it.

Associating unique information about each distributed copy of digital content is called *fingerprinting*, and watermarking is an appropriate solution for that application because it is invisible and inseparable from the content. This type of application is also known as *traitor tracing* because it is useful for monitoring or tracing illegally produced copies of digital work. Also, because watermarking can be used to keep track of multiple transactions that have taken place in the history of the copy of a digital content, the term *transaction tracking* has been used as well.

Digital Watermarking for Content Authentication

Multimedia editing software makes it easy to alter digital content. For example, Figure 7.12 shows three images. The left one is the original,

Figure 7.12. Original image, tampered image, and detection of tampered regions. (Courtesy of D. Kirovski's PowerPoint presentation.)

authentic image. The middle one is the modified version of the original image, and the right one shows the image region that has been tampered with. Because it is so easy to interfere with a digital content, there is a need to be able to verify the integrity and authenticity of the content.

A solution to this problem could be borrowed from cryptography, where a digital signature has been studied as a message authentication method. A digital signature essentially represents some kind of summary of the content. If any part of the content is modified, its summary, the signature, will change, making it possible to detect that some kind of tampering has taken place. One example of digital signature technology being used for image authentication is the trustworthy digital camera described in Reference 29.

Digital signature information needs to be somehow associated and transmitted with a digital content from which it was created. Watermarks can obviously be used to achieve that association by embedding a signature directly into the content. Because watermarks used in the content authentication applications have to be designed to become invalid if even slight modifications of digital content take place, they are called *fragile watermarks*. Fragile watermarks, therefore, can be used to confirm authenticity of a digital content. They can also be used in applications where it is important to figure out how the digital content was modified or which portion of it has been tampered with. For digital images, this can be done by dividing an image into a number of blocks and creating and embedding a fragile watermark into each and every block.

Digital content may undergo lossy compression transformation, such as JPEG image conversion. Although resulting JPEG compressed image still has an authentic content, the image authenticity test based on the fragile watermark described earlier will fail. *Semifragile watermarks* can be used instead. They are designed to survive standard transformations, such as lossy compression, but they will become invalid if a major change, such as the one in Figure 7.12, takes place.

Digital Watermarking for Broadcast Monitoring

Many valuable products are regularly broadcast over the television network: news, movies, sports events, advertisements, and so forth. Broadcast time is very expensive and advertisers may pay hundreds of thousands of dollars for each run of their short commercial that appears during commercial breaks of important movies, series, or sporting events. The ability to bill accurately in this environment is very important. It is important to advertisers, who would like to make sure that they will pay only for the commercials that were actually broadcast. Also, it is important for the performers in those commercials, who would like to collect accurate royalty payments from advertisers.

Broadcast monitoring is usually used to collect information about the content being broadcast, and this information is then used as the bases for billing as well as other purposes. A simple way to do monitoring is to have human observers watch the broadcast and keep track of everything they see. This kind of broadcast monitoring is expensive and it is prone to errors. Automated monitoring is clearly better. There are two categories of automated monitoring systems: passive and active. Passive monitoring systems monitor the content being broadcast and make an attempt to recognize it by comparing it with the known content stored in a database. They are difficult to implement for a couple of reasons. It is difficult to compare broadcast signals against the database content and it is expensive to maintain and manage a large database of content to compare against. Active monitoring systems rely on the additional information that identifies the content and gets the broadcast together with the content. For analog television broadcast, this content identification information can be encoded in the vertical blanking interval (VBI) of the video signal. The problem with this approach is that it is suitable for analog transmission only, and even in that case, it may not be reliable because, in the United States, content distributors do not have to distribute information embedded in the VBI.

A more appropriate solution for active monitoring is based on watermarking. The watermark containing broadcast identification information gets embedded into the content, and the resulting broadcast monitoring solution becomes compatible with broadcast equipment for both digital and analog transmission.

Digital Watermarking for System Enhancement

Digital watermarking can also be used to convey side-channel information with the purpose of enhancing functionality of the system or adding value to the content in which it is embedded. This type of applications, where a device is designed to react to watermark for the benefit of the user, is also referred to as device control applications [11].

An example of an early application of watermarking for system enhancement is described in the Ray Dolby's patent application filed in 1981, where he proposed to make radio devices that would turn Dolby FM noise reduction control system on and off automatically in response to an inaudible signal broadcast within the audio frequency spectrum. Such a signal constitutes a simple watermark, and the proposed radio device was an enhancement compared to the radio devices used at that time, where listeners had to manually turn their radio's Dolby FM decoder on and off.

More recently, Philips and Microsoft have demonstrated an audio watermarking system for music. Basically, as music is played, a microphone on a PDA can capture and digitize the signal, extract the embedded watermark, and based on information encoded in it, identify the song. If a PDA is network connected, the system can link to a database and provide some additional information about the song, including information about how to purchase it.

Another example of a similar application is Digimarc's MediaBridge system. On the content production side, watermarks representing unique identifiers are embedded into images and then printed and distributed in magazines as advertisements. On the user side, an image from a magazine is scanned, the watermark is extracted using the MediaBridge software, and the unique identifier is used to direct a Web browser to an associated Web site.

CONCLUSIONS

In this chapter, we presented an overview of digital watermarking. First, we looked into various watermarking techniques. We presented a general model of the watermarking system and identified its two main components: embedder and detector. Depending on whether the original content was needed for detection, we classified watermarking systems into blind or informed detectors. We drew a parallel between a watermarking system and communications and recognized the possibility of applying various communications system techniques to watermarking. In the overview of watermarking techniques, we introduced various watermarking system solutions for embedding a single bit in different domains: spatial, global, and block DCT, wavelet, and discrete Fourier domains. Those systems were based on blind embedding, where the watermark is both created and modulated independently of the original cover. We then presented improvements that could be achieved if the side information about the original cover is used by the embedder. Those systems have been divided into two groups. The first group represents watermarking systems that use informed embedding. Those systems create watermarks independently of the original cover, but they use the original cover information in the watermark modulation process trying to maximize the watermark energy without increasing perceptual distortion of the watermarked cover. The second group represents watermarking systems that use informed coding. Those systems do not generate watermark independently of the original cover. Instead, they use the original cover information to select one watermark out of a set of available watermarks that creates the least amount of distortion and causes the least interference with the original cover. We then discussed the issues related to embedding multibit payloads, as well as issues related to the evaluation of watermarking systems. Then, we looked at the range of applications that could benefit from applying digital watermarking technology. Protection of intellectual property is very important nowadays because digital multimedia content can be copied and distributed quickly, easily, inexpensively, and with high quality. Watermarking has been accepted as a complementary technology to multimedia encryption, providing some additional level of protection of intellectual property rights. Other applications, such as fingerprinting, content authentication, copy protection, and device control have also been identified.

REFERENCES

1. Bender, W., Gruhl, D., Morimoto, N., and Lu, A., Techniques for data hiding, *IBM Syst. J.*, 35(3&4), 313–336, 1996.
2. Fridrich, J., Robust Bit Extraction from Images, Presented at IEEE International Conference on Multimedia Computing and Systems, 1999, Vol. 2, 536–540.
3. Pitas, I., A Method for Signature Casting on Digital Images, in Proceedings of International Conference on Image Processing, 1996, Vol. 3, pp. 215–218.
4. Wolfgang, R.B. and Delp, E.J., A watermark for Digital Images, in Proceedings of International Conference on Image Processing, 1996, Vol. 3, pp. 219–222.
5. Zeng, W. and Liu, B., On Resolving Rightful Ownerships of Digital Images by Invisible Watermarks, in Proceedings of International Conference on Image Processing, 1997, Vol. 1, pp. 552–555.
6. Cox, I.J., Kilian, J., Leighton, F.T., and Shamoon, T., Secure spread spectrum watermarking for multimedia, *IEEE Trans. Image Process.*, 6(12), 1673–1687, 1997.
7. Zhu, W., Xiong, Z., and Zhang, Y.Q., Multiresolution Watermarking for Images and Video: A unified approach, in Proceedings of 1998 International Conference on Image Processing, 1998, Vol. 1, pp. 465–468.
8. Ramkumar, M., Akansu, A.N., and Alatan, A.A., A Robust Data Hiding Scheme for Images using DFT, in Proceedings of 1999 International Conference on Image Processing, 1999, Vol. 2, pp. 211–215.
9. Ruanaidh, J.J.K.O., Dowling, W.J., and Boland, F.M., Phase Watermarking of Digital Images, in Proceedings of International Conference on Image Processing, 1996, Vol. 3, pp. 239–242.
10. O'Ruanaidh, J.J.K. and Pun, T., Rotation, Scale and Translation Invariant Digital Image Watermarking, in Proceedings of International Conference on Image Processing, 1997, Vol. 1, pp. 536–539.
11. Cox, I.J., Miller, M.L., and Bloom, J.A., *Digital Watermarking*, Morgan Kaufmann, 2001.
12. Hartung, F. and Kutter, M., Multimedia watermarking techniques, *Proc. IEEE*, 87(7), 1079–1107, 1999.

13. Wolfgang, R.B., Podilchuk, C.I., and Delp, E.J., Perceptual watermarks for digital images and video, *Proc. IEEE*, 87(7), 1108–1126, 1999.
14. Costa, M., Writing on dirty paper, *IEEE Trans. Inf. Theory*, 29(3), 439–441, 1983.
15. Eggers, J.J., Su, J.K., and Girod, B., Robustness of a Blind Image Watermarking Scheme, in Proceedings of 2000 International Conference on Image Processing, 2000, Vol. 3, pp. 17–20.
16. Chen, B. and Wornell, G.W., Quantization index modulation: A class of provably good methods for digital watermarking and information embedding, *IEEE Trans. Inf. Theory*, 47(4), 1423–1443, 2000.
17. Koch, E. and Zhao, J., Towards Robust and Hidden Image Copyright Labeling, presented at IEEE Workshop on Nonlinear Signal and Image Processing, 1995.
18. Katzenseisser, S. and Petitcolas, F.A.P., *Information Hiding Techniques for Steganography and Digital Watermarking*, Artech House, Boston, 2000.
19. Kutter, M., and Petitcolas, F.A.P., A Fair Benchmark for Image Watermarking Systems, in Proceedings on Security and Watermarking of Multimedia Contents, SPIE 1999, Vol. 3657, pp. 226–239.
20. Petitcolas, F.A.P., Watermarking schemes evaluation, *IEEE Signal Process. Mag.*, 17(5), 58–64, 2000.
21. Watson, A.B., DCT quantization matrices optimized for individual images, in *Human Vision, Visual Processing, and Digital Display IV*, SPIE, 1993, Vol. 1913, p. 202.
22. Ahumada, A.J., Jr. and Peterson, H.A., Luminance-Model-Based DCT Quantization for Color Image Compression, presented at Human Vision, Visual Processing, and Digital Display III, Proc. SPIE, 1992, Vol. 1666, pp. 365–374.
23. Wang, Z. and Bovik, A.C., A universal image quality index, *IEEE Signal Process. Lett.*, 9(3), 81–84, 2002.
24. Nikolaidis, A., Tsekeridou, S., Tefas, A., and Solachidis, V., A Survey on Watermarking Application Scenarios and Related Attacks, in Proceedings of 2001 International Conference on Image Processing, 2001, Vol. 3, pp. 991–994.
25. Craver, S., Memon, N., Yeo, B.-L., and Yeung, M.M., Resolving rightful ownerships with invisible watermarking techniques: limitations, attacks, and implications, *IEEE J. Selected Areas Commn.*, 16(4), 573–586, 1998.
26. Craver, S., Memon, N., Yeo, B.-L., and Yeung, M.M., On the Invertibility Of Invisible Watermarking Techniques, in Proceedings of International Conference on Image Processing, 1997, Vol. 1, pp. 540–543.
27. Craver, S.A., Wu, M., and Liu, B., What Can We Reasonably Expect from Watermarks?, presented at IEEE Workshop on the Applications of Signal Processing to Audio and Acoustics, 2001, pp. 223–226.
28. Bloom, J.A., Cox, I.J., Kalker, T., Linnartz, J.-P.M.G., Miller, M.L., and Traw, C.B.S., Copy protection for DVD video, *Proc. IEEE*, 87(7), 1267–1276, 1999.
29. Friedman, G.L., The trustworthy digital camera: Restoring credibility to the photographic image, *IEEE Trans. Consumer Electron.*, 39(4), 905–910, 1993.

8

Robust Identification of Audio Using Watermarking and Fingerprinting

Ton Kalker and Jaap Haitsma

INTRODUCTION

There are a large number of (audio) applications where audio identification plays a large role in the feasibility and profitability of the overall systems. One of the better known applications in this context is broadcast monitoring. It refers to the automatic playlist generation of radio, television, or Web broadcasts for, among others, purposes of royalty collection, program verification, advertisement verification, and people metering. Currently, broadcast monitoring is still a manual process, that is, organizations interested in playlist, such as performance rights organizations, currently have "real" people listening to broadcasts and filling out scorecards.

Connected audio is another interesting (consumer) applications where music is somehow connected to additional and supporting information. Using a mobile phone to identify a song is one of these

0-8493-2773-3/05/$0.00+1.50
© 2005 by CRC Press

examples. This business is actually pursued by a number of companies [1,2]. The audio signal in this application is severely degraded due to processing applied by radio stations, FM and AM transmission, the acoustical path between the loudspeaker and the microphone of the mobile phone, speech coding, and, finally, the transmission over the mobile network. Therefore, from a technical point of view, this is a very challenging application. Other examples of connected audio are (car) radios or applications "listening" to the audio streams leaving or entering a soundcard on a personal computer (PC). By pushing an "info" button in the application, the user could be directed to a page on the Internet containing information about the artist, or by pushing a "buy" button, the user would be able to buy the album on the Internet.

Filtering on peer-to-peer (P2P) networks (referring to active intervention in content distribution) is another good example of the usefulness of audio identification. The prime example for filtering technology for file sharing was Napster [3]. Starting in June 1999, users who downloaded the Napster client could share and download a large collection of music for free. Later, as a result of a court case by the music industry, Napster users were forbidden to download copyrighted songs. Therefore, in March 2001, Napster installed an audio filter based on file names to block downloads of copyrighted songs. The filter was not very effective, because users started to intentionally misspell file names. In May 2001, Napster introduced an audio fingerprinting system by Relatable [4], which aimed at filtering out copyrighted material even if it was misspelled. Owing to Napster's closure only 2 months later, the effectiveness of that specific fingerprint system is, to the best of the author's knowledge, not publicly known. In a legal file-sharing service, one could apply a more refined scheme than just filtering out copyrighted material. One could think of a scheme with free music, different kinds of premium music (accessible to those with a proper subscription), and forbidden music. Although, from a consumer standpoint, audio filtering could be viewed as a negative technology, there are also a number of potential benefits to the consumer. For example, organizing music song titles in search results can be done in a consistent way by using reliable metadata obtained through robust identification.

Currently, a number of techniques are available for audio identification to enable applications such as the above. The simplest technique is, without doubt, the explicit addition of labels in headers, such as ID3 tags in MP3 files. Although this technique is easy to implement, both for embedding and detection, it is also extremely fragile. Removal or changing of header information is achieved by unintentional and intentional processing. In particular, most format conversion will completely eradicate any header information. For real robust audio identification, two techniques are currently available. The first of these techniques is referred to as audio watermarking; the second is referred to as audio fingerprinting. The former technique is characterized by *active modifications* of the audio waveform. These modifications, referred to as watermarks, are designed to be imperceptible to the human ear and robust to the quality-preserving processing. An enormous amount of literature on this topic of watermarking has appeared in the last 10 years, and great progress has been made in understanding and designing audio watermarking systems, ranging from perception to robustness to security [5–10]. This chapter will assume that the reader is sufficiently familiar with audio watermarking to allow him to appreciate the analysis on the pros and cons of this technique with respect to the audio fingerprinting technique discussed next.

Audio fingerprinting is characterized by *passive recognition* of audio files. In an enrollment phase, relevant features are derived from an original audio file. In an identification phase, corresponding features are derived from a query audio file for matching against the features derived in the enrollment phase. As such, audio fingerprinting has many similarities with biometrics. In fact, the name *audio fingerprinting* is directly derived from biometric fingerprinting as applied to humans.

One of the goals of this chapter is to introduce the reader to the basic technical aspects of audio fingerprinting. We do so by providing an overview of an audio fingerprinting technology as developed by Philips Research. Second, this chapter provides an analysis on the commonalities and the differences between watermarking and fingerprinting. Its aim is to provide the reader with insights on which identification technology to use in which application. Finally, we discuss some issues for future research.

AUDIO FINGERPRINTING

Audio Fingerprint Definition

An audio fingerprint is defined as (a representation of) a perceptual summary of an audio object. More formally, a fingerprint function F maps an audio object X, consisting of a large number of bits, to a fingerprint $F(X)$ of only a limited number of bits, such that $F(X)$ captures most of the perceptually relevant aspects. Here, we can draw an analogy with well-known cryptographic hash functions. A cryptographic hash function H maps an (usually large) object X to a (usually small) hash value (a.k.a. message digest). A cryptographic hash function allows comparison of two large objects X and Y by just comparing their respective hash values $H(X)$ and $H(Y)$. Strict mathematical equality of the latter pair implies equality of the former, with only a very low probability of error. For a properly designed cryptographic hash function, this probability p equals 2^{-n}, where n equals the number of bits of the hash value. Using cryptographic hash functions, an efficient method exists to check whether or not a particular data item X is contained in a given and large data set $Y = \{Y_i\}$. Instead of storing and comparing with all of the data in Y, it is sufficient to store the set of hash values $\{h_i = H(Y_i)\}$ and to compare $H(X)$ with this set of hash values.

At first, one might think that cryptographic hash functions are a good candidate for fingerprint functions. However, recall that, instead of strict mathematical equality, we are interested in perceptual similarity. For example, an original compact disk (CD) quality version of 'Rolling Stones — Angie' and an MP3 version at 128 Kb/sec sound the same to the human auditory system, but their waveforms can be quite different. Therefore, although the two versions are perceptually similar, they are mathematically quite different. Cryptographic hash functions cannot decide upon perceptual equality of these two versions. Even worse, cryptographic hash functions are typically bit sensitive: a single bit of difference in the original object results in a completely different, totally uncorrelated hash value.

Another valid question the reader might ask is: "Is it not possible to design a robust 'cryptographic' fingerprint function that can decide upon perceptually similarity by computing an appropriate representation of the essential perceptual features?" In other words, can perceptual similarity in one way or another decided by mathematical equality? The question is valid, but the answer is disappointing in the sense that such a modeling of perceptual similarity is fundamentally not possible. To be more precise, it is a well-known fact that perceptual similarity is not transitive and this prevents a solution as implied by the above question. Perceptual similarity of a pair of objects X and Y and of another pair of objects Y and Z does not necessarily imply the perceptual similarity of objects X and Z. However, modeling perceptual similarity by mathematical equality (of fingerprints) would imply such a transitive relationship.

Given the above arguments, we propose to construct a fingerprint function in such a way that perceptual similar audio objects result in similar fingerprints. Furthermore, in order to be able discriminate between different audio objects, there must be a very high probability that dissimilar audio objects result in dissimilar fingerprints. More mathematically, for a properly designed fingerprint function F, there should be a threshold T such that with very high probability $\|F(X) - F(Y)\| \leq T$ if objects X and Y are similar and $\|F(X) - F(Y)\| > T$ when they are dissimilar.

Audio Fingerprinting Parameters

Having a proper definition of an audio fingerprint, we now focus on the different parameters of an audio fingerprint system. The main parameters, in the form of five questions, are given below. Note that some of these parameters (robustness, granularity, and reliability) also have meaning in the context of audio watermarking.

- *Robustness*: Can an audio clip still be identified after severe signal degradation? In order to achieve high robustness, the fingerprint should be based on perceptual features that are invariant (at least to a certain degree) with respect to signal degradations. Preferably, severely degraded audio still leads to very similar fingerprints. The false-negative rate is generally used to express the robustness. A false negative occurs when the fingerprints of perceptually similar audio clips are too different to lead to a positive match.

- *Reliability*: How often is a song incorrectly identified? For example, how often is "Rolling Stones — Angie" identified as "Beatles — Yesterday." The rate at which this occurs is referred to as the false-positive rate.

- *Fingerprint size*: How much storage and bandwidth is needed for a fingerprint? To enable fast searching, fingerprints are usually stored in RAM memory. Therefore, the fingerprint size, usually expressed in bits per second or bits per song, determines, to a large degree, the memory and bandwidth resources that are needed for a fingerprint database server.

- *Granularity*: How many seconds of audio is needed to identify an audio clip? Granularity is a parameter that can depend on the application. In some applications, the whole song can be used for identification, in others, one prefers to identify a song with only a short excerpt of audio.

- *Search speed and scalability*: How long does it take to find a fingerprint in a fingerprint database? What if the database contains thousands and thousands of songs? For the commercial deployment of audio fingerprint systems, search speed and scalability are key parameters. Search speed should be on the order of milliseconds for a database containing over 100,000 songs using only limited computing resources (e.g., a few high-end PCs).

These five basic parameters are strongly interrelated. For instance, a smaller granularity typically implies a reduced reliability in terms of false-positive and false-negative error rates. Also, search speed, in general, profits from fingerprint robustness: A better robustness implies a reduced search space and, therefore, less search effort is required.

In the next section, we give a description of an audio fingerprint description as developed at Philips Research Eindhoven.

PHILIPS AUDIO FINGERPRINTING
Guiding Principles

Audio fingerprints intend to capture the relevant perceptual features of audio. At the same time, extracting and searching fingerprints should be fast and easy, preferably with a small granularity to allow usage in highly demanding applications (e.g., recognition of songs by using a mobile phone). A few fundamental questions have to be addressed before starting the design and implementation of such an audio fingerprinting scheme. The most prominent question to be addressed is: What kind of features are the most suitable? A scan of the existing literature shows that the set of relevant features can be broadly divided into two classes: the class of semantic features and the class of nonsemantic features. Typical elements in the former class are *genre, beats-per-minute,* and *mood.* These types of features usually have a direct interpretation and are actually used to classify music, generate playlists, and more. The latter class consists of features that have a more mathematical nature and are difficult for humans to "read" directly from music. A typical element in this class is *AudioFlatness*, which is proposed in MPEG-7 as an audio descriptor tool [11]. For the work described in this chapter, we have for a number of reasons explicitly chosen to work with nonsemantic features:

1. Semantic features do not always have a clear and unambiguous meaning; that is, personal opinions differ over such classifications. Moreover, semantics may actually change over time. For example, music that was classified as *hard rock* 25 years ago may be viewed as *soft listening* today. This makes mathematical analysis difficult.
2. Semantic features are, in general, more difficult to compute than nonsemantic features.
3. Semantic features are not universally applicable. For example, *beats-per-minute* does not typically apply to classical music.

A second question to be addressed is the representation of fingerprints. One obvious candidate is the representation as a vector of real numbers, where each component expresses the weight of a certain basic perceptual feature. A second option is to stay closer in spirit to cryptographic hash functions and represent digital fingerprints as bit strings. For reasons of reduced search complexity, we have decided for the latter option in this work. The first option would imply a similarity measure involving real-valued additions and subtractions and, depending on the similarity measure, maybe even real multiplications. Fingerprints based on bit representations can be much more easily compared using the Hamming distance (i.e., bit error rate). Given the expected variety of application scenarios, we do not expect a high robustness for each and every bit in such a binary fingerprint. Therefore, in contrast to cryptographic hashes, which typically have a few hundred bits at the most, we will allow fingerprints that have a few thousand bits. Fingerprints containing a large number bits allow reliable identification even if the percentage of nonmatching bits is relatively high.

A final question involves the granularity of fingerprints. In the applications that we envisage, there is no guarantee that the audio files that need to be identified are complete. For example, in broadcast monitoring, *any* interval of 5 sec is a unit of music that has commercial value and, therefore, may need to be identified and recognized. Also, in security applications such as file filtering on a P2P network, one would not wish that deletion of the first few seconds of an audio file would prevent identification. In this work, we therefore adopt the policy of *fingerprint streams* by assigning *subfingerprints* to sufficiently small atomic intervals (referred to as *frames*). These subfingerprints might not be large enough to identify the frames themselves, but a longer interval, containing sufficiently many frames, will allow robust and reliable identification.

Extraction Algorithm

Most fingerprint extraction algorithms are based on the following approach. First, the audio signal is segmented into frames. For every frame, a set of features is computed. Preferably, the features are chosen such that they are invariant (at least to a certain degree) to signal degradations. Features that have been proposed are well-known audio features such as Fourier coefficients [12], Mel Frequency Cepstral coefficients (MFFCs) [13], spectral flatness [11], sharpness [11], Linear Predictive Coding (LPC) coefficients [11], and others. Also, derived quantities such as derivatives, means, and variances of audio features are used. Generally, the extracted features are mapped into a more compact representation by using classification algorithms, such as hidden Markov models [14] or quantization [15]. The compact representation of a single frame will be referred to as a *subfingerprint*. The global fingerprint procedure converts a stream of audio into a stream of subfingerprints. One subfingerprint usually does not contain sufficient data to identify an audio clip. The basic unit that contains sufficient data to identify an audio clip (and therefore determining the granularity) will be referred to as a *fingerprintblock*.

The proposed fingerprint extraction scheme is based on this general streaming approach. It extracts 32-bit subfingerprints for every interval of 11.6 msec. A fingerprint block consists of 256 subsequent subfingerprints, corresponding to a granularity of only 3 sec. An overview of the scheme is

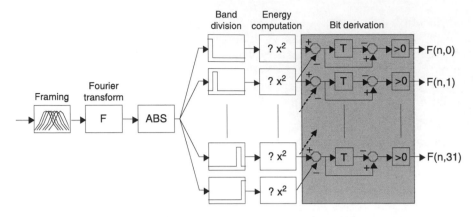

Figure 8.1.   Overview of fingerprint extraction scheme.

shown in Figure 8.1. The audio signal is first segmented into *overlapping* frames. The overlapping frames have a length of 0.37 sec and are weighted by a Hanning window with an overlap factor of 31/32. This strategy results in the extraction of one subfingerprint for every 11.6 msec. In the worst-case scenario, the frame boundaries used during identification are 5.8 msec off with respect to the boundaries used in the database of precomputed fingerprints. The large overlap assures that even in this worst-case scenario, the subfingerprints of the audio clip to be identified are still very similar to the subfingerprints of the same clip in the database. Due to the large overlap, subsequent subfingerprints, have a large similarity and are slowly varying in time. Figure 8.2a is an example of an extracted fingerprint block and the slowly varying character along the time axis.

The most important perceptual audio features live in the frequency domain. Therefore, a spectral representation is computed by performing a Fourier transform on every frame. Due to the sensitivity of the phase of the Fourier transform to different frame boundaries and the fact that the Human Auditory System (HAS) is relatively insensitive to phase, only the absolute value of the spectrum (i.e., the power spectral density) is retained.

In order to extract a 32-bit subfingerprint value for every frame, 33 nonoverlapping frequency bands are selected. These bands lie in the range from 300 to 2000 Hz (the most relevant spectral range for the HAS) and have a logarithmic spacing. The logarithmic spacing is chosen because it is known that the HAS operates on approximately logarithmic bands. Experimentally, it was verified that the sign of energy differences (simultaneously along the time and frequency axes) is a property that is very robust to many kinds of audio processing step. If we denote the

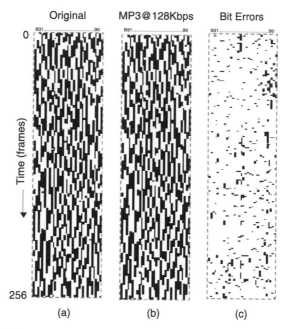

Figure 8.2. (a) Fingerprint block of original music clip, (b) fingerprint block of a compressed version, and (c) the difference between (a) and (b) showing the bit errors in black (bit error rate [BER] = 0.078).

energy of band *m* of frame *n* by $E(n,m)$ and the *m*th bit of the subfingerprint of frame *n* by $F(n,m)$, the bits of the subfingerprint are formally defined as (see also the gray block in Figure 8.1, where T is a delay element):

$$F(n,m) = \begin{cases} 1 & \text{if } E(n,m) - E(n,m+1) - (E(n-1,m) - E(n-1,m+1)) > 0 \\ 0 & \text{if } E(n,m) - E(n,m+1) - (E(n-1,m) - E(n-1,m+1)) \leq 0 \end{cases}$$

$$(8.1)$$

The reason why the sign of energy differences of neighboring bands is robust can be elucidated using the following example. The 33 frequency bands have a width of approximately one-twelfth of an octave, which corresponds to the distance between two semitones. If the excerpt to be fingerprinted contains at a certain time a strong B tone and no C tone, the energy difference between these two bands is much larger than zero. If this excerpt is subsequently subjected to audio processing steps, such as coding or filtering (see Section 4 for a more extensive list and experimental results), the energy difference will most likely change. However, it is very unlikely that the energy difference will change sign, because then the listener will no longer hear a clear B tone.

269

Figure 8.2 is an example of 256 subsequent 32-bit subfingerprints (i.e., a fingerprint block), extracted with the above scheme from a short excerpt of "O Fortuna" by Carl Orff. A 1-bit corresponds to a white pixel and a 0-bit corresponds to a black pixel. Figure 8.2a and Figure 8.2b show a fingerprint block from an original CD and the MP3 compressed (32 Kbps) version of the same excerpt, respectively. Ideally, these two figures should be identical, but due to the compression, some of the bits are retrieved incorrectly. These bit errors, which are used as the *similarity measure* for our fingerprint scheme, are shown in black in Figure 8.2c.

The computing resources needed for the proposed algorithm are limited. Because the algorithm only takes into account frequencies below 2 kHz, the received audio is first downsampled to a mono audio stream with a sampling rate of 5.5 kHz. The subfingerprints are designed such that they are robust against signal degradations. Therefore, very simple downsample filters can be used without introducing any performance degradation. Currently, 16 tap Fourier infrared (FIR) filters are used. The most computationally demanding operation is the Fourier transform of every audio frame. In the downsampled audio signal a frame has a length of 2048 samples and a subfingerprint is calculated every 64 samples. If the Fourier transform is implemented as a fixed-point real-valued fixed Fourier transform (FFT), the fingerprinting algorithm has been shown to run efficiently on portable devices such as a personal digital assistant (PDA) or a mobile phone.

False-Positive Analysis

Two 3-sec audio signals are declared similar if the Hamming distance (i.e., the number of bit errors) between the two derived fingerprint blocks is below a certain threshold T. This threshold value T directly determines the false-positive rate P_f (i.e., the rate at which audio signals are incorrectly declared equal): The smaller T, the smaller the probability P_f will be. On the other hand, a small value of T will negatively affect the false-negative probability P_n (i.e., the probability that two signals are "equal," but not identified as such).

In order to analyze the choice of this threshold T, we assume that the fingerprint extraction process yields random iid (independently and identically distributed) bits. The number of bit errors will then have a binomial distribution (n,p), where n equals the number of bits extracted and p $(=0.5)$ is the probability that a "0" or "1" bit is extracted. Because n $(=8192=32 \times 256)$ is large in our application, the binomial distribution can be approximated by a Normal distribution with a mean $\mu = np$ and standard deviation $\sigma = \sqrt{(np(1-p))}$. Given a fingerprint block F_1, the probability that a randomly selected fingerprint block F_2 has less than $T = \alpha n$ errors with respect to F_1 is given by

$$P_f(\alpha) = \frac{1}{\sqrt{2\pi}} \int_{(1-2\alpha)\sqrt{n}}^{\infty} e^{-x^2/2}\, dx = \frac{1}{2}\mathrm{erfc}\left(\frac{(1-2\alpha)}{\sqrt{2}}\sqrt{n}\right) \qquad (8.2)$$

where α denotes the bit error rate (BER).

However, in practice the subfingerprints have a high correlation along the time axis. This correlation is due not only to the inherent time correlation in audio but also to the large overlap of the frames used in fingerprint extraction. A higher correlation implies a larger standard deviation, as shown by the following argument.

Assume a symmetric binary source with alphabet $\{-1,1\}$ such that the probability that symbol x_i and symbol x_{i+1} are the same and equal to q. Then, one may easily show that

$$E[x_i x_{i+k}] = a^{|k|} \qquad (8.3)$$

where $a = 2 \cdot q - 1$. If the source Z is the exclusive-or of two such sequences X and Y, then Z is symmetric and

$$E[z_i z_{i+k}] = a^{2|k|} \qquad (8.4)$$

For N large, the probability density function of the average $\bar{Z}_N$ over N consecutive samples of Z can be approximately described by a Normal distribution with mean 0 and standard deviation equal to

$$\sqrt{\frac{1 + a^2}{N(1 - a^2)}} \qquad (8.5)$$

Translating the above back to the case of fingerprints bits, a correlation factor a between subsequent fingerprint bits implies an increase in standard deviation for the BER by a factor

$$\sqrt{\frac{1 + a^2}{1 - a^2}} \qquad (8.6)$$

To determine the distribution of the BER with real fingerprint blocks, a fingerprint database of 10,000 songs was generated. Thereafter, the BER of 100,000 randomly selected pairs of fingerprint blocks were determined. The standard deviation of the resulting BER distribution was measured to be 0.0148, approximately three times higher than the 0.0055 one would expect from random independent and identically distributed (i.i.d.) sources.

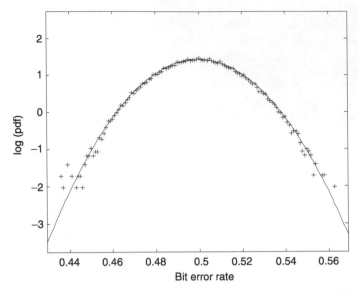

Figure 8.3. Comparison of the probability density function of the BER plotted as + and the normal distribution.

Figure 8.3 shows the log probability density function (pdf) of the measured BER distribution and a Normal distribution with mean of 0.5 and a standard deviation of 0.0148. The pdf of the BER is a close approximation to the Normal distribution. For BERs below 0.45, we observe some outliers, due to insufficient statistics. To incorporate the larger standard deviation of the BER distribution, Equation 8.2 is modified by inclusion of a factor 3:

$$P_f(\alpha) = \frac{1}{2} \text{erfc}\left(\frac{(1-2\alpha)}{3\sqrt{2}}\sqrt{n}\right) \qquad (8.7)$$

The threshold for the BER used during experiments was $\alpha = 0.35$. This means that out of 8192 bits, there must be less than 2867 bits in error in order to decide that the fingerprint blocks originate from the same song. Using Equation 8.7, we arrive at a very low false-positive rate of 3.6×10^{-20}.

Table 8.1 the experimental result for a set of tests based on the SDMI specifications for audio watermark robustness. Almost all the resulting bit error rates are well below the threshold of 0.35, even for global system for mobile communication (GSM) encoding.[1] The only degradations that lead to a BER above threshold are large linear speed changes. Linear speed changes larger then $+2.5\%$ or -2.5% generally result in bit error

---

[1] Recall that a GSM codec is optimized for speech, not for general audio.

Table 8.1. BER for different kinds of signal degradations

Processing	Orff	Sinead	Texas	AC/DC
MP3@128Kbps	0.078	0.085	0.081	0.084
MP3@32Kbps	0.174	0.106	0.096	0.133
Real@20Kbps	0.161	0.138	0.159	0.210
GSM	0.160	0.144	0.168	0.181
GSM C/I = 4dB	0.286	0.247	0.316	0.324
All-pass filtering	0.019	0.015	0.018	0.027
Amp. Compr.	0.052	0.070	0.113	0.073
Equalization	0.048	0.045	0.066	0.062
Echo Addition	0.157	0.148	0.139	0.145
Band Pass Filter	0.028	0.025	0.024	0.038
Time Scale +4%	0.202	0.183	0.200	0.206
Time Scale -4%	0.207	0.174	0.190	0.203
Linear Speed +1%	0.172	0.102	0.132	0.238
Linear Speed -1%	0.243	0.142	0.260	0.196
Linear Speed +4%	0.438	0.467	0.355	0.472
Linear Speed -4%	0.464	0.438	0.470	0.431
Noise Addition	0.009	0.011	0.011	0.036
Resampling	0.000	0.000	0.000	0.000
D/A A/D	0.088	0.061	0.111	0.076

rates higher than 0.35. This is due to misalignment of the framing (temporal misalignment) and spectral scaling (frequency misalignment). Appropriate prescaling (e.g., by exhaustive search) can solve this issue.

Search Algorithm

Finding extracted fingerprints in a fingerprint database is a nontrivial task. Instead of searching for a bit exact fingerprint (easy!), the *most similar* fingerprint needs to be found. We will illustrate this with some numbers based on the proposed fingerprint scheme. Consider a moderate size fingerprint database containing 10,000 songs with an average length of 5 min. This corresponds to approximately 250 million subfingerprints. To identify a fingerprint block originating from an unknown audio clip, we have to find the most similar fingerprint block in the database. In other words, we have to find the position in the 250 million subfingerprints where the bit error rate is minimal. This is, of course, possible by brute force searching. However, this takes 250 million fingerprint block comparisons. Using a modern PC, a rate of approximately of 200,000 fingerprint block comparisons per second can be achieved. Therefore, the total search time for our example will be in the order of 20 min! This shows that brute force searching is not a viable solution for practical applications.

We propose using a more efficient search algorithm. Instead of calculating the BER for every possible position in the database, such as in the brute force search method, it is calculated for a few candidate positions only. These candidates contain with very high probability the best matching position in the database.

In the simple version of the improved search algorithm, candidate positions are generated based on the assumption that it is very likely that at least one subfingerprint has an exact match at the optimal position in the database [14,15]. If this assumption is valid, the only positions that need to be checked are the ones where 1 of the 256 subfingerprints of the fingerprint block query matches perfectly. To verify the validity of the assumption, the plot in Figure 8.4 shows the number of bit errors per subfingerprint for the fingerprints depicted in Figure 8.2. It shows that there is indeed a subfingerprint that does not contain any errors. Actually, 17 out of the 256 subfingerprints are error-free. If we assume that the "original" fingerprint of Figure 8.2a is indeed loaded in the database, its position will be among the selected candidate positions for the "MP3@128Kbps fingerprint" of Figure 8.2b.

The positions in the database where a specific 32-bit subfingerprint is located are retrieved using the lookup table (LUT) of the database architecture of Figure 8.5. The LUT has an entry for all possible 32-bit subfingerprints. Every entry points to a list with pointers to the positions

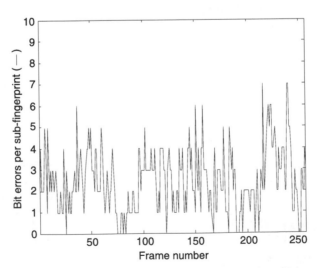

Figure 8.4. Bit errors per subfingerprint for the "MP3@128 Kbps version" of excerpt of "O Fortuna" by Carl Orff.

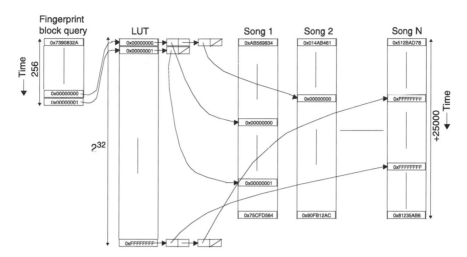

Figure 8.5. Fingerprint database layout.

in the real fingerprint lists where the respective 32-bit subfingerprints are located. In practical systems with limited memory,[2] a LUT containing 2^{32} entries is often not feasible, or not practical, or both. Furthermore, the LUT will be sparsely filled, because only a limited number of songs can reside in the memory. Therefore, in practice, a hash table [16] is used instead of a lookup table.

Let us again do the calculation of the average number of fingerprint block comparisons per identification for a 10,000-song database. Because the database contains approximately 250 million subfingerprints, the average number of positions in a list will be $0.058 (= 250 \times 10^6/2^{32})$. If we assume that all possible subfingerprints are equally likely, the average number of fingerprint comparisons per identification is only 15 $(= 0.058 \times 256)$. However, we observe in practice that, due to the nonuniform distribution of subfingerprints, the number of fingerprint comparisons increases roughly by a factor of 20. On average, 300 comparisons are needed, yielding an average search time of 1.5 msec on a modern PC. The LUT can be implemented in such a way that it has no impact on the search time. At the cost of a LUT, the proposed search algorithm is approximately a factor of 800,000 times faster than the brute force approach.

The observing reader might ask: "But, what if your assumption that one of the subfingerprints is error-free does not hold?" The answer is that the assumption almost always holds for audio signals with "mild" audio signal degradations. However, for heavily degraded signals, the assumption is, indeed, not always valid. An example of a plot of the bit errors per

---

[2]For example a PC with a 32-bit Intel processor has a memory limit of 4 GB.

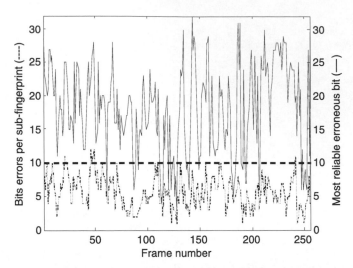

Figure 8.6.   Bit errors per subfingerprint (dotted line) and the reliability of the most reliable erroneous bit (solid line) for the "MP3 @ 32 Kbps version" of "O Fortuna" by Carl Orff.

subfingerprint for a fingerprint block that does not contain any error-free subfingerprints is shown in Figure 8.6. There are, however, subfinger-prints that contain only one error. Therefore, instead of only checking positions in the database where 1 of the 256 subfingerprints occurs, we can also check all the positions where subfingerprints occur that have a Hamming distance of 1 (i.e., 1 toggled bit) with respect to all the 256 subfingerprints. This will result in 33 times more fingerprint comparisons, which is still acceptable. However, if we want to cope with situations that, for example, the minimum number of bit errors per subfingerprint is three (this can occur in the mobile phone application), the number of fingerprint comparisons will increase with a factor of 5489, which leads to unacceptable search times. Note that the observed nonuniformity factor of 20 is decreasing with increasing number of bits being toggled. If, for instance, all 32 bits of the subfingerprints are used for toggling, we end up with the brute force approach again, yielding a multiplication factor of 1. Because randomly toggling bits to generate more candidate positions results very quickly in unacceptable search times, we propose using a different approach that uses soft decoding information; that is, we propose to estimate and use the probability that a fingerprint bit is received correctly.

The subfingerprints are obtained by comparing and thresholding energy differences (see bit derivation block in Figure 8.1). If the energy difference is very close to the threshold, it is reasonably likely that the bit was received incorrectly (an unreliable bit). On the other hand, if the energy difference is much larger than the threshold, the probability of an incorrect bit is low (a reliable bit). By deriving reliability information for every bit of a subfingerprint, it is possible to expand a given fingerprint into a list of probable subfingerprints. By assuming that one of the most probable subfingerprints has an exact match at the optimal position in the database, the fingerprint block can be identified as earlier. The bits are assigned a reliability ranking from 1 to 32, where a 1 denotes the least reliable and a 32 the most reliable bit. This results in a simple way to generate a list of most probable subfingerprints by toggling only the most unreliable bits. More precisely, the list consists of all the subfingerprints that have the N most reliable bits fixed and all the others variable. If the reliability information is perfect, one expects that in the case where a subfingerprint has three bit errors, the bits with reliability 1, 2, and 3 are erroneous. In this case, fingerprint blocks for which the minimum number of bit errors per subfingerprint is three, are identified by generating candidate positions with only eight $(= 2^3)$ subfingerprints per subfingerprint. Compared to the factor 5489 obtained when using all subfingerprints with a Hamming distance of 3 to generate candidate positions, this is an improvement with a factor of approximately 686.

In practice, the reliability information is not perfect (e.g., it happens that a bit with a low reliability is received correctly and vice versa) and, therefore, the improvements are less spectacular, but still significant. This can, for example, be seen from Figure 8.6. The minimum number of bit errors per subfingerprint is one. As already mentioned, the fingerprint blocks are then identified by generating 33 times more candidate positions. Figure 8.6 also contains a plot of the reliability for the most reliable bit that is retrieved erroneously. The reliabilities are derived from the MP3@32Kbps version using the proposed method. We see that the first subfingerprint contains eight errors. These eight bits are not the eight weakest bits because one of the erroneous bits has an assigned reliability of 27. Thus, the *reliability information* is not always *reliable*. However if we consider subfingerprint 130, which has only a single bit error, we see that the assigned reliability of the erroneous bit is 3. Therefore, this fingerprint block would have pointed to a correct location in the fingerprint database when toggling only the three weakest bits. Hence, the song would be identified correctly.

We will finish this subsection by again referring to Figure 8.5 and giving an example of how the proposed search algorithm works. The last extracted subfingerprint of the fingerprint block in Figure 8.5 is 0x00000001. First, the fingerprint block is compared to the positions in the database where subfingerprint 0x00000001 is located. The LUT is pointing to only one position for subfingerprint 0x00000001, a certain position p in song 1. We now calculate the BER between the 256 extracted subfingerprints (the fingerprint block) and the subfingerprint values of song 1 from position $p - 255$ up to position p. If the BER is below the threshold of 0.35, the probability is high that the extracted fingerprint block originates from song 1. However, if this is not the case, either the song is not in the database or the subfingerprint contains an error. Let us assume the latter and that bit 0 is the least reliable bit. The next most probable candidate is then subfingerprint 0x00000000. Still referring to Figure 8.5, subfingerprint 0x00000000 has two possible candidate positions: one in song 1 and one in song 2. If the fingerprint block has a BER below the threshold with respect to the associated fingerprint block in song 1 or 2, then a match will be declared for song 1 or 2, respectively. If neither of the two candidate positions give a below threshold BER, either other probable subfingerprints are used to generate more candidate positions or there is a switch to 1 of the 254 remaining subfingerprints where the process repeats itself. If all 256 subfingerprints and their most probable subfingerprints have been used to generate candidate positions and no match below the threshold has been found, the algorithm decides that it cannot identify the song.

Experimentally, it has been verified that the proposed algorithm performs extremely well in a wide variety of applications.

COMPARING FINGERPRINTS AND WATERMARKS

In this section, we will make explicit the differences between audio watermarks and audio fingerprints. Although some of these observations might seem trivial, in our experience these differences are not always fully appreciated. An explicit consideration will, in our opinion, help the interested reader in making an educated choice for an audio identification technology.

1. Audio watermarking and audio fingerprinting are both signal-processed identification technologies. From their definitions, the most important difference is easily deduced: Watermarking involves (host) signal *modifications*, whereas audio fingerprinting does not. Although watermarks are designed to be imperceptible, there are, nonetheless, differences between original and watermarked versions of a signal. The debate whether or not these differences are perceptible very often remains a point of contention. Practice has shown that for any watermarking technology, audio clips and human ears can be found that will perceive the difference between original and watermarked versions. In some applications, such as archiving, the slightest degradation of the original content is sometimes unacceptable, ruling out audio watermarking as the identification technology of choice. Obviously, this observation does not apply to audio fingerprinting.

2. In the majority of cases, audio watermarking is not part of the content creation process. In particular, all *legacy content* that is currently in the user and consumer domain is not watermarked. This implies that any entity that aims at identifying audio on the basis of watermarks at least needs the cooperation of the content creator or distributor to enable watermark embedding. In other words, watermarking solutions might not always work well in the context of applications with a large amount of legacy content. A typical example is given by broadcast monitoring, where monitoring entities can only track broadcasts on radio stations that actively participate by watermarking all distributed content. Obviously, this observation does not apply to audio fingerprinting: A monitoring organization on the basis of audio fingerprints can work completely independently of broadcasting entities [1,11,12,14,15,17].

3. Audio fingerprints only carry a *single message*, the message of content identity. An audio fingerprint is not able to distinguish between a clip broadcasted from one radio station or another. Watermarks, on the other hand, in general provide a communication channel that is able to carry *multiple messages*. As an example, fingerprinting technology is able to distinguish between copyrighted content and free content, but for distinguishing between copy-once content and copy-no-more content (in a copy protection application), a watermark is needed.

4. A fingerprint application critically depends on a database of templates for linking content to the appropriate metadata. In the majority of applications (e.g., broadcast monitoring), this database of templates is too large to be stored locally. Dedicated hardware and software are required to search fingerprint databases, and client applications need to be *connected* to this database to function properly. Watermarking applications, on the other hand, typically operate in a *stand-alone* mode. For example, the copyright information carried by a watermark in a copy protection application [18] is sufficient to allow a nonconnected device to decide whether or not copying is allowed.

5. Fingerprinting and watermarking technologies both have an inherently statistical component. Error analysis is, therefore, an essential ingredient in establishing robustness and performance. In the case of watermarking, there is a large body of literature on false identification errors. Using well-established mathematical methods and using only simple models of audio, it is possible to derive error rates that correspond well to reality. This is typically not the case for fingerprinting systems, where error rate analysis depends critically on good content models. More precisely, most

audio fingerprint technologies (including the one described in this chapter) assume that fingerprints are uniformly distributed in some high-dimensional template space. Under such an assumption, error analysis is very well possible, but the verification of this assumption is typically only possible by large-scale *experimental verification*. In summary, *mis-identification rates* are well understood analytically for watermarking methods, but can only by verified by large-scale experiments in the case of audio fingerprints. This situation is further complicated by the fact that the notion of perceptual equality is an ambiguous notion that is context and application dependent, even further complicating establishing the performance of an audio fingerprinting system.

6. Browsing through watermarking literature, one finds that security for watermarking always has been an important component. Although the verdict is still is out on whether watermarking can be used as a security technology, similar or in analogy with cryptography, there is at least a large body of literature available on improving the security of (sophisticated) attacks on watermarking systems. Unfortunately, in case of fingerprinting, literature on security is as good as nonexistent. More precisely, the authors are not aware of any work that aims at modifying audio content in an imperceptible manner, such that fingerprint identification is inhibited. The general belief is that such methods, if they exist, are extremely complicated, but tangible evidence is currently lacking.

Apart from the technical issues raised above, there is also an unresolved *legal* issue concerning copyright on audio fingerprints. As is well known in these days of P2P systems (KaZaa, Overnet, and others), music is copyrighted and cannot be freely distributed. Song titles, on the other hand, are not copyrighted and may be shared and distributed at will (e.g., in the form of playlists). The main argument for the latter is that song titles are sufficiently distant from the actual work of art (the song). Audio fingerprints form an interesting middle ground. On the one hand, audio fingerprints are directly derived from audio samples and arguments can be made that audio fingerprints should be treated as *creative art*. On the other hand, fingerprints are such an abstract representation of the audio that reconstruction of the original audio from fingerprints is not possible. In that sense, fingerprints are similar to song titles and, therefore, should not be copyrighted.

CONCLUSIONS

In this chapter, we have introduced the reader to audio fingerprinting as relatively new technology robust identification of audio. The essential ingredients of an audio fingerprinting technology developed at Philips Research are presented. Assuming general knowledge of the reader with watermarking technology, we highlighted the commonalities and differences between the two technologies. We argued that the choice for a particular technology is highly application dependent. Moreover, we raised a number of questions with respect to fingerprinting (error rates, security) that still require further study, hoping that the scientific community will pick them up.

REFERENCES

1. Philips Content Identification, http://www.contentidentification.com, 2003.
2. Shazam, http://www.shazamentertainment.com, 2002.
3. Napster, http://www.napster.com, 2001.
4. Relatable, http://www.relatable.com, 2002.
5. Bassia, P. and Pitas, I., Robust Audio Watermarking in the Time Domain, presented at EUSIPCO, Rodos, Greece, 1998, Vol. 1.
6. Swanson, M.D., Zhu, B., Tewfik, A.H., and Boney, L., Robust audio watermarking using perceptual masking, *Signal Process.*, 66, 337–355, 1998.
7. Jessop, P., The Business Case for Audio Watermarking, presented at IEEE International Conference on Acoustics, Speech and Signal Processing, Phoenix, AZ, 1999, Vol. 4, pp. 2077–2080.
8. Gruhl, D., Lu, A., and Bender, W., Echo Hiding, in Proceedings Information Hiding Workshop, Cambridge, 1996, pp. 293–315.
9. Kirovski, D. and Malvar, H., Robust convert communication over a public audio channel using spread spectrum, in Fourth International Information Hiding Workshop, Pittsburgh, USA, April 2001.
10. Steinebach, M., Lang, A., Dittmann, J., and Petitcolas, F.A.P., Stirmark Benchmark: Audio Watermarking Attacks Based on Lossy Compression, in Proceedings of SPIE Security and Watermarking of Multimedia, San Jose, CA, 2002, pp. 79–90.
11. Allamanche, E., Herre, J., Hellmuth, O., Fröbach, B., and Cremer, M., AudioID: Towards Content-Based Identification of Audio Material. Presented at 100th AES Convention, Amsterdam, 2001.
12. Fragoulis, D., Rousopoulos, G., Panagopoulos, T., Alexiou, C., and Papaodysseus, C., On the automated recognition of seriously distorted musical recordings, *IEEE Trans. Signal Process.*, 49(4), 898–908, 2001.
13. Logan, B., Mel Frequency Cepstral Coefficients for Music Modeling, in Proceedings of the International Symposium on Music Information Retrieval (ISMIR), Plymouth, MA, 2000.
14. Neuschmied, H., Mayer, H., and Battle, E., Identification of Audio Titles on the Internet, in Proceedings of International Conference on Web Delivering of Music, Florence, 2001.
15. Haitsma, J., Kalker, T., and Oostveen, J., Robust Audio Hashing for Content Identification, presented at Content Based Multimedia Indexing 2001, Brescia, Italy, 2001.
16. Cormen, T.H., Leiserson, C.H., and Rivest, R.L., *Introduction to Algorithms*, MIT Press, Cambridge, MA, 1998.
17. Yacast, http://www.yacast.com, 2002.
18. SDMI, http://www.sdmi.org, 2000.

9

High-Capacity Real-Time Audio Watermarking with Perfect Correlation Sequence and Repeated Insertion

Soo-Chang Pei, Yu-Feng Hsu, and Ya-Wen Lu

INTRODUCTION

Because of the maturization of efficient and high-quality audio compression techniques (MPEG I-Layer III or MP3 in brief) and the booming of the internet connection, copyrighted audio productions are spread widely and easily. This enables pirating and illegal usage of unauthorized data and intellectual property problems become serious. To deal with this problem, a sequence of data can be embedded into the audio creation, and in the case of authority ambiguity, only those with the correct key can extract the embedded watermark to declare their ownership.

Digital audio watermarking remains a relatively new area of research compared with digital image and text watermarking. Due to its special features, watermark embedding is implemented in very different ways other than image watermarking.

0-8493-2773-3/05/$0.00+1.50
© 2005 by CRC Press

The simplest and most intuitive way of embedding data into audio signals is replacing the least significant bits (LSBs) of every sample by the binary watermark data [1]. Ideally, the capacity will be the same as the sampling rate that is adopted in the audio signal because 1 bit of watermark data corresponds to exactly one sample in the host audio. However, audible noise may be introduced in the stego audio after the watermark embedding process and it is not as robust as in the most common operations like channel noise, resampling, and so forth. Any manipulation can dramatically remove the watermark. It is possible to increase its robustness by redundancy techniques, but this may reduce the watermarking capacity at the same time. In practice, this method is generally useful only in a closed digital-to-digital environment.

It is also possible to embed the watermark in the phase component of an audio clip. Phase coding is one of the most effective watermarking techniques that can achieve a high perceived signal-to-noise ratio [1]. However, in order to spread the embedded watermark information into the host audio and improve the robustness against geometric attacks, embedding a watermark into the transform coefficients has become a well-known approach. Transforms such as fast Fourier transform (FFT), discrete cosine transform (DCT), and modified DCT (MDCT) are widely used in watermark insertion because of their popularity in audio signal processing, including coding and compression. Recently, as wavelet transform prevails in various fields of application, it has become more and more popular in digital watermarking. In this chapter, two techniques are described as examples in transform domain watermarking.

A watermarking technique using the MDCT coefficients is proposed by Wang [2]. MDCT is widely used in audio coding and the fast MDCT routine is available.

The watermark is embedded into the MDCT-transformed permuted block with embedding locations chosen quite flexibly. A pseudorandom sequence is used as the watermark and embedded by modifying the LSBs of AC coefficients. In the watermark extraction process, the stego audio is first transformed into the MDCT domain and then the watermark is be obtained by the AC coefficients modulo 2.

Embedding watermarks in the MDCT domain is very practical, because MDCT is widely used in audio coding standards. The watermark embedding process can be easily integrated into the existing coding schemes without additional transforms or computation. Therefore, this method introduces the possibility of industrial realization in our everyday life.

Another cepstrum domain watermarking technique is proposed in Reference 3. A pseudorandom binary sequence is used as the watermark and embedded into the cepstrum of the host audio signal. The correlated result in Reference 3 shows that the detection algorithm gives no false alarm and it is very easy to set up a threshold of classifying the stego audio as valid or invalid.

Swanson, Zhu, Tewfik, and Boney proposed a robust and transparent audio watermarking technique using both temporal and frequency masking [4]. Based on the concept that only temporal masking or frequency masking will not be good enough for watermark inaudibility, both masks are used in order to achieve minimum perception distortion in the stego audio. Experimental results show that this technique is both transparent and robust. Even professional listeners can barely distinguish the differences between the host and stego audio. The similarity values prove that the watermark can still survive under cropping, resampling, and MP3 encoding. Therefore, this technique is an effective one under all of the considerations for audio watermarking.

An echo hiding technique is proposed by Gruhl, Lu, and Bender in Reference 5. This idea arises from the requirement of inaudibility. The echo introduced in this method is similar to that introduced when we listen to the music through audio speakers in a room. There is, therefore, no severe degradation in the stego audio because it matches our everyday life experiences. The distortion can be further minimized by carefully and properly choosing the parameters in the watermark embedding process. There are four parameters that can be adjusted when an echo is added to the original audio: initial amplitude, decay rate, "zero" offset, and "one" offset. In a general binary watermark embedding processes, two delay times are used respectively for bit zero or one. The experimental results have revealed that the major two parameters out of the four are the initial amplitude and the offset, where the former one determines the recover rate of the extracted watermark and the latter one controls the inaudibility. Echo hiding, in general, can be a very transparent way of data hiding. At the same time, its robustness can be guaranteed to an extent on a limited range of audio types. Further research is still under development to improve its performance on other types of audio data.

Some other techniques such as spread spectrum watermarking, bit-stream watermarking, and watermarking based on audio content analysis and so forth are also developed and experimented [6–13]. Nowadays, because of the booming of efficient audio coding standards and Internet distribution, audio watermarking has become a more and more important issue and many challenges reside in the watermarking process. There are still many topics to be researched and exploited.

In the above existing watermarking schemes [1–12], the complexity is high and time-consuming, their data hiding capacity are very low and limited, the stored watermark is only a short binary PN sequence such as 64-bit information, it cannot embed a high capacity audio chip, music seqment, or speech data into the host audio data, and the most difficult part is that extracted watermark survive under various attacks and be perceived by human ears with less audio distortion. In the following section, an effective audio watermarking with perfect correlation sequence is proposed; it has very low complexity and can embed high-capacity audio chips as watermark data other than a binary sequence to survive under MP3 compression attacks. Also, it can be implemented in real time by high-speed correlation.

This chapter is organized as follows: Sequences with an autocorrelation function of zero sidelobe such as perfect sequences and uniformly redundant arrays will be introduced in section "Sequences with Autocorrelation of Zero Sidelobe". After that, the proposed watermark embedding and extraction schemes will be presented in section "Proposed Watermarking Technique", with a new audio similarity measurement adopted and explained. In section "Experimental Results", experimental results and discussions will be presented. Finally, the chapter will be concluded in section "Conclusion".

SEQUENCES WITH AUTOCORRELATION OF ZERO SIDELOBE

A sequence with autocorrelation of zero sidelobe is used as the direct sequence in spread spectrum communications. After the signal is correlated with the sequence, it becomes a random signal, just like white noise. If this noise is correlated with the sequence again, the original signal is restored.

Reconstruction using sequences with autocorrelation functions of low sidelobe is mentioned in Reference 13, in which URAs (uniformly redundant arrays) were introduced. Perfect sequences were further developed in Reference 14 as well as other sequences in References 15–21. These two sequences will be used as spreading sequences in our research.

Perfect Sequences

Assume a perfect sequence $s(n)$ with length N, and its periodic sequence $s_p(n)$ with period N. Some properties of perfect sequences are presented in the following subsections.

Synthesis of Perfect Sequences. Each perfect sequence $s_p(n)$ possesses its DFT $S_p(k)$ with constant discrete magnitude. This property is used in perfect sequence synthesis. Combining a constant amplitude frequency spectrum with any odd-symmetrical phase spectrum

$$\psi(N - k) = -\psi(k), \quad \text{for } 0 \le k < N \tag{9.1}$$

can always give a real, perfect sequence by inverse DFT.

Correlation Properties. The autocorrelation function, or the PACF (periodic repeated autocorrelation function), of $s_p(n)$ is given by

$$\varphi(m) = \sum_{n=0}^{N-1} s_p(n)s_p(n + m) \tag{9.2}$$

Then,

$$\varphi(m) = E \cdot \delta(m) = \begin{cases} E, & m = 0 \\ 0, & m \ne 0 \end{cases} \tag{9.3}$$

where the energy E of the sequence is given by

$$E = \sum_{n=0}^{N-1} s_p^2(n) \tag{9.4}$$

and the DFT of $s_p(n)$ is

$$S_p(k) = \sum_{n=0}^{N-1} s_p(n)e^{-j2nk/N}, \quad 0 \le k < N \tag{9.5}$$

By combining Equations 9.1, 9.2, and 9.4, and taking the absolute value of $S_p(k)$, the DFT of the PACF is given by

$$\varphi(m) = E \cdot \delta(m) \leftrightarrow |S_p(k)|^2 = E \tag{9.6}$$

Therefore, the magnitude of the spectrum of a perfect sequence is always the constant $\sqrt{E}$.

For most applications, perfect sequences should possess a good energy efficiency η, as given by

$$\eta = \sum_{n=0}^{N-1} \frac{s_p^2(n)}{N \max(s_p^2(n))} \tag{9.7}$$

Product Theorem. Consider two periodic perfect sequences $s_1(n)$ and $s_2(n)$ whose periods are N_1 and N_2, with N_1 and N_2 relatively prime, and energy efficiencies η_1 and η_2. Then, their product is also a perfect sequence with period $N_1 \cdot N_2$. Also, the energy efficiency of the product sequence is the product of energy efficiencies of the two original sequences; that is,

$$\eta = \eta_1 \cdot \eta_2 \tag{9.8}$$

287

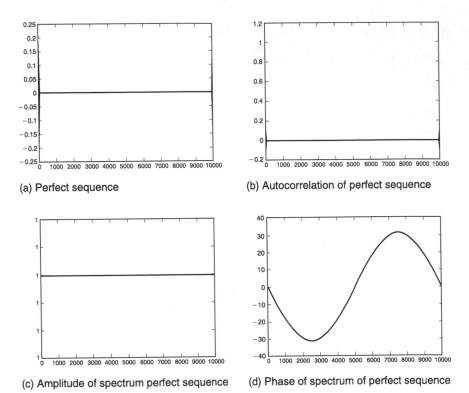

(a) Perfect sequence

(b) Autocorrelation of perfect sequence

(c) Amplitude of spectrum perfect sequence

(d) Phase of spectrum of perfect sequence

Figure 9.1. (a) Perfect sequence of length 10,000, (b) its autocorrelation, (c) amplitude frequency spectrum, and (d) phase frequency spectrum.

This enables the generation of longer perfect sequences based on shorter ones.

Because the only constraint in sequence generation is the odd symmetry of the phase spectrum, arbitrary lengths of perfect sequences are possible. This advantage widens its application when lengths of desired signals are not constant. An example of perfect sequence of length 10,000 is shown in Figure 9.1.

Uniformly Redundant Array

Uniformly redundant arrays as introduced in Reference 13, are some binary matrices with autocorrelation of zero sidelobe. They are first developed to enhance the performance of coded aperture array image processing. A complete URA set consists of a pair of matrices, A and G, where A is the key used in the embedding or scrambling process and G is used in the extraction or restoration process. The synthesis and correlation property are described in the following subsections.

288

Synthesis of URAs. The size of a URA matrix is not arbitrary. Given a URA of dimension r by s, then it must be satisfied that r and s are both prime numbers and $r - s = 2$. The elements in the matrix are denoted as $A(i,j)$, where $i = 0 \sim r - 1$, and $j = 0 \sim s - 1$. URA's are generated as follows [13]:

$$
\begin{aligned}
A(i,j) &= 0 && \text{if } i = 0 \\
&= 1 && \text{if } j = 0, i \neq 0 \\
&= 1 && \text{if } C_r(i)C_s(j) = 1 \\
&= 0 && \text{otherwise}
\end{aligned}
\tag{9.9}
$$

where $C_r(i) = 1$ if there exist an integer x, $1 \leq x < r$ such that $i = x^2 \bmod r$ and -1 otherwise.

C_r can be calculated using a simple method. First, evaluate $i = x^2 \bmod r$ for all x from 1 to $r - 1$; then, the resulting values give the locations (i) in C_r that contain $+1$. Therefore, C_r can be constructed by filling the other terms in it with -1.

The extraction key G is generated by assigning

$$
\begin{aligned}
G(i, j) &= 1 && \text{if } A(i, j) = 1 \\
&= -1 && \text{if } A(i, j) = 0
\end{aligned}
\tag{9.10}
$$

This is used because

$$
\sum_i \sum_j A(i,j) G(i + k, j + p) = \frac{rs + 1}{2} \quad \text{if } k \bmod r = 0 \text{ and } p \bmod s = 0
$$

$$
= 0 \quad \text{otherwise}
\tag{9.11}
$$

Uniformly redundant arrays with size (7,5) and (13,11) are shown in Figure 9.2.

Correlation Property. The circular correlation function of A and G is a two-dimensional (2-D) delta function with the element in the intersection of the first column and the first row proportional to the number of 1's in A, which is the value $(rs + 1)/2$ as indicated in Equation (9.11) and Figure 9.3 and the rest all zeros. For example, the URA of size (7,5) in Figure 9.2a contains in each matrix $(7 \times 5 + 1)/2 = 18$ of 1's; therefore, the circular correlation function in Figure 9.3a is a matrix of all 0's except that the upper-left-most element is 18. It is similar with the URA of size (13,11) in Figure 9.2b and Figure 9.3b.

Because the correlation gain is proportional to the number of 1's in the URA, it is generally preferred to generate URAs of larger dimensions to yield better performance. However, because it is more difficult to find

$$A(7,5) = \begin{Bmatrix} 1 & 0 & 0 & 1 & 1 \\ 1 & 0 & 0 & 1 & 1 \\ 0 & 1 & 1 & 0 & 1 \\ 1 & 0 & 0 & 1 & 1 \\ 0 & 1 & 1 & 0 & 1 \\ 0 & 1 & 1 & 0 & 1 \\ 0 & 0 & 0 & 0 & 0 \end{Bmatrix}$$

(a) URA of size (7,5) — matrix A

$$A(13,11) = \begin{Bmatrix} 1 & 0 & 1 & 1 & 1 & 0 & 0 & 0 & 1 & 0 & 1 \\ 0 & 1 & 0 & 0 & 0 & 1 & 1 & 1 & 0 & 1 & 1 \\ 1 & 0 & 1 & 1 & 1 & 0 & 0 & 0 & 1 & 0 & 1 \\ 1 & 0 & 1 & 1 & 1 & 0 & 0 & 0 & 1 & 0 & 1 \\ 0 & 1 & 0 & 0 & 0 & 1 & 1 & 1 & 0 & 1 & 1 \\ 0 & 1 & 0 & 0 & 0 & 1 & 1 & 1 & 0 & 1 & 1 \\ 0 & 1 & 0 & 0 & 0 & 1 & 1 & 1 & 0 & 1 & 1 \\ 0 & 1 & 0 & 0 & 0 & 1 & 1 & 1 & 0 & 1 & 1 \\ 1 & 0 & 1 & 1 & 1 & 0 & 0 & 0 & 1 & 0 & 1 \\ 1 & 0 & 1 & 1 & 1 & 0 & 0 & 0 & 1 & 0 & 1 \\ 0 & 1 & 0 & 0 & 0 & 1 & 1 & 1 & 0 & 1 & 1 \\ 1 & 0 & 1 & 1 & 1 & 0 & 0 & 0 & 1 & 0 & 1 \\ 0 & 0 & 0 & 0 & 0 & 0 & 0 & 0 & 0 & 0 & 0 \end{Bmatrix}$$

(b) URA of size (13,11) — matrix A

$$G(7,5) = \begin{Bmatrix} 1 & -1 & -1 & 1 & 1 \\ 1 & -1 & -1 & 1 & 1 \\ -1 & 1 & 1 & -1 & 1 \\ 1 & -1 & -1 & 1 & 1 \\ -1 & 1 & 1 & -1 & 1 \\ -1 & 1 & 1 & -1 & 1 \\ -1 & -1 & -1 & -1 & -1 \end{Bmatrix}$$

(c) URA of size (7,5) — matrix G

$$G(13,11) = \begin{Bmatrix} 1 & -1 & 1 & 1 & 1 & -1 & -1 & -1 & 1 & -1 & 1 \\ -1 & 1 & -1 & -1 & -1 & 1 & 1 & 1 & -1 & 1 & 1 \\ 1 & -1 & 1 & 1 & 1 & -1 & -1 & -1 & 1 & -1 & 1 \\ 1 & -1 & 1 & 1 & 1 & -1 & -1 & -1 & 1 & -1 & 1 \\ -1 & 1 & -1 & -1 & -1 & 1 & 1 & 1 & -1 & 1 & 1 \\ -1 & 1 & -1 & -1 & -1 & 1 & 1 & 1 & -1 & 1 & 1 \\ -1 & 1 & -1 & -1 & -1 & 1 & 1 & 1 & -1 & 1 & 1 \\ -1 & 1 & -1 & -1 & -1 & 1 & 1 & 1 & -1 & 1 & 1 \\ 1 & -1 & 1 & 1 & 1 & -1 & -1 & -1 & 1 & -1 & 1 \\ 1 & -1 & 1 & 1 & 1 & -1 & -1 & -1 & 1 & -1 & 1 \\ -1 & 1 & -1 & -1 & -1 & 1 & 1 & 1 & -1 & 1 & 1 \\ 1 & -1 & 1 & 1 & 1 & -1 & -1 & -1 & 1 & -1 & 1 \\ -1 & -1 & -1 & -1 & -1 & -1 & -1 & -1 & -1 & -1 & -1 \end{Bmatrix}$$

(d) URA of size (13,11) — matrix G

Figure 9.2. URA of size (7,5) and (13,11).

$$circorr\{A(7,5), G(7,5)\} = \begin{Bmatrix} 18 & 0 & 0 & 0 & 0 \\ 0 & 0 & 0 & 0 & 0 \\ 0 & 0 & 0 & 0 & 0 \\ 0 & 0 & 0 & 0 & 0 \\ 0 & 0 & 0 & 0 & 0 \\ 0 & 0 & 0 & 0 & 0 \\ 0 & 0 & 0 & 0 & 0 \end{Bmatrix}$$

(a) Circular correlation results of size (7,5) URA

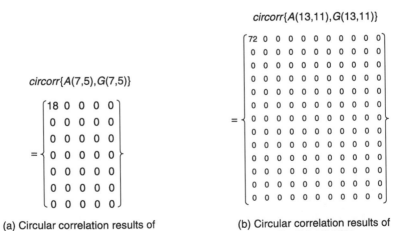

$$circorr\{A(13,11), G(13,11)\} = \begin{Bmatrix} 72 & 0 & 0 & 0 & 0 & 0 & 0 & 0 & 0 & 0 & 0 \\ 0 & 0 & 0 & 0 & 0 & 0 & 0 & 0 & 0 & 0 & 0 \\ 0 & 0 & 0 & 0 & 0 & 0 & 0 & 0 & 0 & 0 & 0 \\ 0 & 0 & 0 & 0 & 0 & 0 & 0 & 0 & 0 & 0 & 0 \\ 0 & 0 & 0 & 0 & 0 & 0 & 0 & 0 & 0 & 0 & 0 \\ 0 & 0 & 0 & 0 & 0 & 0 & 0 & 0 & 0 & 0 & 0 \\ 0 & 0 & 0 & 0 & 0 & 0 & 0 & 0 & 0 & 0 & 0 \\ 0 & 0 & 0 & 0 & 0 & 0 & 0 & 0 & 0 & 0 & 0 \\ 0 & 0 & 0 & 0 & 0 & 0 & 0 & 0 & 0 & 0 & 0 \\ 0 & 0 & 0 & 0 & 0 & 0 & 0 & 0 & 0 & 0 & 0 \\ 0 & 0 & 0 & 0 & 0 & 0 & 0 & 0 & 0 & 0 & 0 \\ 0 & 0 & 0 & 0 & 0 & 0 & 0 & 0 & 0 & 0 & 0 \\ 0 & 0 & 0 & 0 & 0 & 0 & 0 & 0 & 0 & 0 & 0 \end{Bmatrix}$$

(b) Circular correlation results of size (13,11) URA

Figure 9.3. Circular correlations of URA with size (7,5) and (13,11).

pairs of prime numbers with difference 2 when the number gets larger, there may need to be extra computation to determine valid sizes of URAs of larger sizes. The size and the computational loading can be a trade-off.

PROPOSED WATERMARKING TECHNIQUE

Watermark Embedding

The proposed technique makes use of spread spectrum approach and repeated insertion. The watermark W is first correlated with a sequence P, resulting in a noiselike signal I, which is scaled by a factor k and added into the host audio A, producing the stego audio S:

$$S = A + kI = A + k(W \otimes P) \qquad (9.12)$$

where $\otimes$ stands for circular correlation. The scaling factor k is typically 0.1.

The block diagrams of watermark embedding and extraction are shown in Figure 9.4 and Figure 9.5.

When I is added into the host audio A, repeated insertion is adopted; that is, each sample in I is repeatedly added into L consecutive samples of each block with size L in A. This concept is illustrated in Figure 9.6 [22].

Watermark Extraction

The process of watermark extraction is simply the inverse of watermark embedding. Because the scrambled watermark is inserted repeatedly, the average of each repeating block must be computed to determine the original added signal.

It is necessary to refer to the original host audio A when the watermark is to be extracted. The received stego audio S is subtracted from A,

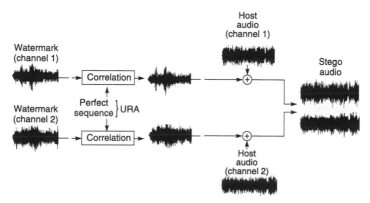

Figure 9.4.   Illustration of watermark embedding.

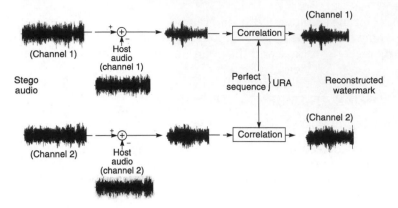

Figure 9.5.   Illustration of watermark extraction.

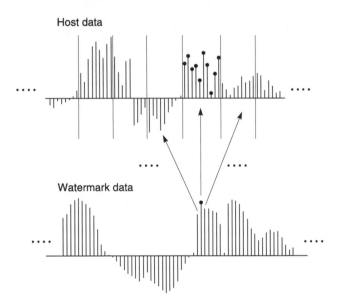

Figure 9.6.   Illustration of repeated insertion.

obtaining the noiselike signal I, which is then correlated with the perfect sequence P to restore the watermark W:

$$
\begin{aligned}
W' &= (S - A) \otimes P \\
&= kI \otimes P \\
&= k(W \otimes P) \otimes P \\
&= kW \otimes (P \otimes P) \\
&= kW \otimes \delta \\
&= kW
\end{aligned}
\tag{9.13}
$$

Audio Similarity Measure

Digital audio signals possess special features that make it improper to calculate the similarities between two signals such as conventionally used in digital signal processing. Traditional objective and quantitative approaches used in digital images, such as mean-square-error (MSE), correlation, or signal-to-noise-ratio (SNR) measurement, can hardly reflect the perceived audio similarity. Two perceptually similar signals may result in large MSE difference. Nowadays, the perceived audio similarity is commonly measured by subjective listening tests. The Mean Opinion Score (MOS) is graded by human beings according to how similar they think the two signals are, after they have listened to them. The score typically ranges from 1 to 5, proportional to the perceived similarity.

In practice, however, an objective measurement is anticipated. Human judgment can vary from person to person; therefore, it is less convincing to express the audio similarity in terms of human judgments. To deal with this problem, several methods of evaluation have been studied [23–27]. Voran [25,26] has developed a new objective measurement to simulate the human perception. In Reference 25, the Measuring Normalizing Block (MNB) is developed, and in Reference 26, the MNB measurement results are compared with other methods and proves to be closest to human perception.

In brief, Voran's method calculates the perceived sound pressure level by converting the spectrum into decibel (dB) scale and preserving only the significant frames based on the energy criterion. A MNB is applied thereafter to compute the perceived similarities in different bands. A vector is obtained after the MNB, and the acoustic distance (AD) is given by properly weighting the elements in the vector. The logistic function of AD is then calculated to convert the AD into another finite-range measurement.

The steps can be summarized as follows:

1. The two signals to be compared are normalized by removing the mean values and normalized to a common root mean square (RMS) level.
2. Each signal is broken into frames, with 50% overlapping. Then, each frame is multiplied by the Hamming window and transformed into the frequency domain by FFT. Only the squared magnitude of samples of DC to Nyquist are retained.
3. Select frames with energy above a given threshold for further consideration. Transform the frequency domain samples into the dB scale by taking logarithm.
4. Apply the frequency MNB (FMNB).

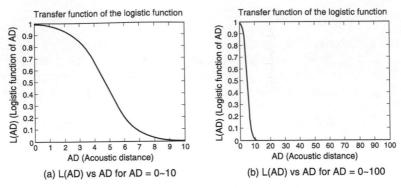

Figure 9.7. Logistic function of ADs from 0 to 10 and 0 to 100.

5. Apply either the time MNB (TMNB) of structure 1 or structure 2.
6. Apply linear combination and logistic function to obtain the AD and the logistic function of AD [L(AD)].

The range of AD is from 0 to infinity, and the range of L(AD) is from 1 to 0.

$$L(AD) = \frac{1}{1 + e^{AD-4.6877}} \tag{9.14}$$

The logistic functions L(AD) for AD = 0–10 and AD = 0–100 are shown in Figure 9.7. If AD closer to 0 and L(AD) closer to 1, the two audio signals are of higher perceptual similarity. For two identical signals, AD is 0 and L(AD) is 0.9909 [25,26].

The correlation of this measurement with subjective test results is compared with other objective test results [26]. Both of the two structures of MNB yield high correlation values of 0.986 and 0.959, whereas the L(BSD) (logistic function of Bark spectral distortion) is only 0.368 and L(ND) (logistic function of noise disturbance) is 0.793, as shown in Figure 9.8.

EXPERIMENTAL RESULTS

In this experiment, watermark robustness is tested against the MPEG I-Layer III (MP3) compression attack. The stego audio is compressed into MP3 format and then decompressed into WAV format again. To increase its robustness, the precompression process is implemented in Figure 9.9. Before the watermark embedding, both the host audio and the watermark clips are compressed and then decompressed to remove any residual data. This precompression process erases the information that is beyond consideration and preserves only the meaningful parts in MP3 encoding, minimizing the impact that MP3 attack may have on the stego audio.

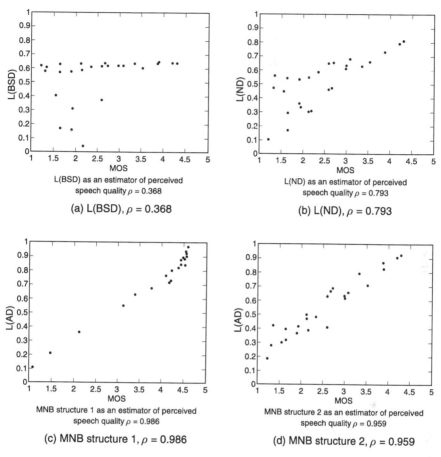

(a) L(BSD), $\rho = 0.368$

(b) L(ND), $\rho = 0.793$

(c) MNB structure 1, $\rho = 0.986$

(d) MNB structure 2, $\rho = 0.959$

Figure 9.8. **Correlation with subjective test results.**

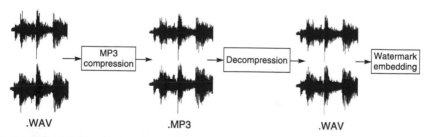

Figure 9.9. **MP3 compression before watermark embedding.**

The precompression is a very common method in both image and audio watermarking to increase the watermark robustness and decrease the degradation of the stego signal when undergoing compression attacks.

MP3 compression not only alters the frequency domain features but also delays the audio clip by a certain amount. If the starting point is

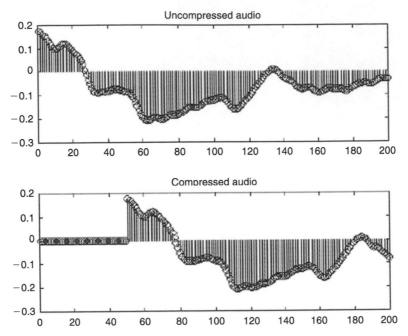

Figure 9.10. Temporal delay introduced by MP3.

delayed, the first sample of the stego audio is no longer the first in the host audio and the resulting difference by subtracting the host audio from the stego audio will be completely wrong. The temporal delay is shown in Figure 9.10. From our experiment, the compressed audio is delayed by 1,058 samples when the audio clip is sampled with a sample rate of 44,100 Hz, equaling 0.024 sec. It implies that if the synchronization is not recovered before watermark extraction, the 1059th sample of stego audio will be subtracted an amount equal to the 1059th sample of the host audio, whereas the desired result is the 1059th sample of stego audio will be subtracted the amount of the first sample of host audio, if the waveform is carefully observed and subtraction is carefully observed. This will damage the extraction process severely because the subtraction is carefully observed and sutraction results will be completely wrong.

There are several methods to overcome this problem. What we use in this research is the simplest one: adding a number, say, 10, of 1's as the starting tag of host audio, as shown in Figure 9.11. In the watermark extraction process, the ten 1's are searched; after we successfully find the ten consecutive 1's in the 1059th to 1068th samples of the stego audio, the 1st to 1058th samples are removed, retaining the data beginning at the 1059th sample. Further computations are done after the synchronization is recovered.

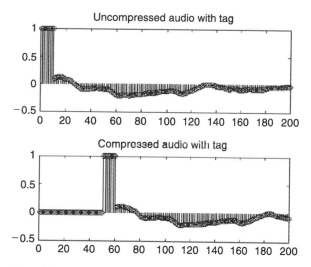

Figure 9.11. Tag added to identify the starting point.

Experimental Audio Clip and PN Binary Data Profiles

Different types of audio clip and pseudo noise (PN) binary data are tested in this research, including music, speech, and binary data. The data profiles are listed in Table 9.1 and Table 9.2. All the audio clips are 16-bit PCM stereo WAV format, except the host ones are of sample rate 44,100 Hz and the watermark ones are of sample rate $44,100/6 = 7,350$ Hz. For simplicity in signal representation, every audio clip is denoted by a

Table 9.1. Audio clips used as host audio (sample rate = 44,100 Hz)

Clip No.	Clip Name	Clip Content	Clip Length
1	PIANO	Piano Solo: "Etude Op.25 No.12, Chopin"	13.531 sec
2	SPMLE	English Speech: "Time Magazine", Male	14.315 sec
3	STREN	String Ensemble: "L'Arlesienne Suite No.2, Bizet"	17.914 sec
4	SPFLE	English Speech: "Time Magazine", Female	17.580 sec

Table 9.2. Audio clips and PN binary data used as watermark (sample rate = 7350 Hz)

Clip No.	Clip Name	Clip Content	Clip Length
1	GUITR	Guitar solo: "Petenera Para Guitarra"	2.900 sec
2	SPFSE	English speech: "from Time Magazine," female	5.538 sec
3	VOILN	Violin Solo: "Hungarian Dances No.1," Brahms	4.232 sec
4	PN sequence	Pseudorandom binary sequence	30,000 bits

string of five capital letters. Also, pseudorandom binary data of 30,000 bits are used as watermarks in the experiments.

Using Perfect Sequences

The perfect sequence is used to scramble the watermark before watermark embedding. There is no single sequence used for all watermark clips. Because arbitrary lengths of perfect sequences can be generated, for different watermark audio clips sequences of lengths corresponding to the watermark audio clip lengths are produced.

The experimental results consist of two parts: stego audio quality and extracted watermark quality. The former represents the watermarking transparency and is measured by the similarity values between the host and the stego audio clips.

The extracted watermark, after the stego audio has gone through MP3 compression and decompression, is compared with the original watermark and the similarity results are also measured. This value stands for the robustness of the watermarking technique against MP3 attack. The MP3 encoding specification tested here is the standard bit rate of 128 Kbps.

The experimental results using perfect sequences are shown in Table 9.3 to Table 9.8. The block size is fixed at 16. Audio clips and PN binary data that are used as watermarks are listed in Table 9.3 to Table 9.6 and Table 9.7 to Table 9.8, respectively. In these tables, the stego audio qualities are displayed in the first row of each table to show its embedded

Table 9.3. Stego audio and extracted watermark quality: music clips in piano music; audio similarity acoustic distance measure using AD: $0-\infty$ and L(AD): 0–1, and higher perceptual similarity means AD $\cong 0$, L(AD) $\cong 1$ (host audio: PIANO, watermark: GUITR)

Perfect Sequence	Watermark : Music		Host : Music	
	Channel 1		Channel 2	
	AD	L(AD)	AD	L(AD)
Stego audio quality	0.3216	0.9875	0.3515	0.9871
MP3	2.2553	0.9193	2.3489	0.9120
Cropping (20%)	0.2448	0.9884	2.0587	0.9327
Downsampling (50%)	0.1789	0.9891	2.0005	0.9363
Echo (delay = 40)	0.3846	0.9867	2.1495	0.9268
Time stretch (2%)	2.7722	0.8716	2.8355	0.8644
Quantization (16 → 8 bits)	1.6150	0.9558	2.0612	0.9325

Table 9.4. Stego audio and extracted watermark quality: music clips in female English speech; audio similarity acoustic distance measure using AD: 0–∞ and L(AD): 0–1, and higher perceptual similarity means $AD \cong 0$, $L(AD) \cong 1$ (host audio: SPMLE, watermark: GUITR)

| Perfect Sequence | Watermark : Music | | Host : Speech | |
| | Channel 1 | | Channel 2 | |
	AD	L(AD)	AD	L(AD)
Stego audio quality	0.5159	0.9848	0.7301	0.9812
MP3	2.6243	0.8873	2.9198	0.8542
Cropping (20%)	0.2121	0.9887	2.0706	0.9320
Downsampling (50%)	0.1550	0.9894	2.0188	0.9352
Echo (delay = 40)	0.3730	0.9868	2.1544	0.9264
Time stretch (2%)	2.7817	0.8706	2.8463	0.8631
Quantization (16 → 8 bits)	2.0568	0.9328	2.3456	0.9123

Table 9.5. Stego audio and extracted watermark quality: English speech in piano music; audio similarity acoustic distance measure using AD: 0–∞ and L(AD): 0–1, and higher perceptual similarity means $AD \cong 0$, $L(AD) \cong 1$ (host audio: PIANO, watermark: SPFSE)

| Perfect Sequence | Watermark : Speech | | Host : Music | |
| | Channel 1 | | Channel 2 | |
	AD	L(AD)	AD	L(AD)
Stego audio quality	0.1909	0.9890	0.1682	0.9892
MP3	4.1408	0.6334	4.1542	0.6303
Cropping (20%)	0.3574	0.9870	0.9173	0.9775
Downsampling (50%)	0.2910	0.9878	0.8842	0.9782
Echo (delay = 40)	0.4964	0.9851	1.0296	0.9749
Time stretch (2%)	4.9556	0.4334	4.9680	0.4304
Quantization (16→8bits)	3.5377	0.7595	3.5444	0.7583

watermark imperceptibility, and the extracted watermark qualities are shown in the lower ones. Different various attacks are experimented to illustrate the robustness of the proposed watermarking scheme.

Using Uniformly Redundant Array

Despite the URA being a 2-D form, that its size is limited, and that its correlation property is valid only for 2-D circular correlations, it is still possible to use URAs in one-dimensional (1-D) signal correlation through a simple dimension conversion process described as follows.

Table 9.6. Stego audio and extracted watermark quality: English speech in female English speech; audio similarity acoustic distance measure using AD: 0–∞ and L(AD): 0–1, and higher perceptual similarity means AD ≅ 0, L(AD) ≅ 1 (host audio: SPMLE, watermark: SPFSE)

Perfect Sequence	Watermark : Speech		Host : Speech	
	Channel 1		Channel 2	
	AD	L(AD)	AD	L(AD)
Stego audio quality	0.2926	0.9878	0.3651	0.9869
MP3	4.2222	0.6143	4.2521	0.6072
Cropping (20%)	0.2350	0.9885	0.8666	0.9786
Downsampling (50%)	0.2147	0.9887	0.8648	0.9786
Echo (delay = 40)	0.4028	0.9864	0.9854	0.9759
Time stretch (2%)	4.9703	0.4298	4.9757	0.4285
Quantization (16→8 bits)	3.8327	0.7016	3.8481	0.6984

Table 9.7. Stego audio quality (watermark: PN data); audio similarity acoustic distance measure using AD: 0–∞ and L(AD): 0–1, and higher perceptual similarity means AD ≅ 0, L(AD) ≅ 1

Host	Channel 1		Channel 2	
	AD	L(AD)	AD	L(AD)
Music (PIANO)	0.9612	0.9765	0.8343	0.9792
Speech (SPMLE)	1.3566	0.9655	1.3195	0.9667

Table 9.8. Extracted watermark quality (watermark: PN data)

Perfect Sequence	Repeat (times)	Music (PIANO) (bits)	Detection Rate	Speech (SPMLE) (bits)	Detection Rate
MP3	5	60/6,000	99%	20/6,000	99.7%
	10	0/3,000	100%	0/3,000	100%
	15	0/2,000	100%	0/2,000	100%
Cropping (20%)	×	3,078/30,000	89.8%	3,079/30,000	89.8%
Downsampling (50%)	×	0/30,000	100%	0/30,000	100%
Echo (delay = 40)	×	34/30,000	99.9%	34/30,000	99.9%
Time stretch (2%)	×	14,951/30,000	50.2%	14,955/30,000	50.2%
Quantization	×	14,858/30,000	50.2%	13,404/30,000	55.4%
(16→8 bits)	5	2,036/6,000	66.1%	2,255/6,000	62.5%
	10	2/3,000	99.9%	16/3,000	99.5%

In our experiment, the audio signal is 1-D with length N and the URA is 2-D with size 43×41, where $N > (43 \times 41) = 1763 = B$. First, the watermark signal is divided into frames of length B, with the number of frames given by

$$\text{Number of frames} = \left\lceil \frac{N}{B} \right\rceil \tag{9.15}$$

where $\lceil \ \rceil$ is the smallest integer that is greater than the number inside the bracket. Second, each frame is converted into a matrix of size 43×41 and 2-D circular correlated with the URA. Finally, by converting the resultant matrix into a 1-D array again, this segment of watermark is successfully scrambled. The total scrambled watermark is obtained by concatenating all of the scrambled frames, and the stego audio is produced by adding the scrambled watermark onto the host audio signal.

In the watermark extraction phase, the above array–matrix conversions take place as well. The signal I_a that is obtained by subtracting the host audio from the attacked stego audio is divided into frames of length B, just like the watermark audio in the embedding process. Similarly, every frame is converted into a matrix, correlated with the URA, and then converted into an array to form the extracted watermark segment. By concatenating all of the segments, the extracted watermark audio clip can be obtained.

The URA of size (43,41) and its autocorrelation function used in our experiment are shown in Figure 9.12 and the usage of 2-D URAs in 1-D audio watermark embedding and extraction described above are illustrated in Figure 9.13 and Figure 9.14.

The similarities of extracted and original watermarks under several attacks are listed in Table 9.9 to Table 9.14. Just as described in the previous subsection, both the stego audio quality and the extracted watermark quality are shown in these tables.

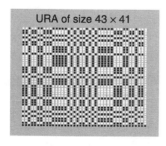

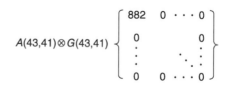

(a) URA of size 43×41 used in watermark embedding

(b) Autocorrelation of URA

Figure 9.12. URA of size (43,41) and its autocorrelation.

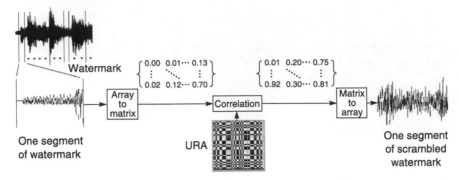

Figure 9.13. Illustration of correlation of 2-D URA and 1-D watermark in watermark embedding.

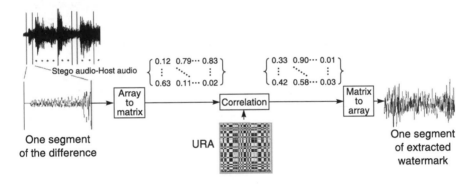

Figure 9.14. Illustration of correlation of 2-D URA and 1-D watermark in watermark extraction.

Table 9.9. Stego audio and extracted watermark quality: music clips in piano music; audio similarity acoustic distance measure using AD: 0–∞ and L(AD): 0–1, and higher perceptual similarity means AD ≅ 0, L(AD) ≅ 1 (host audio: PIANO, watermark: GUITR)

URA	Watermark : Music		Host : Music	
	Channel 1		Channel 2	
	AD	L(AD)	AD	L(AD)
Stego audio quality	0.4142	0.9863	0.4030	0.9864
MP3	2.1438	0.9272	2.0706	0.9320
Cropping (20%)	0.9284	0.9772	0.8398	0.9791
Downsampling (50%)	0.2347	0.9885	0.1492	0.9894
Echo (delay = 40)	2.3018	0.9157	2.2319	0.9210
Time stretch (2%)	2.3392	0.9128	2.3851	0.9091
Quantization (16 → 8 bits)	1.8426	0.9451	1.7457	0.9499

Table 9.10. Stego audio and extracted watermark quality: music clips in female English speech; audio similarity acoustic distance measure using AD: 0–∞ and L(AD): 0–1, and higher perceptual similarity means AD ≅ 0, L(AD) ≅ 1 (host audio: SPMLE, watermark: GUITR)

URA	Watermark : Music		Host : Speech	
	Channel 1		Channel 2	
	AD	L(AD)	AD	L(AD)
Stego audio quality	0.6994	0.9818	0.7428	0.9810
MP3	2.2325	0.9209	2.1569	0.9263
Cropping (20%)	0.9191	0.9774	0.8470	0.9790
Downsampling (50%)	0.1990	0.9889	0.1580	0.9893
Echo (delay = 40)	2.2978	0.9161	2.2329	0.9209
Time stretch (2%)	2.3395	0.9128	2.3842	0.9092
Quantization (16→8 bits)	1.9464	0.9394	1.9909	0.9368

Table 9.11. Stego audio and extracted watermark quality: English speech in piano music; audio similarity acoustic distance measure using AD: 0–∞ and L(AD): 0–1, and higher perceptual similarity means AD ≅ 0, L(AD) ≅ 1 (host audio: PIANO, watermark: SPFSE)

URA	Watermark : Speech		Host : Music	
	Channel 1		Channel 2	
	AD	L(AD)	AD	L(AD)
Stego audio quality	0.2700	0.9881	0.2470	0.9883
MP3	4.1208	0.6380	4.1994	0.6197
Cropping (20%)	3.5394	0.7592	0.9192	0.9774
Downsampling (50%)	3.2272	0.8116	0.2676	0.9881
Echo (delay = 40)	3.7214	0.7244	2.7445	0.8747
Time stretch (2%)	4.3002	0.5957	4.2771	0.6012
Quantization(16→8 bits)	3.8005	0.7083	3.6250	0.7432

Discussions

The above-measured stego audio qualities imply that this technique can guarantee watermark transparency to an extent, but not good enough. It should be the ultimate goal to achieve zero acoustic distances between the stego and host audio clips.

If the extracted watermark qualities under all kinds of combination are carefully investigated, for MP3 attack it is clear that URA outperforms the perfect sequences no matter what types of audio clip are under

Table 9.12. Stego audio and extracted watermark quality: English speech in female English speech; audio similarity acoustic distance measure using AD: 0–∞ and L(AD): 0–1, and higher perceptual similarity means AD $\cong$ 0, L(AD) $\cong$ 1 (host audio: SPMLE, watermark: SPFSE)

URA	Watermark:Speech		Host:Speech	
	Channel 1		Channel 2	
	AD	L(AD)	AD	L(AD)
Stego audio quality	0.4740	0.9854	0.4643	0.9856
MP3	4.2248	0.6137	4.1630	0.6283
Cropping (20%)	0.9584	0.9766	0.9583	0.9766
Downsampling (50%)	0.3172	0.9875	0.3101	0.9876
Echo (delay = 40)	2.7280	0.8765	2.7444	0.8747
Time stretch (2%)	4.2417	0.6097	4.2799	0.6006
Quantization (16→8 bits)	4.0618	0.6516	4.0600	0.6520

Table 9.13. Stego audio quality (Watermark: PN data); audio similarity acoustic distance measure using AD: 0–∞ and L(AD): 0–1, and higher perceptual similarity means AD $\cong$ 0, L(AD) $\cong$ 1

Host	Channel 1		Channel 2	
	AD	L(AD)	AD	L(AD)
Music (PIANO)	0.7357	0.9811	0.6403	0.9828
Speech (SPMLE)	1.1333	0.9722	1.1001	0.9731

Table 9.14. Extracted watermark quality (watermark: PN data)

URA	Repeat (times)	Music (PIANO) (bits)	Detection Rate	Speech (SPMLE) (bits)	Detection Rate
MP3	5	1,013/6,000	84.1%	462/6,000	92.3%
	10	268/3,000	92.1%	99/3,000	96.7%
	15	76/2,000	96.2%	30/2,000	98.5%
Cropping (20%)	×	4,790/30,000	84.0%	4,787/30,000	84.0%
Downsampling (50%)	×	0/30,000	100%	0/30,000	100%
Echo (delay = 40)	×	861/30,000	97.1%	860/30,000	97.1%
Time stretch (2%)	×	14,950/30,000	50.2%	14,953/30,000	50.2%
Quantization	×	136/30,000	99.5%	1,451/30,000	95.2%
(16→8 bits)	5	37/6,000	99.4%	355/6,000	94.1%
	10	9/3,000	99.7%	76/3,000	97.5%

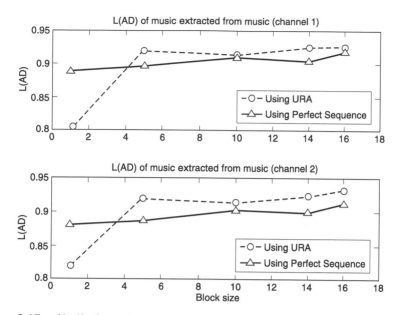

Figure 9.15. Similarity values of music clips extracted from a piano solo.

Table 9.15. Extracted watermark quality using difference block sizes; audio similarity acoustic distance measure using AD: 0–∞ and L(AD): 0–1, and higher perceptual similarity means AD ≅ 0, L(AD) ≅ 1 (host audio: PIANO, watermark: GUITR)

| URA | Watermark : Music | | Host : Music | |
| | Channel 1 | | Channel 2 | |
Block size	AD	L(AD)	AD	L(AD)
1	0.3080	0.7989	3.1766	0.8192
5	2.2381	0.9205	2.2356	0.9207
10	2.3130	0.9149	2.3002	0.9159
14	2.1618	0.9259	2.1793	0.9247
16	2.1438	0.9272	2.0706	0.9320

consideration in addition to nonrepeating schemes (block size = 1). The similarity values of both cases with respect to the repeating block sizes are plotted in Figure 9.15 and given in Table 9.15 and Table 9.16. The larger block size will get the better performance, but by not much. In both cases, the host and watermark combinations can be classified into five categories: "music in music," "speech in music," "speech in speech," "PN data in music," and "PN data in speech." The first combination is more robust against MP3 attack than the latter two. When a music clip is used as a watermark, after the compression attack it can still possess

Table 9.16. Extracted watermark quality: the comparisons of 1-D and 2-D perfect sequences; audio similarity acoustic distance measure using AD: $0-\infty$ and L(AD): 0–1, and higher perceptual similarity means AD $\cong 0$, L(AD) $\cong 1$

Perfect Sequence Watermark_host		Channel 1		Channel 2	
		AD	L(AD)	AD	L(AD)
Music_music P	1-D	2.2553	0.9193	2.3489	0.9120
	2-D	2.0981	0.9302	2.2149	0.9222
Music_speech P	1-D	2.6243	0.8873	2.9198	0.8542
	2-D	2.3117	0.9150	2.4622	0.9025
Speech_music P	1-D	4.1408	0.6334	4.1542	0.6303
	2-D	4.0692	0.6499	4.0832	0.6467
Speech_speech P	1-D	4.2222	0.6143	4.2521	0.6072
	2-D	4.1005	0.6427	4.1018	0.6424

major features of the original watermark that can be recognized by both humans and MNB similarity measurement. In the case of speech watermark clips, however, the results are definitely worse. Because of the frequency distribution and silence durations of speech clips, more information is removed during MP3 encoding than music clips. Therefore, the extracted speech watermark quality is much more degraded. Although the similarity values are not as high as music clips, it should be noted that in the perfect sequence and URA cases, the extracted speech clips are still recognizable. Based on the hearing tests and the MNB measurement, it is appropriate to set a threshold at L(AD) = 0.4 to classify extracted watermarks into two categories of "similar" and "different." Signals with less than a similarity value of 0.4 are generally perceptually very unsimilar. For MP3 attack, URA performs better than perfect sequences. Even speech clips are used as a watermark; perfect sequence and URA can restore the watermark to the extent that L(AD) > 0.4.

Given the same host audio, watermark audio, and scrambling sequence, larger repeating block sizes yield slightly better results in similarity measurement. Moreover, for the perfect sequence case, nonrepeating schemes (block size = 1) also can get acceptable larger similarity values. The major factor that counts in robustness improvement is the efficiency of the scrambling sequence; in the URA case, it is the number of 1's in the matrix.

For embedding PN binary data as watermarks, the input is needed to repeat several times before entering the proposed watermarking scheme; this step is very necessary for improving the robustness and detection rate. In our experiments, if the binary input is repeated 5, 10, or 15 times, the detection rates in Tables 9.8 and Table 9.14 can reach over 94.1% under most various attacks except cropping, time stretch, and quantitation cases. An interesting point needs to be noted: For MP3 attack, the detection rate of binary data is very high, from 99.7% to 100%; for perfect sequence, it is from 92.3% to 98.5%. The perfect sequence's performance is much better than URA for binary data, but is compatible or less than URA for an audio chip.

In addition to the basic difference between binary URA and the real perfect sequence, an additional difference is that the circular correlation of the perfect sequence is 1-D, but the circular correlation of URA is 2-D. By the product theorem in Equation 9.8, we can use two 1-D perfect sequences to form a 2-D perfect sequence whose autocorrelation function is also a delta function. Table 9.17 to Table 9.19 list the performance

Table 9.17. Extracted watermark quality for URA under MP3; audio similarity acoustic distance measure using AD: 0–∞ and L(AD): 0–1, and higher perceptual similarity means AD ≅ 0, L(AD) ≅ 1

URA Watermark_host	Channel 1		Channel 2	
	AD	L(AD)	AD	L(AD)
Music_music	2.1438	0.9272	2.0706	0.9320
Music_speech	2.2325	0.9209	2.1569	0.9263
Speech_music	4.1208	0.6380	4.1994	0.6197
Speech_speech	4.2248	0.6137	4.1630	0.6283

Table 9.18. Extracted watermark quality: comparisons of 1-D and 2-D perfect sequences

Perfect sequence PN data (repeat times)	2-D		1-D	
	Music (bits)	Speech (bits)	Music (bits)	Speech (bits)
Repeat 5	47/6,000 (99.2%)	19/6,000 (99.7%)	60/6,000 (99%)	20/6,000 (99.7%)
Repeat 10	2/3,000 (99.9%)	0/3,000 (100%)	0/3,000 (100%)	0/3,000 (100%)
Repeat 15	0/2,000 (100%)	0/2,000 (100%)	0/2,000 (100%)	0/2,000 (100%)

Table 9.19. Extracted watermark quality for URA under MP3

URA PN data (repeat times)	Music (bits)	Speech (bits)
Repeat 5	1013/6,000 (84.1%)	462/6,000 (92.3%)
Repeat 10	268/3,000 (92.1%)	99/3,000 (96.7%)
Repeat 15	76/2,000 (96.2%)	30/2,000 (98.5%)

comparisons of 1-D and 2-D perfect sequences and 2-D URA under MP3 attack. It is clear that the performances of 2-D are perfect sequence and URA, a little better than 1-D perfect sequence. However we have to pay the price of an extra matrix–array conversion in the 2-D cases.

CONCLUSION

A high-capacity, real-time audio watermarking technique based on the spread spectrum approach is proposed in this Chapter. Sequences with autocorrelation function of zero sidelobe are introduced, investigated, and tested in the experiments. Also, their results under MP3 compression and other attacks are presented in a new objective and quantitative audio similarity measure.

The URA provides better robustness for audio clip watermark embedding than the perfect sequence, but the perfect sequence's performance is much better than URA for binary data. The input repeated insertion for binary data is adopted to improve its robustness and detection rate greatly.

The main contribution of this research is that audio clips and PN data are used as high-capacity watermarks in addition to the commonly used binary sequences. The employment of audio clips as a watermark introduces many more challenges than binary signals, such as the distortion of stego audio after the watermark is embedded, the evaluation of the extracted watermark, and the robustness issues. On the other hand, however, it has pointed another method of digital watermarking. Watermarks can be larger and more meaningful signals other than binary sequences, carrying more information about the author, owner, or the creation, such as the date of origination, the genre, and the company to which it belongs, just like the ID3 tag in the MP3 audio format. It will be of wide application in the future multimedia-and Internet-oriented environments.

REFERENCES

1. Bender W., Gruhl D., Morimoto N., and Lu A., Techniques for data hiding, *IBM Syst. J.*, 35(3&4), 313–336, 1996.
2. Wang Y., A New Watermarking Method of Digital Audio Content for Copyright Protection, in Proceedings of IEEE International Conference on Signal Processing 1998, Vol. 2, pp. 1420–1423.
3. Lee, S.K., and Ho, Y.S., Digital audio watermarking in the cepstrum domain, *IEEE Trans. Consumer Electron.*, 46(3), 744–750, 2000.
4. Swanson, M.D., Zhu, B., Tewfik A.H., and Boney, L., Robust audio watermarking using perceptual masking, *Signal Process.*, 66(3), 337–355, 1998.
5. Gruhl, D., Bender, W., and Lu A., Echo hiding, Information Hiding: 1st International Workshop, Anderson, R.J., Ed., Lecture Notes in Computer Science, Vol. 1174, Springer-Verlag, Berlin, 1996.

6. Kundur, D., Implications for High Capacity Data Hiding in the Presence of Lossy Compression, in Proceedings International Conference on Information Technology: Coding and Computing, 2000, pp.16–21.

7. Magrath, A.J., and Sandler, M.B., Encoding Hidden Data Channels in Sigma Delta Bitstreams, in Proceedings. IEEE International Symposium on Circuits and Systems, 1998, Vol. 1, pp. 385–388.

8. Tilki, J.F., and Beex, A.A., Encoding a Hidden Auxiliary Channel onto a Digital Audio Signal Using Psychoacoustic Masking, in Proceedings IEEE Southeaston 97' Engineering New Century', 1997, pp. 331–333.

9. Furon, T., Moreau, N., and Duhamel, P., Audio Public Key Watermarking Technique, in Proceedings IEEE International Conference on Acoustics, Speech, and Signal Processing, 2000, Vol. 4, pp. 1959–1962.

10. Wu, C.P., Su, P.C., and Kuo, C.C.J., Robust Frequency Domain Audio Watermarking for Copyright Protection, in SPIE 44 Annual Meeting, Advanced Signal Processing Algorithms, Architectures, and Implementations IX (SD41), SPIE, 1999, pp. 387–397.

11. Jessop, P., The Business Case for Audio Watermarking, in Proceedings IEEE International Conference on Acoustics, Speech, and Signal Processing, 1999, Vol. 4, pp. 2077–2078.

12. Swanson, M.D., Zhu, B., and Tewfik, A.H., Current State of the Art, Challenges and Future Directions for Audio Watermarking, presented at IEEE International Conference on Multimedia Computing and Systems, 1999, Vol. 1, pp. 19–24.

13. Fenimore, E.E., and Cannon, T.M., Coded aperture imaging with uniformly redundant arrays, *Appl. Opti.*, 17(3), 1978.

14. Luke, H.D., Sequences and arrays with perfect periodic correlation, *IEEE Trans. Aerospace Electron. Syst.*, 24(3), 287–294, 1988.

15. Luke, H.D., Bomer, L., and Antweiler, M., Perfect binary arrays, *Signal Process.*, 17, 69–80, 1989.

16. Suehiro, N., and Hatori, M., Modulatable orthogonal sequences and their applications to SSMA systems, *IEEE Trans. Inf. Theory*, 34(1), 93–100, 1988.

17. Suehiro, N., Elimination filter for co-channel interference in asynchronous SSMA systems using polyphase modulatable orthogonal sequences, *IEICE Trans. Comm.*, E75-B(6), 1992, pp. 494–498.

18. Suehiro, N., Binary or Quadriphase Signal Design for Approximately Synchronized CDMA Systems Without Detection Sidelobe nor Co-Channel Interference, presented at IEEE International Symposium on Spread Spectrum Techniques and Applications, 1996, Vol. 2, pp. 650–656.

19. Chu, D., Polyphase codes with good periodic correlation properties, *IEEE Trans. Inf. Theory*, 18(4), 531–532, 1972.

20. Bomer, L., and Antweiler, M., Two-Dimensional Binary Arrays With Constant Sidelobe in Their PACF, in Proceedings IEEE International Conference on Acoustics, Speech, and Signal Processing, 1989, Vol. 4, pp. 2768–2771.

21. van Schyndel, R., Tirkel, A.Z., Svalbe, I.D., Hall, T.E., and Osborne, C.F., Algebraic Construction of a New Class of Quasi-orthogonal Arrays in Steganography, SPIE Security and Watermarking of Multimedia Contents, 1999, Vol. 3675, pp. 354–364.

22. Lee, C.H., and Lee, Y.K., An adaptive digital image watermarking technique for copyright protection, *IEEE Trans. Consumer Electron.*, 45(4), 1005–1015, 1999.

23. Caini, C., and Coralli A.V., Performance Evaluation of an Audio Perceptual Subband Coder With Dynamic Bit Allocation, in Proceedings IEEE 13th International Conference on Digital Signal Processing, 1997, Vol. 2, pp. 567–570.

24. Espinoza-Varas, B., and Cherukuri, W.V., Evaluating a Model of Auditory Masking for Applications in Audio Coding, presented at *IEEE ASSP Workshop on Applications of Signal Processing to* Audio and Acoustics, 1995, pp. 195–197.

25. Voran, S., Objective estimation of perceived speech quality, Part I: Development of the measuring normalizing block technique, *IEEE Trans. Speech Audio Process.*, 7(4), 371–382, 1999.
26. Voran, S., Objective estimation of perceived speech quality, Part II: Evaluation of the measuring normalizing block technique, *IEEE Trans. Speech Audio Process.*, 7(4), 383–390, 1999.
27. Noll, P., Wideband speech and audio coding, *IEEE Commn. Mag.*, 31(11), 34–44, 1993.

10
Multidimensional Watermark for Still Image
Parallel Embedding and Detection

Bo Hu

INTRODUCTION

Digital data access and operation became easier because of the rapid evolution of digital technology and the Internet. Security of multimedia data has been a very important issue.

One approach to data security is to use cryptography. However, it should be noted that a cryptosystem restricts access to the data. Every person who wants to access the data should know the key. Once the data are decrypted, the protection of data are invalidated. Unauthorized copying and transmission of the data cannot be prevented.

The digital watermark has been proposed as an effective solution to the copyright protection of multimedia data. Digital watermark is a process of embedding information or signature directly into the media data by making small modifications to them. With the detection of the signature from the watermarked media data, the copyright of the media data can be resolved.

0-8493-2773-3/05/$0.00+1.50
© 2005 by CRC Press

A good watermark should have the following characteristics:

Imperceptibility: The watermark should not change the perceptual characteristics of the original image. If a watermark is easy to detect, it is easy to be attacked, too. In addition, a perceptible watermark will obviously reduce the value of original work.

Robustness: The watermark should be robust as much as possible against attacks and processing such as lossy compression, filtering, scaling, and so forth. Even if one-half of an image is cut, the watermark should be detected.

Statistically invisible: The possession of a large set of images, watermarked by the same watermark, should not help an attacker to determine the watermark.

Support multiple watermarks: Multiple watermarks on the same image are necessary for authorization. At each layer, the dealer can embed its own watermark, which cannot be detected and destroyed by others. This could protect the copyright of the image's owner. With multiple watermarks, it is easy to find who transmitted or copied the image without authorization.

Not using the original data in watermark detection: By comparing the original image and the one with the watermark, the detection possibility could be higher. However, the transmission of the original image may be attacked also, and, the original image cannot been affirmed only through human perceptivity. If the watermark can be detected without the original image, it will be very useful for copyright protection.

Among these characteristics, imperceptibility and robustness, which conflict with each other, are the basic requirements of a digital watermark. Less modification is expected for the sake of imperceptibility. Otherwise, the more watermark power embedded in the image, the more robustness it may have. The trade-off between them has been one of the keys of watermarking.

The methods for image watermarking can be classified into two categories: the spatial domain approach [1] and the transform domain approach. The spatial domain approach has low computation complexity. However, this approach has weak robustness. The watermark can be detected and destroyed easily. By simply letting the less significant bits of each pixel be zero, the watermark will be removed.

Comparing with the spatial domain approach, the methods to embed the watermark in the transform domain, such as discrete cosine transform (DCT) [10], wavelet transform [11], and discrete Fourier transform (DFT) [12], have many advantages. It is hard to detect and destroy. Some perceptual models have been developed:

In order to improve the robustness, the watermark should be embedded in the perceptually significant component. Some researchers use low-frequency coefficients in the DCT domain, including the direct current (DC) component of the image [2,3]. However, this method may result in blockiness, which is sensible to human eyes. It will reduce the quality of image. Also, it is hard to embed multiple watermarks on the DC coefficient.

A feature-based approach is proposed in References 4 and 5, in which, coefficients larger than the threshold value is selected to be watermarked, including the high-frequency coefficient. However, when the image is filtered by a LPF (low-pass filter), the robustness of the watermark may be reduced.

Wu et al. [6] proposed a multilevel data hiding approach based on two-category classification of data embedding. Data hiding in the low band of the DCT can improve the robustness, and high-band data hiding can embed secondary data at a high rate. The high-band watermark proposed by Wu is usually used for image certification rather than copyright protection.

In this chapter, we introduce a multidimensional watermark scenario. The mutual independent watermarks, which are spread spectrum, are embedded in the low band to midband in the DCT domain in parallel, exploiting the properties of the human visual system (HVS) [13]. Without the original image, the watermark can be detected based on the joint probability distribution theory. Gray-scale image watermarking only is studied in this chapter, but the method can be used for color image also.

The arrangement of this chapter is as follows. We present the process of multidimensional watermark embedding in Section "Embedding of Multidimensional Watermarks," then the watermark detection, hypothesis testing, and joint probability distribution are introduced. Experimental results appear in Section "Robustness of the Multidimensional Watermark." Finally, Section "Conclusion" presents the conclusions.

EMBEDDING OF MULTIDIMENSIONAL WATERMARKS

Spread Spectrum Signal

In binary spread spectrum communication, there are two types of modulation: direct sequence (DS) and frequency hopped (FH). Figure 10.1 shows the basic elements of the DS spread spectrum system.

The information stream $a(n)$ with bit rate R is not transmitted directly. Each bit of it will be represented by a binary pseudorandom sequence

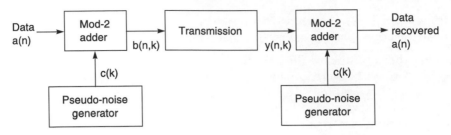

Figure 10.1. Model of the direct sequence spread spectrum system.

$c(k)$ with length $L_c \gg 1$. This is performed by the block named "*Mod-2 adder*," which is defined as

$$b(nT_b + kT_c) = a(nT_b) \oplus c(kT_c), \qquad k = 0, 1, \dots, L_c - 1 \qquad (10.1)$$

where $1/T_b$ is the bit rate R of the information sequence, $1/T_c$ is the bit rate $R' = RL_c$ of the spread spectrum sequence, and $\oplus$ is the "Exclusive OR."

Because $L_c \gg 1$, to transmit $b(n, k)$ in the same period, the bit rate R' is much larger than R and needs more bandwidth. However, the spread spectrum technique is widely used in digital communication because of its many desirable attributes and inherent characteristics, especially in an environment with strong interference. The three main attributes are:

The use of the pseudorandom code to spread the original signal allows easier resolution at the receiver of different multipaths created in the channel.

The spreading follows a technique where the signal is spread into a much larger bandwidth, decreasing the ability for any outside party to detect or jam the channel or the signal.

The last and the most important feature is the forward error correction coding; when done with the spreading and the despreading operations, it provides robustness to noise in the channel.

At the receiver, the received signal $y(n,k)$ is despread with the same pseudorandom sequence $c(k)$ and the transmitted information sequence $a'(n)$ is decided by

$$a'(nT_b) = \begin{cases} 1, & \sum_{K=0}^{L_c-1} y(nT_b + kT_c) \oplus c(kT_c) \geq L_c/2 \\[2ex] 0, & \sum_{K=0}^{L_c-1} y(nT_b + kT_c) \oplus c(kT_c) < L_c/2 \end{cases} \qquad (10.2)$$

If there is no error in transmission, $\sum_{k=0}^{L_c-1} y(nT_b + kT_c) \oplus c(kT_c)$ will be L_c for $a(n) = 1$ or 0 for $a(n) = 0$. The distance between 0 and 1 information

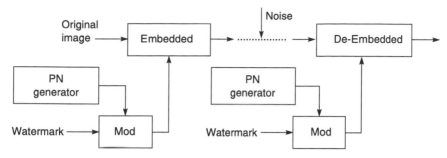

Figure 10.2. Block diagram of a watermark.

bit is much larger than that without spreading. From the theory of communication, it means greater ability for combating the interference. A detailed discussion of spread spectrum signal can be found in the book by Proakis [7].

In 1996, Cox et al. introduced the spread spectrum signal into the digital watermark to improve its robustness [2]. As shown in Figure 10.2, digital watermark can be thought as a spread spectrum communication problem. The watermark information to be sent from the sender to the receiver is spread by a pseudorandom sequence and transmitted through a special channel. The channel is composed of the host image and noise introduced by signal processing or attack on the watermarked image.

Similar to the spread spectrum communication, the watermark information is represented by a random sequence with a much larger length. Then, the power of the spread spectrum sequence could be very small when embedded into the image, but the robustness of the watermark is kept.

In our method, the watermark information is represented by four spread spectrum sequences and embedded into the image independently. By joint detection discussed in Section "Joint Detection of a Multidimensional Watermark," the robust of the watermark is much improved.

Watermark in the DCT Domain Considering HVS

In order to increase the compression ratio, we must take advantage of the redundancy in most image and video signals, including spatial redundancy and temporal redundancy. The transform coding methods, such as DFT, DCT, and wavelet transform, belong to the spatial domain methods. In the transform domain, the energy of signals is mostly located in the low-frequency components. With carefully selected quantization thresholds on different frequency bands and entropy coding, the image or video can be efficiently compressed.

Among them, the DCT based method is widely used, such as in JPEG, MPEG-I, MPEG-II, and so forth, because in its better ability in information packing and low computation load. Therefore, in this chapter, we study the embedding watermark in DCT domain, too.

The original image is first divided into 8×8 pixels blocks. Then, we can perform a DCT on each block, as shown in Equation 10.1:

$$T(u, v) = \sum_{x=0}^{7} \sum_{y=0}^{7} f(x, y) a(u) a(v) \cos\left[\frac{(2x + 1)u\pi}{16}\right] \cos\left[\frac{(2y + 1)v\pi}{16}\right],$$

$$u, v, x, y = 0, 1, \ldots, 7 \tag{10.3}$$

where

$$a(u) = \begin{cases} \sqrt{\frac{1}{8}} & \text{for } u = 0 \\ \sqrt{\frac{1}{4}} & \text{for } u = 1, 2, \ldots, 7 \end{cases} \tag{10.4}$$

and similarly for $a(v) \cdot f(x, y)$ is the value of image pixel (x, y) and $T(u, v)$ is that of the component (u, v) in the DCT domain.

After transform, the frequency components are usually Zig-Zag ordered as shown in Figure 10.3. The DC coefficient is indexed as 0 and the highest frequency component is indexed as 63.

As specified earlier, the watermark should be embedded in the low-frequency band to improve its robustness. However, the low-frequency coefficients generally have much higher power than others. A small change of them will results in a severe degradation of image.

0	1	5	6	14	15	27	28
2	4	7	13	16	26	29	42
3	8	12	17	25	30	41	43
9	11	18	24	31	40	44	53
10	19	23	32	39	45	52	54
20	22	33	38	46	51	55	60
21	34	37	47	50	56	59	61
35	36	48	49	57	58	62	63

Figure 10.3. Frequency components order in the DCT domain.

0	1	5		14	15	27	28
2	4			16	26	29	42
				17	25	30	41	43
9		18	24	31	40	44	53
10	19	23	32	39	45	52	54
20	22	33	38	46	51	55	60
21	34	37	47	50	56	59	61
35	36	48	49	57	58	62	63

Figure 10.4. Watermark location in the DCT matrix

Therefore, we take advantage of the HVS. Because the sensitivity to various types of distortion as a function of image intensity, texture, and motion is different, researchers have developed the Just Noticeable Distortion (JND) profile [14]. With this profile, we choose the coefficients of 3, 6, 7, and 8, which have middle energy power, to embed the watermark rather than the very low-frequency band such as 0, 1, and 2. The coefficients 4 and 5 are not used because in the human perceptual modal, their visual thresholds are too low to embed enough watermarks. Figure 10.4 shows the selection of frequency components into which we may embed a watermark.

Embedding of a Multidimensional Watermark

The architecture of multidimensional watermark embedding is shown in Figure 10.5. The basic steps of embedding the multidimensional watermark into the host image are:

- *Mapping user's message into four-dimensional (4-D) pseudorandom sequences*: The information sequence $s = s_1 s_2 \cdots s_L$, provided by the

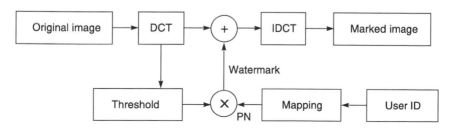

Figure 10.5. Block diagram of watermark embedding.

user, which can be a meaningful signature or registered owner ID, is divided into four subsequences $s'_j, j = 1, 2, \ldots, 4$. Each s'_j determines which pseudorandom sequence is selected and what its initial phase is, from a set of PN sequences. This set of pseudo noise (PN) sequences is generated by the m-sequence. The selected PN sequence is mapped into a bipolar sequence $PN_j(i)$,

$$PN_j(i) = \begin{cases} -1 & \text{if the bit of the } m - \text{sequence is } 0 \\ 1 & \text{if the bit of the } m - \text{sequence is } 1 \end{cases} \quad (10.5)$$

Each $PN_j(i)$ is independent on the others in the set.

The selection of the PN sequence is completed by an irrevisible function $f(\cdot)$, mapping part of s'_j into an index of PN sequences. Here, a rule is needed to avoid the same selection of a PN sequence even if the subsequences s'_i and s'_j, $i \neq j$, are same.

The initial phase is calculted by the rest of the subsquences s'_j. Then, the four independent PN sequences $PN_j(i), j = 1, 2, \ldots, 4$, are obtained and can be used to embedded into the image.

- *Performing image blocking and DCT; obtaining the sequence to be watermarked*: The image is divided into nonoverlapped 8×8 blocks with length of $L_h \times L_v$, wherer L_h is the number of horizontal blocks and L_v is the number of vertical blocks in the image. Then, each block is DCT transformed and organized in zigzag order (shown in Figure 10.3).

As shown in Figure 10.3, the frequency components of all blocks, indexed by $3, 6, 7,$ and 8, are composed of four sequences $I(j)$, $j = 1, 2, \ldots, 4$. The watermark will be embedded into these sequences.

Because each sequence has a length of $L_h \times L_v$, the cycle of pseudorandom sequences should be larger than $L_h \times L_v$.

- *Calculating the corresponding masking thresholds*: Based on measurements of the human eye's sensitivity to different frequencies, an image-independent 8×8 matrix of threshold can be obtained, denoted as $T_f(u, v)$, $u, v = 1, 2, \ldots, 8$.

Considering of the luminance of each image, we can get a more accurate perceptual model:

$$T_f(u, v, b) = T_f(u, v)[X(0, 0, b)/\overline{X(0, 0)}]^\alpha \quad (10.6)$$

$$\overline{X(0, 0)} = \frac{1}{L_h \times L_v} \sum_{b=1}^{L_h \times L_v} X(0, 0, b) \quad (10.7)$$

where $X_{0,0,b}$ is the DC coeffient of block b, $b = 1, 2, \ldots, L_h \times L_v$, and α is the parameter that controls the degree of luminance sensitivity. A value of 0.649 is suggested for α in Reference 8.

From $T_f(u, v, b)$, the threshold vector for sequence $I(j)$, $j = 1, 2, \ldots, 4$, is obtained and denoted by $G_j(b)$, $j = 1, 2, \ldots, 4$, $b = 1, 2, \ldots, L_h \times L_v$. To embed the watermark with the same energy, the masking threshold of the embedding signal must satisfy the condition as follows:

$$G(b) = \min(G_j(b)), \qquad j = 1, \ldots, 4 \tag{10.8}$$

Embedding PN$_j$, $j = 1, 2, \ldots, 4$, into $I_j(i)$, $j = 1, 2, \ldots, 4$

According to the masking threshold, $G(i)$, $i = 1, 2, \ldots, L_h \times L_v$, 4-D watermarks are calculated as

$$W_j(i) = \alpha_j \times \mathrm{PN}_j(i) \times G(i) \tag{10.9}$$

where $j = 1, 2, \ldots, 4$, $i = 1, 2, \ldots, L_h \times L_v$, and

$$\alpha_j = \begin{cases} 0.26 & \text{if mean } (I_j(i)) \geq 0 \\ -0.26 & \text{otherswise} \end{cases} \tag{10.10}$$

It can be proved that the mean value of $W_j(i)$ is zero. Then, the watermark signal is embedded into the host image in the DCT domain in parallel:

$$I'_j(i) = I_j(i) + W_j(i), \qquad i = 1, \ldots, L_v \times L_h, \, j = 1, \ldots, 4 \tag{10.11}$$

By using the masking threshold, the watermarked image is imperceptible.

- *Performing inverted discrete cosine transform (IDCT) and obtaining the watermarked image*: After the watermark being embedded, we perform IDCT as following and gets the watermarked image I':

$$f(x, y) = \sum_{u=0}^{7} \sum_{v=0}^{7} T(u, v) a(x) a(y) \cos\left[\frac{(2u + 1)x\pi}{16}\right] \cos\left[\frac{(2v + 1)y\pi}{16}\right] \tag{10.12}$$

The result of the multidimensional watermark is given in Figure 10.6. Here, the length of **s** is fixed to eight characters. If the user signature is shorter than eight characters, it is made up to eight by default. The mapping function $f(\cdot)$ is defined as follows. First, we denote an integer V_i ($1 \leq V_i \leq 31$) as the corresponding value of character s_i (the characters from "a" to "z" and five resolved characters such as "_," "@," "-," "~," and "'"). Then, the integer value $(V_{2j-1} + V_{2j})/2$ is used to select the pseudorandom sequence PN$_j$ from a set of 10 order m-sequences.

Figure 10.6. The effect of a multidimensional watermark (a) Initial image of Lena; (b) initial image of baboon; (c) Lena with a multidimensional watermark; (d) baboon with multidimensional watermark.

Figure 10.6. Continued.

The original phase of PN$_j$ is determined by the 10-bit binary number,

$$(V_j\%4) \times 256 + (V_{j+1}\%4) \times 64 + (V_{j+2}\%4) \times 16 + (V_{j+3}\%4) \times 4 + V_{j+4}\%4$$

where $i = 1, \ldots, 8$, and $j = 1, \ldots, 4$.

The watermarks are embedded into 512×512 Lena and baboon images in the DCT coefficients with indices 3, 6, 7, and 8. Figure 10.6 show the watermarked images with the signature string "fudan."

Compared with the original one, the modifications of watermark to "Lena" and "Baboon" are to SNR $= 42.0210$ dB and SNR $= 42.8520$ dB additive noise. We can claim that the watermark is imperceptive.

JOINT DETECTION OF A MULTIDIMENSIONAL WATERMARK

The detection of a multidimensional watermark in a still image is an inverse process. In this section, we will discuss how to detect the watermark without an initial image. The block diagram of watermark detection is given in Figure 10.7.

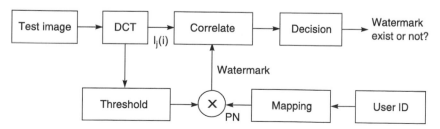

Figure 10.7. Block diagram of watermark embedding.

Similar to the embedding procedure, the following steps should be followed to decide whether the test image contains a watermark,

Mapping user's message into 4-D pseudorandom sequences with the same rule.

Performing image blocking and DCT; obtaining the sequences $I_j'(i)$, $j = 1, 2, \ldots, 4$, which may contain a watermark.

Calculating the corresponding masking thresholds $G'(i)$, $i = 1, 2, \ldots$, $L_h \times L_v$, following Equations 10.6 to Equation 10.8.

Calculating the embedded sequence $W_j'(i)$, $j = 1, 2, \ldots, 4$, $i = 1, 2, \ldots$, $L_h \times L_v$, by Equation 10.9 and Equation 10.10. The $W_j'(i)$ may not be the same one as $W_j(i)$ but will be highly correlated with $W_j(i)$.

Correlating $W_j'(i)$ with $I_j'(i)$, $j = 1, 2, \ldots, 4$, $i = 1, 2, \ldots, L_h \times L_v$. Each correlation output q_j is compared with threshold T_j for judgment. When all of the conditions are satisfied, the claimed watermarks are contained in the test image.

The judgment is based on hypothesis testing and joint probability distribution theories, which can reduce the error judgment probability and improve the reliability.

Hypothesis Testing

In watermark detection, the test DCT coefficient vector $I_j'(i), j = 1, 2, \ldots, 4$, $i = 1, 2, \ldots, L_h \times L_v$, is derived from the test image and correlate with $W_j'(i)$, $j = 1, 2, \ldots, 4$, $i = 1, 2, \ldots, L_h \times L_v$. The correlation output q_j is compared to a threshold T_j. The watermarks detection is accomplished via hypothesis testing:

$$
\begin{aligned}
&H_0: X_j(i) = I_j(i) + N(i) \\
&H_1: X_j(i) = I_j(i) + W_j(i) + N(i)
\end{aligned}
\tag{10.13}
$$

Under hypothesis H_0, the image does not contain the claimed watermark, whereas it does under hypothesis $H_1 \cdot N(i)$ is the interference possibly resulting from signal processing. The correlation detector outputs the test statistics q_j:

$$
q_j = \frac{\sum_{i=1}^{n} Y_j(i)}{V_y \sqrt{n}} = \frac{M_y \sqrt{n}}{V_y}
\tag{10.14}
$$

$$
Y_j(i) = X_j(i) W_j'(i)
\tag{10.15}
$$

where n is $L_h \times L_v$, the size of test vector X_j. M_y and V_y are the mean value and variance of $Y_j(i)$, respectively.

Assume that the sequence $\{Y_j(i)\}$ is stationary and at least l-dependent for a finite positive integer l; then under hypothesis H_0, the statistic q_j follows a Normal distribution $N(0,1)$ when n is large enough. Under hypothesis H_1, q_j follows another distribution $N(m,1)$, where

$$m = \frac{\{E[W_j(i)W_j'(i)] + E[N(i)W_j'(i)]\}\sqrt{n}}{V_y}$$

$$\approx \frac{\sum_{i=1}^{n}[W_j(i)W_j'(i) + N(i)W_j'(i)]}{V_y\sqrt{n}} \tag{10.16}$$

Image processing does not affect the distribution of H_0 basically, but it may reduce the mean value m of q_j and increase its variance under H_1.

The output q_j is then compared to threshold T_j. If $q_j > T_j$, the test image is declared to have been watermarked by the claimed signature s'_j, otherwise it is not.

The distribution of q_j under hypotheses H_0 and H_1 is shown in Figure 10.8. If the two distributions are overlapped, hypothesis testing may have two types of error (Figure 10.8a): Type 1 error P_{err1} is accepting the existence of signature under H_0 (area **B**), and Type 2 error P_{err2} is rejecting the existence of a signature under H_1 (area **A**). The threshold T_j must be the minimum of the total error probability $P_{err} = P_{err1} + P_{err2}$; assuming that H_0 and H_1 are equiprobable, T_j is chosen to be $m/2$. If T_j is chosen to satisfy the condition $P_{err1} < a$, the value of m must larger than $2T_j$ (i.e., $m > 2T_j$). If $T_j = 3.72$ for $a = 0.0001$ under the Normal distribution, to ensure that $P_{err} < a$, m must be larger than 7.44.

If the distribution of q_j under hypotheses H_0 and H_1 is not overlapped (Figure 10.8b), to minimize the error probability P_{err}, the threshold T_j could be much less than $m/2$, and the Type 2 error P_{err2} may decrease mostly.

If the distribution of q_j is overlapped (Figure 10.8a), no matter what threshold value is chosen, the decision error is sure to occur. The only means for reducing P_{err} is to increase the expected value m of q_j. There are two ways to increase m:

1. Increasing the energy of watermark, such as increasing α_j or $G(i)$ in Equation 10.9. However, the quality of the image may decrease and the watermark may be detected easily.
2. Increasing the length of $PN_j(i)$. Because the number of blocks in one image is fixed, we will use the multidimensional watermark for joint detection.

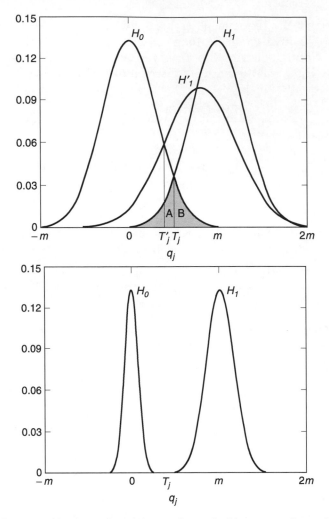

Figure 10.8. Hypothesis testing: (a) overlapped; (b) nonoverlapped.

Joint Detection

For n-dimension random variables, if we have

$$F_{x_1x_2\cdots x_n}(x_1, x_2, \ldots, x_n) = P\{X_1 \leq x_1, X_2 \leq x_2, \ldots, X_n \leq x_n\} \qquad (10.17)$$

then $F_{x_1x_2\ldots x_n}(x_1, x_2, \ldots, x_n)$ is the joint probability distribution (JPD) of n-dimension random variables. If the n-dimension random variables are mutually independent, Equation 10.17 will be

$$P\{X_1 \leq x_1, X_2 \leq x_2, \ldots, X_n \leq x_n\} = P\{X_1 \leq x_1\}P\{X_2 \leq x_2\} \cdots P\{X_n \leq x_n\}$$

$$(10.18)$$

which means

$$F(x_1, x_2, \ldots, x_n) = F(x_1)F(x_2)\cdots F(x_n) \qquad (10.19)$$

As discussed earlier, the test statistics q_j is a random variable with independent identical distribution (i.i.d.). Then, from Equation 10.19, the total Type 1 error probability p_{err1} in the detection of 4-D watermarks is

$$
\begin{aligned}
p_{err1} &= P(x_1 < T_1, x_2 < T_2, \ldots, x_4 < T_4) \\
&= P_{err1}(x_1 < T_1)P_{err1}(x_2 < T_2)\cdots P_{err1}(x_4 < T_4) \qquad (10.20)
\end{aligned}
$$

If we want p_{err1} to be less than 0.0001, the P_{err1} for each q_j needs to be less than $0.0001^{1/4} = 0.1$, that means $T_j = 1.28$ for Normal distribution $N(0,1)$. To ensure that the total error probability of Type 2 is not large than a, the correct detection probability of q_j must satisfy $P_{err2} > (1 - 0.0001)^{1/4} = 0.9999975$; that is, $m > T + 4.05 = 5.33$. Therefore, the requirement of m is decreased; in other words, the error detection probability is lower than the single-dimension watermark when equal watermark energy is embedded.

For the 512×512 Lena image into which we embedded the 4-D watermark, the pdf of the detection output q_1 under hypotheses H_0 and H_1 is shown in Figure 10.9, which obey the Normal distributions $N(0,1)$ and $N(m,1)$. The mean values of q_1 are -0.0015 and 11.9382, respectively.

The four detection outputs are similar; their corresponding detection threshold can be chosen to be the same T. The choice of T will affect the hypothesis testing error probability, Table 10.1 shows the two types of

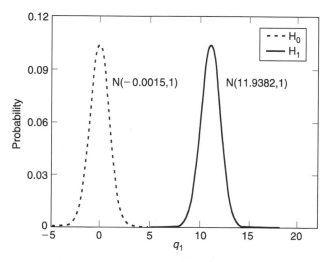

Figure 10.9. **Distribution of detection output q_1.**

TABLE 10.1. Error probability in hypotheses testing

T	Type 1 Error	Type 2 Error
1.0	6.0996e−4	—
1.4	4.0549e−5	—
1.8	1.5471e−6	—
...	...	...
7.8	—	3.6204e−5

error corresponding to different T's; "-" indicates that the error does not occur in simulation. Actually, when the threshold is set to 7.8, the probability of Type 2 error is nearly 3.62041×10^{-5}, when the type 1 error probability is 0.

The effect of a multidimensional watermark is evident. In a single-dimension watermark, we must choose $T = 3.9$ to let P_{err1} be less than 0.0005; however, $T = 1.3$ in 4-D watermark is enough.

In our simulation, 27621 different watermarks are embedded and the simulation results are similar.

ROBUSTNESS OF THE MULTIDIMENSIONAL WATERMARK

In this section, the robustness of the multidimensional watermark is tested for common signal processing such as JPEG compression [15], low-pass filtering, image scaling, multiple watermarking, and so forth.

JPEG Compression

If the test image has been JPEG compressed, the output q will decrease with the JPEG quality factor Q (Q is defined in MATLAB, shown in Figure 10.10) and the testing error probability will increase.

Figure 10.11 shows the detection of a multidimensional watermark compared to the single-dimension watermark. If we use the single-dimension watermark proposed in Reference 9 with threshold $T = 3.9$, the watermark can be detected correctly when the JPEG compressed with factor $Q > 25\%$. While by using 4-D watermark, we can detect the watermark correctly when Q is not less than 15%.

Low-Pass Filtering

A watermark in the low to mid DCT band makes it more robust to low-pass filtering, which degrades the high frequency primarily. A total of 27,621 different watermarked images are low-pass filtered using 2-D Gaussian low-pass filters, whose standard deviation increases

326

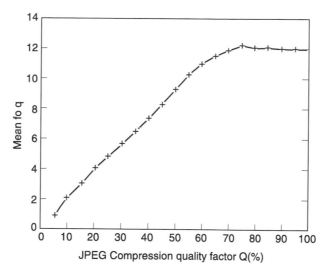

Figure 10.10. **Mean value of *q* as function of the JPEG quality factor Q.**

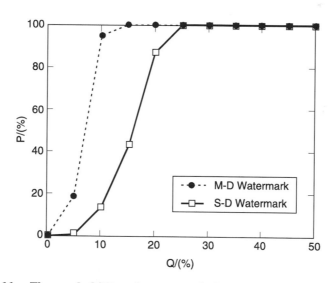

Figure 10.11. **The probability of watermark detection as a function of the JPEG quality factor Q.**

from 0.3 to 5.3. The mean value of q as a function of standard deviation is given in Figure 10.12. The filtered image is shown in Figure 10.13a, where the difference images after filtering is shown in Figure 10.13b, Because the mean value of q is still larger than 7.4, the watermarks are all detected correctly:

$$h_g(x, y) = e^{-(x^2 + y^2)/2\sigma^2} \tag{10.21}$$

327

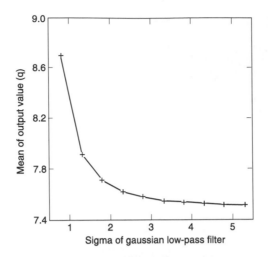

Figure 10.12. The mean value of *q* as a function of the standard deviation.

Figure 10.13. The effect of a Gaussian low-pass filter on a watermarked image: (a) low-pass-filtered watermarked 512×512 Lena image by Gaussion filter $\sigma = 5.3$); (b) the image after filtering.

TABLE 10.2. Image scaling influence on watermark

Geometrical Operation	Mean of q	SNR (dB)
Linear interpolation upscaling (1024 × 1024)		
Bits drawn–out downscaling	11.9382	42.0210
Mean value downscaling	10.4648	41.4656
Bits drawn–out downscaling (256 × 256)		
Bit-repeat upscaling	9.6853	40.9581
Linear interpolation upscaling	9.3468	40.2361

TABLE 10.3. Multiple watermark's influence on Lena

	Mean of q	SNR (dB)
1 watermark	11.9448	42.0210
2 watermarks	10.9527	38.9410
3 watermarks	10.3564	37.1568

Scaling

Before watermark detection, the test image will be changed to its initial size. Image scaling results in noise to the image. As shown in Table 10.2, the effect of detection is knee high to a mosquito.

Multiple Watermarking

Multiple watermarks can be contained in the same image simultaneously. By using spread spectrum signals, the different watermarks are mutually independent. We embed 27,621 different three watermarks in 512 × 512 Lena; Table 10.3 shown the mean SNR multiple watermarks introduced. Three watermarks only introduce 37.1568 dB noise, so multiple watermarks can be detected correctly.

CONCLUSION

Image watermarking for resolving copyright protection is a private watermark; its aim is high robustness, not large capability. Multi-dimensional watermarking in the low to mid-band of DCT coefficients improves the watermark robustness. It is robust to JPEG compression, low-pass filtering, and so forth. Experiments shows that this scenario has high robustness.

REFERENCES

1. Voyatzis, G., Nikolaidis, N., and Pitas, I., Digital Watermarking: an Overview, Eus'98, 1998.

2. Cox, I.J., Kilian, J., Leighton, T., and Shamoon, T., Secure Spread Spectrum Watermarking for Images, Audio and Video, in Proceedings ICIP'96, 1996, vol. 4, pp. 243–246.
3. Huang, J., Shi, Y.Q., and Shi, Y., Embedding image watermarks in DC components, *IEEE Trans. Circuits Syst. Video Technol.*, 10, 974–979, 2000.
4. Craver, S., Memon, N., Yeo, B., and Yeung, M., Can Invisible Watermarks Resolve Rightful Ownerships?, in Proceedings IS&T/SPIE Electronic Publishing, 1994.
5. Zeng, W. and Liu, B., A statistical watermark detection technique without using original images for resolving rightful ownerships of digital images, *IEEE Trans. Image Process.*, 8, 1534–1545, 1999.
6. Wu, M., Yu, H.H., and Gelman, A., Multi-level Data Hiding for Digital Image and Video, in *Photonics East '99 — Multimedia Systems and Applications II*, A.G. Tescher, B. Vasudev, V.M. Bove, Jr., and B. Derryberry, Eds., Proc. SPIE, November 1999, Boston, MA, Vol. 3845, pp. 10–21.
7. Proakis, J.D., *Digital Communications*, 3rd ed., McGraw-Hill, New York, 1995.
8. Peterson, H.A., Ahumada, A.J. Jr., and Watson, A.B., Improved Detection Model for DCT Coefficient Quantization, in Proceedings SPIE Conference on Human Vision. Visual Processing and Digital Display IV, 1993, Vol. 1913, pp. 191–201.
9. Piva, A., Barni, M., Bartolini, F., and Cappellini, V., Threshold Selection for Correlation-Based Watermark Detection, Cost'98, 1998.
10. Hernández, J.R., Amado, M., and Pérez-González, F., DCT-domain watermarking techniques for still images: Detector performance analysis and a new structure, *IEEE Trans. Image Process.*, 9(1), 55–68, 2000.
11. Kundur, D., and Hatzinakos, D., A Robust Digital Image Watermarking Method Using Wavelet-Based Fusion, in Proceedings IEEE International Conference on Image Processing '97, 1997, vol. 1, pp. 544–547.
12. Piva, A., Barni, M., and Bartolini, F., Cappellini, V., Copyright Protection of Digital Images by Means of Frequency Domain Watermarking, SPIE'98, 1998.
13. Podilchuk, C., and Zeng, W., Image adaptive watermarking using visal models, *IEEE J. Selected Areas Commn.*, 16(4), 525–539, 1998.
14. Reininger, R.C., and Gibson, J.D., Distribution of the two-dimensional DCT coefficients for images, *IEEE Trans. Commun.*, 31(6), 835–839, 1983.
15. Wallace, G.K., The JPEG still picture compression standard, *IEEE Trans. Consumer Electron.*, 38(2), 18–34, 1992.

11

Image Watermarking Method Resistant to Geometric Distortions*

Hyungshin Kim

INTRODUCTION

With the advances in digital media technology and proliferation of internet infrastructure, modification and distribution of multimedia data became a simple task. People can copy, modify, and reformat digital media and transmit them over the wireless high-speed Internet with no burden. Companies owning multimedia contents wanted to secure their property from illegal usage. For this purpose, the digital watermark has attracted attention from the market. The digital watermark is the invisible message embedded into the multimedia content. Since its introduction into the field of multimedia security and copyright protection, responses from the researchers were overwhelming.

*Some material in this chapter are from "Rotation, scale, and translation invariant image watermark using higher order spectra," *Opti. Eng.*, 42(2), 2003 by H. S. Kim, Y. J. Baek, and H. K. Lee with the permission of SPIE.

A vast amount of work has been reported in various venues [1]. Initial works were concentrated on perceptual invisibility and distortions by additive noise and JPEG compression. As demonstrated in Reference 2, minor geometric distortions — the image is slightly stretched, sheared, shifted, or rotated by an unnoticeable random amount — can confuse most watermarking algorithms. This minor distortion can be experienced when the image is printed and scanned. This attack is a very powerful attack because it does not affect image quality but does affect the synchronization at the watermark detector. This vulnerability of the previous works against the geometric attack called for new watermarking methods. Since then, many works have been claimed to be robust to geometric distortions. However, until now, not a single algorithm is shown to meet the robustness requirements of the watermarking community.

Many watermarking methods have been reported to be resilient to geometric distortions. One approach is to embed a known template into images along with the watermark [3,4]. The template contains the information of the geometric transform undergone by the image. The transform parameters are estimated from the distorted template. The undistorted watermarked image is recovered using the parameters, and then the watermark can be extracted. This method requires embedding a template in addition to the watermark so that this may reduce image fidelity and watermark capacity.

Another approach is to embed a structured watermark into the image [5]. A repeated pattern is inserted and the autocorrelation function (ACF) is used for detection. The location of the correlation peaks will generate a pattern of peaks. Changes in the pattern due to the geometric distortion can be identified and used for the restoration.

Watermarks that are invariant to geometric distortions have been employed. Rotation, scale, and translation (RST)-invariant watermarks can be designed with the magnitude of the Fourier–Mellin transform of an image [6,7]. Although their watermarks were designed within the RST-invariant domain, they suffer severe implementation difficulty. It is mainly due to the computational complexity and the unstable log-polar mapping during the Fourier–Mellin transform.

Watermarking algorithms using a feature of an image were proposed as the second-generation watermark [8,9]. Because feature vectors of images are invariant to most of the image distortions, they were used as the keys to find the embedding location.

In this chapter, we propose a new watermarking algorithm using a RST-invariant feature of images. However, we use the feature vector as a watermark not as a key. The vector is defined with higher-order spectra (HOS) of the Radon transform of the image. For the use of HOS, we adopt the bispectrum (the third-order spectra) feature, which is known to have invariance to geometrical distortions and signal processing [10].

Our algorithm is different from the previous methods using the Fourier–Mellin transform in that we use the Radon transform, which has less aliasing during embedding than the log-polar mapping. Furthermore, we embed the watermark signal in the Fourier phase spectrum, and this makes our system more difficult to tamper with than the Fourier–Mellin-based methods, in which the Fourier magnitude spectrum is used [11].

This chapter is organized as follows. In section "Watermarking Systems and Geometric Distortions", we start with a closer look at various geometric distortions and their implications in watermarking methods. In section "A RST-Invariant Watermarking Method", we propose a new watermarking method based on an invariant feature vector. Test results are provided in section "Experimental Results". We discuss the results in section "Discussion" and we conclude this chapter in section "Conclusion".

Throughout this chapter, we will focus on two-dimensional (2-D) images. However, most of the arguments can be extended to video as well.

WATERMARKING SYSTEMS AND GEOMETRIC DISTORTIONS

In this section, we describe various geometric distortions that should be considered during watermark design. We start this section with the definition of the watermarking framework. Within the proposed framework, we discuss geometric distortions. Geometric distortions can be classified in two separate global distortions. One is the rigid-body distortion such as RST and the other is the non-rigid-body distortion, which includes shearing, projection, and general geometric distortions. In this section, we will focus only on the rigid-body distortions. We will provide a definition of each distortion with vector notation. Distortions are shown graphically by example figures. We also provide spectral aspects of the distortions. As most of the known watermarking systems are based on the Fourier transform, we will show the corresponding Fourier transform representation to distortions.

Watermarking Framework

Before we look into the geometric distortions that we are interested in, we first define the watermarking system and we will discuss them within this framework. Figure. 11.1 shows the general framework of a water-marking system. We have divided the watermarking system into three subsystems: Insertion, Distribution and use, and Extraction. During insertion, message m is encoded with a private key k. The encoded message w is added into the cover image c to generate a stego image s.

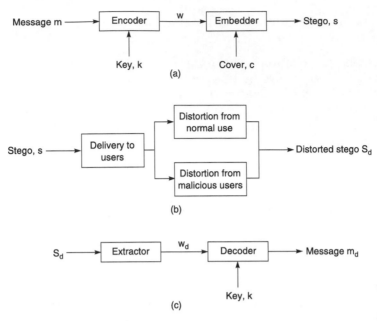

Figure 11.1. A framework of the general watermarking system: (a) embedder; (b) distribution and use; (c) extraction.

The watermarked image s is distributed to a user. Whereas faithful users may inadvertently change their images during their *normal* use of images, malicious users try to remove the watermark and defeat the detector by introducing well-planned distortions into the watermarked image. Among all of the possible distortions on the stego, we will deal only the geometric distortions. A user may try to crop out from s and rotate and resize to put it onto a Web page. A user may print the stego s and digitize the image using a scanner for other uses. These are the typical normal uses that fool the most of the watermark detectors.

The watermark detector should be able to detect the watermark from the distorted stego s_d. As is shown in Figure 11.1, we consider only blind detection, which means that the cover image is not available at the detector. If we are allowed to use the cover, we will be able to correct the distortions added to the cover image by a registration process and our problem will become a trivial case. Hence, we assume blind detection throughout this chapter.

Geometric Distortions

The most common geometric distortions are rotation, scale, and translation and they belong to rigid-body transformation in that objects in the image retain their relative shape and size. In this subsection, we discuss these rigid-body distortions and their impact on the watermark detector. The translated image $i'(x, y)$ can be expressed as

$$i'(x, y) = i(x + t_x, y + t_y) \tag{11.1}$$

A translation of a point $i(x_1, y_1)$ into a new point $i(x_1 + t_x, y_1 + t_y)$ is expressed with vector-space representation as

$$\begin{bmatrix} x_2 \\ y_2 \end{bmatrix} = \begin{bmatrix} x_1 \\ y_1 \end{bmatrix} + \begin{bmatrix} t_x \\ t_y \end{bmatrix} \tag{11.2}$$

Figure 11.2 shows some of the translated images. A watermark from the translated stego image can be detected using 2-D cross-correlation $c(x,y)$ as

$$c(x, y) = \sum_{\text{all } x} \sum_{\text{all } y} w'(x, y) \cdot w(x, y) \tag{11.3}$$

Figure 11.2.   Translated Lena images.

where w' is the distorted watermark after extraction and w is the provided watermark at detector. Using Equation 11.3, the translation can be identified by the location, where $c(x,y)$ shows the peak value. Translation can be most importantly characterized by a phase shift when the translated image is represented into the Fourier transform. This can be shown as

$$i(x + t_x, y + t_y) \Leftrightarrow I(u, v) \exp[j(ut_x + vt_y)] \qquad (11.4)$$

where $I(u, v)$ is the Fourier transform of the image $i(x,y)$. Hence, it is clear that the magnitude of the Fourier transform is invariant to translation in the spatial domain. This translation property of the Fourier transform is very useful, as we can use the Fourier magnitude spectrum whenever we need the translation invariance. Figure 11.3 shows the magnitude spectrum of the original Lena image and the translated Lena image. It verifies that the magnitudes of the Fourier transform of translated image are unchanged after translation.

Scaling of an image with a scaling factor s is expressed as

$$i'(x, y) = i(sx, sy) \qquad (11.5)$$

Note that scaling factors in the row and column directions are the same. It is because we are dealing with rigid-body distortion. Their vector-space representation is given as

$$\begin{bmatrix} x_2 \\ y_2 \end{bmatrix} = \begin{bmatrix} s & 0 \\ 0 & s \end{bmatrix} \begin{bmatrix} x_1 \\ y_1 \end{bmatrix} \qquad (11.6)$$

Figure 11.3. Magnitude Fourier transform of (a) original Lena image and (b) translated Lena image.

From the signal processing's viewpoint, a size change of an image means the change of the sampling grid. For enlarged images, sampling at a finer grid is required, and for reduced images, sampling at a coarse grid is performed. This sampling can be modeled as a reconstruction filter. Although the sampling method may affect the final image quality, it is not our interest in this section. If a watermark is to be correctly extracted from the reduced-sized image, it is usually scaled-up to a standard size. This resizing process is modeled as a spatial low-pass filtering operation and the details of the image will be blurred due to the process. The spatial domain characteristics of the scale operation can be better viewed using the log-polar mapping (LPM). The LPM is defined as the coordination change from the Cartesian coordinate (x,y) into the log-polar coordinate (μ,θ) and this can be achieved as follows:

$$x = e^{\mu} \cos \theta, \quad y = e^{\mu} \sin \theta \qquad (11.7)$$

The sampling grid of the LPM is shown as circles in Figure 11.4. The points are sampled at exponential grids along the radial direction. Figure 11.4b shows the LPM result of the Lena image. The vertical direction corresponds μ and the horizontal direction is the θ-axis. Figure 11.4c and Figure 11.4d show the LPM of two scaled Lena images with different scale ratio.

As expected, the LPM representation of the original Lena is shifted along the radial direction. After simple math, the scale in the Cartesian

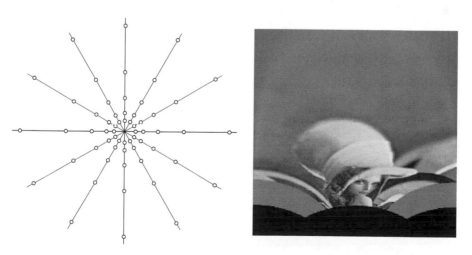

Figure 11.4. Log-polar mapping examples: (a) LPM sampling grid; (b) LPM of Lena; (c) LPM of 50% enlarged Lena; (d) LPM of 50% reduced-size Lena.

Figure 11.4. Continued.

coordinate can be expressed as a translation in the log-polar coordinate and this can be shown as follows:

$$(sx, sy) \Rightarrow (\mu', \theta)$$

$$\mu = \frac{1}{2} \ln(x^2 + y^2)$$

$$\mu' = \mu + \ln s \tag{11.8}$$

For the spectral aspects of the scaled image, we first look at the 1-D signal model and then we simply extend our discussion into 2-D image. The scale-change of an image can be viewed as the sampling rate change in 1-D sequences. For scaling down ($s < 1$), *decimation* is the related operation, and for scaling up ($s > 1$), *reconstruction* is the related 1-D operation. If $x(t)$ is the 1-D signal and its Fourier transform is $X(e^{jw})$, the decimation and reconstruction can be expressed in the spectrum domain as

$$x(t) \Leftrightarrow X(e^{jw})$$

$$x(st) \Leftrightarrow sX(e^{jw/s}) \tag{11.9}$$

The characteristics of scaling in 1-D Fourier transform is explained using a simple 1-D signal model. The Fourier magnitude spectrum of a 1-D signal $x(t)$ with its Nyquist frequency ϖ_N is shown in Figure 11.5a. When we decimate the signal, the spectrum is spread wide, as shown in Figure 11.5b. Note that as the decimation factor increases, the resulting spectrum approaches its Nyquist frequency. If the scaling down factor s

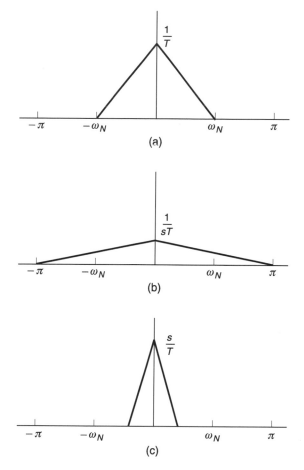

Figure 11.5. Decimation and reconstruction in a 1-D signal: (a) Fourier spectrum of the 1-D signal $x(t)$; (b) Fourier spectrum of a decimated 1-D signal; (c) Fourier spectrum of a scaled up and reconstructed 1-D signal.

becomes very small and, hence, if the resulting image shows aliasing, exact recovery of the original signal is impossible. When we reconstruct a 1-D signal at a finer sampling grid, the spectrum is contracted as shown in Figure 11.5c. The spectrum of the scaled 2-D image is the straightforward extension of the 1-D case and it is expressed as

$$i(sx, sy) \Leftrightarrow s^2 I\left(\frac{u}{s}, \frac{v}{s}\right) \tag{11.10}$$

The frequency spectrum is spread or contracted in the 2-D frequency plane in response to the scale down or scale up, respectively.

An image $i(x, y)$ rotated by $\alpha°$ is expressed as

$$i'(x, y) = i(x \cos \alpha + y \sin \alpha, \; -x \sin \alpha + y \cos \alpha) \tag{11.11}$$

in vector notation,

$$\begin{bmatrix} x_2 \\ y_2 \end{bmatrix} = \begin{bmatrix} \cos \alpha & -\sin \alpha \\ \sin \alpha & \cos \alpha \end{bmatrix} \begin{bmatrix} x_1 \\ y_1 \end{bmatrix} \tag{11.12}$$

The rotation in the spatial domain can be viewed as a translation in the log-polar domain. This can be expressed as

$$i'(\mu, \theta) = i(\mu, \theta + \alpha) \tag{11.13}$$

Figure 11.6 shows the log-polar-mapped images of Lena. The horizontal axis is the angular direction. As the image rotates, the LPM images are circularly rotated along the angular direction.

When an image is rotated, its Fourier spectrum is also rotated. This rotation property of the Fourier transform can be expressed as

$$I'(u, v) = I(u \cos \alpha + v \sin \alpha, \; -u \sin \alpha + v \cos \alpha) \tag{11.14}$$

Figure 11.7a shows the rotated Lena with $\alpha = 30°$ and Figure 11.7b shows its Fourier magnitude spectrum. Note artifacts in the spectrum on the bright "cross". As explained in Reference 12, this phenomenon results from the image boundaries, and any method using the rotation

Figure 11.6. Log-polar-mapped images of (a) original, (b) 92°, (c) 183°, and (d) 275° rotated Lena.

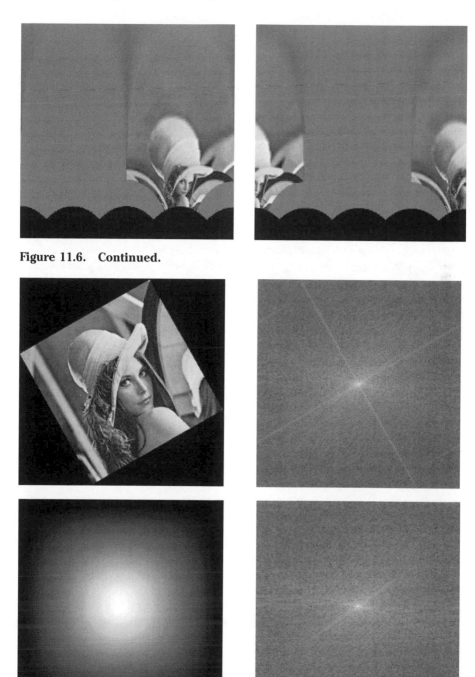

Figure 11.6. **Continued.**

Figure 11.7. **Fourier magnitude characteristics of rotated images: (a) no rotation; (b) after 30° rotation; (c) raised cosine window; (d) after windowing.**

property of the discrete Fourier transform (DFT) will suffer from these. This unwanted "cross" can be reduced by mitigating the strength of the boundary. If a circularly symmetric window pattern with smaller intensities along the image boundary is multiplied to the image, the artifacts can be reduced or removed. Figure 11.7c shows a window generated using the raised-cosine function. In Figure 11.7d, the Fourier spectrum after multiplying the window to the rotated Lena shows reduced intensity of the artifacts.

A RST-INVARIANT WATERMARKING METHOD [13]

In the previous section, we have reviewed the major geometric distortions. In this section, we explain how to implement a watermarking method resistant to such distortions. Our approach uses the invariant feature vector as the watermark and this makes our method robust to various distortions. Spatial and spectral domain understanding of the geometric distortions will be applied during the feature vector design and invariance implementation.

Bispectrum Feature Vector of the Radon Transform

The bispectrum, $B(f_1, f_2)$, of a 1-D deterministic real-valued sequence is defined as

$$B(f_1, f_2) = X(f_1)X(f_2)X^*(f_1 + f_2) \tag{11.15}$$

where $X(f)$ is the discrete-time Fourier transform of the sequence $x(n)$ at the normalized frequency f. By virtue of its symmetry properties, the bispectra of a real signal is uniquely defined in the triangular region of computation, $0 \leq f_2 \leq f_1 \leq f_1 + f_2 \leq 1$.

A 2-D image is decomposed into N 1-D sequences $g(s, \theta)$ using the Radon transform. The Radon transform $g(s, \theta)$ of a 2-D image $i(x, y)$ is defined as its line integral along a line inclined at an angle θ from the y-axis and at a distance s from the origin and it is shown as follows:

$$g(s, \theta) = \int_{-\infty}^{\infty} \int^{\infty} i(x, y)\delta(x \cos \theta + y \sin \theta - s) \, dx \, dy,$$
$$-\infty < s < \infty, 0 \leq \theta < \pi \tag{11.16}$$

Figure 11.8 shows the line integral procedure and an example the Radon transform of Lena. The projection slice theorem [14] states that the Fourier transform of the projection of an image onto a line is the 2-D Fourier transform of the image evaluated along a radial line. From the theorem, we can use 2-D Fourier transform instead of the Radon transform during implementation.

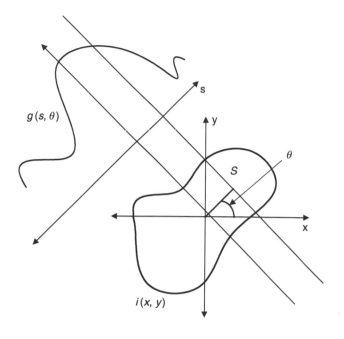

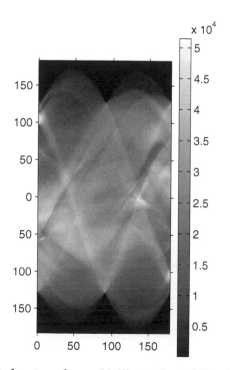

Figure 11.8. **The Radon transform: (a) illustration of line integral; (b) Radon transform of Lena.**

343

A parameter $p(\theta)$ is defined as the phase of the integrated bispectra of a 1-D Radon projection $g(s,\theta)$ along the line of $f_1 = f_2$ and it can be expressed with the polar mapped 2-D DFT, $I(f,\theta)$, as follows using the projection slice theorem:

$$p(\theta) = \angle \left[\int_{f_1=0^+}^{0.5} B(f_1, f_1)\, df_1 \right]$$

$$= \angle \left[\int_{f_1=0^+}^{0.5} I^2(f,\theta) I^*(2f,\theta)\, df \right] \tag{11.17}$$

Although the parameter can be defined along a radial line of slope a, $0 < a \leq 1$ in the bifrequency space, we compute $p(\theta)$ at $a = 1$, where $f_1 = f_2$. In this way, we can avoid interpolation during the computation of $p(\theta)$.

A vector $\mathbf{p}$ of length N is defined as $\mathbf{p} = (p(\theta_2), p(\theta_2), \ldots, p(\theta_N))$. From the properties of the Radon transform and bispectrum parameter $p(\theta)$, $\mathbf{p}$ can be shown to be invariant to geometric distortions.

The translation invariance of the vector $\mathbf{p}$ can be shown using the translation property of the Fourier transform and polar mapping. The Fourier transform of the translated image $i'(x,y)$ becomes Equation 11.4. We change the coordinate from Cartesian to polar coordinate and Equation 11.4 is transformed to

$$I'(f,\theta) = I(f,\theta) e^{jf(x_t \cos \theta + y_t \sin \theta)} \tag{11.18}$$

where $I(f,\theta)$ is the polar-mapped Fourier transform of the original image. From Equation 11.17, the parameter $p'(\theta)$ becomes

$$p'(\theta) = \angle \left[\int_{0^+}^{0.5} [I'(f,\theta)]^2 [I'(2f,\theta)]^*\, df \right]$$

$$= \angle \left[\int_{0^+}^{0.5} I^2(f,\theta) I^*(2f,\theta)\, df \right]$$

$$= p(\theta) \tag{11.19}$$

Because $p'(\theta) = p(\theta)$, the vector $\mathbf{p}$ is invariant to translation.

A scaled image $i'(x,y)$ is expressed as Equation 11.5 and its polar-mapped Fourier transform is shown as

$$I'(f,\theta) = sI\left(\frac{f}{s}, \theta\right) \tag{11.20}$$

Assuming $f/s < 0.5$, the parameter $p'(\theta)$ can be expressed as

$$p'(\theta) = \angle\left[\int_{0^+}^{0.5} s^3 I^2\left(\frac{f}{s},\theta\right)I^*\left(\frac{2f}{s},\theta\right)df\right]$$

$$= \angle\left[\int_{0^+}^{0.5/s} I^2\left(\frac{f}{s},\theta\right)I^*\left(\frac{2f}{s},\theta\right)df\right]$$

$$= \angle\left[\int_{0^+}^{0.5} I^2(f,\theta)I^*(2f,\theta)\,df\right]$$

$$= p(\theta) \tag{11.21}$$

Because $p'(\theta)=p(\theta)$, the vector **p** is invariant to scale.

A rotated image $i'(x,y)$ by $\alpha°$ is expressed as Equation 11.11 and its Fourier transform is shown in Equation 11.14. The polar-mapped Fourier transform of the rotated image $i'(x,y)$ can be expressed as

$$I'(f,\theta) = I(f,\theta - \alpha) \tag{11.22}$$

Then, the parameter $p'(\theta)$ becomes

$$p'(\theta) = \angle\left[\int_{0^+}^{0.5} I^2(f,\theta-\alpha)I^*(2f,\theta-\alpha)\,df\right]$$
$$= p(\theta - \alpha) \tag{11.23}$$

Hence, the vector **p** will be circularly shifted by α.

Watermark System Design

We use a modified feature vector of an image as the watermark. The watermark is embedded by selecting a vector from the set of extracted feature vectors. The chosen feature vector is used as the watermark and the inverted image is used as the watermarked image. The watermarks are generated through an iterative feature modification and verification procedure. This procedure avoids the interpolation errors that can occur during insertion and detection of the watermark. At the detector, the feature vector is estimated from the test image. We use root-mean-square error (RMSE) as our similarity measure instead of the normalized correlation. It is because the feature vectors are not white and the correlation measure cannot produce a peak when they are the same vectors. Hence, we measure the distance between the two vectors using the RMSE function. If the RMSE value is smaller than a threshold, the watermark is detected. The original image is not required at the detector. We define the detector first and an iterative embedder is designed using the detector.

Watermark Detection. Given a possibly corrupted image $i(x, y)$, the $N \times N$, 2-D DFT is computed with zero padding:

$$I(f_1, f_2) = \mathrm{DFT}\{i(x, y)\} \tag{11.24}$$

The $M \times N$ polar map $I_p(f, \theta)$ is created from $I(f_1, f_2)$ along N evenly spaced θ's in $\theta = 0°, \ldots, 180°$ and it is shown as

$$I_p(f, \theta) = I(f \cos \theta, f \sin \theta) \tag{11.25}$$

where

$$f = \sqrt{f_1^2 + f_2^2}$$
$$\theta = \arctan(f_2/f_1) \tag{11.26}$$

The frequency f is uniformly sampled along the radial direction, and bilinear interpolation is used for simplicity. This is equivalent to the 1-D Fourier transform of the Radon transform. The $p(\theta)$ is computed along the columns of $I_p(f,\theta)$ by Equation 11.17 to construct the feature vector $\mathbf{p}$ of length N. The similarity s is defined with the RMSE between the extracted vector $\mathbf{p}$ and the given watermark $\mathbf{w}$ as follows:

$$s(\mathbf{p}, \mathbf{w}) = \sqrt{\frac{1}{N} \sum_{i=1}^{N} [p(\theta_i) - w_i]^2} \tag{11.27}$$

where N is the length of the feature vector. If s is smaller than the detection threshold T, the watermark is detected. Figure 11.9 shows the detection procedure.

Watermark Embedding. Let $i(x, y)$ be a $N \times N$ grays-cale image. We compute the 2-D DFT $I(f_1, f_2)$ of the image. The $M \times N$ polar map $I_p(f, \theta)$ is created from $I(f_1, f_2)$ along N evenly spaced θ's in $\theta = 0°, \ldots, 180°$ as Equation 11.25 and Equation 11.26. The feature vector $\mathbf{p} = (p(\theta_1), p(\theta_2), \ldots, p(\theta_N))$, is computed as Equation 11.17. The watermark signal is embedded by modifying k elements of the vector. From a pseudorandom number generator, the number of modifications, k, and the insertion angles θ_w are determined to select projections for embedding at $\theta_w \in [0° \cdots 180°)$. If we shift all of the phases of a column

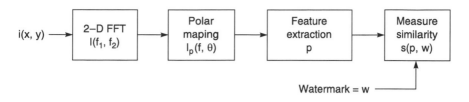

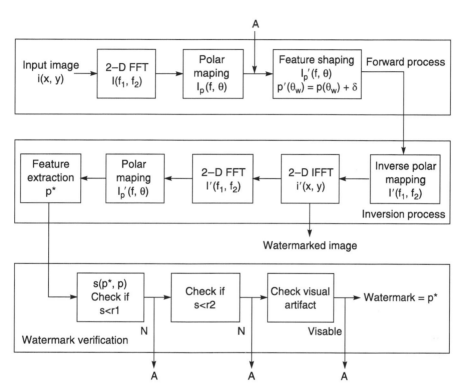

Figure 11.9. Watermarking system: (a) detection procedure; (b) insertion procedure.

θ_w of I_p by δ, we have a modified component $p'(\theta_w)$ as follows:

$$p'(\theta_w) = \angle\left[\int_{f_1=0^+}^{0.5} I_p(f_1)e^{j\delta}I_p(f_1)e^{j\delta}I_p^*(f_1+f_1)e^{-j\delta}\,df_1\right]$$

$$= \angle\left[\int_{f_1=0^+}^{0.5} I_p(f_1)I_p(f_1)I_p^*(f_1+f_1)\,df_1\right] + \delta$$

$$= p(\theta_w) + \delta \tag{11.28}$$

347

After shifting the phases of the selected columns of $I_p(f,\theta)$, we inverse-transform it to have the watermarked image i'.

However, we cannot extract the exact embedded signal at detector. As reported in previous work [6,7], algorithms that modify the Fourier coefficients in polar or log-polar domains experience three problems. First, interpolation at the embedder causes errors at the detector. During polar or log-polar mapping, an interpolation method should be involved because we are dealing with discrete image data. Although we choose a more accurate interpolation function, there will be some errors as long as we are working with discrete images. Second, zero-padding at the detector degrades the embedded signal further. By zero-padding, spectrum resolution is improved, but the interpolation error is increased. Third, the interrelation of the Fourier coefficients in the neighboring angles causes "smearing" of the modified feature values. If we modify a single element $p(\theta)$, it affects other values nearby. In Reference 7, the authors have provided approximation methods to reduce the effects of these errors. Instead of using a similar method, we approach this problem differently. After modifying some elements of the feature vector, the watermarked image that contains the implementation errors is produced by the inverse 2-D DFT. We apply the feature extractor from this watermarked image and use the extracted feature $\mathbf{p}^*$ as the embedded watermark instead of the initially modified feature vector $\mathbf{p}'$. In this way, we can avoid the inversion errors.

However, to guarantee the uniqueness of the watermark and its perceptual invisibility after insertion, we need a validation procedure to use $\mathbf{p}^*$ as a watermark. We empirically measure the maximum noise level $r1$ resulting from geometric distortions with the detector response s as follows:

$$r1 = \max\{s(\mathbf{p}_i, \mathbf{q}_i)|i = 0, \ldots, M\} \qquad (11.29)$$

where $\mathbf{p}_i$ is the feature vector of an image and $\mathbf{q}_i$ is the feature vector of the image after RST distortions. We should adjust the embedding strength so that the distance between the two features from the unmarked and the marked image show a value higher than $r1$. However, the embedding strength cannot be higher than $r2$, which defines the minimum distance between features of the images:

$$r2 = \min\{s(\mathbf{p}_i, \mathbf{p}_j)|i \neq j, \text{ and } i,j = 0, \cdots M\} \qquad (11.30)$$

where $\mathbf{p}_i$ and $\mathbf{p}_j$ are the feature vectors of different images.

We preset $r1$ and $r2$ values empirically. With varying the embedding strength δ, k, and θ_w, it is checked if $r1 < s(\mathbf{p}, \mathbf{p}^*) < r2$. If this condition is satisfied and the embedded signal is unobtrusive, $\mathbf{p}^*$ is accepted as a watermark. We repeat this validation procedure until we get the right result. In this way, we can embed the watermark without exact inversion of the modified signal. Figure 11.9b shows the watermark embedding procedure.

EXPERIMENTAL RESULTS

For valid watermark generation, $r1$ and $r2$ are determined empirically using unmarked images. The similarity s is measured between unmarked test images, and the smallest s is chosen for $r2$. For the determination of $r1$, robustness of the defined feature vector is tested. We used the Stirmark [2] to generate attacked images. Similarly, s is measured between the original image and attacked images. The largest s is chosen for $r1$. For our implementation, we set $r1 = 4.5$ and $r2 = 20$.

Feature vectors are modified with $\delta = 5°-7°$ at randomly selected angles. The number of insertion angles is randomly determined between 1 and 3. A threshold $T = 4.5$ is used for the detection threshold. Watermarks are generated using the iterative procedure described in section "Watermark System Design". During the iteration, parameters are adjusted accordingly. Figure 11.10a shows the watermarked Lena image and Figure 11.10b shows the amplified difference between original and watermarked images. The watermarked image shows a PSNR of 36 dB and the embedded signal is invisible. During the watermark insertion, we maintained the PSNR of the watermarked images higher than 36 dB.

Figure 11.10. Embedding example: (a) watermarked Lena; (b) embedded watermark.

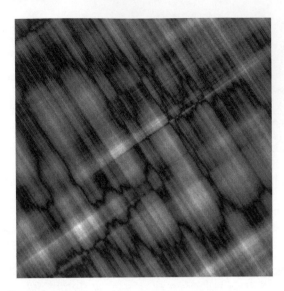

Figure 11.10. Continued.

Table 11.1. Stirmark score

	Proposed Approach	Digimarc	Suresign
Scaling	1.0	0.72	0.95
Small-angle rotation and cropping	0.95	0.94	0.5
Random geometric distortion	0.93	0.33	0
JPEG compression	1.0	0.81	0.95

Two experiments are performed for the demonstration of the robustness. Using images of Lena, a mandrill, and a fishing boat, the watermark detection ratio is measured as implemented by the Stirmark benchmark software. The second experiment is performed to estimate the false-positive (P_{fp}) and false-negative probability (P_{fn}) with 100 images from the Corel image library [14].

Table 11.1 shows the watermark detection ratio using the Stirmark benchmark tool. The ratio 1 means 100% detection success and 0 means the complete detection failure. Against scaling, the watermark is successfully detected in 50% scaling down. For small-angle rotations in the range $-2°$ to $2°$, the watermark is successfully detected without synchronization procedure. It shows that our method outperforms the other commercially available algorithms against geometric attacks. We refer to the results of other methods presented in Reference 12.

Robustness of the watermark against each attack is measured with 100 unmarked images and 100 marked images. We measure the empirical

Table 11.2. **Estimated false-positive and false-negative probabilities**

Distortion	False-Positive Probability	False-Negative Probability
Rotation	3.36×10^{-2}	2.3×10^{-3}
Scaling	3.5×10^{-6}	2.21×10^{-6}
Random geometric attack	7.89×10^{-2}	2.90×10^{-3}
Compression	2.8×10^{-3}	2.2×10^{-20}
Gaussian noise	6.64×10^{-15}	2.85×10^{-3}

probability density function (pdf) of the computed s using a histogram. Although we do not know the exact distribution of s, we approximate the empirical pdf of s to the Gaussian distribution to show a rough estimation of the robustness. The false-positive and false-negative probabilities can be computed using the estimates of mean and variance. The resulting estimated errors are shown in Table 11.2. Random geometric attack performance is the worst with $P_{fp} = 7.89 \times 10^{-2}$ and $P_{fn} = 2.90 \times 10^{-3}$. This attack simulates the print-and-scanning process of images. It applies a minor geometric distortion by an unnoticeable random amount in stretching, shearing, or rotating an image [2]. It shows that our method performs well over the intended attacks. The similarity histograms and receiver operating characteristic (ROC) curves (P_{fp} vs. P_{fn} for several thresholds) are produced for analysis.

Figure 11.11 shows the histogram of s and ROC curve against rotation. Although the rotation by large angle can be detected by cyclically shifting the extracted feature vector, the performance of rotation by a large angle is poor due to the difficulty of interpolation in the Fourier phase spectrum. For this reason, we show the results of rotation by small angles. This problem is discussed in section "Discussions". With $T = 4.5$, P_{fp} is 3.36×10^{-2} and P_{fn} is 2.30×10^{-3}. False-negative probability shows better performance than false-positive probability in this attack. This is because the pdf of the similarity between unmarked images and watermarks has a relatively large variance that resulted into the larger false-positive probability. As P_{fp} and P_{fn} show, our method is robust against rotation by small angles.

Figure 11.12 shows detector performance after scaling distortion. The detection histogram was measured using 50% scaled-down images and 200% scaled-up images. As the histogram in Figure 11.12a shows, the watermarked images show strong resistance to scaling attack. Figure 11.12b shows that P_{fp} is 3.5×10^{-6} and P_{fn} is 2.21×10^{-6}. These values are relatively lower than other attacks and this means that our method performs well with scaling attacks. Our method has strong robustness against scaling attack even after scaling down to 50%.

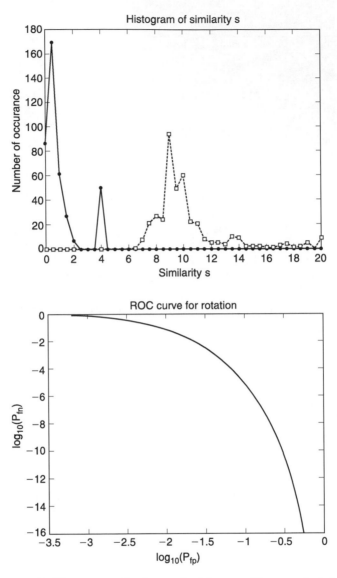

Figure 11.11. (a) Histogram of s (•, marked image; □ unmarked image) and (b) receiver operating characteristic (ROC) curve of unmarked and marked image after rotation ($\pm 0.25°$, $\pm 0.5°$).

Figure 11.13 shows detector performance after random geometric distortion. In Figure 11.13a, the histogram shows large variance in the similarity between watermark and unmarked image. As the result, P_{fp} is 7.89×10^{-2} and P_{fn} is 2.90×10^{-3}, which are relatively large compared with others. Not many previous methods survive this attack and our algorithm works well even with those numbers.

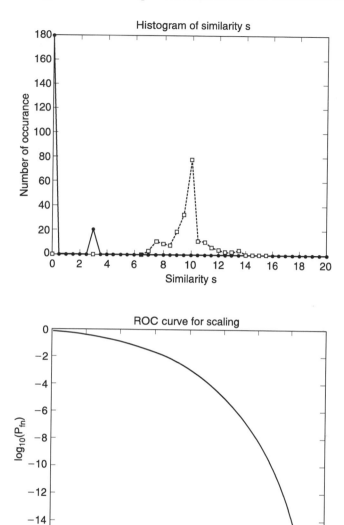

Figure 11.12. (a) **Histogram of** s (•, **marked image;** □ **unmarked image) and** (b) **ROC curve of unmarked and marked image after scaling** ($\times 0.5$, $\times 2.0$).

DISCUSSIONS

Rotation Invariance

Because we use the rotation property of DFT for rotation invariance, we need to employ methods that can compensate for the problems identified in the literature [7,15]. For algorithms that use the Fourier magnitude

353

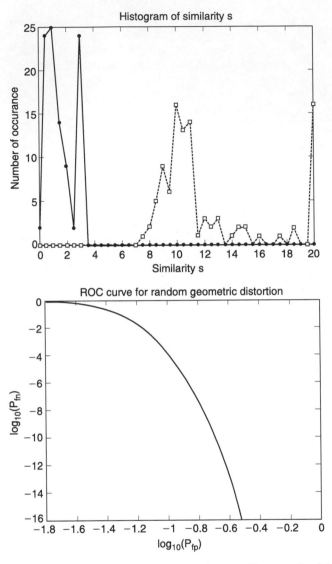

Figure 11.13. (a) **Histogram of** s (•, **marked image;** □ **unmarked image) and** (b) **ROC curve of unmarked and marked image after a random geometric attack.**

spectrum, zero-padding and windowing show the required rotation property. Zero-padding, centered padding, or side padding show no difference. This is because the magnitude spectrum is invariant to the shift resulting from the zero-padding. Symmetric 2-D windowing removes the cross artifact in the frequency domain. Windows such as Hanning, Hamming, and Kaiser reduce the effect of the image boundary, keeping the signal loss low.

For the methods using phase spectrum, such as our algorithm, zero-padding and windowing are not as effective as with the magnitude spectrum. It is because the phase spectrum changes more rapidly than the magnitude spectrum in the frequency domain. Direct interpolation in the frequency domain can solve this problem. The jinc function, which is the circular counterpart of the sinc function in the 2-D spectrum domain, is preferable for the interpolation function.

Complexity

The embedding algorithm requires computation of two fast Fourier transforms (FFTs) (one to embed a watermark signal and one to find the actual watermark) and one inverse FFT (IFFT) (the second FFT does not need to be inversed). Two polar mappings and one inverse polar mapping are required. Computations for the determination of thresholds are not considered, as they are one-time operations. The extraction algorithm requires computation of one FFT and one polar mapping. The embedding algorithm takes 15–20 sec, whereas detection takes 5–7 sec on a Pentium 1 GHz with the Mathworks' Matlab [16] implementation. The polar and inverse polar mapping consumes most of the computation time. The valid watermark embedding was achieved after one or two iteration most of the time.

Capacity

Current implementation works for the zero-bit watermark. With simple modification, we can embed more bits. After the iterative procedure to generate the valid watermarks, we can construct a code book such as "dirty-paper code" [17]. Information is assigned to each valid watermark code during embedding. At the detector, the code book is implemented within the detector. During detection, the detector compares extracted feature with the vectors registered in the codebook. When a measured similarity value reaches a previously determined threshold, it shows the assigned information from the codebook.

Embedding without Exact Inversion

If an embedding function does not have an exact inversion function, the resulting watermarked image will be distorted. This distortion reduces the image fidelity and watermark signal strength. As argued in Reference 7, having an exact inversion is not a necessary condition for the embedding function. Two approaches can be considered. One method is defining a set of invertible vectors and working only with those vectors during the embedding procedure. Although the embedding space is reduced, exact inversion is possible. Another approach is to use a conversion function that maps the embedded watermark and the

355

extracted vector. Our approach belongs to this category. At the detector, after estimation of the watermark, this signal is mapped into the inserted watermark using the conversion function.

CONCLUSION

In this chapter, we have proposed a new RST-invariant watermarking method based on an invariant feature of the image. We have overviewed the properties of RST distortion in various aspects. Based on understanding of geometric distortions, we have designed a watermarking system that is robust against geometric distortions.

A bispectrum feature vector is used as the watermark and this watermark has a strong resilience to RST attacks. This approach shows the potential in using a feature vector as a watermark. An iterative embedding procedure is designed to overcome the problem of inverting the watermarked image. This method can be generalized for other embedding functions that do not have an exact inverse function.

In the experiments, we have shown the comparative Stirmark benchmark performance and the empirical probability density functions with histograms and the ROC curves. Experimental results show that our scheme is robust against the designed attacks. The use of the bispectrum feature as an index for an efficient watermarked image database search may offer new application possibilities. Various embedding techniques and capacity issues for the generic feature-based watermark system should be further researched.

Throughout this chapter, we have looked at the geometric distortions in 2-D images and their impact on watermarking system. Those rigid-body distortions explained in this chapter are only a small fraction of the whole class of geometric distortion. Nonrigid distortions such as shear, projection, and general linear distortions are more difficult and yet to be solved. We believe that there will be no single method that could survive all of the known distortions. Instead, watermark developers should tailor each method according to their application. Only in that way will sufficiently robust solutions be provided into the commercial market.

REFERENCES

1. Hartung, F., and Kutter, M., Multimedia watermarking technique, *Proc. IEEE* 87, 1079, 1999.
2. Petitcolas, F.A.P., Anderson, R.J., and Kuhn, M.G., Attacks on Copyright Marking Systems, in Proceedings 2nd International Workshop on Information Hiding, 1998, p. 218.

3. Pereira, S., and Pun, T., Robust template matching for affine resistant image watermarks, *IEEE Trans. Image Process.*, 9, 1123, 2000.
4. Csurka, G., Deguillaume, F., O'Ruanaidh, J.J.K., and Pun, T., A Bayesian Approach to Affine Transformation Resistant Image and Video Watermarking, in Proceedings 3rd International Workshop on Information Hiding, 1999, p. 315.
5. Kutter, M., Watermarking Resisting to Translation, Rotation, and Scaling, in Proceedings SPIE Multimedia Systems Applications, 1998, p. 423.
6. O'Ruanaidh, J.J.K., and Pun, T., Rotation, scale, and translation invariant spread spectrum digital image watermarking, *Signal Process.*, 66, 303, 1998.
7. Lin, C.Y., Wu, M., Bloom, J.A., Cox, I.J., Miller, M.L., and Lui, Y.M., Rotation, scale, and translation resilient watermarking for images, *IEEE Trans. Image Process.*, 10, 767, 2001.
8. Kutter, M., Bhattacharjee, S.K., and Ebrahimi, T., Towards Second Generation Watermarking Schemes, in Proceedings IEEE International Conference on Image Processing, 1999, p. 320.
9. Guoxiang, S., and Weiwei, W., Image-feature based second generation watermarking in wavelet domain, in Lecture Notes in Computer Science, Vol. 2251, Springes-Verlag, Berlin, p. 2001.
10. Chandran, V., Carswell, B., Boashash, B., and Elgar, S., Pattern recognition using invariants defined from higher order spectra: 2-D image inputs, *IEEE Trans. Image Process.*, 6, 703, 1997.
11. Ruanaidh, J.O., Dowling, W.J., and Boland, F.M., Phase watermarking of digital images, in Proceedings IEEE International Conference on Image Processing, 1996, p.239.
12. Lin, C.Y., Wu, M., Bloom, J.A., Cox, I.J., Miller, M.L., and Lui, Y.M., Rotation, scale, and translation resilient watermarking for images, *IEEE Trans. Image Process.*, 10, 767, 2001.
13. Kim, H.S., Baek, Y.J., and Lee, H.K., Rotation, scale, and translation invariant image watermark using higher order spectra, *Opti. Eng.*, 42, 340, 2003.
14. Jain, A.K., Image reconstruction from projections, in *Fundamentals of Digital Image Processing*, Prentice-Hall, Englewood Cliffs, NJ, 1989, p. 431.
15. Altmann, J., On the digital implementation of the rotation-invariant Fourier–Mellin transform, *J. Inf. Process. Cybern.*, EIK-28(1), 13, 1987.
16. MATLAB, The MathWorks, Inc.
17. Miller, M.L., Watermarking with dirty-paper codes, in Proceedings IEEE International Conference on Image Processing, 1999, p. 538.
18. Corel Corporation, Corel Stock Photo Library 3.

12

Fragile Watermarking for Image Authentication

Ebroul Izquierdo

INTRODUCTION

In an age of pervasive electronic connectivity, hackers, piracy and fraud, authentication and repudiation of digital media is becoming more important than ever. Authentication is the process of verification of the genuineness of an object or entity in order to establish its full or partial conformity with the original master object or entity, its origin, or authorship. In this sense, the authenticity of photographs, paintings, film material, and other artistic achievements of individuals have been preserved, for many years, by recording and transmitting them using analog carriers. Such preservation is based in the fact that the reproduction and processing of analog media is time-consuming, involves a heavy workload, and leads to degradation of the original material. This means that content produced and stored using analog devices has an in-built protection against unintentional changes and malicious manipulations. In fact, conscious changes in analog media are not only difficult, but they can be easily perceived by a human inspector. As a consequence, the authenticity of analog content is inherent to the original master picture. In this electronic age, digital media has become pervasive, completely substituting its analog counterpart. Because affordable image processing tools and fast transmission mechanisms are available everywhere, visual media can be copied accurately, processed, and distributed around the world within seconds. This, in turn, has led to an acute need to protect images and other digital media from fraud and tampering and a heightened awareness of the necessity of automatic tools to establish the authenticity and integrity of digital content. Indeed, creators, legitimate distributors, and end users enjoy the flexibility and user-friendliness of digital processing tools and networks to copy, process, and distribute their content over open digital channels at electronic speed. However, they also need to guarantee that material used or being published at the end of the distribution chain is genuine. Digital content and, specifically, image authentication can be defined as a procedure to verify that a given image is either an identical copy of an original or, at least, it has not been altered in order to convey a different meaning.

In the context of transmission and distribution of genuine digital images across open networks using distributed digital libraries two main threats can be identified:

Masquerade: transformation of an original image into another with similar content but conveying a totally different meaning. In Figure 12.1, the car license plate has been replaced by a different

Figure 12.1. (a) Original image and (b) tampered version.

one. The tampered image may convey a different meaning and could be used, for instance, to confuse evidence in forensic applications.

Modification: The original image is transformed by cropping, swapping areas, replacing portions of the image with content from other images, or applying image transformations to change the original image structure. In Figure 12.1, the right part of the original image has been replaced by a similar area cropped from another image. Although, the meaning may remain the same, the swapped area may contain relevant content.

Measures to deal with these threats have largely found a solution in cryptography, which guarantees the integrity of general content transmission by using digital signatures with secret keys. Cryptographic techniques consist of authentication functions and an authentication protocol that enables the end user to verify the authenticity of a given image. Usually, three types of mechanism can be used to produce authentication functions: full encryption, in which the ciphertext generated from the image file is the authenticator; message authentication codes consisting of a public function and a secret key that enable the authenticator to generate an authentication value; and hash functions, which are public transforms mapping the original content into a fixed-length hash value. Because digital images are bulky in terms of the number of bits needed to represent them, authentication is usually performed using short signatures generated by the owner of the original content. The authentication mechanism is applied to the image signature and the authentication protocol acts on the encrypted digital signature. Using digital signatures, the most popular authentication method is based on the classic protocol introduced by Diffie and Hellman in 1976 [1]. The digital signature is generated by the transmitter depending on secret information shared with the receiver. The digital signature is used to verify the image authenticity, which is endorsed by the sender. Digital signatures can be inserted in the header of an image file assuming that the header, or at least the part of the header where the signature is embedded, remains intact through all transmission stages.

The main drawback of traditional cryptosystems is that they do not permanently associate cryptographic information with the content. Cryptographic techniques do not embed information directly into the message itself, but, rather, hide a message during communication. For several applications, no guarantees can be made by cryptography alone about the redistribution or modification of content once it has passed through the cryptosystem. To provide security using signatures embedded directly in the content, additional methods need to be considered. Techniques that have been proposed to address this problem belong to a more general class of methods known as digital watermarking [2–9].

Although digital watermarking can be used for copyright protection [3–7], data hiding [8], ownership tracing [9], authentication [10–12], and so forth, in this chapter, we are only concerned with watermarking addressing image authentication applications.

CONVENTIONAL WATERMARKING FOR IMAGE AUTHENTICATION

According to the objectives of the targeted authentication application, watermarking techniques can be classified into two broad categories: semifragile and fragile. Semifragile watermarking addresses content verification assuming that some image changes are allowed before, during, or after transmission and the transformed image is still regarded as genuine. Fragile watermarking is used for complete verification assuming the content is untouchable; that is, the image received by the end user has to be exactly the same as the original one.

Semifragile watermarks are designed to discriminate between expected image changes in most cases due to application constraints (e.g., compression to meet bandwidth requirements) and intentional image tampering. These techniques provide an additional functionality for specific application scenarios [11–13]. The objective in this case is to verify that the "essential characteristics" of the original picture have not been altered, even if the bit stream has been subject to changes while traveling the delivery chain, from producer to end user. The underlying assumption in the authentication process is that the meaning of the digital picture is in the abstract representation of the content and not in the low-level digital representation encapsulated in the bit streams. In some cases, manipulations on the bit streams without altering the meaning of content are required by the targeted application. As a consequence, these changes should be considered as tolerable, rendering the changed digital image as authentic. In Figure 12.2, an ideal

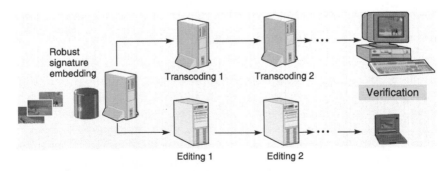

Figure 12.2. Semifragile watermarking for content verification. In an ideal case, the image is endorsed by the producer and authenticated in the last stage by the end user.

watermarking scheme is depicted in which the original image is endorsed by the producer, undergoes several "natural" nonmalicious changes, and is authenticated by the end user only once in the last stage. Unfortunately, it is impossible for an automatic authenticator to know the objective of an arbitrary alteration of the image. For that reason, the ideal universally applicable semifragile watermarking method will remain an open paradigm. However, if the targeted alteration is fixed in the model, authenticators can be designed according to the characteristics of the expected manipulation.

Compression is probably the best example of such changes. To satisfy distribution and broadcasting requirements, storage and transmission should be carried out using compressed digital images. Because lossy compression changes the original image, an authenticator robust to compression should consider lossy encoded digital images as genuine. One representative watermarking technique belonging to this class was reported by Lin and Chang in the late 1990s [14,15]. The proposed authentication approach exploits invariance properties of the discrete cosine transform (DCT) coefficients, namely the relationship between two DCT coefficients in the same position in two different blocks of a JPEG compressed image. It is proven that this relationship remains valid even if these coefficients are quantized by an arbitrary quantization table according to the JPEG compression process. Their technique exploits this invariance property to build an authentication method that can distinguish malicious manipulations from transcoding using JPEG lossy compression.

Fragile watermarking treats any manipulation of the original image as unacceptable and addresses authentication with well-defined tamper-detection properties. Most applications are related to forensic and medical imaging, in which the original content is regarded as essential and no alterations can be tolerated. Another application scenario relates to the content creator — for example, the photographer Alice and her customers — the magazine editor Bob. Alice stores her pictures in an open image database accessible over the Web. Each time Bob downloads a picture from Alice's database to illustrate the stories in his magazine, he expects to be using genuine images. However, because Alice's database and the transmission channel are open, it can happen that a malicious person intercepts transmissions and changes the images for forgeries misrepresenting the original pictures. If Alice's images have been watermarked, Bob can verify the authenticity of received images using a secret verification key shared by Alice and himself. A discussion on other related application scenarios can be found in Reference 13. Figure 12.3 outlines the authentication process using fragile watermarking along the complete delivery chain.

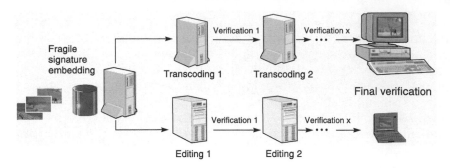

Figure 12.3. Fragile watermarking for complete image verification: the image is authenticated after each editing, transcoding, and transmission stage.

One early technique for image authentication was proposed by Walton in Reference 16. He uses a check-sum built from the seven most significant bits, which is inserted in the least significant bit (LSB) of selected pixels. Similar techniques based on LSB modification were also proposed in References 17 and 18. These LSB techniques have different weaknesses related to the degree of security that can be achieved. A well-known fragile watermarking algorithm was proposed by Yeung and Mintzer in Reference 19. The Yeung–Mintzer algorithm uses the secret key to generate a unique mapping that randomly assigns a binary value to gray levels of the image. This mapping is used to insert a binary logo or signature in the pixel values. Image integrity is inspected by direct comparison between the inserted logo or signature and the decoded binary image. The main advantage of this algorithm is its high localization accuracy derived from the fact that each pixel is individually watermarked. However, the Yeung–Mintzer algorithm is vulnerable to simple attacks, as shown by Fridrich in Reference 13. Another well-known scheme for image authentication was proposed by Wong [20]. It embeds a digital signature extracted from the most significant bits of a block in the image into the LSB of the pixels in the same block. One common feature of these and other techniques from the literature is that authentication signatures are embedded in the image content, either in the pixel or a transform domain, and the security of the schemes resides in a hash or encryption mechanism. Although these authentication methods use watermarking technology to embed signatures in the content, the security aspects are tackled by borrowing from conventional cryptography strategies based on message authentication codes, authentication functions, and hash functions. The technique presented in the next section was reported by Izquierdo and Guerra in Reference 21 and it is essentially different from other authentication methods reported in the literature. It is based on the inherent instability property of inverse ill-posed problems and the fact that slight changes in the input data cause huge changes in any approximate solution. Singular-valued decomposition and other fundamental linear algebra tools are used to construct a highly ill-conditioned matrix interrelating the original image and the watermark signal. This is achieved by exploiting the relation among singular values, the least square solution of linear algebraic equations, and the high instability of linear ill-posed operators.

AN ILL-POSED OPERATOR FOR SECURE IMAGE AUTHENTICATION

The following analysis leads to a secure watermarking approach performed on blocks of pixels extracted from the original image. Each single block is regarded as a generic matrix of intensity values. Likewise, the watermarking signal is considered as a matrix of real positive numbers. A main difference between the proposed watermarking approach and others from the literature is that the watermark signal is embedded in the original image in a different manner, and conventional cryptography strategies based on message authentication codes, authentication functions, or hash functions are avoided. First, the smallest singular value of the matrix to be watermarked is used to artificially create a minimization problem. The solution to this problem involves a least squares approximation of the previously defined ill-posed operator in order to find an unknown parameter. The solution process links the watermark with the image using the underlying ill-posed operator. An image block is considered watermarked by setting its smallest singular value equal to the parameter estimated from the minimization task. Thus, the watermark is spread over the whole image in a subtle but quite complex manner. A major advantage of this scheme is that the distortion induced by the watermarking procedure can be strictly controlled because it depends only on changes in the smallest singular values of each block. The verification procedure solves the same optimization problem and compares the norm of the solution with a large secret number. In this section, this unconventional authentication approach will be described in detail.

Because the watermarking and verification algorithms are based on few theoretical statements, the first part is dedicated to a brief overview of singular-value decomposition of matrices and linear ill-possed operators. In the remainder of the section the authentication technique is elaborated. In the next section, important security issues are discussed. This involves analysis of the algorithm vulnerability to attacks targeting the recovery of some information about the secret authentication key, assuming that the attacker has access to a library of images authenticated with the same key. Furthermore, it is shown that the watermarking scheme is capable of localizing tampering accurately and that it is less vulnerable to the sophisticated vector quantization counterfeiting or the popular cropping attack than techniques using independent blockwise authentication.

Singular-Value Decomposition and Linear Ill-Posed Operators

Digital images are usually stored as $m \times n$ arrays of non-negative scalars, where m is the number of columns and n is the number of lines. Thus, an image I can be regarded as a matrix $I \in \Re^{m \times n}$. Because the analysis leading to the proposed watermarking scheme is performed on blocks of pixels extracted from an image, in this chapter, A denotes a generic matrix of intensity values representing either a complete image I or just a rectangular block of any dimension extracted from I. For the sake of simplicity and without loss of generality, in this section we assume that A is square of dimension $n \times n$.

A fundamental result of linear algebra states that any matrix can be represented as

$$A = USV^T \tag{12.1}$$

where $U = (u_1, \ldots, u_n) \in \Re^{n \times n}$ and $V = (v_1, \ldots, v_n) \in \Re^{n \times n}$. The columns $\{u_k\}$, $k = 1, \ldots, n$, of U are called the *left singular vectors* and form an orthonormal basis; that is, $u_i \cdot u_j = 1$ if $i = j$ and $u_i \cdot u_j = 0$ otherwise. The rows of V^T are the *right singular vectors*, $\{v_k\}, k = 1, \ldots, n$ and also form an orthonormal basis. $S = \text{diag}(s_1(A), \ldots, s_n(A))$ is a diagonal matrix whose diagonal elements are the *singular values* of A. If $\text{rank}(A) = r \leq n$, then $s_k(A) > 0$ for $k = 1, \ldots, r$, $s_k(A) \geq s_{k+1}(A)$ for $k = 1, \ldots, r - 1$, and $s_k(A) = 0$ for $k > r$. One important result derived from the singular-valued decomposition (SVD) is that for $S_l = \text{diag}(s_1(A), \ldots, s_l(A), 0, \ldots, 0)$, the matrix $A_l = US_lV^T$ is the matrix with rank l closest to A. This means that A_l minimizes the sum of the square differences between its elements and the elements of A: $\min_{a_{ij} \in A, \tilde{a}_{ij} \in A_l} \sum_{i,j} |a_{ij} - \tilde{a}_{ij}|^2$. This result can be used to measure image distortions introduced by setting the last singular values of a matrix to zero. A related result applies to distortions induced by changes in the last singular value of A. Assuming that $s_r(A)$ is the last nonzero singular value of A, $\hat{s}_r(A)$ is a real positive number and $\hat{A}$ is the matrix obtained by replacing $s_r(A)$ by $\hat{s}_r(A)$ in Equation 12.1, then

$$\left\| A - \hat{A} \right\|_2 = |s_r(A) - \hat{s}_r(A)| \tag{12.2}$$

where $\| \cdot \|_2$ denotes the $L_2 -$ norm.

In many applications of linear algebra, it is necessary to find a good approximation $\hat{x}$ of an unknown vector $x \in \Re^n$ satisfying the linear equation

$$Ax = b \tag{12.3}$$

for a given right-hand-side vector $b \in \Re^n$. The degree of difficulty in solving Equation 12.3 depends exclusively on the condition number of the matrix A. At first glance, the vector $\hat{x} = A^+b$ seems to be the solution of Equation 12.3. Here, $A^+ = (A^TA)^{-1}A^T$ (i.e., A^+denotes the pseudoinverse of A). However, if A is ill-conditioned or singular $\hat{x} = A^+b$, if it exist at all, is a meaningless poor approximation of x. In fact, the usual error estimates given by $\|x - \hat{x}\| \leq \|A^+\|\|A\hat{x} - b\|$ tell us that the approximation error can grow proportional to the norm of the inverse of A. Because the norm of the inverse of A is proportional to the condition number of A, it is evident that the greater the ill-conditioning of A, the larger the difference between x and $\hat{x}$. Furthermore, the estimation of the inverse of an ill-conditioned matrix is not straightforward. Actually, this is essentially the same problem. Moreover, when A is ill-conditioned, solving Equation 12.3 becomes equivalent to solving the optimization problem $\min_{x \in \Re^n} \|Ax - b\|^2$ for a predefined norm $\|\cdot\|$. It is well known that the L_2 − norm solution of this least squares problem is given by

$$\hat{x} = \sum_{s_i(A) \neq 0} \frac{u_{A_i}^T b}{s_i(A)} v_i \tag{12.4}$$

It becomes evident from Equation 12.3 that errors in either any of the left singular vectors of A or in the right-hand side b are drastically magnified by the smallest singular value of A. As a consequence, the L_2 − norm solution of the least squares problem 12.3 is useless for problems modeled by a linear operator with at least a very small but nonzero singular value. Ill-conditioned linear operators are characterized by the presence of very small singular values. This discussion allows us to formulate the following proposition, which is fundamental for the authentication method introduced in this chapter.

Proposition 1: Let $\tilde{x}$ be the L_2 − norm solution of the least squares problem $\min_{x \in \Re^n} \|\tilde{A}x - \tilde{b}\|_2^2$, with $\tilde{A}$ and $\tilde{b}$ distorted forms of A and b. Then, the difference between the L_2 − norm of x satisfying Equation 12.3 and $\tilde{x}$ becomes large and grows proportional to the inverse of the smallest singular value of $\tilde{A}$.

Watermark Generation

Given an image I of dimensions $m \times n$, a watermark Ω of the same dimensions is built. In its simplest form, Ω may be just an array of randomly generated binary or real numbers. A more attractive procedure to generate Ω uses a small binary logo. Initially, a mosaiclike binary picture L of dimension $m \times n$ is built by tiling the original image with the logo. The watermark is then defined as $\Omega = L \oplus \varpi$, where ϖ is a $m \times n$ array of randomly generated binary numbers and $\oplus$ denotes the bitwise

XOR operator. In either case, no assumption needs to be imposed on the statistical properties of the random number generator. The binary or real numbers used to generate the watermark can follow any probability distribution and is not restricted to Gaussian or uniform. This is because the proposed technique does not rely on statistical analysis for authentication or tamper detection. The only crucial assumption is that Ω depends on a secret key K and it is impossible, or at least extremely hard, for an attacker to generate Ω without knowing K. Observe that K is the seed of the random number generator.

Watermarking Process

Like most methods from the literature, this technique achieves good localization by performing blockwise watermarking. However, the proposed scheme uses block interdependency in order to overcome the vulnerability to vector quantization attacks. In the remainder of this section, the watermarking and verification algorithms are described for single image blocks. The strategy used to break block independency will be presented in the next section.

Blockwise watermarking is performed by partitioning the original image I into l small blocks $A^{(k)}, k = 1, \ldots, l$, of dimensions $p \times q$. Likewise, Ω is partitioned into l blocks $W^{(k)}, k = 1, \ldots, l$, of dimension $p \times q$. For the sake of notation simplicity, the upper index representing the block number will be omitted in the remaining of this section. Without loss of generality, it will be assumed also in the subsequent discussions that the blocks are squares (i.e., $p = q$).

Given a block A, the corresponding watermarked block is defined as the matrix $\hat{A}$ generated according to the following considerations. Initially, singular-value decomposition of A and W is performed to obtain $A = USV^T$ and $W = U_w S_w V_w^T$, respectively. Let $S = \text{diag}(s_1(A), \ldots, s_r(A))$ and $S_w = \text{diag}(s_1(W), \ldots, s_t(W))$ be the nonzero singular values of A and W. The two diagonal matrices $\hat{S} = \text{diag}(s_1(A), \ldots, \hat{s}_r(A))$ and $\hat{S}_w = \text{diag}(s_1(W), \ldots, \hat{s}_t(W))$ are then built by replacing the last nonzero singular value $s_r(A)$ and $s_t(W)$ by two specific real positive numbers $\hat{s}_r(A)$ and $\hat{s}_t(W)$, respectively. Here, it is assumed that the smallest nonzero singular value of A is $s_r(A)$ [i.e., $\text{rank}(A) = r$] and the smallest nonzero singular value of W is $s_t(W)$ [i.e., $\text{rank}(W) = t$]. Using $\hat{S}$, the watermarked block $\hat{A}$ is defined as

$$\hat{A} = U\hat{S}V^T \tag{12.5}$$

Likewise, $\hat{S}_w$ is used to build an ill-conditioned matrix $\hat{W}$ according to

$$\hat{W} = U_w \hat{S}_w V_w^T \tag{12.6}$$

The crucial part of the watermarking process is the choice of the two values $\hat{s}_r(A)$ and $\hat{s}_t(W)$. In fact, the whole watermarking approach depends on the estimation of these parameters. Using Equation 12.5 and Equation 12.6 as the basic relations of the watermarking process, three fundamental conditions are used to constraint the estimation of $\hat{s}_r(A)$ and $\hat{s}_t(W)$.

Fragility: Any change on single or multiple elements of $\hat{A}$ can be detected by the authentication procedure. This condition guarantees the effectiveness of the proposed tamper detection and authentication technique. It is achieved by setting $s_t(W) = \varepsilon$ in Equation 12.6. Choosing $\varepsilon > 0$ as an almost infinitesimal real number, $\hat{W}$ becomes extremely ill-conditioned. Once $\hat{W}$ has been generated, it is linked with $\hat{A}$ via matrix multiplication to obtain the ill-conditioned matrix $B = \hat{A}\hat{W}$. Although $\hat{W}$ is completely defined by ε, $\hat{A}$ still depends on an unknown parameter $\hat{s}_r(A)$. For that reason, B can be regarded as the parametric family of matrices:

$$B(\hat{s}_r) = \hat{A}(\hat{s}_r)\hat{W} \tag{12.7}$$

This parametric family of matrices $B(\hat{s}_r)$ determines the linear ill-posed operator used in the tamper detection procedure according to Proposition 1 and the next two conditions.

Uniqueness: For a predefined large real number N, there exist a unique value $\hat{s}_r(A)$, so that the $L_2 - norm$ solution of the least squares problem

$$\min_{x \in \Re^p} \|Bx - b\|_2^2 \tag{12.8}$$

is N^2. Here, b is an arbitrary vector defining the right-hand side of the linear system to be minimized in Equation 12.8. Although the choice of b does not play any role in the actual watermark embedding or in the fulfillment of the fragility and uniqueness conditions, it plays an important role in breaking the blockwise independence of the watermarking approach. This discussion is elaborated in the next section, where different attacks are considered in more detail.

Imperceptibility: $\hat{s}_r(A)$ is a real positive number satisfying the relation

$$\max(eps, s_r(A) - \delta) \le \hat{s}_r(A) \le s_r(A) + \delta$$

where eps is the machine precision and δ is a scalar used to control the distortion induced by the watermark. Observe that the expression $\max(eps, s_r(A) - \delta)$ defining the lowest bound of the feasible interval guarantees that $\hat{s}_r(A)$ remains nonzero and positive. This condition together with Equation 12.2 allows us to keep the distortion below the user-defined value δ.

Clearly, the proposed scheme heavily relies on a suitable process to find $\hat{s}_r(A)$ obeying these three conditions. Furthermore, there are two arbitrary but fix parameters inherent to the watermarking algorithm: a maximum tolerable distortion δ to control image quality and the almost infinitesimal real number $\varepsilon > 0$ used to impose ill-posedness to the linear operator in Equation 12.8. These parameters can be adjusted according to user requirements or specific applications.

For given input parameters δ and ε, Figure 12.4 outlines the watermarking procedure in four main algorithmic steps.

1. Generation of a K-dependent watermarking matrix W from Ω.
2. Construction of the parametric family of matrices $B(\hat{s}_r)$ defined by Equation 12.7.

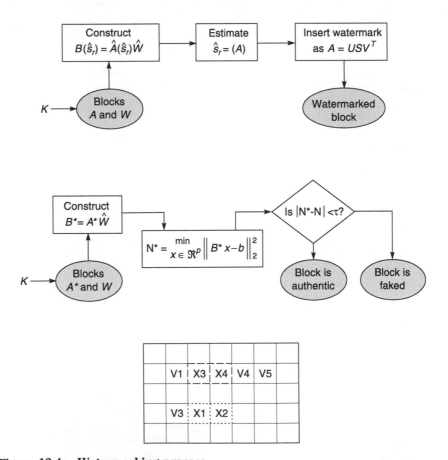

Figure 12.4. Watermarking process.

3. Estimation of the unique parameter $\bar{s}_r(A) \in [\max(eps, s_r(A) - \delta),$ $s_r(A) + \delta] := [H_0, H_1]$, which minimizes the expression

$$\min_{\hat{s}_r \in [H_0, H_1]} \left\{ \sum_{i=1}^{q} \left(\frac{u_{B_i}^T b}{s_i(B(\hat{s}_r))} \right)^2 - N^2 \right\} \qquad (12.9)$$

where u_{B_i} is the ith column of the matrix formed with the right singular vectors of B, $s_i(B)$ are the singular values of B, b is the right-hand-side vector given in Equation 12.8 and N is a large real number.

4. Estimation of the watermarked block $\hat{A} = U\hat{S}V^T$ by setting

$$\hat{S} = \text{diag}(s_1(A), \ldots, s_{r-1}(A), \bar{s}_r(A))$$

The last equation shows how to chose the value $\hat{s}_r(A)$ in Equation 12.5 namely by setting $\hat{s}_r(A) = \bar{s}_r(A)$, where $\bar{s}_r(A)$ is the result of the minimization problem 13.9. Like K, the number N in Equation 12.9 is also secret. Although it can be assumed that N depends on K, or vice versa, higher security is achieved when N and K are chosen independently. Thus, the security of the proposed approach resides in the secrecy of set of keys $\kappa = \{K, N\}$. Obviously, the feasibility and effectiveness of this watermarking procedure is based in the existence and uniqueness of the value $\bar{s}_r(A) \in [H_0, H_1]$ obeying the imperceptibility, fragility, and uniqueness conditions stated earlier. The corresponding analysis to proof the existence of $\bar{s}_r(A)$ is given next.

The feasibility and effectiveness of the proposed watermarking and verification algorithms is based on two specific conditions: the ill-posedness of the linear operator to be minimized in Equation 12.8 or Equation 12.10 and the existence of a unique $\bar{s}_r(A) \in [H_0, H_1]$, minimizing Equation 12.9 for a fix value N. If we show that B is highly ill-conditioned, the validity of the first condition becomes evident. This can be done using the following propositions.

Lemma 1: Let A and W be two square matrices with the same dimension and let $s_k(A)$ and $s_k(W)$ their kth singular values, respectively. Then,

$$s_{i+j-1}(AW) \leq s_i(A)s_j(W) \quad \text{for all integers } i \text{ and } j$$

For the proof of this lemma, the reader is referred to Proposition 2.3.12 of Reference 22.

Proposition 2: Let the smallest singular values of $B = AW$ and W be $s_r(B)$ and $s_t(W)$, respectively; then,

$$s_r(B) \leq s_{r-t+1}(A) \cdot s_t(W) = \varepsilon s_{r-t+1}(A) \quad \text{for } t \leq r \qquad (12.10)$$

Proof: It follows directly from Lemma 1, setting $i = r - t + 1$ and $j = t$.

Because ε is chosen to be very small, Inequality Equation 12.10 guarantees that the smallest singular value of B is also very small and, therefore, extremely ill-conditioned. Usually, the matrices A and W have full rank (i.e., $t = r$). However, it is possible to build counterexamples with $t > r$. Even in such unusual situations, Equation 12.10 can be applied by setting $s_k(W) = 0$ for all $k > r$. Observe that because $\hat{W}$ is artificially constructed, nothing prevents us from setting the required values to zero. As a consequence, the condition $t \leq r$ in Equation 12.10 can be assumed in any case.

In order to prove the existence of $\bar{s}_r(A) \in [H_0, H_1]$, minimizing Equation 12.9 for a fix value N, let us consider the real-valued functions $h(z) : [H_0, H_1] \rightarrow \Re^+$ and $g(z) : [H_0, H_1] \rightarrow \Re^+$ defined as

$$h(z) = s_r(B) \quad \text{and} \quad g(z) = \min_{x \in \Re^p} \|B^*(z)x - b\|_2^2 \qquad (12.11)$$

$h(z)$ can be written as $h(z) = s_r(A(z)\hat{W}) \equiv (h_1 \circ h_2)(z)$, with $h_1(z) = s_r(B(z))$ and $h_2(z) = A(z)\hat{W}$. The two functions h_1 and h_2 are continuous in the interval $[H_0, H_1]$. Hence, $h(z)$ is also continuous in $[H_0, H_1]$. The continuity of $h(z)$ can now be used to prove that $g(z)$ is continuous in $[H_0, H_1]$. Using Equation 12.4, it is straightforward to derive the following expression:

$$g(z) = \sum_{i=1}^{n} \left(\frac{u_{B_i(z)}^T b}{s_i(B(z))} \right)^2 \qquad (12.12)$$

Thus, $g(z)$ is the sum of quotients of continuous functions. Therefore, $g(z)$ is also continuous in $[H_0, H_1]$.

Proposition 3: Consider $h \max = \max(g(z))$ and $h \min = \min(g(z))$. If $N \in [g(h \max), g(h \min)]$, then there exists $\bar{z} \in [H_0, H_1]$ such that $g(\bar{x}) = N$.

Proof: It follows from the continuity of $g(z)$ in $[H_0, H_1]$ and the mean value theorem of continuous functions.

Propositions 2 and 3 guarantee the effectiveness and feasibility of the proposed approach. The underlying operator (Equation 12.8), can be made extremely ill-posed while the norm of its solution is kept equal to N. Furthermore, by selecting $\hat{s}_r(A) \in [H_0, H_1]$, the distortion on the original image remains below the input parameter δ. However, this last property constrains the variation of $\hat{s}_r(A)$ to a very small interval. Because $\hat{s}_r(A)$ depends on N, an important question arises: How does the small interval $[H_0, H_1]$ constrain the set of feasible values N? This question is extremely important because N is a secret key and we are assuming that it is extremely difficult to estimate it. Obviously, the smaller the set of feasible values for N, the easier it is to estimate N and so to mount an attack successfully. This important security aspect will be elaborated in the next subsection.

Verification Procedure

To prove authenticity and detect tampered areas, the receiver of an image I^* needs to test if single blocks A^* have been attacked. It is assumed that the receiver is a trusted party who knows the secret set of keys $\kappa = \{K, N\}$. In addition to ε, a tolerance value τ is used in the verification process. This parameter protects against approximation errors inherent to any numerical process. ε and τ are fix numbers intrinsic to the algorithm and for that reason are of a public nature. As shown in Figure 12.5, most algorithmic steps in the verification procedure coincide with the watermarking steps. Using K, the receiver first generates the watermark W. Next, ε is used to build the matrix $\hat{W}$ by setting $\hat{S}_w = \mathrm{diag}(s_1(W), \ldots, \varepsilon)$ as in Equation 12.6. After that, the ill-conditioned matrix $B^* = A^* \hat{W}$ is built and the solution of the minimization problem

$$\min_{x \in \Re^p} \|B^* x - b\|_2^2 \tag{12.13}$$

is calculated. Once Equation 12.13 has been solved, N^* is defined as the square root of the norm of the vector x minimizing Equation 12.13. The final verification step consists of a simple comparison between N^* and the secret value N. A Boolean response is obtained by thresholding the absolute difference $|N^* - N| = \gamma$. If $\gamma \leq \tau$, A^* is genuine, otherwise A^* is declared a fake.

Because the verification algorithm only requires the norm of the vector x minimizing Equation 12.13, N^* can be estimated directly from

$$(N^*)^2 = \sum_{i=1}^{n} \left(\frac{u_{B_i^*}^T b}{s_i(B^*)} \right)^2 \tag{12.14}$$

where $u_{B_i^*}$ is the ith column of the matrix of the right singular vectors of B^*, and $s_i(B^*)$ are the corresponding singular values [23].

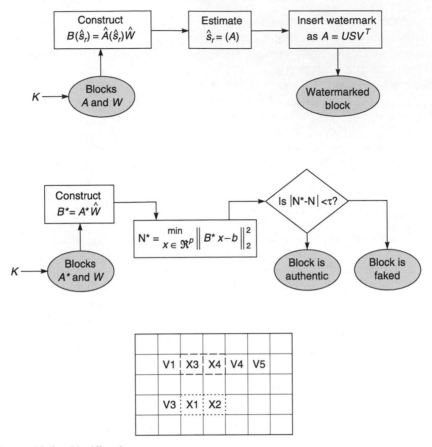

Figure 12.5. Verification process.

According to Proposition 1, if the watermarked block has been manipulated, the difference $|N^* - N|$ becomes large. This behavior is due to ill-conditioning of B^*. Because the smallest singular value of B^* is very close to zero, any modification of $\hat{A}$ will be reflected in B^*. Thus, the norm of the least squares solution of Equation 12.13 will be strongly magnified. Consequently, γ becomes huge and, with certainty, larger than τ.

A fundamental difference between this watermarking approach and others from the literature is that the watermark W is embedded in A by transforming W according to Equation 12.6 and using the result of this operation to estimate $\hat{s}_r(A)$. Consequently, the information contained in the watermark W is first concentrated in $\hat{s}_r(A)$ and then spread in A according to Equation 12.5. A major advantage of this scheme is that the distortion or change in the original image can be strictly controlled using Equation 12.2. Because the distortion induced by the watermarking procedure depends only on changes in the smallest singular values of each block, it is straightforward to control the distortion produced on the watermarked image. The following example illustrates this property of the authentication approach.

Figure 12.6 shows the original test image Lena and the corresponding watermarked image using 8×8 blocks, $\delta = 0.2$, $\varepsilon = 10^{-12}$, and $N = 876,531$. According to Equation 12.2, L_2 − norm distortion is 0.2 and the PSNR rises over 60 dB, a clearly negligible distortion invisible to the human eye. To evaluate the sensitivity of the verification procedure, single pixels are

Figure 12.6. **(a) Original image Lena and (b) watermarked version using blocks of dimension 8×8 and $\delta = 0.2$, and (c) result of the verification procedure after the intensity value of four single pixels have been distorted by a factor of 0.1.**

Figure 12.6. Continued

distorted by very small random perturbations. The image in Figure 12.6a shows the detection result obtained when four pixels of the watermarked image are distorted by adding a random value $\eta \in [0.01, 0.1]$ to the corresponding four intensity values. The distorted pixels correspond to the sampling positions (24, 24), (72, 136), (352, 168), and (264, 336). In Figure 12.6, the detected distorted blocks are shown as black squares enclosed by larger white circles. Although only four single pixels of the image are distorted by small intensity changes, the algorithm detects the presence of the manipulation. This experiment was repeated for many different values of η. The proposed technique can detect very small distortions on single pixels as small as $\eta = 10^{-6}$ when high arithmetic precision is used in the calculations.

The next example shows the behavior of the threshold value τ used in the verification procedure. This parameter depends on the tolerance used in the optimization code for the minimization task according to Equation 12.9 and Equation 12.14. Additional distortions arise from numerical approximations and rounding errors introduced by the operations involved in the different algorithmic steps. These errors and the high instability of the system impede the estimation of the exact value N by solving the underlying minimization problem. Actually, even if an undistorted watermarked block $\hat{A}$ is used in the verification procedure, it cannot be expected that $|\hat{N} - N| = 0$, where $\hat{N}$ is the value obtained from Equation 12.14 using an undistorted watermarked block $\hat{A}$. Although the difference $|N^* - N|$ varies according to the severity of the attack, the $|\hat{N} - N|$ remains almost constant for a fixed block size. For blocks of dimension 8×8, this difference is of the order of 10^{-2} using standard precision and can be reduced to less than 10^{-6} when high arithmetic

Table 12.1. Difference $|N^* - N|$ for selected pixels distorted at levels in the watermarked image Lena

Distortion	$\eta = 0.001$	$\eta = 100$	Tattoo Manipulation
Difference $|N^* - N|$	0.4812e + 000	5.0366e + 005	3.0000e + 004
	2.0011e + 000	2.3448e + 004	3.0000e + 004
	5.1869e + 000	3.0000e + 004	7.5929e + 005
	0.1427e + 000	1.4094e + 003	3.0000e + 004
	1.8862e + 000	2.9563e + 004	3.0000e + 004
	0.2368e + 000	2.4147e + 003	7.9900e + 006
	2.9367e + 000	2.9994e + 004	3.0000e + 004
	0.1990e + 000	2.0223e + 003	3.0000e + 004

precision is used. By setting $\tau = 0.1$, we can be sure that distortions at the subpixel level can be detected using standard arithmetic precision. In Table 12.1, the difference $|N^* - N|$ for selected distortions of the watermarked image are shown for the test image Lena. The second and third columns show the minimum value of $|N^* - N|$ when several randomly selected single pixels were distorted by adding η to their intensity values. Clearly, for negligible distortions ($\eta = 0.001$), the difference $|N^* - N|$ is larger than 0.1. The fourth column of Table 12.1 shows the minimum difference for all tampered blocks of Figure 12.7 (tattoo on Lena's shoulder and inscription on her hat).

The results of selected experiments aimed at detecting intentional manipulations and localization accuracy are presented next. The image at the left of Figure 12.7 shows Lena manipulated with a tattoo in her shoulder and an inscription on her hat in a mirror. The image in the

Figure 12.7. Response of the verification algorithm for a manipulated Lena (left), response of the verification procedure with tampered blocks marked white (middle) and zooms of the areas detected as tampered (right).

Figure 12.8. **Response of the verification algorithm for the tampered image shown in Figure 12.1a.**

middle displays the response of the verification algorithm. In this representation, the detected manipulated blocks are highlighted white. The image at the right of Figure 12.7 shows a zoom of the areas that the verification algorithm fails to authenticate. Figure 12.8 shows the result of the verification procedure applied to the image shown in Figure 12.1a. In this representation, the tampered block is highlighted with a green texture. Clearly, all tampered areas were detected. The localization accuracy is constrained by the size of the block, which is 16×16 in Figure 12.7 and Figure 12.8.

ATTACKS AND COUNTERMEASURES

To evaluate the performance of any authentication technique, two main aspects should be considered: imperceptibility and security. Obviously, watermarking is rendered useless if the quality of the original content is degraded. Because the changes produced by embedding a watermark are, by definition, negligible and invisible to the human eye, quality preservation is a property inherent to any watermarking approach. The security aspect is more complex and relates to the capacity of a given watermarking schemes to detect changes and survive malicious attacks. During the last few years, different attacks and countermeasures have been proposed to either break or increase the security of watermarking-based image authentication algorithms. In this section, some relevant attacks are described and used to analyze the security provided by the authentication technique described in the previous section.

The Vector Quantization Attack

The Achilles' heel of all blockwise independent watermarking techniques is the vector quantization attack introduced by Holliman and Memon

[24]. It belongs to the class of vector quantization counterfeiting and has been proven to defeat any technique targeting localization accuracy by watermarking small image blocks independently. Assuming the availability of a large library of images watermarked with the same key, the attacker builds a vector quantization codebook using blocks extracted from the authenticated images. The image to be faked is then approximated using this codebook. Because each image block in the library is authenticated independently of the other images and blocks, the fake image is obviously accepted as genuine by the watermark detector. Clearly, the quality of the counterfeiting image depends on two factors: the size of the database with available watermarked images and the size of the blocks used in the watermarking process. The more images available, the higher the probability to find better approximations of the fake image. Furthermore, for a given database with a limited number of watermarked images, the quality of a counterfeit can only been improved by reducing the block size. This effect is illustrated in Figure 12.9, where a non-watermarked image (fake) is approximated by the vector

Figure 12.9. **The vector quantization attack: (a) original fake image, (b) counterfeit created using blocks of size 16 × 16, (c) 8 × 8 and (d) 4 × 4.**

Figure 12.9. Continued.

quantization attack using a database containing 496 pictures of cars captured by a conventional surveillance camera. The fake image is shown at the top left, the remaining images show the approximation using blocks of size 16×16, 8×8 and 4×4, respectively. Clearly, the quality of a image approximated using small blocks is high enough even when a reduced database is used.

Several countermeasures to this attack can be found in the literature: increasing the block dimension or using information from surrounding blocks [25], including a block or image index in the signature [26], embedding a unique index and additional image information at multiple image positions [27], or using overlapping block with hierarchical structures [25].

To achieve robustness against these kind of attack, the watermarking approach described in Section 3 does not perform authentication completely independent of other image blocks. As described at the beginning of Section 3, the method partitions the original image into l small blocks $A^{(k)}$, $k = 1, \ldots, l$, of dimension $p \times q$. Likewise, W is also split into l blocks $W^{(k)}$ of dimension $p \times q$. Because the vector quantization attack relies on block-independent watermark embedding, the block independency is broken using the right-hand side b of the linear operator defining the minimization problem (Equation 12.8). Observe that the description given in the previous section do not specify how to chose b. Assuming that the watermarking process is performed sequentially, the block $A^{(k)}$, $k = 2, \ldots, l$, is processed once the blocks $A_w^{(k)}$, $k = 1, \ldots, l - 1$, has been watermarked. To obtain the vector $b^{(k)}$ needed to watermark the current block, a random sequence $Z^{(k)}$ of 1 and -1 is generated using K as the seed. The right-hand side of Equation 12.8 is then defined as

$$b^{(k)} = \begin{cases} A_w^k Z^{(k)} & \text{for } k = 1 \\ A_w^{k-1} Z^{(k)} & \text{else} \end{cases}$$

This simple strategy increases the difficulty of successfully undertaking a vector quantization attack. Basically, using the proposed scheme, the only way to mount this attack is by replacing large image areas containing several authenticated blocks. Even so, the blocks at the border of the swapped area will be recognized as fake blocks.

Swapping Attack

Another related attack consists of swapping blocks of a watermarked image. This attack will be recognized by the algorithm described previously. As shown in Figure 12.10, even by swapping pairs of consecutive blocks that have been watermarked together, the verification procedure detects the changes. In Figure 12.10, the dashed blocks labeled with a **X** have been swapped. Because the first tampered block **X3** was authenticated together with **V3** (and not with **V1**), the verification procedure will mark block **X3** as tampered. In the first pass, block **X4** will be marked as genuine, because the pairs (**X1, X2**) and (**X3, X4**) are watermarked together. However, block **V4** will appear as tampered because it was authenticated together with **X2** (not with **X4**). Consequently, block **X4** will be relabeled as tampered. Finally, block **V5** will appear as authentic, showing that **V4** has to be authentic as well. Thus, at the end of the verification process, only **X3** and **X4** will be labeled as tampered. Likewise, tampering by swapping identically positioned blocks from several authenticated images can be detected.

Cropping

Cropping is probably the simplest form of image manipulation. For instance, the tampered image in Figure 12.1a was created by using "cropped areas" from similar images. Given a cropped image, it is

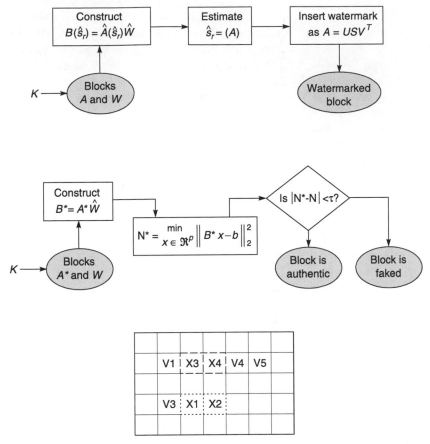

Figure 12.10. **Breaking block independency: swapping pairs of block can be detected.**

desirable to recognize it as part of a genuine image. In most cases, cropping of watermarked images leads to severe desynchronization between the boundaries of the block used in the watermarking process and those used in the verification. As a consequence, the main challenge presented by cropping attacks is to regain synchronization. Conventional strategies are based on search using a "sliding block" to achieve sychronization. This approach is effective when the block dimension is fairly small, as expected when accurate localization is targeted. In the worst case, the number of search positions is equal to the dimension of the used block.

Estimation of the Set of Keys

A watermarking scheme for authentication is secure if an unauthorized party cannot produce any fake image that is recognized as authentic by the verification algorithm or cannot amount a successful attack on either the watermarking or the verification procedures in order to gain information about the secret keys. Furthermore, a watermarking scheme is trustworthy if, in addition, Kerckhoff's principle applies: that is, the security resides only in the secrecy of a key but not in the secrecy of the algorithm. Basically, the watermarking scheme should guarantee that it is impossible or very unlikely to generate counterfeit images or defeat the authentication process without knowing the secret keys. In this context, the remainder of this subsection is dedicated to a discussion related to some important security aspects of the authentication procedure described previously.

The security of this technique resides in a secret set of keys $\kappa = \{K, N\}$. K is the seed of the random number generator and can be selected freely. It could be chosen as a function of N in order to deal with a single key. However, this leads to the question about the possibility of estimating N using the knowledge in the algorithm. If N can be estimated and K depends on N, the full set $\kappa = \{K, N\}$ would be compromised. Fortunately, the range of values that we can use for N is extremely large, making it almost impossible for an attacker to estimate it. Because the distortion introduced by the watermark can be strictly controlled by the distortion coefficient δ, this coefficient defines the feasibility interval $[H_0, H_1]$. Clearly, this interval is very small. Its maximum length does not exceed 2δ, and according to Proposition 3, it defines the range of permissible values for $N \in [g(h\max), g(h\min)]$. Because N should be a large number, to guarantee the security of the algorithm, it is also important to show that the interval of permissible values is also very large. Variations of $z \in [H_0, H_1]$ are reflected in variations of the smallest singular value of B. According to Equation 12.10, the smallest singular value of B is very close to ε. This fact can be used to find an estimate for the interval $[g(h\max), g(h\min)]$. Let us consider the hyperbola $p(z) = C + D/y^2$ with $C = \sum_{i=1}^{r-1} (u_{B_i}^T b / s_i(B))^2$ and $D = (u_{Br}^T b)^2$. Because the variation of $z \in [H_0, H_1]$ determines the variation of $p(z)$, this gives the range for possible values of N. Observe that changes in z also affect C and D, but, actually, the smallest singular value of B is the leading term determining the behavior of $p(z)$. Clearly, $p(z) \to \infty$ if $z \to 0$. Furthermore, p maps very small intervals very close to zero into a very large interval. For instance, if $\varepsilon = 10^{-16}$ and $\delta = 10^{-2}$, then z will approximately vary between the machine precision eps (e.g., 10^{-32} and 10^{-2}). In this case, $[g(h\max), g(h\min)] \approx [10^2, 10^{32}]$. As a consequence, for this particular example, N could be selected from the interval $N \in [10^2, 10^{32}]$. These arguments show that the range of permissible values of N is huge and it would be extremely hard for an attacker to estimate N. Observe that the previous discussion assumes that the two keys N and K depend on each other and K can be inferred from N. However, nothing prevents us from selecting these two keys totally independently of each other. In this case, estimating one of the keys will not be of much use for an attacker and the algorithm appears twice as secure.

Stego Image Attack

It is assumed that the attacker has a single authenticated image and his goal is to produce changes to the image that go undetected or to recover some secret information from the scheme. To do that, the attacker needs to create a new matrix A^* such that $B = A^*W$ satisfies

$$\left| \sum_{i=1}^{n} \left(\frac{u_{B_i}^T b}{s_i(B)} \right)^2 - N^2 \right| \leq \tau \qquad (12.15)$$

Given a matrix $\hat{A} = U\hat{S}V^T$, there exists an infinite number of matrices A^* satisfying relation 12.15. This can be shown by considering the SVD $B = U_B S_B V_B^T$. Because Equation 12.15 does not depend on V_B, this inequality is satisfied by any matrix A^* built as $A^* = B^*W^{-1}$, with $B^* = U_B S_B V^{*T}$ and V^* any orthogonal matrix. Obviously, a matrix A^* constructed in this manner will be recognized as authentic by the proposed algorithm. However, it is almost impossible for an attacker to generate such a A^* because he does not know U_B, S_B, and W and any of these three matrices cannot be estimated from $\hat{A}$, which is the only matrix known by the attacker. The attacker could try to estimate N using Equation 12.15. In this case, he needs to solve a highly nonlinear inequality with $n^2 + 2n + 1$ unknown parameters. Evidently, such an inequality has an infinity number of solutions. Even if the attacker has access to a large number m of watermarked images, an analytical inference of the involved parameters is almost hopeless. In this case, the attacker can build m inequalities, but each new inequality introduces $n^2 + 2n + 1$ additional unknown parameters. The resulting nonlinear system has m nonlinear equations and $m(n^2 + 2n)$ unknowns. This system has, again, an infinite number of solutions and it would be easier just to guess the key rather than to find the right solution.

SUMMARY AND CONCLUSIONS

In this chapter, the fundamentals of watermark-based image authentication were described. The main motivations and the rationale for research, development, and use of this technology were presented. A brief review of state-of-the-art techniques for image authentication was also given. Theoretical and practical aspects related to feasibility, effectiveness, and security were considered and analyzed. The main part of the chapter was dedicated to a detailed description of a technique for fragile image authentication derived from the extreme sensitiveness of linear ill-posed operators to small distortions in the input data. The watermarking process is achieved by linking the watermarked image with a very ill-conditioned matrix derived from a secret key and the original image. The solution of the underlying ill-posed operator is extremely sensitive to changes in the watermarked image. This property is used to verify authenticity and detect tampering of image regions. A novel feature of this approach is that it is based on a nonconventional model. In contrast to other techniques from the literature, it uses linear algebra tools and numerically unstable systems without relying on hash-based algorithms and encryption. Furthermore, the distortion induced by the watermarking process is invisible and fully controllable.

REFERENCES

1. Diffie, W. and Hellman, M.E., New directions in cryptography, *IEEE Trans. Inf. Theory*, 22(6), 644–654, 1976.
2. *Signal Processing*, 66(3), 1998, (special issue on watermarking).
3. *IEEE Transactions on Circuits and Systems of Video Technology*, 13(8), 2003, (special issue on authentication, copyright protection and information hiding).
4. Feng, Y. and Izquierdo, E., Robust Local Watermarking on Salient Image Areas, in Proceedings International Workshop on Digital Watermarking, Seoul, 2002.
5. *Proceedings of the IEEE*, 87(7), 1999, special issue on identification and protection of multimedia information.
6. Cox, I.J., Kilian, J., Leighton, T., and Shamoon, T., Secure spread spectrum watermarking for images, audio and video, *IEEE Trans. Image Process.*, 6(12), 1673–1686, 1997.
7. Wolfgang, R.B., Podilchuk, C., and Deip, E. J., Perceptual watermarks for digital images and video, *Proc. IEEE*, 87(7), 1108–1126, 1999.
8. Ting-Hsu, C. and Ling-Wu, J., Hidden digital watermarks in images, *IEEE Trans. Image Process.*, 8(1), 58–68, 1999.
9. Barni, M., Bartolini, F., Cappellini, V., Lippi, A., and Piva, A., DWT-Based Technique for Spatio-Frequency Masking of Digital Signatures, in *Proceedings SPIE, Security Watermarking Multimedia Contents*, SPIE, 1999, pp. 31–39.
10. Kundur, D. and Hatzinakos, D., Towards a Telltale Watermarking Technique for Tamper Proofing, in Proceedings ICIP, Chicago, 1998.
11. Lin, C.-Y. and Chang, S.-F., Semi-Fragile Watermarking for Authenticating JPEG Visual Content, in Proceedings SPIE, Security and Watermarking of Multimedia Contents, San Jose, CA, SPIE, 2000, pp. 140–151.
12. Wolfgang, R.B. and Delp, E.J., Fragile Watermarking Using the VW2D Watermark, in Proceedings SPIE, Security and Watermarking of Multimedia Contents, San Jose, CA, SPIE, 1999, pp. 204–213.
13. Fridrich, J., Security of Fragile Authentication Watermarks with Localization, in *Proceedings SPIE*, SPIE, 2002.
14. Lin, C.-Y. and Chang, S.-F., A robust image authentication method distinguishing JPEG compression from malicious manipulation, *IEEE Trans. Circuits Syst. Video Technol.*, 11(2), 153–168, 2001.

15. Lin, C.-Y. and Chang, S.-F., A Robust Image Authentication Method Surviving JPEG Lossy Compression, presented at *SPIE Storage and Retrieval of Image/Video Databases*, San Jose, CA, SPIE, 1998.
16. Walton, S., Information authentication for a slippery new age, *Dr. Dobbs J.*, 20(4), 18–26, 1995.
17. van Schyndel, R.G., Tirkel, A.Z., and Osborne, C.F., A Digital Watermark, in Proceedings IEEE International Conference on Image Processing, Austin, TX, 1994, Vol.2, pp. 86–90.
18. Wolfgang, R.B. and Delp, E.J., A Watermark for Digital Images, in Proceedings IEEE International Conference on Image Processing, 1996, Vol.3, pp. 219–222.
19. Yeung, M.M. and Mintzer, F., An Invisible Watermarking Technique for Image Verification, in Proceedings ICIP, Santa Barbara, CA, 1997.
20. Wong, P.W., A Public Key Watermark for Image Verification and Authentication, in Proceedings ICIP, Chicago, 1998.
21. Izquierdo, E., An ill-posed operator for secure image authentication, *IEEE Trans. Circuits and Syst. Video Technol.*, 13(8), 842–852, 2003.
22. Pietsch, A., *Eigenvalues and s-Numbers*, Cambridge University Press, Cambridge, 1997.
23. Golub, G.H. and Van Loan, C.F., *Matrix Computation*, 3rd ed., Johns Hopkins University Press, Baltimore, MO, 1996.
24. Holliman, M. and Memon, N., Counterfeiting attacks on oblivious block-wise independent invisible watermarking schemes, *IEEE Trans. Image Process.*, 9(3), 432–441, 2000.
25. Celik, M.U., Sharma, G., Saber, E., and Tekal, A.M., Hierarchical watermarking for secure image authentication with localization, *IEEE Trans. Image Process.*, 11(6), 585–505, 2002.
26. Coppersmith, D., Mintzer, F., Tresser, C., Wu, C.W., and Yeung, M. M., Fragile Imperceptible Digital Watermark with Privacy Control, in Proceedings SPIE, Security and Watermarking of Multimedia Contents, San Jose, CA, SPIE, 2000, pp. 79–84.
27. Wong, P.W. and Memon, N., Secret and Public Key Authentication Watermarking Schemes that Resist Vector Quantization Attack, in Proceedings SPIE Security and Watermarking of Multimedia Contents, San Jose, CA, SPIE, 2000, pp. 417–427.

13

New Trends and Challenges in Digital Watermarking Technology. Applications for Printed Materials

Zheng Liu

INTRODUCTION

In recent years, digital watermarking has become a very popular topic that interests more and more persons coming from institutes or companies to take part in its research and application developments. A well-known reason for this is that the rapid growth of the Internet and the widespread use of digital contents create an urgent need for the protection of intellectual property. Although there are still many issues to be solved technically and legally before digital watermarking technology can be applied in the real world, more and more digital watermarking products have entered the market for watermarking applications. Moreover, currently, watermarking techniques are expected to be used for more and more kinds of media, such as the printed materials, screen images, cloth materials, and even the images painted on

0-8493-2773-3/05/$0.00+1.50

the wall and floor, which have reached beyond the original scope of digital watermarking (i.e., the digital domain). With the needs of watermarking techniques increased, digital watermarking becomes more fascinating with new challenges.

In this chapter, we will introduce some new trends and challenges in the research and application of digital watermarking for printed materials. The chapter is organized as follows. In Section "Overview of Watermarking Technology," we have a brief overview of the current digital watermarking technology and discuss the corresponding issues. In Section "Watermarking Techniques for Printed Materials," we introduce digital watermarking techniques used for printed materials, which is a very challenging topic in the digital right management (DRM) system. In Section "Extracting Watermark using Mobile Cameras," we introduce watermarking techniques used for printed images by using a mobile camera phone, which may be the most challenging topic in current digital watermarking research and applications. With this techniques, users can simply connect to a Web site with the information extracted from a printed sample image by using a mobile camera phone, where more information about the image can be acquired. Finally, some discussions and conclusions are given in Section "Conclusion."

OVERVIEW OF WATERMARKING TECHNOLOGY

Modern digital watermarking technology has rather short history [1]; it started about 1990 [2,3]. It is reported that there were only 21 publications in the public domain until 1995 [4]. Since the mid-1990s, digital watermarking technology has grown rapidly in research. The number of publications in 1998 reached 103 [4], and the number is more than 1500 by the end of 2003. This trend points out that watermarking research is a growing field and we anticipate its continuing progress in both academic research and industrial applications in the next few years.

We know that the original motivation of digital watermarking technology is to protect the copyright of digital contents, and, theoretically, the related techniques have been researched thoroughly and even have been studied repeatedly. However, the issues of for what digital watermarking technology can achieve and when they can be used practically for copyright protection, authentication, and so on remain unanswered. Therefore, before the discussion of the main topics in this chapter, we should have an overview of the current digital watermarking technology and corresponding issues.

Digital Watermarking Technology

As a traditional watermarking technique, a watermark is usually used as a mark to be implanted transparently into printed materials, such as a

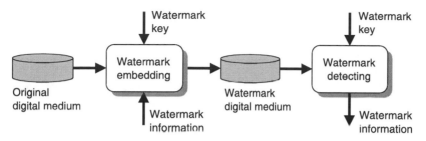

Figure 13.1. Digital watermarking system.

document, paper money, and so forth for the purpose of genuineness certification. In recent years, the digital watermarking technology is used to protect the copyright of digital content by adding watermark information into the content, such as still images, digital video, and digital audio.

Digital Watermarking System. A digital watermarking system is shown in Figure 13.1, in which the watermark can be an image, such as an organization's logo, an author's signature and fingerprint image, or a stream of ASCII codes representing the information about copyright, ownership, and a timestamp. The watermarking system essentially consists of two processes: watermark embedding and watermark detecting. In the embedding process, the watermark information encrypted by a watermark key is embedded into digital content. In the detecting process, the embedded watermark is extracted from the watermarked content and then decrypted by the same key that was used for watermark embedding. The role of a watermark key is to ensure that the embedded watermark is detected by authorized users only. In recent years, there are many methods proposed for digital watermarking technology, which can be divided into two categories: spatial domain watermarking and frequency domain watermarking because, since the watermark is embedded either in the spatial domain or in the frequency domain.

Spatial-Domain Watermarking. Spatial domain watermarking is the method in which the watermark is embedded into pixel values. The least significant bit (LSB) substitution is the main structure of this method. The advantage of spatial domain watermarking is relatively easy to implement, but is weak in geometric signal manipulations such as rotation, scale and, translation (called RST distortion) [5]. The reason for this is that the embedded watermark for effective watermark detection for spatial domain watermarking should be read exactly from the same pixels in which the watermark is embedded and, therefore, a slight geometric change will disturb the order of extracting the watermark and make the watermark functionless. However, if the geometrically distorted

389

image (or signal) can be revised by inverting the distortion such as the methods we will introduce below, the spatial domain watermark can have be as robust as the frequency domain watermarking.

Frequency Domain Watermarking. Frequency domain watermarking is the method in which the watermark information is embedded in the frequency domain. Its major advantage is its robustness to most of the common signal manipulations. The general methods used for signal transformation from the spatial domain to the frequency domain are discrete Fourier transformation (DFT), discrete cosine transformation (DCT), and the discrete wavelet transformation (DWT).

Generalized Scheme for Watermark Embedding. A generalized scheme for watermarking embedding, which can be used both in the spatial domain and the frequency domain, is expressed as

$$X = S(1 + \alpha W) \tag{13.1}$$

where S and X are the original signal and the watermarked signal, respectively, which can be the pixel value in the spatial domain or the element value in the frequency domain. W is the watermark information, which can be represented as a rectangular array of numeric elements, called the watermarking plane. α is a scaling factor to be adjusted between 0 and 1 to provide a good trade-off between imperceptibility and robustness.

Equation 13.1 seems uncomplicated in the academic environment and it may be partly due to this academic simplicity that allows many people coming from different research fields to take part in the watermarking research easily without need of special discipline knowledge, as long as he has a some knowledge of digital signal processing. However, as mentioned earlier, copyright protection by digital watermarking technology does not allow much optimism because there are still many crucial issues to be solved technically and politically before practical applications, and so far, in proposed watermarking methods, any proposal which is robust to most of the common signal manipulations does still not exist.

Applications of Digital Watermarking

Although, digital watermarking technology was originally used to protect the copyright for digital content, they have some other important applications in practie [1,6]. In general, there are three major roles for using digital watermarking technology in practical applications as follows.

Copyright Protection. Copyright protection is a well-known application for digital watermarking technology. There are two important roles for using digital watermarking technology to protect content copyright:

1. To establish the content copyright by embedding the copyright-related information as well as some attached information into digital contents, such as the content ID, the standard time, and so forth
2. To embed digital contents with copy control information (CCI) to indicate the status of the contents, such as "never copy," "one copy allowed" and "copy freely"
3. To embed the digital contents with a unique user ID code to specify the authorized users of the contents

As the first role of digital watermarking is for content copyright protection, with embedding copyright information into digital contents, it is possible to deliver functions for the management of digital content as follows: (1) The copyright holders can verify the ownership of their distributed contents by extracting the copyright information from watermarked content; (2) the owners can distribute their contents on the Internet with a confidence that their contents will not be illegally redistributed; (3) the consumers can assure that the content they have bought or want to buy are legitimate; (4) the possibility of watermark extracted from the contents will discourage those who might wish to redistribute the contents illegally.

With embedding CCI data into digital content, as the second role, the devices of replication can prevent the unauthorized replication of the contents. For example, if a document embedded with CCI data does not specify permission to replicate the document, the scanners and printers will refuse to operate it. Meanwhile, with the unique user ID codes embedded into digital content, as the third role, it is possible to track the use of the content and detect illegal replication of the content by identifying the user ID embedded in the content.

Authentication. Authentication is the second possible application for digital watermarking technology. In general, there are two types of authentication application using digital watermarking.

1. To use digital watermarking for authentication of the genuineness of printed materials, such as identity cards, passports, personal checks, coupons, and so forth.
2. To use digital watermarking for tamper-proofing still images by embedding a fragile watermark into the images. In other words, the embedded watermark should be fragile enough so that even small change of the content will destroy the embedded watermark.

Data Hiding. Data hiding is the third possible application for watermarking technology, by which the contents can be embedded with the information about the content. In data hiding, the objective is to provide supplemental information for using the content rather than to protect the copyright of the content. There are usually two major roles of data hiding:

1. To embed the Web site address concerning the contents, by which users can link the contents, such as printed images in journals, posters, and so forth, to the Web site where the users can find more information about the contents or make some applications on line

2. To embed the information used for retrieving the contents, by which users can retrieve or classify the contents more effectively

Note: Steganography also belongs to the techniques of data hiding, which is an important tool for secret communication. The motivation of steganography is to hide a secret message within the contents, such as a still image, audio signals, and so forth, and then transport it to the other side. However, there is an essential difference between steganography and digital watermarking. In steganography, the main body is the message to be hidden secretly within an image and the image is used as envelop to carry the message, thus this image is usually called as "cover image." To the contrary, in digital watermarking, the main body is the image itself and the embedded information is used for the image; thus, this image is usually called as "host image."

Features of Digital Watermarking

The features of watermarking techniques are usually regarded as the functional requirements a watermarking technique should satisfy for a given application. Obviously, different applications will have different requirements for watermarking techniques. Therefore, it is difficult to have a unique set of requirements that all watermarking techniques must satisfy. In general, there are three basic features that are usually required in the most practical application; they are discussed in the following subsections.

Imperceptibility. Imperceptibility is an essential condition for digital watermarking; that is, the embedded watermark should be imperceptible and the quality of the watermarked signal should be as good as the original one perceptually. This requirement determines that the space in a host signal where the watermark can be embedded with robustness will be limited.

Robustness. Robustness means that the watermark cannot be destroyed unless the image (or signal) is altered to the extent of no value.

Under the condition of imperceptibility, to implement a watermarking scheme with the robustness that can endure signal manipulations as much as possible is always a challenging task in digital watermarking. Therefore, robustness is an important feature for evaluating the performance of a watermarking scheme.

Capability. Capability is also an important feature for evaluating the performance of a watermarking scheme. Under the condition of imperceptibility as well as the requirements of robustness, to embed watermark information as much as possible is a more difficult task in digital watermarking. In many articles, the authors really showed their approach to be more robust with the experimental results. But, in their experiments, the image was usually embedded with only 1-bit information, which is called zero-bit information because it can only say "yes" or "no" to the watermark without more additional information. Needless to say, the proposed method will lose its robustness dramatically if the image is embedded with more information.

As mentioned earlier, therefore, the conditions of imperceptibility, robustness, and capability are limited by each other. In other words, for any watermarking scheme, it is impossible to meet these three requirements simultaneously. Figure 13.2 shows a general performance space corresponding to these three features, where the coordinate axes *I*,

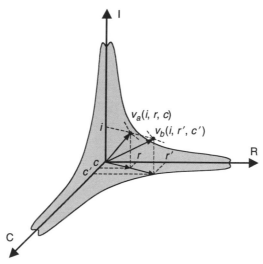

Figure 13.2. The general performance of a watermarking scheme. The coordinates *I*, *R*, and *C* represents the performance of imperceptibility, robustness, and capability, respectively, and the curved surface in gray represents the general performance of a watermarking scheme.

R, and C represent the performance of imperceptibility, robustness, and capability, respectively. The curved surface in gray represents the general performance of a watermarking scheme. Assume that there are two watermarking schemes a and b, which have the general performance expressed in curved surface V_a and V_b, respectively, and V_b is positioned over V_a. The pointer $v_a(i, r, c)$ on the curved surface V_a indicates the general performance of scheme a corresponding to the performance i, r, c. Extending the dotted line $i-v_a$ to form the pointer $v_a(i, r, c)$ on V_a to the pointer $v_b(i, r', c')$ on V_b, method b will have the performance $r' > r$ and $c' > c$. This result means that, given the same imperceptibility, scheme b will have better performance both in robustness and capability compared with the scheme a.

In addition to the above three basic features, there are some other features of watermarking techniques depending on the applications:

1. In general, the embedded watermark is expected to be detectable without using the original signal, because it is usually difficult and even impossible to keep the original signals in advance.

2. In some applications, such as copyright protection, the embedded watermark is usually expected to be inseparable from the watermarked contents, to prevent unauthorized users from reembedding the watermark.

3. In some applications, such as content distribution on the Internet, the embedded watermark is expected to be removable from the watermarked contents in order to renew the watermark information for the distributed contents.

4. In some applications, watermarking schemes are expected to be unopened to the public to prevent unauthorized users from detecting and modifying the embedded watermark.

5. In some applications, the watermarking schemes are required to be opened to the public. Otherwise, the standardization for watermarking technology is impossible.

As mentioned earlier, some of the features required depending on the applications are contradict each other. In other words, it is difficult and even impossible to meet all of the requirements at the same time.

Classifications of Digital Watermarking

Digital watermarking techniques can be classified in a number of ways depending on the applications. Although there are several types of watermarking techniques, in general there are some common ones used in practical applications as follows.

Blind and Nonblind Watermarking. The major difference between blind and nonblind watermarking techniques is that the detection of the embedded watermark from the watermarked signal is accomplished without or with using the original signal. The watermark detection relying on the original signal will be more robust, because with the subtraction of the original signal from the watermarked signal, the embedded watermark can be detected more easily. However, it is usually difficult and even impossible in some applications to have the original signals in advance. The requirements of the original signal greatly limit the practical application of nonblind watermarking techniques. Moreover, when distributed on the Internet, the contents are usually compressed by lossy compression such as JPEG, MP3, and MPEG-1/2/4. It is, therefore, difficult to prepare the original signals compressed with the same parameters that used for the watermarked signal.

In contrast to nonblind watermarking techniques, the blind watermarking techniques do not need the original signals for watermark detection, which makes the method more feasible. Therefore, most of the watermarking products are using blind watermarking techniques. However, the cost of blind watermarking techniques is that the robustness of a watermarking scheme will decrease.

Robust and Fragile Watermarking. The objective of robust watermarking techniques is to embed a watermark into the host signal as robust as possible to endure any possible signal manipulations to prevent unauthorized users from destroying or removing the watermark. Therefore, the robust watermarking techniques are mostly used for copyright protection. To the contrary, the objective of fragile watermarking methods is to embed a watermark into the host signal in such a way that it will be destroyed immediately when that watermarked signal is modified or tampered with. Therefore, the fragile watermarking is often used for tamper-proofing of images.

Reversible and Inseparable Watermarking. Currently, most of the proposed watermarking approaches belong to inseparable watermarking techniques: that is, the watermark is embedded in such a way that it will be difficult for unauthorized users to separate or remove the watermark from the watermarked signal, because only in this way can the watermark be effectively used for copyright protection and authentication. However, in some situations, the watermarked signals are expected to be reversible, that is, the original signal can be retrieved from a watermarked signal. In practice, there are two kinds of application of reversible watermarking. First, when the host signals are valuable data, such as a medical image, military image, artistic work, and so forth, it is usually required that the original signal can be retrieved completely from the watermarked signal, because even the lowest-bit information may have some value that which will be useful to future work. Second, in the applications for contents management, the embedded watermark is usually expected to be reversible in order to renew the embedded information, because the status of contents is changed with the contents distributed on the Internet.

The retrievable watermarking techniques are also a troublesome task, as challenging as the inseparable watermarking techniques, because it is usually required that the original signals be retrieved completely from the watermarked signals. One necessary condition for retrievable watermarking is that the bit information replaced by embedded watermark should be kept within the image, and, theoretically, it is impossible to have such remaining space in an image unless the image has some pixels unused. Moreover, if the retrievable watermark is expected to be robust to general signal manipulations, the task will be more difficult. Currently, unfortunately, this kind of retrievable watermarking is still not available.

Public and Private Watermarking. As indicated by its name, private watermarking techniques allow the embedded watermark to be detected only by authorized users. To the contrary, public watermarking techniques allow the embedded watermark to be detected by the public. In general, private watermarking is more robust than public watermarking, because it will be much easier for attackers to destroy or remove the embedded watermark if the algorithm was opened to public. However, an important condition for copyright protection using watermarking techniques is that the watermark should be embedded by authorized watermarking schemes; otherwise, the embedded watermark will lose its original means (i.e., used as evidence of copyright). In other words, the premise of being an authorized scheme is to open the algorithm first.

Based on the above definitions of public and private watermarking, in a way, the nonblind and blind watermarking belong to the public and private watermarking, respectively. The reason for this is that blind watermarking can be only used for private watermarking; otherwise, it is not necessary to use a watermarked image if anyone has the original one previously.

Major Issues of Digital Watermarking

We have given a brief overview of digital watermarking technology and corresponding issues. Indeed, so far, many interesting methods have been proposed and the technical issues for many kinds of application have been discussed. However, as mentioned earlier, there are still many crucial issues to be solved before the practical application of digital watermarking is possible. Among these issues, those concerned with the copyright protection may be the most crucial ones, because the applications of copyright involve the issues of legality, which not only concerns the content owners but also the users who use the contents. The major issues for the applications of copyright protection are presented in the following subsections.

Limitation of Watermarking Technology. This issue belongs to the technical problems of digital watermarking itself. Because the space in any contents where watermarking techniques can be effectively used is very limited, it is impossible for us to use multiple watermarking techniques for an unrestrained application. Therefore, this crucial condition may determine that those unsolved issues for the copyright protection will be a lengthy topic with controversy.

Administration of Watermarking Technology. This issue belongs to the problems beyond the scope of digital watermarking technology. The watermark does not originally provide any legal information of ownership; that is, without being registered to a trusted agent, the watermark embedded by individuals or some associations could be invalid in law. Especially, by current watermarking techniques, it still difficult to solve the problem of who really watermarked the content first. Therefore, the only way to address this issue is to have a united administration for copyright protection. In other words, a trusted third party is needed to establish a verification service for digital water-marking.

Standardization of Watermarking Technology. This issue belongs to technical problems of digital watermarking as well. In order to have a united administration for the applications of watermarking technology, there should be a worldwide standardization for watermarking techniques as well as the technical issues, such as the JPEG standard for still image compression, the MPEG standard for video signal compression, and the MP3 standard for audio signal compression. However, in order to realize the standardization of watermarking techniques, there are some preconditions:

1. The algorithms of watermarking techniques should be opened for technical verification so as to establish the credibility to the public.
2. The watermark approach should be robust to all possible signal manipulations in practice.

3. The embedded watermark should be reversible for authorized administrators, but should be inseparable for general users.

The first condition (e.g., the threshold in watermark detection, which is usually considered as a criterion to judge if the watermark was embedded) should be determined by the standardization committee, not done ambitiously by the researchers themselves. Of cause, the researchers may assert that the reliability of their threshold can be assured by using some check-code such as cyclic redundancy check (CRC) codes or Bose-Chaudhuri-Hochquenghem (BCH) codes in their algorithm. Even so, the researchers still have the obligation to show the reliability of the check-code to the public. Moreover, as mentioned earlier, the open algorithm will greatly reduce the security of watermarking schemes; that is, once the approach for watermarking is known, it will be much easier for an attacker to destroy the watermark. For the second and third conditions, they are difficult to meet because of the capability of current watermarking techniques.

In order to develop standards for digital watermarking technology, many associations and organizations are working toward standardization of watermarking techniques, such as the Copy Protection Technical Working Group (CPTWG, http://www.cptwg.org), the Digital Audio-Visual Council (DAVIC, http://www.davic.org), the Secure Digital Music Initiative (SDMI, http://www.sdmi.org), and the Japanese Society for Rights of Authors Composers and Publishers (JASRAC, http://www.jasrac.or.jp). Unfortunately, the standardization attempts have either ended without conclusion or have been postponed indefinitely.

Even though digital watermarking technology still has many issues to be resolved for its practical applications, as mentioned earlier, many watermarking technologies have been used commercially. The reason for this is that the potential crisis of data pirating has made customers eager to protect their intelligent copyright by choosing a watermarking technology. Finally, let us use a phenomenon in our daily life to describe the situation in the current research and application of digital watermarking technology. It is well known that everyone should have a key on his door no matter how robust the key will be; otherwise, he will feel uneasy when not home. Meanwhile, in law, it is unnecessary and even impossible to require all keys should have a standard in its robustness. The applications of digital watermarking for copyright protections is just like the "key" on our door; although we cannot assure that this "key" will act as robust as expected in any situations, its important role may be that any possibility of watermark detection will greatly discourage those who might wish to use the contents illegally. Therefore, the applications of watermarking technology will always be challenging topics now and in the future.

WATERMARKING TECHNIQUES FOR PRINTED MATERIALS

About Printed Materials

In recent years, the digital right management (DRM) system has been become a very popular topic, because it promises to offer a secure framework for distributing digital content. Digital watermarking techniques can be used in DRM systems for establishing ownership rights, tracking usage, ensuring authorized access, preventing illegal replication, and facilitating content authentication. However, the watermark embedded in digital images or documents will be lost if they are printed on paper, because the effective scope of digital watermarking techniques is limited to within the digital domain. In other words, the effective scope of DRM systems is limited within the digital domain as well. Therefore, watermarking for printed materials is an important issue for the DRM system, which poses a new challenge and attracts many researchers to this area. In general, there are two kinds of printed material used in the DRM system:

1. Printed images
2. Printed textual images

Printed Images. As indicated by its name, printed images are images printed on paper, such as the paper image, passport photos, coupons, and so forth. There are a number of applications in which digital watermarking techniques can play an important role for printed images:

1. The reason why of some content owners deliver their digital images on the Internet is to provide the images to be used for printed materials. In this application, watermarking techniques are highly necessary to deliver the copyright protection for this kind of image.
2. With the rapid progress in printing and scanning techniques, the easy forgery of passport, ID cards, and coupons becomes a critical social problem, for which watermarking techniques can be used to authenticate the genuineness.
3. With the rapid development of mobile techniques and the rapid growth of the mobile market, many advertising agencies urgently expect that the customers can easily get onto their Web site with the information simply extracted from the printed sample image by using mobile camera phone.

All of the above applications require that watermarking techniques should be robust enough to endure the printing and scanning process.

Printed Textual Images. Printed textual images are the images printed form the text documents or textual images scanned from a paper document. There are several applications where watermarking techniques can be used for printed textual images:

1. Before the wide use of digital office tools, such as Microsoft Word and so on, the paper documents were usually generated by a word processor. To restore them as digital content, the important text documents are usually redigitized by scanning.
2. In general, the text documents need to be printed on paper.

Watermarking for printed textual images can provide an effective protection for text documents against illegally copying, distributing, altering, or forging. However, watermarking for the printed textual image will be more difficult than that for the printed image, because the images for printed textual image are usually the binary image with only two values: black and white.

Watermarking for Printed Images

As in general watermarking techniques, watermarking for printed materials consists of two processes as well: watermark embedding and watermark detection. In the watermark embedding process, the watermark is embedded into a textual image or a text document and then the watermarked image is printed on paper. In the watermark detection process, the watermarked printed material is scanned with a scanner and then the watermark is extracted from the scanned image. Because watermarking for printed materials is a process of digital-to-analog and analog-to-digital transforms, there are a number of problems that will cause the watermark embedded in printed materials to weakened or even be destroyed.

Problems of Watermarking for Printed Images. As described in References 7 and 8, there are two major problems that affect watermark detection on printed images:

1. Geometric distortions
2. Signal distortions

Geometric distortions are invisible to the human eye, which belong the one of most problematic attacks in digital watermarking, because a small distortion, such as rotation or scale, will not cause much change in image quality but will dramatically reduce the robustness of the embedded watermark. Therefore, the issue for geometric distortions is always an active topic in digital watermarking, which attracts many researchers to this area. Generally, the major geometric distortions occurring in printing and scanning are rotation, scale, and translation (RST).

The rotation distortion is caused in the scanning process, because it is difficult to ensure that the paper image is placed on the scanning plane horizontally. The scale distortion is caused by an inconsistency in resolutions between printing and scanning, because it is difficult to have the same resolution both in printing and scanning processes. In general, there are two kinds of scale distortion that are strict and even mortal to the embedded watermark: pixels blur and aspect ratio inconsistency. The pixel blur is caused in the printing process, because the watermarked images are usually printed with a high resolution in dpi (dots per inch) in order to have a high image quality; thus, the watermark information embedded in near pixels will be blurred. The aspect ratio inconsistency is caused in the scanning process by a different ratio between the horizontal and vertical scanning, which is a kind of nonlinear distortion and thus is more difficult to deal with than general linear-scale distortion. The translation distortion is caused in the scanning process as well, because the printed images are usually scanned partially or scanned including the blank margin.

Signal distortions are visible to the human eye, which include the changes of brightness, colors, contrast, and so forth. Contrary to geometric distortions, pixel value distortions will cause a change in image quality, but they will not have much effect on the watermark embedded in the printed images.

Watermarking Techniques for Printed Images. In general, the printed materials, such as books, magazines, and newspapers, are the result of a halftoning process, in which a continuous-tone image is transformed into a halftone image, a binary image of black and white. Therefore, there are two kinds of watermarking technique that can be used for printed images:

1. Watermarking for a continuous-tone image
2. Watermarking for a halftone image

Watermarking for the continuous-tone image belongs to the general watermarking techniques, by which the watermark is embedded into an image in its spatial domain or frequency domain and then the watermarked image is printed on paper by the halftoning process. Therefore, the watermark embedded in the image may be destroyed due to the transformation of a continuous-tone image to a binary image in the halftoning process. To the contrary, watermarking for a halftone image is a technique by which the watermark is embedded in the halftoning process by exploiting the characteristics of the halftone images. Therefore, in principle, watermarking techniques for the halftone image will be more effective in printed images than the general watermarking techniques of the continuous-tone image. However, there are two major disadvantages in watermarking techniques for the halftone image:

1. The methods are sensitivity to geometric distortions such as rotation, small scale, and the presence of stains or scribbles on the printed materials, because the watermark is embedded in the spatial domain.

2. It is usually required that the print resolution be significantly lower than the scanning resolution in order to ensure a high effectiveness in watermark detection. Therefore, the printed images have to reduce their visual quality.

Continuous-Tone Watermarking Techniques

As mentioned earlier, the major problem in watermarking for the printed image is geometric distortion of rotation, scale, and translation (RST), which is a major issue for robust watermarking. Therefore, it is possible to use robust watermarking techniques to deal with the problems in watermarking for printed images. In general, there are two important measures to deal with the problems of geometric distortions in the printing and scanning process:

1. To build a robust watermark with a structure against geometric distortions

2. To accomplish watermarking by RST robust watermarking techniques

Robust Watermark Structure. In a general watermarking scheme, as expressed in Equation 13.1, the watermark is usually embedded into an image pixel by pixel. Therefore, a small geometric distortion will change the order of extracting the watermark and make the watermark functionless, because the watermark should be extracted by locating the pixels in which the watermark is inserted. As pointed by Cox et al. [9], there two important parts for building a robust watermark:

1. The watermark structure
2. The insertion strategy

To build a robust watermark structure, it is usually required that the watermark be built as an integer structure so that if there is local damage in the watermark, it will not have much effect on the total watermark structure. In frequency domain watermarking, a well-known method for building a robust watermark is the spread spectrum technique as described in Reference 9, by which the watermark is generated using pseudo random sequences and then embedding it into the middle range in the frequency domain. The reason for this is that with a signal embedded in the frequency domain, the energy of the signal will be spread uniformly into the spatial domain and, therefore, a location damage of the signal in the spatial domain will not have much effect on the total energy of that signal in the frequency domain. In spatial domain watermarking, it is usual to use a pseudorandom sequence with a period as large as a rectangular array of numeric elements to represent a 1-bit watermark. Therefore, if a small part of the signal is damaged, the watermark plane can still have a good correlation with the original pseudorandom sequence due to the correlation properties of pseudo random sequence.

As an insertion strategy for robust watermark, Cox et al. [9] have proposed embedding the watermark pattern into the perceptually most significant component of signals. This concept is based on the following arguments:

1. The watermark should not be placed in perceptually insignificant regions of the image, since many common signal and geometric processes affect these components.
2. The Significant components have a perceptual capacity that allows watermark insertion without perceptual degradation.
3. Therefore, the problem then becomes how to imperceptibly insert a watermark into perceptually significant components of the signals.

With the above arguments, Cox et al. [9] proposed a method using spread spectrum watermarking techniques, in which the watermark represented using a pseudorandom sequence is embedded into the middle range in the DFT domain. Another well-known scheme for watermark embedding is to embed the watermark according to the masking criterion based on the model of the human visual system (HVS), which exploits the limited dynamic range of the human eye to guarantee that the watermark is embedded imperceptibly with the most robustness. The details about watermarking using the HVS model can be found in References 10–13.

RST Robust Watermarking. Building RST robust watermarking is a very popular topic in recent watermarking research, because to challenge the issues of RST distortion is always a work representing the highest level in digital watermarking. For instant, if you have skimmed the recent journal of *IEEE Transactions on Circuits and Systems for Video Technology* published in August 2003 (Vol. 13, No. 8), you will find that, among the six articles about robust watermarking, there are four articles [17,23,24,26] concerned with the issues of RST distortion. So far, there are a number methods proposed for RST distortion, which can be mainly divided into three types as follows:

1. To embed a template into images along with the watermark
2. To embed the watermark into a domain that is invariant to geometric distortions
3. To synchronize the watermarks by exploiting the features of an image

The idea for the methods of type 1 is to identify what the distortion was and to measure the exact amount of the distortion in order to restore an undistorted watermarked image by inverting the distortion before applying watermark detection. This can be accomplished by embedding an additional template along with the general watermark [14–17]. The template contains the information of geometric transformations undergone by the image and is used to detect the distortion information used for image geometric revisions. With the image geometrically restored from the scanned image, it is possible to extract the general watermark correctly from the geometrically restored image. However, the cost of these methods is a reduction in image fidelity, because it is required to embed the watermark with additional template information.

Some methods have been proposed based on the idea of type 1, in which the watermark was embedded into the mid-frequency range in the DFT domain [14–16] or in the DWT domain [17] in the form of a spread spectrum signal. The template consisted of a number of peaks random arranged in the mid-frequency range in the DFT domain as well. However, some researchers have complained that these templates may be easy to remove.

Instead of using an additional template, Kutter [18] has proposed a method based on an autocorrelation function (ACF) of a specially designed watermark. In the method [18], the watermark is replicated in the image in order to use the autocorrelation of the watermark as a reference point. Voloshynovskiy et al. [19] have proposed a method based on the shift-invariant property of the Fourier transform. In the method [19], the watermark is embedded into a period block allocation in order to recover watermark pattern from geometric distortion.

The idea for the methods of type 2 is that the watermark should be embedded into a domain that is invariant to geometric distortion. Theoretically, the Fourier–Mellin domain is the place that is invariant to RST distortion. O'Ruanaidh and Pun [20] first proposed the watermarking scheme based on the Fourier–Mellin transform and showed that the method can be used to produce watermarks that are resistant to RST distortions. Figure 13.3 is a diagram of a prototype RST-invariant

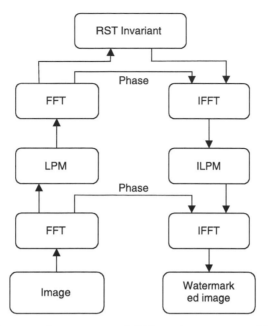

Figure 13.3. Diagram of a prototype RST-invariant watermarking scheme.

watermarking scheme. In the proposed method, the watermark embedding is accomplished in the process as follows:

1. To compute the discrete Fourier transform (DFT) of an image
2. To compute the Fourier–Mellin transform of the Fourier magnitude spectrum, where the Fourier–Mellin transform is a log-polar mapping (LMP) followed by a Fourier transform, which is a RST-invariant domain
3. To embed the watermark into the RST-invariant domain (i.e., the magnitude of the Fourier–Mellin transform)
4. To compute an inverse discrete Fourier transform (IDFT)
5. To compute an inverse Fourier–Mellin transform, an inverse log-polar mapping (ILMP) followed by an IDFT

With the accomplishment of procedures 1 to 5, the watermarked image is accomplished. In the watermark detection process, the watermark is extracted by transforming the watermarked image into the RST-invariant domain. However, there are several problems that greatly reduce the feasibility of the Fourier–Mellin method. First, the method suffers severe implementation difficulty. Second, the watermarked image will have to endure both the LPM and ILPM transforms, which make the image quality unacceptable. For this reason, O'Ruanaidh and Pun [21] have proposed an improved method for using Fourier–Mellin transform, as shown in

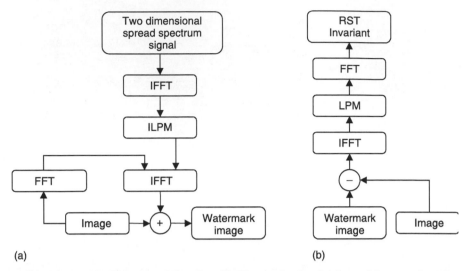

(a)   (b)

Figure 13.4. The improved Fourier–Mellin method, which avoids mapping the original image into the RST-invariant domain: (a) the watermark embedding process; (b) the watermark detecting process.

Figure 13.4. The consideration is very interesting, based on which only the watermark data goes through the ILPM and then it is inserted into the magnitude spectrum of the image. By applying the IDFT to the modified magnitude spectrum, one can get the watermarked image. With the improved method, the image quality will be increased and the implementation will be simplified, because only the watermark signal goes through the ILPM transform. However, the improved method needs the original image for watermark detection.

Other researchers have proposed methods based on the Fourier–Mellin transform in order to improve the Fourier–Mellin method to be more feasible for practical applications. Lin et al. [22] proposed a method for developing a watermark that is invariant to RST distortion based on the Fourier–Mellin transform, in which the watermark is embedded into a one-dimension signal obtained by taking the Fourier transform of the image, resampling the Fourier magnitudes into log-polar coordinates, and then summing a function of those magnitudes along the log-radius axis. The method uses a translation- and scaling-invariant domain, whereas the resistance to rotation is provided by an exhaustive search.

Zheng et al. [23] proposed a method based on the Fourier–Mellin transform, in which the watermark is embedded in the LPMs of the Fourier magnitude spectrum of an image, and uses the phase correlation between the LPM of the original image and the LPM of the watermarked image to calculate the displacement of the watermark position in the LPM domain. The method preserves image quality by avoiding computing the inverse log-polar mapping (ILPM) and it produces a smaller correlation coefficient for original images by using phase correlation to avoid an exhaustive search.

The idea for methods of type 3 is to synchronize the watermark by exploiting the features of an image. Based on the consideration of type 3, Simitopolus et al. [24] have proposed a method in which the amount of geometric distortion is computed by using two generalized Radon transformations; Shim and Jeon [25] have proposed a method that exploits the orientation feature of an image by using two-dimensional (2-D) Gabor kernels; Kim and Lee [26] have proposed a method in which the rotation invariance is achieved by taking the magnitude of the Zernike moments.

Similar to the consideration of type 3, in References 27 and 28, the authors have proposed an interesting concept called the second-generation watermark. Based on the concept, the watermark is generated by using perceptually significant features in the images. The features can be edges, corners, textured areas, or parts in the images with specific characteristics that are invariant to geometric distortions. As defined in Reference 27, for the requirements for watermarking, the features should have the following properties:

1. Invariance to noise (lossy compression, additive, multiplicative noise, etc.)
2. Covariance to geometric transformations (rotation, translation, subsampling, change of aspect ratio, etc.)
3. Localization (cropping the data should not alter the remaining feature points)

Property 1 ensures that only significant features are chosen. Attacks are likely to alter significant features because, otherwise, the commercial value of the data would be lost. Therefore, selecting salient features implies that these features are resistant to noise. Property 2 describes the behavior of the feature if the host data are geometrically distorted. Moderate amounts of geometric modification should not destroy or alter the feature. Property 3 implies that the features should have a well-localized support. The use of such features makes the watermarking scheme resilient to data modifications such as cropping.

As described earlier, the issue of RST robust watermarking is really an active topic with controversy in current watermarking research, and in the near future, more new approaches will be reported. However, what real value practically will a proposed method have and which of the proposed methods will be the best one for RST robust watermarking? If there is not a standard for the evaluation of robustness to RST distortion, it will be quite difficult and even impossible to evaluate their real value or to compare their performances with each other. However, so far, almost all of the proposed methods were only evaluated based on only two features: robustness and imperceptibility.

In practice, as a standard of a robust watermarking with the possibility for application, it is usually required that the watermark capability should have about 64 bits or at least should have multibits rather than a zero-bit watermark, except for the bits for the check-code, such as CRC codes or BCH codes. For instance, the Content ID Forum in Japan (cIDf, http://www.cidf.org) used 64 bits as the content ID for embedding a watermark into contents [1,6]. In STEP2001 (http://www.jasrac.or.jp/ejhp/news/1019.htm), a technical evaluation for audio watermarking was sponsored by the Japanese Society for Rights of Authors Composers and Publishers (JASRAC, http://www.jasrac.or.jp), The samples delivered for evaluation should be embedded with 2 bits of watermark data in a timeframe of no more than 15 sec and 72 bits in one of no more than 30 sec [1,6]. Therefore, if evaluated based on the criteria of robustness, imperceptibility, and watermark capability simultaneously, the methods of type 1 is possibly the most promising for future applications, because it is possible for type 1 methods to select a watermarking scheme with the capability of watermarking according to the application requirements.

Unfortunately, so far, a method that can be accepted worldwide based on the criteria of robustness, imperceptibility, and watermark capability is still not reported. As an example of the application using robust watermarking for the printed image, the solution of Digimarc's Media-Bridge produced by the Digimarc Company [29] has been placed on the market. With the Digimarc MediaBridge, the users can link advertisement images on the pages of a journal directly to a concerned Web page by using a digital camera to extract the Web site information from the images.

Halftone Image Watermarking Techniques

What Is the Halftoning Technique. Halftoning [30] is a traditional technique used to transform continuous-tone images into binary images, which look like the original images when viewed from a distance. Halftoning techniques are widely used in printing books, newspapers, magazines, and so forth, because in the general printing processes, the printed materials can be generated by only two tones: black and white. There are number of methods proposed for halftone image techniques, which can be divided into three categories:

1. Ordered dither method
2. Error diffusion method
3. Other methods

Table 13.1. 8 × 8 Threshold matrix

0	32	8	40	2	34	10	42
48	16	56	24	50	18	58	26
12	44	4	36	14	46	6	38
60	28	52	20	62	30	54	22
3	35	11	43	1	33	9	41
51	19	59	27	49	17	57	25
15	47	7	39	13	45	5	37
63	31	55	23	61	29	53	21

The ordered dither method [31], which is the oldest halftoning technique used in printing, applies a periodic threshold matrix to each image pixel. Table 13.1 shows an example of the threshold matrix with a dimension (period) 8×8 containing the thresholds from 1 to 64, which represents the values of $i/64$ $(i=0\text{–}63)$. Compared with the threshold matrix, the pixel (x, y) can be converted to zero (black) or one (white) as follows:

$$b(x,y) = \begin{cases} 1, & p(x,y) \geq T(x,y) \\ 0, & p(x,y) < T(x,y) \end{cases}$$ (13.2)

where $p(x, y)$ is the input pixel value in the normalized range $[0,1]$ and $b(x, y)$ is the binary image with the values 0 and 1. $T(x, y)$ is the threshold matrix with a period of 8×8.

Because the output pixels are obtained independently, the ordered dither method has the advantage of being computationally inexpensive. However, the drawback of the ordered dither method is that halftones suffer from periodic patterns due to the use of a periodic matrix.

To the contrary, the error diffusion method, first proposed by Floyd and Setinberg in 1976 [32], produces halftones of higher quality than the ordered dither method. The flowchart for the error diffusion method is shown in Figure 13.5.

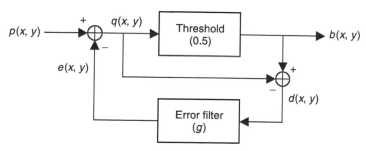

Figure 13.5. The error diffusion process.

In Figure 13.5, $p(x, y)$ denotes the pixels of the continuous-tone image, with the values normalized in range $[0, 1]$, which flows sequentially, and $e(x, y)$ denotes the error diffused from past halftones by the error filter g. Thus, adding the diffused error $e(x, y)$, we have the quantization $q(x, y)$ and the output halftone $b(x, y)$ as follows:

$$q(x,y) = p(x,y) + e(x,y), \tag{13.3}$$

$$b(x,y) = \begin{cases} 1, & q(x,y) \geq 0.5 \\ 0, & q(x,y) < 0.5 \end{cases} \tag{13.4}$$

Here, the halftone error $d(x, y)$ is defined as

$$d(x,y) = b(x,y) - q(x,y) \tag{13.5}$$

Then, the halftone error $d(x, y)$ is diffused to future pixels by the calculation of the error filter g as showed in Table 13.2. Other well-known error diffusion filters are Jarivis–Judice–Ninke [33] and Stucki [34] filters.

However, because halftones are calculated with diffused past errors, the main drawback of the error diffusion method is that it is computationally expensive. Therefore, some other methods have been proposed attempting to retain the good features of error diffusion while keeping the computation simple. More details are provided in References 35–37.

Watermarking Techniques for Halftone Images. So far, a number of methods used for watermarking halftone images have been proposed, which can be divided into the following categories:

1. Watermarking for the ordered dither halftone
2. Watermarking for the error diffusion halftone
3. Other types of watermarking the halftone image

Table 13.2.   The Floyd and Steinberg filter

$p(x-1, y-1)$	$p(x, y-1)$	$p(x+1, y-1)$
Past	Past	Past
$p(x-1, y)$	$d(x, y)$	$e(x+1, y) = d(x, y) \cdot 7/16$
Past	Present	Future
$e(x-1, y+1) = d(x, y) \cdot 3/16$	$e(x, y+1) = d(x, y) \cdot 5/16$	$e(x+1, y+1) = d(x, y) \cdot 1/16$
Future	Future	Future

Watermarking for the ordered dither halftone is a technique by which the watermark information is embedded into the halftone images by exploiting the process of ordered dithering. Therefore, the major advantage of these method is that it is computationally inexpensive. There are still a few watermarking techniques for the ordered dither halftone [38–40], and the basic idea is to embed watermark information using a sequence of two (or more) different threshold matrices in the halftoning process.

Similar to the ordered dithering methods, watermarking for error diffusion halftone embeds watermark information into the halftone image by exploiting the process of error diffusion. There are several methods proposed for error diffusion of halftone images. Pie and Guo [41] have proposed a method in which the error filter g shown in Figure 13.5 was replaced by using two error filters, Jarvis [33] and Stucki [34], to represent 0 and 1 of the watermark information. Their combination will not have much effect on the quality of the halftone images because these two filters are compatible each other. Hsu and Tseng [42] have proposed a method in which the watermark is embedded into the halftone image by adding additional probability condition in the process of error diffusion. Fu and Au [43,44] have proposed a method called DHST (data hiding by self-toggling), in which a set of locations within the images is generated by using a pseudorandom number generator with a known seed. Then, 1 bit of watermark information is embedded in each location by forcing the pixel at the location to be 0 or 1 (normalized), corresponding to black or white. Therefore, compared with the threshold 0.5 in the error diffusion process, the embedded pixel will remain unchanged. However, this method will introduce a visible salt-and-pepper noise into the halftone images due to the random distribution of the watermark information. Moreover, Kim and Afif [45] have proposed an improved method for DHST. However, objective is tamper-proofing for halftone images, not watermarking for printed images.

There are some other types of method used for watermarking halftone images, in which the watermark is a binary logo mark image and it is embedded into the original image imperceptibly, as the general water-marking techniques do and the watermark is extracted by showing the logo mark directly on the printed image, which is done as for the traditional watermark (i.e., the watermarks can be observed on the printed materials under some conditions).

Hsiao et al. [46] have proposed a method for the purpose of anticounterfeiting printed images by using varied screen rulings. In the proposed method, a binary logo mark image is embedded into the original image in the form of varied screen rulings. As for the general watermarking techniques, the watermark embedded in the printed image

411

is invisible perceptively. The embedded logo mark can be observed on the paper if the printed image is replicated by a copy machine. As described in Reference 46, the idea for the proposed method is taking advantage of the limited ability of sampling by copying machines. By viewing a properly handled binary image with varied screen rulings, the HVS can integrate the neighboring halftone dots and perceive a uniform gray-scale image without observing the varied screen rulings. For most copy machines, the resolution is up to approximately 600 dpi. In designing the binary image, part of the logo mark is generated with screen rulings higher than 600 dpi in order to achieve undersampling effects [47], and part of the background is generated with the general screen rulings lower than 600 dpi.

The techniques of hiding a mark in printed materials using varied screen rulings have been widely used in the printing of blank papers. For example, as shown in Figure 13.6, in Japan, paper certificates are generally printed with the background pattern where an invisible mark or some characters, such as "COPY MATTER," or "COPYING PROHIBITED," was hidden. If the certificates are replicated by a copy machine, the hidden mark or characters will appear on the paper documents with the disappearance of the background pattern. Recently, some products using hidden mark techniques, such as the products of Fuji Xerox [48], have been put on the market.

Li et al. [49] have proposed another method using masking techniques called MCWT (multichannel watermarking techniques), which is similar to the method of Hsiao et al. [46]. The proposed method is used to hide a predefined logo mark in an exhibit image by exploiting the characteristics of halftone images. The watermarked image is then printed on paper with the hidden logo for the purpose of copyright authentication. In the MCWT method, the dots of the original image are located in general positions, and the logo mark is embedded into the original image with the dots shifted by four positions of the halftone cells. Figure 13.7 shows the process of the MCWT method, in which the embedded logo mark is

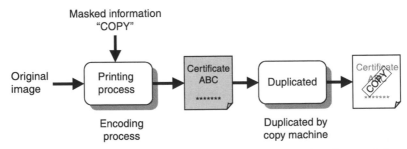

Figure 13.6.   The process of masking information into printed materials.

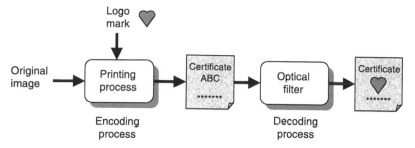

Figure 13.7.   The watermarking process of the MCWT method.

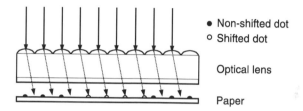

Figure 13.8.   The design of an optical lens for decoding. The incident light will focus on shifting dots through the optical lens at the given angle. (From Li, J.Y., Chou, T.R., and Wang, H.C., presented at IPPR Conference on Computer Vision, Graphics and Image Processing, Kinmen, Taiwan, 2003.)

detected with an optical lens that filters the printed image to exhibit the hidden one.

Figure 13.8 shows the designed optical lens for decoding. The number of lenticules per inch is decided by the screening resolution in the halftoning process. For example, if the screening resolution is 150 dpi, the watermarked image probably will probably be decoded with an optical lens of 75 dpi or 150 dpi. Therefore, the embedded watermark will become apparent as the refracted light is aimed the shifted dots and when the lens is rotated in the appropriate direction.

Watermarking for Printed Textual Images

Roles of Watermarking Printed for Textual Images. The demand for document security is increasing higher in recent years, because with the fast developments in hardcopy techniques, the photocopy infringements of copyright are always important issues concerning publishers. Especially, with the spread of the Internet, an electronic document can be easily sent to other persons by e-mail with far less cost than the hardcopy by copy machines. Therefore, the copyright protection and

413

authentication of electronic documents are important issues for the document DRM system, which includes the following roles:

- *Copyright protection*: To identify and protect copyright ownership
- *Copy control*: To give permission for a legal hardcopies of documents
- *Tracking*: To identify users who replicate the documents illegally
- *Tamper proofing*: To detect the modification made to the document

Watermarking Techniques for Printed Textual Images. To date, most proposed watermarking methods are used for color or gray-scale images in which the pixels have a wide range of values. Therefore, it is possible to embed watermark information into the pixel values, because a small change in pixel values may not be perceptible by the HVS. However, watermarking techniques for printed textual images are quite different from those for the general image, because printed textual images are generated either by textual documents or textual images that have only two kinds of pixel value: black and white. Currently, the proposed watermarking techniques for textual documents or textual images can be divided into the following categories:

1. Watermarking for text documents
2. Watermarking for binary images

Watermarking for Text Documents

To date, a number of methods have been proposed for watermarking text documents [50–56], which can be divided into three types:

1. Line-shift coding
2. Word-shift coding
3. Feature coding

Line-shift coding is a method that alters a document by vertically shifting the locations of text lines to encode the document. This method can be applied either to the format file or to the binary textual images. Watermark detection can be accomplished without the use of original images when the original document has a uniform line space.

Word-shift coding is a method that alters a document by horizontally shifting the locations of words within text lines to the format file or to the binary textual image. This method can be applied to either the format file or the binary textual image. Word-shift coding method has the advantage in that the watermarked documents are least visible because the watermark can be embedded into the interword space by adjusting the white space in the variable spacing. Variable spacing is a general treatment of adjusting white space uniformly distributed within a text line. However, the variable spacing will make the detection more complex; thus, the watermark detection usually requires the original image.

Feature coding is a method in which the image is examined for chosen text features, and the watermark is embedded into a document by changing the text features of selected characters. Watermark detection requires the original image.

As pointed out in Reference 51, among the three methods described above, the line-shift method is likely to be the most easily discernible by readers. However, the line-shift method is most robust in the presence noise. This is because the long length of text lines provides a relatively easily detectable feature. Therefore, the line-shift method is the most promising method for printed textual documents. The word-shift method is less discernible to the reader than the line-shift method because the spacing between adjacent words on a line is often varied to support text justification. The feature coding method is also indiscernible to readers. The method has the advantage that it can accommodate a particularly large number of sanctioned document recipients because there are frequently two or more features available for encoding in each word. Another advantage is that it can be applied simply to image files, which allows encoding to be introduced in the absence of a format file.

Huang and Yan [56] have proposed a method for printed text documents in which the interword spaces of different text lines are modulated by a sinusoidal function and the watermark information is carried by the phase of the sinusoidal function. There are some advantages of modulating interword space by the sinusoidal function:

1. The variation of interword spaces may be less discernible because a sinusoidal wave varies gradually.
2. The periodical property makes the watermark detection much easier and more reliable.

Other authors [57–61] have shown that for some watermarking applications, the properties of the sinusoidal function are very useful for building a robust and reliable digital watermarking.

Watermarking for Binary Images

As mentioned earlier, binary images, such as scanned text, figures, and signatures, are also an important issue that should be taken into account for the document DRM system. In general, it is not difficult to embed watermark information into a digital binary image, which can be scanned from the printed materials. However, the task becomes very challenging if it is expected that the embedded watermark can be extracted from a printed binary image. Note that the techniques used for watermarking a text document, mentioned earlier are still useful for the scanned binary images, but they will be invalid when the images are textual images. In other words, we have to take into account other methods when the scanned binary images are figures, signatures, and so forth. So far, there are some proposed methods that can be used for binary images, including the textual image, figure, and signature, which can be divided into two large categories as follows:

1. Interior watermark embedding method
2. Exterior watermark embedding method

Interior Watermark Embedding Methods. Interior watermark embedding is a method similar to general digital watermarking; that is, the watermark information is embedded within a binary image imperceptibly by exploiting the characteristics of the image. The basic method of embedding data in binary images is to change the values of pixels according to the characteristics of the images. So far, there are mainly three types of approach to embed data in binary images by changing pixel values.

The first type of approach is to embed watermark information by replacing the pixel patterns in a window size of $N \times N$ with previously defined window patterns. For example, Abe and Inoue [62] have used eight window patterns as shown in Figure 13.9 to represent bits 0 and 1.

There are other criteria for the generation of window patterns. In Reference 63, the pixel patterns were changed in a way that to represent

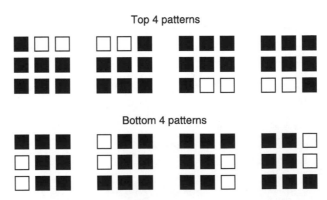

Figure 13.9. Window patterns; the top four patterns represent bit 1 and the lower four patterns represent bit 0.

"0," the total number of black pixels in the window is an even number. Similarly, to represent "1," the total number of black pixels in the window is an odd number. To increase the robustness of the window patterns, the difference of total black pixels can be quantified to be $2kQ$ (for some integer k) to embed "0" and to be $(2k+1)Q$ to embed "1."

The second type of approach is to embed a watermark by changing the features [63], such as the thickness of strokes, curvature, relative positions, and so foth. These approaches may be more suitable for the cases of figures, such as the electric circuit figure, construction figure, and so on. The last type of approach is to embed watermark information into places having structure characteristics. For example, as shown in Figure 13.10, the watermark is embedded into the three corners within the character "F." The basic idea for this method is that some structure characteristics in the binary image, such as the corner of a character or a pattern, are invariant to geometric distortion.

Exterior Watermark Embedding Methods. Exterior watermark embedding is the approach in which the watermark information is embedded into the area outside of the main constituents within an image, such as the background patterns, the blank places, and so forth. Some examples of using exterior watermark embedding methods are:

1. Embed the watermark into the background, a design pattern
2. Embed the watermark into the background, a noise pattern
3. Embed the watermark into a logo mark pattern

Figure 13.11 is an example of embedding a watermark into the design of a coupon [6], where the watermark is embedded by making use of the structure characteristics of the waves in the design, such as the interval of the wave lines and the difference in the form of the waves. The watermark can be extracted by capturing watermarked wave pattern with scanner or digital camera, and then, extracting the watermark from the re-digitalized image.

Figure 13.12 shows an example of printing a watermarked pattern on the back of a coupon. The example in Figure 13.12 was generated as

(a)   (b)

Figure 13.10. Embedding watermark information into the places having structure characteristics: (a) before embedding; (b) after embedding.

Figure 13.11. Example of embedding a watermark into the design of a coupon.

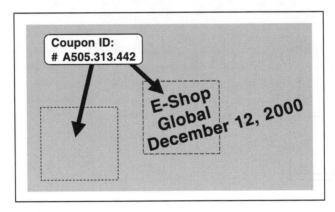

Figure 13.12. An example of printing a watermarked pattern on the back of a coupon.

follows: The watermarked pattern was generated in block size $N \times N$ by using pseudorandom sequences and then the watermarked pattern was printed on the back of the coupon block-by-block repeatedly over all the area of coupon. The watermark can be extracted by capturing the watermarked pattern with a scanner or digital camera and then detecting the watermark from the redigitalized image.

Figure 13.13 shows four designs of "acuaporta," a registered logo mark for the watermark products of M. Ken Co. Ltd. [64], in which the watermark is generated by using a pseudorandom sequence and then the watermarked pattern was placed to the area inside or outside of the logo mark, as showed in Figure 13.13. The watermark can be extracted by capturing the printed logo mark with a scanner or digital camera and then detecting the watermark from the redigitalized image.

Figure 13.13. Four designs of logo mark in which the watermark was embedded.

Compared with interior watermark embedding methods, exterior watermark embedding methods have the advantage that it is robust to the geometric distortion of RST because the detected image is the watermarked pattern. However, this application is limited to some printed materials, such as the certification, personal checks, coupons, and so forth.

EXTRACTING WATERMARK USING MOBILE CAMERAS

About the Current Mobile Market

The mobile phone is one of the most significant technological advances of the 20th century. It has been reported that in the last 10 years, the numbers of mobile phone subscribers have grown from 16 million in 1991 to an astounding figure of 941 million in 2001, and the world mobile phone figure is estimated to grow to about 2.2 billion by 2006 [65]. In Japan, it has been reported that there were 78 million mobile phone subscribers in 2003 [66], a figure equivalent to 58% of the nation's population. In the near future, the mobile phone can play many important roles in daily life than just its use as a phone:

- *Health supervision*: Used to supervise the health of patients or elderly persons by monitoring their vital life signs like breathing, blood pressure, and so forth.
- *Identification card*: Used as a personal ID card, your mobile phone can spend your money for you automatically and can also show your identification, qualification, and capability by replacing the traditional identification card, license, and even passport.
- *Home supervision*: Used as a remote control, your mobile phone can control the electrical equipment in your home and can also supervise home security for you while you are out.

419

- *Multimedia terminal*: Used as a multimedia terminal for playing or displaying multimedia such as radio news and music, television news and programming, and so forth.

With the introduction of the above-described advanced mobile phone services, mobile phones will become an indispensable tool for daily life. Moreover, another very exciting innovation is the mobile camera, by which you can capture a picture and share it with friends and family. In Japan, Sharp and J-Phone [67] introduced the first camera–phone (J-SH04) in November 2000. Now, 3 years later, about 60% of mobile phones in Japan are camera phones and it is expected that in 2005, the market penetration will almost reach 100%. The image quality has skyrocketed from the original 110 K pixels of the J-SH04 to the 1.3 M pixels of SO505i (DoCoMo [68], released June 4, 2003).

With the rapid growth of mobile phone technology and the market, the multimedia messaging service (MMS) has opened new business opportunities for content providers to distribute their contents in the mobile phone market. However, if mobile operators cannot provide a DRM to prevent the illegal redistribution of the contents, the content owners and providers will not have full confidence in exploiting this business opportunity. Therefore, as for the contents distributed on the Internet, content copyright protection and authentication are also urgent issues for the business of content distribution on the mobile Internet. Some of the researchers have sensed the importance and criticality of these issues and concerned discussions have been reported [69]. Certainly, more research on watermarking techniques used to protect the digital copyright for the mobile phone media will be reported in near future. In this section, we will introduce another kind of digital watermarking applications for mobile phone (i.e., extracting the watermark from printed images using a mobile camera).

Extracting a Watermark Using a Mobile Camera

Applications by Mobile Phone. In recent years, with the fast spread of the use of the mobile phone, especially the use of the mobile camera-phones, more and more advertisers are trying to exploit the functions of mobile phones to make it easier for customers to enter their Web site where they can purchase chosen commodities using the information printed in journals or on posters. Currently, there are a number of ways for making commodity application by using the mobile phone:

1. Using the telephone function
2. Using the bar code function
3. Using the distributor code function
4. Using the digital watermarking function

The first method makes use of the telephone function of the mobile phone. The method is accomplished by the following procedure: (1) printing the commodity sample on the paper with the telephone number or a Web site address of a service center where the commodity applications are accepted; (2) connecting to the service center to make the application for the chosen commodity by telephoning or using the Web site address.

The second method makes use of the function of 1-D barcode or 2-D barcode. The 1-D barcode [70] is familiar to most people; it consists basically of a series of parallel lines, or the bar, of varying width representing numbers, letters, and other special characters. The 2-D code [71] is an improvement of the 1-D barcode, such as the QR Code (quick response code), PDF417 code, DataMatrix code, and Maxi code. The method of making use of barcodes is accomplished with the following procedure: (1) printing the commodity sample on the paper with a 1-D/2-D code in which the address of a service center is encoded; (2) capturing the 1-D/2-D code using a mobile camera; (3) decoding the message encoded in the 1-D/2-D code using the decode function of the mobile phones; (4) connecting to a Web site using the decoded address where the customer can find more information or make the application for chosen commodities.

Recently many mobile phone makers have provided the function of decoding 1-D/2-D codes in their products. The KDDI company [72] released, in January 2004, mobile phones equipped with the new EZ Appli (BREW^{TM}) [73], which can read 2-D and 1-D codes printed on magazines, goods, and business cards [74]. NTT DoCoMo Company has also released their newest mobile phone products, the DoCoMo's 505 series, equipped with the i-Appli (i-mode with Java) [75], which can read 2-D and 1-D codes the same as the KDDI does [76]. In the products of both companies, the 2-D codes supported are QR codes as defined in an ISO standard and the 1-D barcodes of JAN codes are also supported. Figure 13.14 shows a sample of the QR code.

Figure 13.14. An example of the QR code.

The third method makes use of the distributor code, a four-digit code representing the distributor's Web site address. The method is accomplished by the following procedure: (1) printing the commodity sample on paper with the distributor code for commodity application; (2) making the application for the chosen commodity by inputting the four-digit distributor code; (3) transforming the distributor code into a corresponding Web site address by using the code-transforming function installed in the hardware or software in the mobile phone; (4) connecting to a Web site where the customer can find more information or make the application for chosen commodities.

The last method makes use of the function of digital watermarking. The method is accomplished by the following procedure: (1) printing the commodity sample on paper in which the concerned Web site address is embedded; (2) capturing the printed sample image using a mobile camera if a customer want to make the application for chosen commodities; (3) extracting the Web site address from the captured image; (4) connecting to a Web site where the customer can find more information or make the application for the chosen commodities. Note that there are usually two types of methods for extracting the Web site address from the captured image in part 3 of the procedure. First, transmitting the captured image to a center and then extracting the Web site address by using a general computer in the center. Second, extracting the Web site address by the computation of mobile phone itself.

Why Do Advertisers Choose Watermarking Techniques. The first method is the most tedious way because it is required to input either the telephone number or the Web site address key by key using the mobile phone. The second method is much simpler compared to the first one, because it is required to capture the image of a 1-D or 2-D barcode by pushing only one button. However, as shown in Figure 13.14, the major disadvantage of using the QR code is its unattractive appearance. In other words, the effectiveness of an elaborately designed commodity sample will be reduced if each sample is printed with a QR code on its side. The third method is an alternative to both the first and second methods, because compared with the first method, the input of a long telephone number or Web site address can be replaced by only four-digit data; compared with the second method, its look is not as unattractive.

The last method is usually considered the best way of using mobile phone to make commodity applications, because with this method, customers can make the applications for their chosen commodities by pushing only one bottom on the mobile phone; meanwhile, it does not require the sample images to be printed with an attached pattern. For example, compare the two business cards shown in Figure 13.15 and Figure 13.16, where the information about the Web site address was

Figure 13.15. An example of a business card on which the photo was embedded with the information about the Web site address of the company.

Figure 13.16. An example of a business card on which the 2-D code contained the information about the Web site address of the company.

inserted in the photo on the card of Figure 13.15 and the 2-D code on the card of Figure 13.16. Needless to say, the card printed with the watermarked face photo is better. Therefore, this is may be the most reason why advertisers choose watermarking technology as the method for applications using mobile phones.

Watermarking Techniques of Using Mobile Cameras

Major Problems for Watermarking Using Mobile Cameras. Similar to the general process of watermarking printed materials, watermarking using a mobile camera involves the processes of printing and image capturing. In the printing process, the image is embedded with the concerned information, such as the Web site address, and then the watermarked image is printed on paper. In the image capturing process, a customer can capture the printed image using a mobile camera and then connect to a Web site with the information extracted from the captured image. Therefore, the major problem of watermarking using a mobile camera is the geometric distortion, such as rotation, scale, and

Figure 13.17. An example of watermarking using a mobile camera; (a) an original image of 512×512 pixels printed at 360 dpi; (b) the image captured using DoCoMo's F505i mobile camera with a size of 480×640 pixels.

translation (RST). Moreover, the geometric distortion occurring in the mobile camera will be much more deteriorated than the geometric distortion in printing and scanning.

Figure 13.17 presents an example of watermarking using a mobile camera. Figure 13.17a is an original image of 512×512 pixels printed at 360 dpi; Figure 13.17b is the image captured by using DoCoMo's F505i mobile camera with 480×640 pixels. As showed in Figure 13.17, in addition to the general RST distortion, there is surface curve distortion occurring near the four image edges. The surface curve distortion is mainly caused by lens distortion, which is more troublesome than the RST distortion.

In addition to the above-mentioned geometric distortions, another major problem is the computation capability of current mobile phones, which may be most troublesome in watermarking using mobile phones because without enough computation capability, any robust watermarking will be ineffective. Currently, the average computation capability of mobile phones is only about $1/n$-hundred of the general PC computer. Moreover, with the additional conditions such as that in DoCoMo's machines, the i-Appli software functions based on the Java 2 platform (J2SE SDK), and the computation speed will decrease further.

Watermarking Scheme. Currently, there are three types of watermarking scheme for watermarking using a mobile camera:

1. Frequency domain watermarking
2. Spatial domain watermarking
3. Design pattern watermarking

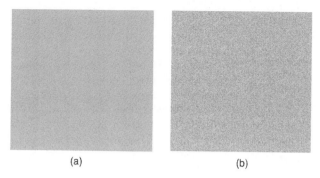

(a)   (b)

Figure 13.18. **Examples of a watermark pattern that are exaggerated both in size and watermark strength for easy observation: (a) a watermark pattern generated by the frequency domain scheme; (b) a watermark pattern generated by the spatial domain scheme.**

Frequency domain watermarking is accomplished in such a way that the watermark plane is generated by a template consisting of number of peaks arranged in the mid-frequency range in the DFT domain. As mentioned earlier, the advantage of frequency domain watermarking is its robustness to RST distortion compared with spatial domain watermarking.

Spatial-domain watermarking is accomplished in such a way that the watermark plane is generated with a noise pattern that consists of a pseudorandom sequence. As mentioned earlier, a noise pattern consisting of a pseudorandom sequence has the autocorrelation property, so a small part of the damage to the pattern will not affect the autocorrelation property of that pattern. The disadvantage of the spatial domain scheme is that the methods are weak in RST distortion because the watermarked pattern and original one are required to have synchronization as consistent as possible. However, as mentioned earlier, with the revision of RST distortion, the spatial domain scheme will have robustness as strong as the frequency domain scheme. Figure 13.18 shows the examples of watermark patterns, which are exaggerated both in size and watermark strength for easy observation. Figure 13.18a is a watermark pattern generated by the frequency domain scheme and Figure 13.18b is a watermark pattern generated by the spatial domain scheme.

Design pattern watermarking is accomplished in such a way that the watermark plane is generated with a design pattern such as that showed in Figure 13.11.

Watermarking System. Currently, there are two types of watermarking system for watermarking using a mobile camera.

1. Local decoding
2. Center decoding

The local decoding type of watermarking system is accomplished by the following procedure: (1) capturing the printed image using a mobile camera; (2) extracting the embedded information by the computation in the mobile phone; (3) connecting the mobile phone to a Web site with the extracted information.

The center decoding type of watermarking system is accomplished by the following procedure: (1) capturing the printed image using a mobile camera; (2) transmitting the captured image to a computer center; (3) extracting the embedded information from the captured image by a general computer in the center; (4) returning the extracted information to the mobile phone; (5) connecting the mobile phone to a Web site with the extracted information.

Compared with the center decoding type, needless to say, the local decoding type is an ideal type, with the merits both in efficiency and effectiveness. However, the major problem for the local decoding type is the computation capability of current mobile phones. Therefore, it is just for this reason that some developers have chosen center decoding for their watermark extraction.

Challenges to Watermarking by Using Mobile Cameras. With the rapid improvement of mobile camera performance in Japan, some makers have begun to challenge the summit of digital watermarking, the watermarking techniques for printed image using mobile camera. However, there are still many problems in their practical application.

On June 17, 2003, Kyodo Printing Company [77] announced that it has succeeded in developing a system for extracting watermark information from the printed image using a mobile camera [78]. In its system, the information about the Web site address is embedded in a design pattern that is thinly spread over the image. The watermark extraction is accomplished by using center decoding. Figure 13.19 shows a sample from the Kyodo Printing Company in which a sample of a zoo map where the block images indicating the animal locations was inserted into the design pattern (Figure 13.19a), a signboard showing how to use a mobile camera to get information about the animals in local places (Figure 13.19b), a enlarged block image where a design pattern can be clearly observed (Figure 13.19c).

On July 7, 2003, NTTGroup [79] announced that it has succeeded in developing a system for extracting watermark information from the printed image using a mobile camera [80]. In its system, the information about the Web site address is embedded into the noise pattern and then the watermarked pattern is inserted as a background pattern. The watermark extraction is accomplished by using center decoding.

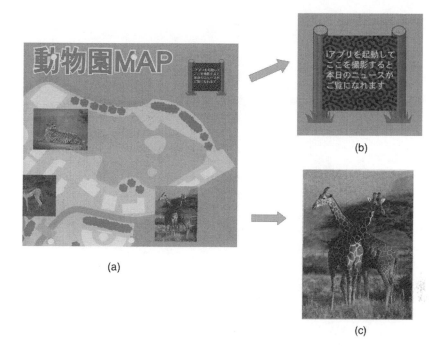

Figure 13.19. A demo image from Kyodo Printing Company. (a) A sample of a zoo map where the block images indicating the animal locations were inserted with the design pattern; (b) a signboard showing how to use a mobile camera to get information about local animals; (c) a enlarged block image in which the design pattern can be clearly observed.

Figure 13.20 presents an example of NTTGroup's extracting watermark information from the printed image using a mobile camera.

Following Kyodo Printing and NTTGroup, on November 6, 2003, M. Ken Co. Ltd. [64] announced that it has succeeded in developing a technique by which the watermark information can be extracted from the printed image by using local decoding; that is, the watermark is extracted directly in the body of the mobile camera-phone [81,82]. In its method, the watermark is embedded using the frequency domain scheme and the watermark extraction can be accomplished within about 5 sec by using DoCoMo's F505i with i-acuaporta, a software developed by M. Ken using i-Appli software. Figure 13.21 presents an example in which a poster of a block image with a white frame was embedded with the information about a Web site address by using M. Ken's watermarking technique.

Comparing the methods developed by the three makers, the major problems of the methods of Kyodo Printing and NTTGroup are that (1) the image quality will be decreased substantially if embedded with

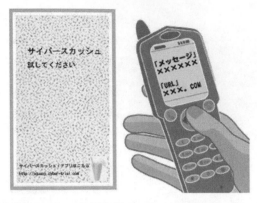

Figure 13.20. An example of NTTGroup's extracting a watermark from a printed image using a mobile camera.

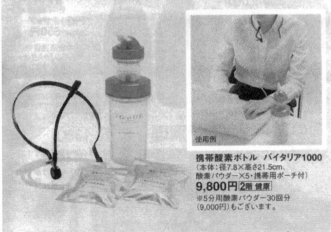

Figure 13.21. An example of the poster in which a block image with a white frame was embedded with the information about a Web site address by using M. Ken's watermarking technique.

Table 13.3. **The decode time for the mobile camera-phone of DoCoMo's 505 series**

Machine	F505i	P505i	D505i	SH505i	SO505i	N505i
Decode time (sec)	4.2	42.2	5.6	4.1	45.1	70.8

the design pattern or noise pattern and (2) the captured images have to be transmitted to a computer center for watermark extraction, which will greatly reduced the effectiveness of using a mobile camera for watermark extraction. Compared with the methods of Kyodo Printing and NTTGroup, the advantages of the M. Ken's method are that (1) the watermark is embedded using the frequency domain scheme, thus it has good image quality as well as high robustness, and (2) the watermark can be extracted in the body of a mobile camera-phone, thus it has the merits both of efficiency and effectiveness.

However, there are still critical problems for the practical application of M. Ken's method. Table 13.3 lists the speed test results of the experiment using M. Ken's method. In Table 13.3, the maximal ratio of the fastest one to the lowest one is about 17. In other words, if it can be accomplished within about 5 sec, watermark decoding using the mobile phone may be an interesting experience for using the mobile camera to capture the image. However, it will be worse if the decoding time is as long as over 1 min. Therefore, unless the computation capability of a mobile phone is developed to about same level as the general computer, it is still a long way before practical application is possible for using a mobile camera to extract watermark.

CONCLUSION

In this chapter, we have introduced new intentions and challenges in the research and application of watermarking technology for printed materials, including watermarking techniques for extracting the watermark from a printed image using mobile camera-phones.

In the second section, we gave a brief overview of the current digital watermarking technology and discussed the corresponding issues, which are currently very popular topics because they are concerned with copyright protection of the digital contents on the Internet, but also they are controversial issues without any final conclusions.

In the third section, we have introduced the watermarking technology used for printed materials, which is an important topic with challenges in the DRM system. Figure 13.22 to Figure 13.24 outline this.

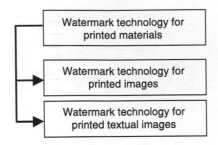

Figure 13.22. The outline of watermarking technology used for printed materials (1).

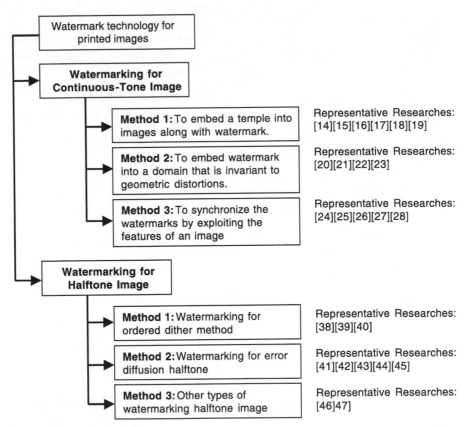

Figure 13.23. The outline of watermarking technology used for printed materials (2).

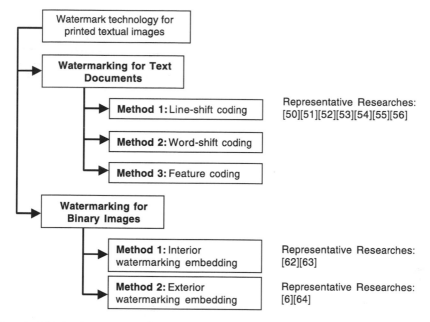

Figure 13.24. The outline of watermarking technology used for printed materials (3).

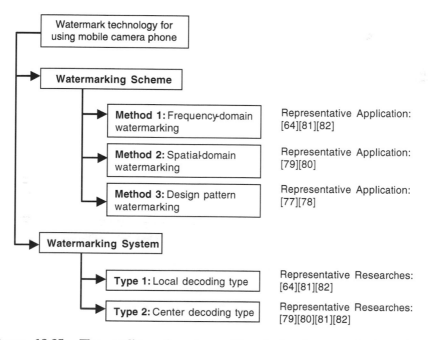

Figure 13.25. The outline of watermarking technology used for printed materials using mobile camera-phones.

In the fourth section, we have introduced watermarking technology used for printed materials using the mobile camera, which is the most challenging task in the current research and application of digital watermarking. Figure 13.25 provides an outline of this.

REFERENCES

1. Liu, Z., Huang, H.C., and Pan, J.S., Digital watermarking — backgrounds, techniques, and industrial applications, *Commn. CCISA*, 10(1), 78, 2004.
2. Tirkel, A.Z., Rankin, G.A., Schyndel, R.M., Ho, V., W.J., Mee, N.R.A., and Osborne, C.F., Electronic water mark, Presented at International Symposium on Digital Image Computing Techniques and Applications, Sydney, Australia, December 8–10, 1993, p. 666.
3. Tirkel, A.Z. and Hall, T.E., A unique watermark for every image, *IEEE Multimedia*, 8(4), 30, 2001.
4. Petitcolas, F.A.P., Anderson, R.J., and Kuhn, M.G., Information hiding — a survey, *Proc. IEEE*, 87(7), 1062, 1999.
5. Lim, Y., Xu, C., and Feng, D.D., Web-Based Image Authentication Using Invisible Fragile Watermark, in Proceedings of the Pan-Sydney Area Workshop on Visual Information Processing 2001, Sydney, 2001, p. 31.
6. Liu, Z. and Inoue, A., Watermark for industrial application, in *Intelligent Watermarking Techniques*, Pan, J.S., Huang, H.C., and Jain, L.C., Eds., World Scientific, Company, Singapore, 2004, chap. 22.
7. Lin, C.Y. and Chang, S.F., Distortion Modeling and Invariant Extraction for Digital Image Print-and-Scan Process, presented at ISMIP 99, Taipei, 1999.
8. Lin, C.Y., Public Watermarking Surviving General Scaling and Ccropping: An Application for Print-and-Scan Process, presented at Multimedia and Security Workshop at ACM Multimedia 99, Orlando, FL, 1999.
9. Cox, I.J., Kilian, J., Leighton, F.T., and Shamoon, T., Secure spread spectrum watermarking for multimedia, *IEEE Trans. Image Process.*, 6(12), 1673, 1997.
10. Swanson, M.D., Zhu, B., and Tewfik, A.H., Transparent Robust Image Watermarking, in Proceedings of ICIP 96, IEEE International Conference on Image Processing, Lausanne, 1996, 211.
11. Delaigle, J.F., Vleeschouwer, C.D., and Macq, B., Psychovisual approach to digital picture watermarking, *J. Electron. Imaging*, 7(3), 628, 1998.
12. Delaigle, J.F., Vleeschouwer, C.D., and Macq, B., Watermarking algorithm based on a human visual model, *Signal Process.: Image Commn.*, 66(3), 319, 1998.
13. Wolfgang, R.B., Podilchuk, C.I., and Delp, D.J., Perceptual watermarks for digital images and video, *Proc. IEEE*, 87(7), 1108, 1999.
14. Pereira, S. and Pun, T., Fast robust template matching for affine resistant image watermarking, Lecture Notes in Computer Science, Vol. 1768, Dresden, 1999, p. 200.
15. Csurka, G., Deguillaume, F., O'Ruanaidh, J.J.K., and Pun, T., A Bayesian approach to affine transformation resistant image and video watermarking, in Lecture Notes in Computer Science, Vol. 1, 1768, Springer-Verlag, Berlin, 1999, p. 270.
16. Pereira, S., O'Ruanaidh, J.J.K., Deguillaume, F., Csurka, G., and Pun, T., Template Based Recovery of Fourier-Based Watermarks Using Log-Polar and Log-Log Maps, in Proceedings of IEEE Multimedia Systems 99, International Conference on Multimedia Computing and Systems, Florence, 1999, Vol. 1, p. 870.
17. Kang, X., Huang, J., Shi, Y.Q., and Lin, Y., A DWT-DFT composite watermarking scheme robust to both affine transform and JPEG compression, *IEEE Trans. Circuits Syst. Video Technol.*, 13(8), 776, 2003.

18. Kutter, M., Watermarking resisting to translation, rotation, and scaling, in *Proceedings of SPIE*, 3528, 1998, 423.
19. Voloshynovskiy, S., Deguillaume, F., and Pun, T., Content Adaptive Watermarking Based on a Stochastic Multiresolution Image Modeling, presented at Tenth European Signal Processing Conference (EUSIPCO'2000), Tampere, Finland, 2000.
20. O'Ruanaidh, J.J.K. and Pun, T., Rotation, Scale and Translation Invariant Digital Image Watermarking, in Proceedings of ICIP 97, IEEE International Conference on Image Processing, Santa Barbara, CA, 1997, p. 536.
21. O'Ruanaidh, J.J.K. and Pun, T., Rotation, scale and translation invariant spread spectrum digital image watermarking, *Signal Process.*, 66(3), 303, 1998.
22. Lin, C.Y., Wu, M., Bloom, A., Cox, I.J., Miller, M.L., and Lui, Y.M., Rotation, scale, and translation resilient watermarking for images, *IEEE Trans. Image Process.*, 10(5), 767, 2001.
23. Zheng, D., Zhao, J., and Saddik, A.E., RST-invariant digital image watermarking based on log-polar mapping and phase correlation, *IEEE Trans. Circuits Syst. Video Technol.*, 13(8), 753, 2003.
24. Simitopoulos, D., Koutsonanos, D.E., and Strintzis, M.G., Robust images watermarking based on generalized radon transformations, *IEEE Trans. Circuits Syst. Video Technol.*, 13(8), 732, 2003.
25. Shim, H.J. and Jeon, B., Rotation, scaling, and translation robust image watermarking using Gabor kernels, Proc. SPIE, 4675, 563, 2002.
26. Kim, H.S. and Lee, H.K., Invariant image watermark using Zernike moments, *IEEE Trans. Circuits Syst. Video Technol.*, 13(8), 766, 2003.
27. Kutter, M., Bhattacharjee, S.K., and Ebrahimi, T., Towards second generation watermarking schemes, in Proceedings of IEEE International Conference on Image Processing, 1999, p. 320.
28. Guoxiang, S. and Weiwei, W., Image-feature based second generation watermarking in wavelet domain, in Lecture Notes in Computer Science, Vol. 2251, Springer-Verlag, Hong Kong, 2001, p. 16.
29. Digimarc Company, http://www.digimarc.com.
30. Ulichney, R., *Digital Halftoning*, MIT Press, Cambridge, MA, 1987.
31. Mese, M. and Vaidyanathan, P.P., Optimized halftoning using dot diffusion and methods for inverse halftoning, *IEEE Trans. Image Process.*, 9(4), 691, 2000.
32. Floyd, R. and Steinberg, L., An adaptive algorithm for spatial greyscale, in *Proc. Soc. Inf. Dis.*, 17(2), 75, 1976.
33. Jarvis, J.F., Judice, C.N., and Ninke, W.H., A survey of techniques for the display of continuous-tone pictures on bilevel displays, *Computer Graph. Image Process.*, 5, 13, 1976.
34. Stucki, P., MECCA—A multiple-Error Correcting Computation Algorithm for Bilevel Image Hardcopy Reproduction, IBM *Research Laboratory Report*, RZ1060, Zurich, 1981.
35. Knuth, D.E., Digital halftones by dot diffusion, *ACM Trans. Graph.*, 6, 245, 1987.
36. Anastassiou, D., Neural net based digital halftoning of images, *Proc. ISCAS*, 1, 507, 1988.
37. Seldowitz, M.A., Allebach, J.P., and Sweeney, D.E., Synthesis of digital holograms by direct binary search, *Appl. Opt.*, 26, 2788, 1987.
38. Baharav, Z. and Shaked, D., Watermarking of dither halftoned images, in *Proc. SPIE Electron. Imaging*, 3657, 307, 1999.
39. Hel-Or, H.Z., Copyright Labeling of Printed Images, in Proceedings of ICIP 2000, Vancouver 2000, Vol. 3, p. 307.
40. Wang, S.G. and Knox, K.T., Embedding digital watermarks in halftone screens, *Proc. SPIE*, 3971, 218, 2000.

41. Pei, S.C. and Guo, J.M., Hybrid pixel-based data hiding and black-based watermarking for error-diffused halftone images, *IEEE Trans. Image Process.*, 13(8), 867, 2003.

42. Hsu, C.Y. and Tseng, C.C., Digital Halftone Image Watermarking Based on Conditional Probability, presented at IPPR Conference on Computer Vision, Graphics and Image Processing, Kinmen, Taiwan, 2003, p. 305.

43. Fu, M.S. and Au, O.C., Data Hiding by Smart Pair Toggling for Halftone Images, in Proceedings of IEEE International Conference Acoustics, Speech and Signal Processing, 2000, Vol. 4, 2318.

44. Fu, M.S. and Au, O.C., Data hiding watermarking for halftone images, *IEEE Trans. on Image Process.*, 11, 477, 2002.

45. Kim, H.Y. and Afif, A., Secure Authentication Watermarking for Binary Images, in Proceeding of Brazilian Symposium on Computer Graphics and Image Processing, 2003, p. 199.

46. Hsiao, P.C., Chen, Y.T., and Wang, H.C., Watermarking a Printed Binary Image with Varied Screen Rulings, presented at IPPR Conference on Computer Vision, Graphics and Image Processing, Kinmen, Taiwan, 2003.

47. Oppenheim, A.V. and Schafer, R.W., *Discrete-Time Signal Processing*, Prentice-Hall, Englewood Cliffs, NJ, 1989.

48. Fuji Xerox: http://www.fujixerox.co.jp/release/2001/1010_TrustMarkingBasic.html.

49. Li, J.Y., Chou, T.R., and Wang, H.C., Multi-channel Watermarking Technique of Printed Images and Its Application to Personalized Stamps, presented at IPPR Conference on Computer Vision, Graphics and Image Processing, Kinmen, Taiwan, 2003.

50. Brassil, J., Low, S., Maxemchuk, N., and O'Gorman, L., Electronic Marking and Identification Techniques to Discourage Document Copying, in Proceedings of INFOCOM'94, 1994, p. 1278.

51. Brassil, J., Low, S., Maxemchuk, N., and O'Gorman, L., Electronic marking and identification techniques to discourage document copying, *IEEE J. Selected Areas Commn.*, 13, 1495, 1995.

52. Brassil, J., Low, S., and Maxemchuk, N., Copyright protection for the electronic distribution of text documents, *Proc. IEEE*, 87(7), 1108, 1999.

53. Low, S., Maxemchuk, N., Brassil, J., and O'Gorman, L., Document Marking and Identification Using Both Line and Word Shifting, in Proceedings of IEEE INFOCOM'95, 1995.

54. Brassil, J., Low, S., Maxemchuk, N., and O'Gorman, L., Hiding Information in Document Images, in Proceedings of the 29th Annual Conference on Information Sciences and Systems, 1995, p. 482.

55. Low, S.H. and Maxemchuk, N.F., Performance comparison of two text marking methods, *IEEE J. Selected Areas Commn.*, 16(4), 8, 1998.

56. Huang, D. and Yan, H., Interword distance changes represented by sine waves for watermarking text images, *IEEE Trans. Circuits Sys. Video Technol.*, 11(12), 1237, 2001.

57. Liu, Z., Kobayashi, Y., Sawato, S., and Inoue, A., Robust audio watermark method using sinusoid patterns based on pseudo-random sequences, *Proc. SPIE*, 5020, 21, 2003.

58. Liu, Z. and Inoue, A., Audio watermarking techniques using sinusoidal patterns based on pseudo-random sequences, *IEEE Trans. Circuits Syst. Video Technol.*, 13(8), 801, 2003.

59. Choi, H., Kim, H., and Kim, T., Robust sinusoidal watermark for images, *Electron. Lett.*, 35(15), 1238, 1999.

60. Chotikakamthorn, N. and Pholsomboon, S., Ring-Shaped Digital Watermark for Rotated and Scaled Images Using Random-Phase Sinusoidal Function, in Proceedings of Electrical and Electronic Technology, TENCON, 2001, p. 321.

61. Petrovic, R., Audio Signal Watermarking Based on Replica Modulation, in Proceedings of TELSIKS 2001, 2001.

62. Abe, Y. and Inoue, K., Digital Watermarking for Bi-level Image, Technical Report 26, Ricoh Company, Ltd., Japan, 2000.
63. Wu, M. and Liu, B., Data hiding in digital binary images, in Proceedings IEEE International Conference on Multimedia and Expo (ICME), New York, 2000.
64. M. Ken Co. Ltd.: http://c4t.jp/ (since January 20, 2004, M. Ken Co. has merged with C4Technology Inc.)
65. *COAI News Bulletin*, 10, May 25, 2002, http://www.coai.com/.
66. NTT DoCoMo Report, March 2003, http://www.nttdocomo.com/presscenter/publications/.
67. J-Phone, http://www.vodafone.jp/scripts/english/top.jsp.
68. DoCoMo, http://www.nttdocomo.co.jp/english/index.shtml.
69. Mobile Content Protection and DRM, Baskerville Communications, 2002.
70. One dimensional code, http://www.barcodehq.com/primer.html.
71. Two dimensional code, http://www.qrcode.com.
72. KDDI, http://www.kddi.com/english/index.html.
73. EZ Appli: BREW™, http://www.qualcomm.com/brew.
74. KDDI EZ Announces New Two-Dimensional Code Reader Application http://www.mobiletechnews.com/info/2003/12/21/025034.html.
75. i-Appli: i-mode with Java, http://www.nttdocomo.com/corebiz/imode/services/iappli.html.
76. NTT DoCoMo Barcode, http://www.nttdocomo.co.jp/p_s/imode/barcode/.
77. Kyodo Printing Company, http://www.kyodoprinting.co.jp/kphome/welcomee.htm.
78. Release, http://www.itmedia.co.jp/mobile/0306/17/n_kyouritu.html.
79. NTTGroup, http://www.ntt.co.jp/index_e.html.
80. Release, http://japan.internet.com/webtech/20030708/5.html.
81. Nikeisanngyou News, Japan, http://ss.nikkei.co.jp/ss/.
82. DIME, Japan, http://www.digital-dime.com/, pp. 82, December 18, 2003.

14

Robust Watermark Detection from Quantized MPEG Video Data

D. Simitopoulos, A. Briassouli, and M. G. Strintzis

INTRODUCTION

The acquisition, representation, distribution, and storage of multimedia data in digital format has led to significant advances in image and video technology. Data can be easily manipulated and distributed and great compression can be achieved while maintaining a high quality of the transmitted data. However, the facility and efficiency with which high-quality digital information can be reproduced and distributed to many users has also created significant problems. Intellectual property rights are violated and digital information can easily be distributed to numerous unauthorized users. This has led to increased interest in the development of new methods for copyright protection. Watermarking has received considerable attention lately, as it provides an effective alternative to methods used in the past for the protection of digital data. The basic principle is to embed information directly into the data, which serve as a cover for that information. A big difference between watermarking and cryptography is that in watermarking, the protected data can still be used, whereas encrypted data cannot. There exist both visible and invisible watermarks, depending on the intended use of the digital data. The former are usually used in preview images in the World Wide Web or in image databases to prevent their commercial use. Systems for copyright protection employ invisible watermarks [1], which are the concern of the present chapter.

A challenging but very realistic problem in watermarking is the imperceptible embedding of a watermark in compressed data and its reliable *blind* detection, where the receiver has no knowledge of the host signal. The imperceptible embedding of a digital signature in multimedia data presents a challenge of its own, because the hidden signal cannot be too weak, or else it cannot be detected. Watermarking of compressed data is of particular interest because in real applications, the multimedia data cannot avoid quantization. However, it is also more difficult, as there is less variability in the data, which leaves fewer possibilities for the imperceptible embedding of a signal. Based on existing methods for watermarking of compressed and uncompressed data, we propose an effective way of imperceptibly hiding a signal in a quantized host.

The detection of the watermark is another very important issue, as it needs to be very reliable in real-life applications. The probability of detection needs to be very high, and the probability of false alarm should remain as low as possible for a real and efficient watermarking scheme. There is a trade-off between the imperceptibility and the detectability of the watermark: The system designer aims to embed the strongest possible signal, to ensure its reliable detection, but at the same time limit its strength to keep it imperceptible.

Perhaps the greatest challenge in watermarking is the fact that many factors, the so-called "attacks," can affect the detectability of the embedded signal. Attacks may be nonmalicious modifications to the protected data, such as compression, scaling, geometric transformations, and warping, many of which are common during the manipulation of digital data. However, there are also malicious attacks, where "pirates" attempt to make unauthorized use of the watermarked data by removing the hidden signal or, more effectively, by making it undetectable. A realistic and reliable watermarking scheme needs to keep the watermark imperceptible and at the same time make it as robust as possible, so it can remain detectable even in the presence of attackers.

Lossy compression can create many difficulties in watermarking systems, but it is absolutely necessary in modern digital applications. It might be intentional, but the quantization employed by the most common compression algorithms is non-malicious, although it weakens the watermark. The effect of quantization in the detection process has been analyzed for some detection systems [2]. Various approaches have already been proposed to improve watermark detection of nonquantized data [3,4]. We present a novel approach for robust detection using the *quantized* domain data of MPEG [5] compressed video frames.

Most watermarking systems aim to extract or detect a watermark without the use of the original host signal and without knowledge of the hidden watermark. They only use the received signal, which may or may not be protected, so they are blind methods. In practice, the "correlation detector" [6,7] is often used, which is optimal for Gaussian data. However, the quantized discrete cosine transform (DCT) domain data does not follow a normal distribution: The Laplacian distribution has been proposed in the literature [8,9] for the modeling of quantized transform domain coefficients. Thus, a statistical watermark detector based on this distribution will perform more efficiently in the compressed domain. The improved detection performance of the proposed system is also expected to have increased robustness to attacks. Indeed, under image blurring and cropping, it is found that the Laplacian detector yields better detection results than the Gaussian correlator in most cases.

The detector performance based on the Laplacian distribution is evaluated in MPEG video data watermarked with an imperceptible watermark that is added to the quantized DCT coefficients. Detection in the compressed domain results in a fast detection method that may be utilized by real-time systems. This is particularly useful for practical applications, because it enables the effective protection of digital data like video, which is of great interest to the industry. It also enables the use of watermarking not just for copyright protection [10] but also for applications where watermark detection modules are incorporated in real-time decoders and players, such as broadcast monitoring [11,12].

IMPERCEPTIBLE WATERMARKING IN THE COMPRESSED DOMAIN

Many watermarking systems are inspired by the spread spectrum modulation schemes used in digital communications in jamming environments [13–15]. The role of the jammer in the watermarking problem is assumed by the attacker who intentionally or unintentionally tries to destroy or extract the embedded watermark [16], whereas the watermark is the hidden information signal and the host signal plays the role of additive noise. Such watermarks can be embedded either in the spatial or in the transform domain, in original or quantized data. Watermark embedding in the quantized transform domain is of particular importance for video watermarking [17–19]. Thus, the proposed method embeds imperceptible watermarks in selected quantized DCT coefficients of the luminance component of MPEG coded I-frames. The embedding may be performed either as part of the MPEG coding of live or recorded video or on precompressed MPEG bit streams by partly decoding (up to the level of the extraction of DCT coefficients) and reencoding them.

The watermarking scheme examined here is a spread spectrum additive system, so the information to be embedded (for the detection problem, only 1 bit) is modulated by a pseudorandom sequence and is multiplied by appropriately chosen scaling factors that ensure both robustness and imperceptibility; the resulting watermark signal is added to the host signal.

Generation of the Spreading Sequence

The pseudorandom spreading sequence S is a zero-mean, unit-variance process with values either $+1$ or -1. The watermark generation procedure is depicted in Figure 14.1. The random number generator is seeded with the result of a hash function to increase the system security. Specifically, the watermark is generated so that even if an attacker finds a watermark sequence that leads to a high correlator output, he still

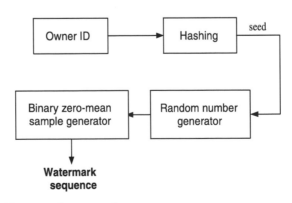

Figure 14.1. Watermark generation.

cannot find a meaningful owner ID that would produce the watermark sequence and, therefore, cannot claim ownership of the data [20].

Perceptual Analysis and Block Classification

The proposed watermark embedding scheme (see Figure 14.2) alters only the quantized AC coefficients $X_Q(m, n)$ of a luminance block (where m and n are indices indicating the position of the current coefficient in an 8×8 DCT block) and leaves chrominance information unaffected. In order to make the watermark imperceptible, a novel method is employed, combining perceptual analysis [21,22] and block classification techniques [23,24]. These are applied in the DCT domain to adaptively select which coefficients are best for watermarking and also to specify the embedding strength.

For the design of the classification mask C, each DCT luminance block is initially classified with respect to its energy distribution to one of five possible classes: *low activity, diagonal edge, horizontal edge, vertical edge,* and *textured block.* The calculations of the energy distribution and the block classification are performed as in Reference 24, returning the class of the block examined. For each block class, the binary classification mask C determines which coefficients are the best candidates for watermarking. Thus,

$$C(m, n) = \begin{cases} 0 & \text{the } (m, n) \text{ coefficient will not be watermarked} \\ 1 & \text{the } (m, n) \text{ coefficient can be watermarked [if } M_Q(m, n) \neq 0] \end{cases}$$

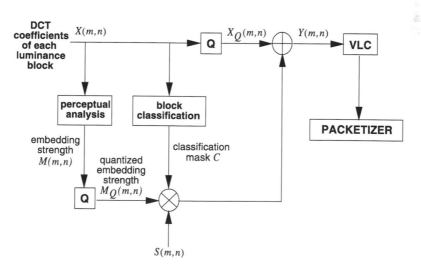

Figure 14.2. **Watermark embedding scheme.**

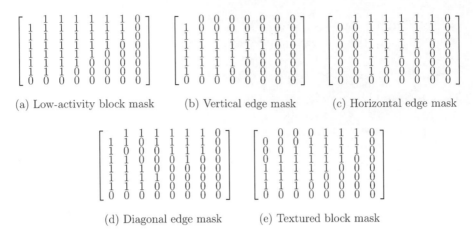

(a) Low-activity block mask   (b) Vertical edge mask   (c) Horizontal edge mask

(d) Diagonal edge mask   (e) Textured block mask

Figure 14.3.   The classification masks that correspond to each one of the five block classes.

where $m, n \in [0, 7]$ and $M_Q(m, n)$ is the quantized embedding strength for the (m, n) coefficient of the block, which is derived from the perceptual analysis. The perceptual analysis that follows the block classification process leads to the final choice of the coefficients that will be watermarked and defines the embedding strength.

Figure 14.3 depicts the mask C for all the block classes. As can be seen, the classification mask for all classes contains "zeros" for all high-frequency AC coefficients. These coefficients are not watermarked because the embedded signal is likely to be eliminated by low-pass filtering or transcoding to lower bitrates. The rest of the zero $C(m, n)$ values in each classification mask (apart from the low-activity block mask) correspond to large DCT coefficients, which are left unwatermarked because their use in the detection process may reduce the detector performance [24].

The perceptual model that is used is a novel adaptation of the perceptual model proposed by Watson [22]. Specifically, a measure $T''(m, n)$ is introduced that determines the maximum Just Noticeable Difference (JND) for each DCT coefficient of a block. This model is then adapted to be applicable to the domain of quantized DCT coefficients.

For 1/16 pixels/degree of visual angle and 48.7 cm viewing distance, the *luminance masking* and the *contrast masking* properties of the human visual system (HVS) for each coefficient of a discrete cosine transformed block are estimated as in Reference 22. The matrices, $T'(m, n)$ (luminance masking) and $T''(m, n)$ (contrast masking) are calculated. Each one of the values of $T'(m, n)$ is compared with the absolute value of each DCT coefficient $|X(m,n)|$: the values of $T''(m,n)$ determine the embedding strength of the watermark only when $|X(m,n)| > T'(m,n)$:

$$M(m,n) = \begin{cases} T''(m,n) & \text{if } |X(m,n)| > T'(m,n) \\ 0 & \text{otherwise} \end{cases}$$

The classification mask C is thus used to determine which quantized coefficients will be watermarked, and the embedding strength values $M(m,n)$ are used to specify the watermark strength, as explained in section "Quantized Domain Embedding."

Quantized Domain Embedding

Following the block classification and perceptual analysis described earlier, two options exist for watermark embedding: The watermark will be added either to the DCT coefficients or to the quantized DCT coefficients [25,26]. If the watermark is added to the DCT coefficients, there is the danger that it may be completely eliminated by the subsequent quantization process if the watermark magnitude is quite small in comparison to the quantizer step. Obviously, in such a case, the value of a quantized DCT coefficient remains unchanged, despite the addition of the watermark to the unquantized data. This may happen to many coefficients, essentially removing a large part of the watermark and leading to an unreliable detector performance [20]. Thus, in the case of video watermarking, it is advantageous to embed the watermark directly in the quantized DCT coefficients. Because the MPEG coding algorithm [5] performs no other lossy operation after quantization (see Figure 14.4), any information embedded after the quantization stage does not run the risk of being eliminated by subsequent processing and the watermark remains intact in the quantized coefficients during the detection process. The quantized DCT coefficients $X_Q(m,n)$ are watermarked as follows (see Figure 14.2):

$$Y(m,n) = X_Q(m,n) + C(m,n)M_Q(m,n)S(m,n) \tag{14.1}$$

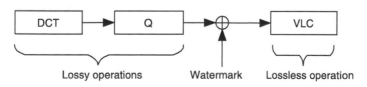

Figure 14.4. MPEG encoding operations.

where $M_Q(m, n)$ is calculated by

$$M_Q(m, n) = \begin{cases} \text{quant}[M(m, n)] & \text{if quant}[M(m, n)] > 1 \\ 1 & \text{if quant}[M(m, n)] \leq 1 \text{ and } M(m, n) \neq 0 \\ 0 & \text{if } M(m, n) = 0 \end{cases}$$

$$(14.2)$$

where quant$[\cdot]$ denotes the quantization function used by the MPEG video coding algorithm. In the sections that follow, we will use the simpler notation

$$W = CM_Q S \qquad (14.3)$$

for the watermark that is finally embedded in the quantized domain data and X, X_Q, and Y for the original DCT coefficients, the quantized DCT coefficients, and the watermarked quantized DCT coefficients, respectively.

The invisibility of the watermark that results through this embedding process can be verified in Figure 14.5. This figure depicts a frame from the standard video sequence "Susie" (MPEG video compliance bit stream), the corresponding watermarked frame, and the difference between the two frames, which is amplified in order to make the modification

Figure 14.5. (a) Original frame from the video sequence Susie, (b) watermarked frame, and (c) amplified difference between the original and the watermarked frames.

444

Figure 14.5.   Continued.

produced by the watermark embedding more visible. In Figure 14.5c, it is clear that in highly textured image areas, the embedded watermark amplitude is higher, but the watermark is still imperceptible.

Various video sequences were watermarked and viewed in order to evaluate the imperceptibility of the watermark embedding method. The viewers were unable to locate any degradation in the quality of the watermarked videos. Table 14.1 presents the mean of the PSNR values

445

TABLE 14.1. Mean PSNR values for the frames of four watermarked video sequences (MPEG-2, 6 Mbits/sec, PAL)

Video Sequence	Mean PSNR for All Video Frames	Mean PSNR for I-Frames Only
flowers	38.6 dB	36.5 dB
Mobile and calendar	33.1 dB	30 dB
Susie	45.6 dB	40.4 dB
Table tennis	35.6 dB	33.2 dB

of all the frames of some commonly used video sequences. In addition, Table 14.1 presents the mean of the peak to noise ratio (PSNR) values of the I-frames (watermarked frames) of each video sequence.

MODELING OF QUANTIZED DCT DOMAIN DATA

It is well known from the literature, that the low- and mid-frequency DCT coefficients carry the most information of an image or video frame. Thus, they are more finely quantized than the high-frequency coefficients, which often vanish after the quantization process. The probability density functions (pdfs) of these coefficients are similar to the Gaussian pdf, as they remain bell-shaped, but their tails are quite heavier [27,28]. This is the reason why the low- and mid-frequency DCT coefficients are often modeled by the heavy-tailed Laplacian, generalized Gaussian, or Cauchy distributions. In the case of quantized data examined here, the DCT coefficients become more discrete-valued, depending, of course, on the degree of quantization. Nevertheless, their heavy-tailed nature is not significantly affected, as we show through statistical fitness tests.

A model often used in the literature [29] with heavier tails than the Normal pdf is that of the Laplacian distribution

$$f_X(x) = \frac{b}{2} \exp(-b|x - \mu|) \tag{14.4}$$

where μ is its mean and b is defined as

$$b^2 = \frac{2}{\text{var}(x)}. \tag{14.5}$$

We focus on the often considered case of the zero-mean Laplacian model, where $\mu = 0$. The Laplacian distribution is frequently used because of its simplicity and accuracy, as well as its mathematical tractability [8,9]. An additional characteristic that makes it particularly

446

useful in practical applications, and especially in video applications, is the fact that its parameter b can be found very easily from the given dataset, whereas the parameters of other statistical models are more difficult to estimate and their computation creates additional overhead.

The Laplacian parameter b for quantized DCT coefficients can be estimated using Equation 14.5, where $b = \sqrt{2/v_Q}$, and v_Q is the variance of the quantized data. Alternative methods for accurately estimating the Laplacian parameter from the quantized data have been presented in the literature [8,9]. In Reference 8 this parameter is estimated by

$$b = \frac{2}{Q}\log t \tag{14.6}$$

where Q is the quantization step size and t is found [8] through

$$t = \frac{u + \sqrt{u^2 - 4}}{2} \tag{14.7}$$

where

$$u = \frac{1 + \sqrt{1 + 16h^2}}{2h}, \quad h = \frac{v_Q}{Q^2} \tag{14.8}$$

The estimate of the Laplacian parameter b from the *quantized* DCT coefficients through Equation 14.6 to Equation 14.8 is more accurate than the estimate using Equation 14.5, as it takes into account the quantization effects [8]. Note that in the case of MPEG-1/2 video, the quantization step is different for every DCT coefficient. To overcome this practical problem, we approximate Q by the mean value of the quantization steps of the coefficients examined, without introducing a significant approximation error. The suitability of the modified Laplacian model is verified by our experiments, which show that the parameter of Equation 14.6 leads to more accurate statistical fitting.

In order to model the low- and mid-frequency quantized DCT coefficients, one can examine their amplitude probability density (APD), $P(|X| > a)$. The APD can be evaluated experimentally, directly from the data by simply counting the values of X for which $|X| > a$. Its theoretical expression can also be derived for every statistical model directly from the corresponding pdf. Thus, the APD for Gaussian data is given by

$$P(|X| > a) = 2Q\left(\frac{a - \mu}{\sigma}\right) \tag{14.9}$$

where μ and σ are the data mean and standard deviation, respectively, and $Q(x)$ is given by

$$Q(x) = \frac{1}{\sqrt{2\pi}} \int_x^\infty e^{-t^2/2}\, dt \qquad (14.10)$$

The Laplacian APD is

$$P(|X| > a) = \begin{cases} 2 - \exp(b(a - \mu)) & \text{for } a < \mu \\ \exp(-b(a - \mu)) & \text{otherwise} \end{cases} \qquad (14.11)$$

To evaluate the accuracy of the Laplacian model for quantized DCT coefficients, the APDs for the quantized DCT domain coefficients of a frame of the standard Susie video sequence are estimated. The parameters for the best-fitting Gaussian model were the mean μ and standard deviation σ of the quantized data. The parameters of the Laplacian pdf are estimated from Equation 14.5 and the quantized data, whereas the parameters of the modified "quantized Laplacian" model are computed using Equation 14.6 to Equation 14.8, and the quantized data. As Figure 14.6 shows, these coefficients indeed exhibit quite heavier tails than the corresponding Gaussian distribution. The Laplacian pdf offers a more accurate fit, as it captures both the mode and the tails of the experimental APD quite accurately. The modified Laplacian model of Equation 14.6 to Equation 14.8 leads to even more accurate fitting of the data than Equation 14.5, as Figure 14.6 shows, because its APD curve practically coincides with the experimental one. Consequently, the quantized Laplacian parameters are used in the sequel.

The suitability of the modified (quantized) Laplacian distribution for our dataset is also examined in case of attacks. In particular, the experimental, Gaussian, and Laplacian APDs for quantized DCT coefficients are estimated under blurring. Figure 14.7 and Figure 14.8 show that the Laplacian model, and particularly the modified Laplacian model of Equation 14.6 to Equation 14.8, is still appropriate for blurring attacks of various degrees, because the pdf of the coefficients remains heavy-tailed. These satisfactory modeling results and the simplicity, and, consequently, the practicality, of the (modified) Laplacian pdf motivate its use for the distribution of the quantized DCT coefficients.

WATERMARK DETECTION

Detection Based on Hypothesis Testing

The known or approximated statistical properties of the DCT coefficients may lead to a blind watermark detection method. Specifically, the water-

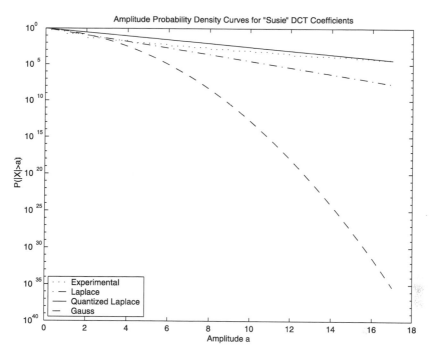

Figure 14.6.   APD for Susie.

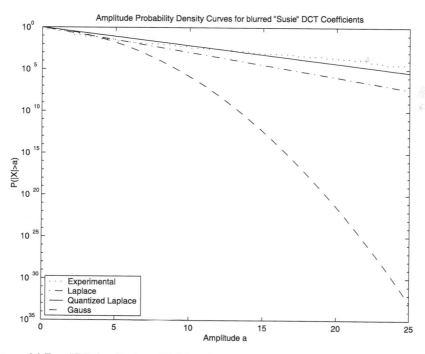

Figure 14.7.   APD for Susie with blurring.

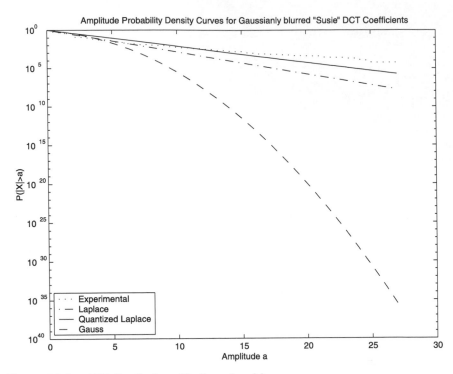

Figure 14.8. APD for Susie with Gaussian blur.

mark detection problem can be formulated as a binary hypothesis test [30], where the two hypotheses concern the existence or nonexistence of a watermark. This technique is *blind* because the test is performed without knowledge of the original, unwatermarked data. The two hypotheses for the test are formulated as follows:

$$H_1: Y = X_Q + W$$
$$H_0: Y = X_Q \tag{14.12}$$

The watermark W is the desired signal or information, whereas the quantized DCT video coefficients X_Q, the host signal for the watermark, play the role of unknown additive noise. The goal of the watermark detector examined here is to verify whether or not a watermark exists in the given data. A basic assumption is that the statistical distribution of the coefficients is not significantly altered by the presence of a watermark, so the pdf of the original data does not change with the embedding of the watermark. This assumption can also be intuitively justified by the fact that the watermarked signal should follow a statistical distribution similar to that of the original signal if it is to remain very similar to it. The

450

decision is then based on the log-likelihood ratio test

$$l(Y) = \ln\left(\frac{f(Y|H_1)}{f(Y|H_0)}\right) \underset{<H_0}{\overset{>H_1}{}} \eta \tag{14.13}$$

The threshold η is often determined by the probability of the false alarm, P_{fa}, leading to the Neyman–Pearson (N-P) detector, which minimizes the probability of missing a watermark with bounded false-alarm probability P_{fa}.

Statistical Detectors

In most current practical image and video watermarking systems, the image and watermark distributions are considered to be Gaussian random processes, which lead to the likelihood ratio

$$l(Y) = \frac{1}{2\sigma^2} \sum (|Y|^2 - |Y - W|^2) \tag{14.4}$$

where σ^2 is the data variance [3,31] and the summation is performed over all coefficient indices. Equation 14.14 reduces to the well-known correlation detector [31]

$$l(Y) = \frac{1}{2\sigma^2} \sum (-W^2 + 2WY) \Rightarrow \sum \frac{WY}{\sigma^2} \underset{<H_0}{\overset{>H_1}{}} \tau \tag{14.15}$$

where

$$\tau = \eta + \frac{1}{2} \sum \frac{W^2}{\sigma^2} \tag{14.16}$$

and η is the threshold of Equation 14.13. This correlator is often used in watermark detection schemes because of the simplicity of its implementation. Also, in cases of blind watermark detection when the statistics of the host signal are unknown, this processor is reasonably robust. However, it is optimal only for Gaussian data and not for the more heavy-tailed probability densities of image DCT coefficients.

Because the quantized DCT coefficients examined here are satisfactorily modeled by the Laplacian distribution as shown in section "Modeling of Quantized DCT-domain Data," a Laplacian detector is expected to be more appropriate for them. The Laplacian detector [31] is derived by simply substituting Equation 14.14 into Equation 14.13, which, for the zero-mean Laplacian pdf, leads to

$$l(Y) = \frac{\sqrt{2}}{\sigma} \sum (|Y| - |Y - W|) \tag{14.17}$$

This likelihood ratio is compared against a threshold η as in Equation 14.13. In the case of N-P hypothesis testing, this threshold is determined by P_{fa}, as will be shown in section "Performance Analysis of Statistical Detectors."

It should be noted that for all of the experiments presented in section "Experimental Results," before applying the detector, the block classification and perceptual analysis procedures are performed as described in section "Imperceptible Watermarking in the Compressed Domain." This is done in order to define the set of the quantized DCT coefficients that are expected to be watermarked and will be used by the detector.

PERFORMANCE ANALYSIS OF STATISTICAL DETECTORS

The performances of the conventional Gaussian and the Laplacian detectors are measured in terms of the detection and error probabilities. In order to analyze the performance of the detector (Equation 14.14), we note that the likelihood ratio consists of the sum of a large number of independent random variables and, thus, it may be approximated by a Normal distribution according to the central limit theorem. The performance of this detector can be measured in terms of the probabilities of detection P_{det} and false alarm P_{fa}, which lead to the receiver operating characteristic (ROC) curves [31]. In order to derive these probabilities, we first need to estimate the mean and variance of the test statistic under H_0 and H_1.

Certain properties of the embedded watermark should be taken into account to enable the determination of the likelihood ratio. More specifically, the watermark value at each position of quantized DCT coefficient is given by $W = CM_QS$ (see Equation 14.3), where M_Q is the quantized watermark strength and S is the corresponding value of the pseudo-random spreading sequence. We consider the case [3] where the pseudorandom sequence takes the equiprobable values $+1$ and -1 with respective probabilities $1/2$, so the watermark is equal to either M_Q or $-M_Q$ for each watermarked quantized DCT coefficient (i.e., each coefficient for which $C = 1$). Note that for the sake of notational simplicity, the coefficient indices m and n have been suppressed.

The input data to the detector is Y, as in Equation 14.12, so $Y = X_Q$ under H_0 and $Y = X_Q + W$ under H_1. Thus, the mean of the likelihood ratio of Equation 14.14 under H_0 is found [2,3] to be

$$E[l(Y|H_0)] = m_0 = -\frac{1}{2\sigma^2}\sum M_Q^2 \qquad (14.18)$$

where the summation is over the set of watermarked pixels. The variance of the likelihood ratio is defined by

$$\text{Var}[l(Y|H_0)] = \sigma_0^2 = [(l(Y|H_0) - m_0)^2] \tag{14.19}$$

so the variance under H_0 is given by

$$\sigma_0^2 = \frac{1}{\sigma^4} \sum M_Q^2 X_Q^2 \tag{14.20}$$

In the case of the Laplacian likelihood ratio of Equation 14.17, the mean and variance under H_0 are similarly found [3] to be

$$m_0 = \sum \frac{\sqrt{2}}{\sigma} \left(|X_Q| - \frac{1}{2} (|X_Q + M_Q| + |X_Q - M_Q|) \right) \tag{14.21}$$

and

$$\sigma_0^2 = \frac{1}{4} \sum \frac{2}{\sigma^2} \left(|X_Q + M_Q| - |X_Q - M_Q| \right)^2 \tag{14.22}$$

It can be easily proven [3] that under H_1, the mean is simply $m_1 = -m_0$ and the variance does not change (i.e., $\sigma_1^2 = \sigma_0^2$). With the mean and variance of the Normally distributed likelihood ratio known, the detection and false-alarm probabilities are respectively given by

$$P_{\text{fa}}(\eta) = Q\left(\frac{\eta - m_0}{\sigma_0}\right), \quad P_{\text{det}}(\eta) = Q\left(\frac{\eta - m_1}{\sigma_1}\right) \tag{14.23}$$

where η is the threshold against which the data are compared and $Q(x)$ is defined in Equation 14.10. For a given P_{fa}, we can compute the required threshold [31] for a watermark to be detected:

$$\eta = m_0 + \sigma_0 Q^{-1}(P_{\text{fa}}) \tag{14.24}$$

By predefining this threshold and setting the signal-to-noise ratio (SNR) equal to m_0^2/σ_0^2 as in Reference 3, we can find the relation between P_{fa} and P_{det}, which leads to the ROC curves:

$$P_{\text{det}} = Q(Q^{-1}(P_{\text{fa}}) - 2\sqrt{\text{SNR}}) \tag{14.25}$$

These curves can be used to measure the performance of the statistical detectors and to compare the detection results of the Gaussian correlator with those of the Laplacian detector. Higher values of the SNR correspond to improved detection performance, so this quantity is what essentially defines and affects the performance of the two detectors.

EXPERIMENTAL RESULTS

Experimental Values of m_0 and σ_0^2

In order to verify the validity of Equation 14.18 to Equation 14.20 for the Gaussian and Laplacian detectors, experiments in which the mean and variance of the log-likelihood ratio are estimated experimentally as well as theoretically are conducted. The watermark is embedded in an I-frame of the standard video sequence Susie in certain quantized DCT coefficients, as described in section "Imperceptible Watermarking in the Compressed Domain." Monte Carlo experiments are carried out by embedding a large number of pseudorandomly generated watermarks (1000 watermarks) in the host data and estimating the log-likelihood ratios of Equation 14.14 and Equation 14.17 in each case. The detection and false-alarm probabilities are also estimated experimentally by this procedure, by comparing the likelihood ratios to a threshold. The threshold for a predefined probability of false alarm is estimated for each detector using Equation 14.24 and the theoretical values of m_0 and σ_0. We set $P_{fa} = 10^{-2}$ for both detectors, which can be reliably estimated through 1000 Monte Carlo runs.

The theoretical values of the likelihood ratio mean and variance are computed using Equation 14.18 to Equation 14.22 for both detectors, whereas their experimental values are obtained directly from the mean and variance of the computed likelihood ratios. As Table 14.2 and Table 14.3 show, the theoretically estimated values of m_0 and σ_0^2 are

TABLE 14.2. **Theoretical and experimental values of the Gaussian likelihood ratio mean and variance for Susie under various attacks**

Data	Theor. m_0	Exp. m_0	Theor. σ_0^2	Exp. σ_0^2
Susie	−10,657	−10,623	66,339	76,679
Susie with blur	−6,050.5	−6,119.7	57,973	55,195
Susie with more blur	−4,652.4	−4,607.6	34,316	27,702
Susie with Gaussian blur	−7,313.4	−7,290	71,277	66,593
Susie with median filter	−5,515	−5,493	114,510	125,140
Subregion "Hair" of Susie (cropping attack)	−75.4	−78.1	420.5	443.5
Subregion "Eye" of Susie (cropping attack)	−316.4	−316.7	2,648.8	2,485.4
Subregion "Lips" of Susie (cropping attack)	−121.3	−120.6	910.1	960.5

TABLE 14.3. Theoretical and experimental values of the Laplacian likelihood ratio mean and variance for Susie under various attacks

Data	Theor. m_0	Exp. m_0	Theor. σ_0^2	Exp. σ_0^2
Susie	−10,215	−10,210	76,457	75,435
Susie with blur	−8,092.4	−80,79.2	7,165	8,131
Susie with more blur	−5,420	−5,407	6,737	6,610
Susie with Gaussian blur	−9,062	−9,077	9,593	9,668
Susie with median filter	−5,044	−5,050	6,259	6,531
Subregion "Hair" of Susie (cropping attack)	−14.3	−15.3	66.9	67.8
Subregion "Eye" of Susie (cropping attack)	−157.8	−156.9	203.9	229.7
Subregion "Lips" of Susie (cropping attack)	−68.9	−69.1	116.5	118.9

very close to the experimental ones for both detectors, thus verifying the results of section "Performance Analysis of Statistical Detectors." Because the experimental results validate the theoretical expressions, it is possible to evaluate the performance of the two detectors theoretically, before actually conducting experiments. Consequently, the suitability of the proposed detection schemes can be predicted *a priori*.

Detector Performance Under Attacks

Filtering Attacks. The proposed quantized Laplacian detector is expected to outperform the conventional Gaussian correlator, as it is based on a more accurate statistical model of the quantized transform domain data. It is also expected to exhibit increased robustness against various image and video processing attacks, either malicious or nonmalicious ones. Experiments are conducted to examine the validity of this expectation in the presence of four common image processing attacks that are applied independently: A blurring filter is applied twice to the host data, degrading it more in the second case, a Gaussian blurring operation is also applied, and, finally, the data are passed through a median filter.

Cropping Attacks. The performances of the two detection schemes are examined under cropping, a very common and usually nonmalicious geometric attack. In a cropping attack, a region of the host data may be removed if it contains information of specific interest. In general, cropping does not necessarily degrade the visual quality of the data, but it creates quite a few problems in watermark detection. When an image is cropped, its origin is shifted and synchronization is lost. We consider that it can be regained by inserting a suitable synchronization signal or by exhaustive search of the image origin, as proposed in References 32 and 33. However, image cropping creates other problems, apart from loss of synchronization. In particular, the detector must extract the watermark or detect its existence using much fewer pixels than when it uses the whole video frame; in other words, a significant amount of information is lost through this procedure. The detector can effectively find the watermark using only the data from the unchanged region, as long as the watermark information is spread all over the image, as in the case of spread spectrum watermarking examined here. This is necessary because if the watermark is embedded only in a small part of the video frame and this area is distorted or removed, the remaining data are not water-marked and it is impossible to find whether or not the original frame contained a watermark.

It must be noted that the reliable detection of a watermark from subregions of an image or video frame can be very useful in various applications, where regions of interest are extracted from the original data. The detection of a watermark after cropping also presents great interest for object-oriented image and video coding standards (JPEG2000, MPEG-4), as it enables the watermarking of separate image or video objects [32]. The performance of the two detection schemes on certain areas of the original video frame is examined to study the effect of cropping. In particular, three subregions from a video frame of the Susie video sequence that present interest are cropped, the "Hair," the "Eyes," and the "Lips" areas (see Figure 14.9 to Figure 14.11).

Experimental Values of m_0 and σ_0^2 under Attacks. The robustness of the detectors and their theoretical expressions are verified experimentally through the application of certain common image processing attacks to the video frame of Susie. The experimental values of the mean and the variance of the likelihood ratio verify their analytical expressions under all of the filtering attacks applied, as shown in Table 14.2 and Table 14.3. The robustness of the theoretical estimates of m_0 and σ_0^2 is also tested under geometrical attacks by applying the two detection schemes to the

Figure 14.9. "Hair" subregion of Susie.

Figure 14.10. **"Eye" subregion of Susie.**

Figure 14.11. **"Lips" subregion of "Susie".**

areas shown in Figure 14.9 to Figure 14.11. In particular, it is verified that the mean and variance of the likelihood ratio can be reliably computed *a priori* from smaller image areas using the theoretical expressions of section "Performance Analysis of Statistical Detectors." Table 14.2 and Table 14.3 show that the analytical expressions still lead to the correct values of the likelihood ratio mean and variance when the watermarked data have been cropped.

The values of the likelihood ratio are plotted in Figure 14.12 and Figure 14.13 for both detectors over 1000 Monte Carlo runs under hypotheses H_1 and H_0 as well as the corresponding threshold for a predefined $P_{fa}(10^{-2})$. Because of the large number of experiments conducted and the similarity of the corresponding plots, we plot the values of the log-likelihood ratio over a set of 1000 Monte Carlo runs carried out only for the "Lips" subregion of Susie. Figure 14.12 and Figure 14.13 show that the theoretical mean of each likelihood ratio is very close to its experimental values under both hypotheses, as it remains between $m_0 \pm \sigma_0$ and $m_1 \pm \sigma_1$ over all 1000 Monte Carlo runs. It is also obvious that for the particular case examined, the watermark will always be detected by both processors and the false-alarm probabilities coincide with their predefined values, approximately 0.01 for $P_{fa} = 10^{-2}$.

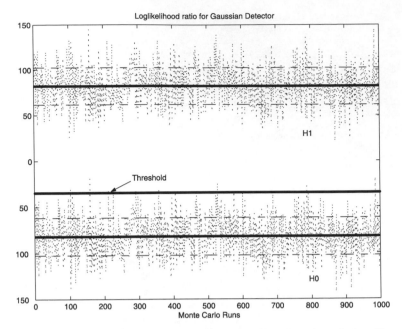

Figure 14.12. Monte Carlo runs for the likelihood ratio of the Gaussian detector for subregion "Lips" of Susie.

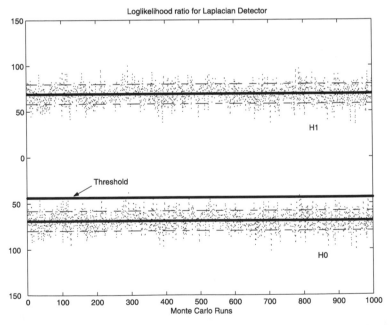

Figure 14.13. Monte Carlo runs for the likelihood ratio of the Laplacian detector for subregion "Lips" of Susie.

Detector SNR and ROC Curves. The detection performance of the Gaussian and the Laplacian schemes are examined under various filtering and geometric attacks. The frame of the standard video sequence Susie is initially blurred slightly. It also undergoes a stronger blurring attack as well as Gaussian blurring. Note that the strength of the watermark relative to the strength of the image is not greatly affected by these attacks. In order to measure the watermark strength quantitatively, we introduce the watermark-to-document Ratio (WDR) [2], using the expression.

$$\text{WDR} = 10 \log\left(\frac{\sigma_w^2}{\sigma_x^2}\right) \tag{14.26}$$

where σ_w^2 is the watermark variance or energy and σ_x^2 is the image variance. For the original data, WDR $\simeq -3$ dB, as Table 14.4 shows, which indicates that the watermark is relatively strong because watermarks are usually characterized by lower WDRs [2]. It must be emphasized that the relatively high strength of the watermark in the quantized domain does not affect its invisibility, as shown in section "Imperceptible Watermarking in the Compressed Domain" and Figure 14.5. This is due to the embedding process, which takes into account the properties of the HVS, the characteristics of each watermarked block, and the effect of quantization on the data, thus leading to the embedding of quite high watermark values that still remain invisible. In addition, it must be noted that the WDR is not an objective measure of the watermark visibility and is used as an indicative estimate of the watermark strength. Finally, Table 14.4 shows that the WDR does not change significantly after the blurring attacks.

As discussed in section "Performance Analysis of Statistical Detectors," the ROC curves and, consequently, the performance of the detectors depend solely on the SNR $= m_0^2/\sigma_0^2$, so this quantity is used to compare the performance of the two detectors in various situations. Table 14.4 depicts the values of the SNR for the two detectors as well as

TABLE 14.4. **Signal to Noise Ratio SNR $= m_0^2/\sigma_0^2$ for Susie under various attacks**

Data	Gaussian SNR (dB)	Laplacian SNR (dB)	WDR (dB)
Susie	32.335	41.350	-2.962
Susie with blur	28.004	39.610	−3.762
Susie with more blur	27.990	36.390	−4.594
Susie with Gaussian blur	28.753	39.325	−3.905
Susie with median filter	24.242	36.100	−5.160

the corresponding values of the WDR. It is obvious that the attacks affect the performance of the two detection schemes, because the values of the SNR became lower after the filtering attacks. The results of Table 14.4 prove that the Laplacian detector yields improved detection results, because it leads to higher SNRs not only for the original frame but also for all the attacked images. As expected, the SNR decreases as the attack strength increases. Table 14.4 shows that the stronger blurring attack leads to lower SNRs than the initial blurring attack. The Gaussian blurring process gives results similar to the initial blurring of the video frame, because it degrades this data as much as the simple blurring process. The most significant reduction of the SNR is caused by the application of the median filter, because the SNR for the Gaussian detector decreases from 32.335 to 24.240 dB and the Laplacian SNR becomes 36.100 dB, whereas its initial value was 41.350 dB.

The performances of the Gaussian and Laplacian detectors are also compared in the case of cropping attacks, where a part of the original video frame was removed. In particular, the detection results on the subregions "Hair" (4×4 macroblocks or $64 \times 64 = 4096$ pixels), "Eye" (6×6 macroblocks or $96 \times 96 = 9216$ pixels), and "Lips" (4×6 macroblocks or $64 \times 96 = 6144$ pixels) are examined. The performances of the two schemes are expected to worsen because fewer pixels are used for the detection process. The entire Susie frame had 704×576 ($= 405,504$) pixels and the subregions examined here contain only 4096–9216 pixels. Indeed, Table 14.5 shows that the SNRs obtained for these subregions are quite lower than those obtained when using the entire Susie frame for watermark detection. Table 14.5 shows that the Laplacian detector does not outperform the conventional Gaussian correlator when the watermark detection takes place in the "Hair" subregion. This can be explained by the fact that this area is smoother, so the corresponding quantized DCT data are more Gaussian-like than the data from the entire video frame. However, the Laplacian detector still outperforms the Gaussian correlator in the "Eye" and "Lips" subregions of Susie. This is expected because these areas contain more details that lead to less

TABLE 14.5 Signal to Noise Ratio SNR= m_0^2/σ_0^2 for Susie under cropping attacks

Data	Gaussian SNR (dB)	Laplacian SNR (dB)	WDR (dB)
Susie	32.335	41.350	−2.962
Subregion "Hair" of Susie	6.507	3.468	−4.990
Subregion "Eyes" of Susie	5.657	5.894	−6.493
Subregion "Lips" of Susie	8.220	10.480	−6.030

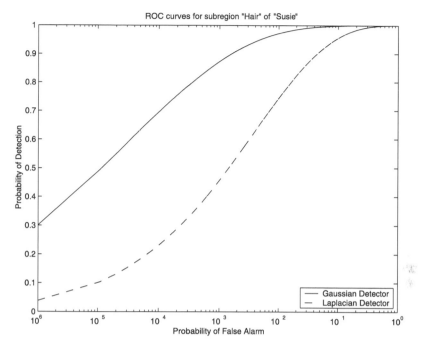

Figure 14.14. **ROC curves for subregion "Hair" of Susie.**

Gaussian-like data, for which the Laplacian detector is more suitable. It has already been mentioned in the literature [3] that the performance and optimality of statistical detectors often depends on the particular characteristics of the data (e.g., more detailed or more textured images or video frames).

Figure 14.14 to Figure 14.16 display the ROC curves for the two detection schemes for these subregions of the Susie video frame. The probability of detection has been plotted against the probability of false alarm for each area that has been cropped from the original frame. Figure 14.14 shows that the Gaussian detector is better than the Laplacian in the subregion "Hair," as was expected, because its SNR was lower than the Laplacian SNR. The ROC curves of Figure 14.15 and Figure 14.16 show that the Laplacian detector leads to higher detection probabilities for the more detailed "Eye" and "Lips" areas, which makes it more suitable for these subregions. The experimental results in the presence of cropping attacks show that both detectors are resistant to such operations and can still detect the watermark even from small areas. A possible explanation for this robustness is that the watermark energy remains quite high in these regions, as the WDRs in Table 14.5 show, allowing the effective detection of the hidden information.

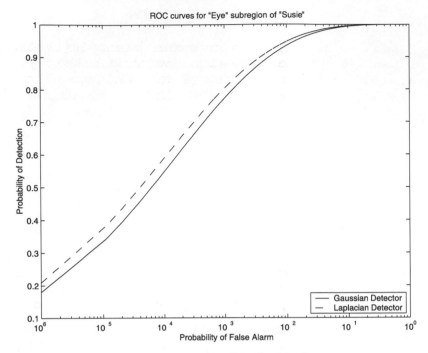

Figure 14.15.   ROC curves for subregion "Eye" of Susie.

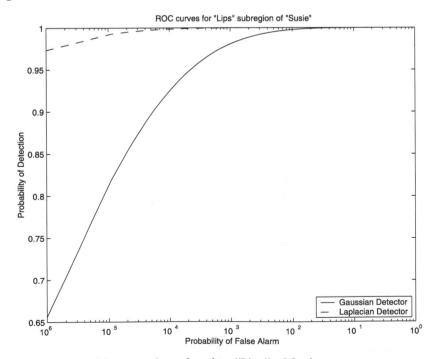

Figure 14.16.   ROC curves for subregion "Lips" of Susie.

CONCLUSIONS

An integrated realistic video watermarking system has been presented that incorporates a novel watermark embedding scheme for compressed domain video data as well as an improved watermark detector for the quantized DCT coefficients of a video frame. A spread spectrum watermark is effectively hidden in the quantized DCT coefficients of a standard video frame, achieving invisibility and quite high watermark energy at the same time. In order to achieve this, both the properties of the human visual system as well as characteristics of the video frame are adopted. The compressed domain transform coefficients are found to follow a more heavy-tailed distribution than the Gaussian model usually used. Thus, they are modeled by a modified version of the Laplacian pdf, which is particularly accurate for quantized DCT data. A statistical detector is then designed based on this improved model of the data. The performance of the Gaussian and quantized Laplacian detectors is analyzed theoretically as well as experimentally. The experimental results verify the theoretical analysis and it is found that the Laplacian scheme, indeed, outperforms the conventional correlator, as expected.

Experiments are carried to measure the robustness of these detection schemes under various attacks. The Laplacian detector performs better than the correlator after blurring or median filtering. The performance of these systems is also examined in the case of cropping. Small regions of interest are removed from the original video frame and the detection process is applied on these data. The experiments prove that both detectors perform quite well under cropping. An interesting point is that although the Laplacian detector in general leads to better performance, the conventional correlator gives better results for smoother subregions of the video frame. However, in general the Laplacian scheme exhibits better overall performance, proving to be robust for the original and also for attacked data.

Thus, the proposed detector can become part of a reliable and robust video watermarking system in the compressed domain. It allows the efficient, invisible embedding of a watermark in the quantized transform domain as well as its reliable detection, even in the presence of cropping attacks, which make it particularly useful for practical systems and object-oriented standards.

REFERENCES

1. Swanson, M.D., Zhu, B., and Tewfik, A. H., Transparent Robust Image Watermarking, in Proceedings IEEE International Conference on Image Processing, Lausanne, 1996, pp. 211–214.
2. Eggers, J. J., and Girod, B., Quantization effects on digital watermarks, *Signal Process.*, 81(2), 239–263, 2001.

3. Hernandez, J.R., Amado, M., and Perez-Gonzalez, F., DCT-domain watermarking techniques for still images: Detector performance analysis and a new structure. *IEEE Trans. Image Process.*, 9, 55–68, 2000.

4. Briassouli, A., and Strintzis, M. G., Locally optimum nonlinearities for DCT watermark detection, *IEEE Trans. Image Process.*, accepted for publication.

5. ISO/IEC 13818-2. Information Technology — Generic coding of moving pictures and asscociated audio: Video, 2000.

6. Barni, M., Bartolini, F., Cappelini, V., and Piva, A., A DCT-domain system for robust image watermarking, *Signal Process.*, 66(3), 357–372, 1998.

7. Proakis, J. G., *Digital Communications*, McGraw-Hill, New York, 1995.

8. Lee, B., Modeling quantization error from quantized Laplacian distributions, *Electron. Lett.*, 36(15), 1270–1271, 2000.

9. Price, J., and Rabbani, M., Dequantization Bias for JPEG Decompression, in Proceedings International Conference on Information Technology: Coding and Computing, 2000, pp. 30–35.

10. Simitopoulos, D., Tsaftaris, S. A., Boulgouris, N. V., and Strintzis, M. G., Digital Watermarking of MPEG-1 and MPEG-2 Multiplexed Streams for Copyright Protection, in Proceedings Second International Workshop on Digital and Computational Video, 2001, pp. 140–147.

11. Busch, C., Funk, W., and Wolthusen, S., Digital watermarking: from concepts to real-time video applications, *IEEE Computer Graphics Applic.*, 19(1), 25–35, 1999.

12. Kalker, T., Depovere, G., Haitsma, J., and Maes, M., A video watermarking system for broadcast monitoring, in Proceedings of SPIE Electronic Imaging 99, Security Watermarking Multimedia Contents, San Jose, CA, 1999, pp. 103–112,

13. Cox, I.J., Kilian, J., Leighton, F.T., and Shamoon, T., Secure spread spectrum perceptual watermarking for images, audio and video, *IEEE Trans. Image Process.*, 6, 1673–1687, 1997.

14. Cox, I.J., Miller, M.L., and McKellips, A., Watermarking as communications with side information, *Proc. IEEE*, 87, 1127–1141, 1999.

15. Hernandez, J.R., Perez-Gonzalez, F., Rodriguez, J.M., and Nieto, G., Performance analysis of a 2D-multipulse modulation scheme for data hiding and watermarking of still images. *IEEE J. Selected. Areas Commn.*, 16, 510–524, 1998.

16. Hartung, F., and Kutter, M., Multimedia watermarking techniques, *Proc. IEEE*, 87, 1079–1107, 1999.

17. Hartung, F., and Girod, B., Digital Watermarking of MPEG-2 Coded Video in the Bitstream Domain, in International Conference on Acoustics, Speech, and Signal Processing, 1997, pp. 2621–2624.

18. Hartung, F., and Girod, B., Watermarking of uncompressed and compressed video, *Signal Process.*, 66, 283–301, 1998.

19. Matsui, K., and Tanaka, K., Video-steganography: How to secretly embed a signature in a picture, in IMA Intellectual Property Project Proceedings, Vol. 1, pp. 187–206, 1994.

20. Zeng, W., and Liu, B., A statistical watermark detection technique without using original images for resolving rightful ownerships of digital images, *IEEE Trans. Image Process.*, 8(11), 1534–1548, 1999.

21. Wolfgang, R. B., Podilchuk, C. I., and Delp, E. J., Perceptual watermarks for digital images and video. *Proc. IEEE*, 87(7), 1108–1126, 1999.

22. Watson, A. B., DCT quantization matrices visually optimized for individual images, in Proceedings SPIE Conference on Human Vision, Visual Processing and Digital Display IV, 1993, pp. 202–216.

23. Rao, K. R., and Hwang, J. J., *Techniques and Standards for Image, Video and Audio Coding*, Prentice-Hall PTR, London, 1996.

24. Chung, T.-Y., Hong, M.-S., Oh, Y.-N., Shin, D.-H., and Park, S.-H., Digital watermarking for copyright protection of MPEG-2 compressed video, *IEEE Trans. Consumer Electron.*, 44(3), 895–901, 1998.

25. Meng, J., and Chang, S.-F., Embedding Visible Video Watermarks in the Compressed Domain, in Proceedings International Conference on Image Processing, 1998, ICIP 98, 1998, Vol. 1, pp. 474–477.

26. Chuhong Fei, F., Kundur, D., and Kwong, R., The Choice of Watermark Domain in the Presence of Compression, in Proceedings International Conference on Information Technology: Coding and Computing, 2001, pp. 79–84.

27. Reininger, R. C., and Gibson, J. D., Distributions of the two-dimensional DCT coefficients for images. *IEEE Trans. Commn.*, 31, 835–839, 1983.

28. Birney, K. A., and Fischer, T. R., On the modeling of DCT and subband image data for compression, *IEEE Trans. Image Process.*, 4, 186–193, 1995.

29. Müller, F., Distribution shape of two-dimensional DCT coefficients of natural images, *Electron. Lett.*, 29, 1935–1936, 1993.

30. Papoulis, A., *Probability, Random Variables, and Stochastic Processes*, 2nd Ed., McGraw-Hill, New York, 1987.

31. Poor, H. V., *An Introduction to Signal Detection and Estimation*, 2nd Ed., Springer-Verlag, New York, 1994.

32. Barni, M., Bartolini, F., and Piva, A., Improved wavelet-based watermarking through pixel-wise masking. *IEEE Trans. Image Process.*, 10, 783–791, 2001.

33. Barni, M., Bartolini, F., De Rossa, A., and Piva, A., A new decoder for the optimum recovery of nonadditive watermarks, *IEEE Trans. Image Process.*, 10, 755–766, 2001.

15

Image Watermarking Robust to Both Geometric Distortion and JPEG Compression

Portions reprinted with permission from X. Kang, J. Huang, Y. Q. Shi and Y. Lin, "A DWT-DFT composite watermarking scheme robust to both affine transform and JPEG compression," *IEEE Trans. on Circuits and Systems for Video Technology*. vol. 13, no. 8, pp. 776–786, Aug., 2003.

Xiangui Kang, Jiwu Huang, and Yun Q. Shi

INTRODUCTION

Digital watermarking has emerged as a potentially effective tool for multimedia copyright protection, authentication, and tamper-proofing [1]. Robustness of watermarking is one of the key issues for some applications, such as intellectual property protection and covert communication. A serious problem constraining some practical exploitations of watermarking technology is the insufficient robustness of existing watermarking algorithms against geometrical distortions such as translation, rotation, scaling, cropping, change of aspect ratio, and shearing.

These geometrical distortions cause the loss of geometric synchronization that is necessary in watermark detection and decoding [2]. Recently, it has become clearer that even a very small geometric distortion may fail watermark detection and, thus, vulnerability to geometric distortion is a major weakness of many watermarking methods [3,4]. Robustness against geometric distortion is known as a difficult issue in watermarking.

There are two different types of solution to geometrical attacks: nonblind and blind [5]. The nonblind solutions find applications in which either the original image and video frames or the watermarked image and video frames before geometric distortion is available.

The existing nonblind approaches to watermarking fall mainly into two categories: point matching [6,7] and motion estimation [8,9]. Johnson et al. [6] proposed recognizing distorted images using salient *feature points* (in fact, 5×5 to 11×11 rectangular regions with very low correlation with any other group of pixels nearby) first and then using the original image to fine-tune image parameters based on *normal flow* (displacement field). In the solution of Braudaway and Minter [7], a set of three or more dispersed reference points is established in both the reference and marked images. An exhaustive search for the best matching between these reference points is conducted to determine the approximate horizontal and vertical position distortions of each pixel in the marked image, thus restoring the geometrically distorted watermarked image. The solution proposed by Davoine et al. [8] is to split the reference image into a set of triangular patches. This mesh of patches then serves as the reference mesh and is kept in memory for a preprocessing step of mark signal retrieval. Loo et al. [9] proposed a registration scheme based on motion estimation between the distorted image and a reference copy. Complex wavelets provide a hierarchical framework for the motion estimation algorithm, and radial basis functions provide the means to correct erroneous motion vectors. Experimental results have shown that the proposed approach can estimate the small distortion quite accurately and allow correct watermark detection. The methods in References 6–9, just as their authors mentioned [8,9] or admitted (in our communications with the authors of References 6 and 7), cannot cope with large distortion such as rotation, scale and translation (RST) and cropping. In Reference 10, an effective, accurate, and efficient nonblind technology that can resist various geometrical distortions, including both large and small geometric distortions, is proposed. A distance measure is introduced between the distorted and undistorted images and video in order to determine the distortion that a watermarked image may experience. Then, the geometric distortion is inversed to regain synchronization. Using multiresolution coarse–fine searching to prune the searching space, the computation of the algorithm is reduced drastically and, hence, possibly implemented in real time. The watermark is robust to geometric distortion combined with JPEG compression with a quality factor as low as 10 (denoted by JPEG 10).

The blind solution, which does not use the original image in watermark extraction, has more applications but is obviously more challenging. Three major approaches to the blind solution have been reported in the literature.

The first approach hides a watermark signal in the invariants of a host signal (invariant with respect to scaling, rotation, shifting, etc.), and is somewhat awkward due to the theoretical and practical difficulties in constructing invariants with respect to combinations of the above-mentioned operations. Specifically, O'Ruanaidh and Pun [11] first proposed a watermarking scheme based on transform invariants by applying the Fourier–Mellin transform to the magnitude spectrum of an original image. In the first step, the Fourier transform of the image is computed. Because shifting in the spatial domain results in a phase shift in the Fourier domain and leaves the magnitude part intact, the magnitude part of the Fourier coefficients is chosen to achieve translation invariance. In the second step, the log-polar mapping (LPM) is applied so that the magnitude of the Fourier transform is changed from Cartesian coordinates to the log-polar grid i.e., the coordinates with logarithm radius and angle axes). Thus, both the scaling and rotation distortions can be converted into horizontal and vertical shifts in the new coordinate. By computing the discrete Fourier transform (DFT) of the log-polar map and keeping the DFT magnitude only, a rotation- and scaling-invariant representation ("strong invariant" domain) is obtained. Taking the Fourier transform of a log-polar map is actually equivalent to computing the Fourier–Mellin transform. The watermark is then embedded into the magnitude of the Fourier–Mellin transform by using the spread spectrum modulation. The watermarked image is constructed by applying inverse log-polar mapping. The phases in both Fourier transforms are not modified, but simply computed with the watermarked magnitudes for the inverse Fourier transforms. In Reference 4, instead of a "strong invariant" domain, the watermark is embedded into the magnitudes of the DFT coefficients resampled by the LPM. The detection process involves a comparison of the watermark with all cyclic shifts of the extracted watermark to cope with rotation. To deal with scaling, the correlation coefficient is selected as the detection metric. Only 1-bit information is hidden in the image. The main drawback is that the watermark cannot resist the general transformations. In addition, because some form of interpolation is always needed when we change the coordinates, LPM and inverse LPM may cause a loss of image quality from sampling.

The second approach exploits the self-reference principle based on an autocorrelation function (ACF) or the Fourier magnitude spectrum of a periodical watermark [12,13]. In Reference 12, Kutter introduced the idea of self-reference systems that embeds the watermark several times at shifted locations. The watermark becomes a reference of itself, making the synchronization possible without using original information, simply using the relative position of the marks. The watermark pattern is replicated in the image in order to create four repetitions of the same watermark pattern such that the experienced generalized geometrical transformation can be detected by applying autocorrelation to the investigated image. The four patterns are shifted copies of each other. The initial watermark pattern is a two-dimensional (2-D) random number array. The second watermark pattern is then formed by horizontally shifting the first pattern by x_δ columns periodically. The third watermark pattern is formed by vertically shifting y_δ rows periodically. Finally, the fourth pattern is formed by shifting the first pattern by x_δ columns y_δ rows periodically. The four watermark patterns are embedded in an orthogonal way. The first watermark pattern is embedded at locations with odd rows and odd columns. The second pattern is embedded in locations with odd rows and even columns. The third pattern is embedded in locations with even rows and odd columns. The fourth pattern is embedded in locations with even rows and even columns. In the watermark recovery process, a prediction of the embedded watermark based on a prediction filter is computed. Then, the ACF is computed for this prediction. The multiple embedding of the watermark results in additional autocorrelation peaks. Nine peaks can be detected in the ACF. The center peak represents the energy of the filtered image, whereas the other eight peaks, which are symmetric around the center owing to the symmetric structure of the autocorrelation, are generated by the four embedded patterns. By comparing the location of the extracted peaks with their expected location, we can determine the affine distortion applied to the image. The distortion can then be inverted. Experimental results showed that the algorithm could resist generalized geometrical transformations. However, the watermark embedded in spatial domain is vulnerable to the lossy coding scheme such as JPEG compression. In Reference 13, the use of a periodical block allocation of a watermark pattern for recovering from geometrical distortions is proposed. The message m is encoded using some Error-correcting coding (ECC), encrypted, mixed with the reference watermark, and allocated into a block, depending on the secret key k. This block is then upsampled and flipped, and the resulting macroblock is tiled up to the complete image size. In watermark recovery, a maximum *a posteriori* probability (MAP) estimate is used to estimate the watermark. The periodical watermark having a discrete magnitude spectrum makes it possible to obtain a regular grid of reference points that can be employed for recovering from general affine transformations. However, it is noted that the watermark estimation is key independent: The periodical watermark results in an underlying regular grid. Hence, the watermark may be detected and may be destroyed [14,15]. It has been proposed recently that the resulting watermark can be slightly predistorted in such a way that this problem may be resolved [16]. In Reference 16, it is reported that the watermark is robust against local or nonlinear geometrical distortions, such as random bending attack.

The third approach utilizes an additional template (e.g., a cross [17] or a sinusoid [18]). Bender et al. [17] suggested a scheme in which multiple cross shapes are embedded into the image (e.g., by least significant bit [LSB] plane manipulation). Any geometrical transformation applied to the image will reflect in the shape and position of the embedded crosses. This information can be used to determine the affine transformation. The drawback of this scheme is its low robustness toward noise, such as compression noise. Fleet and Heger [18] proposed embedding sinusoidal signals in the spatial domain of the image. These sinusoids act as a grid, providing a coordinate frame for the image. The sinusoids appear as peaks in the frequency domain and can then be used to determine the geometrical distortions. In Reference 19, Pereira and Pun proposed an affine resistant image watermark in the DFT domain. The informative watermark (the message to be conveyed to the detector and receiver) together with a template are embedded in the middle-frequency components in the DFT domain. The template consists of intentionally embedded peaks in the Fourier spectrum; that is, the watermark embedder casts extra peaks in some locations of the Fourier spectrum. The locations of the peaks are predefined or selected by a secret key. The strength is adaptively determined by using local statistics of the Fourier spectrum. In watermark detection, all local maxima are extracted by using small windows. Although resistant to affine transform, the watermark generated with the scheme is not robust enough. In particular, it is not robust to JPEG compression, with the quality factor below 75. Although some significant progress has been made recently, these new schemes are normally not robust to JPEG compression at the same time, as shown below.

In this chapter, a new blind image watermarking algorithm robust against both affine transformations and JPEG compression is proposed. The proposed discrete wavelet transform (DWT)–DFT composite watermarking scheme embeds a message and a training sequence in the LL_4 subband of the DWT domain, and it embeds a template in the magnitude spectrum of the DFT domain. The watermarking incorporates DWT and DFT, concatenated coding of direct sequence spread spectrum (DSSS) and Bose-Chaudhuri-Hochquenghem (BCH), 2-D interleaving, resynchronization based on the template, and training sequence. Experimental results have demonstrated that the watermark is robust against both affine transformations and JPEG compression when the quality factor is as low as 10.

The rest of this chapter is organized as follows. Watermark embedding is introduced in section "Watermark Embedding." Section "Watermark Extraction with Resynchronization" describes the watermark extraction based on resynchronization. The experimental results are presented in section "Experimental Results." Conclusion is drawn and future research is discussed in the last section.

WATERMARK EMBEDDING

This algorithm achieves enhanced robustness by improving the embedding strategy and watermark structure and using a new effective synchronization technique. DWT is playing an increasingly important role in watermarking, due to its good spatial-frequency characteristics and its wide applications in the image and video coding standards. According to Cox et al. [20] and Huang et al. [21], the watermark should be embedded in the DC and low-frequency AC coefficients in the DCT domain because of their large perceptual capacity. The strategy can be extended to the DWT domain [22]. We embed an informative watermark into the LL_4 subband in the DWT domain to make it more robust while keeping the watermark invisible. When the marked image undergoes affine transformation

$$\left(\begin{bmatrix} x' \\ y' \end{bmatrix} = B \begin{bmatrix} x \\ y \end{bmatrix} + \begin{bmatrix} t_x \\ t_y \end{bmatrix} \right),$$

the matrix B can be determined by using a template as reference. The template is embedded into the middle-frequency components in the magnitude spectrum to avoid interfering with the informative watermark. To determine the translation parameter, we embed a training sequence in the DWT domain. To survive all kinds of attack, we use the concatenated coding of the BCH and DSSS method to encode the message m $\{m_i; i = 1, \ldots, L, m_i \in \{0,1\}\}$ ($L = 60$ in our work). To cope with bursts of errors possibly occurring with a watermark, a newly developed 2-D interleaving [23,24] is exploited. The watermark embedding process is shown in Figure 15.1.

The Message Encoding and Training Sequence Embedding in the DWT Domain

The watermark embedding in DWT domain is implemented through the following procedures (Figure 15.1):

- *DWT decomposition*: By using Daubechies 9/7 biorthogonal wavelet filters, we apply a four-level DWT to an input image $f(x,y)$ ($512 \times 512 \times 8$ bits, in our work), generating 12 subbands of high frequency (LH_i, HL_i, HH_i, $i = 1$–4) and one low-frequency subband (LL_4).

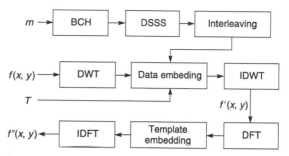

Figure 15.1. The watermark embedding process. (From X. Kang et al., *IEEE Trans. on Circuits and Systems for Video Technology*. vol. 13, no. 8, pp. 776–786, Aug., 2003. With permission.)

- *Message encoding*: The message m is first encoded using BCH (72, 60) to obtain the message m_c of length $L_c = 72$. Then, each bit m_{ci} of m_c is DSSS encoded using an N_1-bit bipolar pseudo noise sequence (PN) $p = \{p_j; j = 1,\ldots,N_1\}$, where "1" is coded spreadly as $\{+1 \times p_j; j = 1,\ldots,N_1\}$ and "0" as $\{-1 \times p_j; j = 1,\ldots,N_1\}$, thus obtaining a binary string W:

$$m_{ci} \xrightarrow{\text{DSSS coding}} W_i\{w_{ij}; w_{ij} \in \{-1, +1\}, 1 \le j < N_1\}, \quad 1 \le i < L_c$$

- *Training sequence*: The training sequence T $\{T_n; n = 1,\ldots,63\}$, $T_n \in \{-1, 1\}$, which is a key-based sequence, should be distributed all over the image in order to survive all kinds of attack, especially cropping. To this purpose, in our work, the training sequence T (63 bits) is embedded in row 16 and column 16 of the LL_4 subband (see Figure 15.2). It is noted that a different combination of row and column corresponding to the important image portion can also be chosen. The informative watermark W is 2-D interleaved and embedded in the leftover portion of the LL_4 subband (Figure 15.2).

  In implementation, we allocate the 63 bits of the training sequence T into row 16 and column 16 of a 32×32 2-D array; we then deinterleave [24] it, resulting in a new 32×32 array. The binary string W is embedded into the remaining portion of the above-mentioned array. By applying the 2-D interleaving technique [24] to this array, we obtain another 2-D array. Scanning this 2-D array, say, row by row, we convert it into a 1-D array X.

- *Data embedding*: Coefficients of the LL_4 subband are scanned in the same way as in the embedding, resulting in a 1-D array, denoted by C. We adopt quantization-based embedding Equation (15.1) to embed the binary data X into C to obtain C' [25]. Here, α is a parameter related to the watermark embedding strength, $C(i)$ and $C'(i)$ denote the amplitude of the ith element in C and C', respec-

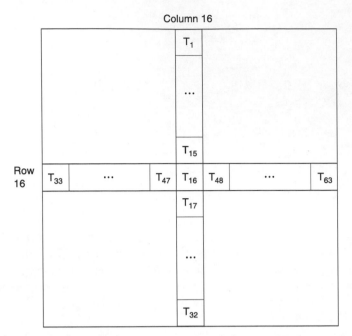

Figure 15.2. Training data-set. (From X. Kang et al., *IEEE Trans. on Circuits and Systems for Video Technology*. vol. 13, no. 8, pp. 776–786, Aug., 2003. With permission.)

tively. The quantizer $q(\cdot)$ is a uniform, scalar quantization function of step size α, and $q(x) = k\alpha + 0.5\alpha$, $k = \lfloor x/\alpha \rfloor$, $k \in Z$, and $\lfloor \cdot \rfloor$ denotes the *floor* operation. Equation (15.1) indicates that the proposed embedding method tries to output a $C'(i)$ value that is closest to $C(i)$ and whose corresponding embedding bit value equals x_i (Figure 15.3).

$$
\begin{cases}
C'(i) = q\left(C(i) - \dfrac{1}{4}\alpha\right) + \dfrac{1}{4}\alpha & \text{if } x_i = +1 \\[2mm]
C'(i) = q\left(C(i) + \dfrac{1}{4}\alpha\right) - \tfrac{1}{4}\alpha & \text{if } x_i = -1
\end{cases}
\tag{15.1}
$$

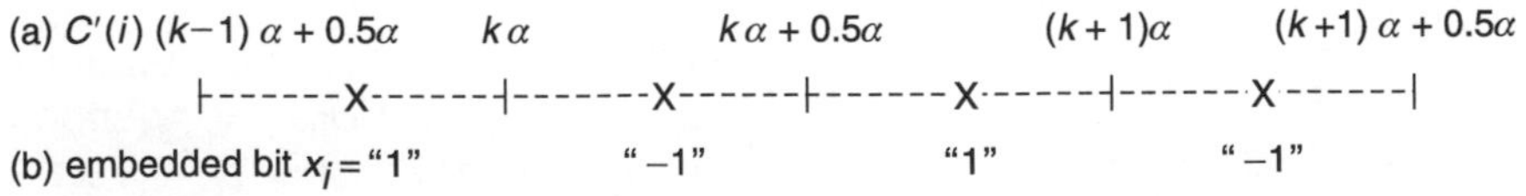

(a) $C'(i)$ $(k-1)\,\alpha + 0.5\alpha$ $k\,\alpha$ $k\alpha + 0.5\alpha$ $(k+1)\alpha$ $(k+1)\,\alpha + 0.5\alpha$

├------X------┼-------X------├------X------┼------X------┤

(b) embedded bit $x_i =$ "1"   "−1"   "1"   "−1"

Figure 15.3. Graphical illustration of data embedding: (a) "x" indicates a possible $C'(i)$ value after one bit is embedded; (b) the corresponding embedded bit.

Note that the difference between $C(i)$ and $C'(i)$ is between -0.5α and $+0.5\alpha$. If $x_i = -1$, $C'(i) \bmod \alpha = 0.25S$. If $x_i = +1$, $C'(i) \bmod \alpha = 0.75S$. Therefore, in the extraction, if the extracted coefficient $C^*(i)$ has $C^*(i) \bmod \alpha > \alpha/2$, then the recovered binary bit $x_i^* = +1$; Otherwise, $x_i^* = -1$ (Equation 15.2):

$$
x_i^* = \begin{cases} +1 & C^*(i) \bmod \alpha > \alpha/2 \\ \\ -1 & \text{otherwise} \end{cases} \tag{15.2}
$$

In order to extract an embedded bit correctly, the absolute error (introduced by image distortion) between $C^*(i)$ and $C'(i)$ must be less than 0.25α. The parameter α can be chosen so as to make a good compromise between the contending requirements of imperceptibility and robustness. We chose $\alpha = 90$ in our work. Performing the inverse DWT on the modified image, we obtain a watermarked image $f'(x, y)$.

This training sequence helps to achieve synchronization against translation possibly applied to stegoimage. If the correlation coefficient between the training sequence T and the test sequence S obtained from a test image satisfies the condition $\rho_{T,S} = (1/63) \sum_{n=1}^{63} (T_n S_n) \geq \text{thresh}_1$, we regard S matched to T and consider synchronization achieved. Here, we can calculate the corresponding probability of false positive (false synchronization) as $H_{\text{fp}} = (1/2^{63}) \sum_{k=63-e}^{63} \binom{63}{k}$, where $e = \text{round}\,((63/2)(1 - \text{thresh}_1))$, and round $(\cdot)$ means taking the nearest integer. In our work, we choose $\text{thresh}_1 = 0.56$ (empirically value); thus $H_{\text{fp}} = 5.56 \times 10^{-6}$, which may be sufficiently low for many applications.

- *IDWT*: Performing the inverse DWT (IDWT) on the modified DWT coefficients, we produce the watermarked image $f'(x, y)$. The peak signal-to-noise ratio (PSNR) of thus generated marked images $f'(x, y)$ vs. the original image is higher than 42.7 dB.

Template Embedding in the DFT Domain

Because the DWT coefficients are not invariant under geometric transformation, to resist the affine transform, we embed a template in the DFT domain of the watermarked image (inspired by Reference 19). Before embedding the template, the image $f'(x, y)$ is padded with zeros to a size of 1024×1024 in order to have the required high resolution. Then, the fast Fourier transform (FFT) is applied. The template embedding is as follows.

A 14-point template uniformly distributed along two lines (refer to Figure 15.4) is embedded, 7 points of each line in the upper half-plane in the DFT domain at angles θ_1 and θ_2 with radii varying between

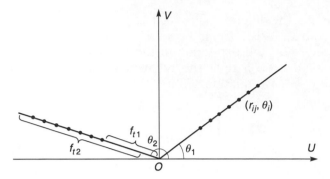

Figure 15.4. Template embedding. (From X. Kang et al., *IEEE Trans. on Circuits and Systems for Video Technology.* vol. 13, no. 8, pp. 776–786, Aug., 2003. With permission.)

f_{t1} and f_{t2}. The angles θ_i and radii r_{ij}, where $i = 1, 2, j = 1,\ldots,7$, may be chosen pseudorandomly as determined by a key. We require at least two lines in order to resolve ambiguities arising from the symmetry of the magnitude of the DFT, and we choose to use only two lines because adding more lines increases dramatically the computational cost of template detection. We found empirically that seven points per line are enough to lower the false-positive probability to a satisfactory level during detection. However, to achieve more robustness against JPEG compression than the technique reported in Reference 19, a lower-frequency band, say, $f_{t1} = 200$ and $f_{t2} = 305$, is used for embedding the template. This corresponds to 0.2 and 0.3, respectively, in the normalized frequency, which is lower than the band of 0.35–0.37 used in Reference 19. Because we do not embed the informative watermark in the magnitude spectrum of DFT domain, to be more robust to JPEG compression, a larger strength of the template points is chosen than in Reference 19. Concretely, instead of the local average value plus two times the standard deviation [19], we use the local average value of DFT points plus five times standard deviation. According to our experimental results, this higher-strength and lower-frequency band has little effect on the invisibility of the embedded watermark (refer to Figure 15.5 and Figure 15.6).

Correspondingly, another set of 14 points are embedded in the lower half-plane to fulfill the symmetry constraint.

To calculate the inverse FFT, we obtain the DWT–DFT composite watermarked image $f''(x,y)$. The PSNR of $f''(x,y)$ vs. the original image is 42.5 dB, which is reduced by 0.2 dB compared with the PSNR of $f'(x,y)$ vs. the original image due to the template embedding. The experimental results demonstrate that the embedded data are perceptually invisible.

Figure 15.5. Original images: (a) Baboon, (b) Lena, and (c) Plane. (From X. Kang et al., *IEEE Trans. on Circuits and Systems for Video Technology.* vol. 13, no. 8, pp. 776–786, Aug., 2003. With permission.)

Figure 15.6. The watermarked images with PSNR > 42.5 dB. (From X. Kang et al., *IEEE Trans. on Circuits and Systems for Video Technology.* vol. 13, no. 8, pp. 776–786, Aug., 2003. With permission.)

WATERMARK EXTRACTION WITH RESYNCHRONIZATION

In order to extract the hidden information, we extract a data sequence S $\{S_n; n = 1, \ldots, 63\}$ in row 16 and column 16 in the DWT LL_4 subband of the to-be-checked image $g(x, y)$, which is rescaled to the size of the original image at first, in our work, 512×512. (We assume that the size of the original image is known to the detector.) If, $\rho_{T,S} \geq \text{thresh}_1$, we can then extract the informative watermark and recover the message from the LL_4 subband directly. Otherwise, we need to resynchronize the hidden data before extracting the informative watermark.

In order to resynchronize the hidden data after geometric distortion, we restore affine transform according to the template embedded and then restore translation using the training sequence. Therefore, the procedure of information extraction is divided into three phases: template detection, translation registration, and decoding.

Template Detection

We first detect the template embedded in the DFT domain. By comparing the detected template with the originally embedded template, we can determine the affine transformation possibly applied to the test image. To avoid high computational complexity, we propose an effective method to estimate the affine transformation matrix.

A linear transform applied in spatial domain results in a corresponding linear transform in the DFT domain; that is, if a linear transform B is applied to an image in the spatial domain,

$$\begin{bmatrix} x \\ y \end{bmatrix} \rightarrow B \begin{bmatrix} x \\ y \end{bmatrix}$$

(15.3)

then, correspondingly, the following transform takes place in the DFT domain [19]:

$$\begin{bmatrix} u \\ v \end{bmatrix} \rightarrow (B^{-1})^T \begin{bmatrix} u \\ v \end{bmatrix}$$

(15.4)

The template detection is conducted in the following way, which is basically the same as in Reference 19, except it is a more efficient way to estimate the linear transform matrix. First, apply a Bartlett window to the to-be-checked image $g(x, y)$ to produce the filtered image Iw. Calculate the FFT of the image padded with zero to the size of 1024×1024. A higher resolution in the FFT is expected to result in a more accurate estimation of the undergone transformations. However, as we increase the amount of zeropadding, we also increase the volume of the calculations. We find that using a size of 1024×1024 yields a suitable compromise. Then, extract and record the positions of all the local peaks (p_{ui}, p_{vi}) in the image. Sort the peaks by angle and divide them into N_b equally spaced bins based on the angle.

For each of the equally spaced bins, search for a K where $K_{\min} < K < K_{\max}$ such that at least N_m local peaks having radial coordinate r_{li}, where $i \in 1, \ldots, N_b$, match the peaks in the original template having radial coordinate r_{Tj} along line j where $j \in 1, 2$. Here, by matching it is meant that $|r_{li} - K r_{Tj}| < \text{thresh}_2$. If at least N_m points match, we store the set of matched points. (In our work, we chose $K_{\min} = 0.5$ and $K_{\max} = 2$. The corresponding scaling factor considered is, hence, between 2 and 0.5).

From all of the sets of matched points, choose one set matching to template line 1 and another to template line 2 such that the angle between two sets deviates from the difference between θ_1 and θ_2, as shown in Figure 15.4, within a threshold θ_{diff}. Calculate a transformation matrix A such that the mean square error (MSE) defined in Equation (15.5) is minimized:

$$
\text{MSE} = \frac{1}{\text{nummatches}} \left\| A \begin{bmatrix} u_{11} & v_{11} \\ \vdots & \vdots \\ u_{1l_1} & v_{1l_1} \\ u_{21} & v_{21} \\ \vdots & \vdots \\ u_{2l_2} & v_{2l_2} \end{bmatrix}^T - \begin{bmatrix} u'_{11} & v'_{11} \\ \vdots & \vdots \\ u'_{1l_1} & v'_{1l_1} \\ u'_{21} & v'_{21} \\ \vdots & \vdots \\ u'_{2l_2} & v'_{2l_2} \end{bmatrix}^T \right\|^2 \tag{15.5}
$$

where (u, v) with subscripts represent a peak point's coordinate, (u', v') with subscripts represent the original (known) templates coordinate, nummatches $= l_1 + l_2$ is the number of matching points. We note that A is a 2×2 transformation matrix $\begin{bmatrix} a & c \\ b & d \end{bmatrix}$ and the matrix inside $\| \cdots \|$ is of $2 \times (l_1 + l_2)$. The notation $\| \cdots \|^2$ denotes the sum of all of these squared error elements in the error matrix. The rows contain the errors in estimating the u and v from the original (known) template positions u' and v' after applying the linear transformation.

In the following, we propose a method to estimate the transformation matrix A. It is noted that Equation (15.6) links the set of matched points $(u_{i_1j_1}, v_{i_1j_1})$ and the set of the original template points $(u'_{i_1j_1}, v'_{i_1j_1})$ with

$i_1 = 1, 2, j_1 = 1, 2, \ldots, l_1$ or l_2, $N_m \le l_1$ or $l_2 \le 7$.

$$
\begin{bmatrix}
u_{11} & v_{11} & 0 & 0 \\
\vdots & \vdots & \vdots & \vdots \\
u_{1l_1} & v_{1l_1} & 0 & 0 \\
u_{21} & v_{21} & 0 & 0 \\
\vdots & \vdots & \vdots & \vdots \\
u_{2l_2} & v_{2l_2} & 0 & 0 \\
0 & 0 & u_{11} & v_{11} \\
\vdots & \vdots & \vdots & \vdots \\
0 & 0 & u_{1l_1} & v_{1l_1} \\
0 & 0 & u_{21} & v_{21} \\
\vdots & \vdots & \vdots & \vdots \\
0 & 0 & u_{2l_2} & v_{2l_2}
\end{bmatrix}
\begin{bmatrix} a \\ b \\ c \\ d \end{bmatrix}
=
\begin{bmatrix}
u'_{11} \\
\vdots \\
u'_{1l_1} \\
u'_{21} \\
\vdots \\
u'_{2l_2} \\
v'_{11} \\
\vdots \\
v'_{1l_1} \\
v'_{21} \\
\vdots \\
v'_{2l_2}
\end{bmatrix}
\tag{15.6}
$$

Equation (15.6) can be rewritten in the following matrix–vector format:

$$MA_c = N \tag{15.7}$$

or

$$
\begin{bmatrix} M_1 & 0 \\ 0 & M_1 \end{bmatrix}
\begin{bmatrix} A_1 \\ A_2 \end{bmatrix}
=
\begin{bmatrix} N_1 \\ N_2 \end{bmatrix}
\tag{15.8}
$$

We seek A that minimizes $\|E\|^2 = \|N - MA_c\|^2$; the solution that satisfies the requirement is [26]

$$M_1^T M_1 A_1 = M_1^T N_1 \tag{15.9}$$

$$M_1^T M_1 A_2 = M_1^T N_2. \tag{15.10}$$

Because the matrix $M_1^T M_1$ is a positive definite 2×2 matrix, $M_1^T N_1$ and $M_1^T N_2$ are 2×1 matrices, the above two linear equation systems can be easily solved for $A_1 = (M_1^T M_1)^{-1} M_1^T N_1$ and $A_2 = (M_1^T M_1)^{-1} M_1^T N_2$. Thus, we obtain the candidate of transformation matrix

$$
A = \begin{bmatrix} A_1^T \\ A_2^T \end{bmatrix},
$$

and the corresponding MSE according to Equation (15.5) is:

$$\text{MSE} = \frac{1}{\text{nummatches}} \|\boldsymbol{E}\|^2$$

$$= \frac{1}{\text{nummatches}} (\|\boldsymbol{N_1} - \boldsymbol{M_1 A_1}\|^2 + \|\boldsymbol{N_2} - \boldsymbol{M_1 A_2}\|^2) \tag{15.11}$$

Because we only work on the upper half-plane, in order to resolve possible ambiguities, we add 180° to the angles in the sets of matched points corresponding to line 1 of the template (either line can be used), we then repeat the previous step.

Finally, choose the $\boldsymbol{A}$ that results in the smallest MSE. If the minimized error is larger than the detection threshold T_d, we conclude that no watermark was embedded in the image. Otherwise, we proceed to decoding. According to Equation (15.3) and Figure (15.4), we can obtain the linear transform matrix $\boldsymbol{B}$, described at the beginning of Section "Template Detection," and $\boldsymbol{B} = \boldsymbol{A}^T$. Applying the inverse linear transform, $\boldsymbol{B}^{-1}$, to the image $g(x, y)$, we obtain an image $g'(x, y)$, which has corrected the applied linear transform. One example is shown in Figure 15.7a and Figure 15.7b.

Translation Registration and Decoding

Assume that the linear transform corrected image $g'(x, y)$ has size $M \times N$. Padding $g'(x, y)$ with 0's to the size 512×512 generates the image $I(x, y)$ (Figure 15.7c).

One way to restore translation is to search by bruteforce for the largest correlation coefficient between the training sequence $\boldsymbol{T}$ and the data sequence $\boldsymbol{S}$ extracted from the DWT coefficients in row 16 and column 16 of the LL_4 subband corresponding to the following set of all possible translated images. (Note that we construct $\boldsymbol{S}$ sequence in the way as we embed the $\boldsymbol{T}$ sequence in the LL_4 sunband; refer to Figure 15.2.)

$$I_t(x, y) = I((x - t_x) \bmod 512, (y - t_y) \bmod 512);$$

$$\left\{ -\frac{1}{2}(512 - M) \le t_x < \frac{1}{2}(512 - M); \ -\frac{1}{2}(512 - N) \le t_y < \frac{1}{2}(512 - N) \right\} \tag{15.12}$$

where t_x and t_y are the translation parameters in the spatial domain. This method demands a heavy computational load when $(512 - M) > 16$ and $(512 - N) > 16$. We dramatically reduce the required computational load by performing DWT for at most 256 cases according to the dyadic nature of the DWT; that is, if an image is translated by $16x_{t1}$ rows and $16y_{t1}$ columns $(x_{t1}, y_{t1} \in Z)$, then the LL_4 subband coefficients of the image are translated by x_{t1} rows and y_{t1} columns accordingly. This property is

(a) (b)

(c) (d)

Figure 15.7. Resynchronization: (a) the to-be-checked image $g(x, y)$, which is 512×512 and experienced a rotation of $10°$, scaling, translation, cropping, and JPEG compression with a quality factor of 50; (b) the image $g'(x, y)$, which is 504×504 and has been recovered from the linear transform applied; (c) the image $I(x, y)$, which has been padded with 0's to the size 512×512; (d) the resynchronized image $g^*(x, y)$, which is 512×512 and has been padded with the mean gray-scale value of the image $g(x, y)$. The embedded message was finally recovered without error. (From X. Kang et al., *IEEE Trans. on Circuits and Systems for Video Technology.* vol. 13, no. 8, pp. 776–786, Aug., 2003. With permission.)

utilized to efficiently handle translation synchronization in our algorithm; that is, we have $t_x = 16x_{t1} + x_t$ and $t_y = 16y_{t1} + y_t$, where $-8 \leq x_t$ and $y_t < +8$. In each of the 256 pairs of (x_t, y_t), we perform DWT on the translated image $I_t(x, y)$, generating the LL_4 coefficients, denoted by $LL_{4t}(x, y)$. We then perform translations on $LL_{4t}(x, y)$:

$$LL'_{4t}(x, y) = LL_{4t}((x - x_{t1}) \bmod 32, \ (y - y_{t1}) \bmod 32);$$
$$\{-T_1 \leq x_{t1} < T_1; \ -T_2 \leq y_{t1} < T_2\}$$

(15.13)

where x_{t1} and y_{t1} are the translation parameters in the LL_4 subband, $T_1 = \text{round}(1/2(512 - M)/16)$ and $T_2 = \text{round}(1/2(512 - N)/16)$. Each time, we extract the data sequence $\boldsymbol{S}$ in row 16 and column 16 in the $LL'_{4t}(x, y)$.

The synchronization is achieved when $\rho_{T,S} \geq \text{thresh}_1$ or $\rho_{T,S}$ is largest. For example, for the image in Figure 15.7c, the maximum correlation coefficient $\rho_{T,S}(=0.87)$ is achieved when $x_t = -3$, $y_t = -4$, and $x_{t1} = 0$, $y_{t1} = 0$. Finally, we obtain the translation parameters ($t_x = 16 \times x_{t1} + x_t$, $t_y = 16 \times y_{t1} + y_t$).

After restoring affine transform and translation and padding with the mean gray-scale value of the image $g(x, y)$, we obtain the resynchronized image $g^*(x, y)$ (Figure 15.7d). The LL_4 coefficients of $g^*(x, y)$ are scanned in the same way as in data embedding, resulting in a 1-D array, C^*. The extracted hidden binary data, denoted by $X^* = \{x_i^*\}$, are extracted as in Equation (15.2).

Deinterleaving [24] the 32×32 2-D array, constructed from X^*, we can obtain the recovered binary data W^*. We segment W^* by N_1 bits per sequence and correlate the obtained sequence with the original PN sequence p. If the correlation value is larger than zero, the recovered bit is "1," otherwise "0." The binary bit sequence b can thus be recovered.

The recovered bit sequence b is now BCH decoded. In our work, we use BCH(72, 60). Hence, if there are fewer than five errors, the message m will be recovered without error; otherwise, the embedded message cannot be recovered correctly.

EXPERIMENTAL RESULTS

We have tested the proposed algorithm on images shown in Figure 15.5 and Figure 15.8. The results are reported in Table 15.1. In our work,

Figure 15.8. Some original images used in our test. (From X. Kang et al., *IEEE Trans. on Circuits and Systems for Video Technology.* vol. 13, no. 8, pp. 776–786, Aug., 2003. With permission.)

Figure 15.8. Continued. (From X. Kang et al., *IEEE Trans. on Circuits and Systems for Video Technology.* vol. 13, no. 8, pp. 776–786, Aug., 2003. With permission.)

Figure 15.8. Continued. (From X. Kang et al., *IEEE Trans. on Circuits and Systems for Video Technology*. vol. 13, no. 8, pp. 776–786, Aug., 2003. With permission.)

Table 15.1. Experimental results with StirMark 3.1. (From X. Kang et al., *IEEE Trans. on Circuits and Systems for Video Technology*. vol. 13, no. 8, pp. 776–786, Aug., 2003. With permission.)

	Lena	Baboon	Plane	Boat	Drop	Pepper	Lake	Bridge
StirMark functions								
JPEG 10~100	1	1	1	1	1	1	1	1
Scaling	1	1	1	1	1	1	1	1
Jitter	1	1	1	1	1	1	1	1
Cropping_25	1	1	1	1	1	1	1	1
Aspect ratio	1	1	1	1	1	1	1	1
Rotation (autocrop, scale)	1	1	1	1	1	1	1	1
General linear transform	1	1	1	1	1	1	1	1
Shearing	1	1	1	1	1	1	1	1
Gauss filtering	1	1	1	1	1	1	1	1
Sharpening	1	1	1	1	1	1	1	1
FMLR	1	1	1	1	1	1	1	1
2×2 median_filter	1	0	1	0	1	1	1	0
3×3 median_filter	1	0	1	0	1	0	0	0
4×4 median_filter	0	0	0	0	1	0	0	0
Random bending	0	0	0	0	0	0	0	0

we chose $L = 60$, $N_1 = 11$, $N_b = 180$, $N_m = 5$, $\text{thresh}_1 = 0.56$, $\text{thresh}_2 = 0.002$, and the detection threshold $T_d = 1.0 \times 10^{-6}$. The PSNRs of the marked images are higher than 42.5dB (Figure 15.6 and Figure 15.9). The watermarks are perceptually invisible. The watermark embedding takes

Figure 15.9. The watermarked images with PSNR > 42.5 dB. (From X. Kang et al., *IEEE Trans. on Circuits and Systems for Video Technology.* vol. 13, no. 8, pp. 776–786, Aug., 2003. With permission.)

Figure 15.9.   Continued. (From X. Kang et al., *IEEE Trans. on Circuits and Systems for Video Technology.* vol. 13, no. 8, pp. 776–786, Aug., 2003. With permission.)

less than 4 sec, whereas the extraction takes about 2–38 sec on a Pentium PC of 1.7 GHz using C language.

Figure 15.10 shows a marked Lena image that has undergone JPEG compression with a quality factor of 50 (JPEG_50) in addition to general linear transform (a StirMark test function: linear_1.010_0.013_0.009_1.011; Figure 15.10a) or rotation 30° (autocrop, autoscale; Figure 15.10b). In both cases, the embedded message (60 information bits) can be recovered with no error. This demonstrates that our watermarking method is able to resist both affine transforms and JPEG compression. Table 15.1 shows more test results with our proposed algorithms by using StirMark 3.1. In Table 15.1, "1" represents that the embedded 60-bit message can

(a)   (b)

Figure 15.10.   Watermarked images that have undergone JPEG_50 in addition to an affine transform (StirMark function): (a) JPEG_50 + linear_1.010_0.013_ 0.009_1.011; (b) JPEG_50 + rotation_scale_30.00. (From X. Kang et al., *IEEE Trans. on Circuits and Systems for Video Technology*. vol. 13, no. 8, pp. 776– 786, Aug., 2003. With permission.)

be recovered successfully, whereas "0" means the embedded message cannot be recovered successfully. It is observed that the watermark is robust against Gaussian filtering, sharpening, FMLR (frequency mode Laplacian removal), rotation (autocrop, autoscale), aspect ratio variations, scaling, jitter attack (random removal of rows and columns), general linear transform, and shearing. In all of these cases, the embedded message can be recovered. We can also see that the watermark can effectively resist JPEG compression and cropping. It is noted that our algorithm can recover the embedded message for JPEG compression with the quality factor as low as 10. The watermark is recovered when up to 65% of the image has been cropped. The watermark can also resist the combination of RST (rotation, scaling, translation) and cropping with JPEG_50 (Figure 15.7 and Figure 15.10).

We have also tested the proposed algorithm using different wavelet filters, such as orthogonal wavelet filters Daubechies-N (N = 1∼10) and other bi-orthogonal wavelet filters (for example, Daubechies 5/3 wavelet filter). The similar results have been obtained.

CONCLUSIONS

In this section, we propose a DWT–DFT composite watermarking scheme that is robust to affine transforms and JPEG compression simultaneously. The watermarking scheme embeds a template in magnitude spectrum in the DFT domain to resist affine transform and uses a training sequence embedded in the DWT domain to achieve synchronization against translation. By using the dyadic property of the DWT, the number of the DWT implementation is dramatically reduced, hence lowering the computational complexity. A new method to estimate the affine transform matrix, expressed in Equation (15.9) and Equation (15.10), is proposed. It can reduce the high computational complexity required in the iterative computation. The proposed watermarking scheme can successfully resist almost all of the affine-transform-related test functions in StirMark 3.1 and JPEG compression with a quality factor as low as 10.

However, the robustness of watermarking against median filtering and random bending needs to be improved. According to our latest work, robustness against median filtering may be improved by applying the adaptive receiving technique [27]. In Reference 27, instead of the fixed interval representing the hidden binary data "+1" in $(0.5S, S)$ and representing "−1" in $(0, 0.5S)$ just as presented in this chapter, the intervals representing the hidden binary data bit "+1" and "−1" are adjusted adaptively according to the responsive distribution of the embedded training sequence. Experimental results show that the proposed watermarking is robust to median filtering (including 2×2, 3×3, 5×5, 7×7 median filtering). Robustness against median filtering can be further improved by increasing the strength of the informative watermark via adaptive embedding based on perceptual masking [13,15,21,28–30].

To the best of our knowledge, the watermarking scheme that is capable of extracting a hidden watermark signal with no error from the marked image, which has been attacked by the randomization and bending, a test function in StriMark 3.1, remains open. This issue is currently under our investigation. It appears that increasing the embedding strength is able to overcome this difficulty. The watermarking schemes with a larger bit rate to resist the randomization and bending and other attacks such as print and scanning are other future research subjects.

Moreover, it is noted that the template embedded in the DFT domain may be removed by the attacker [31]. This issue needs to be addressed in the future.

The proposed blind resynchronization technique is presented with respect to a DWT-based marking algorithm in this chapter. However, it can also be applied to other marking algorithms, including DCT-based ones.

ACKNOWLEDGMENT

The work on this chapter was partially supported by NSF of china (60325208, 60133020, 60172067), NSF of Guangdong (04205407), Foundation of Education, Ministry of China, New Jersey Commission of Science and Technology via NJWINS.

REFERENCES

1. Hartung, F. and Kutter, M., Multimedia watermarking techniques, *Proc. IEEE*, 87(7), 1079–1107, 1999.
2. Deguillaume, F., Voloshynovskiy, S., and Pun, T., A Method for the Estimation and Recovering from General Affine Transforms in Digital Watermarking applications, in Proceedings of the SPIE: Security and Watermarking of Multimedia Contents IV, San Jose, CA, 2002, pp. 313–322.
3. Petitcolas, F.A.P., Anderson, R.J., and Kuhn, M.G., Attacks on copyright marking systems, Proceedings of the 2nd Information Hiding Workshop, Lecture Notes in Computer Science, (D. Aucsmith Ed.) vol. 1525, Springer-Verlag, Berlin. pp. 218–238, Portland, 1998.
4. Lin, C.-Y., Wu, M., Bloom, J.A., Cox, I.J., Miller, M.L., and Lui, Y.-M., Rotation, scale, and translation resilient watermarking for images, *IEEE Trans. Image Process.*, 10(5), 767–782, 2001.
5. Dugelay, J.-L. and Petitcolas, F.A.P., Possible Counter-attackers Against Random Geometric Distortions, in Proceedings of the SPIE: Security and Watermarking of Multimedia Contents II, San Jose, CA, 2000, pp. 358–370.
6. Johnson, N.F., Duric, Z., and Jajodia, S., Recovery of watermarks from distorted images, in Proceedings of the 3rd Information Hiding Workshop, Lecture Notes in Computer Science (A. Pfitzman Ed.) Vol. 1768, Springer-Verlag, Berlin, 1999, pp. 318–332.
7. Braudaway, G.W. and Minter, F., Automatic Recovery of Invisible Image Watermarks from Geometrically Distorted Images, in Proceedings of the SPIE: Security and Watermarking of Multimedia Contents I, San Jose, CA, 2000, pp. 74–81.
8. Davoine, F., Bas, P., Hébert, P.-A., and Chassery, J.-M., Watermarking et résistance aux déformations géométriques, in Cinquièmes journées d'études et d'échanges sur la compression et la représentation des signaux audiovisuals (CORESA'99), Sophia-Antipolis, France, 1999.
9. Loo, P. and Kingsbury, N., Motion Estimation Based Registration of Geometrically Distorted Images for Watermark Recovery, in Proceedings of the SPIE Security and Watermarking of Multimedia Contents III, CA, 2001, pp. 606–617.
10. Kang, X., Huang, J., and Shi, Y.Q., An image watermarking algorithm robust to geometric distortion, in Proceedings of the International Workshop on Digital Watermarking 2002 (IWDW2002), Lecture Notes in Computer Science (F.A.P. Petitcolas and H.J. Kim Eds.), Vol. 2613, Springer-Verlag, Heidelberg, 2002, pp. 212–223.
11. Ruanaidh, J.J.K.O. and Pun, T., Rotation, scale and translation invariant spread spectrum digital image watermarking, *Signal Process.*, 66(3), 303–317, 1998.
12. Kutter, M., Watermarking Resistance to Translation, Rotation, and Scaling, in Proceedings of the SPIE: Multimedia Systems Applications, 1998, pp. 423–431.
13. Voloshynovskiy, S., Deguillaume, F., and Pun, T., Content adaptive watermarking based on a stochastic multiresolution image modeling, in Proceedings of Tenth European Signal Processing Conference (EUSIPCO'2000), Tampere, Finland, 2000.
14. Shim, H.J. and Jeon, B., Rotation, Scaling, and Translation Robust Image Watermarking Using Gabor Kernels, in Proceedings of the SPIE: Security and Watermarking of Multimedia Contents IV, San Jose, CA, 2002, pp. 563–571.
15. Voloshynovskiy, S., Herrigel, A., Baumgärtner, N., and Pun, T., Generalized Watermark Attack Based on Watermark Estimation and Perceptual Remodulation, in Proceedings of the SPIE: Electronic Imaging 2000, Security and Watermarking of Multimedia Content II, San Jose, CA, 2000, pp. 358–370.
16. Voloshynovskiy, S., Deguillaume F., and Pun, T., Multibit Digital Watermarking Robust Against Local Nonlinear Geometrical Distortions, in Proceedings of IEEE International Conference on Image Processing, Thessaloniki, Greece, 2001, Vol. 3, pp. 999–1002.

17. Bender, W., Gruhl, D., and Morimoto, N., Techniques for data hiding, *IBM Syst. J.*, 35, 313–337, 1996.
18. Fleet, D.J. and Heger, D.J., Embedding invisible information in color images, in Proceedings of IEEE International Conference on Image Processing, Santa Barbara, CA, 1997, Vol. 1, pp. 532–535.
19. Pereira, S. and Pun, T., Robust template matching for affine resistant image watermarks, *IEEE Trans. Image Process.*, 9(6), 1123–1129, 2000.
20. Cox, I.J., Killian, J., Leighton, F.T., and Shamoon, T., Secure spread spectrum watermarking for multimedia, *IEEE Trans. Image Process.*, 6(12), 1673–1687, 1997.
21. Huang, J., Shi, Y.Q., and Shi, Y., Embedding image watermarks in DC components, *IEEE Trans. Circuits Syst. Video Technol.*, 10(6), 974–979, 2000.
22. Huang, D., Liu, J., and Huang, J., An embedding strategy and algorithm for image watermarking in DWT domain, *J. Software*, 13(7), 1290–1297, 2002.
23. Shi, Y.Q., Ni, Z., Ansari, N., and Huang, J., 2-D and 3-D Successive Packing Interleaving Techniques and Their Applications to Image and Video Data Hiding, presented at IEEE International Symposium on Circuits and Systems 2003, Bangkok, Thailand, 2003, Vol. 2, pp. 924–927.
24. Shi, Y.Q. and Zhang, X.M., A new two-dimensional interleaving technique using successive packing, *IEEE Trans. Circuits Syst. I*, 49(6), 779–789, 2002.
25. Chen, B. and Wornell, G.W., Quantization index modulation: A class of provably good methods for digital watermarking and information embedding, *IEEE Trans. Inf. Theory*, 47(4), 1423–1443, 2001.
26. Stewart, G.W., *Introduction to Matrix Computations*, Academic Press, New York, 1973.
27. Kang, X., Huang, J., Shi, Y.Q., and Zhu, J., Robust watermarking with adaptive receiving, Proceedings of 2nd International Workshop on Digital Watermarking, Lecture Notes in Computer Science (T. Kalker, I. J. Cox, Y. M. Ro Eds.), Vol. 2989, Springer-Verlag, Heidelberg, pp. 396–407.
28. Podilchuk, C.I. and Zeng, W., Image-adaptive watermarking using visual models, *IEEE Trans. Selected Areas in Commn.*, 16(4), 525–539, 1998.
29. Huang, J. and Shi, Y.Q., An adaptive image watermarking scheme based on visual masking, *IEE Electron. Lett.*, 34(8), 748–750, 1998.
30. Voloshynovskiy, S., Herrigel, A., Baumgärtner, N., and Pun, T., A stochastic approach to content adaptive digital image watermarking, in Proceedings of the 3rd Information Hiding, Lecture Notes in Computer Science, (A. Pfitzman Ed.) Vol. 1768, Springer-Verlag, Berlin, 1999, pp. 212–236.
31. Voloshynovskiy, S., Pereira, S., Iquise, V., and Pun, T., Attack modeling: Towards a second generation benchmark, *Signal Process.*, 81(6), 1177–1214, 2001.

16
Reversible Watermarks Using a Difference Expansion

Adnan M. Alattar

INTRODUCTION

Watermarking valuable and sensitive images such as artworks and military and medical images presents a major challenge to most watermarking algorithms. First, such applications may require the embedding of several kilobytes of data, but most robust watermarking algorithms can embed only several hundred bits of data. Second, the watermarking process usually introduces a slight but irreversible degradation in the original image. This degradation may reduce the aesthetic and monetary values of artwork, and it may cause the loss of significant artifacts in military and medical images. These artifacts may be crucial for an accurate diagnosis from the medical images or for an accurate analysis of the military images. Just as importantly, the degradation may introduce new, misleading artifacts.

The demands of the aforementioned applications can be met by reversible watermarking techniques. Unlike their robust counterparts, reversible watermarking techniques are fragile and employ an embedding process that is completely reversible. Furthermore, some of these techniques allow the embedding of about a hundred kilobytes of data

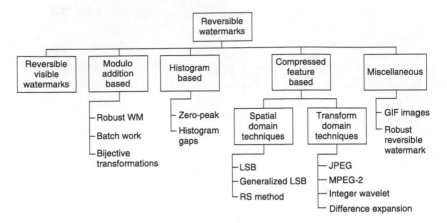

Figure 16.1.   Classification of reversible watermarks.

in a 512×512 image. The noise introduced by these methods can be removed and the original image can be completely restored from the watermarked image alone. These techniques are extremely sensitive to the smallest change in the image, which makes them inherently suitable for image authentication. The nature of the payload data varies according to the intended application, which may include image authentication, copyright protection, data hiding, secret communication, and so forth.

As shown in Figure 16.1, reversible watermarking techniques can be classified into four main classes: visible watermarks, modulo addition based, histogram based, compressed feature based, and miscellaneous. These classes are briefly discussed in the next section. However, the remainder of this chapter is then dedicated to an important technique from the compressed-feature-based class. This technique uses the difference expansion of a generalized reversible integer transform, which allows the technique to provide a higher embedding capacity at a lower noise level than all of the other published techniques.

BACKGROUND

Several reversible watermarking algorithms that allow the complete restoration of the original image from a watermarked image have been developed [1–22]. As stated in the Introduction, Figure 16.1 presents a classification of these algorithms, which can be classified into four main classes and some miscellaneous algorithms. In this section, we will briefly discuss the four main classes.

The first class includes one of the earliest of these algorithms, known as a reversible visible watermarking algorithm. This algorithm adds a watermark to the image in the form of a semivisible pattern [1,2]. The algorithm lets the user restore the original image by subtracting the embedded pattern from a "teaser" image using a secret key and a special "vaccine" program.

The second class of algorithms uses modulo arithmetic and additive, nonadaptive, robust watermarking techniques [3,5,7], such as the phase dispersion [4] and batch work [6] techniques, to embed a short message that can be completely removed from the original image. The length of the message is less than 100 bits, and the early version of this technique suffers from a "salt-and-pepper" noise caused by the wraparound effect of the modulo arithmetic.

The third class of reversible watermark algorithms is based on histogram modification. These algorithms systematically create gaps in the histogram. Then, they use the missing gray levels to embed the watermark data [18,20]. Decoding may require overhead information that must also be embedded in the image. The embedding capacity of this method is highly dependent on the nature of the image. Ni et al. [20] reported about 0.06 bits per pixel, whereas Least et al. [18] reported 0.06–0.6 bits per pixel for their technique that creates multiple gaps. They both reported high image quality with peak signal-to-noise ratios (PSNR) greater than 45 dB.

The fourth class of reversible watermarking algorithms is based on replacing an insignificant image feature with a pseudorandomized feature to create the embedded image. An insignificant image feature is any feature that can be changed without affecting the visual quality of the image in a discernable way. Several examples are given later. To ensure reversibility, the original feature is compressed losslessly and used with the payload to construct a composite bit stream that represents the pseudorandomized feature. To restore the original image, the pseudor-andomized feature is first extracted from the embedded image. Then, the original feature is retrieved from the composite bit stream using decompression. Finally, the randomized feature is replaced with the retrieved original feature. The capacity of this technique highly depends on the choice of the image feature. If this feature is highly compressible, a high capacity can be obtained. This technique can be applied to both the spatial domain and the transform domain. In the remainder of this section, we briefly discuss several examples of compressed-feature-based techniques.

Fridrich et al. [8] has proposed the use of one of the least-significant-bit (LSB) planes of the spatial image as the insignificant image feature. They compress the selected LSB plane and append to it a 128-bit image hash that can be used to authenticate the image. They then replace the selected LSB plane with the resulting composite bit stream. Celik et al. [12] extended Fridrich et al.'s technique and proposed using a general-ized LSB embedding to increase the capacity and decrease the distortion.

In his technique, Celik et al. first quantize each pixel with an L-step uniform quantizer and use the quantization noise as the insignificant image feature. Then, they losslessly compress the quantization noise using the CALC algorithm and append the payload to the resulting bitstream. They compute the L-ary representation of the results, then add each digit to one of the quantized pixel values. With the Lena image, Celik et al. reported an embedding capacity as high as 0.68 bits/pixel at 31.9 dB.

Later, Fridrich et al. [10] proposed a technique to improve the capacity of their original method. They called this technique the RS-embedding technique. They used a discrimination function to classify the image vectors into three groups: regular, singular, and unusable. The insignificant image feature, in this case, is a binary map consisting of binary flags indicating the locations of the regular and singular vectors. This map is losslessly compressed, and a new image feature is composed from the compressed bit stream and a payload. Whenever necessary, they used a flipping function to change a usable vector from one type to another to match the new image feature. With the Lena image, Fridrich et al. reported an embedding capacity as high as 0.137 bits/pixel at 35.32 dB.

Fridrich et al. also applied their original technique to JPEG compressed images [10,11] in two approaches. In the first approach, they used an insignificant image feature that consisted of the LSB of prespecified discrete cosine transform (DCT) coefficients collected from the mid-band of all the blocks. They losslessly compressed these LSBs and appended a payload to the resulting bit stream to form a composite bit stream. They used the bits from the composite bitstream to replace the original LSBs of the prespecified DCT coefficients. They reported an embedding capacity of 0.061 bits/pixel with 25.2 dB for the JPEG compressed Lena with a 50% quality factor. They also reported the same capacity with a 38.6 PSNR for the JPEG compressed Lena with a 90% quality factor.

In their second approach, Fridrich et al. first preprocessed the bit stream by multiplying each DCT coefficient by 2, saving the parity of the quantization factor in the LSB of the corresponding new DCT coefficient, and then dividing each quantization factor by 2. Finally, they used the LSB of the new DCT coefficients as the image feature, which they compressed and appended to the payload. They reported an embedding capacity of 0.061 bits/pixel with a 40.7 PSNR for the JPEG compressed Lena with a 50% quality factor. They also reported the same capacity with a 44.3 PSNR for the JPEG compressed Lena with a 90% quality factor.

Xuan et al. selected an insignificant image feature from the integer wavelet domain [19]. To avoid overflow and underflow, they first preprocessed the image and slightly compressed (losslessly) its histogram. They recorded all of the modifications that they introduced and added them to the payload. They selected the fourth bitplane of the HH, LH, and HL wavelet subbands as the insignificant image feature. They losslessly compressed this feature, appended the payload data to it, and then replaced the original feature with the resulting data. When the inverse wavelet transform is computed, a small shift in the image mean may occur. This drawback prompted the authors to apply a small circular rotation to the histogram to restore the original value of the mean. The authors reported a capacity of 0.32 bits/pixel at 36.64 dB for the Lena image.

Tian introduced a high-capacity, reversible watermarking technique that uses a simple integer wavelet transform. This transform computes the average of and the difference between each adjacent pair of pixels [15]. He recognized that many of the pairs in the image have very small difference coefficients that can be doubled and the LSB of the results can be randomized without noticeably affecting the image quality or causing an overflow or underflow in the spatial domain. Therefore, he called this group the *expandable* group. He devised his algorithm to record the locations of the expandable pairs and then double their difference coefficients. He also recognized that the slightest change (randomizing their LSBs) in the difference coefficients of some pairs would cause an overflow or underflow in the spatial domain. He further recognized that the LSB bit of the difference coefficient of each of the remaining pairs can be changed without causing overflow or underflow. He called this group the *changeable* group.

To allow for the restoration of the original image, Tian first collected the LSBs of the changeable group and included them with the payload to form a composite bit stream. He then compressed the information that indicated the locations of the expandable pairs and appended it to the composite bit stream. Finally, he replaced the LSBs of the difference coefficients of all expandable and changeable pairs with a bit from the composite bit stream. This, in essence, changes the mix of the expandable and changeable pairs, which can be considered here as the insignificant image feature. Tian reported a very high capacity of 0.84 bits/pixel for Lena at 31.24 dB, which is higher than any other technique in the literature.

Tian's technique is a special case of a family of techniques based on a generalized reversible integer transform. Due to its importance, we dedicate the rest of this chapter to the discussion of this family of techniques. We use the generalized transform to introduce similar but more efficient techniques that operate on vectors of dimensions higher than 2. We begin the next section by introducing the generalized reversible integer transform (GRIT) with the dyad, triplet, and quad vectors as examples [23]. In section "The Difference Expansion and Embedding," we introduce the difference expansion and the LSB embedding in the GRIT domain. In section "Algorithm for Reversible Watermark," we discuss the

497

generalized embedding and recovery algorithms. In section "Payload Size," we consider the size of the payload that can be embedded using the GRIT. In section "Data Rate Controller," we present an algorithm to adjust the internal thresholds of the algorithm to embed the desired payload size into a host image. In section "Recursive and Cross-Color Embedding," we discuss the ideas of recursive embedding and embedding across color components to embed more data into a host image. In section "Experimental Results," we present simulation results of the algorithm using triplet and quad vectors, and we compare it to the result for dyads. Finally, in the last section, we present a summary and conclusions.

THE GENERALIZED REVERSIBLE INTEGER TRANSFORM

In this section, we introduce the GRIT, which is the base of a family of high-capacity, reversible watermarking algorithms. We begin with the definition of a vector; then, we give the transform pair, prove its reversibility, and give some examples of GRIT that will be used later in the simulation.

Vector Definition

Let the vector $u = (u_0, u_1, \ldots, u_{N-1})^T$ be formed from N pixel values chosen from N different locations within the same color component according to a predetermined order. This order may serve as a security key. The simplest way to form this vector is to consider every set of $a \times b$ adjacent pixel values, as shown in Figure 16.2, as a vector. If w and h are the width and the height of the host image, respectively, then $1 \leq a \leq h$, $1 \leq b \leq w$, and $a + b \neq 2$.

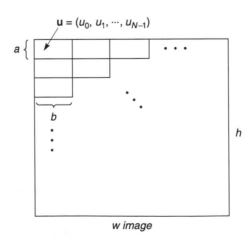

Figure 16.2. **Vector configuration in an image.**

For simplicity, we require that each color component be treated independently and, hence, have its own set of vectors. Also, we require that vectors do not overlap each other (i.e., each pixel exists in only one vector). These requirements may be removed at the expense of complicating the watermarking algorithm due to the extra caution required to determine the processing order of the overlapped vectors.

The GRIT

Theorem: If $\mathbf{D}$ is an $N \times N$ full-rank matrix with an inverse $\mathbf{D}^{-1}$, $\mathbf{u}$ is an $N \times 1$ integer column vector, and $\mathbf{Du} = [a\, d_1\, d_2 \cdots d_{N-1}]^T$, where a is the weighted average value of the elements of $\mathbf{u}$ and $d_1, d_2, \ldots, d_{N-1}$ are independent pairwise differences between the elements of $\mathbf{u}$, then $\mathbf{v} = \lfloor \mathbf{Du} \rfloor$ and $\mathbf{u} = \lceil \mathbf{D}^{-1}\mathbf{v} \rceil$ form a GRIT pair, where $\lceil \cdot \rceil$ indicates the ceiling function (i.e., round up to nearest integer) and $\lfloor \cdot \rfloor$ indicates the floor function (i.e., round down to the nearest integer).

Proof: To satisfy $\mathbf{DD}^{-1} = I$, each element of the first column of $\mathbf{D}^{-1}$ must be 1. This result is required because the inner product of the first row of $\mathbf{D}$ and the first column of $\mathbf{D}^{-1}$ must be 1, and the inner product of each of the remaining $N-1$ difference rows of $\mathbf{D}$ and the first column of $\mathbf{D}^{-1}$ must generate a 0. Therefore, $\mathbf{D}^{-1}$ can be written as

$$\mathbf{D}^{-1} = \begin{bmatrix} 1 & \alpha_{0,1} & \alpha_{0,2} & \cdots & \alpha_{0,N-2} & \alpha_{0,N-1} \\ 1 & \alpha_{1,1} & \alpha_{1,2} & \cdots & \alpha_{1,2} & \alpha_{1,2} \\ \vdots & \vdots & \vdots & \vdots & & \\ 1 & \alpha_{N-2,1} & \alpha_{N-2,2} & \cdots & \alpha_{N-2,N-2} & \alpha_{N-1,N-1} \\ 1 & \alpha_{N-1,1} & \alpha_{N-1,2} & \cdots & \alpha_{N-1,N-2} & \alpha_{N-1,N-1} \end{bmatrix}$$

Because the d_i's are integers, $\mathbf{v} = \lfloor \mathbf{Du} \rfloor = [\lfloor a \rfloor\, d_1\, d_2 \cdots d_{N-1}]^T$ and $\mathbf{u} = \mathbf{D}^{-1}\mathbf{Du}$, each element of u can be expressed explicitly as $u_i = a + \sum_{j=1}^{N-1} \alpha_{j,i} d_j$. Hence, $\sum_{j=1}^{N-1} \alpha_{j,i} d_j = u_i - a$. Now, an approximation of u can be written as $\hat{\mathbf{u}} = \mathbf{D}^{-1}\mathbf{v}$. Each element of this approximation can be expressed explicitly as $\hat{u}_i = \lfloor a \rfloor + \sum_{j=1}^{N-1} \alpha_{j,i} d_j$. Substituting $\sum_{j=1}^{N-1} \alpha_{j,i} d_j = u_i - a$ in the previous equation gives $\hat{u}_i = u_i + (\lfloor a \rfloor - a)$. However, for $-1 < \lfloor a \rfloor - a < 0$ $\lceil \hat{u}_i \rceil = \lceil u_i + (\lfloor a \rfloor - a) \rceil = u_i$, because u_i is an integer. Hence,

$$\mathbf{v} = \lfloor \mathbf{Du} \rfloor, \tag{16.1a}$$

$$\mathbf{u} = \lceil \mathbf{D}^{-1}\mathbf{v} \rceil, \tag{16.1b}$$

define a GRIT.

The above transform forms a family of transforms, where different matrices lead to different transforms. In the next subsections, we will discuss the GRIT transforms for vectors of size 2×1, 3×1, and 4×1, which we will refer to as dyad, triplet, and quad vectors, respectively.

Dyad-Based GRIT

A dyad-based GRIT can be easily derived from the transform pair given in section "The GRIT" by setting $N=2$ and selecting a proper 2×2 **D** matrix. The transform used by Tian [15] can be obtained by setting

$$\mathbf{D} = \begin{bmatrix} a_0/c & a_1/c \\ 1 & -1 \end{bmatrix}$$

where $c = a_0 + a_1$. The inverse matrix is

$$\mathbf{D}^{-1} = \begin{bmatrix} 1 & -a_1/c \\ 1 & a_0/c \end{bmatrix}$$

In this case, the transform of the dyad $\mathbf{u} = (u_0, u_1)^T$ is the dyad $\mathbf{v} = (v_0, v_1)^T$, whose coefficients are given by

$$v_0 = \left\lfloor \frac{a_0 u_0 + a_1 u_1}{a_0 + a_1} \right\rfloor$$

$$v_1 = u_0 - u_1 \tag{16.2a}$$

$$u_0 = \left\lceil v_0 + \frac{a_1 v_1}{a_0 + a_1} \right\rceil$$

$$u_1 = \left\lceil v_0 - \frac{a_0 v_1}{a_0 + a_1} \right\rceil \tag{16.2b}$$

Because v_0 is an integer,

$$u_0 = v_0 + \left\lfloor \frac{a_1 v_1}{a_0 + a_1} + \frac{a_0 + a_1 - 1}{a_0 + a_1} \right\rfloor$$

$$u_1 = v_0 - \left\lfloor \frac{a_0 v_1}{a_0 + a_1} \right\rfloor \tag{16.2c}$$

The above equations indicate that the first coefficient of the transformed dyad is the integer representation of the weighted average of the elements of the original dyad. The other coefficient is the difference between the second and the first elements of the original vector.

500

For $a_0 = a_1 = 1$, the transform pair becomes

$$v_0 = \left\lfloor \frac{u_0 + u_1}{2} \right\rfloor$$
$$v_1 = u_0 - u_1$$

(16.2d)

$$u_0 = v_0 + \left\lfloor \frac{v_1 + 1}{2} \right\rfloor$$
$$u_1 = v_0 - \left\lfloor \frac{v_1}{2} \right\rfloor$$

(16.2e)

which is identical to the transform pair used by Tian [15]. Similar dyad-based transforms can also be derived by changing the values of a_0 and a_1 and subtracting u_0 from u_1 instead of subtracting u_1 from u_0.

Triplet-Based GRIT

A triplet-based GRIT can be easily derived from the transform pair given in section "The GRIT" by setting $N=3$ and selecting a proper 3×3 **D** matrix. The transform used by Alattar [16] can be obtained by setting

$$\mathbf{D} = \begin{bmatrix} a_0/c & a_1/c & a_2/c \\ -1 & 1 & 0 \\ -1 & 0 & 1 \end{bmatrix}$$

where $c = a_0 + a_1 + a_2$. The inverse matrix is

$$\mathbf{D}^{-1} = \begin{bmatrix} 1 & -a_1/c & -a_2/c \\ 1 & (a_0 + a_2)/c & -a_2/c \\ 1 & -a_1/c & (a_0 + a_1)/c \end{bmatrix}$$

In this case, the transform of the triplet $\mathbf{u} = (u_0, u_1, u_2)^T$ is the triplet $\mathbf{v} = (v_0, v_1, v_2)^T$, whose coefficients are given by

$$v_0 = \left\lfloor \frac{a_0 u_0 + a_1 u_1 + a_2 u_2}{a_0 + a_1 + a_2} \right\rfloor$$
$$v_1 = u_1 - u_0$$
$$v_2 = u_2 - u_0$$

(16.3a)

$$u_0 = \left\lfloor v_0 - \frac{a_1 v_1}{a_0 + a_1 + a_2} - \frac{a_2 v_2}{a_0 + a_1 + a_2} \right\rfloor$$
$$u_1 = \left\lfloor v_0 + \frac{(a_0 + a_2) v_1}{a_0 + a_1 + a_2} - \frac{a_2 v_2}{a_0 + a_1 + a_2} \right\rfloor$$
$$u_2 = \left\lfloor v_0 - \frac{a_1 v_1}{a_0 + a_1 + a_2} + \frac{(a_0 + a_1) v_2}{a_0 + a_1 + a_2} \right\rfloor$$

(16.3b)

Because v_0 is an integer,

$$u_0 = v_0 - \left\lfloor \frac{a_1 v_1 + a_2 v_2}{a_0 + a_1 + a_2} \right\rfloor$$

$$u_1 = v_0 - \left\lfloor \frac{a_2 v_2 - (a_0 + a_2) v_1}{a_0 + a_1 + a_2} \right\rfloor \qquad (16.3c)$$

$$u_2 = v_0 - \left\lfloor \frac{a_1 v_1 - (a_0 + a_1) v_2}{a_0 + a_1 + a_2} \right\rfloor$$

The values of u_1 and u_2 can also be calculated easily after calculating the value of u_0 using the equations

$$u_1 = v_1 + u_0$$
$$u_2 = v_2 + u_0 \qquad (16.3d)$$

For $a_0 = a_1 = a_2 = 1$, the transform pair becomes

$$v_0 = \left\lfloor \frac{u_0 + u_1 + u_2}{3} \right\rfloor$$

$$v_1 = u_1 - u_0 \qquad (16.3e)$$

$$v_2 = u_2 - u_0$$

$$u_0 = v_0 - \left\lfloor \frac{v_1 + v_2}{3} \right\rfloor$$

$$u_1 = v_1 + u_0 \qquad (16.3f)$$

$$u_2 = v_2 + u_0$$

However, for $a_0 = 2$ and $a_1 = a_2 = 1$, the transform pair becomes

$$v_0 = \left\lfloor \frac{2u_0 + u_1 + u_2}{4} \right\rfloor$$

$$v_1 = u_1 - u_0 \qquad (16.3g)$$

$$v_2 = u_2 - u_0$$

$$u_0 = v_0 - \left\lfloor \frac{v_1 + v_2}{4} \right\rfloor$$

$$u_1 = v_1 + u_0 \qquad (16.3h)$$

$$u_2 = v_2 + u_0$$

which is identical to the transform pair used by Alattar [16].

502

If we think of u_0, u_1, and u_2 as the green, blue, and red components of a colored image, respectively, and we think of v_0, v_1, and v_2 as the Y, U, and V components in the YUV color space, then the above transform can be written using the RGB and YUV notation as follows:

$$Y = \left\lfloor \frac{R + 2G_0 + B}{4} \right\rfloor$$

$$U = R - G$$

$$V = B - G \tag{16.3i}$$

$$G = Y - \left\lfloor \frac{U + V}{4} \right\rfloor$$

$$R = U + G \tag{16.3j}$$

$$B = V + G$$

which is identical to the reversible component transform proposed in JPEG2000 for color conversion from RGB to YUV [9].

Similar triplet-based transforms can also be derived by changing the values of a_0, a_1, and a_2 and changing the way the pairwise differences is computed from u_0, u_1, and u_2.

Quad-Based GRIT

A quad-based GRIT can be easily derived from the transform pair given in section "The GRIT" by setting $N=4$ and selecting a proper 4×4 **D** matrix. The transform used by Alattar [17] can be obtained by setting

$$\mathbf{D} = \begin{bmatrix} a_0/c & a_1/c & a_2/c & a_3/c \\ -1 & 1 & 0 & 0 \\ -1 & 0 & 1 & 0 \\ -1 & 0 & 0 & 1 \end{bmatrix}$$

where $c = \sum_{i=0}^{3} a_i$. The inverse matrix is

$$\mathbf{D}^{-1} = \begin{bmatrix} 1 & -a_1/c & -a_2/c & -a_3/c \\ 1 & (a_0 + a_2 + a_3)/c & -a_2/c & -a_3/c \\ 1 & -a_1/c & (a_0 + a_1 + a_3)/c & -a_3/c \\ 1 & -a_1/c & -a_2/c & (a_0 + a_1 + a_2)/c \end{bmatrix}$$

In this case, the transform of the quad $\mathbf{u} = (u_0, u_1, u_2, u_3)^T$ is the quad $\mathbf{v} = (v_0, v_1, v_2, v_3)^T$, whose coefficients are given by

$$v_0 = \left\lfloor \frac{a_0 u_0 + a_1 u_1 + a_2 u_2 + a_3 u_3}{a_0 + a_1 + a_2 + a_3} \right\rfloor$$

$$v_1 = u_1 - u_0$$

$$v_2 = u_2 - u_0 \qquad (16.4a)$$

$$v_3 = u_3 - u_0$$

$$u_0 = v_0 - \left\lfloor \frac{a_1 v_1 + a_2 v_2 + a_3 v_3}{a_0 + a_1 + a_2 + a_3} \right\rfloor$$

$$u_1 = v_0 - \left\lfloor \frac{-(a_0 + a_2 + a_3)v_1 + a_2 v_2 + a_3 v_3}{a_0 + a_1 + a_2 + a_3} \right\rfloor$$

$$\qquad (16.4b)$$

$$u_2 = v_0 - \left\lfloor \frac{a_1 v_1 - (a_0 + a_1 + a_3)v_2 + a_3 v_3}{a_0 + a_1 + a_2 + a_3} \right\rfloor$$

$$u_3 = v_0 - \left\lfloor \frac{a_1 v_1 + a_2 v_2 - (a_0 + a_1 + a_2)v_3}{a_0 + a_1 + a_2 + a_3} \right\rfloor$$

The values of u_1 and u_2 can also be calculated easily after calculating the value of u_0 using the equations

$$u_1 = v_1 + u_0$$

$$u_2 = v_2 + u_0 \qquad (16.4c)$$

$$u_3 = v_3 + u_0$$

For $a_0 = a_1 = a_2 = a_3 = 1$, the transform pair becomes

$$v_0 = \left\lfloor \frac{u_0 + u_1 + u_2 + u_3}{4} \right\rfloor$$

$$v_1 = u_1 - u_0$$

$$v_2 = u_2 - u_0 \qquad (16.4d)$$

$$v_3 = u_3 - u_0$$

$$u_0 = v_0 - \left\lfloor \frac{v_1 + v_2 + v_3}{4} \right\rfloor$$

$$u_1 = v_1 + u_0$$

$$u_2 = v_2 + u_0 \qquad (16.4e)$$

$$u_3 = v_3 + u_0$$

A different transform for the quad can be obtained by rearranging the elements of the matrix $\mathbf{D}$ such that u_0 is subtracted from u_1, u_1 is subtracted from u_2, and u_2 is subtracted from u_3. In this case,

$$\mathbf{D} = \begin{bmatrix} a_0/c & a_1/c & a_2/c & a_3/c \\ -1 & 1 & 0 & 0 \\ 0 & -1 & 1 & 0 \\ 0 & 0 & -1 & 1 \end{bmatrix}$$

where $c = \sum_{i=0}^{3} a_i$,

$$\mathbf{D}^{-1} = \begin{bmatrix} 1 & -(c-a_0)/c & -(a_2+a_3)/c & -a_3/c \\ 1 & a_0/c & -(a_2+a_3)/c & -a_3/c \\ 1 & a_0/c & (a_0+a_1)/c & -a_3/c \\ 1 & a_0/c & (a_0+a_1)/c & (c-a_3)/c \end{bmatrix}$$

In this case, the transform can be written as

$$v_0 = \left\lfloor \frac{a_0 u_0 + a_1 u_1 + a_2 u_2 + a_3 u_3}{a_0 + a_1 + a_2 + a_3} \right\rfloor$$

$$v_1 = u_1 - u_0$$

$$v_2 = u_2 - u_1 \qquad\qquad (16.4\text{f})$$

$$v_3 = u_3 - u_2$$

$$u_0 = v_0 - \left\lfloor \frac{(a_1 + a_2 + a_3)v_1 + (a_2 + a_3)v_2 + a_3 v_3}{a_0 + a_1 + a_2 + a_3} \right\rfloor$$

$$u_1 = v_1 + u_0$$

$$u_2 = v_2 + u_1 \qquad\qquad (16.4\text{g})$$

$$u_3 = v_3 + u_2$$

THE DIFFERENCE EXPANSION AND EMBEDDING

In this section, we explain the expansion of the difference coefficients of the GRIT, explain the LSB embedding, and give formal definitions of expandable and changeable vectors. These processes and definitions are necessary for the embedding algorithm that will be explained in section "Algorithm for Reversible Watermark."

The Difference Expansion

The difference expansion of the vector $\mathbf{u} = (u_0, u_1, \ldots, u_{N-1})^T$ is computed by first calculating the GRIT transform $\mathbf{v} = f(\mathbf{u})$, then modifying

$\mathbf{v} = (v_0, v_1, \ldots, v_{N-1})^T$ according to the following equations:

$$\tilde{v}_0 = v_0$$
$$\tilde{v}_1 = 2 \times v_1$$
$$\vdots$$
$$\tilde{v}_{N-1} = 2 \times v_{N-1}$$

(16.5)

Finally, the inverse transform $\tilde{\mathbf{u}} = f^{-1}(\tilde{\mathbf{v}})$ is computed to obtain the expanded vector. Hence, this expansion can be written as $\tilde{\mathbf{u}} = f^{-1}(d(f(\mathbf{u})))$, where $d(\cdot)$ is a function that denotes the expansion of Equation 16.5.

It should be noted here that the above difference expansion only changes the differences of the vector $\mathbf{u}$. Each of $\tilde{v}_1, \tilde{v}_2, \ldots, \tilde{v}_{N-1}$ is a 1-bit left-shifted version of the original value $v_1, v_2, \ldots, v_{N-1}$, respectively, but potentially with a different LSB. The weighted average v_0 of $\mathbf{u}$ remains unchanged.

The LSB Embedding in the GRIT Domain

The LSB embedding of the binary bits $b_1, b_2, \ldots, b_{N-1} \in \{0, 1\}$ in the vector $\mathbf{u}$ is a two-step process. It first transforms the vector $\mathbf{u}$ with the GRIT transform and then embeds the bits according to the following equations:

$$\hat{v}_0 = v_0$$
$$\hat{v}_1 = 2 \times \left\lfloor \frac{v_1}{2} \right\rfloor + b_1$$
$$\hat{v}_2 = 2 \times \left\lfloor \frac{v_2}{2} \right\rfloor + b_2$$
$$\vdots$$
$$\hat{v}_{N-1} = 2 \times \left\lfloor \frac{v_{N-1}}{2} \right\rfloor + b_{N-1}$$

(16.6)

Again, the above modification only changes the differences of the vector $\mathbf{u}$. $\hat{v}_1, \hat{v}_2, \ldots, \hat{v}_{N-1}$ are the same as the original $v_1, v_2, \ldots, v_{N-1}$, but potentially with different LSBs. The weighted average v_0 of $\mathbf{u}$ remains unchanged. This LSB embedding can be denoted as $\tilde{\mathbf{u}} = f^{-1}(e(f(\mathbf{u})))$, where $e(\cdot)$ is the LSB embedding operation described by Equation 16.6.

Definition 1: Expandable

The vector $\mathbf{u} = (u_0, u_1, \ldots, u_{N-1})^T$ is said to be *expandable* with respect to the GRIT pair $f(\cdot)$ and $f^{-1}(\cdot)$, if its transform can be expanded by the difference expansion $d(\cdot)$ to produce $\widetilde{\mathbf{v}} = d(f(\mathbf{u}))$, which can be embedded by the LSB embedding function $e(\cdot)$ with arbitrary bits $b_1, b_2, \ldots, b_{N-1} \in \{0, 1\}$ to produce $\widehat{\widetilde{\mathbf{v}}} = e(\widetilde{\mathbf{v}})$ without causing overflow or underflow in the inverse GRIT $\widehat{\mathbf{u}} = f^{-1}(\widehat{\widetilde{\mathbf{v}}})$.

Definition 2: Changeable

The vector $\mathbf{u} = (u_0, u_1, \ldots, u_{N-1})^T$ is said to be *changeable* with respect to the GRIT transform pair $f(\cdot)$ and $f^{-1}(\cdot)$, if its transform can be embedded by the LSB embedding function $e(\cdot)$ with arbitrary bits $b_1, b_2, \ldots, b_{N-1} \in \{0, 1\}$ to produce $\widehat{\widetilde{\mathbf{v}}} = e(\widetilde{\mathbf{v}})$ without causing overflow or underflow in the inverse GRIT $\widehat{\mathbf{u}} = f^{-1}(\widehat{\widetilde{\mathbf{v}}})$.

It should be noted here that an expandable vector is also a changeable vector, and it remains changeable even after expanding the difference coefficients of its GRIT and replacing their LSBs with arbitrary bits. Similarly, a changeable vector remains changeable even after changing the LSB of each of the difference coefficients of its GRIT. This property is very important for the reversible watermarking algorithm discussed in the next section.

ALGORITHM FOR REVERSIBLE WATERMARK

The following assumptions and explanations of notations are necessary for a precise description of the reversible watermarking algorithm of this section.

Let $I(i, j, k)$ be an RGB image where i and j are the two spatial indices and k is the color component index, and assume that the following:

1. The pixel values in the red component, $I(i, j, 0)$, are arranged into the set of $N \times 1$ vectors $U_R = \{\mathbf{u}_l^R, l = 1, \ldots, L\}$ using the security key K_R.

2. The pixel values in the green component, $I(i, j, 1)$, are arranged into the set of $N \times 1$ vectors $U_G = \{\mathbf{u}_h^G, h = 1, \ldots, H\}$ using the security key K_G.

3. The pixel values in the blue component, $I(i, j, 2)$, are arranged into the set of $N \times 1$ vectors $U_B = \{\mathbf{u}_p^B, p = 1, \ldots, P\}$ using the security key K_B.

Although it is not necessary, usually all color components in the image have the same dimensions and are transformed using the same GRIT. This makes the number of vectors in the sets U_R, U_G, and U_B be the same (i.e., $L = H = P$). Also, let the set $U = \{\mathbf{u}_r, r = 1, \ldots, R\}$ represent any of the above set of vectors U_R, U_G, and U_B and let K represent its associated security key. In addition, let $V = \{\mathbf{v}_r, r = 1, \ldots, R\}$ be the transformation of U under the GRIT function $f(\cdot)$ [i.e., $V = f(U)$ and $U = f^{-1}(V)$].

Finally, note that $\mathbf{u}_r = (u_0, u_1, \ldots, u_{N-1})^T$ and its GRIT vector is $\mathbf{v}_r = (v_0, v_1, \ldots, v_{N-1})^T$.

The vectors in U now can be classified into three groups according to the definitions given in sections "Definition 1: Expandable" and "Definition 2: Changeable." The first group, S_1, contains all expandable vectors whose $v_1 \leq T_1, v_2 \leq T_2, \ldots, v_{N-1} \leq T_{N-1}$, where $T_1, T_2, \ldots, T_{N-1}$ are predefined thresholds. The second group, S_2, contains all changeable vectors that are not in S_1. The third group, S_3, contains the rest of the vectors (not changeable). Also, let S_4 denote all changeable vectors (i.e., $S_4 = S_1 \cup S_2$).

Let us now identify the vectors of S_1 using a binary location map, M, whose entries are 1's and 0's, where the 1 symbol indicates the S_1 vectors and the 0 symbol indicates the S_2 or S_3 vectors. Depending on how the vectors are formed, the location map can be one dimensional (1-D) to or two dimensional (2-D). For example, if vectors are formed from 2×2 adjacent pixels, the location map forms a binary image that has one-half the number of rows and one-half the number of columns as the original image. However, if a random key is used to identify the locations of the entries of each vector, then the location map is a binary stream of 1's and 0's. In this case, the security key and an indexing table are needed to map the 0's and 1's in this stream to the actual locations in the image. Such a table must be predefined and assumed to be known to both the embedder and the reader.

Embedding a Reversible Watermark

The embedding algorithm can be summarized using the steps below.

For every $U \in \{U_R, U_G, U_B\}$, do the following:

1. Form the set of vectors U from the image $I(i, j, k)$ using the security key K.
2. Calculate V using the forward GRIT $f(\cdot)$ (see Equation 16.1a).
3. Use definitions 1 and 2 in sections "Definition 1: Expandable" and "Definition 2: Changeable," respectively, to divide U into the sets S_1, S_2, and S_3.
4. Form the location map, M; then, compress it using a lossless compression algorithm, such as joint bi-level image experts group (JBIG) or an arithmetic compression algorithm, to produce sub-bit stream B_1. Append a unique identifier, end-of-stream (EOS), symbol to B_1 to identify the end of B_1. The EOS is optional because the decompression process during image restoration can be stopped once M is completely restored.
5. Extract the LSBs of $v_1, v_2, \ldots, v_{N-1}$ of each vector in S_2. Concatenate these bits to form sub-bit-stream B_2. One may choose to losslessly

compress these LSBs to reduce their size; however, not much of a gain should be expected, because these bits usually have high entropy.

6. Assume the watermark to be embedded forms a sub-bit-stream B_3 and concatenate sub-bit-streams B_1, B_2, and B_3, to form the bit stream B.

7. Sequence through the member vectors of S_1 and S_2 as they occur in the image and through the bits of the bit stream B in their natural order. For S_1, expand the vectors as described in Equation 16.5 and embed the bit stream bits using the LSB embedding described in Equation 16.6. For S_2, embed the bit stream bits directly in the difference coefficients using the LSB embedding described in Equation 16.6. The values of $b_1, b_2, \ldots, b_{N-1}$ are taken sequentially from the bit stream.

8. Calculate the inverse GRIT of the resulting vectors using $f^{-1}(\cdot)$ (see Equation 16.1b) to produce the watermarked S_1^w and S_2^w.

9. Replace the pixel values in the image, $I(i,j,k)$, with the corresponding values from the watermarked vectors in S_1^w and S_2^w to produce the watermarked image $I^w(i,j,k)$.

It should be noted here that the size of bit stream B must be less than or equal to $N-1$ times the size of the set S_4. To meet this condition, the values of the threshold $T_1, T_2, \ldots, T_{N-1}$ must be set properly. Also, note that the algorithm is not limited to RGB images. Using the RGB space in the previous discussion was merely for illustration purposes, and using the algorithm with other types of color images is straightforward.

Reading a Watermark and Restoring the Original Image

To read the watermark and restore the original image, the steps below must be followed.

For every $U \in \{U_R, U_G, U_B\}$, do the following:

1. Form the set of vectors U from the image $I^w(i,j,k)$ using the security key K.

2. Calculate V using the forward GRIT, $f(\cdot)$ (see Equation 16.1a).

3. Use definition 2 of section "Definition 2: Changeable" to divide the vectors in U into changeable and nonchangeable vectors. Let $\hat{S}_4$ contains the changeable vectors and S_3 contain the nonchangeable vectors. $\hat{S}_4$ has the same vectors as S_4, which was constructed during embedding, but the values of the entities in each vector may be different. Similarly, S_3 is the same set constructed during embedding because it contains nonchangeable vectors.

4. Extract the LSBs of $\tilde{v}_1, \tilde{v}_2, \ldots, \tilde{v}_{N-1}$ of each vector in $\hat{S}_4$ and concatenate them to form the bit stream B, which is identical to that formed during embedding.

5. Identify the EOS symbol and extract sub-bit-stream B_1. Then, decompress B_1 to restore the location map M, and, hence, identify the member vectors of the set S_1 (expandable vectors). Collect these vectors into set $\hat{S}_1$.

6. Identify the member vectors of S_2. They are the members of $\hat{S}_4$ that are not members of $\hat{S}_1$. Form the set $\hat{S}_2 = \hat{S}_4 - \hat{S}_1$.

7. Sequence through the member vectors of $\hat{S}_1$ and $\hat{S}_2$ as they occur in the image and through the bits of the bit stream B in their natural order after discarding the bits of B_1. For $\hat{S}_1$, restore the original values of $v_1, v_2, \ldots, v_{N-1}$ as follows:

$$v_1 = \left\lfloor \frac{\tilde{v}_1}{2} \right\rfloor, \quad v_2 = \left\lfloor \frac{\tilde{v}_2}{2} \right\rfloor, \ldots, \quad v_{N-1} = \left\lfloor \frac{\tilde{v}_{N-1}}{2} \right\rfloor \tag{16.7}$$

8. For $\hat{S}_2$, restore the original values of $v_1, v_2, \ldots, v_{N-1}$ according to Equation 16.6. The values of $b_1, b_2, \ldots, b_{N-1}$ are taken sequentially from the bit stream.

9. Calculate the inverse GRIT of the resulting vectors using $f^{-1}(\cdot)$ (see Equation 16.1b) to restore the original S_1 and S_2.

10. Replace the pixel values in the image $I^w(i,j,k)$ with the corresponding values from the restored vectors in S_1 and S_2 to restore the original image $I(i,j,k)$.

11. Discard all the bits in the bit stream B, which were used to restore the original image. Form the sub-bit-stream B_3 from the remaining bits. Read the payload using the watermark contained in B_3.

PAYLOAD SIZE

To be able to embed data into the host image, the size of the bit stream B must be less than or equal to $N-1$ times the size of the set S_4. This means that

$$\|S_1\| + \|S_2\| \geq \frac{\|B_1\| + \|B_2\| + \|B_3\|}{N - 1} \tag{16.8}$$

where $\|x\|$ indicates number of elements in x. However, $\|B_2\| = (N - 1) \|S_2\|$; hence, Equation 16.8 can be reduced to

$$\|B_3\| \leq (N - 1)\| S_1 \| - \|B_1\| \tag{16.9}$$

For Tian's algorithm, the bit stream size is $\|B_3\| \leq \|S_1\| - \|B_1\|$, which can be obtained from Equation 16.9 by setting $N=2$.

Equation 16.9 indicates that the size of the payload that can be embedded into a given image depends on the number of expandable vectors that can be selected for embedding and on how well their location map can be compressed.

With a $w \times h$ host image, the algorithm would generate $(w \times h)/N$ vectors per color component. Only a portion, $\alpha (0 \leq \alpha \leq 1)$, of these vectors can be selected for embedding [i.e., $\|S_1\| = \alpha(w \times h)/N$]. Also, the algorithm would generate a binary map, M, containing $(w \times h)/N$ bits. This map can be compressed losslessly by a factor β $(0 \leq \beta \leq 1)$. This means that $\|B_1\| = \beta(w \times h)/N$. Using Equation 16.9, the potential payload size (in bits) becomes

$$\|B_3\| \leq (N-1)\alpha\frac{w \times h}{N} - \beta\frac{w \times h}{N}$$
$$\leq \left(\frac{N-1}{N}\alpha - \frac{1}{N}\beta\right) \times w \times h \qquad (16.10)$$

Equation 16.10 indicates that the algorithm is effective when N and the number of selected expandable vectors are reasonably large. In this case, it does not matter if the binary map, M, is difficult to compress (because its size is very small). However, when each vector is formed from N consecutive pixels (rowwise or columnwise) in the image and when N is large, the number of expandable vectors may decrease substantially; consequently, the values of the thresholds $T_1, T_2, \ldots, T_{N-1}$ must be increased to maintain the same number of selected expandable vectors. This increase causes a decrease in the quality of the embedded image. Such a decrease can be ignored by many applications because the embedding process is reversible and the original image can be obtained at any time. In this case, the algorithm becomes more suitable for low-signal-to-noise-ratio (SNR) embedding than for high-SNR embedding. To maximize $\|B_1\|$ for high-SNR embedding, either N must be kept relatively small or each vector must be formed from adjacent pixels in the 2-D area in the image. The quad ($N=4$) structure given in the next section satisfies both requirements simultaneously.

The maximum payload size can be achieved when N is extremely large ($N \approx N-1$) and all vectors in the image are expandable ($\alpha = 1$). The binary map, in this case, will be extremely compressible ($\beta \approx 0$) because it contains no zeros. Substituting these values of N, α, and β in Equation 16.10, we find that the maximum possible payload size equals the area of the image. Hence, the maximum capacity of this algorithm is 1 bit/pixel per color component.

When $\alpha \leq \beta/(N-1)$, the payload size in Equation 16.10 becomes negative. In this case, nothing can be embedded into the image.

This scenario is less likely to happen with natural images. Most lossless compression algorithms can achieve a $2:1$ compression ratio easily (i.e., $\beta = 1/2$). In this case, α must be greater than $1/2(N-1)$ to be able to embed a nonzero payload. This ratio can be satisfied easily when $N > 2$.

For Tian's algorithm, where $N = 2$, the payload size becomes

$$\|B_3\| \leq \left(\frac{\alpha}{2} - \frac{\beta}{2}\right) \times w \times h \tag{16.11}$$

Equation 16.11 suggests that the ratio of selected expandable pairs, α, has to be much higher than the achievable compression ratio, β, for Tian's algorithm to be effective. When the compression ratio of the binary map is more than the ratio of selected expandable pairs, Tian's algorithm cannot embed anything into the host image. However, because Tian uses pairs of pixels as vectors, the correlation of the pixels in each pair is expected to be very high in natural images. This correlation makes the pair easier to satisfy smaller thresholds and, hence, to produce a large portion of selected expandable pairs. The main drawback of Tian's algorithm is the size of the binary map. To almost double the amount of data that can be embedded into the host image, Tian applies his algorithm rowwise, then columnwise.

DATA RATE CONTROLLER

For a given vector size, N, the payload size that can be embedded into an image and the quality of the resulting image is solely determined by the host image itself and by the value of the thresholds used. However, most practical applications require the embedding of a fixed-size payload regardless of the nature of the host image. Hence, an automatic data rate controller is necessary to adjust the value of the thresholds properly and to compensate for the effect of the host image. The simple iterative feedback system depicted in Figure 16.3 can be used for this purpose.

If $T(n) = [T_1(n), T_2(n), \ldots, T_{N-1}(n)]$ is the threshold's vector at the nth iteration and if C is the desired payload length, then the following proportional feedback controller can be used:

$$T(n) = T(n-1) - \lambda(C - \|B_3\|)T(n-1) \tag{16.12}$$

where $0 < \lambda < 1$ is a constant that controls the speed of convergence. $T(0)$ is a preset value that reflects the relative weights between the entities of the vector used in the difference expansion transform.

512

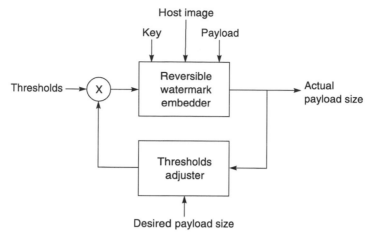

Figure 16.3. **Feedback system for adjusting the thresholds.**

RECURSIVE AND CROSS-COLOR EMBEDDING

Recursive Embedding

Applying the algorithm recursively as in Figure 16.4 can increase its hiding capacity. This recursive application is possible because the proposed watermark embedding is reversible, which means that the input image can be recovered exactly after embedding. However, the difference between the original image and the embedded image increases with every application of the algorithm. At some point, this difference becomes unacceptable for the intended application. However, because the original image always can be recovered exactly, most applications have a high tolerance to this error.

One way to reduce the error when the algorithm is applied recursively is to use permutations of the elements of the input vector, which is

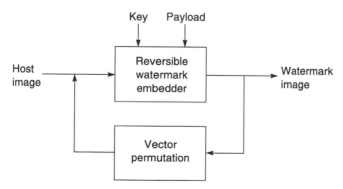

Figure 16.4. **Recursive embedding of the reversible watermark.**

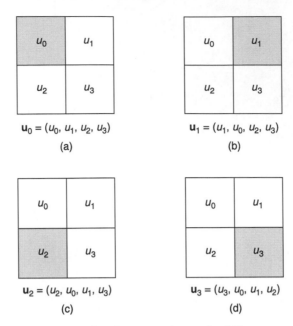

Figure 16.5. Quad permutation for recursive embedding.

depicted in Figure 16.5 for quad vectors. Figure16.5 suggests four different quad structures, each of which can be used in a different iteration, for a total of four iterations. The vectors $\mathbf{u}_0$, $\mathbf{u}_1$, $\mathbf{u}_2$, and $\mathbf{u}_3$ are different permutations of the same vector $\mathbf{u}$. For $\mathbf{u}_0$, the GRIT is performed based on u_0, so the closer u_0 is to u_1, u_2, and u_3, the smaller the difference is and, hence, the smaller the embedding error is. Similarly, for $\mathbf{u}_1$, $\mathbf{u}_2$, and $\mathbf{u}_3$, the GRIT will be based on u_1, u_2, and u_3 components, respectively. This use of permutations lets the algorithm exploit the correlation within a quad completely.

Cross-Color Embedding

To hide even more data, the algorithm can be applied across color components after it is applied independently to each color component. In this case, the vector $\mathbf{u}$ contains the color components (R, G, B) of each pixel arranged in a predefined order. The GRIT for the cross-color arrangement is given in Equation 16.3i and Equation 16.3j.

Although, the spirit of the payload size analysis of section "Payload Size" applies to the cross-color vectors, the results must be slightly modified to reflect the fact that the number of vectors, in this case, equals the area of the location map, which equals the area of the original image. Hence,

$$\|B_3\| = 2\|S_1\| + \|B_1\|$$
$$\|B_3\| = (2\alpha - \beta) \times w \times h$$

(16.13)

514

EXPERIMENTAL RESULTS

Tian [15] implemented a special case of the algorithm we detailed in section "Algorithm for Reversible Watermark" for the dyads vector when $a_0 = a_1 = 1$. However, we implemented the general form of the algorithm when $a_0 = a_1 = , \ldots, = a_{N-1} = 1$ and tested it with spatial triplets, spatial quads, cross-color triplets, and cross-color quads. In all cases, we used a random binary sequence derived from a uniformly distributed noise as a watermark signal. We tested the algorithm with the 512×512 RGB images: Lena, Baboon, and Fruits. In all of the experiments, we set $T_1 = T_2 = , \ldots, = T_{N-1} = C$ and adjusted the value of C to produce the desired peak SNR (PSNR). We used a payload that consists of pure text obtained from a typical text document.

Spatial Triplets

A spatial triplet is a 1×3 or 3×1 vector formed from three consecutive pixel values in the same color component rowwise or columnwise, respectively. We applied the algorithm recursively to each color component: first to the columns and then to the rows. The payload size embedded into each of the test images (all color components) is plotted against the PSNRs of the resulting watermarked image in Figure 16.6. The plot indicates that the achievable embedding capacity depends on the nature of the image itself. Some images can bear more bits with lower distortion in the sense of PSNR than others. Images with many low-frequency contents and high correlation, like Lena and Fruits, produce

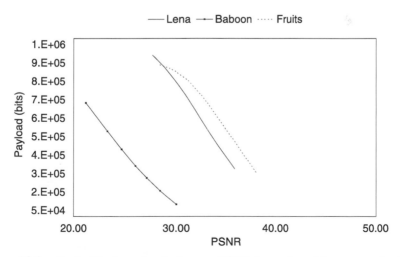

Figure 16.6. **Embedded payload size vs. PSNR for colored images embedded using spatial triplet-based algorithm.**

515

more expandable triplets with lower distortion (in the PSNR sense) than high-frequency images, such as Baboon, and, hence, can carry more watermark data at higher PSNRs.

The proposed algorithm performs slightly better with Fruits than with Lena. With Fruits, the algorithm is able to embed 858 kbits (3.27 bits/pixel) with an image quality of 28.52 dB. The algorithm is also able to embed 288 kbits (1.10 bits/pixel) with the reasonably high image quality of 37.94 dB. Nevertheless, the performance of the algorithm is lower with Baboon than with Lena or Fruits. With Baboon, the algorithm is able to embed 656 kbits (2.5 bits/pixel) at 21.2 dB and 115 kbits (0.44 bits/pixel) at 30.14 dB.

The visual quality of the watermarked images is shown in Figure 16.7 and Figure 16.8 for Lena and Baboon, respectively, embedded at very low, low, and medium PSNRs. In general, the embedded images can hardly be distinguished from the original. However, a sharpening effect can be observed when the original and the embedded images are displayed

Figure 16.7. Lena embedded using the spatial triplet-based algorithm: (a) original, (b) 27.76 dB embedded with 910,802 bits (3.47 bits/pixel), (c) 31.44 dB embedded with 660,542 bits (2.52 bits/pixel), (d) 35.80 dB embedded with 305,182 bits (1.16 bits/pixel).

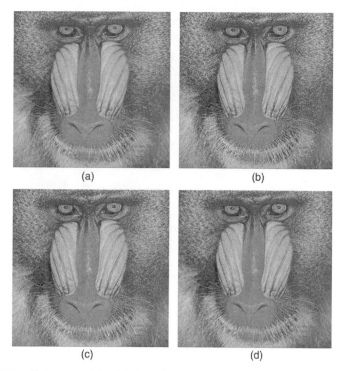

Figure 16.8. Baboon embedded using the spatial triplet-based algorithm: (a) original, (b) 21.20 dB embedded with 656,296 bits (2.50 bits/pixel), (c) 24.74 dB embedded with 408,720 bits (1.56 bits/pixel), (d) 30.14 dB embedded with 115,026 bits (0.44 bits/pixel).

alternatively. This effect is more noticeable at lower PSNRs than at higher PSNRs. It is also more noticeable for a high-frequency image, such as Baboon, than for Lena and Fruits.

Spatial Quads

A spatial quad was assembled from 2×2 adjacent pixels in the same color component as shown in Figure 16.5a. We applied the algorithm to each color component independently. The payload size embedded into each of the test images (all color components) is plotted against the PSNR in Figure 16.9. Again, the plot indicates that the achievable embedding capacity depends on the nature of the image itself. The algorithm performs with Fruits and Lena much better than Baboon, and it performs slightly better with Fruits than with Lena. With Fruits, the algorithm is able to embed 508 kbits (1.94 bits/pixel), with an image quality of 33.59 dB. It is also able to embed 193 kbits (0.74 bits/pixel) with the high image quality of 43.58 dB. Nevertheless, with Baboon, the algorithm is able to

517

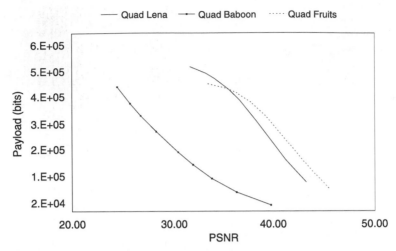

Figure 16.9. Embedded payload size vs. PSNR for colored images embedded using the spatial quad-based algorithm.

embed 482 kbits (1.84 bits/pixel) at 24.73 dB and 87 kbits (0.33 bits/pixel) at 36.6 dB.

The visual quality of the watermarked image is shown in Figure 16.10 and Figure 16.11 for Lena and Baboon embedded at low, medium, and high SNRs. In general, the quality of the embedded images is better than that obtained by the algorithm using spatial triplets. Also, the sharpening effect is less noticeable.

Figure 16.12 combines Figure 16.9 and Figure 16.6. Figure 16.12 reveals that the spatial quad-based and the spatial triplet-based algorithms seem to have different operation ranges with some overlap. At higher PSNRs, the spatial triplet-based algorithm was unable to generate many results, but it can be observed from the tendency of the curves that the spatial quad-based algorithm seems to have superior performance compared to the spatial triplet-based algorithm. This result is because 2×2 spatial quads have a higher correlation than 1×3 spatial triplets and because the single location map used by the spatial quad-based algorithm is smaller than each of the two location maps used by the spatial triplet-based algorithm (one location map for each pass).

On the other hand, although the spatial quad-based algorithm was unable to generate many results at lower PSNRs, it can be observed from the tendency of the curves that the spatial triplet-based algorithm seems to have superior performance. This behavior is attributed to the fact that the spatial triplet-based algorithm is applied to the image twice (rowwise and columnwise), whereas the quad-based algorithm is applied only once.

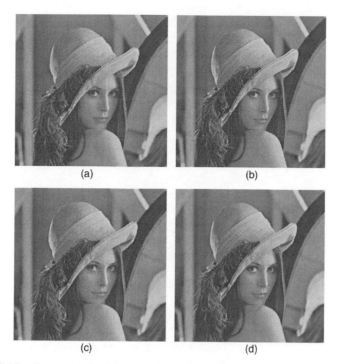

Figure 16.10. Lena embedded using the spatial quad-based algorithm: (a) original, (b) 31.78 dB embedded with 569,317 bits (2.17 bits/pixel), (c) 37.34 dB embedded with 410,520 bits (1.57 bits/pixel), (d) 44.56 dB embedded with 90,443 bits (0.34 bits/pixel).

As shown in Figure 16.13, when the quad-based algorithm was applied to the image twice (as described in section "Embedding a Reversible Watermark"), about a 100-bit increase in the data to be hidden was observed over the spatial triplet-based algorithm when the PSNR was kept constant, and about a 2-dB increase in the image quality was observed over the spatial triplet-based algorithm when the amount of data to be hidden was kept constant.

Cross-Color Embedding

The cross-color triplets were formed from the RGB values, and the GRIT is applied to it in two different ways. In the first way, we used equal weighting in the GRIT (i.e., $a_0 = a_1 = a_2 = 1$). In the second way, we used different weightings (i.e., $a_0 = a_2 = 1$ and $a_1 = 2$), as described in section "Cross-Color Embedding" under "Recursive and Cross-Color Embedding."

Figure 16.14 and Figure 16.15 plot the size of the payload embedded into each of the test images against PSNR for the equal-weighting and different-weighting cases, respectively. Both figures show that the achievable payload size and the PSNR using cross-color vectors are much

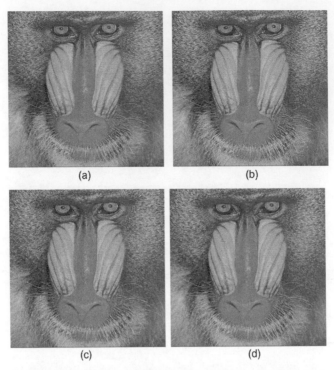

Figure 16.11. **Baboon embedded using the spatial quad-based algorithm: (a) original, (b) 24.73 dB embedded with 481,624 bits (1.84 bits/pixel), (c) 30.19 dB embedded with 258,053 bits (0.98 bits/pixel), (d) 40.00 dB embedded with 39,829 bits (0.15 bits/pixel).**

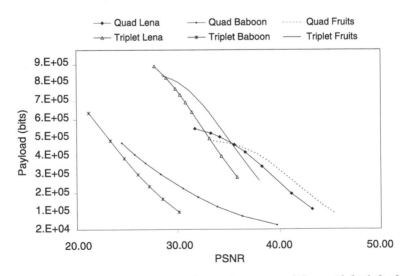

Figure 16.12. **Comparison between the performance of the spatial triplet-based algorithm and the spatial quad-based algorithm applied to the image once.**

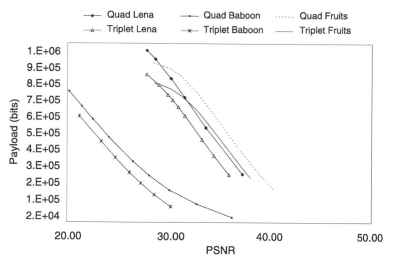

Figure 16.13. **Comparison between the performance of the spatial triplet-based algorithm and the spatial quad-based algorithm applied to the image twice.**

lower than those using spatial vectors. Hence, for a given PSNR level, it is better to use spatial vectors than cross-color vectors.

Also, Figure 16.14 and Figure 16.15 clearly show that the cross-color algorithm with equal weighting has almost the same performance as the cross-color algorithm with different weightings with all test images except Lena at PSNR greater than 30. Although the equal-weighting

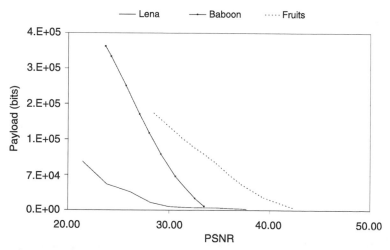

Figure 16.14. **Embedded payload size vs. PSNR for colored images embedded using cross-spectral with equal-weighting GRITs.**

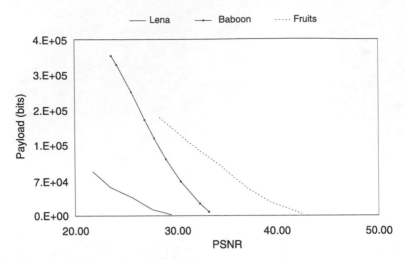

Figure 16.15.   Embedded payload size vs. PSNR for colored images embedded using cross-spectral with different-weighting GRITs.

algorithm was able to embed small payloads at these higher PSNRs, the different-weighting algorithm was not.

Upon closer inspection of the Lena image, we noted that the blue channel of Lena is very close to the green channel. Also, upon further inspection of the equal-weighting and different-weighting GRIT transforms, we noted that when the red or blue channel is close in value to the green channel, the dynamic range of G after expansion according to Equation 16.15 becomes wider for the different-weighting transform than for the equal-weighting transform. Hence, in this case, the equal-weighting GRIT algorithm has the potential of producing more expandable vectors and a location map of less entropy than the different-weighting GRIT algorithm. Indeed, this was the case with the Lena image, as can be seen in Figure 16.16 and Figure 16.17.

Comparison with Other Algorithms in the Literature

We also compared the performance of the proposed algorithm with that of Tian's described in Reference 15 using gray-scale Lena and Barbara images. Recall that Tian's algorithm uses spatial pairs rather than spatial triplets and spatial quads. The results are plotted in Figure 16.18 for the spatial triplet-based algorithm and in Figure 16.19 for the spatial quad-based algorithm. As expected, Figure 16.18 indicates that our spatial triplet-based algorithm outperforms Tian's at low PSNRs, but Tian's algorithm outperforms ours at high PSNRs. In contrast, Figure 16.19 indicates that our spatial quad-based algorithm outperforms Tian's at PSNRs higher than 35 dB, but Tian's algorithm marginally outperforms

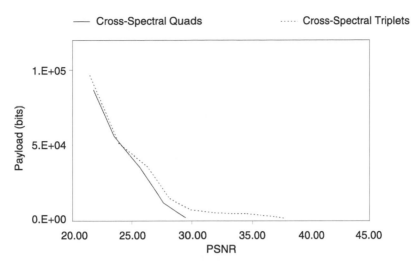

Figure 16.16. Payload size vs. PSNR for Lena colored image using cross-spectral with equal-weighting and different-weighting GRIT transforms.

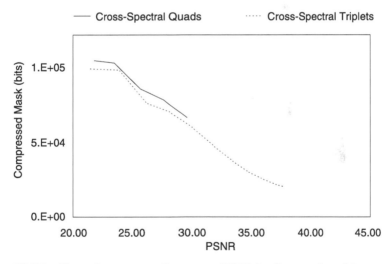

Figure 16.17. Size of compressed map vs. PSNR for Lena colored image using cross-spectral with equal-weighting and different-weighting GRIT transforms.

ours at lower PSNRs. Moreover, our spatial quad-based algorithm is applied once to the image data, whereas Tian's is applied twice: The first time, it is applied columnwise, and the second time, it is applied, rowwise. Having to apply the algorithm only once makes our algorithm more efficient.

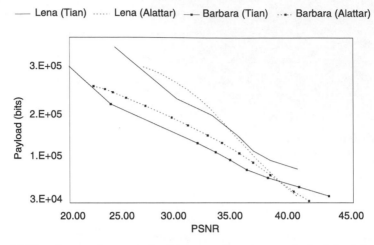

Figure 16.18. Comparison results between the proposed spatial triplet-based algorithm and Tian's using gray-scale images.

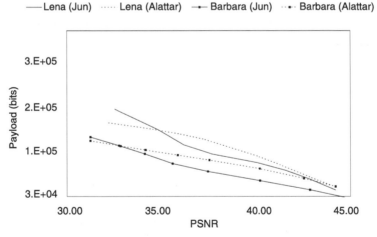

Figure 16.19. Comparison results between the proposed spatial quad-based algorithm and Tian's using gray-scale images.

We also compared our proposed algorithm with that of Celik [12] using gray-scale Lena and Barbara images. The results are plotted in Figure 16.20 for the spatial triplet-based algorithm and in Figure 16.21 for the spatial quad-based algorithm. Figure 16.20 indicates that our spatial triplet-based algorithm also outperforms Celik's at low PSNRs, but our algorithm has similar performance to Celik's at high PSNRs. In contrast, Figure 16.21 indicates that our quad-based algorithm is superior to Celik's at almost all PSNRs.

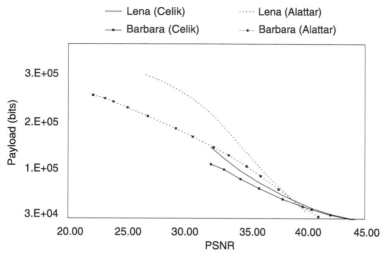

Figure 16.20. Comparison results between the proposed spatial triplet-based algorithm and Celik's using gray-scale images.

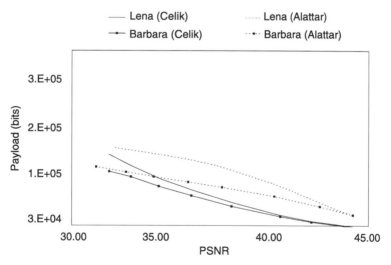

Figure 16.21. Comparison results between the proposed spatial quad-based algorithm and Celik's using gray-scale images.

SUMMARY

In this chapter, we described a family of reversible watermarking algorithms that has very high capacity and causes low distortion in the image. This family is based on the expansion of the difference coefficients of a GRIT of vectors of an arbitrary size. Test results of the spatial dyad-based, triplet-based, and quad-based algorithms indicate that the amount of data that one can embed into an image depends highly on the nature of the image. Test results also indicated that the performance of the quad-based algorithm is superior to that of the dyad-based algorithm, and it is also superior to that of the triplet-based algorithm at higher PSNRs. These results also show that applying the algorithm across the color components has inferior performance to applying the algorithm spatially; hence, cascading cross-color with spatial applications would be useful only when there is a need to hide a large amount of data without regard to the quality of the watermarked image.

ACKNOWLEDGMENTS

The author thanks Ammon Gustafson, Joel Mayer, Tony Rodriguez, Kyle Smith, and John Stach at Digimarc Corporation for their helpful discussion and feedback.

REFERENCES

1. Morimoto, N., Digital watermarking technology with practical applications, *Inf. Sci.*, 2(pt.4), 107–111, 1999.
2. Mintzer, F., Lotspiech, J., and Morimoto, N., Safeguarding digital library contents and users: digital watermarking. *D-lib Mag.*[Online], 1997; Available at http://www.Dlib.org.
3. Honsinger, C. W., Jones, P. W., Rabbani, M., and Stoffel, J. C., Lossless Recovery of an Original Image Containing Embedded Data, U.S. Patent 6,278,791, 2001.
4. Honsinger, C. and Rabbani, M., Data Embedding Using Phase Dispersion, presented at PICS 2000: Image Processing, Image Quality, Image Capture, Systems Conference, Portland, OR, 2000, pp. 264–268.
5. Macq, B., Lossless Multiresolution Transform for Image Authenticating Watermark, in Proceedings of EUSIPCO, Tampere, Finland, 2000.
6. Bender, W., Gruhl, D., Morimoto, N., and Lu, A., Techniques for data hiding, *IBM Syst. J.*, 35(3–4), 313–336, 1996.
7. De Vleeschouwer, C., Delaigle, J. F., and Macq, B., Circular Interpretation of Histogram for Reversible Watermarking, in Proceedings of IEEE 4th Workshop on Multimedia Signal Processing, 2001.
8. Fridrich, J., Goljan, M., and Du, R., Invertible Authentication, in Proceeding of SPIE Photonics West, Security and Watermarking of Multimedia Contents III, San Jose, CA, 2001, pp. 197–208.
9. Taubman, D. and Marcellin, M., *JPEG2000: Image Compression Fundamentals, Standards, and Practice,* Kluwer Academic, Boston, 2002, pp. 422–423.
10. Fridrich, J., Goljan, M., and Du, R., Lossless data embedding — New paradigm in digital watermarking, *EURASIP J. Appl. Signal Process.*, 2002(2), 185–196, 2002.
11. Fridrich, J., Goljan, M., and Du, R., Lossless Data Embedding for All Image Formats, in Proceedings of SPIE Photonics West, Electronic Imaging 2002, Security and Watermarking of Multimedia Contents, San Jose, 2002, pp. 572–583.
12. Celik, M. U., Sharma, G., Tekalp, A. M., and Saber, E., Reversible Data Hiding, in Proceedings of the IEEE International Conference on Image Processing, 2002, Vol. II, pp. 157–160.

13. Wu, X., Lossless compression of continuous-tone images via context selection, quantization, and modeling, *IEEE Trans. Image Process.*, 6 (5), 656–664,1997.
14. Xuan, G., Zhu, J., Chen, J., Shi, Y. Q., Ni, Z., and Su, W., Distortionless data hiding based on integer wavelet transform, *IEE Electron. Lett.*, 38, (2), 1646–1648, 2002.
15. Tian, J., Reversible data embedding using a difference expansion, *IEEE Trans. Circuit Syst. Video Technol.*, 12(8), 890–896, 2003.
16. Alattar, A. M., Reversible Watermark Using Difference Expansion of Triplets, in Proceedings of the 2003 IEEE International Conference on Image Processing, ICIP'2003, Barcelona, 2003, pp. 501–504.
17. Alattar, A. M., Reversible watermark using the difference expansion of a generalized integer transform, *IEEE Trans. Image Process.*, 13(8), 1147–1156, 2004.
18. Least, A. V., Veen, M., and Bruekers, F., Reversible Image Watermarking, in Proceedings of the IEEE International Conference on Image Processing, 2003, Vol. 2, pp. 731–734.
19. Xuan, G., Chen, J., Zhun, J., and Shi, Y., Lossless Image Digital Watermarking Based on Integer Wavelet and Histogram Adjustment, presented at International Conference on Diagnostic Imaging and Analysis, ICDIA '02, August 18–20, 2002, Shangai, China, pp. 60–65.
20. Ni, Z., Shi, Y. Q., Ansari, N., and Su, W., Reversible Data Hiding, Proceedings of the 2003 International Symposium on Circuits and Systems, 25–28 May 2003, Vol. 2, pp. 912–915.
21. Kalker, T. and Willems, F.M., Capacity Bounds and Code Constructions for Reversible Data-Hiding, in Proceedings of Electronic Imaging 2003, Security and Watermarking of Multimedia Contents V, Santa Clara, CA, 2003.
22. Kalker, T. and Willems, F. M., Capacity Bounds and Constructions for Reversible Data-Hiding, in Proceedings of the International Conference on Digital Signal Processing, 2002, pp. 71–76.
23. Stach, J. and Alattar, A.M., A high capacity invertible data hiding algorithm using a generalized reversible integer transform, IS&T/SPIE's 16th International Symposium on Electronic Imaging, January 2004, Vol. 5306, pp. 386–396.

Part IV
Multimedia Data Hiding, Fingerprinting, and Authentication

17

Lossless
Data Hiding
Fundamentals, Algorithms, and Applications

Yun Q. Shi, Guorong Xuan, and Wei Su

INTRODUCTION

Data hiding has recently been proposed as a promising technique for the purpose of information assurance, authentication, fingerprint, security, data mining, copyright protection, and so forth. By data hiding, pieces of information represented by some data are embedded in a cover medium in such a way that the resultant medium, often referred to as marked medium or stego medium, is perceived with no difference from the original cover medium. Many data hiding algorithms have been proposed in the past several years. As will be shown, in most cases, the cover medium will experience some permanent distortion due to data hiding and cannot be inverted back to the original medium. In the analogy to the classification of data compression algorithms, the vast majority of data hiding algorithms can be referred to as lossy data hiding.

The fact that most of the current data hiding algorithms reported in the literature are not lossless can be shown as follows. For instance, with the most popularly utilized spread spectrum watermarking techniques,

0-8493-2773-3/05/$0.00+1.50
© 2005 by CRC Press

either in the discrete cosine transform (DCT) domain [1] or block 8×8 DCT domain [2], round-off error and truncation error may take place during data embedding. As a result, there is no way to invert the marked media back to the original without distortion. With the popularly used least significant bit (LSB) plane embedding method, the bits in the LSB are replaced according to the data to be embedded and the bit replacement is *not-memorized*. Consequently, the LSB method is not reversible. With another group of frequently used watermarking techniques, called quantization index modulation (QIM) [3], *quantization* error makes lossless data hiding impossible.

In some applications, such as in the fields of law enforcement, medical image systems, in addition to perceptual transparency, it is desired to reverse the marked media back to the original cover media after the hidden data are retrieved because of some legal consideration. In scientific research, military, and some other fields, the original media are very difficult and expensive to obtain. Therefore, it is also desired to have the original media inverted back after the hidden data are extracted. The marking techniques satisfying this requirement are referred to as *reversible*, *lossless*, *distortion-free*, or *invertible* data hiding techniques. For the same consideration mentioned earlier, we prefer to use the term "lossless data hiding."

Some lossless marking techniques have been reported in the literature over the last a few years. In section "Three Categories of Lossless Data Hiding Algorighms," we classify all of these existing algorithms into three categories. The fundamental principles, mechanisms, merits, drawbacks and applications of these algorithm are presented. Some future research issues are discussed in section "Some Future Research Issues." Conclusions are made in section "Conclusion."

THREE CATEGORIES OF LOSSLESS DATA HIDING ALGORTIHMS

In this section, all existing lossless data hiding algorithms published in the literature up to today are classified into the following three categories: those for fragile authentication, those for high embedding capacity, and those for semifragile authentication. The fundamental principle of each algorithm is described and its performance is commented.

Those for Fragile Authentication

The first several lossless data hiding algorithms developed at the early stage belong to this category. Because fragile authentication does not need much data to be embedded in a cover medium, the embedding capacity in this category is not large, normally between 1 kbits to 2 kbits. For a typical 512×512 gray-scale image, this capacity is equivalent to a data hiding rate from 0.0038 bits/pixel (bpp) to 0.0076 bpp.

Barton's patent in 2000 [4] may represent the earliest lossless data hiding algortihm. His algorithm was developed for authentication of digital media, including JPEG and MPEG coded image and videos. The main idea is to losslessly compress the bits to be overlayed and thus leave space for the embedding authentication bit string. No specific performance has been reported.

Honsinger et al.'s patent in 2001 [5] described the second lossless data hiding technique used for fragile authentication. Their method is carried out in the image spatial domain by using modulo 256 addition. In the embedding, $Iw = (I + W) \bmod 256$, where Iw denotes the marked image, I is an original image, and W is the payload derived from the hash function of the original image. On the authentication side, the payload W can be extracted from the marked image by subtracting the payload from the marked image, thus losslessly recovering the original image. By using modulo 256 addition, the issue of overflow or underflow is avoided. Here, by overflow or underflow, it is meant that gray-scale values either exceed its upper bound (*overflow*) or its lower bound (*underflow*). For instance, for an 8-bit gray image, its gray scale ranges from 0 to 255. The overflow refers to the gray scale exceeding 255, whereas the underflow refers to below 0. It is clear that either case will destroy losslessness (i.e., reversibility). Therefore, this issue is often a critical issue in lossless data hiding. Using modulo 256 addition can avoid overflow and underflow on the one hand. On the other hand, however, the marked image may suffer from the salt-and-pepper noise [6] during possible gray-scale flipping over between 0 and 255 in either direction due to the operation of modulo 256 addition. The effect caused by salt-and-pepper noise will be discussed further later in this chapter.

Fridrich's group has explored a deep investigation on lossless data hiding techniques and developed a few algorithms. Their first algorithm [7] is developed in the spatial domain, which losslessly compresses the bitplanes to leave room for data embedding in the spatial domain. In order to leave sufficient room for data embedding, it needs to compress the high-level bitplane, which usually leads to visual quality degradation. One example is given in Figure 17.1. There, the hash value of the original $512 \times 512 \times 8$ Elain image, generated by the MP5 program, represented by a string of 128 bits, is embedded in the fifth LSB by using this algorithm. The artifacts are observed in the marked image, which are due to the perturbation caused by data manipulation on the fifth LSB.

In addition, they described two reversible data hiding techniques [8] for a lossy compressed JPEG image. The first technique is based on lossless compression of biased bit streams derived from the quantized JPEG coefficients. The second technique modifies the quantization matrix to enable lossless embedding of 1 bit per DCT coefficient. In addition, Fridrich's group extended the idea of lossless authentication to MPEG-2 video [9].

Figure 17.1. Elaine image: (a) original and (b) marked with noticeable artifacts.

Those for High Embedding Capacity

All the above-mentioned techniques aim at fragile authentication, instead of data hiding. As a result, the amount of hidden data is rather limited. Hence, Goljan et al. [10] presented the first lossless data hiding technique that is suitable for the purpose of data embedding. The details are described as follows. The pixels in an image are grouped into nonoverlapped blocks, each consisting of a number of adjacent pixels. For instance, it could be a horizontal block consisting of four consecutive pixels. A discrimination function that can capture the smoothness of the groups is established to classify the blocks into three different categories, Regular, Singular and Unusable. An invertible operation F can be applied to groups; that is, it can map a block from one category to another as $F(R)=S$, $F(S)=R$, and $F(U)=U$. It is invertible because applying it to a block twice produces the original block. Hence, this invertible operation is called *flipping F*. An example of the invertible operation F can be the permutation between 0 and 1, 2 and 3, 3 and 4, and so on. This is equivalent to flipping the LSB. Another example is the permutation between 0 and 2, 1 and 3, 4 and 6, and so on (i.e., flipping the second LSB). Apparently, the *strength* of the latter flipping is stronger than the former. The principle to achieve lossless data embedding lies in that there is a bias between the number of regular blocks and that of singular blocks for most images. This is equivalent to saying that there is a redundancy and some space can be created by lossless compression. Together with some proper bookkeeping scheme, one can have lossless data hiding.

The proposed algorithm first scan a cover image block by block, resulting in a what is called *RS*-vector formed by representing, say, an *R*-block by binary 1 and an *S*-block by binary 0 with the *U* groups simply skipped. Then, the algorithm losslessly compresses this *RS*-vector — as an overhead for bookkeeping usage in reconstruction of the original image late. By assigning binary 1 and 0 to R and S blocks, respectively, 1 bit can be embedded into each R or S block. If the bit to be embedded does match the type of block under consideration, the flipping operation F is applied to the block to obtain a match. The to-be-embedded data consist of the overhead and the watermark signal (payload). In data extraction, the algorithm scans the marked image in the same manner as in the data embedding. From the resultant *RS*-vector, the embedded data can be extracted. The overhead portion will be used to reconstruct the original image, whereas the remaining portion is the payload.

Although it is novel and successful in reversible data hiding with a large embedding capacity, the amount of data that can be hidden by this technique is still not large enough for some applications such as covert communications. From what is reported in Reference 10, the estimated embedding capacity ranges from 0.022 to 0.17 bpp when the embedding strength is 6 and the Peak signal-to-noise ratio (PSNR) of the marked image versus the original image is about 3606 dB. Note that the embedding strength 6 is rather high and there are some block artifacts in the marked image generated with this embedding strength. On the one hand, this embedding capacity is much higher than that in the first category discussed in the previous subsection. On the other hand, however, it may be not high enough for some applications. This limited embedding capacity is expected because each block can at most embed 1 bit, U blocks cannot accommodate data, and the overhead is necessary for reconstruction of the original image. Another problem with this method is that when the embedding strength increases, the embedding capacity will increase, at the same time, the visual quality will drop. Often, block artifacts will take place at this circumstance, thus causing visual quality of marked image to decrease.

Xuan et al. proposed a high-capacity lossless data hiding technique based on integer wavelet transform (IWT) [11]. IWT is used in the algorithm to ensure the lossless forward wavelet transform and inverse wavelet transform. After IWT, the bias between binary 1 and 0 in the middle and high bitplanes of IWT coefficients of high-frequency subbands becomes much larger than that in the spatial domain. Although it is observed that the higher the bitplane, the larger the bias, only proper middle bitplanes of these coefficients are selected for losslessly compression in order to maintain a high visual quality of marked images. Histogram modification is used in this algorithm to prevent the overflow and underflow resulting in the perturbation of the bitplanes of the IWT coefficients in data embedding. A block diagram of data embedding is contained in Figure 17.2. Owing to the superior decorrelation capability of integer wavelet transformation, a larger embedding capacity has been achieved, compared with that in Reference 10. Further improvements — in particular, a clever bookkeeping scheme [12] — lead to a higher embedding capacity and better visual quality than that in Reference 11. Some experimental results are presented late in this section.

Ni et al. [13] proposed a new lossless data hiding technique based on the histogram modification. This algorithm utilizes what are known as zero or minimum points of image histogram and modifies the pixel value to embed data. In the image histogram, the algorithm first finds a *zero point* (no pixel assumes that gray value) and a *peak point* (a maximum number of pixels assumes that gray value). Then, the algorithm shifts the histogram between the zero point and the peak point toward the zero point by one unit and leaves the histogram near the peak point empty.

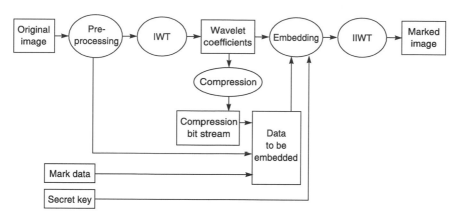

Figure 17.2. Data embedding block diagram of the IWT-based lossless data hiding algorithm (IIWT = inverse IWT).

Finally, the gray scale of the peak point is either replaced by that of its immediately neighboring point or kept intact to embed binary 1 and 0, respectively. This algorithm has a quite large embedding capacity, ranging from 0.019 to 0.31 bpp, while keeping a very high visual quality for all images (the PSNR of marked images vs. original images is guaranteed to be higher than 48 dB). This is indeed a peculiar advantage of this method. In addition, this algorithm is simple and can be implemented rather quickly.

Celik et al. presented a high-capacity, low-distortion reversible data hiding technique [14]. In the embedding phase, the host signal is quantized and the residual is obtained. The algorithm adopts the lossless image compression algorithm, with the quantized values as side information, to efficiently compress the quantization residuals to create a high capacity for the payload data. The compressed residual and payload data are concatenated and embedded into the host signal via the generalized LSB modification method. The experimental results show that the PSNR and capacity are satisfactory.

Tian recently presented a new high-capacity, reversible data embedding algorithm in Reference 15. In the algorithm, two techniques are employed (difference expansion and generalized LSB embedding) to achieve a very high embedding capacity while keep the distortion low. The main idea of this technique is described next. For a pair of pixel values x and y, the algorithm first computes the integer average l and difference h of x and y, where $h = x - y$. Then, h is shifted to the left-hand size by 1 bit and the to-be-embedded bit b is appended into the LSB. This is equivalent to $h' = 2 \times h + b$, where h' denotes the expanded difference, which explains the term "difference expansion." Finally, the new x and y

Table 17.1. **Some experimental results of IWT-based method and difference expansion method**

	PSNR	IWT-Based Method	Difference Expansion Method
Lena	35 dB	0.50 bpp	0.65 bpp
	44 dB	0.15 bpp	0.15 bpp
Barbara	35 dB	0.45 bpp	0.35 bpp
	44 dB	0.12 bpp	0.12 bpp
Mpic2	35 dB	0.70 bpp	Cannot work in this case
	44 dB	0.30 bpp	0.40 bpp

are calculated based on the new difference values h' and the original integer average value l. In this way, the marked image is obtained. To avoid overflow and underflow, the algorithm only embeds data into the pixel pairs that will not lead to overflow and underflow. Therefore, a two-dimensional binary bookkeeping image is losslessly compressed and embedded as overhead.

It has been reported in Reference 15 that the embedding capacity achieved by the difference expansion method is higher than that achieved in both References 10 and 14. Our experimental works demonstrate that both the difference expansion method [15] and the IWT-based method [12] have achieved similar high embedding capacity. Some experimental results are listed in Table 17.1.

In summary, several lossless data hiding algorithms having large embedding capacity have been presented and discussed in this sub-section. The difference expansion method [15] and the IWT-based method [12] may have achieved the highest embedding capacity in this category, whereas the histogram-modification-based method [13] may have kept the highest guaranteed visual quality of marked images in terms of PSNR among all of the existing methods.

Those for Semifragile Authentication

For multimedia, content-based authentication makes more sense than representation-based authentication. This is because the former, often called semifragile authentication, allows some incidental modification, say, compression within a reasonable extent, whereas the latter, called fragile authentication, does not allow it. For instance, when an image goes through JPEG compression with a high quality factor, the content of this image is considered unchanged from the common sense. Hence, it makes sense to claim this compressed image as authentic. For the purpose of semifragile authentication, we need lossless data hiding algorithms that are robust to compression, also referred to as robust lossless data hiding in this chapter. This can be illustrated in the following scenario. In a newly proposed JPEG2000 image authentication framework [16], both fragile and semifragile authentications are included. Within the semifragile authentication, both cases of lossy and lossless compression are considered. In this framework, some features corresponding to an image below a prespecified compression ratio are first identified. By "corresponding," it is meant that these features will remain as long as the compression applied to the image is below this prespecified compression ratio. The digital signature of these features is losslessly embedded into the image. Then, in the verification stage, if the marked image has not been changed at all, the hidden signature can be extracted and the original image can be recovered. The matching between the extracted signature and the signature generated from the reconstructed (*leftover*) image renders the image authentic. If the marked image has been compressed with a compression ratio below the prespecified one, the original image cannot be recovered due to the lossy compression applied, but the hidden signature can still be recovered without error and verify the compressed image as authentic. Obviously, if the marked image goes through a compression with the compression ratio higher than the prespecified ratio, it will render the image unauthentic. A robust lossless data hiding algorithm is indeed necessary for this framework. In general, robust lossless data hiding can be utilized in the lossy environment.

De Vleeschouwer et al. [17] proposed a lossless data hiding algorithm based on patchwork theory [18], which has a certain robustness against JPEG lossy compression. This is the only existing robust lossless data hiding algorithm against JPEG compression. In this algorithm, each hidden bit is associated with a group of pixels (e.g., a block in an image). Each group is pseudorandomly divided into two subsets of equal number of pixels (i.e., zones A and B). The histogram of each zone is mapped into a circle in the following way: Each point on the circle is indexed by the corresponding luminance, and the number of pixels assuming the luminance will be the weight of the point. One can then determine the *mass* center of each zone. It is observed that in the most cases, the vectors pointing from the circle center to the mass center of zones A and B are *close* (almost equal) to each other because the pixels of zones A and B are highly correlated for most images. Considering a group of pixels rotating these two vectors in two opposite directions by a small quantity, say, rotating the vector of zone A counter clockwise and rotating the vector of zone B clockwise, allows for embedding binary 1, whereas rotating the vector of zone A clockwise and the vector of zone B counterclockwise embeds a binary 0. As to the pixel values, rotation of the vector corresponds to a shift of luminance. An embedding diagram is shown in Figure 17.3. In Figure 17.3a and Figure 17.3b, the vectors of zone A and zone B are shown, whereas in Figure 17.3c and Figure 17.3d, these two vectors are rotated in opposite directions, as described earlier, to embed a binary 1. In data extraction, the angles of the mass center vectors of both zone A and zone B vs. the horizontal direction are first calculated, and the difference between these two angles are then determined. A positive difference represents a binary 1, whereas the negative difference represents a binary 0.

One major element of this algorithm is that it is based on the patchwork theory; that is, within each zone, the mass center vector's orientation is determined by all of the pixels within this zone. Consequently, the algorithm is robust to image compression to certain extent. Another major element of this algorithm lies in that it uses modulo 256 addition to avoid overflow and underflow, thus achieving lossless data hiding. This can be easily observed from Figure 17.3. Consequently, as pointed out in section "Those for Fragile Authentication," this algorithm

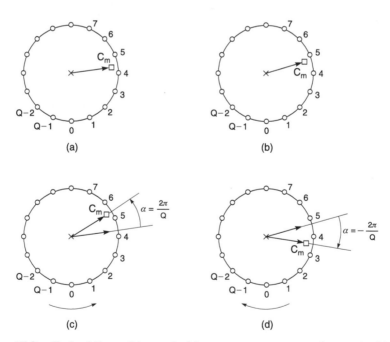

Figure 17.3. Embedding a binary 1: (a) mass center vector of zone A, (b) mass center vector of zone B, (c) counterclockwise rotated *mass center vector* of zone A, (d) clockwise rotated mass center vector of zone B. (From Y. Q. Shi et al., Proceedings of IEEE International Symposium on Circuits and Systems, Vol. II, pp. 33–36, Vancouver, Canada, May 2004. With permission.)

suffers from the salt-and-pepper noise. More investigation in this regard is presented in section "Some Future Research Issues."

SOME FUTURE RESEARCH ISSUES

As pointed in section "Three Categories of Lossless Data Hiding Algorithms," De Vleeschouwer et al.'s method is the only existing lossless data hiding technique robust to high-quality JPEG compression. Hence, it can be used for semifragile authentication. However, our extensive investigation reveals that it has some drawbacks that prevented it from practical usages.

First, some marked images suffer from salt-and-pepper noise because the algorithm utilizes modulo 256 addition; that is, in implementing modulo 256 addition, a very bright pixel with a large gray value close to 255 will be possibly changed to a very dark pixel with a small gray value close to 0, and vice versa. The salt-and-pepper noise becomes severe for some images, where more dark and bright pixels are contained. A typical category of this type of images is medical images. We tried eight medical images and found that five of the eight images suffered from severe salt-and-pepper noise, and other three resulted in some salt-and-pepper noise. Figure 17.4 and Figure 17.5 give such examples of severe and less severe salt-and-pepper noise, respectively. Note that the salt-and-pepper noise becomes so "dense" in Figure 17.4 that the original name of salt-and-pepper appears not to be appropriate. In Figure 17.5, the salt-and-pepper noise manifests itself in the four letters on the four sides of the image.

Not only for medical images but also for color images, the salt-and-pepper noise may be severe. We have applied De Vleeschouwer et al.'s

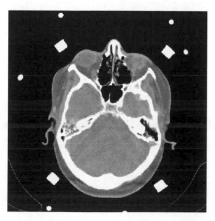

Figure 17.4. Medical picture 1: (a) original and (b) marked with severe salt-and-pepper noise.

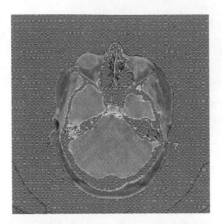

Figure 17.4. Continued.

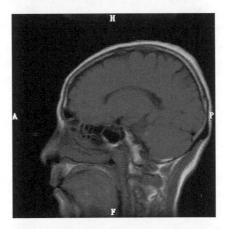

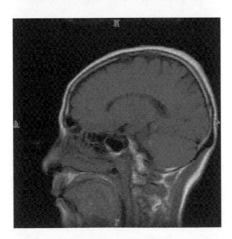

Figure 17.5. Medical picture 3: (a) original and (b) marked with some salt-and-pepper noise (the letters on the four sides have become blurred).

method to eight JPEG2000 test color images. There are four among the eight images that suffer from severe salt-and-pepper noise, whereas the other four suffer from less severe salt-and-pepper noise. Figure 17.6 presents an example of the severe case. This is a JPEG2000 test image. The data for authentication are embedded into the red color component. We can observe severe salt-and-pepper noise, which manifests itself as severe color distortion. Specifically, the color of half of the woman's hair has become dark red and that of most of the palm of her right hand

Figure 17.6. Woman image (N1A): (a) original and (b) marked with severe color distortion due to severe salt-and-pepper noise (the color of half of her hair has become dark red and that of most of palm of her right hand has become green).

Figure 17.7. Cáfe image (N2A): (a) original and (b) marked with severe color distortion due to severe salt-and-pepper noise (the color of the table surfaces has turned to yellow).

has become green. Figure 17.7 provides another such an example, where the salt-and-pepper noise has turned the white coffee tables yellow.

Second, the marked images do not have a high enough PSNR. Table 17.2 contains test results for the eight medical images. The PSNR of marked images is as low as 26 dB (as 476 information bits are embedded in image of $512 \times 512 \times 8$). Note that the salt-and-pepper noise exists in each of

Table 17.2. Test results of eight medical images

	PSNR (dB)	Robustness (bpp)	Salt-and-pepper Noise
Mpic 1	9.28	1.0	Severe
Mpic 2	4.73	2.0	Severe
Mpic 3	26.38	0.8	Severe
Mpic 4	26.49	0.6	Severe
Mpic 5	26.49	0.6	Severe
Mpic 6	5.60	1.6	Severe
Mpic 7	9.64	0.8	Severe
Mpic 8	5.93	2.8	Severe

Table 17.3. Test results with eight JP2000 color images

	PSNR (dB)	Robustness (bpp)	Salt-and-pepper Noise
N1A	17.73	0.8	Severe
N2A	17.73	2.2	Severe
N3A	23.73	0.6	Severe
N4A	19.67	1.2	Severe
N5A	17.28	1.2	Severe
N6A	23.99	0.6	Severe
N7A	20.66	1.4	Severe
N8A	14.32	1.4	Severe

eight marked medical images. When severe salt-and-pepper noise exists, the PSNR drops to below 10 dB.

Table 17.3 contains test results for the eight JPEG2000 test images. The PSNR of marked image vs. the original image is less than 30 dB (as 1412 information bits are embedded into a color image of $1536 \times 1920 \times 24$). Note that the salt-and-pepper noise again exists in each of the eight test images. When severe salt-and-pepper noise takes place, the PSNR can be as low as less than 20 dB.

Third, it is noted that in both Table 17.2 and Table 17.3, there is an item called robustness (in the units of bpp). This means that the hidden data can be error-freely retrieved when the data rate of the image compressed images is above this quantity. We noted, for instance, in Table 17.3, that the robustness of the JPEG2000 test image N6A corresponds to 0.6 bpp, meaning that when the marked N6A goes through a compression, resulting in a data rate above 0.6 bpp, the hidden data by using De Vleeschouwer et al.'s method will be able to recover without error.

However, this robustness may not be strong enough for JPEG2000 compressed images. The reason is that JPEG2000, due to its superiority, can compress an image to a data rate on the order of 0.0xx bpp, i.e., less than 0.1 bpp, while the compressed image is still of good quality for some applications. Hence, a robustness of 0.6 bpp is not enough for semiauthentication of JPEG2000 compressed images.

Therefore, it is necessary to develop new robust lossless data hiding technologies that (1) they do not use modulo 256 addition, (2) consequently, they will not generate the annoying salt-and-pepper noises and the marked images will have a high PSNR vs. the original image, and (3) the robust lossless data hiding techniques are robust enough so that they can work for semifragile authentication of JPEG2000 compressed images.

From the above investigation, it can be concluded that *all lossless data hiding algorithms based on modulo 256 addition to resolve the overflow and underflow issue, say, in References 5 and 17, cannot be applied to real applications*. This type of approach should be avoided.

CONCLUSION

This chapter presents a thorough investigation on the development of all existing lossless data hiding techniques. These techniques are classified into three categories. The principles, merits, and drawbacks of each category are discussed. The future research issues on robust lossless data hiding algorithms, which may find wide applications in semifragile authentication of JPEG2000 compressed images, are addressed. It is expected that lossless data hiding may open a new door to link two groups of data: cover media data and to-be-embedded data. This will find wide applications in information assurance such as authentication, secure medical data system, and intellectual property protection, to name a few.

ACKNOWLEDGMENT

The work on this chapter was partially supported by New Jersey Commission of Science and Technology via New Jersey Center of Wireless Networking and Internet Security (NJWINS).

REFERENCES

1. Cox, I.J., Kilian, J., Leighton, T., and Shamoon, T., Secure spread spectrum watermarking for multimedia, *IEEE Trans. Image Process.*, 6(12), 1673–1687, 1997.
2. Huang, J. and Shi, Y.Q., An adaptive image watermarking scheme based on visual masking, *Electron. Lett.*, 34(8), 748–750, 1998.
3. Chen, B. and Wornell, G.W. Quantization index modulation: a class of provably good methods for digital watermarking and information embedding, *IEEE Trans. Inf. Theory*, 47(4), 1423–1443, 2001.

4. Barton, J.M., Method and Apparatus for Embedding Authentication Information Within Digital Data, U.S. Patent 5,646,997, 1997.
5. Honsinger, C.W., Jones, P., Rabbani, M., and Stoffel, J.C., Lossless Recovery of an Original Image Containing Embedded Data, U.S. Patent 6,278,791, 2001.
6. Gonzalez, R.C. and Woods, R.E., *Digital Image Processing*, Prentice-Hall, Englewood Cliffs, NJ, 2001.
7. Fridrich, J., Goljan, M., and Du, R., Invertible Authentication, in Proceedings of the SPIE, Security and Watermarking of Multimedia Contents, San Jose, CA, 2001, pp. 197–208.
8. Fridrich, J., Goljan, M., and Du, R., Invertible Authentication Watermark for JPEG Images, presented at ITCC 2001, Las Vegas, 2001, pp. 223–227.
9. Fridrich, J. and Du, R., Lossless Authentication of MPEG-2 Video, in Proceedings of ICIP 2002, Rochester, NY, 2002.
10. Goljan, M., Fridrich, J., and Du, R., Distortion-Free Data Embedding, in Proceedings of 4th Information Hiding Workshop, Pittsburgh, PA, 2001, pp. 27–41.
11. Xuan, G., Zhu, J., Chen, J., Shi, Y.Q., Ni, Z., and Su, W., Distortionless data hiding based on integer wavelet transform, *IEE Electron. lett.*, 38(25), 1646–1648, 2002.
12. Xuan, G., Shi, Y.Q., Ni, Z.C., Chen, J., Yang, C., Zhen, Y., and Zhen, J., High Capacity Lossless Data Hiding Based on Integer Wavelet Transform, presented at the IEEE International Symposium on Circuits and Systems, Vancouver, Canada, 2004.
13. Ni, Z., Shi, Y.Q., Ansari, N., and Su, W., Reversible Data Hiding, presented at the IEEE International Symposium on Circuits and Systems, Bangkok, 2003.
14. Celik, M., Sharma, G., Tekalp, A.M., and Saber, E., Reversible Data Hiding, in Proceedings of the International Conference on Image Processing 2002, Rochester, NY, 2002.
15. Tian, J., Reversible data embedding using a difference expansion, *IEEE Trans. Circuits Syst. Video Technol.*, 13(8), 890–896, 2003.
16. Sun, Q., Lin, X., and Shi, Y.Q., A Unified Authentication Framework for JPEG2000 Images, ISO/IEC JTC1/SC29/WG1 N2946, 2003.
17. De Vleeschouwer, C., Delaigle, J.F., and Macq, B., Circular interpretation of bijective transformations in lossless watermarking for media asset management, *IEEE Trans. Multimedia*, 5(1), 97–105, March 2003.
18. Bender, W., Gruhl, D., Mprimoto, N., and Lu, A., Techniques for data hiding, *IBM Syst. J.*, 35(3&4), 313–336, 1996.

18
Attacking Multimedia Protection Systems via Blind Pattern Matching

Darko Kirovski and Jean-Luc Dugelay

THE TARGET

Significantly increased levels of multimedia piracy over the last decade have put the movie and music industry under pressure to deploy a standardized antipiracy technology for multimedia content. Initiatives, such as the Secure Digital Music Initiative (SDMI) [1] and the Digital Versatile Disk (DVD) Copy Control Association (CCA) [2], have been established to develop open technology specifications that protect the playing, storing, and distributing of digital music and video. Although the demand for deploying and standardizing a digital rights management (DRM) technology is strong from both media studios as well as information technology companies, technically, a DRM system that can provide a cryptographic level of multimedia protection has yet to be developed.

The problem of ensuring copyright of multimedia at the client side lies in the fact that traditional data protection technologies such as encryption or scrambling cannot be applied exclusively, as they are prone to digital copying or analog rerecording (commonly referred to as the "analog hole"). Once decrypted or unscrambled, the plaintext content can be either copied in its digital format or digitized from the system's analog output via an analog to digital (A/D) converter. The moment that the adversary obtains a plaintext digital copy of the multimedia clip, its copyright owners, at least technically, lose control over the content's distribution. Thus, almost all modern copyright protection mechanisms tend to rely to a certain extent on watermarks — imperceptive and secret marks hidden in host signals. Two types of protection system have evolved over the past decade: content screening and fingerprinting.

Content Screening

In a typical content screening scenario, a copyright owner protects her distribution rights simply by hiding a unique and secret watermark in her multimedia clips. Whereas both the original multimedia content as well as the key used to generate the secret must be safely guarded by the owner, the marked copy can be distributed using a public communication channel such as the Internet. In general, the marked content can be distributed in plaintext over the communication channel. The client's media player searches the distributed content for hidden information without the presence of the original clip. We refer to such watermark detection as "blind." If the secret mark is detected, the player must verify, prior to playback, whether it has a license to play the content. Only when the license is authentic and valid, the media player may play the protected clip. By default, unmarked content is considered unprotected and is played without any barriers. Hence, a content screening system consists of two subsystems: a watermark detector and a DRM agent that handles license management via standard crypto-graphic tools. An example of a DRM agent is Microsoft's Media Player 9 DRM system [3]. A content screening scenario is illustrated in Figure 18.1.

A content screening system that relies on watermarks must fulfill several requirements. First, watermarks must be *imperceptive* to the human auditory system (HAS). For example, in the case of music, marked audio clips should be perceptually indistinguishable from the original audio signals by people with extraordinary auditory senses — commonly called "golden ears." An additional requirement that relates to system security is that the embedded secret is *robust* to attacks; that is, it cannot be removed from the multimedia clip without the knowledge of the secret

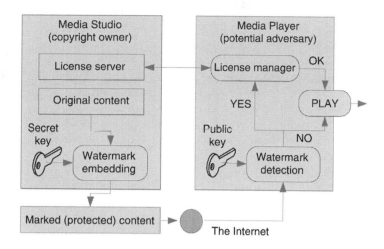

Figure 18.1. Block diagram of a typical content-screening scenario.

key used to generate the watermark. Next, detection must be *reliable* in
the sense that watermark detectors must achieve a low likelihood of false
alarms, both positive (detecting a watermark in an unmarked content)
and negative (not detecting a watermark despite its existence in the clip),
even under attack.

The key requirement for a content screening system is *BORE resistance*. BORE is the acronym for "break once, run everywhere," a typical
vulnerability of content screening systems. Note that if the key used to
detect the watermark is the same secret key used to mark the original
clips, by breaking one player, the adversary gains access to the master
secret key and, as a consequence, can mark or remove marks from
content at will. In order to be BORE resistant, a content screening secret
must deploy either a public key content screening system, much like
existing public key cryptosystems (e.g., the Rivest–Shamir–Adleman
system [RSA] [4]), or use tamper-proof hardware and software. Neither
of the two goals have been achieved to date despite strong efforts
by the academic and industrial research community. For example,
certain progress on public key watermarking has been done by Kirovski
et al. [5].

Fingerprinting

In a typical scenario that uses multimedia marking for forensic purposes,
illustrated in Figure 18.2, studios create a uniquely marked content copy
for each individual user request. User-specific distinct watermarks are

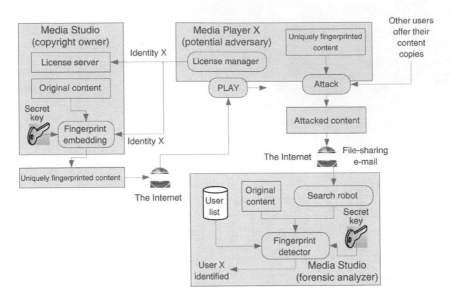

Figure 18.2.   Block diagram of a typical forensic marking scenario.

commonly denoted as fingerprints. The fingerprinted copy is securely distributed to the user who plays the content using a media player that is unmodified compared to modern media players. If the user decides to illegally distribute this content, the media studios deploy search robots to find content copies on the Internet. Illegally distributed content is retrieved, and based on the known user database as well as the original clip, media studios deploy powerful forensic analysis tools that identify the pirate.

Clearly, imperceptiveness, robustness, and reliability remain as key requirements for fingerprints. However, the attack model in this scenario changes. A major problem for fingerprinting systems is the *collusion attack*. To launch such an attack, an adversarial clique of malicious users colludes their copies in order to create a copy that is statistically clean of any fingerprint traces (possibly the original) or a copy that incriminates another innocent user. Collusion resistance for multimedia content is typically low. The best constructive fingerprint codes have been proposed by Boneh and Shaw, with a collusion resistance proportional to the fourth root of content size [6]. Because of this deficiency, multimedia protection systems based on forensic analysis are commonly restricted to small distribution lists. Finally, one of the most devastating problems for fingerprinting systems is, surprisingly, successful identity theft. An adversary with a stolen identity can purchase a multimedia clip and then illegally distribute it, leaving multimedia studios without a target for legal action.

Attacking Arbitrary Content Protection Systems

For content protection to perform securely, both fingerprints and water-marks must be robustly embedded in host signals. Their detection must survive an arbitrary signal processing attack that preserves perceptual fidelity of the target content. In this chapter, we focus our attention to blind pattern-matching attacks. In addition, with certain loss of generality, in the remainder of this chapter we consider attacking only watermarks (i.e., content screening systems). Because forensic detectors use knowledge of the original content, their detection ability is superior to "blind detectors"; however, we deem that due to exceptionally low collusion resistance, such systems are likely to be broken using collusion as the main attack principle.

Blind pattern matching is a new breed of signal processing attacks that does not use any or uses very limited knowledge of the protection algorithms. In related literature, it has been referred to as the swap attack, dewatermaking, and the replacement attack. The goal of the attack is to irreversibly remove watermarks by pseudorandomly permuting perceptually similar signal blocks that are tainted with distinct watermarks. Several adversarial mechanisms surveyed in section "Short Survey of Attacks on Content Protection Systems," have been largely successful in setting up robustness benchmarks for watermarking technologies. However, none of these attack technologies, which do not rely on having access to the watermark detector, remove watermarks without any hope that an irreversible or preventing action is possible.

Historically, the first ideas behind this class of attacks were proposed simultaneously and independently in two research articles [7,8], for images and audio, respectively. Both attacks were replacing blocks of the marked signal with perceptually similar blocks obtained from the same signal elsewhere. Whereas Dugelay et al. used basic geometric transformations to improve their search space in Reference 7, Kirovski and Petitcolas proposed a new version of their initial blind-pattern-matching attack, which computed the replacement signal block as a linear combination of other similar signal blocks [9]. In this chapter, we review the results presented in these articles and pose open research problems that may intrigue the reader to pursue solving them.

SHORT SURVEY OF ATTACKS ON CONTENT PROTECTION SYSTEMS

The main target of an adversarial attack on a content protection system is finding a signal processing primitive that removes the watermark or prevents a detector to find it, or both. We survey the related work in attacking watermarking systems according to these seemingly two classes of attack.

The first class, watermark removal engines, targets estimating the watermark and then subtracting its value from the multimedia clip in hope to remove it. One of the first attacks in this category was the *sensitivity attack* proposed by Linnartz and van Dijk [10]. The sensitivity attack iteratively and accurately concludes bits of the watermark while treating the watermark detector as a black box. This attack assumes that the adversary has access to the watermark detector, which is the case in the content screening scenario. Initially, the adversary performs a binary search for the amplitude of the random noise, which, when mixed with the marked content, causes the detector to be on the verge of positive or negative detection. Next, the adversary iteratively repeats the following procedure. For a certain sample of the marked content, the adversary adds a positive value to it and observes whether the decision of the watermark detector changes toward positive or negative. If it is positive, the adversary concludes that the sign of the added watermark for that sample is positive and vice versa. In time linear with respect to object size, the adversary can accurately estimate the embedded watermark following these steps. Kalker et al. extended the original results on this attack to show that randomized decision thresholding does not prevent the adversary from estimating the watermark [11]. The main conclusion that stems from the sensitivity attack is that watermark estimation is possible even in the case when tamper-proof hardware is available.

Another watermark removal mechanism is the *watermark estimation attack*, which commonly applies a Wiener filter to obtain an optimal estimate of the sign of watermark samples. For instance, Langelaar et al. showed that 3×3 median filtering gives a good approximation of original pictures in the case for which they have been watermarked using the spread spectrum [12]. So far, estimation attacks have introduced fairly strong blurring effects, but recent work based on maximum *a posteriori* watermark estimation and remodulation has given promising results [13,14]. Attacks in this category have been reviewed and analyzed also in References 5, 15, and 16.

To date, several mechanisms that belong to the second class of attacks, watermark desynchronization engines, have been largely successful in setting up robustness benchmarks for watermarking technologies. Early *desynchronization attacks*, such as random geometric distortions [17–19], relied on the fact that most watermarking algorithms are based on some form of correlation, which requires good alignment properties. Breaking this alignment usually prevents reliable detection. Although robustness to this type of attack is not fully solved in the case of images, there already an audio data hiding algorithm that can cope with desynchronization attacks already [20,21]. The technology presented in Reference 20 is a good example of a watermarking system that opens doors to another type of an attack, such as the estimation attack in this case [15,16], while providing robustness to one attack class.

In summary, whereas watermark removal attacks have certain limitations such as access to a watermark detector or upper bounded estimation ability, desynchronization attacks do not provably remove watermarks; hence, certain hope that they can be recovered remains. To the contrary, blind-pattern-matching attacks cannot be prevented and their aim is to irreversibly remove significant portion of the embedded watermarks.

RATIONALE BEHIND THE BLIND PATTERN-MATCHING ATTACK

The key to the success of the blind-pattern-matching (BPM) attack is the existence of repetitive content across one or several multimedia clips. The BPM attack simply replaces one block of the multimedia signal with another, a perceptually similar block extracted or derived from other blocks found in locations that are not tainted with the same watermark. For most types of multimedia, finding such blocks is a challenging but doable task. Redundancy in multimedia is an undeniable fact that stems from the way movies and music are produced. For example, within a common movie scene, its background and encompassed objects experience minor geometric transformations significantly more frequently than changes in appearance. Similarly, repetition is often a principal part of composing music and is a natural consequence of the fact that distinct instruments, voices, and tones are used to create a soundtrack. Thus, it is likely to find similar blocks of musical content within a single musical piece, an album of songs from a single author, or in instrument solos.

Assuming a highly repetitive multimedia content, which is the case with most music and video, the BPM attack adds a nearly marginal noise to the marked content. Although the noise appears to be zero-mean, independent and identically distributed (i.i.d.) Gaussian, it wipes out an enormous percentage of the watermark correlation. An additive and truly Gaussian noise of equivalent variance would not alter the watermark correlation beyond its expected statistics [22]. The surprising effect of the BPM attack stems from the redundancy that exists in multimedia content. By using this information, one can recreate the protected multimedia while being marginally dependent on the originally embedded watermark.

The rationale behind the attack is simple. A given watermarked block $\mathbf{x} + \mathbf{w} \in \{\mathbb{R}\}^N, \mathbf{w} \in \{\pm 1\}^N$ of N samples represents a point in the N-dimensional space. With no loss of generality,[1] the original content can be modeled as a zero-mean, Gaussian random variable of i.i.d. samples $\mathbf{x} = \mathcal{N}(0, \sigma_x)^N$. Based on the statistics of spread spectrum watermarking [22], by adding additive white Gaussian noise $\mathbf{n} = \mathcal{N}(\mu, \sigma)^N$ to the marked

[1]Due to the central limit theorem.

content, the expected normalized correlation value of $E[(\mathbf{x} + \mathbf{w} + \mathbf{n}) \cdot \mathbf{w}] = E[(\mathbf{x} + \mathbf{w}) \cdot \mathbf{w}] = 1$ remains intact for any mean μ and variance σ^2. Operator $(\cdot)$ is defined as a normalized inner product of two vectors: $\mathbf{x} \cdot \mathbf{y} = N^{-1} \sum_{i=1}^{N} x_i y_i$. We denote all points $\mathbf{y} \in \{\mathbb{R}\}^N$ in a ball $Y(\varepsilon)$: $\{\|\mathbf{y} - (\mathbf{x} + \mathbf{w})\| \leq \varepsilon\}$ at minimal Euclidean distance $\varepsilon \geq 1$ from $\mathbf{x} + \mathbf{w}$ as "perceptually valid" points. The goal of the attacker is to drive $\mathbf{x} + \mathbf{w}$ with some noise pattern $\mathbf{n}_a$ into a point of attack $\mathbf{z} = \mathbf{x} + \mathbf{w} + \mathbf{n}_a \in Y(\varepsilon)$, which is not correlated with $\mathbf{w}$ (i.e., $E[\mathbf{z} \cdot \mathbf{w}] = 0$). Although there are many points in Y that are not correlated with $\mathbf{w}$ — for example, $\mathbf{x}$ is, by definition, one of them, as $E[\mathbf{x} \cdot \mathbf{w}] = 0$ — it is difficult to find in a single trial an $\mathbf{n}_a$ that cancels out the effect of $\mathbf{w}$. For example, random $\mathbf{n}_a = \mathbf{n}$ provably does not affect $E[(\mathbf{x} + \mathbf{w} + \mathbf{n}_a) \cdot \mathbf{w}]$.

There are three standard approaches for finding $\mathbf{n}_a$; all of them informally reviewed in section "Short Survey of Attacks on Content Protection Systems." The first one is to use the estimation attack $\mathbf{n}_a = \text{sign}(\mathbf{x} + \mathbf{w})$, which has limited success if $\mathbf{w}$ is not redundantly embedded [5]. If the watermark detector is available to the adversary, as in the case of content screening, she can launch the sensitivity attack: a multitrial search for $\mathbf{n}_a$ that quickly finds an optimal attack vector to achieve its goal [11]. Finally, during the desynchronization attack, the adversary randomly bends the space dimensions to inflict difficulty positioning $\mathbf{w}$ in the same direction as it had when it was embedded into $\mathbf{x}$ [19,21]. This type of an attack can be prevented using redundancy while embedding $\mathbf{w}$ [20]; however, such a solution is prone to the estimation attack [15].

The BPM attack assumes that multimedia is repetitive enough so that for a given point $\mathbf{x} + \mathbf{w}$ of sufficient dimensionality N, there exists at least one other point $\mathbf{a}$ within the same media clip that is within the ball of interest $Y(\varepsilon)$. By definition, if $\mathbf{a}$ is not tainted with $\mathbf{w}$, replacing $\mathbf{x} + \mathbf{w}$ with $\mathbf{a}$, intuitively, removes the targeted correlation. However, if $\mathbf{a} = \mathbf{x} + \mathbf{w}$, then $E[\mathbf{a} \cdot \mathbf{w}] = 1$. With the increase of N, this case can be made arbitrarily unlikely. Although perceptual repetitiveness in audio and, in particular, video is significant, the above assumption is still strong, as it is unlikely to expect that for any high-dimensional point in the media clip there exists a redundant one.

To address this issue, the BPM attack creates a in two steps. First, it finds a set $\mathbf{B}$ of K points $\mathbf{B} = \{\mathbf{b}_1, \ldots, \mathbf{b}_K\}$ in the media clip closest to $\mathbf{x} + \mathbf{w}$ and, then, it computes a least square approximation of $\mathbf{x} + \mathbf{w}$ using a linear combination of transformed vectors from $\mathbf{B}$. The list of transformations T is dependent on the type of multimedia. For example, images as two-dimensional objects provide a richer space for transforms such as rotation, scaling, bending, and normalization [7], as opposed to audio where perceptual masking, pitch shifts, and band filtering can be applied. The resulting approximation is the replacement vector $\mathbf{a}$.

For large N, it is safe to assume that none of the vectors in **B** as well as their considered transforms are correlated with **w**. However, the correlation of their linear combination with **w** increases with the increase of K. Theoretic analysis of this process is exceedingly difficult, as it is hard to model content similarity due to its diverse nature. To address this issue, the BPM attack assumes that there is a "safe distance" $\alpha \leq \|\mathbf{a} - (\mathbf{x} + \mathbf{w})\|$ that **a** needs to have with respect to $\mathbf{x} + \mathbf{w}$ in order to be noncorrelated. This is a model that has proven to yield solid results for small K [7–9].

We expand upon this similarity model by assuming that $\mathbf{a} = \mathbf{x} + \mathbf{d} + \mathbf{v}$, where **d** is the similarity noise in the original content modeled as a zero-mean, normal random variable of i.i.d. samples $\mathbf{d} = \mathcal{N}(0, \sigma_d)$ and $\mathbf{v} \in \{\pm 1\}^N$ is the watermark added to the original block $\mathbf{x} + \mathbf{d}$.

Theorem 1: Computing $E[\mathbf{a} \cdot \mathbf{w}]$. Assuming mutual independence of **x**, **d**, **v**, and **w**, if $\|\mathbf{a} - (\mathbf{x} + \mathbf{w})\| = \alpha$, then

$$2E[\mathbf{a} \cdot \mathbf{w}] = 2 + \sigma_d^2 - \alpha \qquad (18.1)$$

Proof: From $\|\mathbf{a} - (\mathbf{x} + \mathbf{w})\| = \alpha$ and $E[\mathbf{x} \cdot \mathbf{w}] = 0$, we conclude that $2E[\mathbf{a} \cdot \mathbf{w}] = 1 + \|\mathbf{a} - \mathbf{x}\| - \alpha$.   From $\|\mathbf{a} - \mathbf{x}\| = \|\mathbf{d}\| + 1$,   we   derive Equation 18.1.   ∎

From Theorem 1, we conclude that if we want to drive $E[\mathbf{a} \cdot \mathbf{w}]$ to a small value, the minimal distance α must be driven close to $2 + \sigma_d^2$. Needless to say, the efficacy of the attack is highly determined by the similarity metric σ_d^2, which represents the variance of the difference among most similar blocks in a given media clip. By using the least square approximation of $\mathbf{x} + \mathbf{w}$ using transforms of similar blocks from the media clip, we effectively reduce σ_d at the cost of increasing the correlation of **a** and $\mathbf{x} + \mathbf{w}$. Careful execution of this process is the key to the success of the BPM attack.

LOGISTICS OF THE BLIND PATTERN MATCHING ATTACK

The BPM attack is not limited to a type of content or to a particular watermarking algorithm. For example, systems that modulate secrets using a spread spectrum [22] or quantization index modulation (QIM) [23] are all prone to this attack. For brevity, the analysis of the attack in this chapter is restricted to direct sequence spread spectrum watermarks. Readers are encouraged to review an article by Kirovski and Petitcolas, who demonstrate that an implementation of QIM can be successfully attacked using BPM [24].

In order to launch the attack successfully, the adversary does not need to know the details of the watermarking codec. This assumption is convenient for the adversary compared to the required knowledge that other attacks, mentioned in section "Short Survey of Attacks on Content Protection Systems," mandate.

Given a signal $\hat{\mathbf{x}} \in \{\mathbb{R}\}^M$ and a corresponding watermark $\hat{\mathbf{w}} \in \{\pm 1\}^M$, the BPM attack performs the following steps:

   I.   Partition $\hat{\mathbf{x}} + \hat{\mathbf{w}}$ into overlapping blocks $\mathbf{x} + \mathbf{w}$ of length N,

  II.   For each block $\mathbf{x} + \mathbf{w}$, find a set $\mathbf{B}$ of K perceptually most similar blocks in $\hat{\mathbf{x}} + \hat{\mathbf{w}}$ that do not overlap $\mathbf{x} + \mathbf{w}$,

 III.   Compute the replacement block a as a least square linear approximation of $\mathbf{x} + \mathbf{w}$ using transformations of blocks from $\mathbf{B}$,

 IV.   Replace $\mathbf{x} + \mathbf{w}$ with $\mathbf{a}$.

Attack Trade-offs

Considering the issues related to the BPM attack and presented in section "Rationale Behind the Blind Pattern-Matching Attack," we identify several important trade-off decisions that the adversary needs to make before applying the attack. The trade-offs, TO.1 to TO.4, reflect on the following important performance metrics: reduction in correlation, distortion, and speed.

TO.1.   *Point dimensionality.* N has a profound effect on $E[\mathbf{a} \cdot \mathbf{w}]$. By increasing N, the adversary reduces the likelihood that vectors in $\mathbf{B}$, as well as their linear combinations, are correlated with $\mathbf{w}$. On the other hand, significantly increased N reduces the expectation on the cardinality of $\mathbf{B}$ as it increases σ_d, thus reducing attack effectiveness. One heuristic is that N should be maximized for a given clip so as to still produce "perceptually valid" matches.

TO.2.   *Selection of α.* An increased α improves all aspects of attack performance except distortion. This parameter should be maximized for a desired perceptual quality.

TO.3.   *Size M of the lookup media library.* $\hat{\mathbf{x}} + \hat{\mathbf{w}}$ determines the complexity of the attack and can significantly improve σ_d.

TO.4.   *Selection of the set of transform functions.* T increases the search space and reduces σ_d at a significant increase in attack complexity. A BPM attack on images or video is significantly more likely to benefit from such transforms due to the nature of redundancy that occurs in such media [7]. A BPM attack on audio signals benefits less from a diverse T, as most redundancies in audio can be identified without any transforms. One transform that can be applied to an audio BPM with significant impact is band-pass filtering.

ATTACK STEPS
Step I: Signal Partitioning

For improved perceptual quality of the resulting multimedia clip, the protected signal $\mathbf{z} = \hat{\mathbf{x}} + \hat{\mathbf{w}}$ is partitioned into a set of blocks $\Pi = \{\mathbf{p}_1, \ldots, \mathbf{p}_P\}$, where each block $\mathbf{p}_i = \{h_j z_{1+(i-1)N/2+j}, j = 1, \ldots, N\}$ overlaps its neighbors and is windowed with an analysis windowing function $\mathbf{h} \in \{\mathbb{R}\}^N$ that yields perfect reconstruction with its synthesis counterpart. With no loss of generality, we assume that $\hat{\mathbf{x}} + \hat{\mathbf{w}}$ is an one-dimensional signal.

Step II: Search for the Substitution Base

Finding perceptually similar blocks of certain music or video content is a challenging and computationally expensive task. For a given set of transforms $T = \{\tau_1, \ldots, \tau_{|T|}\}$ and for each point $\mathbf{p}_i$ in the multimedia clip, we want to find a set $\mathbf{B}_i$ of K best-matched blocks in $\{\mathbf{z}, \tau_1(\mathbf{z}), \ldots, \tau_{|T|}(\mathbf{z})\}$ denoted as $\mathbf{B}_i = \{\mathbf{b}_1, \ldots, \mathbf{b}_K\}$, with individual points denoted as $\mathbf{b}_j = \mathbf{h} \cdot \{z_{s_j}, \ldots, z_{s_j+N-1}\}$ or $\mathbf{b}_j = \mathbf{h} \cdot \tau_k(\{z_{s_j}, \ldots, z_{s_j+N-1}\})$ if a transform is used, where s_j indexes the location of $\mathbf{b}_j$ in $\mathbf{z}$ or its transform $\tau_k(\mathbf{z})$. Before we define the search process, we adopt a normalized and squared Euclidean distance between two N-dimensional points $\mathbf{a}$ and $\mathbf{b}$ as a similarity metric:

$$\phi(\mathbf{a}, \mathbf{b}) = \frac{\sum_{k=1}^{N}[a_k - b_k]^2}{\sigma_a \sigma_b N} \tag{18.2}$$

where σ_a and σ_b are standard deviations of vectors $\mathbf{a}$ and $\mathbf{b}$, respectively. Because maximized normalized correlation corresponds to a minimal Euclidean distance in L^2, the search for top K matches in $\mathbf{z}$ against each $\mathbf{p}_i$ can be sped up as follows. We first compute the normalized block convolution of the complex conjugate of $\mathbf{p}_i$ with respect to $\mathbf{z}$. This can be done rather fast using the fast Fourier transform (FFT) and the overlap-add fast convolution method [25]. The same method can be used to efficiently compute the standard deviation of every point $\mathbf{p}_j$ in $\mathbf{z}$. The complexity of this step is $\mathcal{O}(M \log_2 N)$ assuming N is a power of 2. The top K correlated blocks in $\mathbf{z}$ and its transforms that do not overlap $\mathbf{p}_i$ constitute the substitution base $\mathbf{B}_i$ for $\mathbf{p}_i$.

Step III: Computing the Replacement

This step of the algorithm is crucial, as it resolves the trade-offs related to the selection of α and the inherent σ_d — the two most important metrics of the BPM attack. First, we review one restriction of the attack.

We restrict that each sample a_i of the replacement block **a** is at a "safe" and "perceptually valid" distance from the sample p_i it is replacing in **p**. More formally,

$$(\forall a_i \in \mathbf{a}) \quad \alpha \le |a_i - p_i| \le \varepsilon \tag{18.3}$$

Bounds α and ε define the "safe" and "perceptually valid" distance, respectively, that a sample of the replacement block must have. In the remainder of this subsection, we review several algorithms for computing a such that the above constraints are satisfied.

Algorithm A0 computes the replacement block **a** as follows. We assume that it is given a set **B** of K points such that for each point $\mathbf{b}_i \in \mathbf{B}$, the following relation holds: $\alpha \le |\mathbf{b}_i - \mathbf{p}| \le \varepsilon$. The replacement block **a** is computed from the selected blocks in **B** such that its similarity with respect to **p** is maximized. More formally, we construct a matrix $\mathbf{s} \in \{\mathbb{R}\}^{K \times N}$, where each row of this matrix represents one block from **B**. We aim to compute a vector λ such that $\|\mathbf{s}\lambda - \mathbf{p}\|$ is minimized. The least square solution to this set of linear equations, commonly called pseudoinverse of **s**, equals $\lambda = (\mathbf{s}^T\mathbf{s})^{-1}\mathbf{s}^T\mathbf{p}$. A temporary replacement block $\mathbf{a}'$ is now computed as $\mathbf{a}' = \mathbf{s}\lambda$. Per sample a'_j, three cases can occur:

1. It satisfies the predefined restriction [e.g., $\alpha \le \phi(p_j, a'_j) \le \varepsilon$], in which case, the replacement sample equals $a_j = a'_j$.
2. Sample a'_j is too similar to p_j [e.g., $\phi(p_j, a'_j) < \alpha$], in which case, a'_j and a corresponding sample t_j from a randomly chosen block $\mathbf{t} \in \mathbf{B}$ such that $\phi(t_j, a'_j) > \alpha$ are mixed to create the replacement sample as $a_j = (1 - q)t_j + qa'_j$, where q is such that $\phi(p_j, a_j) = \alpha$. The mixing parameter q enforces the desired similarity $\phi(p_j, a_j) = \alpha$ if $q = (\alpha + p_j - t_j)/(a'_j - t_j)$.
3. Sample a'_j is perceptually not similar to p_j [e.g., $\phi(p_j, a'_j) > \varepsilon$], in which case, Algorithm A0 sets $a_j = p_j$ to minimize the noise due to the attack.

Algorithm A1 has a different, randomized strategy in computing the replacement block **a** to address better the trade-off between the size of **B** and the correlation of the computed replacement **a** with respect to the watermark contained in **p**. It computes a in several steps. In the first step, it generates c random and distinct subsets of r blocks from **B**. We denote these subsets by $\mathbf{S}_1$ through $\mathbf{S}_c$. For each of these subsets, Algorithm A1 computes the least square approximation of **p**. More formally, for a given $\mathbf{S}_i = \{\mathbf{s}_1, \ldots, \mathbf{s}_r\}$, we are seeking for λ_i such that

$\|\mathbf{S}_i \lambda_i - \mathbf{p}\|$ is minimized. Optimal λ_i is computed simply as

$$\lambda_i = (\mathbf{S}_i^T \mathbf{S}_i)^{-1} \mathbf{S}_i^T \mathbf{p} \tag{18.4}$$

which yields the replacement candidate vector $\mathbf{a}_i = \mathbf{S}_i \lambda_i$. Samples from $\mathbf{a}_i$ can be categorized into two categories: ones that satisfy Equation 18.3 and the ones that do not.

To resolve this issue, we introduce a binary coverage matrix $\mathbf{q}_i = \{0, 1\}^N$ associated with each $\mathbf{S}_i$. We set $q_{i,j} = 1$ if sample $a_{i,j}$ satisfies Equation 18.3 and vice versa. We define as effective dimensionality $\bar{N}(\mathbf{a}_i)$ the number of samples a given replacement $\mathbf{a}_i$ covers $\bar{N}(\mathbf{a}_i) = \sum_{j=1}^{N} q_{i,j}$. Heuristically, we are already driven by the assumption that the larger the effective dimensionality, the stronger the effect of the attack on the resulting correlation. Hence, we can model the goal of our replacement algorithm as a combinatorial optimization problem. Algorithm A1 aims to cover as many samples as possible from $\mathbf{p}$ using as few vectors as possible from the set $\mathbf{A} = \{\mathbf{a}_1, \ldots, \mathbf{a}_c\}$. This problem is better known as minimum cover and is NP-hard [26]. To obtain a good quality solution for this problem, Algorithm A1 uses a greedy heuristic, which iteratively selects vectors from $\mathbf{A}$ to cover the maximum number of remaining uncovered samples. Just as in the case of Algirithm A0, there may be samples that cannot be covered by any $\mathbf{a}_i$ — their values are set to the corresponding values of the marked content $\mathbf{z}$ in order to minimize distortion.

Parameters K, c, and r strongly influence the performance of Algorithm A1. By increasing r, we reduce σ_d at the cost of stronger correlation of each $\mathbf{a}_i$ and $\mathbf{z}$. It has been shown empirically that solid results can be obtained by using relatively small r, usually on the order of $r \in [1, 20]$ [27]. Once r is set, the average effective dimensionality $\bar{N}$ in $\mathbf{A}$ can be computed. The higher the $\bar{N}$, the more candidate trials c Algorithm A1 can afford to test. Again empirically, it has been shown that solid results can be achieved with $c \approx \bar{N}$ [27]. Finally, the size of the substitution database should be kept large at $K \sim \bar{N}$.

Algorithm A2 is a significantly slower, but still randomized, version of Algorithm A1 and Algorithm A0. It is based on the observation that the blocks in $\mathbf{A}$ are highly redundant because they are searched using a common criterion (Equation 18.2). Using only these blocks in linear combinations restricts significantly the search space. A better, but still not optimal strategy in representing $\mathbf{p}$ as accurately as possible using a constant number of blocks from $\mathbf{z}$ is to use a variant of Gram–Schmidt orthonormalization (GSO) [28].

Hence, Algorithm A2 iteratively performs the following process. Using the search procedure already described in section "Step II: Search for the Substitution Base," Algorithm A2 finds the first $\mathbf{A} = \mathbf{a}_1 = \mathbf{B}_1$ with $K = 1$. The most similar point $\mathbf{a}_1$ is subtracted from $\mathbf{p}$ as $\mathbf{p} - \lambda_1 \mathbf{a}_1$, where λ_1 is a scalar equal to the normalized correlation of $\lambda_1 = \mathbf{p} \cdot \mathbf{a}_1 / (\|\mathbf{p}\| \|\mathbf{a}_1\|)$. In the subsequent iteration, Algorithm A2 computes the similarity of $\mathbf{p} - \lambda_1 \mathbf{a}_1$ with $\mathbf{z}$, finds the best match $\mathbf{a}_2$, and subtracts it from the remainder as $\mathbf{p} - \lambda_1 \mathbf{a}_1 - \lambda_2 \mathbf{a}_2$, where $\lambda_2 = (\mathbf{p} - \lambda_1 \mathbf{a}_1) \cdot \mathbf{a}_1 / (\|\mathbf{p} - \lambda_1 \mathbf{a}_1\| \|\mathbf{a}_1\|)$. This procedure is iterated while $\|\mathbf{p} - \sum \lambda_i \mathbf{a}_i\| > \alpha$.

The above version of Algorithm A2 has the problem that not all samples of the final replacement $\mathbf{a} = \sum \lambda_i \mathbf{a}_i$ obey the constraint in Equation 18.3. In order to address this problem, after each iteration, Algorithm A2 discards samples that satisfy Equation 18.3 in the subsequent iterations. In addition, Algorithm A2 considers the top K similar blocks in each iteration because it is not case that the closest point provably has the highest effective dimensionality. These two adjustments marry Algorithm A1 with GSO to best describe Algorithm A2. Similarly, one can advertise Algorithm A1 as a low-cost version of GSO because it does not need to perform the similarity search in each iteration. Finally, because none of the presented algorithms is optimal, we encourage the reader to try new mechanisms when deploying the basic idea behind the BPM attack.

Step IV: Block Substitution

In the final step, each block $\mathbf{p}$ of the original watermarked signal is replaced with the corresponding computed replacement $\mathbf{a}$ to create the output media clip.

BLIND-PATTERN-MATCHING ATTACK FOR AUDIO SIGNALS

In this section, we demonstrate how the generic principles behind the BPM attack can be applied against an audio watermarking technology. We first describe how an audio signal is partitioned and preprocessed for improved perceptual pattern matching. Next, we analyze the similarity function we used for our experiments. The effect of the BPM attack on direct sequence spread spectrum watermark detection is presented in the following subsections.

Audio Processing for the BPM Attack

Because most psychoacoustic models operate in the frequency spectrum [29], we launch the BPM attack in the logarithmic (dB) frequency domain. The set of signal blocks Π is filtered using a modulated complex lapped transform (MCLT) [29]. The MCLT is a 2× oversampled discrete

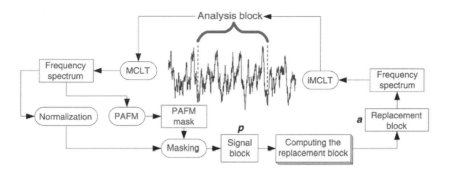

Figure 18.3. Block diagram of the signal processing primitives performed as preprocessing and postprocessing to the BPM attack.

Fourier transform (DFT) filter bank, used in conjunction with analysis and synthesis windows that provide perfect reconstruction. We consider MCLT analysis blocks with 2048 transform coefficients and an $\eta = 0.5$ overlap. Each block of coefficients is normalized and psychoacoustically masked using an off-the-shelf masking model [29]. Similarity is explored exclusively in the audible part of the frequency subband where watermarks are hidden. In this case, we bound this subband within 200 Hz and 7 kHz [20]. Figure 18.3 illustrates the signal processing primitives used to prepare blocks of audio content for substitution.

Watermark length is assumed to be greater than 1 sec. In addition, we assume that watermark chips may be replicated along the time axis at most for one second[2] [20]. Thus, we restrict that for a given block its potential substitution blocks are *not* searched within 1 sec.

In the experiments presented in this section, we considered the following five audio clips:

clip 1: Ace of Base, *Ultimate Dance Party 1999*, Cruel Summer (Blazin' Rhythm Remix)

clip 2: Steely Dan, *Gaucho*, Babylon Sisters

clip 3: Pink Floyd, *The Wall*, Comfortably Numb

clip 4: Dave Matthews Band, *Crash*, Crash Into Me

clip 5: Unidentified classical piece, produced by SONY Ent., selected because of exceptional perceptual randomness (available upon request from the authors)

---

[2]A higher level of redundancy may enable effective watermark estimation.

Analysis of the Similarity Function

We performed several experiments in order to evaluate the effectiveness of the BPM attack. The first set of experiments aimed at quantifying similarity between blocks of several audio clips marked with spread spectrum watermarks with a magnitude of $\delta = 1$ dB. Figure 18.4 shows the values of the similarity function $\phi(\mathbf{b}_i, \mathbf{b}_j)$ for five 2048-long MCLT blocks at positions $i = \{122, 127, 132, 137, 142\}$ against a database of 240 blocks $j = \{1, \ldots, 240\}$ within clip 1. We observe that throughout the database, four different pairs of blocks (circled in Figure 18.4) are found as similar below 4 dB to the pair of blocks with indices 127 and 137. In this experiment, similarity is computed in the MCLT domain. All similar pairs of blocks preserve the same index distance as the target pair. This points to the fact that, in many, cases content similarity is not a result of coïncidence, but a consequence of repetitive musical content.

Figure 18.5 illustrates the probability that for a given 2048-long MCLT block $\mathbf{b}_i$, there exists another block $\mathbf{b}_j$ within the same audio clip that is within $\phi(\mathbf{b}_i, \mathbf{b}_j) \in [x - 0.25, x + 0.25\}$ dB, where x is a real number. For our benchmark set of five distinctly different musical pieces, we conclude that the average $\phi(\mathbf{b}_i, \mathbf{b}_j)$ for two randomly selected blocks within an

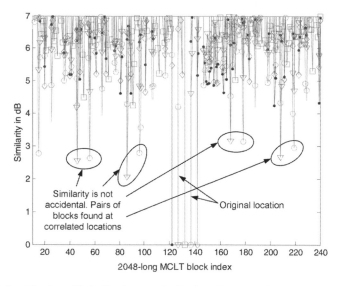

Figure 18.4. Music self-similarity: a similarity diagram for five different 2048-long MCLT blocks taken from clip 1 with a substitution database of 240 MCLT blocks taken from the same clip. Zero-similarity denotes equality. The abscissa x denotes the index of a particular MCLT block. The ordinate denotes the similarity $\phi(x, \mathbf{b}_i)$ of the corresponding block x with respect to the selected five MCLT blocks with indices $\mathbf{b}_i|i = \{122, 127, 132, 137, 142\}$. Here, similarity is computed in the MCLT domain.

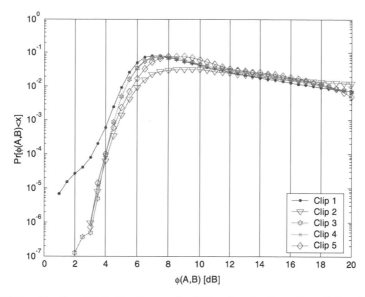

Figure 18.5. Music self-similarity: probability density function of the similarity function $\phi(b_i, b_j)$ within an audio clip for our benchmark of five audio clips. A certain value x on the abscissa represents a histogram bin from $x - 0.25$ to $x + 0.25$ dB.

audio clip is in the range of 6–8 dB. The probability of finding a similar block should rise proportionally to the size of the substitution database, especially if it consists of clips of the same genre or performer. Finally, note that electronically generated music (in our benchmark, clip 1) is significantly more likely to contain perceptually correlated blocks than music performed by humans.

In the following experiment, we demonstrate the similarity exhibited by the search process presented in section "Step II: Search for the Substitution Base." Figure 18.6 illustrates the normalized correlation of a selected point $\mathbf{p}_i = \{z_{560001}, \ldots, z_{564096}\}$ taken from the first million samples of clip 1, which constitutes the substitution database $\mathbf{z}$. Note that perfect correlation is achieved only for the original block. However, within the database $\mathbf{z}$, there are 10 other blocks which retain correlation better than 0.7. Clearly, all of them are solid candidates for building the substitution set $\mathbf{B}$. Four of them with correlation exceeding 0.9 can be considered as direct substitutes. Finally, one can observe that the locations of similar blocks in Figure 18.6 are correlated with the clip's beat. Hence, it would be interesting to demonstrate whether the search for similarity can benefit from an automated beat detector.

A new set of experiments explores the distortion that a simple BPM–Algorithm A0 attack introduces. We consider three cases. In the first case, in the left-hand side of Figure 18.7, we present the probability that the

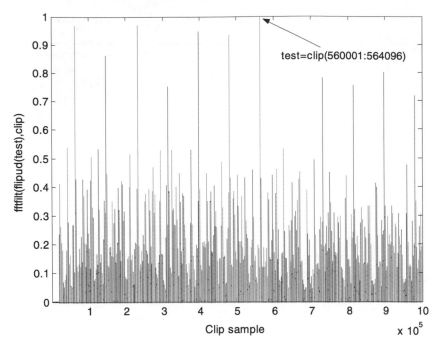

Figure 18.6. Music self-similarity: result of the search process for the source point $p_i = \{z_{560001}, \ldots, z_{564096}\}$ within the first million samples of clip 1. The search was executed using the following MATLAB script: `test = clip(560001:564096); plot(fftfilt(flipud(test),clip)./(fftfilt(ones(4096,1),clip)*norm(test)))`.

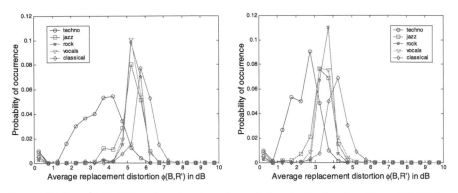

Figure 18.7. Probability density function of the similarity function $\phi(B, R')$ for two different cases: $K = 1$ (left) and $K = 10$ (right).

replacement block $\mathbf{a}'$ is at distance $\phi(\mathbf{b}, \mathbf{a}')$ if $\mathbf{a}'$ equals the most similar block found in the substitution database (e.g., $K = 1$). The right-hand side of Figure 18.7 presents the same metric for the case when $K = 10$ and $\mathbf{a}'$ is computed as described in section "Step III: Computing the Replacement."

Table 18.1. Improvement in signal distortion due to the BPM attack as parameter K is increased

$\phi(\mathbf{b}, \mathbf{a}')$	clip 1	clip 2	clip 3	clip 4	clip 5	Average $\phi_K(\mathbf{b}, \mathbf{a}') - \phi_{K=1}(\mathbf{b}, \mathbf{a}')$
$K = 1$	3.5714	5.2059	5.3774	5.3816	5.8963	N/A
$K = 10$	2.3690	3.2528	3.5321	3.5193	4.0536	1.741
$K = 20$	2.2666	3.0576	3.3792	3.3664	3.7968	1.914
$K = 30$	2.2059	2.9255	3.3061	3.2762	3.5613	2.032
$K = 50$	2.1284	2.6595	3.1702	3.1209	3.0635	2.253
$K = 100$	1.9512	2.1331	2.8631	2.7439	1.8719	2.775

Note: Results are reported on the dB scale. Average effective block dimensionality is $N \approx 400$.

Finally, Table 18.1 quantifies the improvement in the average distortion $\phi(\mathbf{b}, \mathbf{a}')$ as K increases from 1 to 100. We conclude that the BPM attack in our experimental setup induces between 1.5–3 dB distortion noise with respect to the marked copy.

Effect of the Attack on Watermark Detection

In order to evaluate the effect of the BPM–Algorithm A0 attack on direct sequence spread spectrum (DSSS) watermarks, we conducted another set of experiments. We used DSSS sequences with a $\delta = 1$dB amplitude that spread over 240 consecutive 2048-long MCLT blocks (approximately 11 sec long), where only the audible frequency magnitudes in the 2- to 7-kHz subband were marked. We did not use chip replication, as its effect on watermark detection is orthogonal with respect to the BPM attack.

Figure 18.8 shows how normalized correlation of a spread spectrum watermark detector is affected by the increase of the parameter K. During the attack, we replaced each target block $\mathbf{p}$ with its computed replacement block $\mathbf{a}$ following the recipe presented in section "Step III: Computing the Replacement." In Figure 18.8, we show two results. First, we show the average normalized correlation value (left ordinate) across 10 different tests for watermark detection within marked content (curves marked WM) and within marked content attacked with our attack for several values of $K = \{1, 10, 20, 30, 50, 100\}$ (curves marked RA). Second, we show on the right ordinate the signal distortion caused by the BPM attack: the minimal, average, and maximal distortion across all five audio clips. We can conclude from the diagram that for small values of K, its increase results in greatly improved distortion metrics, whereas for large values of K, the computed replacement vectors are too similar with respect to the target blocks, which results in a lesser effect on the normalized correlation.

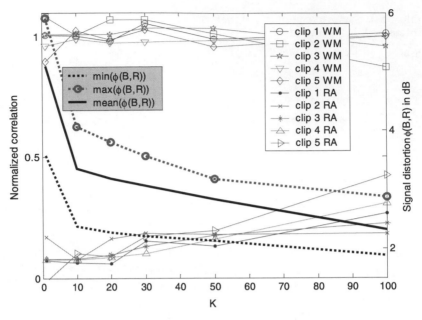

Figure 18.8. Response of a DSSS watermark detector to the BPM attack. The abscissa quantifies the change in parameter K from 1 to 100 for a fixed watermark amplitude of $\delta = 1\text{dB}$. The left ordinate shows the increase of the normalized correlation as K increases. The results are obtained for five full songs in different genres. The right ordinate shows the corresponding minimal, maximal, and average distortions with respect to the set of benchmark clips due to the BPM attack.

The power of the BPM attack is most notably observed by comparing the effect of adding a white Gaussian noise (AWGN) pattern $\mathbf{n} = \mathcal{N}(0, \sigma_n)$ of a certain standard deviation $\sigma_n \in \{2, \dots, 3\}\text{dB}$ to a BPM attack of equivalent distortion. Whereas the dramatic effect of replacement can be observed in Figure 18.8, AWGN does not affect the correlation detector. Finally, additive noise of 2–3 dB in the 2- to 7-kHz subband is a relatively tolerable modification.

Distortion introduced by the BPM attack is linearly proportional to the watermark amplitude δ because $\alpha = c(1) \cdot \delta$ from Theorem 1. Clearly, with the increase of δ, the search process becomes harder because of increased entropy of the blocks' contents. On the other hand, empirical studies have concluded that the watermark amplitude affects the reduction of the normalized correlation minimally [24]. Although stronger watermarks may sound like a solution to the BPM attack, high watermark amplitudes cannot be accepted because of two reasons. First, the requirement for high-fidelity marked content and, second, strong watermarks can be efficiently estimated using an optimal watermark estimator [5]; that is, the estimate $\mathbf{v} = \text{sign}(\mathbf{x} + \mathbf{w})$ makes one error per bit, $\varepsilon = \Pr[v_i \neq w_i] = \frac{1}{2}\mathrm{erfc}(\sigma_x/\delta\sqrt{2})$ exponentially proportional to the watermark amplitude δ.

BLIND-PATTERN-MATCHING ATTACK FOR IMAGES

The BPM attack for still images exploits self-similarity within a given image or a library of images. Self-similarity can be seen as a particular kind of redundancy. Usually, correlation between neighboring pixels is taken into account. With self-similarities, it is the correlation between different parts (more or less spaced) of the image that is of interest. This idea has already been used with success for fractal compression [30].

The basic idea of the attack consists in substituting some parts of the picture with some other parts of itself (or even from an external codebook), which are perceptually similar. This process is depicted in Figure 18.9 and explained in the next subsection. The objective is to approximate the watermarked signal while keeping clear the cover signal. Even though self-similarity can be realized in various transform domains (DCT [31], wavelets), we restrict our presentation here to an implementation of the BPM attack in the spatial domain.

Image Processing for the BPM Attack

In the spatial domain, the original image is scanned block by block. Those blocks are labeled *range blocks* (block R_i) and have a given dimension $n \times n$. Each block R_i is then associated with another block D_i, which looks similar (modulo a pool of possible photometric and geometric transformations) according to a root mean square (RMS) metric defined by

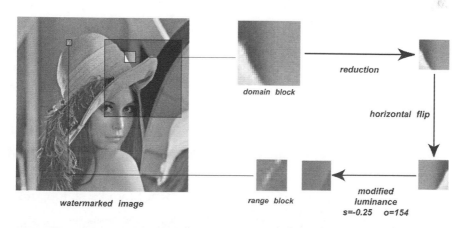

Figure 18.9. **Illustration of the process of searching for similar blocks within the original marked image and its transforms.**

$$\text{RMS}\ (f,g) = \frac{1}{n} \sqrt{\sum_{x=1}^{n} \sum_{y=1}^{n} \left(f(x,y) - g(x,y) \right)^2} \qquad (18.5)$$

The block $\mathbf{D_i}$ is labeled the *domain block* and is searched in a codebook containing Q blocks $\mathbf{Q_j}$. Those blocks may be blocks from the same image or from an external database. In practice, for a given range block $\mathbf{R_i}$, a window is randomly selected in the image. The blocks belonging to this window provide the codebook. Each block $\mathbf{Q_j}$ is scaled if needed in order to match the dimensions of the range block $\mathbf{R_i}$. A set of $|T|$ geometrically transformed blocks $\tau_k(\mathbf{Q_j})$ is then built (identity, four flips, three rotations). For each transformed block $\tau_k(\mathbf{Q_j})$, the photometric scaling $\mathbf{s}$ and offset $\mathbf{o}$ is computed by minimizing the error between the transformed block $g = \tau_k(\mathbf{Q_i})$ and the range block $f = \mathbf{R_i}$ by the least mean square method:

$$R = \sum_{x=1}^{n} \sum_{y=1}^{n} \left(\mathbf{s} \cdot g(x,y) + \mathbf{o} - f(x,y) \right)^2 \qquad (18.6)$$

Eventually, the transformed block $\mathbf{s} \cdot \tau_k(\mathbf{Q_j}) + \mathbf{o}$, which has the lowest RMS distance with the range block $\mathbf{R_i}$, is found and the corresponding block $\mathbf{Q_j}$ will be the domain block $\mathbf{D_i}$ associated with the range block $\mathbf{R_i}$. Because the two blocks $\mathbf{R_i}$ and $\mathbf{D_i}$ look similar, we can substitute $\mathbf{R_i}$ with the transformed version of $\mathbf{D_i}$.

In order to satisfy the restriction from Equation 18.3, for each range block we search for the transformed block $\mathbf{s} \cdot \tau_k(\mathbf{Q_j}) + \mathbf{o}$ that has the lowest RMS distance with the range block $\mathbf{R_i}$ *above the threshold* α. Here we use a new strategy for determining the value of the α threshold. In order to have an image-dependent threshold, α is chosen in such a way that a given percentage p of the range blocks is not optimally substituted. As a result, two search iterations are needed. In the first iteration, the threshold is set to zero and the cumulative histogram of the errors between the range blocks and the domain blocks is built. The adaptive threshold is then determined in order to interfere with $p\%$ of the substitutions during the second iteration.

The new algorithm is likely to introduce visible artifacts. In order to prevent this effect, two constraints have been added:

- Only a given part of the domain block is substituted with the range block. In our case, we used a circular mask inscribed in the block.
- Overlapping range blocks have been used. Consequently, specific care must be taken during the reconstruction. A simple substitution

is not any more pertinent. Instead, the domain blocks are accumulated in a temporary image and, at the end, each pixel value is divided by the number of blocks that contribute to the value of this pixel.

Effect of the Attack on Watermark Detection

This attack has been tested with three publicly available watermarking tools that offer roughly the same capacity (a few bits). A wide range of color images have been tested, although we only report the results with Lena in this chapter. Moreover, we made the assumption that the attacker knows in which color channel, the watermark is embedded.

In the first experiment, the BPM attack has been tested against the **D********* watermarking algorithm. We find out that around 60% of block associations need to be disturbed in order to remove the watermark in all the tested images. The image quality, as it can be seen in Figure 18.10, which presents the original, the marked, and the attacked image, is relatively undisturbed compared to the original and the marked version. Visually, one can notice that only the textured areas are slightly affected. The peak signal-to-noise ratio (PSNR) and weighted PSNR (wPSNR)[3] are equal to 40.32 dB and 53.90 dB, respectively, between the original image and its watermarked version, whereas they equal 35.67 dB, and 51.54 dB, respectively, between the watermarked image and its attacked version. As a result, the BPM attack on this watermarking technology can be considered successful.

In a second experiment, the **S***I**** watermarking algorithm has been put to the test. The watermark is strongly embedded in the **B** channel of the RGB color space [32]. In order to remove the watermark, we need to disturb 99% of the block associations. This results in a strong degradation of the blue channel. However, this degradation is quite invisible because the human eye is less sensible to the blue channel, as can be seen in Figure 18.11, which shows the luminance of the attacked image. The PSNR and wPSNR equal 49.05 dB and 59.73 dB, respectively, between the original image and its watermarked version, whereas they equal 46.52 dB and 59.24 dB, respectively, between the watermarked image and its attacked version. Once again, the BPM attack is a success.

In the last experiment, we tested **S***S*****. The watermark seems to be mainly embedded in the channel **Y** of the color space YUV. It has been determined experimentally that 92% of the block associations have to be disturbed in order to remove the watermark [7]. This results in strong visible artifacts, as can be seen in Figure 18.12. At least for the moment,

---

[3]The PSNR and the wPSNR are computed using the Y channel only of the YUV color space.

Original   Marked   Attacked

Figure 18.10. Attack against D*******. The figure illustrates, from left to right, the original image, the image marked with D******* watermarking software, and the image produced by the BPM attack.

Original   Marked   Attacked

Figure 18.11. Attack against S***I**. The figure illustrates, from left to right, the original image, the image marked with S***I** watermarking software, and the image produced by the BPM attack.

Original   Marked   Attacked

Figure 18.12. Attack against S***S***. The figure illustrates, from left to right, the original image, the image marked with S***S*** watermarking software, and the image produced by the BPM attack.

the simple version of the BPM attack is a failure of this watermarking algorithm.

The presented BPM attack for images has been partially integrated into the Stirmark benchmark v4.0 [17]. The reader is encouraged to download the software and explore new possibilities with the provided tools.

CONCLUDING REMARKS

For any watermarking technology and any type of content, one powerful attack is to rerecord the original content (e.g., perform again the music or capture the image of the same original visual scene). The BPM attack emulates this "perfect" attack using a computing system: The BPM attack aims at replacing small pieces of the marked content with perceptually similar but unmarked substitution blocks created using a library of multimedia content. The hope is that the substitutions have little correlation with the original embedded mark. In this chapter, we presented several versions of the basic BPM attack applied to both audio and images. Richer media such as video is likely to provide significantly higher level of redundancy in its content, which is the ultimate prerequisite for a successful BPM attack.

For those readers who hope to provide watermarking technologies robust with respect to the BPM attack, we identify two possible prevention strategies against it. For example, a data hiding primitive may identify unique parts of the content at watermark embedding time and mark only these blocks. However, this process increases dramatically the complexity of the embedding process while providing little guarantee that similar content cannot be found elsewhere. In the case of spread spectrum watermarks, increased detector sensitivity may enable watermark detection at lower thresholds. Unfortunately, such a solution comes at the expense of having longer watermarks that are prohibitive because of significantly lowered robustness with respect to classic desynchronization attacks such as Stirmark [17].

REFERENCES

1. The Secure Digital Music Initiative, http://www.sdmi.org.
2. The DVD Copy Control Association, http://www.dvdcca.org.
3. Architecture of Windows Media Rights Manager, http://www.microsoft.com/windows/windowsmedia/wm7/drm/architecture.aspx.
4. Rivest, R.L., Shamir, A., and Adleman, L.A., A method for obtaining digital signatures and public-key cryptosystems, *Commn. ACM*, 21(2), 120–126, 1978.
5. Kirovski, D., Malvar, H., and Yacobi, Y., A dual watermarking and fingerprinting system, *ACM Multimedia*, 2002, pp. 372–381.

6. Boneh, D. and Shaw, J., Collusion-secure fingerprinting for digital data, *IEEE Trans. Inf. Theory*, 44(5), 1897–1905, 1998.
7. Rey, C., Doeer, G., Dugelay, J.-L., and Csurka, G., Toward Generic Image Dewatermarking, presented at IEEE International Conference on Image Processing, 2002.
8. Kirovski, D. and Petitcolas, F.A.P., Blind Pattern Matching Attack on Audio Watermarking Systems, presented at IEEE International Conference on Acoustics, Speech, and Signal Processing, 2002.
9. Kirovski, D. and Petitcolas, F.A.P., Replacement Attack on Arbitrary Watermarking Systems, presented at ACM Workshop on Digital Rights Management, 2002.
10. Linnartz, J.P. and van Dijk, M., Analysis of the Sensitivity Attack Against Electronic Watermarks in Images, presented at Information Hiding Workshop, 1998.
11. Kalker, T., Linnartz, J., and van Dijk, M., Watermark Estimation Through Detector Analysis, presented at IEEE International Conference on Image Processing, 1998, pp. 425–429.
12. Langelaar, G.C., Lagendijk, R.L., and Biemond, J., Removing Spatial Spread Spectrum Watermarks by Non-linear Filtering, presented at European Signal Processing Conference, 1998, pp. 2281–2284.
13. Kutter, M., Voloshynovskiy, S., and Herrigel, A., The Watermark Copy Attack, presented at Security and Watermarking of Multimedia Contents II, SPIE, 2000, pp. 371–380.
14. Voloshynovskiy, S., Pereira, S., Herrigel, A., Baumgartner, N., and Pun, T., Generalised watermarking attack based on watermark estimation and perceptual remodulation, presented at Security and Watermarking of Multimedia Contents II, SPIE, 2000, pp. 358–370.
15. Mihcak, M.K., Kesal, M., and Venkatesan, R., Crypto-analysis on Direct Sequence Spread Spectrum Methods for Signal Watermarking and Estimation Attacks, presented at Information Hiding Workshop, 2002.
16. Kirovski, D. and Malvar, H., Spread spectrum watermarking of audio signals, *IEEE Trans. Signal Process.*, 51(4), 1020–1033, 2003.
17. Petitcolas, F.A.P., Anderson, R.J., and Kuhn, M.G., Attacks on Copyright Marking Systems, presented at Information Hiding Workshop, 1998, pp. 218–238, available at http://www.cl.cam.ac.uk/~fapp2/watermarking.
18. Kutter, M. and Petitcolas, F.A.P., A Fair Benchmark for Image Watermarking Systems, presented at Security and Watermarking of Multimedia Contents, SPIE, 1999, pp. 226–239.
19. Anderson, R.J. and Petitcolas, F.A.P., On the limits of steganography, *IEEE J. Selected Areas Commn.*, 16, 474–481, 1998.
20. Kirovski, D. and Malvar, H., Robust Covert Communication Over a Public Audio Channel Using Spread Spectrum, presented at Information Hiding Workshop, 2001, pp. 354–68.
21. Briassouli, A. and Moulin, P., Detection-theoretic Anaysis of Warping Attacks in Spread-Spectrum Watermarking, presented at IEEE International Conference on Acoustics, Speech, and Signal Processing, 2003.
22. Cox, I.J., Kilian, J., Leighton, T., and Shamoon, T., A Secure, Robust Watermark for Multimedia, presented at Information Hiding Workshop, 1996, 183–206.
23. Chen, B. and Wornell, G.W., Quantization index modulation: a class of provably good methods for digital watermarking and information embedding, *IEEE Trans. Inf. Theory*, 47(4), 1423–1443, 2001.
24. Kirovski, D. and Petitcolas, F.A.P., Blind pattern matching attack on watermarking systems, *IEEE Trans. Signal Process.*, 51(4), 1045–1053, 2003.
25. Oppenheim, A.V. and Schafer, R.W., *Discrete-Time Signal Processing*, Prentice-Hall, Englewood Cliffs, NJ, 1989.

26. Garey, M.R. and Johnson, D.S., *Computers and Intractability*, W.H. Freeman, San Francisco, 1979.
27. Kirovski, D. and Landau, Z., Randomizing the Replacement Attack, presented at IEEE International Conference on Acoustics, Speech, and Signal Processing, 2004.
28. Cohen, H., A Course in Computational Algebraic Number Theory, Springer-Verlag, New York, 1993.
29. Malvar, H., A Modulated Complex Lapped Transform and Its Application to Audio Processing, presented at IEEE International Conference on Acoustics, Speech, and Signal Processing, 1999.
30. Fisher, Y., Fractal Image Compression: Theory and Application, Springer-Verlag, New York, 1995.
31. Barthel, K.-U., Schüttemeyer, J., and Noll, P., A New Image Coding Technique Unifying Fractal and Transform Coding, presented at IEEE International Conference on Image Processing, 1994.
32. Kutter, M., Jordan, F., and Bossen, F., Digital Signature of Color Images Using Amplitude Modulation, in Proceedings of Electronic Imaging, 1997.

19

Digital Media Fingerprinting
Techniques and Trends

William Luh and Deepa Kundur

INTRODUCTION

The ease at which digital data can be exactly reproduced has made piracy, the illegal distribution of content, attractive to pirates. As illegal copies of digital data, such as video, and audio proliferate over the Internet, an emerging interest in protecting intellectual property and copyright material has surfaced. One such method of protecting copyright material is called *digital fingerprinting*. Although other means of digital data security exist, fingerprinting aims to deter pirates from distributing illegal copies, rather than actively preventing them from doing so. Fingerprinting is a method of embedding a unique, inconspicuous serial number (fingerprint) into every copy of digital data that would be legally sold. The buyer of a legal copy is discouraged from distributing illegal copies, as these illegal copies can always be traced back to the owner via the fingerprint. In this sense, fingerprinting is a *passive* form of security, meaning that it is effective after an attack has been applied, as opposed to *active* forms of security, such as encryption, which is effective from the point it is applied to when decryption takes place.

In fingerprinting a given dataset, all legal copies of the digital data are "similar," with the exception of the unique imperceptible fingerprints; that is, all copies of the digital data appear to be visually or audibly indifferent. A coalition of pirates can, therefore, exploit this weakness, by comparing their digital data looking for differences and then possibly detecting as well as rendering the fingerprint unreadable. Such an attack is known as *collusion*. One goal of fingerprinting is to ensure that some part of the fingerprint is capable of surviving a collusion attack, so as to identify at least one of the pirates.

For multimedia, an additional extra level of protection is required. In general, multimedia can withstand some amount of (single) user-generated distortion, such that the modifications are imperceptible to the enduser. A coalition of pirates might add noise to the multimedia in addition to their collusion attack. Hence, a fingerprinting scheme for multimedia should also be robust to some amount of user-generated distortion. Examples of user-generated distortions are additive white Gaussian noise, linear filtering such as blurring with Gaussian or Laplacian point spread functions, JPEG compression, geometric distortions such as cropping and resizing, and many other attacks found in References 1 and 2. Fingerprinting has the goal of traceability; hence, fingerprinting for multimedia should be robust to both collusion as well as user-generated distortions.

The scope of this chapter encapsulates symmetric fingerprinting as well as fingerprinting in a broadcast channel environment, which involves sending one encrypted copy of the multimedia to several end users, where symmetric fingerprinting at decryption takes place. A formal presentation of symmetric fingerprinting as well as fingerprinting in a broadcast channel environment is presented, and a survey of selected articles follows.

THE DIGITAL FINGERPRINTING PROBLEM

The problem of digital fingerprinting consists of many components that can be formulated separately. Figure 19.1 shows a breakdown of the overall problem into smaller problems, presented as the leaves of the tree diagram. The problem of designing a fingerprinting scheme consists of designing a fingerprinting code and a watermarking scheme used to embed the code into multimedia as discussed in the next sub-sections. Fingerprinting in a broadcast channel environment is described in a later section.

Fingerprinting Code

A fingerprinting code consists of a codebook and a tracing algorithm.

Definition 1: A *codebook* is a set $\Gamma = \{\gamma^1, \gamma^2, \ldots, \gamma^M\} \subseteq \Sigma^l := \{s_1 s_2 \cdots s_l | s_i \in \Sigma\}$, of M codewords of length l, over some finite alphabet Σ. Any subset $\Gamma' \subset \Gamma$ is also a valid codebook. Also, γ^i can be written as $\gamma^i = \gamma_1^i \gamma_2^i \cdots \gamma_l^i$.

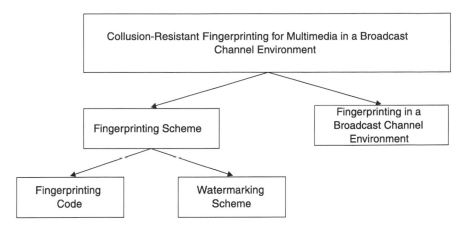

Figure 19.1. **Components of the digital fingerprinting problem.**

Definition 2: A *tracing algorithm* A is a function $\Sigma^l \mapsto 2^{\Sigma^l} \setminus \{\emptyset\}$, where 2^{Σ^l} is the power set of Σ^l.

In a fingerprinting scheme, each codeword from Γ is assigned to a different user. The goal of a malicious coalition of users is to combine their codewords, producing a new codeword η, such that η cannot be traced back to the coalition.

Definition 3: A *successful collusion attack* by a coalition of M users with codewords in Γ is a function Z such that $\eta = Z(\Gamma) \in \Sigma^l$ and $A(\eta) \notin 2^{\Gamma} \setminus \{\emptyset\}$.

While the adversaries of fingerprinting must design a successful attack function Z, the advocates of fingerprinting must design a codebook, and a tracing algorithm to counter collusion attacks.

Definition 4: The triple (Γ, A, Θ) consisting of codebook Γ of cardinality M, tracing algorithm A, and a family of attack functions Θ is said to be *M-collusion resistant to* Θ with ε error if the following properties are satisfied: *Given* $\eta = Z(\Gamma')$ for $Z \in \Theta$ and $\Gamma' \subseteq \Gamma, \Pr\left[A(\eta) \in 2^{\Gamma'} \setminus \{\emptyset\}\right] > 1 - \varepsilon$.

Note, Definition 4 also resists framing of innocent users, as the tracing algorithm A will produce a subset of codewords from Γ', the codebook belonging to the colluders. A typical family of attack functions is the family of Marking Assumption attacks [3], which will be discussed in section "Fingerprinting under the Marking Assumption."

Once the fingerprinting code is designed, a means to imperceptibly embed it into multimedia is necessary. The technique of embedding codewords imperceptibly into multimedia is called *modulation and water-marking*, as defined in the next subsection.

Watermarking Scheme

A digital watermark is an imperceptible mark, such as a logo, that is embedded into multimedia. To quantify the meaning of impercepti-bility, a metric $d(\cdot, \cdot) \geq 0$ is defined to measure the similarity between any two multimedia c and $\tilde{c}_i$. Let $T > 0$ be a threshold variable such that $d(c, \tilde{c}_i) \gg T$ implies c and $\tilde{c}_i$ are not visually similar and $d(c, \tilde{c}_i) < T$ implies they are visually similar. For example, $d(c, \tilde{c}_i)$ can represent the Euclidean distance, or mean square error (MSE), between c and $\tilde{c}_i$. The threshold T is a factor of the human visual system for images and video or a factor of the human audio system for audio.

A watermarking algorithm takes a host multimedia, a watermark, a key, and embeds the watermark into the host in an imperceptible manner.

Definition 5: A *watermarking algorithm* is a function $\omega : C \times W \times L \rightarrow C$, where C is a set of multimedia, W is a set of watermarks, and L is a set of keys. In addition, for any $c \in C$, $w \in W$, and $l \in L$, $\omega(c, w, l)$ is the watermarked multimedia, such that $d(\omega(c, w, l), c) < T$.

Watermarks are extracted from watermarked multimedia via an extraction function X.

Definition 6: A *blind watermark extraction function X*, is a function $C \times L \mapsto W$, such that $X(\omega(c, w, l), l) = w$.

In addition to the imperceptibility requirements of a watermarking algorithm, $\omega(c, w, l)$ should also be robust to some attacks.

Definition 7: A *feasible attack F* on watermarked multimedia, given w, but not l, is a function $C \mapsto C$, such that $d(F(\omega(c, w, l)), c) < T$.

The attack defined in Definition 7 allows the attacker to have knowledge of the watermarking algorithm but not the key, which is known as Kerckhoff's Principle in cryptography [4]. In addition, the attack should have imperceptible effects on the watermarked multimedia. A blind watermarking scheme is said to be robust to a particular attack if the extraction function is able to extract the watermark after that particular attack.

Definition 8: A blind watermarking scheme that is strictly robust to a feasible attack F is a *triple* (ω, X, F), such that $X(F(\omega(c, w, l)), l) = w$. A blind watermarking scheme that is τ-robust to a feasible attack F is a *triple* (ω, X, F), such that $X(F(\omega(c, w, l)), l) = \tilde{w}$ and $d(\tilde{w}, w) < \tau$.

An example of a feasible attack is additive white Gaussian noise (AWGN). Here, F is AWGN with noise power σ^2, defined as

$$F(\omega(c, w, l), l) = \omega(c, w, l) + n \qquad (19.1)$$

n is a noisy video whose frames are Gaussian random processes, with a constant power spectral density (PSD) at σ^2. The attack in Equation 19.1 is made feasible by controlling σ^2.

In order to integrate the fingerprinting code into a watermarking scheme, a modulation function is required to map the codewords into unique watermarks. A demodulation function is used to map watermarks back to codewords.

Definition 9: A modulation function $m : \Gamma \to W$ is a *bijection*, and its inverse function is the demodulation function $m^{-1} = B : W \to \Gamma$.

Definition 9 ties together the fingerprinting code and the watermarking scheme, thus completing the problem formulation of the fingerprinting scheme. The modulation and watermarking process, along with its inverse operations, are depicted in Figure 19.2. In the next subsection, we discuss possible attacks for which we must effectively design fingerprints.

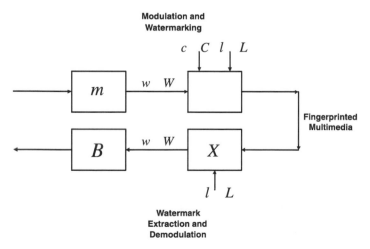

Figure 19.2. Modulation and watermarking and inverse operations.

Attacks

As discussed in the previous subsections, attacks on the fingerprinted multimedia can be classified as either single-user attacks or collusion attacks directly on the multimedia.

Single-User Attacks. Feasible attacks on multimedia as defined in Definition 7 are now presented. These attacks involve one copy of the multimedia in question and can be categorized into *unintentional attacks* and *intentional attacks* [2]. Unintentional attacks are those that occur due to bandwidth constraints, such as lossy copying and transcoding (i.e., compression, change in frame rate, format conversion, conversion in display format). Intentional attacks are those user-generated attacks that aim to remove the watermark or fingerprint in the multimedia. Intentional attacks on video can be categorized into *single-frame attacks* and *statistical attacks* [2]. Single-frame attacks can be categorized into *signal processing attacks* (i.e., band-pass filtering, adaptive Wiener denoising [2], etc.) and *desynchronizing attacks* (i.e., affine transformations, scaling, cropping, etc.) [2]. Statistical attacks for video are sometimes also called collusion. However, there is only one copy of the video in question and the term arises from the fact that consecutive frames in the video are used together to remove the watermark or fingerprint. A simple statistical attack on video is to average a small set of consecutive frames in hopes that this will remove the watermark. A more sophisticated statistical attack is to first estimate the watermark in each individual frame and then average the estimated watermarks to obtain a final estimation, which is then subtracted from each frame.

Section "Multimedia Collision" presents collusion attacks directly on multimedia, by comparing several copies of the same video with different fingerprints embedded. The attacks involve combining matching frames from each copies as opposed to consecutive frames from one copy.

General Notation. Although we focus at times on video fingerprinting to elucidate concepts, much of our insights can be generalized to other types of digital media. Let c represent a plaintext video. The variable $\hat{c}$ is the encrypted version of c (denoted by the caret), and $\tilde{c}_i$ is a plaintext version of c (denoted by the tilde) with an imperceptible fingerprint for User i denoted by a subscript i. For example, the codeword $\gamma^i \in \Gamma$ is modulated and watermarked into c. Furthermore, the video can be broken down according to frame and pixel index, $c^j(x,y)$, where the superscript j denotes the frame number and (x,y) denotes the pixel index. In the next subsection, collusion applied directly on the video will be introduced, and $\underline{c}$ (using the underscore) is the notation used to denote a video obtained by collusion via a set of fingerprinted videos $\{\tilde{c}_i\}$. At times, a probabilistic approach is required, and the pixels in a video can

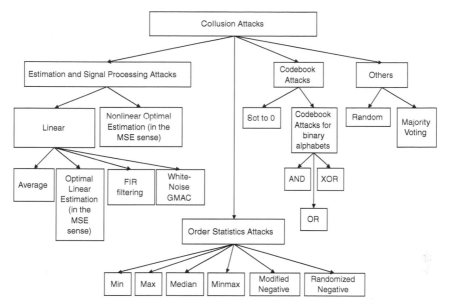

Figure 19.3.   Collusion attacks.

be regarded as a random variable. A random variable is denoted by the uppercase version [i.e., $\tilde{C}_i^j(x, y)$]. Random variables are required when the probability density function is used; otherwise, a particular sample or outcome denoted by the usual lowercase is used.

Multimedia Collusion. Up to this point, collusion has only been addressed for codebooks as in Definition 3. This subsection will give examples of attacks that can be applied to the codebook and attacks that are applied directly to the multimedia. Figure 19.3 presents an example of some collusion attacks. It is assumed User 1 to User M collude without loss of generality, where $M \leq |\Gamma|$.

In general, attacks that are strictly applied to codebooks are not practically useful for multimedia fingerprinting because they do not effectively model collusion directly applied to media files; however, they are discussed because these attacks can serve as a basis for designing collusion-resistant codes, as in References 5–7. If Σ is binary, such as $\Sigma = \{0, 1\}$, any binary operator such as AND, OR, and XOR can be used as attacks. In the Set to 0 attack, $\eta = \eta_1\eta_2\cdots\eta_l$ is the codeword created by colluders, where $\eta_i = 0$ if $\gamma_i^j \neq \gamma_i^k$ for some $j \neq k$, otherwise $\eta_i = \gamma_i^j$ where $\gamma_i^1 = \gamma_i^2 = \cdots = \gamma_i^M$. This attack sets any differing bits (in the case of a binary alphabet) to 0 while preserving bits that are the same across all users. This attack can be modified so that differing bits are always set to one alphabet that appears in the colluders' codewords.

Majority Voting, which can also be a codebook attack, is defined in Equation 19.2 using the MODE operation from statistics, which chooses the most common value from a set:

$$\underline{c}^j(x, y) = \text{MODE}\left(\tilde{c}_1^j(x, y), \tilde{c}_2^j(x, y), \ldots, \tilde{c}_M^j(x, y)\right) \tag{19.2}$$

The Random attack selects pixels from the fingerprinted multimedia with equal probability. Equation 19.3 defines the Random attack using the UNIFORM-RANDOM operation, which selects the indices 1 to M with equal probability:

$$\underline{c}^j(x, y) = \tilde{c}_{\text{UNIFORM-RANDOM}(1, 2, \ldots, M)}^j(x, y) \tag{19.3}$$

From the Estimation and Signal Processing Attacks category, the Non-linear Optimal Estimation (in the MSE sense), given frames in $\{\tilde{c}_i\}_{i=1}^M$, is

$$\underline{c}^j(x, y) = \text{E}\left[c^j(x, y) | \tilde{c}_1^j(x, y), \tilde{c}_2^j(x, y), \ldots, \tilde{c}_M^j(x, y)\right]$$
$$= h\left(\tilde{c}_1^j(x, y), \tilde{c}_2^j(x, y), \ldots, \tilde{c}_M^j(x, y)\right)$$

where $\underline{c}^j(x, y)$ is an estimate of the jth frame at the (x,y)th pixel of the respective frame and pixel in the video c with no fingerprint, and $h(\tilde{c}_1^j(x, y), \tilde{c}_2^j(x, y), \ldots, \tilde{c}_M^j(x, y))$ is given in

$$h\left(\tilde{c}_1^j(x, y), \tilde{c}_2^j(x, y), \ldots, \tilde{c}_M^j(x, y)\right)$$
$$= \int_{-\infty}^{+\infty} z \frac{f_{C^j(x,y), \tilde{C}_1^j(x,y), \tilde{C}_2^j(x,y), \ldots, \tilde{C}_M^j(x,y)}\left(z, \tilde{c}_1^j(x, y), \tilde{c}_2^j(x, y), \ldots, \tilde{c}_M^j(x, y)\right)}{f_{\tilde{C}_1^j(x,y), \tilde{C}_2^j(x,y), \ldots, \tilde{C}_M^j(x,y)}\left(\tilde{c}_1^j(x, y), \tilde{c}_2^j(x, y), \ldots, \tilde{c}_M^j(x, y)\right)} dz \tag{19.4}$$

$f_{C^j(x,y), \tilde{C}_1^j(x,y), \tilde{C}_2^j(x,y), \ldots, \tilde{C}_M^j(x,y)}$ is the joint probability density function (pdf) of the jth frame, (x,y)th pixel random variables of c, and $\{\tilde{c}_i\}_{i=1}^M$. $f_{\tilde{C}_1^j(x,y), \tilde{C}_2^j(x,y), \ldots, \tilde{C}_M^j(x,y)}$ is the joint of the jth frame, (x, y)th pixel random variables $\{\tilde{c}_i\}_{i=1}^M$. To be practical, Equation 19.4 requires that the pdfs be somehow estimated or known. Colluders must therefore resort to more practical techniques, such as those found in the Linear category. The simplest Linear attack averages the set of fingerprinted multimedia, as in

$$\underline{c}^j(x, y) = \frac{1}{M} \sum_{i=1}^M \tilde{c}_i^j(x, y) \tag{19.5}$$

As noted in Reference 5, Equation 19.5 is fair; all members of the coalition contribute equally, hence no member is at a higher risk of being caught than his colleagues.

Another linear attack that is more optimal in the MSE sense than Equation 19.5 is given by

$$\underline{c}^j(x,y) = \sum_{i=1}^{M} \alpha_i \left(\tilde{c}_i^j(x,y) - E\left[\tilde{C}_i^j(x,y) \right] \right) + E[C^j(x,y)] \qquad (19.6)$$

The weights α_i in Equation 19.6 are obtained by solving

$$E\left[\left(C^j(x,y) - E[C^j(x,y)] - \sum_{i=1}^{M} \alpha_i \left(\tilde{C}_i^j(x,y) - E\left[\tilde{C}_i^j(x,y) \right] \right) \right) \right.$$

$$\left. \left(\tilde{C}_k^j(x,y) - E\left[\tilde{C}_k^j(x,y) \right] \right) \right] = 0 \qquad (19.7)$$

for $k = 1, 2, \ldots, M$.

Although Equation 19.7 may seem intimidating, it is a linear equation of M equations (for $k = 1, 2, \ldots, M$) and M unknowns (α_i, for $i = 1, 2, \ldots, M$). The expected values are more easily estimated than the pdfs in Equation 19.4. The Fourier infrared (FIR) filtering attack averages $\{\tilde{c}^j(x,y)\}_{j=1}^{M}$, then applies a FIR filter as in Equation 19.8.

$$\underline{c}^j = h(x,y) \otimes \frac{1}{M} \sum_{i=1}^{M} \tilde{c}_i^j(x,y) \qquad (19.8)$$

$h(x,y)$ is an FIR two-dimensional (2-D) spatial filter and $\otimes$ is the 2-D convolution operator. The goal of Equation 19.8 is to further attenuate the fingerprint by filtering the average. Additive noise can be added to Equation 19.8, as in the Gaussian medium access channel (GMAC) [8].

The Order Statistic attacks found in Reference 9, consist of the min, max, median, minmax, modified negative, and randomized negative defined in Equations 19.9 to Equation 19.14 respectively:

$$\underline{c}^j(x,y) = \min\left(\tilde{c}_1^j(x,y), \tilde{c}_2^j(x,y), \ldots, \tilde{c}_M^j(x,y) \right) \qquad (19.9)$$

$$\underline{c}^j(x,y) = \max\left(\tilde{c}_1^j(x,y), \tilde{c}_2^j(x,y), \ldots, \tilde{c}_M^j(x,y) \right) \qquad (19.10)$$

$$\underline{c}^j(x,y) = \text{median}\left(\tilde{c}_1^j(x,y), \tilde{c}_2^j(x,y), \ldots, \tilde{c}_M^j(x,y) \right) \qquad (19.11)$$

$$\underline{c}^j(x,y) = \frac{1}{2} \left(\min\left(\tilde{c}_i^j(x,y) \right) + \max\left(\tilde{c}_i^j(x,y) \right) \right) \qquad (19.12)$$

$$\underline{c}^j(x,y) = \min\left(\tilde{c}_i^j(x,y) \right) + \max\left(\tilde{c}_i^j(x,y) \right) - \text{median}\left(\tilde{c}_i^j(x,y) \right) \qquad (19.13)$$

$$\underline{c}^j(x,y) = \begin{cases} \min\left(\tilde{c}_i^j(x,y)\right) & \text{with probability } p \\ \max\left(\tilde{c}_i^j(x,y)\right) & \text{with probability } 1-p \end{cases} \quad (19.14)$$

The Digital Fingerprinting Problem

The above subsections formulated the problem piece by piece. The complete problem is stated in this subsection. The goal is to design a transmitter (Γ, m, ω) and a receiver (A, B, X). (Γ, A) is M-collusion resistant with ε error to common attack families such as the family of Marking Assumption attacks. In addition, the entire system should be robust to higher-level attacks on the multimedia, such as those in Figure 19.3. The triplet (ω, X, F) should be τ robust to common feasible watermarking attacks, such as $F = \text{AWGN}$, for $\sigma^2 < \tau$.

PREVIOUS WORK ON FINGERPRINTING

The fingerprinting schemes discussed are known as symmetric fingerprinting. In symmetric fingerprinting, the buyer has no protection against a dishonest merchant whose intent is to frame the innocent buyer. This problem led to a solution known as asymmetric fingerprinting [10], allowing the buyer to avoid being framed by dishonest merchants. The major drawback in this scheme is the amount of personal information that the buyer needs to disclose to merchants. The solution to this problem is known as anonymous fingerprinting [11], where there is a third party devoted to collecting and keeping personal information private. These two fingerprinting schemes are protocol heavy, and some existing symmetric fingerprinting schemes can be modified to adopt these protocols, hence inheriting their frame-proof and privacy advantages, respectively. The reader is referred to the relevant references for further information. The focus of this chapter is on algorithmic solutions to digital fingerprinting.

Symmetric fingerprinting can also be classified by where in the content distribution chain the fingerprint is embedded. The majority of symmetric fingerprinting schemes embed the fingerprint at the end nodes (source and destination) and directly into the media content. Methods for embedding a fingerprint into multimedia can be found in the watermarking literature. However, some fingerprinting schemes, such as traitor tracing [12], do not embed the fingerprint in the actual content and may even make use of intermediate network components to mark the media. The following subsections will delve into specific instances of symmetric fingerprinting. We begin by discussing symmetric fingerprinting methods that make use of intermediate nodes. The advantage of these techniques is that they are more scalable (i.e., make use of less bandwidth) as the number of fingerprinted copies to be transmitted grow. However, to be effective, the intermediate nodes such as routers must be trustworthy and, hence, secure. We then discuss symmetric methods that assume that the fingerprint is embedded directly into the media and only required participation from the end nodes. The reader should note that, for the most part, the techniques are described using the same or similar notation and variables of the original articles.

Symmetric Fingerprinting through Intermediaries

In watercasting by Brown et al. [13], watermarked data packets are selectively dropped to form a fingerprint and, hence, provide an audit trail associating a media clip to a particular user. An active network topology for multicast communications is assumed in which routers are used for the fingerprinting process. The *depth* of the associated multicast distribution tree is assumed to be known and is denoted d. Instead of transmitting n distinct watermarked copies of the media from the source for each of the n users, m (where $d << m < n$) copies are transmitted, providing an immediate bandwidth and complexity savings from the standard approach of embedding the fingerprint at the source and making multicast efficiency difficult to exploit.

The original host media is packetized at the source. For each packet, m slightly different fingerprinted copies, called a transmission group, are generated such that their perceptual media quality is equivalent at any receiver. Each of the packets in the transmission group are encrypted and simultaneously transmitted into the multicast network. When the transmission group reaches the first router, the router pseudorandomly selects one of the m distinct packets to drop and forwards the others. Similarly, as the remaining packets continue downstream along the multicast distribution tree, the other intermediate routers also selectively drop packets. At each "last hop" router (which communicates directly with a receiver), only one packet is pseudorandomly forwarded. Brown et al. discussed that if the pseudorandom packet dropping selection is keyed with the position of the associated router, then the identity of the dropped packets within the overall media data stream serves as an effective *route marker*. This marker can be employed along with a secure *timestamp* associated with the data stream to identify the user to whom the media was transmitted.

The advantage of watercasting is its scalability with respect to bandwidth; multicast communications can be employed and little packet flooding is required. The challenges include making the system robust to unintentional packet dropping due to network congestion (setting $m \gg d$ can account for packet loss, but will trade off bandwidth), the vulnerability to router compromise that may interfere with the fingerprinting process, the overhead associated with adding packet selection and dropping functions into active routers, and high logging requirements because the state of the network during the entire session must be saved. Furthermore, to determine the user associated with a media clip, the entire session and communication transmission events must be resimulated in order to identify the exact route taken by the media stream.

Another approach called *watermarking multicast video with a hierarchy of intermediaries (WHIM)* [14] by Judge and Ammar is a scalable fingerprinting and encryption system that distributes the fingerprinting process to intermediate entities called *intermediaries*. The authors argue that watercasting provides security from the source to the last hop router, which may result in an ambiguous identification of users if there is more than one receiver connected to a last hop router. In contrast, WHIM provides the additional protection between each last hop intermediary and the associate receivers. Their system makes use of an overlay network in order to implement incremental embedding of the fingerprint over the intermediaries such that each receiver obtains a uniquely fingerprinted copy of the media.

The WHIM architecture has two components. The first part is the WHIM backbone, which is comprised of a set of secure servers that performs the function of incremental embedding of unique intermediary IDs to create an audit trail. Because of the hierarchical nature of the intermediaries, the overall fingerprint received by a user is unique and secure [14]. The second part is the WHIM Last Hop, which works between the last hop (or "leaf") intermediaries and their associated receivers. This system fingerprints a unique ID for each user, such as the IP address, and ensures secure transmission.

Thus, overall, the WHIM system can be employed to trace user's location in the overlay network while ensuring bandwidth efficiency because the traffic is multicast. The main disadvantage of the system is that each intermediary must perform a decryption, fingerprint generation and embedding, and reencryption, which introduces a high overhead, especially for applications involving video distribution. In addition to the computational load, the system is more vulnerable to attacks that compromise the intermediary nodes. Neither watercasting nor WHIM specify the embedding methodology required for fingerprinting, but state that it must be effective and robust.

In contrast, Wang et al. [15] focused on the application of tracing leaks of intelligence images. For this application, the authors argued that the lack of lossy compression on intelligence images coupled with their controlled distribution to a limited number of users reduces the robustness requirements of the embedding process. The embedding does not have to be robust to attacks such as JPEG nor does it have to be robust to collusion attacks involving hundreds or even tens of fingerprinted copies.

Their system is based on a controlled tree distribution structure for images in which the root node distributes the host image to a number of intermediate nodes that each fingerprint the received image and pass the result onto other intermediate nodes that, in turn, also embed a fingerprint. The fingerprinted images propagate down the tree, with additional fingerprints being embedded along the way up to the leaf nodes that represent the users of the data. The fingerprinted copies at the user nodes are kept and all other intermediate copies are destroyed in the process for added security.

At each distributing node, the fingerprints are securely generated by the receiving nodes, and the specific discrete wavelet transform used to embed the print and details of the insertion process is randomly selected by the distributing node. The embedding process occurs at a separate location of any node in the distribution tree, which then passes on the fingerprinted copy to the rightful receiver and destroys any copy. This gives the system an asymmetric quality for added protection because both the distributor and the receiver are needed to generate the fingerprinted image and neither party can frame the other. Although the properties of the approach are appropriate for intelligence image distribution, as discussed by the authors it is not suited for other media distribution applications such as video-on-demand due to robustness issues.

Symmetric Fingerprinting via Data Embedding

In contrast to the previous subsection, this subsection surveys fingerprinting that is performed at the end nodes by embedding the data directly into the media or decrypting certain parts of the encrypted media to render a fingerprinted copy.

Limitations. It is with no surprise that with a sufficient number of colluders, any embedding-based fingerprinting scheme can be defeated. In fact, under the framework of Reference 16 using the Euclidean distance as a measure of fidelity, at most $O(\sqrt{n/\ln n})$ colluders are required to render the fingerprinting scheme useless. Here, n represents the size of the data (i.e., number of pixels in an image). However, these findings do not necessarily suggest that the fingerprinting research exists in vein. For example, in Reference 6, subgroup-based constructions of anticollusion codes are used to group buyers suspected of likely colluding. For small groups, this implies that the colluders either have to recruit more colluders or purchase additional copies. Hence, the goal of fingerprinting is to make it more costly, but not necessarily impossible, for colluders.

589

Fingerprinting under the Marking Assumption. One of the very first articles analyzing the feasibility of collusion-secure codes is presented by Boneh and Shaw in Reference 3. A mark in digital data is a position that can take on a finite number of states, including an unreadable state. A collection of marks is defined to be a fingerprint. A mark is said to be detectable by a coalition of colluders if at least one mark differs in state from the other copies. The marking assumption states that only the detectable marks can be altered by colluders, whereas the undetectable marks cannot be altered.

Under the marking assumption, there are no totally collusion-resistant fingerprinting codes; that is, there exists a nonzero probability of missed detection, or probability of false alarm. However, this probability can be made arbitrarily small with the fingerprinting code (Γ, A) introduced here. For $|\Gamma| = n$, construct the $n \times (n - 1)$ matrix

$$
\begin{bmatrix}
1 & 1 & 1 & \cdots & 1 \\
0 & 1 & 1 & \cdots & 1 \\
0 & 0 & 1 & \cdots & 1 \\
0 & 0 & 0 & \ddots & \vdots \\
\vdots & \vdots & \vdots & \ddots & 1 \\
0 & 0 & 0 & \cdots & 0
\end{bmatrix}
$$

and each column is replicated $d = 2n^2 \log(2n/\varepsilon)$ times. A random permutation of the columns of the matrix is then performed. The n rows of this resulting matrix make up Γ and are assigned to n different users. The code is able to resist collusion up to n users with a false-alarm probability of ε/n.

The detection process is rather involved. Let B_m be the set of bit positions in which the first m users see a 1 and the rest see a 0, and the cardinality of B_m is d. Let $R_s = B_{s-1} \cup B_s$ for $s \in \{2, 3, \ldots, n-1\}$. Let weight$(x)$ be the number of 1's in x, where x is a binary string. Let $x_{\hat{S}}$ be the bits in x at positions given by the numbers in the set S. The tracing algorithm A operates on a binary string x produced by colluders:

1. If weight $(x_{\hat{B}_1}) > 0$, then output "User 1 is guilty."
2. If weight $(x_{\hat{B}_{n-1}}) < d$, then output "User n is guilty."
3. For all $s = 2$ to $n - 1$ do:

    Let $k = $ weight $(x_{\hat{R}_s})$.

    If

$$
\text{weight}\left(x|_{B_{s-1}}\right) < \frac{k}{2} - \sqrt{\frac{k}{2} \log \frac{2n}{\varepsilon}}
$$

    then output "User s is guilty."

Note that the unreadable marks are set to 0. Unreadable marks are those states that are neither 0 nor 1 after demodulation B.

The idea behind (Γ, A) is now explained. Suppose that User s is innocent. Then $\gamma^s_{|R_s} = 00 \cdots 0011 \cdots 11$. It is assumed that for large d, a random permutation of $\gamma^s_{|R_s}$ will result in the 1's being evenly distributed in both halves of $\gamma^s_{|R_s}$; that is, $\text{weight}\left(\gamma^s_{B_{s-1}}\right) \approx \text{weight}\left(\gamma^s_{B_s}\right)$. In addition, it is assumed that the attack does not change this distribution. Therefore, if the 1's are not evenly distributed in $x_{\hat{R}_s}$, then s is guilty, which is essentially the step 3 in A. The first two steps are easy to see. If User 1 is innocent, then marks in B_1 are undetectable and cannot be changed by the marking assumption to frame User 1. The same reasoning applies for User n.

A modulation and watermarking scheme is not described in this chapter. However, the following content distribution approach is abstractly presented. A plaintext object P of length L is partitioned in to l pieces. Each piece is embedded with a 0 and a 1, creating a pair of new pieces for each of the original L pieces. Each of the new $2L$ pieces is encrypted with a separate private key, and the $2L$ encrypted pieces are distributed to users. Each user is given a set of private keys that can only decrypt one of the two pairs. Figure 19.4 depicts the process of partitioning, embedding, and encrypting. Recall that each key is, in fact, a fingerprint, and given that the fingerprints are properly designed, keyholders that collude to create another key can be traced.

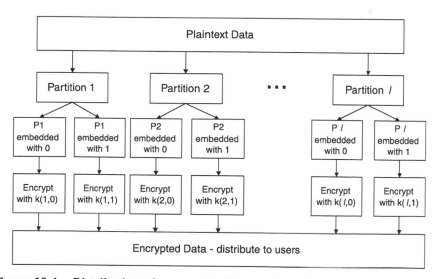

Figure 19.4.   Distribution of encrypted data — a broadcast channel approach. At the user end, each user is given a set of keys that will only decrypt the data embedded with a 0 or a 1. The keys are labeled so that the user end knows which of the ones to decrypt.

The codebook (Γ, A) is created from the binary alphabet $\Sigma = \{0, 1\}$, although the theory applies to q-ary alphabets. Other researchers have implementations for the q-ary alphabet as well as methods for embedding in multimedia, such as images [17]. The work of Reference 3 has also been extended to anonymous fingerprinting [18].

Projective Geometric Codes. In Reference 19, Dittmann et al. described a fingerprint as a collection of marking positions that are either marked with 1 or not marked, being equivalent to 0. The idea is to construct fingerprint codewords that intersect with other fingerprint codewords in unique ways. Assuming that the unique intersection point cannot be changed, as in the marking assumption, it will determine all of the colluders.

The concept of unique intersections has a nice geometric interpretation. For fingerprinting codes that can detect at most two colluders, the codewords that make up Γ can be represented by the edges on the triangle in Figure 19.5. Any two users have a unique intersection point — being a vertex of the triangle. Even if the users remove the detectable marks, the intersection will remain intact, revealing the identities of the two colluders. If the colluders do not remove all of the detectable marks (i.e. some leftover edge), then the leftover edge can be used to detect the colluders. Hence it is in the best interest of colluders to remove detectable marks. A possible attack that can cripple this system is to remove leftover edges but leave the vertices intact. As will be seen later, when the geometric shapes live in higher dimensions and the colluders do not know where a vertex may be embedded, it is difficult to generate this attack.

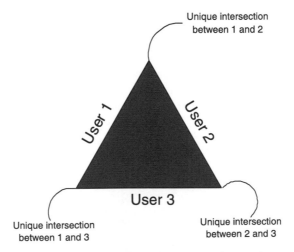

Figure 19.5. **Geometric interpretation of two-collusion-resistant codewords.**

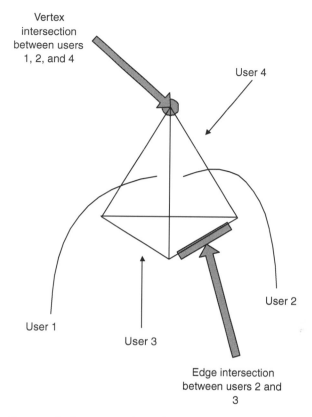

Figure 19.6. **Geometric Interpretation of three-collusion-resistant codewords.**

Figure 19.6 depicts a tetrahedron, where the sides represent the code-words that can detect at most three colluders. The four sides (planes) represent codewords for four users. When two users collude, they share a unique edge. When three users collude, they share a unique vertex.

For codewords that can detect at most four colluders, a geometric shape in four-dimensions is used. In general, codewords that can detect at most n colluders require shapes in n dimensions, or $O(n)$ bases are required. The hyperplanes of the higher-dimension "hypertetrahedron" represents the codewords. These hyperplanes are derived from finite projective spaces PG(d, q), where d is the dimension that the hyper-tetrahedron lives in and $q+1$ is the number of points on a hyperplane, where q is a prime number. PG(d, q) is constructed from a vector space of dimension $d+1$ over the Galois field GF(q). The details of this construction can be found in References 20 and 21.

The tracing algorithm A detects the marked points that have not been removed, and it determines the largest projective subspace spanned by these points. In the theory of projective geometry, hyperplanes can be described by a set of points that satisfy a linear equation. The coefficients of the linear equation are formed from any point in the subspace; hence, by choosing any point of the subspace and then solving the equation, a set of hyperplanes corresponding to fingerprints can be generated.

The modulation function m transforms the vectors of PG(d, q) to a binary string. The watermarking function W consists of the following steps:

1. The luminance channel of the image is split into 8×8 blocks, and a discrete cosine transform (DCT) is performed on each block.
2. The blocks are quantized with the quantization matrix given in Reference 19.
3. Each 8×8 quantized block will have one bit of the fingerprint binary string embedded into it. If the bit is a 1, a pattern is added to the block. If the bit is a 0, no pattern is added. Note that a different pattern is used for each block; however, all copies use the same block-pattern scheme.
4. The inverse operations of steps 2 and 1 are performed, resulting in a fingerprinted image.

The fingerprint is also embedded repeatedly, as redundancy adds an extra level of robustness. The demodulation function B requires the original unfingerprinted image. Basically, the 8×8 blocks are tested to see whether it is a 1 or 0.

Anticollusion Fingerprinting for Multimedia. The idea of using the unique intersection of codes to determine colluders is also used by Trappe et al. in References 5 and 6. Instead of taking a geometric approach, Γ is designed using the balanced incomplete block design (BIBD) from combinatorial design theory [22–24]. The drawback of the BIBD codes is that they are only resilient to the AND operator, which is not always an effective model of collusion.

The work of References 5 and 6 also examines various modulation formats for fingerprint embedding, and it identifies coded modulation, a form of combining orthogonal basis signals, as suitable for fingerprinting. The anti collusion codes (ACCs) are then used as the weights for combining orthogonal basis signals as follows:

$$\vec{w}_j = m(b_{1j}b_{2j}\cdots b_{vj}) = \sum_{i=1}^{v} b_{ij}\vec{u}_i \qquad (19.15)$$

In Equation 19.15, $\vec{w}_j$ is the watermark created via modulation for user j, $\{\vec{u}_i\}_{i=1}^{v}$ is a set of v orthogonal basis signals, and $b_{1j}b_{2j}\cdots b_{vj}$ is an ACC for user j, where $b_{ij} \in \{\pm 1\}$. ACCs have the property that the length $v = O(pn)$ and provides collusion resistance for up to n colluders. This is an improvement over the projective geometric codes requiring $O(n)$ bases. The watermark $\vec{w}_j$ is embedded into the multimedia using robust watermarking techniques such as in Reference 25.

The collusion attack is an equally weighted average of all copies. This attack is used because it is fair for all colluders [5]. Although this scheme is not the optimal estimate of the original host, all colluders have the same vulnerability of being caught. According to References 5 and 6 this attack has the effect of operating on the codewords using the AND operator. This assumption is usually only true for two colluders. However, by adjusting the threshold used in detection, the assumption can accommodate more colluders with a small probability of error (probability of false alarms and missed detections). Three detection processes are presented. The simplest detection scheme is known as hard detection, where the Z-statistic given in Equation 19.16 is compared to a fixed threshold

$$T(i) = \frac{\vec{y}^T \vec{u}_i}{\sqrt{\sigma^2 \left\| \vec{u}_i \right\|^2}} \tag{19.16}$$

$\vec{y}$ is the received vector that has been attacked by collusion and AWGN with variance $\sigma^2 \cdot T(i)$ is compared to a threshold τ_e that is determined empirically from simulations. The output is

$$T(i) \geq \tau_e, \quad r_i = 1$$
$$T(i) < \tau_e, \quad r_i = 0 \tag{19.17}$$

where $r = r_1 r_2 \cdots r_v$ is a bit string uniquely identifying the colluders and r should be the bit string resulting from the AND of the colluders' codewords. τ_e is chosen to minimize the probability of false alarms and the probability of missed detections, via simulation. The drawback of this scheme is that the empirically determined threshold τ_e is different for varying numbers of colluders; hence, simulations would be needed to select τ_e for more than two colluders.

In adaptive detection [7], the likelihood of $\vec{\Phi}$ is defined to be the conditional pdf

$$f\left(T(1), T(2), \ldots, T(v) \mid \vec{\Phi}\right) = N\left(\frac{\alpha_1}{K} B \vec{\Phi}, I_v\right)$$

where $\alpha_1 = \alpha \sqrt{\|\vec{u}\|^2 / \sigma_d^2}$ and K is the number of 1's in $\vec{\Phi}$. This is also the PDF of the Z-statistics under the alternate hypothesis that collusion has taken place. $\vec{\Phi}$ is a vector such that if $\Phi_j = 1$, then the jth user is guilty. The goal is to estimate $\vec{\Phi}$. The algorithm starts by assuming that all users are guilty; that is, $\vec{\Phi}_{est} = (1, 1, \ldots, 1)$. For each iteration, the number of 1's in $\vec{\Phi}_{est}$ is reduced by an element-wise multiplication with the jth row in the code matrix C, such that the $T(j)$'s are in nonascending order. The likelihood $\vec{\Phi}_{est}$ is compared to that of the previous iteration. If the likelihood is greater, then the iterations continue. If the likelihood is smaller, then the previous estimate is used. The advantage of adaptive detection over hard detection is that there is no need to empirically determine a fixed threshold. An improvement to the adaptive detection algorithm is to estimate $\vec{\Phi}_{est}$ at each stage using the likelihood function. This algorithm, termed sequential detection, reigns supreme over the two above-described algorithms in terms of lower probability of false alarm and lower probability of missed detection. However the algorithm is more computationally intensive than the other two algorithms.

Fingerprinting in a Broadcast Channel Environment

In most fingerprinting schemes described in section "Symmetric Fingerprinting via Data Embedding," unique fingerprints are embedded into every copy of digital data before distribution to buyers. This is impractical for on-demand applications such as live-streaming video. The computational burden at the source, as well as the excessive bandwidth usage of embedding different fingerprints, and hence transmitting different copies of a video across a network, motivate a more clever design. Instead of embedding the fingerprint at the source, the fingerprints can be embedded at the user end during decryption of an encrypted video, thereby eliminating both the-above-mentioned problems. Such a scheme is referred to as fingerprinting in a broadcast channel environment.

There are other motivating factors behind fingerprinting at the decryption stage. The most significant are discussed in this sub-section. First, tamper-proof hardware, such as pay-TV decoders, may not be needed, as the receiver is a private key decryptor. All receivers are identical, with the exception of a unique key assigned to each user. In many real-time, on-demand applications, where multimedia is being streamed to viewers, it is important to keep the computational cost of embedding fingerprints at a minimum. Even if the application in question is not real time or on-demand, the time and cost of embedding fingerprints at the source may be too high to bear. For example, with users worldwide buying music online, several dedicated servers are required to embed fingerprints into the music before it is sent to the buyer. It would be easier to give the user a key and then have an executable file decrypt and embed the fingerprint at the user end.

A formal presentation is given next, outlining the criteria needed for fingerprinting in a broadcast channel environment. Some novel ideas for solving this problem are also provided.

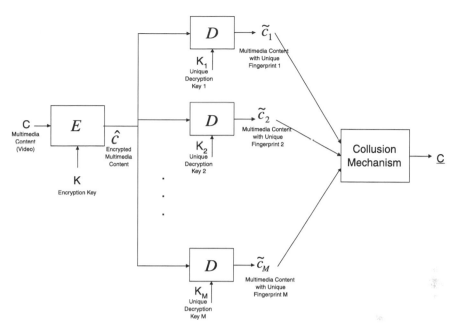

Figure 19.7. Problem formulation of fingerprinting in a broadcast channel environment.

Fingerprinting in a Broadcast Channel Environment Problem Formulation. We next consider how fingerprinting is integrated into a broadcast channel environment. Let c represent the multimedia to be distributed to buyers. In particular, c is a set of N frames, $c = \{c^1, c^2, \ldots, c^N\}$, that represent a perfectly compressed video. Let E represent an encryption function that maps from a set of videos C, and a set of keys K, to a set of encrypted versions of c, $\hat{C}$, such that any $\hat{c} \in \hat{C}$ cannot be understood by eavesdroppers without the appropriate key.

A particular $\hat{c}$ is sent to M receivers who have legally purchased the video content of c. Each receiver has unique keys, $K_1, K_2, \ldots, K_M$. A function D maps $\hat{C}$ and $\{K_1, K_2, \ldots, K_M\}$ to the set $\tilde{C}$, where $\tilde{c}_i \in \tilde{C}$ is a fingerprinted version of c, with a unique fingerprint for receiver i. This scheme is depicted in Figure 19.7. Referring to the notations introduced in Figure 19.7, the following requirements are necessary:

1. Scrambled video signal: $d(c, \hat{c}) \gg T$. The encrypted video $\hat{c}$ does not resemble the unencrypted video c, which means that eavesdroppers will be watching an unintelligible video.

2. Unique fingerprinted videos: $\forall i \neq j, d(\tilde{c}_i, \tilde{c}_j) \neq 0$ and $d(c, \tilde{c}_i) < T$. $\tilde{c}_i$ should also contain codewords that are collusion resistant (Definition 4) and the watermarking scheme should be robust common feasible attacks found in the watermarking literature.

3. Encryption security: Without keys $K_1, K_2, \ldots, K_M$, an eavesdropper, or the function D, cannot derive a $\bar{c}$ given $\hat{c}$, such that $d(c, \bar{c}) < T$.

4. Frame-proof: It is impossible to create any set of keys $\{K_i\} \subset \{K_1, K_2, \ldots, K_M\}$, given another set of keys $\{K_j\} \subset \{K_1, K_2, \ldots, K_M\}$, where $\{K_i\} \cap \{K_j\} = \emptyset$.

The overall problem is tied together by noting that the function D performs decryption as well as modulation and watermarking (as in Figure 19.2) of the fingerprints. Collusion resistance of the keys is addressed by criteria 3 and 4, which state that only appropriate keys can be used to decrypt an encrypted video, and given a set of keys, it is impossible to create different keys that are appropriate.

A notation that encapsulates the process of modulation, watermarking, collusion, feasible attacks, extraction, and demodulation is now introduced. Suppose that $\tilde{c}_1, \tilde{c}_2, \ldots, \tilde{c}_M$ (in Figure 19.7) are fingerprinted with codewords from Γ. Then, define $x = B(X(F(\underline{c}), l))$, where $\underline{c}$ is the video after collusion, as in Figure 19.7, and x is the result of detecting a fingerprint from a signal that is the result of collusion and a feasible attack. Γ is called the underlying codebook of x and is denoted $\Gamma \longrightarrow x$. If Γ was not used, then the notation is $\Gamma \overset{\text{not}}{\longrightarrow} x$. This notation is useful, as it ties together the processes from the transmitter, through the attack channel, to the receiver.

The Chameleon Cipher. In Figure 19.4, the broadcasting scheme used approximately twice the bandwidth of the original plaintext data. The Chameleon Cipher [26] allows the distributor to send one copy of the encrypted data to several end users, whereupon the encrypted data are decrypted and, at the same time, a fingerprint is embedded.

An example construct of the Chameleon Cipher is now presented. Suppose that B is a key consisting of 2^{16} 64-bit words, and A is a key or seed for a pseudorandom number generator. The key A is applied to the pseudorandom number generator to select four words from B, which are then exclusive ORed together, creating the keystream, which is then exclusive ORed with a 64-bit plaintext, resulting in the ciphertext. When the ciphertext is decrypted, a key C is used, where C is the almost the same as B, with the exception of some changed bits, called the marks. The marks can be placed in the least-significant-bit positions; hence, upon decryption, the result is similar to the original plaintext, with the exception that the least significant bits may be different. These differences are the fingerprints, and if enough marks are placed in C, the probability of any two users having a common marked bit will be high. The Chameleon Cipher uses a random approach instead of a systematic codebook, however, the same idea can be applied using a systematic codebook.

The advantage of the Chameleon Cipher is that bandwidth usage is efficient. The disadvantage is that because fingerprinting is performed at the user end, the user knows where the fingerprints are (i.e., the least significant bits) and, hence, the user can readily remove these fingerprints. Even if another key is employed to place these fingerprints in places other than the least significant bit, given knowledge of the key and algorithm, the user can still deduce where the fingerprints are.

Some ways around this is to either apply a black-box philosophy, where the joint fingerprinting and decryption mechanism has a component whose operations are unknown to the user, or the user is given a tamper-proof smart card containing a key that must be authenticated; hence, the user cannot create a fraudulent key of her own. Furthermore, the key stored in the smart card cannot be deduced by the user, unless expensive hardware is designed to read this smart card. In the next section, such a scheme is proposed.

FRAMEWORK TOWARD A SOLUTION FOR FINGERPRINTING IN A BROADCAST CHANNEL ENVIRONMENT

This section presents some criteria and design methods for achieving a solution for fingerprinting in a broadcast channel environment.

Figures of Merit

Some traits that make a fingerprinting scheme strong are as follows:

1. Code efficiency: It is desirable that the fingerprinting codewords $\gamma^i \in \Gamma$ be short, while being resilient to a large coalition of pirates; that is, let $M = |\Gamma|$, and $l = \text{length}(\gamma^i)$, then it is desirable to make l/M as small as possible while satisfying the collusion-resistant criteria. Short codes are attractive because they can be embedded into the media several times, making them more robust.

2. Encryption–decryption efficiency: In Figure 19.7, $\hat{c}$ should be computed from c in polynomial time. In addition, the size of $\hat{c}$ should be close to that of c. In Figure 19.4, $\text{length}(\hat{c}) = 2 \times \text{length}(c)$. This is undesirable, as twice the bandwidth is required to transmit the encrypted video.

3. Low probability of false alarms: It should be difficult for pirates to generate digital data with fingerprints that frame innocent buyers; that is,

$$\Pr\left[A(x) \in 2^{\Gamma} \setminus \{\emptyset\} | \Gamma \xrightarrow{\text{not}} x\right] < \varepsilon \quad \text{for some small } \varepsilon$$

4. High probability of detection: After an attack, some portion of the fingerprint should survive to identify at least one pirate; that is,

$$\Pr[A(x) \in 2^{\Gamma} \setminus \{\emptyset\} | \Gamma \longrightarrow x] < 1 - \varepsilon \quad \text{for some small } \varepsilon.$$

Smart Card Fingerprinting System

This subsection proposes a framework for fingerprinting in a broadcast channel environment that minimizes criterion 2 of Section "Figures of Merit," such that $\text{length}(\hat{c}) = \text{length}(c)$.

Suppose (K, γ^i, l) is encrypted with a private key (in the context of Public Key Cryptography) at the distributor, then, (K, γ^i, l) and its encrypted version are stored on a tamper-proof smart card, resulting in K_i, where K and K_i are as in Figure 19.7, $\gamma^i \in \Gamma$ is a codeword, and $l \in L$ is the watermarking key. The decryption function D first authenticates K_i using the distributor's public key, and then decrypts $\hat{c}$. The fingerprint is embedded using the watermarking key l. To increase security and make it more of a problem for pirates to break, (K, γ^i, l) can be changed, say, every 4 months, and the subscriber would receive a new smart card at the time.

The use of smart card decouples D from the unique keys, allowing manufacturers to mass produce generic D's, and another plant may be responsible for creating the unique smart cards. This cuts costs, as production efficiency is increased. The smart card fingerprinting system is depicted in Figure 19.8.

The security in this scheme requires tamper-proof hardware, as well as tamper-proof smart cards. The keys are stored in the smart card and assumed to be unreadable by an average user. Attackers who are willing to spend more money can execute the following attacks. In the first expensive attack, the user builds a machine to read the keys on the smart card. Once the keys and the watermarking algorithm are known, the

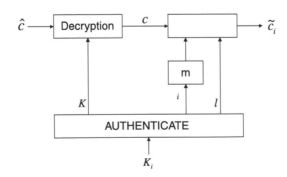

Figure 19.8. Smart card fingerprinting system.

locations of the fingerprints can be located and removed. In the second expensive attack, the user builds D without the watermarking component, thus the resulting $\tilde{c}_i = c$ in Figure 19.8. The third expensive attack is to create fraudulent smart cards. A malicious user cannot easily create fraudulent smart cards unless the public key is retrieved from D and a machine for creating these smart cards is built. Once the public key is retrieved and a machine for creating these smart cards is built, the attacker can generate a key, say u_1, and use the public key to "decrypt" this user-generated key, creating u_2. Now, the user has a user-generated key, u_1, representing the encrypted version of (K, γ^i, l), and a "decrypted" user-generated key, u_2, representing the plaintext (K, γ^i, l). Of course, u_1 has to be generated, such that u_2 contains the actual K, as this is needed to decrypt $\hat{c}$. Additional computation time for searching for a u_1 that satisfies this is required. The attacker would also have to be able to read a genuine smart card to retrieve K. Now, once u_1 and u_2 are generated, the machine for creating smart cards will create a smartcard containing these fraudulent keys. The authentication mechanism in D will apply the public key to the user-generated key, u_1; the result will match u_2 and the authentication test will pass with these fraudulent keys.

FINAL REMARKS

This chapter provides an introduction to the area of modern digital media fingerprinting. The digital fingerprinting problem is explicitly defined, common attacks are described, and three classes of digital fingerprinting techniques are identified. Several influential methods in the literature are summarized and analyzed. The chapter concludes with some novel insights into the use of chameleonlike ciphers for bandwidth efficient combined media decryption and fingerprinting.

From our discussion, it is clear that the design of a good digital fingerprinting method requires the consideration of advanced code development, signal processing, statistical and security analysis, and a little ingenuity or cleverness on the part of the designer. No proposed technique is absolutely robust to all possible attacks. In fact, we believe that given the constraints on the digital fingerprinting problem described here, it is impossible to develop such a method. However, as a passive form of security, digital fingerprinting needs only to be good enough to discourage the widespread propagation of such an attack; that is, overcoming the fingerprinting processes needs to be more expensive than the cost of the digital media content.

We focus, in this chapter, on digital fingerprinting for multimedia; in fact, the majority of literature in the area targets such applications. However, it is the belief of the authors that digital fingerprinting technology has potential for a broad number of other applications such as future sensor networks and computer forensics. Future work will focus on these emerging challenges.

REFERENCES

1. Cox, I. J., Kilian, J., Leighton, T. F., and Shamoon, T., Secure spread spectrum watermarking for multimedia, *IEEE Trans. Image Process.*, 6, 1673, 1997.
2. Deguillaume, F., Csurka, G., and Pun, T., Countermeasures for unintentional and intentional video watermarking attacks, in *IS&T/SPIE's 12th Annual Symposium, Electronic Imaging 2000: Security and Watermarking of Multimedia Content II*, 3971, Wong, P. W. and Delp, E. J., Eds., SPIE Proceedings, Vol. 3971, SPIE, 2000.
3. Boneh, D. and Shaw, J., Collusion-secure fingerprinting for digital data, *IEEE Trans.Inf. Theory*, 44, 1897, 1998.
4. Stinson, D. R., *Cryptography Theory and Practice*, CRC Press, Boca Raton, FL, 1995, p. 24.
5. Trappe, W., Wu, M., Wang, Z. J., and Liu, K. J. R., Anti-collusion fingerprinting for multimedia, *IEEE Trans. Signal Process.*, 51, 1069, 2003.
6. Trappe, W., Wu, M., and Liu, K. J. R., Collusion-resistant fingerprinting for multimedia, presented at IEEE International Conference on Acoustics, Speech, and Signal Processing (ICAPSS '02), Orlando, FL, 2002, Vol. 4, p. 3309.
7. Trappe, W., Wu, M., and Liu, K.J.R., Anti-collusion Fingerprinting for Multimedia. Technical Research Report, Institute for Systems Research, University of Maryland, Baltimore, 2002.
8. Su, J. K., Eggers, J. J., and Girod, B., Capacity of Digital Watermarks Subjected to an Optimal Collusion Attack, presented European Signal Processing Conference (EUSIPCO 2000), Tampere, Finland, 2000.
9. Zhao, H., Wu, M., Wang, Z. J., and Liu K. J. R., Nonlinear Collusion Attacks on Independent Fingerprints for Multimedia, in Proceedings of IEEE International Conference on Acoustics, Speech, and Signal Processing (ICASSP '03), Hong Kong, 2003.
10. Pfitzmann, B. and Schunter, M., Asymmetric fingerprinting, in *Advances in Cryptology — EUROCRYPT '96, International Conference on the Theory and Application of Cryptographic Techniques*, Maurer, U. M., Ed., Lecture Notes in Computer Science, Vol. 1070, Springer-Verlag, Berlin, 1996, p. 84.
11. Pfitzmann, B. and Waidner, M., Anonymous Fingerprinting, *in* Advances in cryptology — EUROCRYPT '97, International Conference on the Theory and Application of Cryptographic Techniques, Fumy, W., Ed., Lecture Notes in Computer Science, Vol. 1233, Springer-Verlag, Berlin, 1997, p. 88.
12. Chor B., Fiat, A., Naor, M., and Pinkas, B., Tracing traitors, *IEEE Trans. Inf. Theory*, 46, 893, 2000.
13. Brown, I., Perkins, C., and Crowcroft, J., Watercasting: Distributed watermarking of multicast media, in *First International Workshop on Networked Group Communication '99*, Risa, L. and Fdida, S., Eds., Lecture Notes in Computer Science, Vol. 1736, Springer-Verlag, Berlin, 1999, p. 286.
14. Judge, P. and Ammar, M., WHIM: Watermarking multicast video with a hierarchy of intermediaries, *IEEE Commn. Mag.*, 39, 699, 2002.
15. Wang, Y., Doherty, J. F., and Van Dyck, R., E., A watermarking algorithm for fingerprinting intelligence images, in *Proceedings of Conference on Information Sciences and Systems*, Baltimore, MD, 2001.
16. Ergun, F., Kilian, J., and Kumar, R., A note on the limits of collusion-resistant watermarks, in *Advances in Cryptology — EUROCRYPT '99, International Conference on the Theory and Application of Cryptographic Techniques*, Stern, J., Ed., Lecture Notes in Computer Science, Vol. 1592, Springer-Verlag, Berlin, 1999, p. 140.

17. Safavi-Naini, R. and Wang, Y., Collusion secure q-ary fingerprinting for perceptual content, in *Security and Privacy in Digital Rights Management ACM CCS-8 Workshop DRM 2001*, Sander, T., Ed., Lecture Notes in Computer Science, Vol. 2320, Springer-Verlag, Berlin, 2002, p. 57.
18. Biehl, I. and Mayer, B., Cryptographic methods for collusion-secure fingerprinting of digital data, *Computers Electr. Eng.*, 28, 59, 2002.
19. Dittmann, J., Behr, A., Stabenau, M., Schmitt, P., Schwenk, J., and Ueberberg, J., Combining digital watermarks and collusion secure fingerprints for digital images, *SPIE J. Electron. Imaging*, 9, 456, 2000.
20. Beautelspacher, A. and Rosenbaum, U., *Projective Geometry*, Cambridge University Press, Cambridge, 1998.
21. Hirschfeld, J. W. P., *Projective Geometries over Finite Fields*, Oxford University Press, Oxford, 1998.
22. Dinitz, J. H. and Stinson, D. R., *Contemporary Design Theory: A Collection of Surveys*, John Wiley & Sons, New York, 1992.
23. Linder, C. C. and Rodger, C. A., *Design Theory*, CRC Press, Boia Raton, FL, 1997.
24. Colbourn, C. J. and Dinitz, J. H., *The CRC Handbook of Combinatorial Designs*, CRC Press, Boca Raton, FL, 1996.
25. Podilchuk, C. and Zeng, W., Image adaptive watermarking using visual models, *IEEE J. Selected Areas Commn.*, 16, 525, 1998.
26. Anderson, R. and Manifavas, C., Chameleon — A new kind of stream cipher, in *Fast Software Encryption 4th International Workshop*, Biham, E., Ed., Lecture Notes in Computer Science, Vol. 1267, Springer-Verlag, 1997, p. 107.

20

Scalable Image and Video Authentication*

Dimitrios Skraparlis

INTRODUCTION

Because digital multimedia data are constantly increasing in quantity and availability, concerns are being expressed regarding the authenticity (by means of origin authentication) and integrity of multimedia data; the ease and efficiency of tampering with digital media has created a need for media authentication systems based on cryptographic techniques. In other words, the trustworthiness of digital media is in question.

Digital media are a key element to the consumer electronics industry. As technology penetrates more and more into everyday life, security has began to emerge as a primary concern and media authentication technology will definitely play an important role in the success of present and future multimedia services. In any case, data origin and data integrity verification would enhance the quality of multimedia services.

Traditional authentication methods (cyclic redundancy checks (CRC) checks or cryptographic checksums and digital signatures) are not fit for modern and future multimedia standards. Modern coding standards introduce scalability and tend to be more content-oriented. Therefore, it would be necessary to devise image and video authentication methods

*© 2003 IEEE. Reprinted, with permission, from Skraparlis D., Design of an efficient authentication method for modern image and video, *IEEE Transactions on Consumer Electronics*, May 2003, pp. 417–426, Vol. 49, issue 2.

that could offer scalability, regard the media content, be based at multimedia standards (and not be custom solutions, which are usually unsuccessful), and also be relatively simple to apply. So far, no authentication mechanism has been able to address all of these properties.

The main contributions of this chapter are the research for a technique to authenticate scalable media efficiently, the introduction of scalable content-based digital signatures, and a proposal of such a signature scheme based on stream signature techniques and a modern image and video coding scheme (JPEG2000 for image and MPEG4 for video).

The applicability of such a scheme to real-world scenarios is further enhanced by the proposal of combining the scheme with special digital signature schemes, such as forward secure and group signatures. Application scenarios regard the security in modern multimedia standards or other applications such as reliable storage in unreliable entities. Examples are the transmission, distribution, or storage of digital photographic images and video on the Internet, press agencies, and similar organizations, reliable storage entities aiding in copyright and ownership disputes, and evidential cameras and motion pictures production and distribution.

The analysis of the authentication method will start in sections "Authentication System Considerations" and "Design of the Authentication Method" with an overview of the design considerations that will form a solid framework for the proposed technique. Prior art will then be reviewed in section "Prior Art," which will present fundamental techniques related to multimedia authentication. The complete proposed method will be presented in section "New Directions in Authentication Method Design." The analysis is finally complemented by the Appendix, which clarifies the concept of authentication and further justifies the design considerations. Throughout this chapter, emphasis will be given to the theory and general design rather than actual results of the implemented method, as actual implementation details will rely on the application and one of the author's intentions was to demonstrate a concise theoretical treatment of the subject.

AUTHENTICATION SYSTEM CONSIDERATIONS

In brief, a practical authentication system should provide the following:

Asymmetric security/source authentication: Asymmetric techniques based, for example, in public key cryptography provide data source and data integrity authentication. Symmetric techniques such as MACs (Message Authentication Codes) and symmetric encryption do not provide source authentication, because the same key is shared between two entities and impersonation is, therefore, possible.

Offline security: Offline security means that the creator does not need to be available for every authentication query. This is a fact in complex media distribution chains that contain one-way distribution links from the producer to the consumer (e.g., offline availability of compact disk (CD) and digital versatile disk (DVD) distributed works).

Open-system security: It is wise to address an open system such as the Internet for possible applications due to its popularity and constant growth. Furthermore, if the method achieves reasonable open-system security, then it could complement closed-systems security as a more secure mechanism.

Applied security: The term "applied security" implies that no custom cryptographic primitives and mechanisms are used (e.g., use public key cryptography infrastructures [PKIs]). PKI's security comes from years of experience in security engineering. In any case, it is much easier and economical to base an authentication mechanism on PKIs than establish a new custom authentication infrastructure that will probably fail to establish itself in the consumer electronics market.

For supplementary analysis on the authentication system design choices, please refer to the Appendix.

DESIGN OF THE AUTHENTICATION METHOD

In the following subsections, the design steps for an authentication method based on the considerations of the previous section will be presented.

How to Do Source and Integrity Authentication

A method to do source and integrity authentication for data that are transmitted through an insecure channel is by sending a cryptographically secure digest of the message (e.g., the output of a hash function) through a secure channel (Figure 20.1).

The channel could either be a secure physical line or a virtual channel provided by a cryptographic technique such as a digital signature. Verification is performed by comparing the authentic message digest with a calculated digest of the received message (Figure 20.2); if the two values are equal, then the message is authentic; that is, it comes from its reputed source. It should be noted that, as it will be clarified in the Appendix, the concept's source authentication and integrity authentication are inseparable.

There also exist methods to do source and integrity authentication using encryption and they are briefly described in the Appendix.

607

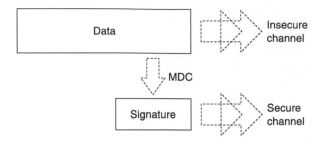

Figure 20.1. How to do source and integrity authentication without the need for feedback. MDC stands for message digest code (a hash function).

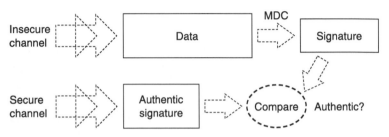

Figure 20.2. Source and integrity authentication verification with no need for feedback.

However, as encryption in a system is optional and encryption in general does not always provide authentication, such methods will not be discussed further in this chapter.

Application to Multimedia Data Formats

The method of the previous subsection, although essential, is not sufficient for still images and video data. Two problems exist. First, we should devise a way to incorporate the signature in the media file. Second, we should construct an authentic or secure channel for the signature's transmission.

For the first problem, possible solutions are watermarking and labeling. Watermarking is obtrusive, as it alters the original media. The method of watermarking is analyzed in more detail in section "Watermarking." Labeling (or captioning) is unobtrusive; that is, it does not alter the original media; Signature data accompany the media file as an authentication label or tag (see section "Labeling (or Captioning)"). Labeling is the preferred technique in this chapter; please refer to section "Watermarking" for justification.

The second problem, the problem of the secure channel of Figure 20.1 and Figure 20.2, can be solved by mathematical methods that provide an authentic channel, such as MACs and digital signatures.

Principles and Requirements of Modern Coding Standards that the Designer of Modern Systems Should Consider

Modern coding standards were designed with new features in mind. Some of these features are progressive transmission or scalability and some features that relate to content-based encoding such as region-of-interest coding and random codestream access and processing. In this chapter, we will work mostly on the JPEG2000 standard [1]. Some introductory work on JPEG2000 are presented in References 2–6.

Scalability. Scalability enables users of different bandwidths or rights to all receive the same content in different qualities. It is also possible to transmit an image or video progressively, by progressive size enlargement or progressive quality enhancement (image is gradually rendered more accurate). Some applications of scalability are the following: multiquality network services, layered multicasting, and database browsing.

Content Based Multimedia. The representation of visual information has a fundamental difference compared to other types of data: Data are considered as a whole (i.e., every bit change is considered significant), whereas the representation of images provides information not directly linked to the binary representation of data. Every image holds information (meaning) that is associated with the human cognitive representations of the world. Modern coding standards such as JPEG2000 and MPEG-4 demonstrate a trend toward content encoding.

The contents of sections "Scalability" and "Content-Based Multimedia" constitute the new directions of research that are presented in section "New Directions in Authentication Method Design."

Design of the Authentication Method Summary

To summarize, the design methodology for the authentication system is the following. For authentication via securely providing authentication information, the designer needs to:

- Provide an authentic channel (by the use of cryptography).
- Find a way to include labels in the data. Also make sure that it is efficient and that an unexpected termination in data transmission does not prove catastrophic for the authentication mechanism (such a desired property would also benefit multiscale data authentication). Image coding methods should therefore be investigated.
- Investigate the "content-based properties" of the label embedding method (i.e., how the position of authentication labels in the bit stream is related to the image content).

In addition, various design considerations in this chapter are as follows:

- Do not create a custom image coding method, but work on a standard.
- Make sure that the cryptographic part of the final method is efficient but as secure as desired for multimedia.

However, before tackling with how to deal with the above, we will review prior art.

PRIOR ART

In the following subsections, we will review prior art in the form of either commercial applications or in the open literature. These will form an overview of prior art in the subject and present the shortcomings and inefficiencies of the methods, creating the need for more complete or advanced techniques. Moreover, some of the prior art techniques will form a base for the novel advances presented in section "New Directions in Authentication Method Design."

Traditional (General Purpose and Multimedia) Authentication Systems

The general problem of data integrity authentication has traditionally been addressed by commercial picture authentication systems and image authentication cameras, which are based on common digital signature standards.

However, as digital signatures have proven to be rather inefficient for multimedia data, and hash and sign methods (section "Hash and Sign Method") are still deemed inefficient and inappropriate for streams and multicasting, other techniques based on stream signatures have been presented recently in the literature and are analyzed in the subsection.

Efficient Data Authentication Techniques: Stream Signatures

Stream signatures aim to provide efficient source and data authentication for scalable data. For example, in modern image and video coding standards such as JPEG2000 and MPEG-4, some receivers opt for not receiving the multimedia data at the maximum rate (best quality). In that case, traditional authentication systems utterly fail, because verification is only possible when the media has been fully received.

Hash and Sign Method. The simplest but most inefficient stream signature is the hash and sign method (see Figure 20.3), which is used in commercial authentication systems. The bit stream is first divided into blocks of equal size and then the hash value of each block is signed by

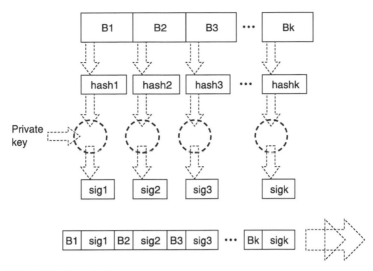

Figure 20.3. **Hash and sign signature.**

a digital signature algorithm (DSA). The signatures are then inserted between each block in the bit stream. The receiver calculates the signatures for each received block and compares them to the received signatures. If they match, the content is authenticated.

The performance cost of hash and sign is one digital signature calculation per block. That could be severely decreased without significant security compromised by introducing hash trees. The idea is to transfer the properties of a single digital signature to all of the blocks through hash trees or chain mechanisms.

Tree-Based Method. The tree-based method can be constructed in several ways, either the standard tree chaining method [7, 8, §13.4.1] or subinstances of the tree chaining method (e.g. star chaining method in the signed Java Applets technology).

As one can see in Figure 20.4, every node in the tree contains the hash of its children. During transmission, every packet p has overhead equal to the size of hash values that constitute the path in the tree that leads to the packet. For data comprising k packets, the amount of hash values n_p that must precede each packet for successful authentication is either $\log_2(k)$, $\sum_{p=1}^{k} n_p = k \log_2(k)$, when packet-loss robustness is required, or $0 \le n_p \le \log_2(k)$, $\sum_{p=1}^{k} n_p = k$, when a minimum number of transmitted packets is required but by adding complexity to the receiver.

Offline or Recursive Method. The recursive stream authentication method of Reference [9] has less overhead than the tree methods, requiring less buffering at the receiver end. Looking at Figure 20.5, block

611

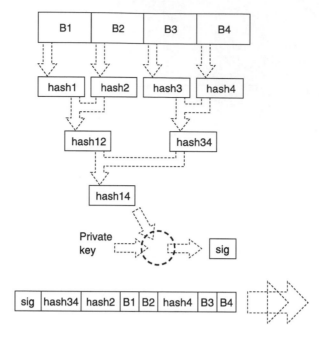

Figure 20.4. **Example of a hash authentication tree method as a stream signature.**

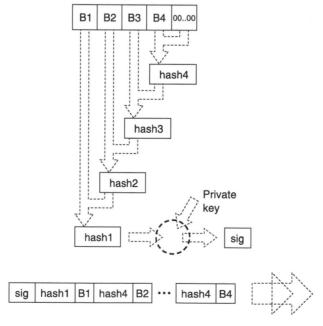

Figure 20.5. **An example of the "offline" or recursive stream signature.**

processing is performed backward (from the end to the first block in the data). The recipient of the data verifies the digital signature sent at the start and then calculates and compares the hash for each block recursively, as shown in the Figure 20.5.

Other Methods. There has been research on stream signatures for streams that suffer from packet losses [10–13], but they will not be considered here. Other methods for embedding stream signatures into the file format do not offer offline security (as described in section "Authentication System Considerations"), as they are 1-time signatures [10,14–16].

A stream signature cannot unfold its power unless we devise a way to embed the overhead packets in the data, as has been previously described in the design considerations of section "Design of the Authentication Method Summary." Stream signatures have been demonstrated to work in MPEG video streams [9] but have not previously been introduced in modern coding standards such as JPEG2000 and MPEG-4, although data embedding methods for these standards exist (section "Labeling (or Captioning)"). This fact led the author to the concept of multiscale image labeling, which will be analysed in section "Scalability in an Authentication Technique: Multiscale Image Labeling."

Medium Oriented Authentication Techniques

Watermarking. The basic idea behind watermarking technology is the slight import of information in the media, taking advantage of the imperfections of the human visual system. There also exist visible watermarking technique, see Reference 17 as an example. An embedded watermark theoretically cannot be removed or modified without visible degradation of the medium. Watermarking is classified among information hiding methods. A bibliography for this subject is given in Reference 18.

Reasons for not choosing the technique of watermarking are the following:

- Watermarking techniques have not yet been convincing for their effectiveness. First-generation watermarking (and practically all of the commercial solutions) has proven to be exceptionally weak on simple attacks, a fact that renders their legal credit easily questionable. The watermarking techniques currently under research are not convincing for their reliability and the results are rather disappointing. The greatest problem, in general, is synchronization of the watermarking detector, which can easily be lost by very simple and also slight or imperceptible image transformations [19,20].

- An important characteristic of watermarking is that watermarking techniques do modify the data, although these modifications are supposed to be imperceptible to the consumer; Quality degradation and modification of the original artistic medium may dissatisfy the candidate recipients of this technology — artists and consumers. Also, certain applications may strictly not allow any modification of the data, as the example of evidential images.
- The complexity and the computational power demand of state-of-the-art watermarking techniques are deemed to be unreasonable; this is a weak point when targeting consumer electronics' and portable devices' technology. Watermarking is probably better suited in the subject of intellectual property, rather than casual data integrity control.

It would also be possible for a security solution to utilize other methods of integrity authentication combined with watermarking. For an example of combining watermarking and labeling, see Reference 21 (refer to section "Labeling (or Captioning)" for the description of labeling).

Labeling (or Captioning). By using labeling, which is the insertion of authentication tags in the image or video data in a way that is transparent to a noncompliant decoder, offline source and integrity authentication is possible. Similar to traditional integrity schemes, the absence of the labels implies uncertainty of the image or video integrity. To support the credibility of the labeling method, several labeling schemes for state-of-the-art coding standards will be introduced in the following subsections.

Labeling in JPEG2000. There exist three possible methods to do labeling in the JPEG2000 coding standard:

- *Component extension.* In component extension, authentication tags are introduced as an extra component in the image, for certain progression orders that allow the multiplexing of the extra component with the other components in the image bit stream. It should be noted that this method is probably not compliant with the basic JPEG2000 standard.
- *After coding termination.* In Reference 22, a method that inserts data into a JPEG2000 codestream is presented. The method uses a detail of the JPEG2000 encoder: After each coding pass termination (between entropy coding and rate allocation), it is possible to insert data without affecting a normal JPEG2000-compliant decoder. More information for the encoder can be found in the standard [23, Annex C: §3.4, p. 90: Annex D, §4.1, p. 99] and in Reference 24. Behind the encoder lies the EBCOT algorithm [25].

- *Multiplexing (general method).* The JPEG2000 standard offers freedom in how to construct the bitstream. This fact allows the multiplexing of the authentication component with the basic image components (as in the component extension method) and additionally allows controlling its frequency of occurrence in the final bit stream with precision. This would, however, raise the complexity of an encoder. A similar technique is opting for using a more general file format, as follows in the next subsections.

Labeling in the jpm and jpx File Formats. The jpx format is described as an extension in JPEG2000–ISO/IEC FCD15444-2, Annex L: extended file format syntax. It allows the insertion of XML descriptors and MPEG-7 metadata. It also incorporates an Intellectual Property Rights (IPR) box and a Digital Signature Box that uses certain hash and digital signature algorithms in order to protect any data part in the .jpx file.

The term "box" refers to a binary sequence that contains objects and has the general form |size|type|contents|. Unknown types of box are to be ignored by normal decoders (according to the standard's L.10 directive), so there is the possibility of creating custom boxes without losing compatibility. This, in fact, enables one to create more diverse labeling methods, such us the one presented in this chapter.

The .jpm (ISO/IEC FCD15444-6) standard is about the transmission of images through the Internet and it also allows the use of XML and IPR boxes. In both of these formats, JPEG2000 codestreams in jpc and jp2 format can be inserted as boxes (named Media Data box and Contiguous Codestream box in the .jpm format).

The incorporation of labeling methods in the jpx and jpm standards is very simple and possibly more robust than using the JPEG2000 labeling methods described earlier.

Labeling in Motion-JPEG2000 and MPEG-4. The JPEG2000, Motion-JPEG2000, MPEG-4, and QuickTime standards have some common characteristics: They all share the same file format structure that comprises objects. Boxes in Motion-JPEG2000 are called atoms in QuickTime and MPEG-4 and they are formatted as |size|type|contents|. Furthermore, unknown objects are ignored, a directive that allows the definition of objects specifically designed for our authentication application. The reader is referred to References 26 and 27.

Content-based Multimedia

Advanced data integrity techniques include content-based signatures. Content-based digital signatures are data integrity techniques applied to the data content, such as low-level image descriptors (edges [28] with the popular canny edge detector [29], textures, histograms [30],

DCT coefficients as feature codes [31], and wavelet coefficient heuristics [32]) or semantic nonautomatic descriptions (e.g., MPEG-7). A digital signature is applied not to the image bit stream, but to the image's content description, thereby authenticating the content of the image. Commercial automatic content-authentication systems typically claim to be highly robust. However, as cited by state-of-the-art research in computer vision, all automated content extraction methods are unreliable and perform satisfactorily only in custom and well-defined conditions. Therefore, human-assisted algorithms would ensure performance of content-based signatures.

Modern coding standards could provide the framework for higher-level human-assisted image descriptors. In modern coding standards, the creator can influence the coding of the image or video. This will be supported in more detail in section "The Example of JPEG2000." This idea is similar to the method of Perceptual Image Hashing with a noninversive compression technique presented in Reference 33. In addition, scalability is a new important trend in image and video coding standards, but commercial content-based authentication systems do not address scalability; so there is a need to optimize content-based signature methods with scalability robustness, a fact that can be tackled with modern scalable coding standards, leading to the term "scalable content-based digital signature" (section "The Example of JPEG2000").

NEW DIRECTIONS IN AUTHENTICATION METHOD DESIGN

Traditional (and commercial) media authentication systems are not designed for multiscale media and are, therefore, inappropriate for multicasting and other applications of scalability. Second, they are usually not content-oriented or they did not consider this trend during their design.

Scalability in an Authentication Technique: Multiscale Image Labeling

To address scalability, each label should only be dependent on its corresponding layer (scale). A simple solution to the scalability of digital media is to insert a label containing a digital signature between each data part of the media, like the "hash and sign" method. However, this solution is inefficient. This can be solved by incorporating research on more advanced stream digital signatures (section "Efficient Data Authentication Techniques: Stream Signatures"), that have the benefit that the cryptographic properties of a single digital signature can be spread over multiple data with low additional computational cost.

What Is the Relationship Between Labeling and the Image Content?

State-of-the-art media coding techniques take steps toward encoding the media content.

The Example of JPEG2000. In JPEG2000, it is possible to ignore designing a content extraction algorithm, because the encoder indirectly provides a version of such an algorithm. More specifically, at JPEG2000:

- Signing the local maxima on the multiscale layers [34] is equivalent to authenticating the image edges [35]. Local maxima remain at high compression rates.

- The basic layer (LL) of the multiscale image is the most important layer and is encoded (and therefore authenticated) first.

- The image bit stream can be reallocated so that most of the image's information is transmitted first and the not-so-important details follow. Human intervention is therefore introduced in deciding which image context or information is important and which is not, a task that is yet impossible for the automatic content extraction tools. JPEG2000 includes ROI (region of interest) coding and, in general, provides the freedom to precisely choose the position of every encoded packet in the final bit stream, for every layer up to providing lossless coding for any image.

If one considers the scalability property of modern coding standards, then one may come up with the term "scalable content-based digital signatures," which provide content authentication of scalable versions of images and video.

Discussion of JPEG2000's Ability for Content Oriented Encoding. One could object that JPEG2000 does not really encode the image content. In addition, content encoding implies that there exist subjective factors that influence image encoding. A good example is that content encoding could prove catastrophic to artistic image encoding; different people perceive the image in completely different ways that no automatic or human-assisted content description operation can ever predict.

In that case, we could say that the encoded image is not content-oriented but, instead, that it is encoded hierarchically in a content-oriented way. This means that, on the one hand, the image is coded losslessly (no information is being left out), but, on the other hand, the encoded image's packet hierarchy is carefully chosen so that the most probably important image elements are encoded first.

To conclude, JPEG2000's design enables us to do integrity authentication of the scalable bit stream by using scalable labeling and at the same time establish a connection to the original image's content. This fact enables the characterization of the labeled encoded JPEG2000 image as a "self-contained content-based digital signature." This type of digital signature is very useful in reliable multimedia storage in insecure systems (section "Application Scenarios").

The Complete Authentication Method

The image authentication mechanism can take two forms. One is the self-contained autonomous content based signature and the other is the authentication information targeted for embedding in specific code-streams. Examples of each type will be given as application scenarios in section "Application Scenarios." The implementation of the authentication method and related issues can be found in Reference 36.

Choosing Parameters and Primitives: Stream Signature, Labeling Technique, Digital Signature Technique. Some important factors for choosing cryptographic primitives are the proven security of the primitive (old, well-tested, and enhanced primitives like Rivest-Shami-Alderman (RSA) may qualify for this), the primitive's throughput (it is important to work with fast algorithms), its availability internationally (possible imposed patents or restrictions are undesirable), and the implementation cost (hardware and software availability — depends on the design of the system). Another very important factor is key length, which should be chosen so that it could protect the data for a reasonable amount of time; in the case of multimedia, this should probably be at least a decade, although it is very difficult to foresee technological advances that will affect the security of the mechanism.

Based on the above, a brief selection procedure will result in the following choices:

- *Stream signature*. The efficiency of the authentication method clearly depends in the type of stream signature. The offline stream signature combined with secure hash algorithm (SHA-1) as a hash primitive and RSA of at least 1024 bits key length (and perhaps 2048 bits for more security) is very efficient compared to other constructions like "hash and sign," as has already been analyzed in section "Efficient Data Authentication Techniques: Stream Signatures."
- *Labeling method*. jpm, jpx for image and MPEG-4 and Quicktime and Motion-JPEG2000 for video are the most elegant file formats that allow the configuration of boxes as labels. However, a design should be oriented to a popular format, but this wide acceptance cannot be foreknown for modern standards at the time of writing.

An example of the complete authentication signature for the low-speed Internet is presented in Figure 20.6. In general, the length of the blocks that the stream signature operates on should actually be selected according to the application needs; the key factor would be the data rate, thus providing low overhead for the signature bits in each case.

Optimizing the Stream Signature Technique for Scalable Video. A modification to stream signatures that exists since the creation of authentication

| 3.194 | 50.352 | 63.292 | 159.953 |

Figure 20.6. These successive instances of the JPEG2000 image have 160-bit authentication tags embedded after each coding termination. Numbers show the file size in bytes after the insertion of each tag.

trees [8, §15.2.1(iv)] is to apply the stream signature to streams where each individual packet could contain a substream. As an example, authentication trees could contain other authentication trees as leaves in the tree. This technique allows the designer of video (and, of course, image) authentication techniques to optimize for scalability:

- In video coding, there exists two-dimensional scalability: time scalability along with quality scalability. Instead of calculating one stream signature (which includes one digital signature) per time interval (e.g., per fixed number of frames), it is possible to expand the stream signature so that it includes blocks of substreams. For example, looking at Figure 20.7, B1 is frame 1, B2 is frame 2, and so on. Each frame's quality is addressable and time scalability

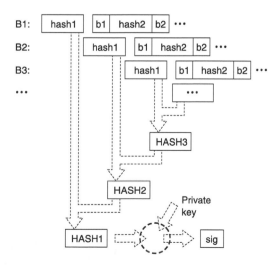

Figure 20.7. Optimized stream signature for scalable video.

is maintained, as each frame can be uniquely decoded and authenticated.

- The technique can be applied similarly to general object-based coding standards like jpx. The stream signature operates on groupings of objects (or boxes) so that a single digital signature operates to each group ("sig" operating in the figure's objects; Figure 20.7) while preserving scalability in each individual object (B1, B2,... in Figure 20.7).

Choice of Digital Signature Schemes for Adaptation to Real Models. Some unpopular digital signature schemes have been devised in the past that could greatly enhance an authentication method's applicability to real-world scenarios.

- *Group signatures.* Group signatures are a relevantly recent concept, introduced in Reference 37 in 1991. Group signatures can be used to conceal an organization's infrastructure (i.e., issue a single public key for the whole organization). Their description is as follows: A member of the group can sign anonymously on behalf of the group. In case of a dispute, the identity of the creator of the signature can be revealed only by a designated entity, the group manager. Nobody (even the group manager) can attribute a signature's origin to someone else or sign on behalf of another person of the group.

  Several group signature techniques have been presented in the literature, designed for small and for large groups. Some relevant publications are References 38 and 39. An application example of group signatures can be seen in Figure 20.9.

- *Forward secure signatures.* The application of forward-secure signatures has the pleasant property that the validity of past signatures is maintained even if the present secret or private key is disclosed. Another property of forward-secure signatures is that it is not possible to forge signatures of past time intervals in cases of secret or private key disclosure. This property is very important, especially if we think that the problem of revealing secret or private key exists in applications but no satisfactory solutions has been given: For example, in Reference 40, the possible problem of a photographic camera being stolen or withheld, along with all the implications of this (key extraction by camera tampering and forging of later signatures), is discussed, however without suggesting any solution other than avoiding this situation.

  In general, it is desired and possible to construct forward-secure signatures based on, for example, RSA. Various techniques have been presented in the literature [41–44].

Application Scenarios

The application examples that follow are directly related to the general applications of scalability.

Image and Video Transmission on the Internet. Looking at Figure 20.8, a lossless image or video is gradually transmitted from the creator to another user up to a desired quality level, being authenticated at the same time. Please note that the public key could also be included in the final file, provided that a digital signature certificate is also included in the data. That is, of course, only if trust could be attributed to the entity responsible for keeping the certificates authentic.

Commercial Case: Secure Transmission and Distribution of Media. A commercial multimedia distribution scheme is presented and described in Figure 20.9. The scheme is complemented by utilizing group signatures. This scheme links the creator of the content to the distributor and then to all of the clients while preserving the uniqueness of the creator's signature, as no entity can forge a group signature. The scheme is straightforward, as there is no need to generate another version of the image or video for each customer. Bearing in mind that the original work is scalable, each customer can receive a quality-degraded part (according to bandwidth, requirements, subscription status, etc.) or even the lossless-complete version by simply cutting down the size or rate of the distributed media.

Reliable Storage of Multimedia Data in Unreliable Storage Entities. Thinking in a similar fashion as in Figure 20.9, the original creator of an image or video could store the work (with the embedded authentication signature) in a server (the distributor in the figure), which can later be accessed by another user or client. An important feature is that, due to the

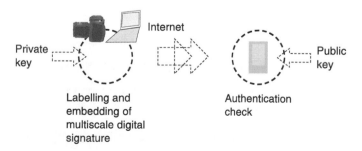

Figure 20.8. Simple case of image or video transmission on the Internet (identical to Figure 20.1 and Figure 20.2).

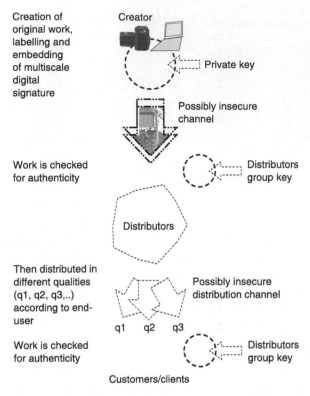

Creation of original work, labelling and embedding of multiscale digital signature

Creator

Private key

Possibly insecure channel

Work is checked for authenticity

Distributors group key

Distributors

Then distributed in different qualities (q1, q2, q3,..) according to end-user

q1   q2   q3

Possibly insecure distribution channel

Work is checked for authenticity

Distributors group key

Customers/clients

Figure 20.9. Image or video original content creation and commercial distribution. The distributor could be a press agency.

presence of a secure channel provided by the digital signature, the storage entity can be unreliable, as no attacker could tamper the image without detection (based on a digital signature's high cryptographic strength). It is therefore possible to relax the requirements for a very secure server that is trusted to keep the integrity of the data stored. (In unreliable storage entities, it is possible that one could regenerate the signature for an original work and claim rights. This can be solved by the use of timestamping during the original work's creation.)

In addition, users can access the media files from the server at different qualities and speeds according to their bandwidth and access rights. This is a main feature of modern coding standards — scalability — so there is no need to store many files for every image or video. Furthermore, authentication for any user is straightforward, backed up by the public key mechanism.

The same concept could be used in human-assisted authentication queries: An image database that would hold a version of the content-oriented digital signatures (i.e., the original work with the embedded authentication signatures up to a certain bit stream size) could be used by a human operator to authenticate the content of a downgraded image in question.

Credibility Problems of the Digital Photographic Image and Digital Video. The introduction of the technique suggested in this chapter in the hardware part of a complete solution could enhance the reliability of an image or video authentication system. This is, of course, not a new idea. Commercial authentication photographic cameras and image authentication systems already exist.

However, do such authentication systems have any credibility? The answer is not yet clear, as the suggestion of hardware digital signature generation transfers the problem to a new domain — the reliability of security hardware — which has been questioned [45].

CONCLUSIONS AND FURTHER WORK

To recapitulate, an authentication method under various considerations has been presented. This method allows the source and integrity authentication of scalable images and scalable video sources in the form of the modern coding standards JPEG2000 and MPEG-4. The use of labeling has been exploited instead of modifications of the medium such as watermarking because of its efficiency and practicality. The above could all be summed up in a "scalable content-based digital signature" concept that has many practical applications.

Finally, it has been found that recent advances in cryptography present new potential dimensions to the manipulation of digital multimedia and could complement the application of the proposed method in real-world scenarios.

Further work on this subject could include investigations of techniques for packet-loss tolerance. A complete system would also involve the combination of all the special digital signature methods in a single primitive.

Another subject would be the investigation of the scalable authentication method in complete DRM systems, an area of current great interest.

ACKNOWLEDGMENT

The author would like to thank Associate Prof. Dimitrios Mitrakos of Aristotle University of Thessaloniki, Department of Electrical and Computer Engineering for his encouragement and support during the early version of this work, as found in Reference 36.

APPENDIX: MULTIMEDIA SECURITY AND THE IMPORTANCE OF AUTHENTICATION

Two important concepts related to the security of digital media are:

- Source authentication (or data origin authentication)
- Integrity authentication (or message authentication)

These two concepts are bound under the term "authentication." Source authentication assures for data origin (that data actually originated from its reputed source) and message authentication provides for the message or media integrity. These two concepts are, by definition, inseparable: The tampering of data has the automatic effect that the tampered data originated from another source (the entity responsible for the modification), whereas claims about the integrity of data originating from an undefined source cannot be held [8, §9.6.1]. In the next subsections, we will clarify the concepts related to the security of digital media and establish a better understanding of security mechanisms and the importance of authentication in the overall design of a system.

Source and Integrity Authentication Should Not Be Confused with Entity Authentication

Source and message integrity authentication and entity authentication are different concepts: Entity authentication (or identification) takes place during an online authentication protocol and its validity is confined to that transaction only.

Entity authentication can be found to be insufficient for large distribution systems, as the complexity of the system can vary and it would, therefore, be necessary to attribute trust to or set up authentication mechanisms for all of the entities that comprise the distribution system (starting from the creator of the content, all the way to all of the distributors and redistributors of multimedia content in the distribution chain).

Source and integrity (message) authentication and entity authentication contribute to classify the authentication systems as either online or offline. However, they are not the absolute factors that classify the system as, for example, there exist both online and offline message authentication techniques. Online and offline methods have been briefly discussed in section "Authentication System Considerations."

Why Authentication and Not Encryption?

Although techniques that provide message authentication by utilizing encryption exist, it is a mistake to assume that encryption automatically

624

does message authentication. See Reference 8 (§9.6.5) for details. Examples of methods that provide authentication by using encryption are the following:

- Embedding the hash value before encryption. Careful consideration should be given to the choice of cryptographic algorithms.
- Embedding a secret authenticator prior to encryption.

REFERENCES

1. ISO/IEC JTC 1/SC 29/WG 1, coding of still pictures. JPEG 2000 Part I Final Committee Draft Version 1.0, ISO/IEC FCD15444-1: 2000 (V1.0, 16 March 2000).
2. Tutorial on JPEG2000 and overview of the technology, http://jj2000.epfl.ch/jj_tutorials/index.html.
3. Rabbani, M. and Santa-Cruz, D., The JPEG 2000 Still-Image Compression Standard, presented at International Conference in Image Processing (ICIP), Thessaloniki, Greece, 2001.
4. Christopoulos, C., Skodras, A., and Ebrahimi, T., The JPEG2000 still image coding system: An overview, *IEEE Trans. Consumer Electron.*, 46(4), 1103–1127, 2000.
5. Special edition on JPEG 2000, Ebrahimi, T., Christopoulos, C., and Leed, D.T., Eds., *Signal Processing: Image Communication, Science Direct*, 17(1), January 2002.
6. Adams, M., The JPEG-2000 still image compression standard, ISO/IEC JTC 1/SC 29/WG 1 N 2412, December 2002.
7. Merkle, R., A certified digital signature, in *Advances in Cryptology, Crypto 89*, Brassard, G., Ed., Lecture Notes in Computer Science, Vol. 435, Springer-Verlag, New York, 1989, pp. 218–238.
8. Menezes, A., van Oorschot, P., and Vanstone, S., *Handbook of Applied Cryptography*. CRC Press, Boca Raton, FL, 1997.
9. Gennaro, N. and Rohatgi, P., How to Sign Digital Streams, presented at Crypto 97, 1997.
10. Perrig, A., et al., Efficient Authentication and Signing of Multicast Streams over Lossy Channels, presented at IEEE Symposium on Security and Privacy 2000, pp. 56–73.
11. Golle P. and Modadugu, N., Authenticating Streamed Data in the Presence of Random Packet Loss, presented at ISOC Network and Distributed System Security Symposium, 2001, pp. 13–22.
12. Wong, C. and Lam, S., Digital signatures for flows and multicasts, *IEEE/ACM Trans. on Networking*, 7(4), 502–513, 1999.
13. Miner, S. and Staddon, J., Graph-Based Authentication of Digital Streams, presented at IEEE Symposium on Security and Privacy, 2001.
14. Anderson, R., et al., A new family of authentication protocols, *ACM Operating Systems Review*, 1998, Vol. 32, no. 4. ACM, pp. 9–20.
15. Rohatgi, P., A Compact and Fast Hybrid Signature Scheme for Multicast Packet Authentication, in Proceedings of 6th ACM Conference on Computer and Communication Security, 1999.
16. Perrig, A., et al., Efficient and Secure Source Authentication for Multicast, presented at Network and Distributed System Security Symposium, NDSS '01, 2001.
17. Mintzer, F., Lotspiech, J., and Morimoto, N., Safeguarding digital library contents and users. Digital watermarking. *D-Lib on-line Mag.*, 1997, http://www.dlib.org.
18. Anderson, R. and Petitcolas, F., Information Hiding: An Annotated Bibliography, http://www.cl.cam.ac.uk/~fapp2/steganography/bibliography/.
19. Petitcolas, F., Anderson, R., and Kuhn, M., Attacks on copyright marking systems, *Second Workshop on Information Hiding*, Lecture Notes in Computer Science, Aucsmith, D., Ed., 1998, Vol. 1525, pp. 218–238.

20. Petitcolas, F., Weakness of existing watermarking schemes. stirmark — samples, http://www.petitcolas.net/fabien/watermarking/stirmark31/samples.html.
21. Herrigel, A., An Image Security Solution for JPEG 2000, presented at *Seminar on Imaging Security and JPEG2000*, Vancouver, 1999.
22. Grosbois, R., Gerbelot, P., and Ebrahimi, T., Authentication and access control in the JPEG 2000 compressed domain, *Applications of Digital Image Processing XXIV*, SPIE 46th Annual Meeting Proceedings, San Diego, July 29–August 3, Vol. 4472, SPIE, 2001, pp. 95–104.
23. JPEG 2000 Related Publications, http://jj2000.epfl.ch/jj_publications/index.html.
24. Taubman, D., et al., Embedded Block Coding in JPEG2000, in *Proceedings of the IEEE International Conference on Image Processing (ICIP)*, September 2000, Vol. 2, pp. 33–36.
25. Taubman, D., High performance scalable image compression with EBCOT, *IEEE Trans. Image Process.*, 9(7), 1151–1170, 2000.
26. Houchin, J. and Singer, D., File format technology in JPEG 2000 enables flexible use of still and motion sequences, *Signal Process.: Image Commn.*, 17, 131–144, 2002.
27. ISO/IEC JTC 1/SC 29/WG 1, coding of still pictures. ISO/IEC 15444-3 (JPEG 2000, Part 3) Committee Draft.
28. Dittmann, J., Steinmetz, A., and Steinmetz, R., Content based digital signature for motion pictures authentication and content-fragile watermarking, in *Proceedings of the IEEE International Conference on Multimedia Computing and Systems (ICMCS)*, IEEE Computer Society, New York, 1999, Vol. II, pp. 209–213.
29. Canny, J., A computational approach to edge detection, *IEEE Trans. Pattern Anal. Mach. Intell.*, 8, 679–698, 1986.
30. Chang, S. and Schneider, M., A Robust Content Based Digital Signature for Image Authentication, presented at IEEE International Conference on Image Processing (ICIP), Lausanne, 1996.
31. Lin, C. and Chang, S., Generating Robust Digital Signature for Image/Video Authentication, presented at ACM Multimedia and Security Workshop, Bristol, 1998.
32. Lu, C. and Liao, H., Structural Digital Signature for Image Authentication: An Incidental Distortion Resistant Scheme, presented at ACM Multimedia Workshop, 2000, pp. 115–118.
33. Mihçak, M. and Venkatesan, R., New Iterative Geometric Methods for Robust Perceptual Image Hashing, presented at ACM Workshop on Security and Privacy in Digital Rights Management, Philadelphia, 2001.
34. Mallat, S., *A Wavelet Tour of Signal Processing*, Academic Press, San Diego, CA, 1998.
35. Mallat, S. and Zhong, S., Characterization of signals from multiscale edges, *IEEE Trans. Pattern Anal. Mach. Intell.*, 14, 710–732, 1992.
36. Skraparlis, D., Methods for Safeguarding Authenticity and Integrity of Digital Image and Video During Reception, Transmission and Distribution via Communication Networks, Diploma thesis, Aristotle University of Thessaloniki, 2002 (in Greek).
37. Chaum, D. and van Heyst, E., Group signatures, in *Advances in Cryptology*, EUROCRYPT'91, Lecture Notes in Computer Science Vol. 547, Springer-Verlag, Berlin, 1991, pp. 257–265.
38. Ateniese, G., et al., A practical and provably secure coalition-resistant group signature scheme, in *Avances in Cryptology CRYPTO 2000*, Lecture Notes in Computer Science, Vol. 1880, Springer-Verlag, 2000, pp. 255–270.
39. Camenisch, J. and Damgård, I., Verifiable Encryption, Group Encryption, and Their Applications to Separable Group Signatures and Signature Sharing Schemes, presented at *Asiacrypt 2000*, 2000, pp. 331–345.
40. Eastman Kodak Whitepaper. Understanding and Integrating Kodak Picture Authentication Cameras, 2000.

41. Krawczyk, H., Simple Forward-Secure Signatures from Any Signature Scheme, presented at ACM Conference on Computer and Communications Security, 2000, pp. 108–115.
42. Anderson, R., Invited Lecture, presented at Fourth Annual Conference on Computer and Communications Security, ACM, 1997.
43. Bellare, M. and Miner, S., A Forward-Secure Digital Signature Scheme, presented at *Crypto 99*, 1999, pp. 431–448.
44. Itkis, G. and Reyzin, L., Forward-Secure Signatures with Optimal Signing and Verifying, presented at Crypto 01, 2001, pp. 332–354.
45. Anderson, R. and Kuhn, M., Tamper Resistance — A Cautionary Note, in The Second USENIX Workshop on Electronic Commerce Proceedings, Oakland, CA, 1996, pp. 1–11, http://www.usenix.org

21

Signature-Based Media Authentication

Qibin Sun and Shih-Fu Chang

In other chapters of this book, we know that (semifragile) watermarking solutions work well in protecting the integrity of multimedia content. Here, we are interested in those authentication methods that protect both content integrity and source identity. In cryptography, digital signature techniques are the natural tools for addressing data authentication issues. To date, many signature-based authentication technologies particularly Public Key Infrastructure (PKI) have been incorporated into the international standards (e.g., X.509) and state-of-arts network protocols (e.g., SSL, Secure Socket Layer) for the purposes of data integrity and source identification. It would be great if we could extend digital signature schemes from data authentication to multimedia authentication. First, the system security of a digital signature scheme, which is a very important issue in authentication, has been well studied in cryptography. Second, the current security protocols in a network, which work for data exchange or streaming, do not need to redesign for multimedia exchange or streaming. Therefore, in this chapter, we focus on the discussion of authenticating multimedia content based on digital signature schemes, although digital watermarking techniques might be employed in some specific schemes.

0-8493-2773-3/05/$0.00+1.50
© 2005 by CRC Press

Considering that the content in this chapter closely relates to the techniques from multimedia processing and analysis, security and cryptography, communication and coding, and so forth, we will start by introducing the basic concepts and unifying the technical terms. More detailed descriptions on the state-of-arts signature based authentication methods are categorized and presented in sections "Complete Authentication," "Content Authentication," and "A Unified Signature-Based Authentication Framework for JPEG2000 Images," respectively. Section "System Performance Evaluation" discusses how to evaluate the performance of multimedia authentication systems. Section "Summary and Future Work" summarizes this chapter and discusses future work.

INTRODUCTION
Basic Concepts and Definitions

Multimedia authentication is a relatively new research area compared to other "traditional" research areas such as multimedia compression. Different researchers with different research backgrounds may have different understandings of the term "authentication." For example, people from multimedia watermarking usually use the term "authentication" to refer to content integrity protection; people from biometrics may use the term to refer to source identification, verification, and so forth. To help the reader, we first introduce some concepts and definitions borrowed from cryptography [1,2] in which the digital signature techniques have been well studied for traditional computer systems. We then extend them for multimedia applications.

Date authentication: Data authentication is a process determined by the authorized receivers, and perhaps the arbiters, that the particular data were most probably sent by the authorized transmitter and have not subsequently been altered or substituted. Data authentication usually associates with data integrity and nonrepudiation (i.e., source identification) because these issues are very often related to each other: Data that have been altered effectively should have a new source; and if a source cannot be determined, then the question of alteration cannot be settled (without reference to the original source).

Data integrity: Data integrity means that the receiver can verify that the data have not been altered by even 1-bit during the transmission. The attacker should not be able to substitute false data for the real data.

Nonrepudiation: Nonrepudiation, also called *source identification*, means the data sender should not be able to falsely deny the fact that he sent the data.

Typical methods for providing data authentication are digital signature schemes (DSS) together with *a* data or message authentication code (DAC or MAC), which actually is a type of keyed one-way hash functions.

One-way hash function: The one-way hash function or crypto hash is a hash function that works only in one direction. A good one-way hash function should have the following requirements: (1) Given a message *m* and a hash function *H*, it should be easy and fast to compute the hash $h = H(m)$; (2) given *h*, it is hard to compute *m* such that $h = H(m)$ (i.e., the hash function should be one way); (3) given *m*, it is hard to find another data m' such that $H(m') = H(m)$ (i.e., collision-free). Such a crypto hash guarantees that even a 1-bit change in the data will result in a totally different hash value. To differentiate it from other kinds of hash functions, we introduce a *crypto hash*, which will be throughout this chapter. Typical crypto hash functions include MD5 (128 bits) and SHA-1 (160 bits) [1,2].

Digital signature schemes: The digital signature scheme (DSS) is a bit string that associates the data with some originating entity. It includes signature generation (signing) procedure and signature verification procedure. Public key DSS is a very common technology and has been adopted as the international standards for data authentication [1,2]. Public key DSS uses the sender's private key to generate the signature and his public key to verify to a recipient the integrity of data and identity of the sender of the data. Typical public key DSS include the well-known Rivest-Shamir-Adleman algorithm (RSA) and digital signature algorithm (DSA) [1,2]. Figure 21.1 shows the block diagram of RSA and DSA digital signature schemes. Given an arbitrary length of data, a short fixed-length hash value is obtained by a crypto hash operation (e.g., 128 bits by MD5 or 160 bits by SHA-1). The signature is generated using the sender's private key to sign on the hash value (a standard length of private key, public key, and the generated signature is 1024 bits). The original data associated with its signature is then sent to the intended recipients. Later, the recipient can verify whether his received data were altered and the data were really sent from the sender, by using the sender's public key to authenticate the validity of the attached signature. The final authentication result is then drawn from a bit-by-bit comparison between two hash values (one is decrypted from the signature and the other is obtained by rehashing the received data; refer to Figure 21.1) by the criterion: Even if for 1-bit difference, the received data will be deemed unauthentic.

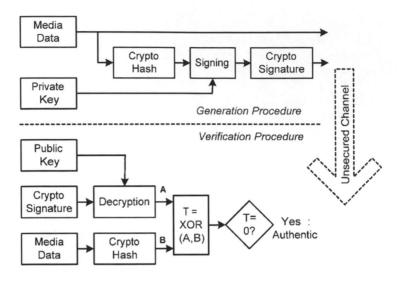

Figure 21.1. Block diagram of crypto signature schemes.

To differentiate the original public key digital signature schemes in cryptography from their modified schemes in multimedia security, we name them *crypto signatures*.

Data vs. content: Given a specific type of multimedia (e.g., image), multimedia data refer to the exact representation (e.g., binary bit stream). Multimedia content refers to its understanding or its meaning on the same representation. For example, before and after lossy compression, multimedia data should be different while multimedia content could be the same.

Content authentication: For content authentication, as long as the meaning of multimedia content is not changed, it should be considered as authentic, because the meaning of multimedia is based on their content instead of their data. Such requirements come from real applications where manipulations on the data without changing the meaning of content are considered as acceptable. Lossy compression and transcoding are examples. In addition to the distortions from malicious attacks, these acceptable manipulations will also introduce the distortions into the multimedia content.

Incidental distortion and intentional distortion: Content authentication has to deal with two types of distortion. One is the distortion introduced by the malicious content modifications (e.g., content meaning alteration) from attackers (hereafter we refer this type of distortions as intentional distortion); the other is the distortion from real applications, where some manipulation of content (e.g., lossy compression or transcoding) has to be considered allowable during the process of media transmission and storage (hereafter we refer to this type of distortion as incidental distortion). Content authentication has to tolerate all incidental distortions and reject any intentional distortions.

Content-based multimedia authentication has many potential applications. Consider the case of a police station transmitting a suspect's image to their officers' mobile terminals in order to identify a criminal. The policemen need to have guarantees that the image they received was indeed from the station. The police station also wishes to ensure that the sent image was not altered during transmission. However, due to different models of mobile terminals used by the policemen and different bandwidth conditions between the station and each policeman, the authentic image will undergo some manipulations such as lossy compression or format conversion before reaching the policemen's mobile terminals. The image still needs to be securely authenticated in such a case. Therefore, we have two requirements to consider. On the one hand, the scheme must be *secure* enough to prevent any attacked content from passing the authentication test (sensitive to intentional distortions). On the other hand, the scheme also needs to be *robust* enough to accept an image that has undergone some acceptable manipulations (insensitive to incidental distortions). The authentication robustness to the incidental distortions is the main difference between data authentication and content authentication because no robustness is required for data authentication (any bit change is not allowable).

It would be interesting to put all authentication-related technologies into one graph (Figure 21.2) in terms of three functions: integrity protection, source identification, and robustness. Ideally from a signed handwritten signature on the paper we (the trained experts) could identify who is the signer even under the case that the handwritten signatures vary a lot; but we cannot tell whether his signed paper content was altered or not because no any link exists between his signature and the paper content he signed. Digital watermarking could achieve a good capability of integrity protection at a high robustness level; but most of the watermarking schemes are unable to securely identify the source (i.e., the true and undeniable watermark embedder) because a symmetric key is used for both watermark embedding and watermark extraction. Crypto signature schemes can achieve both integrity protection and source identification at a very high security level; but its robustness level is fragile, i.e., any bit change will result in a failure of authentication. Media signature for content authentication, which covers all these three functions in a tunable manner, is the topic we mainly discuss in this chapter.

633

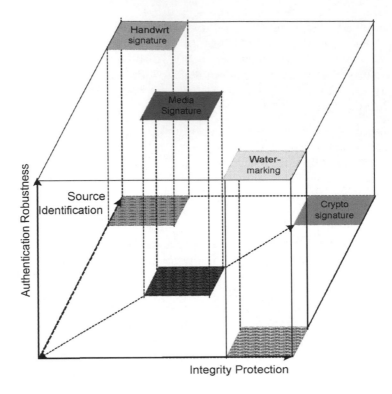

Figure 21.2. Comparison of various authentication related technologies.

Signature-Based Content Authentication

As we described in the previous subsection, data authentication provides the authentication result in such a way that even the data changed by 1 bit will be claimed as unauthentic (i.e., the data to be verified are either authentic or unauthentic). Such a criterion motivated the researchers to create many crypto signature schemes whose system security level is very high and can even be proved mathematically. DSA and RSA are two good crypto signature schemes adopted by the international standards [1,2].

The requirement on a certain level of authentication robustness is the main difference between data authentication and content authentication. Such a difference makes authenticating multimedia content much more challenging and complex than authenticating multimedia data [3,4].

- A good media signature is required to be robust to *all* incidental distortions and sensitive to *any* intentional distortions. However, different incidental distortions come from different defined acceptable content manipulations. Defining acceptable content manipulations is application dependent.

- Given a specific application, it is very hard to formalize a clear boundary between incidental distortions and intentional distortions (i.e., differentiating the acceptable content manipulations clearly from the malicious content modifications). Consequently, instead of presenting the authentication result precisely as either authentic or unauthentic, content authentication may have to say "don't know" in some circumstances, as shown in Figure 21.3. It also means that a certain false acceptance rate and false rejection rate are normal in the case of content authentication.

- An existing gray area in the authentication results implies that the system security is not as good as data authentication, although we could argue that the portion of system security impaired by system robustness could be compensated for by the contextual property of the media content. For example, if the attacker is to modify a black pixel to the white one in a natural image, the case that its surrounding pixels are all black is more likely to be detected than that its surrounding pixels are all white pixels.

Based on the robustness level of authentication and the distortions introduced into the content during content signing, we categorize state-of-the-art techniques based on signature-based authentication as follows (Figure 21.4). Different types of signature-based authentication methods

(a) Data authentication (b) Content authentication

Figure 21.3. Presentation of authentication results.

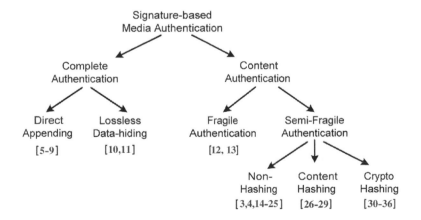

Figure 21.4. Categories of signature-based authentication.

are targeted for different applications:

Complete authentication: Complete authentication refers to directly applying the crypto signature scheme to multimedia data authentication. It works in a way that the multimedia data to be sent to the recipient is either in its original form [5–9] or able to be exactly recovered at the receiver site [10,11]. We call it complete authentication because the case of any bit changing in its original form will be deemed as unauthentic. The generated signature is sent to the recipient either by appending it to the data or by inserting it into the data in a lossless way. Note that complete authentication can not only authenticate the stable multimedia data but also authenticate the streaming multimedia data, even slightly robust to packet loss.

Content authentication: Content authentication includes fragile authentication and semifragile (sometimes it is also called semirobust) authentication. We call it content authentication because both methods need to "distort" the multimedia data to achieve the authentication.

Fragile authentication: Fragile authentication refers to those methods that authenticates the multimedia content via modifying a small portion of multimedia content, but in a fragile way (i.e., any bit changing in the protected multimedia content is not allowed). Typical solutions are to embed the generated signature into the content via fragile watermarking [12,13]. In other words, it does not have any authentication robustness, although it still needs to modify the multimedia data.

Semifragile authentication: Semifragile authentication refers to those methods that are robust to the predefined content manipulations such as lossy compression or transcoding. To achieve a certain level of robustness, the actual data to be authenticated are "distorted" versions of the original multimedia data (e.g., the extracted feature). Its typical approaches include nonhashing methods (i.e., crypto hashing is not applicable for those methods) [3,4,14–25], content hashing methods (i.e., some proprietary methods mapping the content into a short data sequence) [26–29], and crypto hashing methods [30–36].

COMPLETE AUTHENTICATION

Complete authentication refers to directly applying the crypto signature scheme to multimedia data authentication, as shown in Figure 21.1. A detailed description on how to generate and verify the crypto signature can be found in Reference 2 (chap. 20).

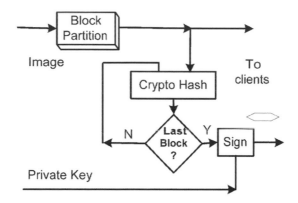

Figure 21.5. Signature generation by successively hashing blocks.

Friedman first proposed to incorporate the crypto signature schemes into digital camera to make the captured images authentic [5]: the trustworthy digital camera. Later, Quisquater et al. proposed an improved solution by partitioning the image into a number of blocks and then signing on the last hash value obtained by successively crypto hashing block by block [6], as shown in Figure 21.5. It brings some authentication flexibilities for imaging applications such as image editing.

In the direction of increasing the authentication flexibility, Gennaro and Rohatgi extend crypto signature schemes from signing static data to signing steaming data and proposed a tree-based authentication structure to increase the efficiency of signature generation [7]. Yu [8] further extended the schemes in Reference 7 to the applications of scalable multimedia streaming. Again, all of these methods do not have any robustness; that is, they cannot tolerate any single bit loss, for authentication. In order to tolerate packet loss, which frequently occurs during data transmission especially over an unreliable channel, Park et al. proposed an interesting authentication work based on the concept of Error Correcting Coding (ECC) [9]. Their basic idea is illustrated in Figure 21.6. Naturally, a MAC has to be appended to its corresponding packet for stream authentication, as shown in the upper part of Figure 21.6. The authentication can be executed after receiving the last packet. Such a scheme cannot deal with the packets loss during the transmission because the signature was generated based on all hash values from all packets. To overcome this problem, a straightforward solution is to add some redundancies by attaching several hashes from other packets into the current transmitting packet. If the current packet (e.g., N) is lost, its hash still can be obtained from other packets (e.g., $N+m$). The verification on the whole stream can still be

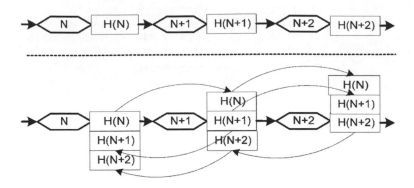

Figure 21.6. Data streaming authentication robust to packet loss.

executed. Obviously, such solutions obtain a certain robustness by paying for the extra transmission cost. In Reference 9, they proposed an optimized solution that balances the robustness and transmission cost. Although it achieves a certain robustness, we still classify it as complete authentication because all of the received data cannot tolerate any bit errors.

Instead of storing and transmitting the generated signature separately, the generated signature can also be inserted back into the multimedia content via lossless data hiding [10,11]. The basic procedure is as follows:

Signing

- Generate the crypto signature from the whole set of multimedia data.
- Embed the signature back into the multimedia content in a lossless way (watermarking).
- Send this watermarked multimedia content to the recipient.

Verifying

- Receive the watermarked content from the sender.
- Extract the signature and recover the original form of the multimedia content.
- Verify the original form of the multimedia content based on the crypto signature scheme.

One example for lossless watermarking is to compress the least significant bit (LSB) portion of the multimedia data by a lossless data compression algorithm such as ZIP and then insert the signature into the free space earned by this lossless data compression algorithm. Interested readers can refer to Chapter 17 of this handbook for more details on how to embed the data in a lossless way.

CONTENT AUTHENTICATION

Content authentication includes fragile authentication and semifragile (sometimes it is called semirobust) authentication. We call it content authentication because both methods need to modify the multimedia data to achieve the authentication, although the meaning of "modification" is different for fragile and semifragile authentication. For fragile authentication, the content needs to be chopped some portions to store the signatures; for semifragile authentication, the content needs to be abstracted to a level to tolerate the incidental distortions.

Fragile Authentication

Fragile authentication refers to those methods that authenticate the multimedia content via modifying a small portion of multimedia content, but in a fragile way (i.e., any bit changing in the protected multimedia content is not allowed). Typical solutions can be depicted as follows [12,13].

Signing

- Set all LSB portions to zero to obtain the protected data.
- Generate the crypto signature by signing on the whole set of protected data.
- Store the signature into the LSB portion to obtain the watermarked multimedia content.
- Send this watermarked multimedia content to the recipient.

Verifying

- Receive the watermarked content from the sender.
- Extract the signature out from the LSB portion of the received multimedia content.
- Regenerate the protected data by setting all LSB portions to zero.
- Verify the protected data based on the selected crypto signature scheme.

Both complete authentication and fragile authentication work on the fragile level; that is, they do not allow any bit change in their protected data. However, complete authentication protects the whole set of multimedia data (i.e., the authentication process will not distort any multimedia data and the users can obtain the original form of their wanted multimedia data), whereas, fragile authentication only protects the multimedia data, excluding the LSB watermarked portion (i.e., the authentication process will distort a small portion of multimedia data permanently and the users can only obtain the watermarked version of the original multimedia data, although the introduced distortion is invisible).

Semifragile Authentication

Semifragile authentication refers to those methods that are robust to pre-defined content manipulations such as lossy compression or transcoding. Usually, it authenticates an abstracted version of the original multimedia data to achieve a certain degree of robustness. Its typical approaches include nonhashing methods (i.e., crypto hashing is not applicable for those methods) [3,4,14–25], content hashing methods (i.e., some proprietary methods mapping the content into a short data sequence) [26–29], and crypto hashing methods (i.e., crypto hashing is incorporated into the scheme) [30–36].

Nonhashing. Originating from the ideas of crypto signature schemes as described earlier, Schneider and Chang [14] and other researchers proposed using some typical content-based measures as the selected features for generating content signature, by assuming that those features are insensitive to incidental distortions but sensitive to intentional distortions. Those features include the histogram map [14], the edge or corner map [16,17], moments [19], a compressed version of the content, and more [20–25]. Considering that applying a lossy compression such as JPEG should be deemed an acceptable manipulation in most applications, Lin and Chang [15] discovered a mathematically invariant relationship between two coefficients in a block pair before and after JPEG compression and used it as the selected feature. Similarly, Lu and Liao [18] presented a structural signature solution for image authentication by identifying the stable relationship of a parent–child pair of coefficients in the wavelet domain.

Referring to Figure 21.7, we can see that the module of the "crypto hash" in Figure 21.1 has been replaced with the module of "feature extraction" here in order to tolerate some incidental distortions. The replacement is applied because the acceptable manipulations will cause changes to the content features, although the changes may be small compared to content-altering attacks. Such "allowable" changes to the content features make the features non-crypto-hashing. (Any minor changes to the features may cause a significant difference in the hashed code due to the nature of the crypto hashing methods such as MD5 and SHA-1.) Accordingly, as a result of the incompatibility with crypto hashing, the generated signature size is proportional to the size of the content, which is usually very large. In typical digital signature algorithms, the signature signing is more computational than signature verifying; this will also result in a time-consuming signing process because the size of the formed signature is much greater than 1024 bits [1,2] and it has to be broken into small pieces (less than 1024 bits) for signing. On the other hand, no crypto hashing on selected features will make the decision of authenticity, which is usually based on

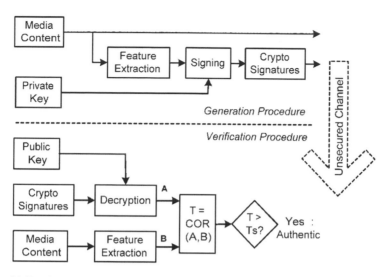

Figure 21.7. Semifragile-signature-based authentication (non-crypto-hashing).

comparison of the feature distance (between the original one decrypted from the signature and the one extracted from the received image) against a threshold value, which is hard to determine in practice and brings some potential security risks.

Content Hashing. Because directly applying crypto hashing to images (features) seems infeasible and feature-based correlation approaches still do not resolve the issues such as signature size and security risks, other researchers have already been working on designing their own hash functions: *robust hash* or *content hash* (see Figure 21.8). Differing from crypto hash functions, content hashing aims to generate two hash codes with a short hamming distance if one image is a corrupted version of another image by incidental distortions. In other words, if two images or two image blocks are visually similar, their corresponding hash codes should be close in terms of hamming distance. For example, in the Fridrich Goljan [26] solution, the hash function is designed to return 50 bits for each 64×64 image block by projecting this 64×64 image block onto a set of 50 orthogonal random patterns with the same size (i.e., 64×64) generated by a secret key. The final hash value of the image is then obtained by concatenating the hash codes from all 64×64 image blocks.

Xie et al. [27] proposed a content hash solution for image authentication called Approximate Message Authentication Codes (AMACs). The AMAC is actually a probabilistic checksum calculated by applying a series of random XOR operations followed by two rounds of majority

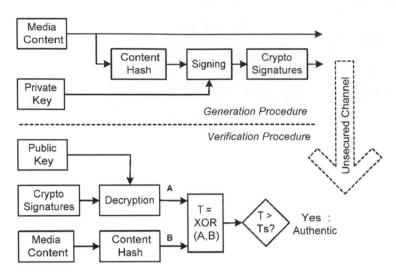

Figure 21.8. **Semifragile-signature-based authentication (content hashing).**

voting to a given message. The similarity between two messages can be measured by the hamming distance of their AMACs. The length of an AMAC is typically around 80–400 bits. Venkatesan et al. [28] also proposed a solution for generating content hash for an image. First, the image is randomly tiled and the wavelet transform is applied to each tile independently. Some statistics measures such as mean and variance are calculated in each wavelet domain. Second, those obtained measures are quantized using a random quantizer (i.e., the quantization step size is random) to increase security against attacks. Third, the quantized statistics measures in each tile are decoded by a predefined error correction coding scheme. Finally, the content hash value of the image is obtained by concatenating the ECC decoder outputs of all tiles. Kailasanathan et al. [29] proposed another compression-tolerant discrete cosine transform (DCT)-based image hash method, which is robust to lossy compression (e.g., JPEG). Conceptually, they obtained the hash value by an intersection of all subsets compressed with different JPEG quality factors.

However, some limitations also exist for this type of content-hash-based scheme. Due to a short representation of generated content hash, it is very hard for content hash to differentiate incidental distortions from intentional distortions, especially when the intentional distortions are the results of attacks to only part of the image. Consequently, it is very difficult to set a proper threshold for making the authentication decision. Furthermore, this scheme also lacks the ability to locate the content modifications if the authentication fails.

Crypto Hashing. After studying the non-crypto-hashing and content-hashing signature schemes for semifragile multimedia authentication, we can summarize here that a good semifragile digital signature scheme should satisfy the following requirements. First, the size of the signature signed by a semifragile scheme should be small enough to reduce the computational complexity. Ideally, the size is comparable to or the same as that which is signed by a crypto signature scheme. Second, such a scheme should be able to locate the attacked portion of the image because that would easily prove the authentication result. Finally, and most importantly, a proper content-based-invariant feature measure should be selected in such a way that it is sensitive to malicious attacks (intentional distortions) and robust to acceptable manipulations (incidental distortions). In addition, the feature should be able to characterize the local property of an image because most attacks only act on part of the image (e.g., only changing the digits in a check image not the whole image); feature selection is obviously application dependent because different applications have different definitions of incidental distortions as well as intentional distortions. Therefore, defining acceptable manipulations and unallowable modifications of the image content is the first step in designing a good semifragile authentication system.

To achieve the requirements described, another type of signature-based approaches is proposed [30–35]. Crypto hashing was incorporated into the proposed approaches to fix the length of the generated signature and enhance system security. Figure 21.9 illustrates the whole framework which comprise three main modules. Signature

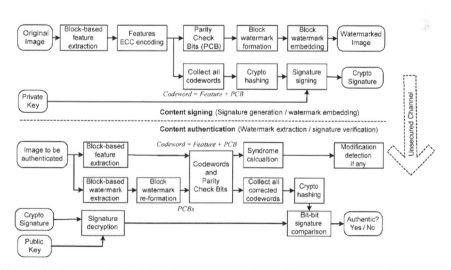

Figure 21.9.   Crypto-hashing-signature-based semifragile authentication.

generation and verification modules are mainly employed for the role of content signing and authentication. Watermark embedding and extraction modules are only used for storing signatures. ECC is employed to tackle the incidental distortions.

The procedure of content signing (the upper part in Figure 21.9) can be depicted as follows. The input original image is first partitioned into nonoverlapping blocks. A transform such as DCT is usually needed for each block. Block-based-invariant features are then extracted and mapped onto binary values if the subsequent ECC scheme is binary. After ECC encoding, their corresponding parity check bits (PCBs) of a each block-based-feature set can be obtained, taking PCBs as the seed of a watermark to form the block-based watermark. One necessary condition in selecting a watermarking scheme is that the embedded watermark should be robust enough for extraction from received images under acceptable manipulations. Therefore, a simple method for generating watermark data is to use ECC again: PCBs are encoded by another ECC scheme and then the ECC-encoded output is embedded as watermark data. The watermark data for each block are embedded either into the same block or into a different block. In the meantime, all codewords (features together with their corresponding PCBs) are concatenated and hashed by typical cryptographic hashing function such as MD5. Finally, the content owner's private key is used to sign the hashed value to generate the robust crypto signature.

The content signing algorithm is further described as follows.

System setup

- Content owner requests a pair of keys (private key and public key) from the PKI authority.

- Select an adequate ECC scheme (N, K, D) given domain-specific acceptable manipulations. Here, N is the length of output encoded message, K is the length of original message, and D is the error correction capability.

- Select another ECC scheme (N', K', D') for watermark formation as described above (optional).

Input

Original image to be signed I_o

Begin

- Partition image into nonoverlapping blocks $(1, \ldots, B)$.

- **For** block 1 **to** block B, **Do**

    Conduct block-based transform such as DCT.

Extract invariant feature.

Map each feature set into one or more binary messages each of which has length K.

ECC encode each binary vector to obtain its codeword W (N bits) and PCBs (N–K bits).

1. Take PCBs is used as watermark
   Embed watermark into selected block
   Inverse transform to obtain watermarked image I_w

2. Collect codewords from all blocks W $(1...B)$. Concatenate them to form a single-bit sequence Z

End

Hash the concatenated codeword sequence Z to obtain $H(Z)$

Sign on $H(Z)$ by the owner's private key to obtain the signature S

End

Output

- Watermarked image I_w
- Content-based crypto signature S

As described earlier only the PCBs are embedded as watermarks and are used later in the verification stage for correcting potential changes in the signatures (i.e., content features). As shown in the example ECC in Table 21.1, we can see that the first four bits in a codeword are from the original message bits. Assume that we want to protect the message 0001; its corresponding codeword is 0001111.

TABLE 21.1. (7, 4) Hamming code

Message	Codeword Message	PCB	Message	Codeword Message	PCB
0 0 0 0	0 0 0 0	0 0 0	1 0 0 0	1 0 0 0	0 1 1
0 0 0 1	0 0 0 1	1 1 1	1 0 0 1	1 0 0 1	1 0 0
0 0 1 0	0 0 1 0	1 1 0	1 0 1 0	1 0 1 0	1 0 1
0 0 1 1	0 0 1 1	0 0 1	1 0 1 1	1 0 1 1	0 1 0
0 1 0 0	0 1 0 0	1 0 1	1 1 0 0	1 1 0 0	1 1 0
0 1 0 1	0 1 0 1	0 1 0	1 1 0 1	1 1 0 1	0 0 1
0 1 1 0	0 1 1 0	0 1 1	1 1 1 0	1 1 1 0	0 0 0
0 1 1 1	0 1 1 1	1 0 0	1 1 1 1	1 1 1 1	1 1 1

We only need to use the last 3 bits (PCBs) to form the MAC and use it as watermark data. Later, assume that we receive a message like 0011 (1-bit change compared to the original message 0001). By checking with its original parity check value 111, we can detect and correct the code 0011 back to 0001 and obtain its correct corresponding codeword 0001111 again. It is clear that by using a simple ECC scheme, the system robustness is improved. It is likely that minor changes caused by acceptable manipulations (e.g., lossy compression or codec implementation variation) can be corrected by the ECC method. However, the use of a ECC also brings some security concerns. Because the mapping from messages to PCBs is a multi-to-one mapping, the reverse mapping is not unique. In the example shown in Table 21.1, one PCB is shared by two messages. It results in some security problems. For example, given the original message 0001, its PCB is 111. We can replace the original message with a faked one: 1111; its corresponding PCB also is not affected — still 111. This case will become worse in practice, as the length of the message (i.e., extracted feature) usually is mush longer than that of PCBs. Such a security problem is solved by cryptographically hashing all codewords instead of only hashing message bits or PCBs. Let us recheck Table 21.1. We can see that, although given one PCB, there are multiple messages sharing the PCB. However, their corresponding codewords are different (0001111 and 1111111, respectively). In other words, each syndrome (message and PCB), is uniquely defined. Any change in the message or the PCBs will make the syndrome different. Given the uniquely defined concatenated codeword sequence, we can apply cryptographic hashing (e.g., MD5) to the codeword sequence and form a much shorter output (about a few hundreds bits).

Refer to Figure 21.9 (lower part), to authenticate a received image content, in addition to the image itself, two other pieces of information are needed: the signature associated with the image (transmitted through external channels or as embedded watermarks) and the content owner's public key. The image is processed in the same way as feature generation; decompose image into blocks, to extract features for each block, to form finite-length messages. From the embedded watermarks, we also extract the PCBs corresponding to messages of each block. Note that messages are computed from the received image content, whereas the PCBs are recovered from the watermarks that are generated and embedded at the source site. After we combine the messages and the corresponding PCBs to form codewords, the whole authentication decision could be made orderly. First, we calculate the syndrome block by block to see whether there exists any blocks whose codewords are uncorrectable. If yes, then we could claim that the image is unauthentic. Second, assume that all codewords are correctable; we replace those erroneous codewords with their corrected ones.

Then, we repeat the same process at the source site: concatenate all corrected codewords into a global sequence and cryptographically hash the result sequence. By using the owner's public key, the authenticator can decrypt the hashed value that is generated at the source site. The final authentication result is then concluded by bit-by-bit comparison between these two hashed sets: If there is any single bit different, the authenticator will report that the image is unacceptable ("unauthentic").

It is interesting and important to understand the interplay between the decision based on the block-based syndrome calculation and the crypto signature. The syndrome can be used to detect any unrecoverable changes in a block. However, because we do not transmit the entire codeword, there exist changes of a block that cannot be detected (as for the case 0001111 vs. 1111111 discussed earlier) because of their normal syndrome calculation. However, such changes will be detected by the crypto signature, because the hashed value is generated by using the entire codewords, not just the PCBs. Therefore, there exist such possibilities: The image is deemed as unauthentic because of inconsistence between hashed sets while we are unable to indicate the locations of attacks because there are no uncorrectable codewords found. In such case, we still claim that the image is unauthentic, although we are not able to indicate the exact alternation locations.

The above-described framework has been successfully applied to JPEG authentication [31,35], JPEG2000 authentication [32], MPEG-4 object [33], MPEG-1/2 [34], and so forth. Their defined acceptable manipulations, selected feature, and watermarking methods are summarized in Table 21.2.

It is interesting to revisit the issue of authentication robust to packet loss here. Recall that such an issue could be solved by complete authentication [9]. Figure 21.10 illustrates the typical procedure of image transmission over a lossy channel (i.e., some data will be lost during transmission). The original image is encoded together with some error-resilient techniques, block by block. The compressed bit stream is then packetized for transmission. Assume that the transmission channel is unreliable; some data packets will be lost before reaching the receiver (e.g., packets corresponding to block 5). Such loss will be known at the receiver end either by the transmission protocols (e.g., Real-Time Protocol [RTP]) or by the content-based error detection techniques. The corrupted image could be approximately recovered by error concealment techniques before displaying or further processing. Refer to Figure 21.10, at the receiver end; the authentication could be done either at point X or point Y, depending on applications. If we want to authenticate the corrupted image without error concealment, then we should conduct

TABLE 21.2. Crypto-hashing-based semifragile signature schemes

Media	Acceptable Manipulations	Malicious Attacks	Features	Watermarking
JPEG [31]	Lossy compression, transcoding	Content copy–paste and cutting, etc.	DCT DC/AC coefficients	DCT quantization
JPEG2000 [32]	Lossy compression, rate transcoding	Content copy–paste and cutting, etc.	Wavelet coefficients	Bitplane modification
MPEG-4/object [33]	Lossy compression, object translation/scaling/rotation, segmentation errors	Background or object replacement, partial object modification, etc.	MPEG-7 object descriptor	FFT grouping and labeling
MPEG-1/2 [34]	Lossy compression, frame resizing, and frame dropping	New frame insertion, content modification within frame, etc.	DCT DC/AC coefficients	DCT grouping and quantization
JPEG [35]	Lossy compression, noise, packet loss	Content copy–paste and cutting, etc.	Preprocessed DCT DC/AC coefficients	DCT quantization

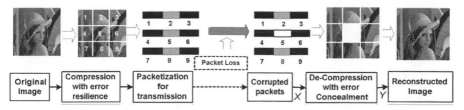

Figure 21.10. Image transmission over an unreliable channel.

the authentication at point X, which has been achieved by complete authentication [9]; otherwise, we should select point Y for authentication. The advantage for authenticating at point Y is that the concealed image is still securely reusable for other applications without redoing error concealment.

The solution of authenticating at point Y was proposed in Reference 35. Figure 21.11 presents some test examples done in Reference 35. Figure 21.11a is the original image with a size of 768×512. Figure 21.11b is the signed watermarked image compressed with JPEG QF (quality factor) 8, the robustness of authentication and watermarking is set to JPEG QF 7. Note that the QF for authentication robustness is lower than the QF for compression. Figure 21.11c is the damaged image of Figure 21.11b due to packet loss. (The bit error rate [BER] is 3×10^{-4}.) It can still pass the authentication after concealing the errors. Figure 21.11d is the attacked image on the damaged image in Figure 21.11c (one window in the image center was filled with its surrounding image content). Figure 21.11e is the recovered image obtained by applying the error concealment techniques in Figure 21.11d. It cannot pass the authentication and Figure 21.11f shows the detected attacked location. Due to packet loss and its subsequent error concealment, some false detected locations occur and are shown in Figure 21.11f.

A UNIFIED SIGNATURE-BASED AUTHENTICATION FRAMEWORK FOR JPEG2000 IMAGES

In this section, we describe a signature-based authentication framework for JPEG2000 images [36]. The proposed framework integrates the complete authentication and semifragile authentication together to form a scalable authentication scheme, targeting various applications because different applications may have different requirements on authentication robustness.

Recently, the Joint Photographic Experts Group issued a new still-image compression standard called JPEG2000 [37]. Compared to JPEG compression based on DCT, this new standard adopts wavelet transform

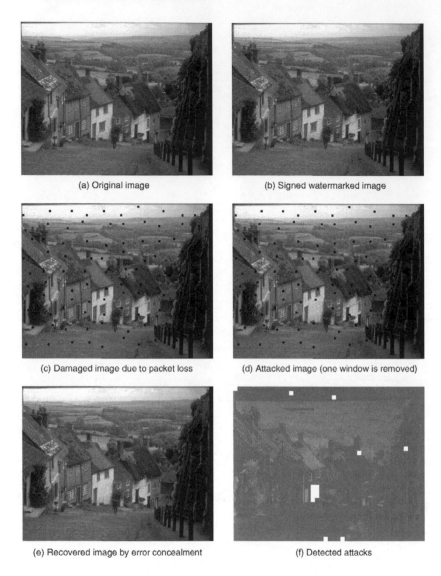

(a) Original image

(b) Signed watermarked image

(c) Damaged image due to packet loss

(d) Attacked image (one window is removed)

(e) Recovered image by error concealment

(f) Detected attacks

Figure 21.11. Authentication examples robust to packet loss.

(WT) to obtain better energy compaction for the image. Therefore, it can achieve improved compression efficiency, especially for a bit-rate compression lower than 0.25 bpp (bits per pixel). By adopting multiple-resolution representation and an embedded bit stream, decoding can be flexibly achieved in two progressive ways: resolution scalability and signal-to-noise (SNR) scalability. Some other rich features provided by JPEG2000 include region-of-interest (ROI) coding, tiling, error resilience, random codestream access, and so forth. Figure 21.12 illustrates the

650

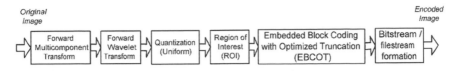

Figure 21.12. A brief diagram of JPEG2000 encoding.

main encoding procedure of JPEG2000. Given a color image, after forward multicomponent transformation (i.e., color transformation), it will be decomposed into different resolution levels and subbands by forward wavelet transformation. It is then quantized by a uniform scalar quantizer (Standard Part I). An adaptive binary arithmetic coder will start encoding all quantized WT coefficients from the most significant bit (MSB) bitplane to the least significant bit (LSB) bitplane. The final bit stream will depend on the predefined compression bit rate and the progression order. The last two steps are also called EBCOT (Embedded Block Coding with Optimized Truncation). Through-EBCOT, the user can quantitatively define the intended compression ratio for his images in terms of the bit rate.

Because of the novel features in JPEG2000, JPEG2000 has received significant attention and support from various applications such as mobile communication, medical imaging, digital photography, digital library, and e-commerce, and so forth. Such applications demand content-integrity protection and source identification (i.e., a secure and robust authentication solution for JPEG2000 compatible images). However, flexible and efficient coding strategies for JPEG2000 also pose some new challenges. For example, JPEG2000 is known as "encode once and decode many times." This means that once a JPEG2000 codestream is formed at a given compression bit rate, any other new codestream whose bit rate is lower than the given compression bit rate can be simply obtained by just truncating this formed codestream from its end. Furthermore, JPEG2000 is the first international image standard that integrates lossy and lossless compression into one codestream. Hence, it requires that the proposed authentication solution be able to align with these flexibilities.

Some potential applications of such JPEG2000 authentication framework could be:

- *Content streaming*: protecting the integrity of the streamed content under given authentication bit rate. The streaming could be done in several ways such as streaming the content into a buffer with bit rate A, later streaming it into the client with bit rate B, and so on. As long as all of the streamed bit rates in terms of the original file size are greater than the said authentication bit rate,

651

the streamed content should be protected against unauthorized modifications.

- *Content transformation in different domains*: Given the authentication bit rate, the content to be protected may undergo some practical transformations among different domains such as digital-to-analog and analog-to-digital. By using the ECC scheme, the transformed content should be protected against unauthorized modifications as long as the bit rate of the transformed content is still greater than authentication bit rate.

Based on these JPEG2000 targeting applications, we identify a list of acceptable manipulations for the proposed solution, which may not be complete:

- *Multicycle compression.* Typically, this involves reencoding a compressed image to a lower bit rate. The reencoding may be repeated multiple times. In practice, there is a minimal requirement for image quality. Thus, one may set a minimal acceptable bit rate beyond which the image will be considered unauthentic. Such manipulations will unavoidably introduce some incidental distortions. Assume that we want a 1-bpp JPEG2000 compressed image. One example is that the original image repeatedly undergoes JPEG2000 compression; that is, compressing the original image into 1 bpp and decompressing it to an 8-bit raw image (one cycle), compressing it again and again at 1 bpp. Furthermore, we know that JPEG2000 provides much flexibility in compressing an image to a targeted compression bit rate, such as directly encoding from raw image or truncating or parsing from a compressed codestream. Another example of obtaining a 1-bpp JPEG2000 compressed image is to compress the original image with full, 4 bpp, and 2 bpp, respectively and then truncating to the targeted bit rate of 1 bpp.
- *Format or codec variations.* Differences may exist between different implementations of JPEG2000 codec by different companies. Such differences can be due to different accuracies of representation in the domains of (quantized) WT, color transformation, and the pixel domain.
- *Watermarking.* Image data are "manipulated" when authentication feature codes are embedded back into the image. Such a manipulation should not cause the resulting image to be considered unauthentic.

In Reference 36, we proposed a unified authentication system that can protect a JPEG2000 image with different robustness modes: complete and semifragile fragile (lossy and lossless). The whole framework is compliant with PKI: After image signing, the signature together with the watermarked image is sent to the recipients. At the receiver site, the recipient can verify the authenticity of the image by using the image sender's public key and the signature. The complete authentication mode is straightforward by employing the traditional crypto signature scheme. Lossy and lossless modes are robust against the predefined image manipulations such as image format conversion or transcoding. In addition, lossy and lossless modes can allocate the attacked area if the image is maliciously manipulated. Moreover, the lossless mode is able to recover the original image after image verification if no incidental distortion is introduced. Finally, the authentication strength could be quantitatively specified by the parameter lowest authentication bit rate (LABR). It means that all data or content of the image above the LABR will be protected. Thus, it will offer users convenience. It is worth mentioning that the proposed solution is also compatible with other JPEG2000 features such as JPEG2000 codestream transcoding (i.e., parsing and truncation), ROI coding, and so forth. This facilitates the proposed solution for more JPEG2000 applications. Figure 21.13 is the diagram of the proposed system.

In Figure 21.13, the proposed solution can work on three modes: complete authentication mode (i.e., traditional signature scheme), semifragile authentication mode with lossy watermarking, and semifragile authentication mode with lossless watermarking. Different levels of the authentication robustness provided by the three authentication modes, together with the authentication strength, which can be quantitatively specified by the LABR parameter, forms a scalable authentication scheme. In addition, the protected areas can also be defined in terms

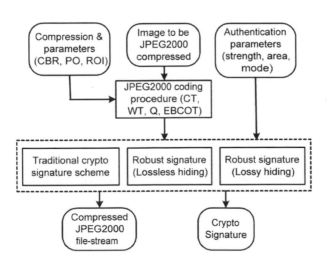

Figure 21.13. Diagram of the proposed authentication framework for JPEG2000 images.

of component, tile, resolution, subband, layer, codeblock, and even packet (all are used for representing JPEG2000 images with different granularities).

Complete Authentication Mode

The complete authentication mode is selected for protecting JPEG2000 compressed codestreams. The signing and verifying operations are quite straightforward. Figure 21.14 shows its signing procedure. During signing, the original image is encoded as per normal. While the codestream is formulated, its protected parts, as specified by LABR and other parameters, are extracted and fed to crypto hashing and signing operation. As result, a crypto signature is generated. The verification procedure is the inverse procedure of signing, as shown in Figure 21.15. During verifying, while the codestream is parsed during decoding, its protected part, as specified by LABR and other parameters, is extracted and fed to the traditional hashing and verifying operation, which

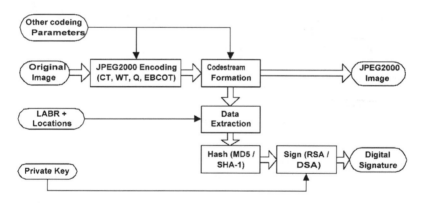

Figure 21.14. Complete authentication mode for JPEG2000 images (signing).

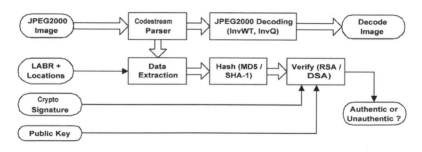

Figure 21.15. Complete authentication mode for JPEG2000 images (verifying).

returns the verification result: Even a 1-bit change in the protected part will be deemed as unauthentic.

Semifragile Content Authentication (Lossy)

The semifragile content authentication (lossy) mode is usually selected for those applications demanding for more robustness such as wireless communication. It is the most robust mode in the three authentication modes.

The proposed signing procedure is shown in Figure 21.16. In the scheme, the signature generation and verification modules are mainly employed for content signing and authentication. Watermark embedding and extraction modules are only used for storing ECC check information. Instead of directly sending an original image to recipients, we only pass them the watermarked copy associated with one signed digital signature whose length is usually very short (e.g., 1024 bits if signed by DSA or RSA). In Figure 21.16, among the four inputs — original image, JPEG2000 compression bit rate (CBR) b, LABR a, and the image sender's private key, CBR is the mandatory input for compressing images into the JPEG2000 format. In addition to the original image, only two inputs are needed to generate the JPEG2000 image signature in a content-based way: the private key and the LABR a. If a JPEG2000 image signature is generated with the LABR value a, a new image with the CBR b will be authentic as long as b is greater than a and the new image is derived from defined acceptable manipulations or transcoded (by parsing or truncation) from a compressed JPEG2000 image.

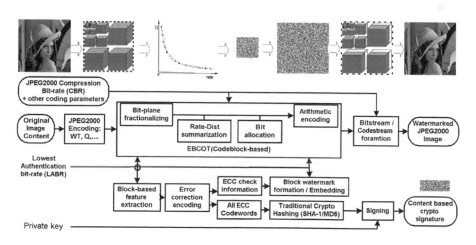

Figure 21.16. **Content authentication (lossy) mode for JPEG2000 images (signing).**

The original image first undergoes color transformation, wavelet transformation, and quantization, which are all basic procedures in JPEG2000 encoding [37]. EBCOT is then employed for bitplane fractionalizing or encoding and optimal bit-rate control. We extract content-based features from the available fractionalized bitplanes by assuming the image is encoded above LABR. Feature values from each codeblock are thresholded and ECC coded to generate corresponding PCBs. Then, we take PCBs as the seed to form the block-based watermark. One necessary condition in watermarking is that the embedded watermark should be robust enough for extraction from received images under acceptable manipulations. Because incidental changes to the embedded watermarks might occur, we apply another ECC scheme again to encode the PCB data before they are embedded. The watermark data for each block are embedded into either the same block or a different block. The watermark embedding location is also determined based on the LABR value. Note that only the PCB data (not including the feature codes) are embedded in the above watermarking process. All codewords (features together with their corresponding PCBs) from all resolution levels and all subbands are concatenated and the resulting bit sequence is hashed by a cryptographic hashing function such as MD5 or SHA-1. The generated semi-fragile crypto hash value can then be signed using the content sender's private key to form the crypto signature. Differing from a data-based signature scheme in which the original data are sent to the recipients associated with its signature, in the proposed solution we send out the watermarked image to the recipients instead of sending the original image.

We select EBCOT [37] to generate robust content-based features based on the following considerations. First, EBCOT is the last processing unit prior to forming the final compressed bitstream in JPEG2000. It means that all possible distortions have been introduced before running EBCOT in the encoding procedure and no distortion is introduced afterward. If we are to authenticate the encoding output or output of directly truncating the compressed bitstream, the extracted features will not be distorted. Second, EBCOT expands the scalability of image by fractionalizing the bitplane. The significant bitplanes of EBCOT represent closely the image content and, thus, it is hard to alter the image while intentionally keeping the same significant bitplanes. Third, in JPEG2000, EBCOT is the engine of bit-rate control and provides exact information about specific passes of data to be included in the final bitstream given a target bit rate. Such information allow us to specify the invariant layers of data and quantitatively select an authenticable level.

EBCOT provides a finer scalability of image content resulted from multiple pass encoding on the codeblock bitplanes (i.e., fractionalizing each bitplane into three subbitplanes: significant pass, magnitude refinement pass and cleanup pass based on predefined contextual models). The coding summaries for each codeblock after EBCOT include feasible truncation points, their corresponding compressed size (rate), and estimated distortions (rate-distortion curve). The target compression bit rate is achieved by globally scanning the contributions from all codeblocks and optimally selecting truncating points for each codeblock (using the Lagrange multiplier method). One important property is worth noting — passes included in the bit stream at a rate (say a) are always included in the bit stream at a higher rate ($\geq a$). Such a property is illustrated in Figure 21.2, in which no curves cross the curves that correspond to passes included at different compression bit rates. Such a property is important for our proposed solution in order to obtain invariant feature sets. Based on the above observation, we select two measures as invariant features directly from EBCOT: One is the state of passes (i.e., the fractionalized bitplanes) of MSBs and the other is the estimated distortion associated with each pass. The estimated distortion is a measure of the change of "1" in a given bitplane [37]. However, from the experiments [32], there are still some small perturbations introduced to the features by the defined acceptable manipulations. This is why we employ ECC to tame such distortions in order to generate stable crypto hash values and signature.

Refer to Figure 21.17. To authenticate the received image content, in addition to the image, two other pieces of information are needed: the signature associated with the image (transmitted through external channels or as embedded watermarks) and the content sender's public key. The image is processed in the same way as content signing: decompose and quantize image into blocks, to extract features for each block. (Note that here we assume that the JPEG2000 image has been decoded into a raw image and we are authenticating this raw image given LABR. If the image is still JPEG2000 compressed, the features and watermarks

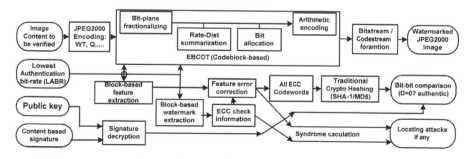

Figure 21.17. Complete authentication (lossy) mode for JPEG2000 images (verifying).

can also be obtained in the EBCOT domain from the JPEG2000 decoding procedure.) From these embedded watermarks, we extract the PCB data generated at the source site. Note that the features are computed from the received image, whereas the PCBs are recovered from the watermarks that are generated and embedded at the source site. After we combine the features and the corresponding PCBs to form codewords, the whole content verification decision could be made orderly. First, we calculate the syndrome of the codeword block by block to see whether any blocks exist whose codewords are uncorrectable. If yes, then we claim that the image is unauthentic and use the above ECC checking process to display alteration locations. If all codewords are correctable (i.e., errors in any feature code are correctable by its PCB), we repeat the same process at the source site: concatenate all corrected codewords into a global sequence and cryptographically hash the result sequence. The final verification result is then concluded through a bit-by-bit comparison between these two hashed sets (i.e., one is this new generated one and the other is decrypted from the associated signature by the obtained public key): If any single bit differs, the verifier will deem the image unacceptable ("unauthentic").

Semifragile Content Authentication (Lossless)

The semifragile content authentication (lossless) mode is usually selected for medical or remote-imaging-related applications where lossless recovery of the original image is required. The signing operation is very similar to lossy signing operation (Figure 21.16). The only difference lies in the watermark embedding module. The lossless watermarking method used is novel [38], which does not generate annoying salt-and-pepper noise, and robust against image compression. The codeblock whose size is usually 64×64 is further divided into 8×8 blocks called patches. The coefficients in a patch are split into two subsets. Then, we calculate the difference value α, which is defined as the arithmetic average of differences of coefficients in two respective subsets. Because in a patch the coefficients are highly correlated, the difference value α is expected to be very close to zero. Furthermore, it has certain robustness against incidental distortions because α is based on all coefficients in the patch. Each patch is embedded with 1 bit. If 1 is to be embedded, we shift the difference value α to the right or left beyond a threshold by adding or subtracting a fixed number from each coefficients within one subset. If 0 is to be embedded, the patch is intact. There is a chance that the value α is originally beyond the threshold and a bit of binary 0 is to be embedded. In this case, we shift the value α further beyond the threshold and rely on the ECC to correct the bit error, because the watermark bits are ECC encoded again before being embedded.

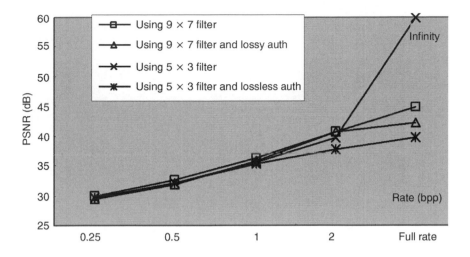

Figure 21.18. Image quality comparison between compressed and compressed plus signed JPEG2000 images.

The verifying operation is also similar to the lossy one, with the exception of watermark extraction. The code block is divided in patches and difference value α of each patch is calculated in the same way as lossless sign. For each patch, if value α is beyond the threshold, a bit of "1" is extracted and the difference value is shifted back to its original position, which means that original coefficients are recovered. If the value α is inside the threshold, a bit of "0" is extracted and nothing needs to be done. Finally, an ECC correction is applied on the extracted bit sequence to get the correct watermark bits.

More detailed test results are given in Reference 36. Figure 21.18 and Figure 21.19 compare the image quality and file size before and after signing. The image is encoded with a 9×7 filter (with and without lossy watermark) and a 5×3 filter (with and without lossless watermark), respectively. We can see that the image quality drops slightly with a watermark embedded and no significant difference between the image sizes.

In summary, the main contributions of the proposed solution could be listed as follows.

- JPEG2000 images can be quantitatively authenticated in terms of the LABR. By utilizing the JPEG2000 coding scheme, the proposed solution can quantitatively authenticate images specified by the LABR. It will offer users more convenience and achieve a good trade-off among system robustness, security, and image quality. It is noted that to control the authentication strength, the

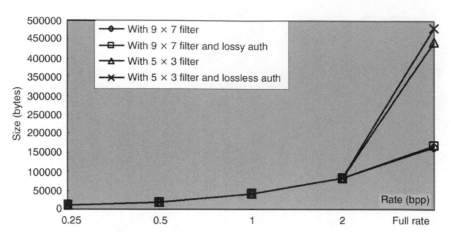

Figure 21.19. **File size comparison between compressed and compressed plus signed JPEG2000 images.**

state-of-the-art content based authentication systems (signature based or watermarking based) are mainly based on the following techniques: empirical trials, (R)MSE measure associated with human vision system (HVS), or quantization measure. Although the latter is "implicitly" related to bit-rate allocation, it is still not convenient because it is not easy to set the quantization step size properly under various types of image content, especially for ordinary users. JPEG2000 images can be flexibly authenticated in three modes: complete authentication, semifragile authentication with lossy watermarking, and semifragile authentication with lossless watermarking. The image can be signed or verified with difference granularities: codestream level, quantized wavelet coefficient level, and wavelet coefficient level. In addition, the protected area can be specified via different representation levels used for JPEG2000 coding, such as ROI, component, tile, resolution, subband, precinct, codeblock, and packet. All of these authentication flexibilities are well aligned with the JPEG2000 coding procedure (Part 1). ECC is employed to tame incidental distortions, which integrates the signature-based solution and the watermarking-based solution into one unified scheme in a semifragile way. Thus, we obtain only one crypto hash or signature per image regardless of its size, in a semifragile way. As we can see later, ECC is used in a novel way in our proposed solution. ECC's message comes from the extracted features of the image. ECC's PCBs are the seeds of watermarks and will be embedded back into the image. Such novel usage of ECC will bring the system the ability to correct some possible errors in its extracted features without increasing system payload: The code-block-based watermarks (PCBs) could be used for locating malicious modifications and the final content-based signature is obtained by cryptographically hashing all corresponding codewords (message + PCB) to make sure that no catastrophic security holes exist.

- The proposed solution can be incorporated into various security protocols (symmetric and asymmetric): By using the ECC scheme on extracted features, we can obtain stable crypto hash results, thus gaining more room and flexibility in designing authentication protocols in real applications. For example, by using PKI, the content signer and content verifier can own different keys. Working under PKI will also let the proposed solution be easily incorporated into current data-based authentication platforms.

- The proposed solution is fully compatible with the JPEG2000 encoder and decoder (Part 1). Based on the description given earlier, we can see that all features are directly extracted from EBCOT and the generated watermarks are also embedded back in the procedure of EBCOT. The proposed solution is fully compatible with JPEG2000 codec and could efficiently co-work with the procedure of JPEG2000 image compression and decompression.

SYSTEM PERFORMANCE EVALUATION

The performance of a media authentication system mainly relates to its system security and system robustness. For the cases of complete authentication, and content-based fragile authentication, which adopt traditional crypto signature schemes, their system security should be equivalent to the traditional crypto signature whose security is assumed to be very secure (it cannot be broken based on the computation capability of current computers); their system robustness is also the same as the traditional crypto signature schemes, which is fragile (even a 1-bit change is not allowed). Therefore, our discussion in this section on system performance evaluation only focuses on semifragile content authentication.

System Security Evaluation

System security plays a vital role in an authentication system. From our previous description of noncrypto-hashing methods, content hashing methods, and crypto hashing methods, we can see that three main modules affect system security (i.e., feature extraction, hashing, and ECC). If we denote the performance of system security in terms of the probabilities of the system being crashed (i.e., given an image, the probabilities of finding another image that can pass the

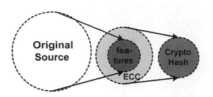

Figure 21.20.   Illustration of system security analysis.

signature verification, under the same parameters). Therefore, the system security can be presented by three probabilities: P_F in the feature extraction module, P_E in the ECC module, and P_C in hashing. Assume that they are mutually independent and very small; the overall system security could be drawn as

$$P = 1 - (1 - P_F)(1 - P_E)(1 - P_C) \approx P_F + P_E + P_C$$

In fact, P_F and P_E impair the system security in different ways, as shown in Figure 21.20. A good feature descriptor should represent the original source as close as possible. Differing from feature extraction, which functions by "removing" redundancy from original source, ECC functions by "adding" redundancy in order to tolerate incidental distortions. Thus, a good feature extraction method and a proper ECC scheme are the key factors in system security. More detailed analysis is given below.

Hash. Typically, two types of hashing function are used in semifragile authentication system: crypto hashing and content hashing. The security of crypto hashing is known to be very secure based on the computation capability of current computers. For example, the security of a 160-bit SHA-1 hash value is about [1,2] $P_C \approx 2^{-80}$, under a birthday attack[1] assumption. Therefore we ignore the security risk of crypto hashing.

Usually content hashing is very robust to the acceptable manipulations. However, a robust content hashing is also robust to the local attacks. Analysis on the security of content hashing is a little complicated because it relates to the following issues. The first one is the feature selected for content hashing: A good feature representation is the first step to design a secure content hashing scheme. The second one is the length of the generated content hash code because it implicitly means how detailed the content is represented given the same feature

---

[1]A typical brute force attack against crypto hashing: An adversary would like to find two random messages, M and M', such that $H(M)=H(M')$. It is named the birthday attack because it is analogous to finding two people with the same birthday [1].

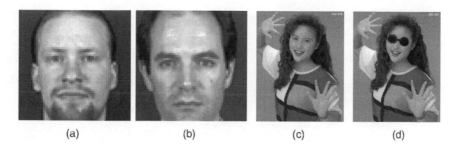

(a)   (b)   (c)   (d)

Figure 21.21. **(a) Original image; (b) attacked image of (a). (c) Original image; (d) attacked image of (c).**

representation. Finally and most importantly, the authentication decision of content hash is based on the (Hamming) distance between the two sets of hash codes and it is hard to differentiate the local attacks from the acceptable manipulations.

Feature Extraction. Again, the purpose of attacking an image authentication system is to make the faked image pass the verification. At the content level, the forgery should alter the image in such a way that its semantic meaning will be changed; for instance, using the face [39] shown in Figure 21.21b to pass the authentication whose signature is generated from the face image shown in Figure 21.21a. Note that the usual attacks may only try to modify parts of the image rather than to take whole total different image for verification by assuming that the authentication system is appropriately designed. For example, Figure 21.21d is the attacked image in Figure 21.21c, where only parts of image altered.

We now evaluate some typical features selected for image authentication.

Histogram. The histogram of an image represents the relative frequency of occurrence of the various gray levels in the image. It is actually a global measure of an image and is unable to characterize the local changes in the image. Therefore, it is too robust to be used as the feature for image authentication in most cases. For example, Figure 21.22a and Figure 21.22b show the histograms of face images in Figure 21.21a and Figure 21.21b, respectively. We can see that their histograms look very similar although they are different faces. Such similarity of Figure 21.22b still fails in passing the authentication of Figure 21.22a. We can adopt a technique called "histogram warping" to further shape Figure 21.22b until it passes the authentication. As a result, the face image shown in Figure 21.22c has a histogram exactly the same as that of

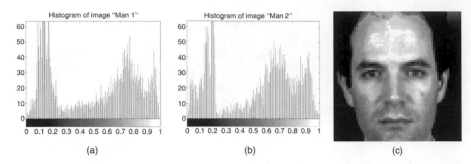

(a)   (b)   (c)

Figure 21.22. The histograms of face images and the attacked image. (a) Histogram of Figure 21.21a; (b) histogram of Figure 21.21b; (c) the modified image of Figure 21.21b but with the same histogram as Figure 21.21a.

(a) Original image   (b) Modified image by adding some clouds

Figure 21.23. The extracted feature points (corner) are used for authentication.

Figure 21.21a. Therefore, the histogram alone is not suitable to be the feature for content authentication.

Edge and Corner. The corner points in an image are defined as the maxima of absolute curvature of image edges that are above a preset threshold value. Image corner points are usually used in computer visions to find the correspondences between two images. Taking image edges or corners as the features for image authentication can be found in some publications such as References 16 and 17. It can characterize the local significances in the image, as shown in Figure 21.23a. However, it usually fails in detecting some gradual changes, as shown in Figure 21.23b, where we did some modifications on Figure 21.23a by adding more clouds on the upper-right site. From their detected corners, we can see that their corner points are almost the same. Therefore, the edge or corner is not suitable for authenticating those attacks, such as gradual changing part of the image.

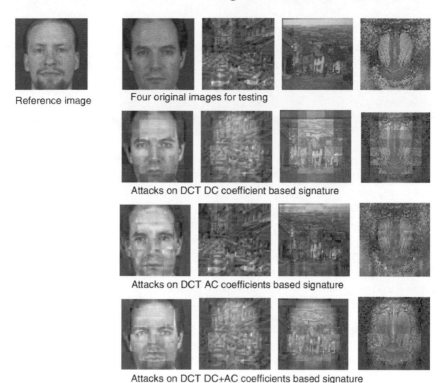

Reference image   Four original images for testing

Attacks on DCT DC coefficient based signature

Attacks on DCT AC coefficients based signature

Attacks on DCT DC+AC coefficients based signature

Figure 21.24. DCT test under attacks.

DCT Coefficients. Because a discrete cosine transform (DCT) is used for JPEG lossy compression, DCT coefficients are also considered to be the feature for image authentication, as usually lossy compression is required as acceptable for authentication. However, how to select DCT coefficients and how many DCT coefficients should be taken as the representative feature of the image still need to be carefully designed, especially when the feature needs to be embedded back to image as watermarks because the watermark capacity also affects the feature selection. Figure 21.24 presents a visual illustration on these issues.

In Figure 21.24 (first row), the image on the left is taken as the reference and the other four images on the right are taken as the attack images. Our purpose is to modify these four images until they can pass the authentication whose signature is based on the reference image. We illustrated three examples for every 8×8 DCT block: DC coefficients only (the second row), three AC coefficients (the third row), and DC plus two AC coefficients (the fourth row). Visually, DC plus AC coefficients are the best selection and, in practice, use as many DCT coefficients as possible.

Compressed Version of the Original Content. Some authentication systems directly employed the compressed version of the original content as the feature for authentication [22–24,32]. In such a case, we argue that the following assumption is valid:

Assumption 1: Lossy compression such as JPEG2000 is the approximate representation of its original source in terms of an optimal rate-distortion measure.

Intuitively, it implies that no security gap exists between the original source and its compressed version under a given authentication strength. Therefore, we would argue that if an authentication system could make use of all compressed information (e.g., all quantized coefficients in all blocks) as the feature, it should be deemed the same as having made use of all original source. In other words, if an attacker intends to modify the content in the spaces between the original source and its compressed version, this attack should not be considered harmful to this content (i.e., cause the meaning of content change) because, eventually, the attacked content will be discarded by its lossy compression.

Note that in the proposed solution for the JPEG2000 image [36], we have used almost all compressed information (e.g., the state of significant pass and the estimated distortions among different passes) above LABR for generating hash. Therefore, we could argue that P_F [36] is negligible (small enough) under Assumption 1.

Generic Framework For System Security Evaluation. Considering the varieties of features selected for authentication, originating from the brute force attack experiences in cryptography [1,2], a similar evaluation framework for content authentication could also be drawn, as shown in Figure 21.25.

Refer to Figure 21.25; given a reference image and an attacked image, we can keep their similarity score increasing by randomly perturbing the attacked image and recording it in a discarding-keeping manner (i.e., if the trial makes the score up, then records it, and discards it otherwise). Such an evaluation framework does not need any prior knowledge of the authentication design (i.e., no information about feature selection) and have been widely used in information security and biometrics systems.

Error Correction Coding. A (N, K, D) code [40] is a code of N-bit codewords, where K is the number of message digits (the extracted feature set in this chapter) and D is the minimum distance between codewords. $N - K = P$ is the number of parity check digits (i.e., checksum or check information). It should be noted that an ECC with rate K/N can correct

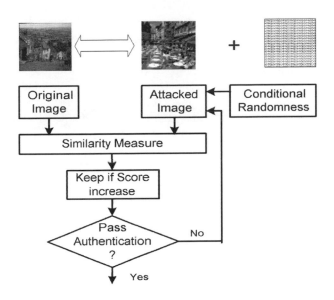

Figure 21.25. Generic framework for security evaluation.

up to $T = (D - 1)/2$ errors. Here, we are only interested in binary BCH ECC (i.e., Bose–Chaudhuri–Hochquenghem Code). BCH code is a multilevel, cyclic, error-correcting, variable-length digital code used to correct errors of up to approximately 25% of the total number of digits. Like other linear ECC codes, the first K-bit digits in its N-bit codeword are exactly the same as its original K-bit message bits. This is the main motivation for selecting the BCH code: In the procedure of content signing, the output (codeword) from the ECC encoder can still be separated into two parts: the original message and its corresponding PCBs. We can then embed the PCB into the image as a watermark. In the procedure of content authentication, the authenticity can be verified by checking the syndrome of the merged codeword (message comes from the feature set extracted from the image to be verified and PCB comes from the watermark extracted from the same image).

To check how the ECC scheme affects the system security (i.e., P_E), we have the following lemma [32]:

Lemma 1: Let H_C be our proposed content hash scheme based on an ECC scheme (N, K, D) with error correction ability t [e.g., $t = (D - 1)/2$]. For any K' that satisfies $\|N - N'\| \leq t$, we have $H_C(N) = H_C(N')$. It means that in a properly selected ECC scheme, all corrupted codewords will be deemed as authentic as long as they are still within the error correction capability. Hence a proper ECC scheme could make P_E negligible.

Clearly, ECC diminishes system security in some sense, as ECC does provide the property of fuzziness. This goes back again to the issue of how to select a proper ECC scheme to balance between system robustness and security. However, we have to accept the fact that ECC does introduce some security risk into the system. Refer to Table 21.1; for instance, if the formed codeword is 0000000, seven codes will be considered as acceptable in hash verification (1000000, 0100000, 0010000, 0001000, 0000100, 0000010, and 0000001) because they are all within the designed error correction capability (correcting 1-bit error) and will be corrected as 0000000 in the procedure of content authentication. Now, the issue is whether taking these seven codes as acceptable is secure? This also requires that a well-designed application-oriented ECC scheme under predefined acceptable manipulations plays a key role in semifragile content-based hash function.

The last concern related to security is watermarking used in References 30–36. In such a case, we do not need to pay more attention to watermarking security because watermarking only functions for storing part of the ECC information.

In practice, the threats from attacks are also closely related to other factors, such as the attackers' knowledge background, and some counter-measures have been addressed and proposed by researchers in cryptography, which are beyond the scope in this chapter.

System Robustness Evaluation

To quantitatively evaluate system performance on robustness, two important measures used are false acceptance rate (FAR) and false rejection rate (FRR). The FRR is the percentage of failure among attempts in generating its derived signature from the authentic image, averaged over all authentic images in a population O. The FAR is the percentage of success among attempts in generating its derived signature from the forged image, averaged over all forged images in a population M. Faced with some practical difficulties when we conduct testing, such as manually forging a number of meaningful images, FAR and FRR may be measured in an alternative way; that is, the FAR and FRR can be derived based on block-based evaluation not image-based evaluation. Given a pair of images to be tested (one is original watermarked and the other one is its fake version), assume that there are a total of N blocks in the pair of images, M blocks are modified, and O blocks remain unchanged; $N = M + O$. We then pass the pair of these images (e.g., watermarked image and modified watermarked image) through various kinds of acceptable manipulation such as multiple-cycle compression, trans-coding, and so forth. FAR and FRR will be derived based on checking all of these distorted images block by block. Assuming that after

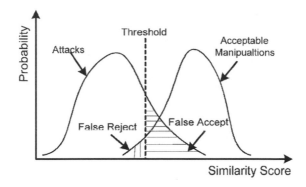

Figure 21.26. Evaluation of system robustness.

distorting images, M' among M blocks can pass the authentication and O' among O cannot pass the authentication. FAR and FRR could be obtained as

$$\text{FRR} = \frac{O'}{O} \quad \text{and} \quad \text{FAR} = \frac{M'}{M}$$

In practice, the system performance could also be described by a receiver operating characteristic (ROC) curve indicating the operating point (threshold) for various applications with different robustness requirements, as shown in Figure 21.26.

Actually, system security and system robustness are mutually dependent. A secure authentication system may result in a low authentication robustness or a high FRR, and a robust authentication system may unavoidably have a high FAR, which means that some attacked content could pass the authentication.

Before we end this section, we tabulate all signature-based authentication methods described earlier (Table 21.3), which is an extension of Figure 21.2 but only focusing on multimedia content.

SUMMARY AND FUTURE WORK

In this chapter, we introduced a list of state-of-the-art signature-based authentication schemes for multimedia applications. Based on whether the original multimedia data can be obtained at the receiver site, they are classified into complete authentication and content authentication. Content authentication can be further classified into fragile authentication and semifragile authentication based on the robustness of authentication. We then focus on discussing various semifragile signature-based authentication techniques. In summary, we argue that a good semifragile authentication system should not only be secure

TABLE 21.3. Comparison of various authentication methods

	Content Integrity	Source Identification	Robustness	Signature size	Security	Attack localization
Complete authentication	Yes	Yes	No	1024 bits	Very strong	No
Fragile authentication	Yes	Yes	No	0	Very strong	Yes
Semifragile (watermarking)	Yes	No	Yes	0	Weak	Yes
Semifragile (nonhashing)	Yes	Yes	Yes	Very large	Some	Some
Semifragile (content hashing)	Yes	Yes	Some	Large	Some	No
Semifragile (crypto hashing)	Yes	Yes	Yes	1024 bits	Strong	Yes

enough to the malicious attacks but also be robust enough to the acceptable manipulations. Such a good system should be application dependent.

The demands of adaptively streaming the media in a pervasive environment pose new requirements as well as new challenges on the semifragile media authentication system design. A pervasive environment may contain many heterogeneous devices with varying computing resources and connected through different channels. Therefore, in such an environment, traditional security solutions, which only provide yes or no answers (i.e., either authentic or unauthentic), cannot work well because different devices and channels may have different required security levels due to their limited computing power or their targeted applications. Therefore, it requires a scalable, robust, and secure authentication framework.

REFERENCES

1. Menezes, A., Oorschot, P., and Vanstone, S., *Handbook of Applied Cryptography*, CRC Press, Boca Raton, FL, 1996.
2. Schneier, B., *Applied Cryptography*, John Wiley & Sons, New York, 1996.
3. Lin, C.-Y. and Chang, S.-F., Issues and Solutions for Authenticating MPEG Video, presented at SPIE International Conference on Security and Watermarking of Multimedia Contents, San Jose, 1999, Vol. 3657, pp. 54–65.
4. Wu, C.-W., On the design of content-based multimedia authentication systems, *IEEE Trans. Multimedia*, 4, 385–393, 2002.

5. Friedman, G. L., The trustworthy digital camera: Restoring credibility to the photographic image, *IEEE Trans. Consumer Electron.*, 39(4), 905–910, 1993.

6. Quisquater, J.-J., Macq, B., Joye, M., Degand, N., and Bernard, A., Practical Solution to Authentication of Images with a Secure Camera, presented at SPIE International Conference on Storage and Retrieval for Image and Video Databases V, San Jose, 1997, Vol. 3022, pp. 290–297.

7. Gennaro, R. and Rohatgi, P., How to Sign Digital Streams, in Proceedings of CRYPTO 97, Santa Barbara, CA, 1997.

8. Yu, H., Scalable Multimedia Authentication, in Proceeding of the 4th IEEE Pacific-Rim Conference on Multimedia (PCM)'03, Singapore, 2003.

9. Park, J. M., Chong, E. K. P., and Siegel, H. J., Efficient Multicast Packet Authentication Using Signature Amortization, in Proceedings of the IEEE Symposium on Security and Privacy, 2002, pp. 227–240.

10. Goljan, M., Fridrich, J., and Du, R., Distortion-free Data Embedding, in Proceedings of 4th Information Hiding Workshop, 2001, pp. 27–41.

11. Shi, Y. Q., Chap. 7, this volume.

12. Yeung, M. M. and Mintzer, F., An Invisible Watermarking Technique for Image Verification, in Proceeding of International Conference on Image Processing'97, 1997.

13. Wong, P. W. and Memon, N., Secret and public key image watermarking schemes for image authentication and ownership verification, *IEEE Trans. Image Process.*, 10(10), 1593–1601, 2001.

14. Schneider, M. and Chang, S.-F., A Robust Content-based Digital Signature for Image Authentication, in Proceedings of International Conference on Image Processing'96, 1996.

15. Lin, C.-Y. and Chang, S.-F., A robust image authentication method distinguishing JPEG compression from malicious manipulation, *IEEE Trans. Circuits Systems Video Technol.*, 11(2), 153–168, 2001.

16. Dittmann, J., Steinmetz, A., and Steinmetz, R., Content-based Digital Signature for Motion Pictures Authentication and Content-Fragile Watermarking, in Proceedings of the IEEE International Conference on Multimedia Computing and Systems, 1999, Vol. II, pp. 209–213.

17. Queluz, M. P., Authentication of digital image and video: generic models and a new contribution, *Signal Process.: Image Commn.*, 16, 461–475, 2001.

18. Lu, C. S. and Liao, H.-Y.M., Structural digital signature for image authentication: an incidental distortion resistant scheme, *IEEE Trans. Multimedia*, 5(2), 161–173, 2003.

19. Tzeng, C.-H. and Tsai, W.-H., A New Technique for Authentication of Image/Video for Multimedia Applications, in Proceedings of ACM Multimedia Workshop'01, Canada, 2001.

20. Wu, C.-P. and Kuo, C.-C. J., Comparison of Two Speech Content Authentication Approaches, presented at SPIE International Symposium on Electronic Imaging 2002, San Jose, CA, 2002.

21. Sun, Q., Zhong, D., Chang, S.-F., and Narasimhalu, A., VQ-based Digital Signature Scheme for Multimedia Content, presented at SPIE International Conference on Security and Watermarking of Multimedia Contents II, San Jose, CA, 2000, Vol. 3971, pp. 404–416.

22. Bhattacharjee, S. and Kutter, M., Compression Tolerant Image Authentication, in Proceedings of the IEEE International Conference on Image Processing'98, Chicago, 1998.

23. Lou, D.-C. and Liu, J.-L., Fault resilient and compression tolerant digital signature for image authentication, *IEEE Trans. Consumer Electron.*, 46(1), 31–39, 2000.

24. Chang, E.-C., Kankanhalli, M. S., Guan, X., Huang, Z., and Wu, Y., Image authentication using content based compression, *ACM Multimedia Syst.*, 9(2), 121–130, 2003.

25. Memon, N. D., Vora, P., and Yeung, M., Distortion Bound Authentication Techniques, presented at SPIE International Conference on Security and Watermarking of Multimedia Contents II, San Jose, CA, 2000, Vol. 3971, pp. 164–175.

26. Fridrich, J. and Goljan, M., Robust Hash Functions for Digital Watermarking, in Proceedings of IEEE International Conference on Information Technology — Coding and Computing'00, Las Vegas, 2000.

27. Xie, L., Arce, G. R., and Graveman, R. F., Approximate image message authentication codes, *IEEE Trans. Multimedia*, 3(2), 242–252, 2001.

28. Venkatesan, R., Koon, S.-M., Jakubowski, M.H., and Moulin, P., Robust Image Hashing, in Proceedings of IEEE International Conference on Image Processing'00, Vancouver, 2000.

29. Kailasanathan, C., Naini, R. S., and Ogunbona, P., Compression Tolerant DCT Based Image Hash, in Proceedings of the 23rd International Conference on Distributed Computing Systems Workshops (ICDCSW'03), Rhode Island, 2003.

30. Sun, Q., Chang, S.-F., and Maeno, K., A New Semi-fragile Image Authentication Framework Combining ECC and PKI Infrastructure, in Proceeding of International Symposium on Circuits and Systems'02, Phoenix, 2002.

31. Sun, Q., Tian, Q., and Chang, S.-F., A Robust and Secure Media Signature Scheme for JPEG Images, in Proceeding of Workshop on Multimedia Signal Processing'02, Virgin Island, 2002.

32. Sun, Q., and Chang, S.-F., A quantitative semi-fragile JPEG2000 image authentication solution, *IEEE Trans. Multimedia*, in press.

33. He, D., Sun, Q., and Tian, Q., An Object Based Authentication System, *EURASIP J. Appl. Signal Process.*, in press.

34. Sun, Q., He, D., Zhang, Z., and Tian, Q., A Robust and Secure Approach to Scalable Video Authentication, in Proceedings of International Conference on Multimedia and Expo'03, 2003.

35. Sun, Q., Ye, S., Lin, C.-Y., and Chang, S.-F., A crypto signature scheme for image authentication over wireless channel, *Int. J. Image Graphics*, 5(1), 1–14, 2005.

36. Zhang, Z., Qiu, G., Sun, Q., Lin, X., Ni, Z., and Shi, Y., A unified authentication framework for JPEG2000, ISO/IEC JTC 1/SC 29/WG1, N3074, and N3107, 2003.

37. JPEG2000 Part 1: Core coding system, ISO/IEC 15444-1:2000.

38. Ni, Z. and Shi, Y. Q., A Novel Lossless Data Hiding Algorithm and Its Application in Authentication of JPEG2000 Images, *Technical Report*, New Jersey Institute of Technology, 2003.

39. Face database: http://www.ee.surrey.ac.uk/Research/VSSP/xm2vtsdb, 2002.

40. Lin, S. and Costello, D. J., *Error Control Coding: Fundamentals and Applications*, Prentice-Hall, Englewood Cliffs, NJ, 1983.

Part V
Applications

22
Digital Watermarking Framework
Applications, Parameters, and Requirements

Ken Levy and Tony Rodriguez

INTRODUCTION

Digital watermarks are digital data that are embedded into content and may survive analog conversion and standard processing. Ideally, the watermark data are not perceptible to the human eye and ear, but can be read by computers. Digital watermarks, as a class of techniques, are capable of being embedded in any content, including images, text, audio, and video, on any media format, including analog and digital.

Digital watermark detection is based on statistical mathematics (i.e., probability). In addition, there are numerous digital watermarking techniques. As such, for a framework chapter like this, many descriptions include terms such as "usually" or "likely" due to the probabilistic nature of detection as well as generalizing the numerous techniques.

Importantly, digital watermarks are traditionally part of a larger system and have to be analyzed in terms of that system. This chapter reviews a framework that includes digital watermark classifications, applications,

0-8493-2773-3/05/$0.00+1.50
© 2005 by CRC Press

important algorithm parameters, the requirements for applications in terms of these parameters, and workflow.

The goals are twofold:

1. Help technology and solution providers design appropriate watermark algorithms and systems
2. Aid potential customers in understanding the applicability of technology and solutions to their markets

DIGITAL WATERMARKING CLASSIFICATIONS

Digital watermarks traditionally carry as payload either one or both of the following types of data:

- Local data
- Persistent identifier that links to a database

The local data can control the actions of the equipment that detected the digital watermark or has value to the user without requiring a remote database. The persistent identifier links the content to a database, usually remote, which may contain any data related to that content, such as information about the content, content owner, distributor, recipient, rights, similar content, URL, and so forth.

Digital watermarking algorithms can also be classified as robust and fragile. Although there are other types, such as semifragile, tamper-evident, and invertible, this chapter is limited to the base types of robust and fragile to simplify our discussion. A robust digital watermark should survive standard processing of the content and malicious attacks up to the point where the content loses its value (economic or otherwise) as dictated by the specific application and system. It can, for example, be used for local control and identification.

A fragile digital watermark, on the other hand, is intended to be brittle in the face of a specific transformation. The presence or absence of the watermark can be taken as evidence that the content has been altered. For some applications, this can also be achieved by using appropriate fingerprinting or hashing techniques in conjunction with a robust watermark. Regardless of the implementation, these techniques are traditionally used when the desire is to determine if the content has been manipulated. In many cases, it is desirable to employ both robust and fragile watermarks in the same content: one to act as a persistent identifier and for local control and the other as an indicator that the content has been modified or has gone through a specific transformation.

Another classification is based on the detection design parameter in watermarking algorithms, whether they are designed to do blind detection or informed detection. Blind detection is the ability of the algorithm to find and decode the watermark without access to the original, unmarked content. Informed detection implies that the detector has access to the original content to aid in finding and decoding the watermark. For purposes of this discussion, the application definitions that follow assume the ability to do blind detection unless otherwise stated, such as for forensic tracking.

A final and independent classification is based on the secrecy of the watermarking key. Most watermarking algorithms use some form of a secret in the form of a key, where the same key is required by the embedder to embed and the detector to decode the watermark. For many digital watermarking methods, this key is usually related to a unique key-based scrambling or randomization of the message structure, the mapping of the message structure to elements or features of the host content in which they are embedded, and unique format of the message structure. Independent of the algorithm used, this key either can be shared in the form of a public detection infrastructure, where the detector (and hence the key in a secure or obfuscated form) is made widely available or can be closely guarded and only divulged in a limited fashion within the constraints of other security mechanisms (e.g., physical security dongles). The later approach is referred to as a private detector infrastructure. For security, there are obvious reasons for reducing the exposure of the key and ease of detecting the watermark, hence implying that the private detector route provides additional security to the overall system. As applications are defined within the model, a distinction will be made between the two approaches to detector deployment. There is also research on watermarking algorithms that use different keys for embedding and detection, similar to public key encryption.

DIGITAL WATERMARKING APPLICATIONS

There are numerous applications to help protect, manage, and enhance content. Other references [1,2] have presented useful application definitions. This work tackles application definitions from the standpoint of market evolution as well as distinct values, although the applications are not mutually exclusive. For example, forensic tracking stands on its own, although it can also be considered part of a Digital Access Management (DAM) or Digital Rights Management (DRM) system.

Figure 22.1 shows an overview of Digital Watermarking (DWM) applications, which are described in detail afterward. The applications are presented in terms of a general evolution (in clockwise fashion in the figure). Every application is applicable to all content types, such as image, text, audio, and video.

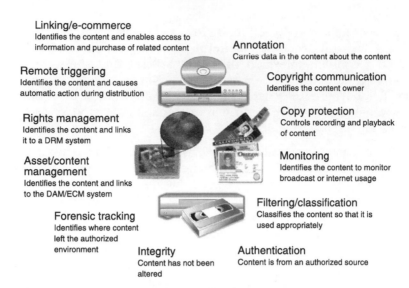

Figure 22.1. Digital watermarking applications.

Annotation

Annotation refers to hiding information, usually about the content, in the content. This approach may be more robust than using headers because annotations are part of the content and can take less space than headers because the information is part of the content. Most robust techniques do not have the data capacity for annotations, but invertible techniques can work perfectly for annotations. An example involves embedding a person's medical information in an x-ray with an invertible technique so the X-ray can be read in nonmodified form even after embedding.

Copyright Communication

Content often circulates anonymously, without identification of the owner or an easy means to contact the owner or distributor to obtain rights for use. Digital watermarks enable copyright holders to communicate their ownership, usually with a public detector, thereby helping to protect their content from unauthorized use, enabling infringement detection and promoting licensing. The watermark payload carries a persistent copyright owner identifier that can be linked to information about the content owner and copyright information in a linked database. For example, photographs can be embedded with the photographer owners' ID to determine whether two photos were taken from a similar location at a similar time, or that one is an edited copy of another. The same can occur with video, such as TV news.

Copy Protection

Digital watermarks enable a means to embed copy and play control instructions within the content. These instructions might indicate that playout is allowed, that a single copy can be made, or that no copies may be made. Woven into the content, these instructions are present even through conversions from digital to analog form, and back. Compliant devices, such as digital video recorders, can use this embedded information to determine whether copying or playing is permitted, thus establishing a copy protection system and guarding against unauthorized duplication. The digital watermark action happens at the local machine; thus, no remote database is usually required.

For example, the digital watermark could carry the copy control information (CCI), including copy control not asserted, copy-once, copy-no-more (state after copy-once content is copied), and copy-never in content, such as for digital versatile disk (DVD) audio and video. Recorders can detect and act appropriately on the CCI state (i.e., not reproduce copy-never and copy-no-more content). Even players can restrict playback of nonencrypted content with CCI bits because encoding rules can show that this content is not legitimate.

Monitoring: Broadcast and Internet

Broadcast monitoring enables content owners and distributors to track broadcast dissemination of their content. Broadcast content is embedded with a persistent content identifier that is unique and, optionally, distributor and/or date and time information. Detectors are placed in major markets, where broadcasts are received and processed. The digital watermark is decoded and used to reference a database, resulting in reports to the owner or distributor that the content played in the given market, at a given time, and whether it played to full length (for audio and video). The value is providing usage and license compliance information, advertising clearance verification, and detection of unauthorized use. Content usage data can help determine trends, such as which content is popular, and used to enhance content creation and distribution. A related database links the content identification to the content owner, and the distributor identification to the content aggregator (including networks and radio groups), or service provider for broadcast video, as well as distributor or retailer for recorded media.

Internet monitoring allows the content owner or distributor to track use of the content over the Internet. Web spiders crawl the Internet, especially known pirate or partner sites, and send content to detectors. The digital watermark is decoded and used to reference a database, resulting in reports for the content owner or distributor providing usage information and locating illegitimate content. The digital watermark carries the content identification, and the database links the content identification to the content owner and licensing information.

Monitoring differs from forensic tracking in that the recipient, such as a distributor, is not identified in the payload, but by where the detector finds the content with the digital watermark.

An example of broadcast monitoring involves using digital watermarks embedded in news stories, ads, and promotions, and a detector infrastructure monitoring radio and TV stations to report which news stories, ads and promotions are used and when, where, and for how long they are aired. The report is accessible in minutes to days and can be used for content usage demographics as well as compliance reporting. An example of Internet monitoring involves embedding a content ID in a digital photograph presented on the owner's Web site, discovering the photograph on an inappropriate Web site, and sending a report to the content owner. This can lead to the photograph being removed, or, more beneficially, properly licensed. Both of these examples potentially increase revenues for the content owners and distributors.

Filtering and Classification

Digital watermarks enable content to be identified, classified, and filtered. Therefore, systems are enabled to selectively filter potentially inappropriate content, such as corporations or parents desiring to restrict viewing of pornographic or other objectionable material. The digital watermark carries the classification codes or identifies the content and links to a remote database with the classification codes. The classification code can be used by the local machine to filter content or the content identification can be used to classify the content in a related database. For example, an adult bit or ratings bits can be embedded in images, audio, and video to aid Web filters in properly blocking adult content.

Authentication

Digital watermarks can provide authentication by verifying that the content is genuine and from an authorized source. The digital watermark identifies the source or owner of the content, usually in a private system. The system can recognize the private watermark on the local machine or link the content owner to a private database for authentication. For example, surveillance video recorders can be embedded with an ID that links it to that specific video recorder. Additionally, ID cards can be embedded with the authorized jurisdiction.

Integrity

Digital watermarks can provide integrity by verifying that the content has not been altered. The presence of the digital watermark and the continuity of the watermark can help ensure that the content has not been altered, usually in a private system. For example, surveillance video can have a digital watermark that can be used to determine if and where the video was modified. Additionally, ID cards can be embedded so that the information cannot be modified, such as stopping photo swapping.

Forensic Tracking

Digital watermarks enable content owners or service providers to track where content left the authorized distribution path. The digital watermark identifies the authorized recipient of the content. Thus, illegitimate content can be tracked to the last authorized recipient. The detector and database are private and managed by the content owner or rendering device (e.g., set-top box, DVD player, etc.) manufacturer. The detector can work in a standard mode that attempts to detect the watermark without user input. Alternatively, the detector can work in an interactive mode where the user (i.e., inspector) can help detect the watermark, such as by providing original content (i.e., informed detection) and estimating the distortion. This database contains the contact information of the user of the receiving device or software and should be protected from privacy.

A few examples of forensic tracking include the following. Prereleased songs or CDs can be embedded with a forensic ID linked to the recipient, such as the music critic or radio station, via a signed contract. The forensic ID can be detected when the song is found in an inappropriate location, such as on file-sharing networks, to determine the source of the leak. The same can happen with movies on DVD and VHS for academy awards, for example. Finally, this system can be extended to downloaded audio and video to consumers via click-through licenses. In this example, the service provider's account ID for the user can be embedded in the downloaded content and the database that links account IDs to users can require a subpoena to access.

Digital Asset Management and Electronic Content Management

Digital watermarks can be used as a persistent media asset tag, acting as keys into a digital asset management (DAM) or electronic content management (ECM) system. In this fashion, any piece of tagged content can lead back to the original, stored in the DAM and ECM management systems. Tagged content can also link to metadata in the DAM and

ECM systems, such as keywords, licensing data, author, and so forth. Such a link can be used when passing content between enterprises or asset management systems to retrieve the related information from remote systems. Digital watermarking extends DAM and ECM solutions outside compliant internal systems and can speed entering metadata into local DAM and ECM systems. The digital watermark carries the content identification and, possibly, distributor identification. The related database links the content identification to the content owner and to the content information in the DAM system.

For example, a digital watermark embedded with a content and version ID in an image of a branded item, such as a car, can be used to determine if the appropriate image (e.g., correct model and year car) is available on a partner's Web site. Similarly, branded audio and video can be verified to be the correct version and in the correct location.

Digital Rights Management

Digital watermarks identify the content and link to appropriate usage rules and billing information in conjunction with a digital rights management (DRM) system. In other words, the digital watermark identifies the content and links to its rights, such as specified by rights languages. This can make it easier for the mass market to purchase legitimate content, rather than use illegitimate content for free. Digital watermarking enables DRM systems to connect content outside the DRM back to the DRM (i.e., a marketing agent for the DRM). More specifically, digital watermarking allows content to pass through the analog domain and over legacy equipment while still carrying information to be reintegrated with the DRM system. The digital watermark contains the content identification and, optionally, distributor identification. The related database links the content identification to the content owner, usage rules, and billing information, and the distributor identification to the method of distribution. Enforcement of rights and identification of infringement can be assisted with every other application, especially copy protection and forensic tracking.

For example, a content ID embedded in video can be used to identify video coming into a set-top box or TV and determine the proper rules, content protection, and billing. The video could be coming from an analog source, such as existing home TV cabling or broadcast TV, or nonencrypted source, such as broadcast digital TV. For content with both a content ID and CCI bits, if the player can detect and link the content ID to the DRM, the DRM could be used instead of the copy protection bits, thus enabling viewing and purchasing by the user rather than blocking it. In the end, this DRM system produces additional revenue for the content owner and service provider.

Remote Triggering

Digital watermarks identify the content and cause an automatic response during the distribution of the content. The watermark can link to unique databases at any individual sites where the watermark is detected, thus creating endless possibilities for the functions that it may enable. For example, video received in a cable head-end with an embedded content ID can be used to link the video to the proper advertisements or interactive TV content, possibly specific to that cable system and head-end location.

Linking and E-commerce

Digital watermarks enable access to information about the content and purchase of the content and related content. The digital watermark includes the content identification and, possibly, distributor identification. A database links the content identification to the content owner and related content and information, and the distributor identification to the method of distribution. Linking and e-commerce can also enhance security, as the digital watermark is a benefit, and attacked content that lacks the digital watermark (as well as being inferior quality due to the attack) is less valuable.

An example involves linking printed materials to related content via a content ID embedded in the printed materials and read by a camera on a mobile phone, such as linking an ad for a new music release to a song sample. This is beneficial because it is difficult to type a URL into a mobile phone. Similarly, songs on the radio can have digital watermarks with content IDs that are detected in mobile devices and used to identify the song title, artist, and so forth, purchase that song and similar songs, or even provide concert information by that artist in the current location.

SIX DWM ALGORITHM PARAMETERS

The framework includes six DWM algorithm parameters: perceptibility, performance, robustness, reliability, payload size, and granularity. They are all interrelated. For example, DWM algorithms are usually more robust when the payload size is reduced or granularity is made coarser. The challenge in commercializing watermarking algorithms is to find the optimal set of trade-offs for a given application and validating performance against customer requirements [3]. By quantifying performance of various DWM algorithms using these parameters and comparing the results against the high-level application requirements provided later in Table 22.1, one can better understand applicability of a given DWM technique implementation.

Perceptibility

Perceptibility refers to the likelihood of visibility artifacts in images, text and video, or audible artifacts for audio due to the digital watermark. Perceptibility is subjective; thus, it is hard to define and includes descriptions like "not visible to the average user" or "doesn't degrade the quality of content." Most important is that the digital watermark does not reduce the quality of the content, which is sometimes different than not being noticeable. However, due to the difficulty of measuring quality, in some instances, watermarks are quantified as Just Noticeable Difference (JND) [4]. Techniques to reduce visibility and audibility leverage modeling of the human visual or auditory system (HVS or HAS), respectively, and can range in complexity from the simple Contrast Sensitivity Curve-based algorithms [5] to full on Attention Models [6] that attempt to predict what portion of the content is deemed most valuable to the user. The likelihood of perceptible artifacts can typically be decreased by increasing computational complexity, decreasing robustness, allowing more false positives, decreasing payload size or coarser granularity — and vice versa.

Performance

Performance refers to the computational complexity of the algorithm for the embedder and detector. It can be described in terms of CPU cycles and memory requirements for general-purpose processors, gates and memory for ASIC designs, or the mixture of the two. Depending on the application, performance of the embedder or detector may be more important. For example, in copy protection systems that use play control, a detector must be located in every consumer device (players and recorders.) As such, the performance of the detector must be very good.

Many times, the embedder or detector can share cycles with other subsystems, such as content compression (e.g., JPEG and MPEG) systems because most watermark algorithms are based on perceptual models and correlation, as are most compression technologies.

Allowing increased computational complexity of the embedder or detector results in reduced likelihood of perceptibility, increased robustness, fewer false positives, increased payload size, or finer granularity — and vice versa.

Robustness

Robustness refers to the ability of the digital watermark to produce an accurate payload at the detector. The content may have been transformed by the workflow or user, or include a malicious attack to purposely remove or modify the watermark. In other words, robustness provides false negatives for each (or lack of) transformation.

In many cases, a "standard" detector may not be able to detect the watermark, but after further processing, such as providing the original content or more computational power, the watermark payload can be accurately read. This is critical for forensic tracking, where, for certain very valuable content, the owner may desire this extra analysis.

Robustness can usually be increased when allowing increased likelihood of perceptibility, increased computational complexity, more false positives, decreased payload size, or coarser granularity — and vice versa.

Reliability and False Positives

Reliability determines how likely the wrong payload is detected (a.k.a. false positive). There are two types of false positive: (1) detecting the wrong payload when nothing is embedded and (2) detecting the wrong payload when another payload is embedded. Many applications require false positives to be in the range of 10^{-9} to 10^{-12} because false positives result in a much worse consumer experience than lack of detection (i.e., robustness or false negative) for those systems. Because false positives are so few, they are hard to measure experimentally.

Allowing more false positives usually causes a reduced likelihood of perceptibility, reduced computational complexity, increased robustness, increased payload size, or finer granularity — and vice versa.

Payload Size

The payload size refers to the number of bits that the digital watermark can carry as its message for a fixed amount of content. It does not include error correction or synchronization bits, for example. Common payload sizes are 8 to 64 bits, as watermarks usually include local control information or identifiers that link to additional information in a remote database. Reversible watermarks and other less robust watermarks can carry kilobytes of information.

Allowing a smaller payload usually causes reduced likelihood of perceptibility, reduced computational complexity, increased robustness, fewer false positives, or finer granularity — and vice versa.

Granularity

Granularity refers to the amount of content required to carry the digital watermark payload. A smaller amount of content to carry the payload is described as finer granularity, and a larger amount of content to carry the

685

payload is described as coarser granularity. For images, granularity refers to the number of pixels or size in print which robustly and accurately carries the payload. For video, it usually refers to the number of frames or seconds of video, but can refer to part of a frame. For audio, it refers to the number of seconds of audio.

Because digital watermarking is statistical in nature, the system may sometimes detect accurate payloads in finer amounts than the specified granularity, but, on average, a certain granularity is needed. In addition, because just a second of video is needed to detect the payload, the location of the beginning or end of the payload can be located with higher precision using additional statistical measures.

Granularity can be directly balanced with payload size. For example, 32 bits per second of video is equivalent to 64 bits per 2 sec of video, where a 32-bit ID is repeated every second in video or a 64-bit ID is repeated every 2 sec. In addition, allowing coarser granularity usually causes reduced likelihood of perceptibility, reduced computational complexity for the detector, increased robustness, or fewer false positives — and vice versa.

Parameter Evaluation

Perceptibility is best measured with subjective tests, although quantitative approaches may be used to aid in analysis (Watson [7], mean squared error, etc.). Performance and robustness are best measured with objective or quantitative experiments. False positives, payload size, and granularity are usually set by the algorithm to provide the desired perceptibility, performance, and robustness — based on an understanding of the nature of the content to be watermarked. Payload size and granularity are usually fixed by the algorithm and do not need to be measured, whereas false positives can be measured. False positives are usually measured theoretically because they are very small (like 10–12) and cannot usually be measured experimentally. False positives can change after content transformations.

The MPEG-21 group is currently working on a technical report about the evaluation of persistent association technology, including digital watermarks [8] through analysis of similar parameters.

APPLICATION AND PARAMETER REQUIREMENTS

The requirements for applications in terms of the above six parameters can be used to compare to a technology's capabilities. An exemplar table is provided in Table 22.1. It is a useful overview of how to use the applications and parameters. However, it is also very general

TABLE 22.1. Requirements for applications in terms of parameters

	Perceptibility	Performance	Robustness	Reliability (FP)	Payload	Granularity
Annotations	Pro	Embed — any Detect — any	Low	Very	kilobytes	N/A
Copyright Communications	Pro	Embed — any Detect — fast	Very	Very	32 bits	Fine
Copy Protection	Pro	Embed — any Detect — very fast	Extremely	Extremely	8 bits	Fine
Monitoring	Consumer	Embed — any Detect — very fast	Very	Very	32 bits	Fine
Filtering	Consumer	Embed — any Detect — very fast	Very	Very	8 bits	Fine
Authentication	Pro	Embed — any Detect — very fast	Extremely	Extremely	32 bits	Fine
Integrity	Pro	Embed — any Detect — very fast	Low	Extremely	32 bits	Very fine
Forensic Tracking	Pro/consumer	Embed — very fast Detect — any	Extremely	Extremely	32 bits	Coarse
DAM	Pro	Embed — any Detect — very fast	Very	Very	32 bits	Fine
DRM	Pro	Embed — any Detect — very fast	Extremely	Very	32 bits	Fine
Remote Triggering	Consumer	Embed — any Detect — any	Very	Very	32 bits	Fine
e-Commerce	Consumer	Embed — any Detect — very fast	Extremely	Very	32 bits	Fine

Perceptibility Key: Pro = acceptable to professional; Consumer = acceptable to consumer. Performance Key: Any = the faster the better. False Positive (FP) Key: extremely = 10–9; very = 10–6; not very = 10–3 (approximate values).

with many assumptions. The surrounding system and requirements should be evaluated to refine Table 22.1 for the specific system being analyzed. For example, integrity requires extreme robustness and only coarse granularity when determining that a photo has not been switched in a printed ID card. However, Table 22.1 refers to integrity as identifying pixels that have been modified, such as required in surveillance video, which requires fine granularity but does not need to be very robust.

In general, Table 22.1 demonstrates that the requirements for most applications include a robust DWM and computationally efficient detector, whereas for forensic tracking, the application requires a very robust DWM with coarser granularity (i.e., more content per payload) and a very computationally efficient embedder. Because forensic tracking allows increased robustness in trade for coarser granularity, there are many watermarking algorithms that are applicable to this growing application today.

WORKFLOW

Workflow is an essential part of a digital watermarking system and framework. The watermarking solution must cause minimal disturbances on the workflow for the customer to adopt the solution because changes in workflow can be more expensive for the customer than the watermarking solution. The requirements for the applications in Table 22.1 are chosen to minimize workflow problems. For example, in forensic tracking, because each piece of content can require a unique ID at the time of distribution, the embedder must be very efficient (as shown in Table 22.1) to not cause workflow troubles.

However, workflow is extremely dependent on system details and cannot be generalized. For example, determining whether the input to the detector is SDI, MPEG-2, or analog video can be critical, but it is highly system dependent. As such, workflow is not considered further within this chapter.

SUMMARY

In summary, this framework includes classifications, definitions of distinct but related watermark applications, six important watermark algorithm parameters, requirements for applications in terms of these parameters, and workflow. This digital watermark framework is helpful to technology and solution providers in designing an appropriate watermark algorithm and solution, and to customers for evaluating the watermark technology and related solutions.

REFERENCES

1. Cox, I., Miller, M., and Bloom, J., *Digital Watermarking*, Academic Press, San Diego, CA, 2002, pp. 12–26.
2. Arnold, M., Schmucker, M., and Wolthusen, S., *Techniques and Applications of Digital Watermarking and Content Protection*, Artech House, Norwood, MA, 2003, pp. 39–53.
3. Rodriguez, T. and Cushman D., Optimized selection of benchmark test parameters for image watermark algorithms based on Taguchi methods and corresponding influence on design decisions for real-world applications, in *Proceedings of Security and Watermarking of Multimedia Contents*, Ping Wah Wong, P.W. and. Delp, E.J. eds., The Society for Imaging Science and Technology and the International Society for Optical Engineering (SPIE), Bellingham, Washington, 2003, pp. 215–228.
4. Engeldrum, P., *Psychometric Scaling*, Imcotek Press, Winchester, MA, 2000, pp. 55–86.
5. Bartens, P., *Contrast Sensitivity of the Human Eye and Its Effects on Image Quality*, SPIE Press, Bellingham, Washington, 1999, pp. 137–157.
6. Wil Osberger, Tektronix, personal communication, 2001.
7. Watson, A.B., Image data compression having minimum perceptual error. U.S. patent 5,629,780.
8. ISO/IEC 21000-11, "Information Technology – Multimedia framework (MPEG-21) — Part 11: Evalutation Tools for Pesistend Association Technologies", submitted for publicaton.

23

Digital Rights Management for Consumer Devices

Jeffrey Lotspiech

INTRODUCTION

The ongoing digital revolution has presented both a threat and a promise to the entertainment industry. On one hand, digital technologies promise inexpensive consumer devices, convenient distribution and duplication, and flexible new business models. On the other hand, digital copies are perfect copies, and the industry risks losing substantial revenue because of casual end-user copying and redistribution. The entertainment industry and the technology companies that service it have been attacking this problem with increasing sophistication. This chapter outlines the history of this effort and concludes with some discussion of a possible future direction.

Digital rights management in consumer electronics devices, to date, is about restricting unauthorized copies. As such, it has been the subject of much controversy and misunderstanding. "Bits want to be free," is John Perry Barlow's catchy phrase [1], and others, even cryptographers such has Bruce Schneier, have joined the chorus. Schneier [2] argues that any technology that tries to prevent the copying of bits is doomed to failure. This is probably true, but misses the point. Bits can be freely copied, moved, and exchanged, but it is possible that the bits themselves are encrypted. The devices that have the keys necessary to decrypt the

0-8493-2773-3/05/$0.00+1.50

bits may be constrained by rules to refuse to decrypt the bits if they find them in the wrong place.

Therefore, modern copy protection is fundamentally about rules, not about technology. The rules may be enforced by law, but the trend has been to encode them instead in voluntary technology licenses. In other words, the manufacturer of a consumer device, wanting to allow a certain type of content to be played, signs a license to obtain the secrets (usually encryption keys) necessary to play the content. In return, the license constrains what functions his device might offer and calls out specific penalties for breaches.

Many copy protection technologies have the technical means to exclude devices from the system whose manufacturers have broken the rules they agreed to in the license. Curiously, however, licenses often have forgone this remedy, depending, instead, on the legal enforcement rather than technical revocation. Technical revocation has been reserved to cases of unlicensed circumvention devices, whose makers have somehow obtained the secrets without signing the license.

"What prevents a playing device, which obviously has the decryption key, from turning around and making an unauthorized copy in the clear?" is a question cryptographers, who think they have found the fatal flaw in a given protection scheme, often ask. The answer is, "Nothing technical, but the manufacturer building such a device would be violating his license and would be liable to civil action and damages." It is the license, not the technology, that offers the fundamental protection; the technology is just the hook that provides something to license. It is no accident that, in a technology I am intimately familiar with, Content Protection for Recordable Media (CPRM), the average specification runs about 40 pages, whereas the average license plus attachments runs about 90.

For the rest of this chapter, I will focus on the technical aspects of specific content protection schemes. However, remember that these technologies are always offered in the context of a license, and the license defines the rules that ultimately shape the consumer experience.

MACROVISION[TM] [1]

Soon after video cassette recorders (VCRs) hit the market in the late 1970s, the race was on for a technology that would prevent consumers simply copying prerecorded movies for their friends. As early as 1979,

[1]MacroVision is a trademark of MacroVision, Incorporated; additional information available at http://www.macrovision.com.

tapes appeared on the market that had techniques to make them difficult to copy. Unfortunately for consumers, the same techniques that made a tape difficult to copy often made it difficult to view on some televisions. In the late 1980s, John Ryan embarked on a systematic study of the differences between a VCR's input circuit and a television's tuner.

Ryan's first successful technology, which he called MacroVision, inserted a 4-μsec high-amplitude noise pulse in the vertical blanking interval of a prerecorded tape. The vertical blanking interval is the period during each frame of a TV broadcast in which the TV swings its picture tube's guns back to the top of the picture to start the next frame. Thus, no picture data are broadcast during that interval, and the TV ignores it. On the other hand, VCRs must record the vertical blanking interval, and the automatic gain circuit of many VCRs can be thrown off by that artificially loud signal. As a result, the VCR may not correctly record hundreds of lines of valid picture data that follow the vertical blanking interval. A recording made of a MacroVision-protected program usually has an objectionable band of low-intensity rolling across the picture.

Unfortunately for the movie studios, not all VCRs are sensitive to this first version of MacroVision, either coincidentally or deliberately. Also, "black boxes" have appeared in the market that strip out the noise pulse on its way to the VCR. Therefore, MacroVision Inc. created a new technology based on modifications to the 11-cycle color burst signal that is present at the beginning of every line of the picture. The company claims that 85% of all VCRs are sensitive to the combination of MacroVision I and MacroVision II. Although this is not enormously effective, unlike the other technologies I will describe, MacroVision (1) works in the analog domain and (2) requires no secrets and no device license, because it is based on fundamental differences between TVs and VCRs.

SERIAL COPY MANAGEMENT SYSTEM

In the late 1980s, the consumer electronic industry had developed a new technology called Digital Audio Tape (DAT). DAT offered consumers the capability to make a perfect digital copy of a music compact disk (CD). It offered the music industry nothing. Prerecorded music CDs are mass produced by a lithographic process analogous to a printing press's process. No magnetic media could possibly achieve the same economy of scale, so CDs would always be preferred to DATs for releasing prerecorded music. DAT was strictly a copying technology.

An alarmed music industry reached a compromise with the consumer electronics industry, and this compromise was enacted in law in the United States as the Audio Home Recording Act of 1992. There were three main features of the act: First, because some uses of DAT would certainly be infringing legitimate copyrights, small royalties would be imposed on blank media, to be paid to the music industry, to compensate them for the infringement. Second, it established it was *not* an infringement for consumers to make copies for "space shifting." In other words, it is not an infringement for consumers to make a copy of music they have purchased to play it on a different device, such as in their car. Third, it had a technical aspect: It required digital audio recording devices to implement the Serial Copy Management System (SCMS).

The Serial Copy Management System is the heart of simplicity. Two bits in a header are used to encode three states: "copy freely," "copy once," or "copy never." If a compliant device makes a copy of content marked "copy once," it must mark the copy "copy never." Of course, it cannot record "copy never" content.

As the hearings about the proposed law were being held, a third industry started to become alarmed. The computer industry, at that time worried about hard disks, felt the royalty on blank media was inherently unfair. Why should someone saving a spreadsheet, for example, end up paying the music industry for the privilege? The industry argued that computer media be exempt from the legislation, and it won.

Since the act was passed, DAT has not been very successful, at least not in the United States, and two unanticipated computer-related technological advances have made the computer exemption, at the very least, ironic. First, effective music compression technologies, such as MP3, were invented, so that even a moderately sized hard disk could store a huge music library. Second, the cost of CD recording technology went from tens of dollars per disk to tens of cents per disk. The impact of all this on the record industry has been substantial. The Recording Industry Association of America has stated that the industry loses $4.2 billion per year due to different forms of "piracy," but this figure includes traditional sources such as counterfeit recordings and is not entirely the effect of home computers.

The Audio Home Recording Act's exemption on computer media leads to other anomalies. Consumer electronics companies have begun to offer purpose-built music CD copying devices. Because these devices are not exempt from the Audio Home Recording Act, they must use special, and more expensive, CD blank disks than those used by computer software to perform the same function. The price of the special blank disk includes the royalty to the music industry. Likewise, the devices honor the SCMS bits. Although some music-copying computer programs also honor SCMS bits, they are probably not legally obligated to, and doing so puts them at a competitive disadvantage.

CONTENT SCRAMBLING SYSTEM

At the end of 1994, the movie industry and the consumer electronics industry were poised to introduce a new technology, Digital Versatile Disk (DVD), which promised a superior consumer experience for prerecorded movies compared to magnetic tapes. This promise has been more than fulfilled; DVDs are the most successful new consumer technology ever introduced, in terms of rate of adoption: more successful than music CDs, more successful than cell phones.

The movie industry, of course, was concerned about rampant copying of the digital DVD movies. They had agreed with the consumer electronics industry that they would ask Congress to extend the law so that SCMS would be compulsory with the new technology. Because there is no strong "space shifting" argument with movies, the new disks could be marked "copy never" at the source. Unlike music CDs, however, DVDs were designed from the beginning to be computer compatible, so they knew that it would be necessary to get the computer industry to agree with this proposed law. They did not anticipate any problem; after all, checking a couple of bits was a trivial function in a device as powerful as a personal computer.

Universally, however, the computer companies that got involved in this issue reacted negatively to the proposal. Their arguments focused on two main problems. First, they argued that computers were general-purpose devices and designers might inadvertently provide a circumvention path while building an entirely different function. For example, you could imagine a medical image system using this new high-density DVD technology to store medical images. If the designers were ignorant of the movie industry and ignored the SCMS bits on the DVDs, the system could be viewed as a circumvention device, with criminal penalties to the manufacturer. Second, they argued that SCMS was a totally ineffective system on the personal computer (PC); any 12-year-old Basic programmer could figure out how to flip the SCMS bits to make copies.

Regardless of the merits of the first argument, the second argument got the attention of the movie industry. After much discussion, the three industries (movie, computer, and consumer electronics) agreed upon some basic principles that still govern content protection technologies today:

1. If content owners want their content protected, they must encrypt it at the source.

2. It is a voluntary choice whether a manufacturer supports that protection scheme or not.
3. Because the content is encrypted, it is not possible for a manufacturer to inadvertently create a circumvention device; circumvention requires overt action.

Based on these principles, in 1996 engineers from Matsushita and Toshiba designed an encryption system for DVD video they called the Content Scrambling System (CSS). Because of concerns about United States' and especially Japan's export laws on cryptography, it was based on a 40-bit key size. Even with 1996 personal computers, a 40-bit key could be found by exhaustive search with less than a day of processing. Even worse, the encryption algorithm they chose is actually invertible, meaning it was not a true encryption algorithm. It is, like its name implied, simply a way to scramble the data. All of the weaknesses of CSS were known by insiders before the system was ever deployed. As a result, the details of the system have always been confidential. The system was deployed in time for Christmas 1996.

The Content Scrambling System works by assigning each device manufacturer a single key. There are 400 different manufacturer keys, and the disk key (the key that fundamentally protects the content on the disk) is "encrypted" by each one of them. The encryptions are placed in the lead-in area on the DVD disk. Each device goes to its manufacturer's particular spot in the lead-in area, then uses its manufacturer's key to decrypt the disk key. From the disk key, it can decrypt individual title keys and decrypt the content itself.

For 3 years, CSS went along swimmingly. In November 1999, however, it was reported that a 16-year-old in Norway found a global secret: one of the manufacturer's keys. In spite of the glaring weaknesses in the cryptography of the system, the break was apparently achieved by reverse engineering a poorly protected software implementation. It was also clearly helped by information that had been circulating for many months on some Linux newsgroups, apparently from anonymous employees of licensees who had access to the confidential specification. Once the break was announced and studied, the cryptographic weaknesses became apparent, and within weeks, the remaining 399 manufacturers' keys had been published on the Internet.

It is easy for cryptographers to criticize the CSS technology. Of course, it is now completely broken; a few minutes search on the Web can get you a program that will allow you to copy DVDs. The technology certainly could have been improved, but only at the cost of missing the critical 1996 Christmas season. Had the entire deployment, and therefore the movie industry's enormous DVD revenues, been moved back a year,

I believe that the industry would have lost more money than they are ever going to lose due to the break. Because the manufacturer's keys act as global shared secrets in the system, an irrevocable break was inevitable in any event.

CONTENT PROTECTION ON RECORDABLE MEDIA[2]

Most of the people who were working on DVD copy protection in 1996 — and I was among them — agreed that the one fundamental weakness of CSS, the manufacturer's key, was unavoidable. It was recognized that if a manufacturer's key was ever compromised, and there were large numbers of devices already using the key, the system would be broken. There would be no way to revoke the compromised key without causing unacceptable consumer dissatisfaction. What else could be done? Because the system needed to play disks worldwide across all manufacturers' players and the players had to work in a disconnected mode, it seemed inevitable that the system had to be based on shared secrets.

It turned out that this intuition was wrong. A few years before, Amos Fiat and Moni Naor had written a theoretical paper [3] in which they asked the question, "Can a key management system be devised if the parties involved have only a one-way communication?" The answer was "yes," and they provided a concrete example. They were thinking of the case of broadcasting encrypted programs to many receivers, so they called their approach "broadcast encryption." It turns out that the one-way property is also essential for protecting physical media: A recorder makes a recording, and years later a player needs to play it back. There may be no opportunity for the player and recorder to get together and share a cryptographic key.

In 1997, Kevin McCurley and I at IBM became attracted to broadcast encryption, not because of its applicability to protecting physical media — we had not made the connection, either — but because it needed substantially less device overhead than public key cryptography. We devised an improvement over the Fiat–Naor method and described it to several companies. A couple of months later, Brendan Traw from Intel came back and said, "you know, this would solve the problem of DVD recording, don't you?" He was right, of course, and with that, the technology called Content Protection for Recordable Media (CPRM) was born. IBM and Intel were later joined by Matsushita and Toshiba in developing and licensing this technology.

Like any broadcast-encryption-type technology, CPRM works by having a block of data, the *media key block*,[3] prerecorded on the blank

---

[2]Addition information available at http://www.4centity.com.
[3]Fiat and Naor called it the "session key block."

media. Every CPRM device in the world has a unique set of keys, called *device keys*. Two devices may have some keys in common, but no two devices will have exactly the same set. The devices process the media key block using their device keys. Each device follows a unique path through the media key block, but they all end up with the same answer, called the *media key*. The media key is the key that fundamentally protects the content on the media. If a circumvention device appears in the world, new media key blocks can be produced so that that particular device's path is blocked — it does not calculate the same media key as the rest of the compliant devices. It cannot be used to copy a recording if it has been made on the new media. Thus, CPRM solves the problem that plagued CSS: If a 16-year-old in Norway discovers a set of keys, the hacked program he or she produces will not work on newly released media. The value of the hack becomes less and less as time passes.

Figure 23.1 illustrates a media key block. For each type of protected media, the CPRM license agency has created a matrix of keys. The matrices are always 16 columns wide, but the number of rows depends on the type of media (from 500 to 24,000 rows). For each device, a manufacturer gets 16 device keys, 1 in each column, shown as small rectangles. In the meantime, the license agency is producing media key blocks for blank media. The media key blocks contain the encryption of the media key, over and over again, at each position in the matrix. The media key is picked at random and is different for each media key block.

When the device processes a media key block, it can use any of its device keys. As of early 2004, no circumvention devices have yet appeared, so any device keys will work. Eventually, however, circumvention devices may appear. The compromised keys they are using are shown as X's. Those positions in the matrix will not contain the correct encryption of the media key. If by bad luck, an innocent device may have a key or two that is compromised, it just uses one of its other keys.

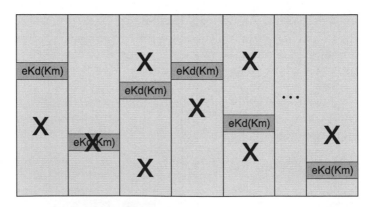

Figure 23.1. The CPRM media key block.

Unless a large fraction of the matrix is covered with X's, an innocent device has an extremely high probability that it will have at least one uncompromised key. If that large a part of the matrix is compromised, the system will be effectively broken. This will require hundreds to tens of thousands of independent attacks, depending on the size of the matrix.

It turns out to be prohibitively expensive, for most types of media, to produce a new media key block for each piece of blank media. Instead, the media key blocks are massed produced (in the case of DVD record-able blanks, up to one million blanks can use the same media key block). However, a unique serial number has been burned into the physical media, and the devices use it as part of the cryptographic calculation. In particular, they perform a cryptographic hash, combining the media key and the serial number. The result is then treated as a key, called the *media unique key,* which, with very high probability, is unique for the particular piece of media. Even though the serial number is freely readable by any device, only compliant devices can calculate the media unique key, because only a compliant device can calculate the media key correctly.

Just like CSS, CPRM adds one more level of indirection before the actual key that encrypts the content is obtained. That key is called the *title key,* and there can be many title keys on a single piece of media. The title key is randomly picked by the recorder, which then encrypts that title key with the media unique key. It places the encrypted content and the encrypted title key on the media. A compliant player can play back the encrypted content by reversing the steps: After calculating the media unique key, the player reads and decrypts the encrypted title key. Then, using the title key, the player can decrypt and play the content itself.

The title key calculation is slightly more complicated on some types of CPRM media: Before encrypting the title key, the recorder exclusive-ors it with the Copy Control Information (CCI). The CCI can be as simple as our old friend, the two bits from SCMS, but it can also contain more information (e.g., the number of copies allowed). Now, if the 12-year-old Basic programmers flip the CCI bits to obtain more copy permission, it does not make it available. They can change the bits, but the compliant players will no longer calculate the title keys correctly, and the content has been rendered useless.

Content Protection on Recordable Media was intended to allow record-ers to make copies that could not be further copied; it was designed for recordable media. The same principles, however, are obviously applicable to prerecorded content. An offshoot of CPRM, called Content Protection for Prerecorded Media (CPPM), has been developed by the same companies. In CPPM, the idea of having a unique serial number per piece of media has been replaced by the idea of having a serial number (called an *album ID*) for each particular release. Other than that, the two technologies are identical. CPPM is used to protect DVD audio, a successor technology to DVD video. To date, DVD audio has been a high-resolution format of interest only to audiophiles, but because of its strong copy protection capability, the music industry is looking at it as a replacement to music CDs.

Both CPRM and CPPM have been remarkably successful. As of year end 2003, over 200 million devices have been licensed. CPRM has been employed on DVD recordables (DVD-RAM, DVD-R, and DVD-RW), on the Secure Digital Memory Card (SD Card[4]), and on secure CompactFlash.[5] To be honest, however, a great deal of the success of CPRM has been due to the success of the SD Card, and the success of the SD Card has everything to do with its price, performance, and convenient form factor, and little to do with its inherent copy protection feature.

DIVX

After examining a successful copy protection technology like CPRM, it might provide some balance to consider a copy protection scheme that failed in the market. Divx was a technology developed by Circuit City together with some Hollywood-based investors. It was an innovative idea, with some interesting consumer and distribution advantages that the company seemed singularly unable to articulate. Fundamentally, Divx was a rental replacement technology. Built on top of the DVD technology, the Divx player would allow the consumer to play the disk for a 48-h period. If the user wanted to play it again later or if he had obtained the disk from a friend who had already used the 48-h period, that was permitted, but it would require an additional payment.

Divx was implemented by requiring the player to have a phone connection to a clearinghouse and periodically to call and report usage. If the consumer blocked the phone calls, the player would eventually refuse to play disks. A consumer might be able to obtain a few weeks of free play by that trick, but it would be a one-time attack.

What were the advantages of the Divx technology? It had the potential to revolutionize the way the movie rental business works. Today, the most difficult aspect of renting movies, from the point of view of the outlet offering them, is maintaining inventory. In other words, the outlets must keep track of the returned movies. Most retail stores shy away from this complication. Divx disks, however, were designed not to be

[4]SD is a trademark of the SD Card Association.
[5]CompactFlash is a trademark of the CompactFlash Association.

returned. The store could manage them like any other retail item. Convenience stores were the ideal retail outlets for Divx disks.

The consumer value proposition for Divx came indirectly from the retail advantage. If the disks were available in many retail outlets, the consumer would not have to make a special trip to rent a movie, and no trip at all to return it. Because the 48-h period did not begin until the disk was inserted in the player, a consumer buying a quart of milk on Tuesday, for example, could rent a movie on impulse for Saturday night. If it turned out that she was busy on that Saturday, she could play the movie the next Saturday.

The Divx company seemed unable to exploit the obvious advantages of their technology. Reasoning, presumably, that consumers would be wary of renting movies from unfamiliar sources, they focused, unsuccessfully, on traditional movie rental outlets such as Blockbuster.[6] Blockbuster must have seen Divx as a major threat. Divx also seemed unable to explain its value proposition to consumers. The San Francisco Bay Area was one of their test markets, so I observed their early advertisements. Their first advertising slogan was, "Build your movie library at $5 a disc!" Divx was many things, but it *absolutely* was not that. The $5 price was a rental fee.

Opponents of content protection technologies have often cited the failure of Divx as an example of consumer revolt against overly restrictive content protection technologies. This appears to be wishful thinking. Although Divx players were sold in only about 10% of the retail consumer electronics outlets in the United States, they accounted for 20% of the sales of DVD players. This remained true up until the moment the company went out of business. It can be argued, then, that consumers preferred Divx players two-to-one over conventional DVD players. Instead the failure of Divx appears to be a failure to correctly exploit a good idea. Or perhaps, it was an idea ahead of its time.

DIGITAL TRANSMISSION CONTENT PROTECTION[7]

Up to now, I have focused in this chapter on content protection for physical media. That is only part of the problem. At the same time that the industry was developing DVD video, they recognized that some of the high quality of the digital content could be lost if it had to be connected to a TV over a conventional analog coax connection. The digital bus called Firewire (IEEE 1394) was designed to carry compressed digital TV streams with perfect fidelity. If the streams were transmitted in

[6]Blockbuster is a trademark of Blockbuster, Inc.
[7]Additional information available at http://www.dtcp.com.

the clear, however, hackers would have an easy circumvention path to get the copy-protected movies.

Five companies (Intel, Toshiba, Sony, Hitachi, and Matsushita) developed a technology to solve this problem. Called Digital Transmission Content Protection (DTCP), it is available on Firewire, on the Universal Serial Bus, and, most recently, on any transmission technology that supports Transmission Control Protocol/Internet Protocol (TCP/IP).

Unlike physical media, transmission media allow the two devices to have a conversation. Therefore, conventional public key cryptography can be applied. Each device has a public key and its corresponding private key. It has a certificate, signed by the licensing agency, that reveals its public key. For example, let us say a DVD player wants to output to a TV over DTCP. The two devices exchange their certificates, and each device verifies that the other's certificate is correctly signed. Furthermore, they both consult a certificate revocation list (called a System Renewal Message in DTCP) to verify that other device has not been revoked.

They also agree on a cryptographic key, using a technique called the Diffie–Hellman protocol. Diffie–Hellman is perhaps the best known public key method, and is based on two principles:

1. $(g^x)^y = (g^y)^x$.
2. There is no efficient way to calculate x if you are given the remainder of g^x after dividing by some number (such as a large prime), even if you know g. Thus, if g and x are large numbers, finding x is an intractable calculation.

All the devices in the system agree upon a number g and a prime P; these are wired in at the factory. One device picks a random x and transmits the remainder of g^x after dividing by P; the other device picks a random y and transmits the remainder of g^y after dividing by P. The first device raises g^y to the x power; the second device raises g^x to the y power. Thus, they both calculate the remainder of g^{xy} after dividing by P, and this becomes the common key in which they encrypt other keys. In effect, they have established a secret value while having a completely public conversation.

In DTCP, each device signs its Diffie–Hellman messages with the private key that corresponds to the public key in its certificate; this is called an *authenticated* Diffie–Hellman protocol. Second, the signatures in the messages and the values in the Diffie–Hellman protocol are not integers, but points on an elliptic curve. Although a discussion of this *elliptical curve cryptography* is too detailed for this chapter, it does not change the underlying Diffie–Hellman mathematical principle. Instead, it allows the calculations to use smaller numbers than would otherwise have been needed to prevent exhaustive search.

This is a very standard cryptographic approach. However, some consumer electronics companies felt that the calculations involved in public key cryptography were too onerous for their low-end devices. A compromise was reached. Devices that could only handle "copy once" content — which is to say, recorders — were allowed to establish the bus key in a simpler way, called *restricted authentication*. The argument was that "copy once" content was less valuable than "copy never" content, so its cryptographic protection did not need to be as strong. Devices that could handle "copy never" content (e.g., TVs, DVD players, set-top boxes) had to implement both types of authentication.

Once a recorder makes a recording of "copy once" content, the recording can no longer be copied. However, if the recording was marked "copy never," that would require the recorders to have the full authentication to play back the recordings they made, which was what the designers were trying to avoid. Instead, the recording is marked "copy no more," a state that is logically the same as "copy never," but one in which the receiving device does not demand full authentication.

The DTCP restricted authentication is superficially similar to CPRM's media key block. Like CPRM, it is based on a matrix of values. The license agency has produced this matrix. Each source device (player, set-top box) is given three rows from the matrix; each sink device (recorder, TV) is given three columns. As sketched in Figure 23.2, the devices are not given the actual values in their rows or columns, instead they are given a

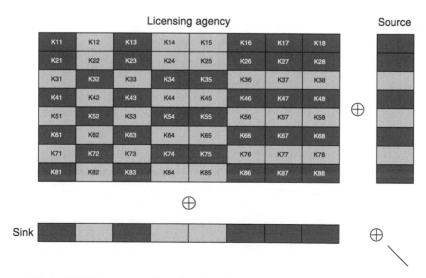

Figure 23.2. DTCP restricted authentication.

summary, which is referred to as their keys. The summary is the element-by-element exclusive-or of the selected rows or columns. For example, the source device is given a key for each row in the matrix, and that value is the exclusive-or of its selected columns at that row.

A source and a sink agree upon a common key as follows: The source device tells the sink which columns it is using. Because the sink device has a key for every column, it selects the keys from the three sink's columns and exclusive-ors them together. Concurrently, the sink device tells the source which rows it is using. The source device exclusive-ors the three keys that correspond to those rows. In effect, each device is calculating the exclusive-or of the nine values they have in common, shown in black, although they are each doing it in a different way and neither of them know the actual nine values. That result of the calculation is then the common key for the two devices.

The amount of revocation that can be achieved with such a system is limited, even though there are two matrices, one for "copy once" transfers, and one for "copy no more" transfers. The DTCP matrix is 12×12; this means that there are only 220 possibilities each for source devices and sink devices. If a particular set of rows or columns needs to be revoked, there will be many innocent devices that are using the same set. Also, each compromised set gives the attackers an equation in 3 unknowns, and after 12 such compromises, they will have 12 equations in 12 unknowns and be able to calculate the entire matrix. The well-publicized "break" of DTCP was nothing more than an observation of this obvious truth, which had been long admitted by the designers of the system. The restricted authentication mode of DTCP is weak, but it is arguably sufficient for the lower-value content that it is intended to protect. The full authentication mode is a sound public-key-based system.

HIGH DEFINITION CONTENT PROTECTION[8]

Digital Transmission-Content Protection was intended to protect compressed video data, especially on the Firewire bus. There is a second digital bus in common use on high-definition televisions that conveys uncompressed video. That bus is called the Digital Video Interface (DVI) and was originally designed for computer monitors. Points in the system that contain uncompressed video are substantially less vulnerable to attack than points that contain compressed data. In order to record uncompressed video, the recorder must have a huge recording capacity, both in terms of bandwidth and total storage, or must be prepared to compress on-the-fly. Either solution has its problems and costs.

[8]Additional information available at http://www.digital-cp.com.

Nonetheless, there were concerns about the DVI bus not having any protection. In 1999, Intel defined the High Definition Content Protection (HDCP) technology to protect the DVI bus. HDCP's keys use the same scheme used in DTCP's restricted authentication mode. Also, the scheme uses scrambling, not encryption, to actually protect the data. So, HDCP is not a particularly strong protection scheme, but the data it is protecting are not particularly vulnerable. Because HDCP can be inexpensively implemented in the device, it appears as though the designers have achieved a reasonable balance.

BROADCAST FLAG

As was alluded to earlier, the licenses for the technologies we have described are often more involved than the specifications. The licenses contain numerous rules describing what the manufacturers can and cannot do. These are called *compliance rules*. However, they also contain rules that constrain how the content owners can use the technology. These are *encoding rules*. The consumer electronics and computer companies that develop these technologies make more money the more copying that is allowed, up to a point. After all, the companies make recorders, hard disks, computers, and so forth that are used for copying. However, it is in no one's interest, not even consumers', if the content owners' worst fears are realized, and rampant copying destroys the entertainment industry.

The encoding rules in the DTCP license, which are also mimicked in the other licenses that want to be compatible, are consumer-oriented:

1. Over-the-air digital broadcasts must be marked "copy freely."
2. Cable-only programs may be marked no more restrictive than "copy once." However, if a cable signal is also part of an over-the-air broadcast, it must be marked "copy freely." For example, 99% of the cable viewers in the United States receive WTBS through their cable system; however, the signal is broadcast over-the-air in Atlanta, so it must be "copy freely."
3. Pay-per-view or video-on-demand programs may be marked "copy never."
4. Temporarily recording a program (called a "transient image") for implementing consumer convenience functions such as rewind and fast forward does not count as a copy; even a "copy never" program may be recorded for this purpose. The content owners control how long the transient image may exist, but they may not set the period to less than 90 min.

These rules were basically acceptable to television and movie studios. However, they had some real concerns about the "copy freely" content remaining in the clear, where it could be widely redistributed on the Internet. They were concerned that in the new high-bandwidth world, consumers might stream a digital over-the-air broadcast out over the Internet in real time. This could destroy the value of regional advertisements, which are a substantial source of revenue for local television stations. They were also concerned that file-sharing networks or casual personal Web sites might save all of the programs from a popular television series as they are broadcast, destroying the value of the reruns of the series in syndication.

The studios were willing to accept DTCP's relatively liberal encoding rules if the technology companies could solve the problem of Internet redistribution for them. The DTCP group came up with the idea called Encryption Plus Non-assertion (EPN). The video content is marked "copy freely," but it is also encrypted. The compliance rules of the DTCP license prohibit devices from putting such content on the Internet. Of course, they are also prohibited from releasing the content in the clear so that downstream, noncompliant devices could put it on the Internet for them.

Incidentally, EPN gave those of us working on CPRM fits. Although DTCP had separate bits for encryption and copy control information, we did not; we had made the simplifying assumption for DVD video recording that encryption was synonymous with "copy no more." So technically, EPN content in CPRM is marked "copy no more." However, there is an 8-byte field in the content itself that says, in effect, "I don't really mean that, it is actually 'copy freely'." This 8-byte value is a cryptographic value called a Message Authentication Code (MAC) and is different for each piece of content and media. It can only be correctly calculated by a device that can calculate the media unique key. In other words, it can only be calculated by a compliant device.

Concurrently with the work being done by the various content protection technology groups, the Motion Pictures Expert Group (MPEG) had defined a field, called the *broadcast flag*, which could be used to denote that a given digital broadcast should have EPN protection. The content companies asked the Federal Communication Commission (FCC) to pass a rule requiring digital tuners to respond to the broadcast flag with an approved content protection scheme. Observe that this approach is completely at odds with the agreements that had been hammered out in the original DVD discussions: Content should be protected at the source, and downstream participation should be strictly voluntary. However, to apply this principle in this case would require the public air waves to carry encrypted signals. This is a political hot potato that few wanted to touch.

In November 2003, the FCC passed a rule requiring digital receivers manufactured after July 2005 to respond correctly to the broadcast flag.

This rule has been controversial. At least some of the opposition, however, seems to be coming from consumer rights groups who are under the misapprehension that the broadcast flag somehow restricts copying. EPN allows unlimited physical copies; if there is any doubt, ask anyone in the CPRM group about the hoops they had to jump through to make that possible. The broadcast flag and EPN is strictly about restricting the Internet redistribution of digital television broadcasts.

ADVANCED BROADCAST ENCRYPTION

Compared to public key cryptography, broadcast encryption is a younger technology, and remarkable strides have been made in recent years. The original Fiat–Naor and the later CPRM matrix approaches have the property that they have only a finite amount of revocation. Although this finite number may be quite large and, therefore, the chance of defeating the system by running out of revocation capability is negligible, this theoretical limitation has been overcome. In 1997 and 1998, two independent groups, one led by Debbie Wallner [4] and one led by C. K. Wong [5] devised a scheme based on a tree of keys instead of a matrix. The tree approach allowed unlimited revocation.

The tree approach also solved a problem that we in CPRM called the "evil manufacturer problem." If a fraudulent manufacturer signs a license and begins using keys to build circumvention devices, he does substantially more damage to CPRM's revocation capability than a hacker compromising individual devices, in spite of a clever key assignment algorithm to mitigate this exact problem. In a tree, it is no more expensive to revoke an entire branch (i.e., an entire manufacturer) than it is to revoke an individual device.

With the invention of the tree approach, the state of the art of broadcast encryption had advanced so that it had equivalent revocation capability to public key cryptography, but it still required substantially more space to convey that information. In other words, the two-way handshake of public key cryptography allowed a concise encoding of the revocation information, whereas the one-way transfer of broadcast encryption needed a much larger media key block (e.g., $100 \times$ larger) to achieve the same amount of revocation. In 2001, however, Dalit Naor and Moni Naor (among others) at IBM broke through this barrier and invented a tree-based scheme [6] that was equivalent in space to a public key certificate revocation list.

In the Naor–Naor scheme, which they called *subset difference*, each device has almost every key in the tree. The tree is carefully designed so it is always possible for you to find a single key that every device in the system knows, except the device you are trying to revoke. What if you are trying to revoke two devices? The answer is that there is not a single tree. There are literally billions of different independent trees. Thus, revoking a list of devices in the subset-difference scheme requires you to first identify the trees you want to use (the "subsets") and then identify the single device being revoked in each tree (the "differences"). You then encrypt the media key repeatedly in the keys that the revoked devices do not know. Like a public key certificate revocation list, the amount of data that needs to be transmitted is proportional to the number of devices you are revoking and, unlike all previous broadcast encryption schemes, independent of the total number of devices in the system.

My description of the subset-difference scheme so far has implied that the device needs billions of keys — clearly impractical. However, an earlier paper by Canetti and others [7] had provided an answer: It is not necessary for the keys to be truly independent. If the keys that are closer to the leaves of the trees are one-way functions of the keys that are closer to the root, devices can calculate almost all the keys they need on-the-fly, from a few keys they have stored. With this principle, subset-difference devices need to store only hundreds of keys, instead of billions. Although storing 250 keys, for example, is eminently practical, it requires noticeably more storage than the 16 keys that the original CPRM technology required. In 2001, however, Havely and Shamir observed [8] that subset-difference's "independent" trees could, in fact, be related to each other in a similar way to the way the keys within the tree were. They called their modification *layered subset difference*, and with that improvement, the number of keys required in the device was comparable to the previous schemes that required much large media key blocks.

So, the state of the art of broadcast encryption has advanced remarkably in the 10 years since it was invented. These advances stand ready to protect the new generation of high-definition DVDs, which will probably hit the market in late 2005. With effectively infinite revocation capability, those new disks will not be at the mercy of the first 16-year-old who discovers a key. However, the promise of broadcast encryption is even greater. Essential for applications like physical media that require its one-way property, broadcast encryption no longer pays any penalty in two-way applications compared to public key cryptography. Because it has some interesting consumer advantages compared to public key cryptography that I will cover in the next section, it seems inevitable that more and more content protection schemes will be based on broadcast encryption.

PERSONAL DIGITAL DOMAIN

The technologies we have discussed so far are focused on restricting the number of physical copies that can be made from a given media source. Even the transmission-oriented technologies such as DTCP and HDCP have this as their end goal; they provide a secure way in which the copy restrictions can be communicated and restrict the transfer of content to compliant devices. Recently, however, a different approach has gotten the attention of both industry and consumer groups: Instead of restricting the number of physical copies of a particular content item, it restricts the number of devices that are allowed to play the item. This idea has come to be called the *personal digital domain*.[9] The idea is that each household's extended devices — not just the devices actually in the home, but in-car devices, portable devices, and perhaps even office personal workstations — would comprise the household's personal digital domain.

The consumer promise of the personal digital domain is enticing. Freed from having to count copies, consumers could make unlimited copies for backup and for the convenience of playback at various places and times. However, this promise could be lost if setting up the personal digital domain is an administrative headache or perceived to be an intrusion on privacy. Those working on this concept in IBM, including myself, became convinced that for the personal digital domain to succeed, it must be essentially "plug and play." Devices must form a personal digital domain simply by being connected, and they must do this autonomously, without external registration, except under the most unusual circumstances.

There are several technical approaches being proposed for the personal digital domain. None has yet been adopted by the consumer electronics industry. Basically, the technical problem resolves itself into having a group of devices agree upon a common cryptographic key. This is a classic application of public key cryptography, and all the solutions save one have used public key techniques. The one exception is the one with which I am involved. In 2001, Florian Pestoni, Dalit Naor, and I, buoyed by the recent substantial improvements in broadcast encryption, devised a personal digital domain scheme based on broadcast encryption. We called our solution the *xCP Cluster Protocol*.[10]

Broadcast encryption brings with it several desirable properties, not the least of which is that it needs substantially less device overhead than public key cryptography. However, it also turns out to be the ideal cryptographic primitive when you are building a system that works autonomously and anonymously. Public key cryptography is fundamentally based on *identity*. Parties first exchange their certificates and then enter a protocol in which they each prove that they have the private key that corresponds to the public key in that certificate. In broadcast

---

[9]Also called the "authorized domain" in some circles.
[10]From Extensible Content Protection, our name for a family of content protection technologies based on broadcast encryption, including CPRM and CPPM.

encryption, identity is irrelevant. A device can safely encrypt content based on a media key block, never knowing which device might finally end up decrypting it. However, the first device is confident that this unknown second device is compliant, because it must be to correctly process the media key block.

The fundamental property of any content protection is compliance, not identity. A content protection scheme needs to guarantee that all the devices are playing by the rules. If the scheme is based on public key cryptography, it achieves compliance as a side effect of determining identity. There is a list of those devices that are not playing by the rules, and the scheme needs the identity of the devices in the system to make sure that none of them are on the list. In effect, there is a level of indirection in achieving compliance. A broadcast-encryption-based scheme is more direct: The media key block enables compliant devices explicitly, by allowing them, and no one else, to calculate the key.

What does this all mean for the personal digital domain? At the risk of overgeneralizing, the public key approach suggests systems that are based on the registration and administration of devices, whereas the broadcast-encryption approach suggests systems where the devices, freed from the task of having to explicitly establish the compliance of the whole system, self-organize into personal digital domains. That is not a hard and fast rule; in both cases, one could design a system that bucks this trend.

Assuming that registration of devices with an external authorization center would lead to an awkward and intrusive consumer experience, what other alternatives are there? Basically, there are two: The personal digital domain can be based on physical proximity or the domain can be approximated simply by limiting the number of devices that can be in a single domain. Using physical proximity has both technical and philosophical problems. Technically, network protocols like TCP/IP are deliberately designed to be transparent to distance. They work the same if the devices are around the room or around the world. Philosophically, shouldn't your vacation home be part of the same personal domain as your main home?

That leaves limitations on the number of devices as the best technique for establishing the personal domain. As long as there is an easy exception mechanism (undoubtedly requiring an external connection) to handle those gadget-happy homes that have far more than the normal number of devices, I think this approach is eminently workable. However, it almost immediately suggests attacks where the attackers try to fool the autonomous devices into creating an artificially large domain. Most of the attacks resolve themselves into the attack I call the "dormitory snowball."

In this attack, the attackers form a domain of two devices and then separate them. To each, they introduce one more device. The domain now contains four devices, but no single device knows about more than two others. The process is repeated with the new devices. The domain ends up with many more devices than it should, perhaps even an unlimited number, depending on how devices are allowed to leave the domain.

This attack is thwarted in the xCP protocol by making the domain key be based, in part,[11] on the list of devices in the domain. xCP calls this list the *authorization table*, and each device has its own copy of it. If two devices have a different authorization table, they cannot successfully share content. Thus, in the dormitory attack, the different overlapped clusters effectively operate as different domains. Of course, it is possible for two devices to have different authorization tables for perfectly benign reasons; for example, a device was powered off or disconnected when a new device joined the domain. In the benign cases, however, one authorization table is always a subset of the other, whereas in the attack scenario, they must remain disjoint (or else the number of devices in the table gets too large). It is easy for compliant devices to detect and automatically update their authorization table in the benign case.

To maintain this important superset and subset property, devices leaving the cluster are not simply deleted from the authorization table. They remain in the table, but they are marked as deleted. Furthermore, they must acknowledge that they have been deleted. The rule is that a deleted device can no longer play any content in the domain. Because all devices are compliant, they play by that rule.

What if the user prevents the device from learning that it has left the domain? Then, it continues to count as one of the devices the domain. Suppose a consumer wants to sell a device on eBay,[12] together with his library of content, but retain a copy of his library for his own use. In effect, he is selling unauthorized copies of his content. There is nothing in our protocol that absolutely prevents this. However, the sold device cannot be deleted from the seller's domain, or else the content would be useless. Therefore, the device must be sold together with an authorization table that contains all of the other seller's devices. This can greatly reduce the value of the device to the buyer. The buyer, to use the content, must have his other devices join this domain, which is artificially large. If, as a result, he needs an external authorization to successfully build the new domain, the authorization center can observe that the

---

[11]The other values that are used to calculate the domain key are a domain identifier and, of course, a media key from a media key block. The latter guarantees that only compliant devices can calculate the domain key.
[12]eBay is a trademark of eBay Inc.

domain has an unusually large number of devices that are not active, and guess what has happened. At that point, the center has several possible courses of action, including refusing to allow the domain to grow. I recommend, however, that the center should offer the consumer in this state the chance to make things right in a way that is not too onerous to the consumer.

Note also that attacks of this type are inherently self-limiting. After one or two such attacks, the size of the authorization tables becomes a serious problem to the attackers. However, we allow the possibility of these attacks because we want normal consumers, who are not attempting to traffic in unauthorized content, to have a completely seamless experience. Ideally, the personal digital domain content protection should be completely transparent to the average consumer.

CONCLUSIONS

In this chapter, I have sketched a history of content protection on consumer electronics devices, together with a brief description of each. Over time, improvements in the underlying cryptography have made the content protection schemes increasingly effective. Perhaps fortuitously, the same underlying cryptographic improvements also offer the consumer the promise of increased privacy, transparency, and flexibility. Rather than cloaking the consumer in a content protection straightjacket, it seems that the newest technologies have the opposite potential: The consumer should be able to do more things with entertainment content than he or she has ever been able to do it in the past. This flexibility can be achieved without substantial inroads to the entertainment industry's business. In fact, the new digital modes, offering new business opportunities and new consumer value, can be a boon to the industry.

REFERENCES

1. Barlow, J.P., The economy of ideas, *Wired*, Vol. 203, March 1994.
2. Schneier, B., The Futility of Digital Copy Prevention, http://schneier.com, May 15, 2001.
3. Fiat, A. and Naor, M. Broadcast encryption, in *Advances in Cryptology — Crypto '93*, Lecture Notes in Computer Science Vol. 773, Springer-Verlag, Berlin, 1994, pp. 480–491.
4. Wallner, D.M., Harder, H.J., and Agee, R.C., Key management for multicast: issues and architectures, IETF draft wallner-key, July 1997. ftp://ftp.ietf.org/internet-drafts/draft-wallner-key-arch-01.txt.
5. Wong, C.K., Gouda, M., and Lam, S., Secure Group Communications Using Key Graphs, in Proceedings of ACM SIGCOMM'98, 1998, p. 68.
6. Naor, D., Naor M., and Lotspiech, J., Revocation and tracing routines for stateless receivers, in *Advances in Cryptology — Crypto '2001*, Lecture Notes in Computer Science Vol. 2139, Springer-Verlag, Berlin, 2001, p. 41.

7. Canetti, R., et al., Multicast Security: A Taxonomy and Some Efficient Constructions, in Proceedings of IEEE INFOCOMM '99, 1999, Vol. 2, p. 708.
8. Halevy, D. and Shamir, A., The LSD broadcast encryption scheme, in *Advances in Cryptology — Crypto '2002*, Lecture Notes in Computer Science, Springer-Verlag, Berlin, 2002.

24

Adult Image Filtering for Internet Safety

Huicheng Zheng, Mohamed Daoudi,
Christophe Tombelle, and Chabane Djeraba

INTRODUCTION

Internet filters allow schools, libraries, companies, and personal computer users to manage the information that final users can access. Filters may be needed for legal, ethical, or productivity reasons. In the context of legal reason, many new laws require schools and libraries to enforce the policy of Internet safety that protects against minors and adults access to unsuitable visual content that are obscene, child pornography, or harmful to minors. More generally:

- Parents can check their home computer for evidence of inappropriate usage by children, as well as protect their children from files that may exist there already.
- Teachers can use the content filters to make sure no improper documents have compromised school and university computer systems. So, schools need to limit access to the Web sites so children will not be exposed to objectionable or inappropriate documents.
- Companies can use the content filters to make sure their systems are not contaminated with adult or otherwise unsuitable documents and to determine if files were downloaded during office hours.

0-8493-2773-3/05/$0.00+1.50
© 2005 by CRC Press

Internet surfing can cut into a company's productivity and place a strain on network services. That is why a company may want to filter access to sites that employees might visit during working hours like Amazon and certain magazine publisher sites. Technical staff can track down the reasons for excess Internet bandwidth and hard drive usage.

To deal with the content filtering, three notions are necessary: image information retrieval, text categorization, and image categorization. Even though these three notions are important to efficient content filtering, we will focus in the scope of the chapter on image content filtering. We plan to consider all of these notions in future experiments.

Why Is Content Filtering a Multimedia Information Retrieval Notion?

Multimedia information retrieval [1] deals with the representation, storage, organization of, and access to multimedia information ingredients. The representation and organization of the multimedia information ingredients should provide the user with easy access to the information in which he is interested. Unfortunately, characterization of the user information requirement is not a simple problem. Consider, for instance, the following hypothetical user information need in the context of the World Wide Web (or just the Web): Find all multimedia documents containing visual descriptions of researches that are maintained by a country, and participate in an European project. To be relevant, result documents [2] must include Web pages including images and video of research and the coordinates of the research team head.

Clearly, this full description of the user visual information need cannot be used directly to request information using the current interfaces of Web search engines. Instead, the user must first adapt this visual information need into a query that can be processed by the search engine. In its most common form, this adaptation meets a set of keywords, which summarizes the description of the user visual information need. Given the user query, the key goal of a multimedia information retrieval system is to retrieve information that might be useful or relevant to the user.

When people refer to filtering, they often really mean information retrieval. The filtering corresponds to the Boolean filter in multimedia information retrieval [3]. So, decisions would concern every multimedia document regarding whether or not to visualize it to the user who is searching the information. Initially, a profile describing the user's information [4] needs is set up to facilitate such decision making; this profile may be modified over the long term through the use of user models. These models are based on a person's behavior — decisions, reading behaviors, and so on — that may change the original profile. Both information retrieval and information filtering attempt to maximize the good material that a person sees (that which is likely to be appropriate to the information problem at hand) and minimize the bad material.

The difference between multimedia information retrieval and multimedia information filtering is that multimedia information retrieval is intended to support users who are actively retrieving for multimedia information, as in Internet searching. Multimedia information retrieval typically assumes a dynamic or relatively active behavior of users, because they interact with the search engine by specifying queries. On the basis of the first results, users interact again with the search engine on the basis of his feedback that are useful for the further queries. By contrast, multimedia information filtering supports people in the passive monitoring for desired information. It is typically understood to be concerned with an active incoming stream of information objects.

In content filtering, we deal with the same problems as in multimedia information retrieval. For example, in the online side of multimedia information retrieval, the accurate specification of the queries is not simple and rough, because users look for information they do not know, and it is probably inappropriate to ask them to specify accurately what they do not know. The representation of multimedia information content requires interpretations [5] by a human or machine indexer. The problem is that anyone's interpretation of a particular text or medium is likely to be different from any indexer else's, and even different for the same indexer at different times. As our state of knowledge or problems change, our understanding of a text and medium changes. Everyone has experienced the situation of finding a document not relevant at some point but highly relevant later on, perhaps for a different problem or perhaps because we are different. Studies showed that when people are asked to represent an information ingredient, even if they are highly trained in using the same indexing language, they might achieve as much as only 60–70% consistency in tasks such as assigning descriptors. We will never achieve "ideal" information retrieval — that is, all the relevant documents and only the relevant documents, or precisely that one thing that a person wants. Because of these roughs, the retrieval process, is also inherently rough and probabilistic.

The understanding of soft- and hard-porn Web sites is subjective, and, therefore, representation is necessarily inconsistent. Algorithms for representing images do give consistent representations. However, they give one interpretation of the text and image, out of a great variety of possible meanings, depending on the interpreter. Language is ambiguous in many ways: polysemy, synonymity, and so on, and the context matters greatly in the interpretation.

The meta-language used to describe information objects, or linguistic objects, often is construed to be exactly the same as the textual language.

However, they are not the same. The similarity of the two languages has led to some confusion. In information retrieval, it has led to the idea that the words in the text represent the important concepts and, therefore, can be used to represent what the text is about. The confusion extends to image retrieval, because images can be ambiguous in at least as many ways as can language. Furthermore, there is no universal meta-language for describing images. People who are interested in images for advertising purposes have different ways to talk and think about them than do art historians, even though they may be searching for the same images. The lack of a common meta-language for images means that we need to think of special terms for images in special circumstances.

In attempting to prevent children from getting harmful material, it is possible to make approximations and give helpful direction. However, in the end, that is the most that we can hope for. It is not a question of preventing someone from getting inappropriate material but, rather, of supporting the person in not getting it. At least part of the public policy concern is children who are actively trying to get pornography, and it is unreasonable to suppose that information retrieval techniques will be useful in achieving the goal of preventing them from doing so.

There are a variety of users. The user might be a concerned parent or manager who suspects that something bad is happening. However, mistakes are inevitable and we need to figure out some way to deal with that. It is difficult to tell what anything means, and usually we get it wrong. Generally, we want to design the tools so that getting it wrong is not as much of a nuisance as it otherwise might be.

Why Is Content Filtering a Text Categorization Notion?

Categorization is a problem that cognitive psychologists have dealt with for many years [6]. In the process of categorization of electronic documents, categories are typically used as a means of organizing and getting an overview of the information in a collection of several documents. Text categorization is the sorting of text into groups, such as pornography, hate speech, violence, and unobjectionable content. A text categorizer looks at a Web page and decides into which of these groups a piece of text should fall.

Automatic categorization of electronic documents is dependent on the structure of the documents — documents can be partitioned into categories based on, for example, differences of their layouts [7]. Keywords or phrases are often very useful for categorizing electronic documents, but only to a certain degree. The physical appearance of a document is not enough for a good categorization of the document: The documents in one domain (e.g., the domain "tax forms") may have a well-known and predefined structure whereas in other domains (e.g., e-mail letters), they may have less structure and, as is typical for e-mail, be more "anarchic" by nature.

Content filtering is a categorization task because the categories are decisions, such as "allow," "allow but warn," or "block." We either want to allow access to a Web page, allow access but also give a warning, or block access. Another way to frame the problem is to say that the categories are different types of content, such as news, sex education, pornography, or home pages. Depending on which category we put the page in, we will take different actions. For example, we want to block pornography and give access to news.

There are two general and basic principles for creating categories: cognitive economy and perceived world structure [6]. The principle of cognitive economy means that the function of categories is to provide maximum information with the least cognitive effort. The principle of perceived world structure means that the perceived world is not an unstructured set of arbitrary or unpredictable attributes. The attributes that an individual will perceive and thus use for categorization are determined by the needs of the individual. These needs change over time and with the physical and social environment. In other words, a system for automatic text categorization should in some way "know" both the type of text and the type of user. The maximum information with least cognitive effort is achieved if categories map the perceived world structure as closely as possible [6].

Searching and retrieving information can be treated as a process in which query information is mapped onto the categories (the structure of the filing system) in order to retrieve a specific document. Relatively unstructured documents, as in the case of e-mail letters, and unstructured queries might require some transformation on the query information.

One of the goals in the author's current research is to find a means for automatically measuring something similar to the typicality of an electronic document (images), so that the typicality can be used for measuring dissimilarity between a pair of documents or between a document and a summary of category members. The typicality of a letter has to be defined — a combination of information retrieval and simple natural language processing techniques may be feasible to use to extract important information from letters.

Text categorization can be seen as the basis for organization and retrieval of information. Automatic categorization may be used to organize relatively unstructured documents and make it easier to use other methods of creating and presenting the documents (e.g., associative methods such as hypertext, which are more natural to a user) [8].

Adding something like measurement of typicality to an individual letter may give e-mail a new "sense." This might make the (manual or automatic) categorization of letters easier, hopefully more natural, and also more like categorization of letters in traditional postal mail.

Why Is Content Filtering an Image Categorization Notion?

In the problem of image categorization, we assume that a fixed set of classes or categories has been defined and that each image belongs to one category. We are given a set of training images whose category membership is known. In addition, we have a set of unlabeled images and we want to assign each of these images into one of the categories. For an automatic approach to this problem to be feasible, one needs to assume that images that are in a way similar will fall into the same category and that this underlying notion of similarity can be captured automatically using a suitable representation of images and some learning algorithm. In addition, before applying machine learning techniques to images in order to learn models, we need to represent images with structures with which known machine learning algorithms can actually work. Thus, the problem of image categorization bears a strong relationship to image retrieval, where the notions of representation and similarity are also of great importance. One way of applying results from image retrieval to the problem of image categorization would be to take image descriptors, such as histograms or autocorrelograms, and use any of the numerous machine learning techniques that can work with image descriptors. Alternatively, one could take an arbitrary similarity measure from image retrieval, no matter what the underlying representation, and use it in combination with a machine learning algorithm that can work with arbitrary similarity or distance measures, such as nearest neighbors or, possibly, the generalized kernel variety of support vector machines [9].

Determining whether a picture is pornographic is difficult for several reasons. First, an image looks different from different angles and in different lights. When color and texture change, images look different. People can change their appearance by moving their heads around. So, we apparently have different images, but these images support the same semantic.

The approach we proposed is situated in continuity of image categorization.

Multimedia Filtering Approaches

It is important to note that no filtering method is perfect. The Web changes minute by minute. Web URLs are fleeting. Companies buy and sell domain names. Web hosting companies are bought by other companies. URLs that use to point to safe content can now point to different content. Filtering products need to be constantly updated or they lose their effectiveness. Nothing can replace proper training and adult supervision. The following are some of the methods that are used to restrict access to Web files:

- Restrict access by file types. Do not allow viewing of graphics, sound, or movie files. Browsers can be set not to display these type of files. Some browsers do not allow the viewing of these file types.

- Restrict access based on words found on the page. If restricted words are found on the page, access to the page will be blocked. A problem with this method is that the word patterns used to block questionable sites also end up blocking sites that probably should not be blocked.

- Restrict access to Web sites by blocking IP addresses. Questionable URLs (internet protocols (IP) addresses) are added to a list or database. The blocking can be to deny access or only allow access if the URL is in the database or list. Rating services and some blocking software use this method. Some software offer updates to their list of questionable sites for a yearly renewal fee. The problem with this type is that a third party decides what should or should not be filtered. Many of these companies do not fully share there blocking criteria.

- Restrict access to Web pages by using a rating system. Rating systems provide a method of assigning values to page content. The system would allow the content to be rated suitable or unsuitable for children. These systems depend on webmasters embedding certain meta-tags in their Web pages. In this way, the content is rated by the content creator. The process is usually handled by the webmaster answering questions on a Web page and the rating system site e-mails the appropriate meta-tags to the webmaster who inserts the meta-tags into his Web pages. Some common rating systems are RSACi (http://www.rsac.org/homepage.asp), ICRA, Weburbia, and SafeSurf. The greatest drawback to this system is the number of sites that have voluntarily rated their Web pages. Although these rated sites number in the hundreds of thousands, there are millions of Web sites that have not adopted any self-rating system. Many educational institutes have thousands of Web pages they would have to rate. The labor costs associated with applying a rating to each Web page makes it a huge investment of educational resources. Access can be restricted by using a PICSRules file (http://www.w3.org/PICS/). A PICSRules file is a text file that uses PICS rules to define what sites can be viewed by a browser. ICRA provides a free template for building your own filter.

Project Interconnect has a PICSRules file you can download and change to suit your needs. PICSRules files have an extension of .prf (profile). The downside of this method is as the file grows larger, it takes more time to look up addresses and slows down the Web responses. It is not that easy for novices to figure out how to allow or block sites.

- Restrict access by redirecting requests to a local Hosts file. A Hosts file resides on many computer platforms (Mac, Windows, Linux) and can be used by the computer to resolve a Web address. A Web address like www.yahoo.com needs to be resolved to its IP number. If the IP number is found in your Hosts file, your computer stops looking and goes to the site listed in your hosts file. If it is not in the Hosts file, a DNS (domain name server) computer is queried and your request is sent to that IP address. Because the Hosts file is checked first, it provides a method that can be used to block sites that have objectionable content, or are used for advertisements, or any other site that you choose to block. A drawback is if the URL is already an IP number, the Hosts file is not queried and the URL cannot be blocked by the Hosts file. Like the PICSRules file, as the Hosts file gets larger, Web response slows down.

Paper Objective

This chapter is aimed at the detection of adult images appearing in the Internet. Indeed, images are an essential part of today's World Wide Web. The statistics of more than 4 million HTML Web pages reveal that 70.1% of webpages contain images and that on average there are about 18.8% images per HTML Webpage [10]. These images are mostly used to make attractive Web contents or to add graphical items to mostly textual content, such as navigational arrows.

However, images are also contributing to harmful (e.g., pornographic) or even illegal (e.g., pedophilic) Internet content. Therefore effective filtering of images is of paramount importance in an Internet filtering solution.

Protecting children from harmful content on the Internet, such as pornography and violence, is increasingly a research topic of concern. Fleck et al. [11] detects naked people with an algorithm involving a skin filter and a human figure grouper. The WIPE system [12] uses Daubechies wavelets, moment analysis, and histogram indexing to provide semantically meaningful feature vector matching. Jones and Rehg [13] propose techniques for skin color detection and simple features for adult images detection. Bosson et al. [14] propose a pornographic image detection system that is also based on skin detection and the multilayer perception (MLP) classifier.

In our adult image filter, the first step is skin detection. We build a model with Maximum Entropy Modeling (MaxEnt) for the skin distribution of the input color image. This model imposes constraints on color gradients of neighboring pixels. We use the Bethe tree approximation to eradicate the parameter estimation and the Belief Propagation (BP) algorithm to obtain an exact and fast solution. Our model outperforms the baseline model in Reference 13 in terms of pixel classification performance. Based on the output of skin detection, features including global ones and local ones are extracted. We train a MLP classifier on 5084 patterns from the training set. In the testing phase, the MLP classifier takes a quick decision on the pattern in one pass. The adult image filter takes 0.08 sec for an inquiry 256×256 image. Compared with 6 min in Reference 11 and 10 sec per image in Reference 12, our system is more practical. Plenty of experimental results on 5084 photographs show stimulating performance.

The chapter is organized as follows. In section "The Structure of the System," we give a brief overview of the system. An adult image filter is presented in sections "Skin Detection" and "Adult Image Detection." Section "Experimental Results" will present the results obtained by our adult image detector. Section "Conclusion" summarizes the chapter.

THE STRUCTURE OF THE SYSTEM

Our object is blocking images that are objectionable, including adult images and harmful symbols. The system structure is shown in Figure 24.1. After an unknown image is captured, it is first classified by a classifier discriminating natural images from artificial images. If an image is classified as a natural image, it is passed to an adult image filter or a symbol filter. In this chapter, we will present our adult image filter.

SKIN DETECTION

Methodology and Notations

Skin detection is of the paramount importance in the detection of adult images. We build a MaxEnt model for this task. MaxEnt is a method for inferring models from a dataset. See Reference 15 for the underlying philosophy. It works as follows: (1) choose relevant features, (2) compute their histograms on the training set, (3) write down the MaxEnt model within the ones that have the feature histograms as observed on the training set, (4) estimate the parameters of the model, and (5) use the model for classification. This plan has been successfully completed for several tasks related to speech recognition and language processing. See, for example, Reference 16 and the references therein. In these

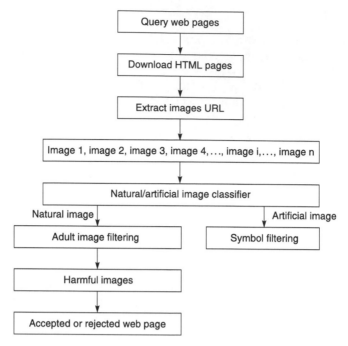

Figure 24.1.   The structure of the system.

applications, the underlying graph of the model is a line graph or even a tree, but in all cases it has no loops. When working with images, the graph is the pixel lattice. It, indeed, has many loops.

A breakthrough appeared with the work in Reference 17 on texture simulation where steps 1 to 4 were performed for images and step 5 was replaced by simulation.

We adapt to skin detection as follows: In step 1, we specialize in colors for two adjacent pixels given "skinness."

We choose RGB color space in our approach. In practice, we know from References 13 and 14 that the choice of color space is not critical given a histogram-based representation of the color distribution and enough training data. In step 2, we compute the histogram of these features in the Compaq manually segmented database. Models for step 3 are then easily obtained. In step 4, we use the Bethe tree approximation; see Reference 18. It consists in approximating locally the pixel lattice by a tree. The parameters of the MaxEnt models are then expressed analytically as functions of the histograms of the features. This is a particularity of our features. In step 5, we pursue the approximation in step 4: We use the BP algorithm (see Reference 19), which is exact in a tree graph but only approximate in loopy graphs.

Indeed, one of us had already witnessed in a different context that a tree approximation to loopy graph might lead to effective algorithms; see Reference 20.

Let us fix the notations. The set of pixels of an image is S. Let $C = \{0, \ldots, 255\}^3$. The color of a pixel $s \in S$ is x_s, $x_s \in C$. The "skinness" of a pixel s, is y_s, with $y_s = 1$ if s is a skin pixel and $y_s = 0$ if not. The set of neighbors of s is denoted as $v(s)$. $\langle s, t \rangle$ denotes a pair of neighboring pixels. We use a four-neighbor system here. The color image, which is the vector of color pixels, is denoted by x and the binary image made up of the y_s's is denoted by y.

From the user's point of view, the useful information is contained in the one-pixel marginal of the posterior; that is, for each pixel, the quantity $p(y_s = 1 | x)$, quantifying the belief of skiness at pixel s. Bayesian analysis tell us that whatever cost function the user might think of, all that is needed is the joint distribution $p(x, y)$. In practice $p(x, y)$ is unknown. Instead, we have the model the segmented Compaq Database [13]. It is a collection of samples

$$\{(x^1, y^1), \ldots, (x^n, y^n)\}$$

where, for each $1 \leq i \leq n$, x^i is a color image and y^i is the associated binary skinness image. We assume that the samples are independent of each other with distribution $p(x, y)$. The collection of samples is referred to later as the training data. Probabilities are estimated using classical empirical estimators and are denoted with the letter q.

First-Order Model

Our MaxEnt model respects the two-pixel marginal of the joint distribution. We define the following constraints:

$$C: \forall\ s \in S, \forall \in V(s), x_s \in C, \forall x_t \in C,$$

$$\forall y_s \in \{0, 1\}, \forall y_t \in \{0, 1\}$$

$$p(x_s, x_t, y_s, y_t) = q(x_s, x_t, y_s, y_t)$$

The quantity $q(x_s, x_t, y_s, y_t)$ is the proportion of times we observe the values (x_s, x_t, y_s, y_t) for a couple of neighboring pixels, regardless of the orientation of the pixels s and t in the training set.

Using Lagrange multipliers, the solution to the MaxEnt problem under $C(x)$ is then the following Gibbs distribution:

$$p(x, y) \approx \prod_{\langle s, t \rangle} \lambda(x_s, x_t, y_s, y_t)$$

Figure 24.2. **The output of skin detection is a gray-scale skin map with the gray level indicating the probability of skin.**

where $\lambda(x_s, x_t, y_s, y_t) > 0$ are parameters that should be set up to satisfy the constraints. This model is refered as the first-order model (FOM). Parameter estimation in the context of MaxEnt is still an active research subject, especially in situations where the likelihood function cannot be computed for a given value of the parameters. This is the case here. We use the Bethe tree approximation to deal with parameter estimation and build a model TFOM (tree approximation of FOM). We use the BP algorithm to obtain an exact and fast solution. The detailed description can be found in our previous work [21].

Our TFOM model outperforms the baseline model in Reference 13, as shown in Reference 21 in the context of skin pixel detection rate and false-positive rate. The output of skin detection is a *skin map* indicating the probabilities of skin on pixels. In Figure 24.2, skin maps of different kinds of people are shown.

Vezhnevets et al. [22] recently compared some most widely used skin detection techniques and conclude that our skin detection algorithm [21] gives the best performance in terms of pixel classification rates.

ADULT IMAGE DETECTION

There are propositions for high-level features based on grouping of skin regions [11], but the actual application need a high-speed solution; therefore, along with References 12 and 13, we are interested in trying simpler features. We calculate two fit ellipses for each skin map: the global fit ellipse (GFE) and the local fit ellipse (LFE). The GFE is computed on the whole skin map, whereas the LFE only on the largest skin region in the skin map (see Figure 24.3).

We extract nine features from the input image, most of them are based on fit ellipses. The first three features are global; they are (1) the average skin probability of the whole image, (2) the average skin probability inside the GFE, and (3) the number of skin regions in the image. The other

Figure 24.3. The original input image (left); the global fit ellipse on the skin map (middle); the local fit ellipse on the skin map (right).

six features are computed on the largest skin region of the input image: (1) distance from the centroid of the largest skin region to the center of the image, (2) angle of the major axis of the LFE from the horizontal axis, (3) ratio of the minor axis to the major axis of the LFE, (4) ratio of the area of the LFE to that of the image, (5) average skin probability inside the LFE, and (6) average skin probability outside the LFE.

Evidence from Reference 14 shows that the MLP classifier offers a statistically significant performance over several other approaches, such as the generalized linear model, the k-nearest-neighbor classifier, and the support vector machine. In this chapter, we adopt the MLP classifier. We train the MLP classifier on 5084 patterns from the training set. In the testing phase, the MLP classifier intakes a quick decision on the pattern in one pass, and the output is a number $o_p \in [0, 1]$, corresponding to the degree of adult. One can set a proper threshold to get the binary decision.

EXPERIMENTAL RESULTS

All experiments are made using the following protocol. The database contains 10,168 photographs, which are imported from the Compaq Database and the Poesia Database. It is split into two equal parts randomly, with 1,297 adult photographs and 3,787 other photographs in each part. One part is used as the training set and the other one, the test set, is left aside for the receiver operating characteristics curve computation. The performance figures are initially correlated into a confusion matrix (Table 24.1), where the letters are defined as follows:

- A: harmless URLs that represent URLs that have been judged harmless by the enduser
- B: harmful URLs that represent URLs that have been judged harmful by the enduser

TABLE 24.1. Evaluation Confusion Matrix

Actual	Predicted		
	Harmless (Accepted) Pages	Harmful (Blocked) Pages	Total
Harmless pages	a	b	A
Harmful pages	c	d	B
Total	C	D	

- C: accepted URLs that represent URLs that have been judged harmless by the POESIA filter
- D: blocked URLs that represent URLs that have been judged harmful by the POESIA filter

Calculation of Evaluation Measures

The results will be presented using the standard measure from Information Retrieval: Precision, Recall, and F-Measure:

- **Precision** is the proportion of pages predicted correctly given the total number of pages predicted in that category.
- **Recall** is the proportion of pages predicted correctly given the total number of pages belonging to that category.
- **F-Measure** is the most commonly used measure for balancing precision and recall. In the standard used here, it provides equal importance to Precision and Recall and is calculated as $(2 \times \text{Precision} \times \text{Recall})/(\text{Precision} + \text{Recall})$. For harmless pages, $\text{Precision} = a/C$ and $\text{Recall} = a/A$; for harmful pages, $\text{Precision} = d/D$ and $\text{Recall} = d/B$.

Blocking Effectiveness and Overblocking

In terms of filtering, the simplest interpretation of the performance of the POESIA system can be seen in terms of:

- *effectiveness* at blocking harmful pages (i.e., maximizing the number of harmful pages that are blocked). This is indicated by the Recall value for harmful pages (i.e., d/B).
- *overblocking* of harmless pages (i.e., minimizing the number of harmless pages that are blocked). This is indicated by a high Recall measure for harmless pages. The overblocking value is generally represented as the inverse of the Recall value for harmless pages [i.e., $1 - (a/A)$].

Graphical Interpretation (ROC)

For image filtering, a ROC (Receiver Operating Characteristics) is also used to show its performance as a tradeoff between the true-positive rate (i.e., effectiveness or Recall value for harmful pages) and the false-positive rate (i.e., overblocking or Recall measure for harmless pages).

One part of image collection is used for the ROC curve computation. In Figure 24.4, we can observe that if we accept 20% of false positives, then we can detect around 91% of pornographic images. In this case, we obtain the results in Table 24.2.

The image filter for the identification of pornographic images provides an effectiveness of 0.91 with a overblocking value of 0.2. These values are worse than those seen in the text filters; however, the image filter is only categorizing a single image rather than all of the image content found on a given Web page.

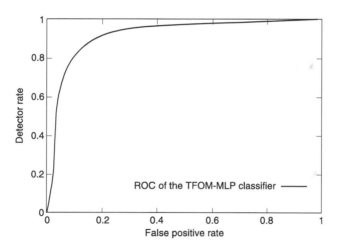

Figure 24.4. ROC curve of adult detection.

TABLE 24.2. Results using harmful filter

Actual	Predicted Harmful	Harmless	Total
Harmful	910	90	1000
Harmless	200	800	1000
Total	1110	890	2000
Precision	0.82	0.90	
Recall	0.91	0.80	
F-Measure	0.86	0.85	

$o_p=0.0$ $o_p=0.0$ $o_p=0.00119$ $o_p=0.0$

Figure 24.5. **Experimental results on nonadult images. Below the images are the associated outputs of the MLP.**

Figure 24.6. **A misclassified images and the associated outputs of the MLP.**

The elapsed time of image filtering is about 0.18 sec per image. Compared with 6 min in Reference 11 and 10 sec per image in Reference 12, our system is more practical. Figure 24.5 shows some examples of nonadults images with the corresponding output neural networks.

However, there are also some cases where this detector does not work well. In Figure 24.6, several such examples are presented. The first adult image is not detected because the skin appears almost white due to overexposure. We see that most of the skin is not detected on the skin map. The second adult image contains two connected large frames. The LFE of this image will then be very large, and the average skin probability inside this LFE will be very small. The third image is benign, but it is detected adult because the toy dog takes a skinlike color and the average skin probabilities inside the GFE and the LFE are very high. The fourth image is a portrait but decided adult because it exposes a lot of skin and even the hair and the clothes take skinlike colors. We believe that skin detection based solely on color information cannot do much more, so perhaps some other types of information are needed to improve the adult image detection performance. For example, some kind of face detector could be implemented to improve the results. However, generally, adult images in Web pages tend to appear together and are surrounded by text, which could be an important clue for the adult content detector.

CONCLUSION

This work aims at filtering objectionable images appearing in the Internet. Our main contributions are as follows:

1. In skin detection, build a FOM that introduces constraints on color gradients of neighboring pixels and use the Bethe tree approximation to eradicate parameter estimation. The BP algorithm is used to accelerate the process. This model improves the performance in a previous work [13].

2. By using simple features based on skin map output by skin detection and MLP classifier, our adult image filter can give a promising performance.

3. Our adult image filter is more practical compared with those in References 11 and 12 in terms of processing speed.

To improve the performance of our filters, we can use a face detector in the adult image filter. Research in face detection has progressed significantly. The best systems recognize 90% of faces, with about 5% false positives. This is good performance and getting much better. In 3–5 years, the computer vision community will have many good face-detection methods. This might help in identifying pornography, because skin with a face is currently more of a problem than skin without a face. Face-detection technology probably can be applied to very specific body parts; text and image data and connectivity information also will help. However, distinguishing semantic concepts such as hard-core from soft-core pornography will remain difficult. These terms are rough, and the relationships between visual features of images and these semantic concepts are not evident. In general, images tend to appear together and are surrounded by text in Web pages, so combining with text analysis could improve the performance of our image filters.

REFERENCES

1. van Rijsbergen. C. J., *Information Retrieval,* Butterworths, Stoneham, MA, 1979.
2. Salton G. and McGill, M. J., *Introduction to Modern Information Retrieval*, McGraw-Hill, New York, 1983.
3. Frakes, W. B. and Baeza-Yates, R., *Information Retrieval: Data Structures & Algorithms*, Prentice-Hall, Englewood Cliffs, NJ, 1992.
4. Lesk, M., *Books, Bytes, and Bucks*, Morgan Kaufmann, San Francisco, 1997.
5. Strzalkowski, T., Ed., *Natural Language Information Retrieval*, Kluwer Academic, Boston, 1999.
6. Rosch, E., Principles of categorization, in *Cognition and Categorization*, Rosch, E. and Lloyd, B. B., Eds., Lawrence Erlbaum Associates, Hillsdale, NJ, 1978, pp. 27–48.
7. Spitz, L. and Dengel, A., Eds., *Document Analysis Systems*, World Scientific, Singapore, 1995.
8. Jonassen, D. H., Semantic network elicitation: tools for structuring hypertext, in *Hypertext: State of the Art*, McAleese, R. and Green, C., Eds., Intellect Limited, Oxford, 1990.

9. Mangasarian, O. L., Generalized support vector machines, in *Advances in Large Margin Classifiers*, Smola, A. J., Bartlett, P. J., Schölkopf, B., and Schuurmans, D., Eds., MIT Press, Cambridge, MA, 2000, pp. 135–146.
10. Starynkevitch, B., et al., POESIA Software Architecture Definition Document, http://www.poesia-filter.org/pdf/Deliverable 3 1.pdf, Deliverable 3.1:7–9, 2002.
11. Fleck, M.M., et al, Finding naked people, in *Proceedings of the European Conference on Computer Vision*, Buxton, B. and Cipolla, R., Eds., Springer-Verlag, Berlin, 1996, Vol. 2, pp. 593–602.
12. Wang, J. Z., et al., System for screening objectionable images, *Computer Commn.*, (21)15, 1355–1360, 1998.
13. Jones, M.J. and Rehg, J.M., Statistical color models with application to skin detection, in Proceedings of the Conference on Computer Vision and Pattern Recognition, 1999, Vol. 1, pp. 1274–1280.
14. Bosson, A., Cawley, G.C., Chian, Y., and Harvey, R., Non-retrieval: Blocking pornographic images, in *Proceedings of the International Conference on the Challenge of Image and Video Retrieval*, Lecture Notes in Computer Science, Vol. 2383, Springer-Verlag, London, 2002, pp. 50–60.
15. Jaynes, E., Probability Theory: The Logic of Science, http://omega.albany.edu:8008/JaynesBook, 1995.
16. Berger, A., Pietra, S.D., and Pietra, V.D., A maximum entropy approach to natural language processing, *Comput. Linguist.*, 22, 39–71, 1996.
17. Zhu, S., Wu, Y., and Mumford, D., Filters, Random fields and maximum entropy (Frame): Towards a unified theory for texture modeling, *Int. J. Computer Vis.*, 27(2), 107–126, 1998.
18. Wu, C. and Doerschuk, P.C., Tree approximations to Markov random fields, *IEEE Trans. Pattern Anal.*, 17, 391–402, 1995.
19. Yedida, J.S., Freeman, W.T., and Weiss, Y., Understanding Belief Propagation and its Generalisations, Technical Report TR-2001-22, Mitsubishi Research Laboratories, 2002.
20. Geman, D. and Jedynak, B., An active testing model for tracking roads in satellite images, *IEEE Trans. Pattern Anal.*, 18(1), 1–14, 1996.
21. Jedynak, B., Zheng, H., and Daoudi, M., Statistical Models for Skin Detection, presented at IEEE Workshop on Statistical Analysis in Computer Vision, in conjunction with CVPR 2003, Madison, WI, 2003.
22. Vezjnevets, V., Sazonov, V., and Andreeva, A., A survey on pixel-based skin color detection techniques, presented at Graphicon 2003, 13th International Conference on the Computer Graphics and Vision, Moscow, 2003.

25

Combined Indexing and Watermarking of 3-D Models Using the Generalized 3-D Radon Transforms

*Petros Daras, Dimitrios Zarpalas,
Dimitrios Tzovaras, Dimitrios Simitopoulos,
and Michael G. Strintzis*

INTRODUCTION

Increasingly in the last decade, improved modeling tools and scanning mechanisms as well as the World Wide Web are enabling access to and widespread distribution of high-quality three-dimensional (3-D) models. Thus, companies or copyright owners who present or sell their 3-D models are facing copyright-related problems.

Watermarking techniques have long been used for the provision of robust copyright protection of multimedia material [1,2] as well as for multimedia annotation, with indexing and labeling information [3]. In this chapter, we propose a methodology for watermarking as a means to content-based indexing and retrieval of 3-D models stored in databases. A short review of current trends in 3-D content-based search and retrieval and in 3-D model watermarking is first presented in the following.

3-D Content-Based Search and Retrieval

Determining the similarity between 3-D objects is a challenging task in content-based recognition, retrieval, clustering, and classification. Its main applications have traditionally been in computer vision, mechanical engineering, education, e-commerce, entertainment, and molecular biology. Such applications are expected to expand into a variety of other fields in the near future due to the increased processing power of today's computers and the demand for a better representation of virtual worlds and scientific data.

Many groups worldwide are currently investigating and proposing new techniques for 3-D model search and retrieval. Specifically, Zhang and Chen [4] for this purpose used features such as volume–surface ratio, moment invariants, and Fourier transform coefficients. They improve the retrieval performance by an active learning phase in which a human annotator assigns attributes such as "airplane," "car," and so on to a number of sample models. A descriptor based on the 3-D discrete Fourier transform (DFT) was introduced by Vranic and Saupe [5]. Kazhdan [6] described a reflective symmetry descriptor associating a measure of reflective symmetry to every plane through the model's centroid. Ohbuchi et al. [7] employed shape histograms that are discretely parameterized along the principal axes of inertia of the model. The three shape histograms used are the moment of inertia about the axis, the average distance from the surface to the axis, and the variance of the distance from the surface to the axis. Osada et al. [8] introduced and compared shape distributions, which measure properties based on distance, angle, area, and volume measurements between random surface points. They evaluated the similarity between the objects using a metric that measures distances between distributions. In their experiments, the "D2" function, which measures the distance between two random surface points, was most effective. However, in most cases, these statistical methods are not discriminating enough to make subtle distinctions between classes of shape.

In Reference 9, a fast querying-by-3-D-model approach, was presented that utilized both volume-based (the binary 3-D shape mask) and edge-based (the set of paths outlining the shape of the 3-D object) descriptors.

The descriptors chosen seem to mimic the basic criteria that humans use for the same purpose. In Reference 10, another approach to 3-D shape comparison and retrieval for arbitrary objects described by 3-D polyhedral models was proposed. The signature of an object was represented as a weighted point set that represents the salient points of the object. These weighted point sets were compared using a variation of the "Earth Mover's Distance" [11].

3-D Model Watermarking

The main challenge in 3-D model watermarking is the robustness of the watermarking techniques against several types of attack that do not substantially degrade the model's visual quality:

- Rotation, translation, and uniform scaling
- Points reordering
- Remeshing
- Polygon simplification
- Mesh smoothing operations
- Cropping
- Local deformations

Due to the multitude of applications as well as the high variability of the above attacks, the determination of a complete generic solution robust to the above attacks is impossible. Instead, different types of application impose different requirements and call for different watermarking techniques, ranging from slight modifications of existing algorithms to completely innovative methods.

Although much work has been done on watermarking of 1-D and 2-D multimedia content, only a few methods have been proposed for the watermarking of 3-D models. In Reference 12, three watermarking algorithms for 3-D models were proposed: the triangle similarity quadruple embedding algorithm, the tetrahedral volume ratio embedding algorithm, and the mesh density pattern embedding algorithm, which provide many useful insights into mesh watermarking. Benedens proposed several watermarking algorithms: the watermarking that is robust to mesh simplification and affine transform [2,13,14], high-capacity watermarking [15], and a combined watermarking system [16]. Praun et al. [17] presented a technique for embedding secret watermarks using a spread spectrum technique. In Reference 18, the proposed method is based on distributing the information over the entire model via vertices scrambling.

In this chapter, a novel technique for 3-D model indexing and water-marking, based on a generalized Radon transform, is described. The proposed transform is a variation of the radial integration transform (RIT) [19], namely the cylindrical integration transform (CIT), which integrates the 3-D model's information on cylinders that begins from its center of mass. The comparison of two 3-D models is achieved after proper positioning and alignment of the models. After solving this "pose estimation" problem, a set of descriptor vectors, completely invariant to translation, rotation, and scaling, is extracted from the reference 3-D model. At the same time a watermarking technique is used in order to embed a specific model identifier, which represents the 3-D model's descriptors, in the vertices of the 3-D model via a modification of their location. This identifier links the model to its descriptor vector, which is extracted only once and stored in a database. Every time this model is used as a query model, the watermark detection algorithm is applied in order to retrieve the corresponding descriptor vector, which can be further used in a matching algorithm. Similarity measures are then computed for the specific descriptors and introduced into a 3-D model-matching algorithm.

The overall method is characterized by the following highly desirable properties concerning the 3-D content-based search:

1. Invariance with respect to translation, rotation, and scaling of a 3-D object
2. No need for model preprocessing, in terms of model's fixing degeneracies (inner planes, single surfaces, collinear triangles)
3. Efficient feature extraction and very fast implementation because matching involves simple comparison of vectors

Further, the proposed watermarking method is robust to geometric distortions such as translation, rotation, and uniform scaling. Additionally, the method is robust against the "points reordering" attack. However, the method is vulnerable to attacks such as mesh smoothing operations, cropping, and local deformations because such operations change the shape of a 3-D model, which then cannot be used as a query model.

The rest of the chapter is organized as follows. Section "The Generalized 3-D Radon Transform" describes mathematically the proposed transform. In section "3-D Model Prepossessing," the necessary preprocessing steps are presented. Section "Content-Based Search and Retrieval" presents in detail the proposed descriptor extraction method, whereas in section "Watermarking for Data Hiding," the detailed watermarking procedure is presented. In section "Matching Algorithm," the matching algorithm used is described. Experimental results evaluating the proposed method both in terms of watermarking for data hiding and content-based retrieval performance are presented in section "Experimental Results." Finally, conclusions are drawn in section "Conclusions."

THE GENERALIZED 3-D RADON TRANSFORM

Let M be a 3-D model and $f(x)$, $x = \{x, y, z\}$, be the volumetric binary function of M, which is defined as

$$f(x) = \begin{cases} 1 & \text{when } x \text{ lies within the 3-D model's volume} \\ 0 & \text{otherwise} \end{cases}$$

Let also η be the unit vector in $\Re^3$ and l a real number. The 3-D generalized Radon transform $R_f(\eta, l)$ [20] is a function that associates to each pair (η, l) the integral of $f(x)$ on the curve $C(\eta, l) = \{x | \psi(x; \eta, l) = 0\}$, where $\psi(x; \eta, l)$ denotes the transformation curve:

$$R_f(\eta, l) = \int_{x \in C(\eta, l)} f(x) \, dx \tag{25.1}$$

The 3-D Radon transform $R_f(\eta, l)$ [21] of the $f(x)$, is produced by Equation 25.1 and it is a function that associates to each pair (η, l) the integral of $f(x)$ on the plane $\Pi(\eta, l) = \{x | x^T \cdot \eta = l\}$, where the superscript T indicates vector transposition. This plane is normal to the direction η and at a distance l to the origin:

$$R_f(\eta, l) = \int_{x \in \Pi(\eta, l)} f(x) \, dx \tag{25.2}$$

The radial integration transform $\text{RIT}_f(\eta)$ [19] of a 3-D model's binary function $f(x)$ is produced by Equation 25.1, and it is a function that associates to each η the integral of $f(x)$ on the line $L(\eta) = \{x | (x/|x|) = \eta\}$ passing through the origin:

$$\text{RIT}_f(\eta) = \int_{x \in L(\eta)} f(x) \, dx \tag{25.3}$$

The CIT is a slight modification of the RIT where the cylinder $\text{CYL}(\eta)$ is used instead of the line $L(\eta)$. The radius of the cylinder is Th and its axis

737

Figure 25.1.   The cylindrical integration transform.

is the line $L(\eta)$. Thus, CIT is given by

$$\text{CIT}_f(\eta) = \int_{x \in \text{CYL}(\eta)} f(x) \, dx \tag{25.4}$$

where $x \in \text{CYL}(\eta)$ simply means $\sqrt{|x|^2 - |x \cdot \eta|^2} \leq Th$.

The discrete form of CIT, which will be used for the actual extraction of the shape descriptors, is given by

$$\text{CIT}(\eta_i) = \sum_{x_j \in \text{CYL}(\eta_i)} f(x_j), \qquad i = 1, \ldots, N_{\text{CYL}}, \quad j = 1, \ldots, J \tag{25.5}$$

where N_{CYL} is the total number of cylinders and J is the total number of points x_j. An illustration of CIT is given in Figure 25.1, where the red dots indicate the points x_j, the green line segments indicate the lines $L(\eta_i)$, and the yellow cylinders $\text{CYL}(\eta_i)$ indicate the cylindrical integration area.

3-D MODEL PREPROCESSING

A 3-D model M is composed of a set of vertices $\mathbf{V}$ and a set of connections between the vertices. Each vertex $\mathbf{v}_i$ has three coordinates in the Cartesian space, $\mathbf{v}_i = \{x_i, y_i, z_i\}$. Before applying the proposed transform, a canonical position and orientation is estimated for each 3-D model using two steps:

1. *Model rotation and translation.* Let $\mathbf{Q}$ be the class of vectors for all pairs of vertices of the 3-D model. The vector $\mathbf{q}_1$ is calculated, where $|\mathbf{q}_1| = \max\{|\mathbf{q}| : \mathbf{q} \in \mathbf{Q}\}$. Further, the most distant vertex $\mathbf{v}_d$ from $\mathbf{q}_1$ and its projection $\mathbf{O}'$ to $\mathbf{q}_1$ are found. Then, the vector $\mathbf{q}_2 = \overrightarrow{\mathbf{O}'\mathbf{v}_d}$ is formed. The point $\mathbf{O}' = \{\bar{x}, \bar{y}, \bar{z}\}$ is the new origin of the model. The model is translated so that the new origin coincides with the old origin:

$$\acute{x}_i = x_i - \bar{x}, \qquad \acute{y}_i = y_i - \bar{y}, \qquad \acute{z}_i = z_i - \bar{z} \qquad (25.6)$$

where x_i, y_i, and z_i are the coordinates of the vertex $\mathbf{v}_i$ and $\acute{x}_i$, $\acute{y}_i$, and $\acute{z}_i$ are the coordinates of the translated vertex $\acute{\mathbf{v}}_i$. In this way translation invariance is accomplished. Finally, the model is rotated so that $\mathbf{q}_1$ coincides with the z-axis and $\mathbf{q}_2$ coincides with the x-axis. Rotation invariance follows.

2. *Model scaling.* In order to achieve scaling invariance, the maximum distance $d_{\max}$ between the center of mass and the most distant vertex, is calculated. Then, the model is scaled so that $d_{\max} = 1$. At this point, scaling invariance is also accomplished.

The translated, rotated, and scaled model is then placed into a bounding sphere with radius $R_a = d_{\max}$.

CONTENT-BASED SEARCH AND RETRIEVAL

The bounding sphere is partitioned in equal cube-shaped voxels $\mathbf{u}$ with centers v. Let $\mathbf{U}$ be the set of all voxels inside the bounding sphere and $\mathbf{U}_1 \subseteq \mathbf{U}$, be the set of all voxels belonging to the bounding sphere and lying inside M. Then, the discrete binary volume function $\hat{f}(v)$ of M is defined as

$$\hat{f}(v) = \begin{cases} 1 & \text{when } \mathbf{u} \in \mathbf{U}_1 \\ 0 & \text{otherwise} \end{cases}$$

In Figure 25.2, a model and its discrete binary volume function $\hat{f}(v)$ are illustrated, where the red dots indicate the centers of the voxels in $\mathbf{U}_1$.

CIT-Based Descriptors

After proper positioning of the model M, Equation 25.5 is applied to $\hat{f}(v)$, producing the CIT vector with elements $\text{CIT}_f(\eta_i)$, where $i \in S_C = \{1, \ldots, N_{\text{CYL}}\}$ and η_i form the lines $L(\eta_i)$. Each η_i begins from the origin $\mathbf{O}'$ and ends at a point $\mathbf{P}_i$, which lies on the bounding unit sphere and

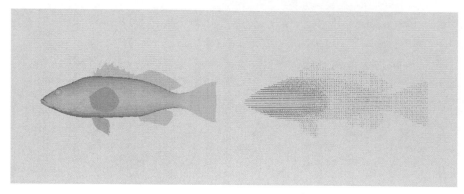

Figure 25.2.   Model and voxels inside the model $[\hat{f}(v)]$.

can be expressed using spherical coordinates as $\boldsymbol{\eta}_i = [\cos\phi_i \sin\theta_i,$ $\sin\phi_i \sin\theta_i, \sin\theta_i]$, where ϕ_i is the azimuthal angle and θ_i is the polar angle of the vector $\boldsymbol{\eta}_i$. Thus,

$$\text{CIT}_f(\boldsymbol{\eta}_i) = \text{CIT}_f(\theta_i, \phi_i) \tag{25.7}$$

In order to obtain a more compact representation of the information contained in the CIT vector, a clustering of $\boldsymbol{\eta}_i$ where $i \in S_C$, is performed. A cluster is defined as $\text{Cluster}(k) = \{\boldsymbol{\eta}_i | \|\boldsymbol{\eta}_k - \boldsymbol{\eta}_i\| \le d_c\}$, where d_c is a pre-selected threshold, $k \in S_{R1} \subset S_R$, $S_{R1} = \{1, \ldots, N_{\text{cluster}}\}$, and N_{cluster} is the total number of clusters. For each cluster, a single characteristic value is calculated as the sum of the CIT values weighted with the sigmoid function:

$$W(d; \beta, d_c) = 1 - \frac{1}{1 + \exp^{-\beta(d-d_c)}} \tag{25.8}$$

where the parameter β influences the sharpness of the function and $d = |\boldsymbol{\eta}_k - \boldsymbol{\eta}_i|$. In this way, the CIT feature vector becomes

$$\mathbf{u}_{\text{CIT}}(k) = \sum_{i=1}^{N_{\text{CYL}}} \text{CIT}_f(\boldsymbol{\eta}_i) W(|\boldsymbol{\eta}_k - \boldsymbol{\eta}_i|; \beta, d_c) \tag{25.9}$$

Enhanced CIT-Based Descriptors

The large and important amount of information contained in the $\text{CIT}_f(\boldsymbol{\eta})$, or in terms of spherical coordinates $\text{CIT}_f(\theta, \phi)$, can be further exploited in order to enhance the CIT-based descriptor vectors. For this reason, an approach similar to the one introduced in Reference 22, namely the trace transform, was followed. The trace transform consists of tracing an image (2-D function) with straight lines along which certain functionals of the image are calculated. According to Reference 22, a set of invariant functionals are applied to $\text{CIT}_f(\theta, \phi)$ in order to produce a new descriptor vector $\mathbf{u}_{\text{EnCIT}}$, which represents a set of features of $\text{CIT}_f(\theta, \phi)$.

The most suitable set of functionals for the proposed application [22] is the following:

$$F_1(g) = \max\{g(t_i)\}, \quad i = 1, \ldots, N \tag{25.10}$$

$$F_2(g) = \sum_{i=1}^{N} |g'(t_i)| \tag{25.11}$$

$$F_3(g) = \sum_{i=1}^{N} g(t_i) \tag{25.12}$$

$$F_4(g) = \max\{g(t_i)\} - \min\{g(t_i)\}, \quad i = 1, \ldots, N \tag{25.13}$$

where g is a differentiable function, g' is its derivative, t_i, $i = 1, \ldots, N$ are sample points for g, and N is their total number.

The goal is the gradual reduction of the dimensions of $\text{CIT}_f(\theta, \phi)$, so as to produce a compact representation, which could be, for example, a single number, the descriptor. For this reason, we define

$$G_k(\phi) = F_k[\text{CIT}_f(\theta, \phi)], \quad k = 1, 2, 3, 4 \tag{25.14}$$

and

$$G_k(\theta) = F_k[\text{CIT}_f(\theta, \phi)], \quad k = 1, 2, 3, 4 \tag{25.15}$$

and choose as descriptors

$$A_{kj} = F_j[G_k(\theta)], \quad k, j = 1, 2, 3, 4 \tag{25.16}$$

and

$$B_{kj} = F_j[G_k(\phi)], \quad k, j = 1, 2, 3, 4 \tag{25.17}$$

A set of $N_{\text{EnCIT}} = 32$ descriptor values A_{kj} and B_{kj} is produced. The enhanced CIT-based descriptor vector is defined by

$$\mathbf{u}_{\text{EnCIT}}(i) = \{A_{kj}, B_{kj}\} \tag{25.18}$$

where $k, j = 1, 2, 3, 4$ and $i = 1, \ldots, N_{\text{EnCIT}}$.

This procedure is repeated for every 3-D model contained in a database and the extracted $\mathbf{u}_{\text{EnCIT}}(i)$ descriptor vector is stored along with the corresponding model.

WATERMARKING FOR DATA HIDING

The proposed transform (CIT) is also used in the watermarking procedure, in order to produce the unique identifiers that link a model M
with its descriptor vector $\mathbf{u}_{\mathrm{EnCIT}}$. As was mentioned in section "3-D Model
Prepossessing," a 3-D model M is composed of a set of vertices
$\mathbf{V} = \{\mathbf{v}_1, \mathbf{v}_2, \ldots, \mathbf{v}_{N_M}\}$, where N_M is the total number of vertices of the
model. A function $h(\mathbf{v})$ is defined for each M as

$$h(\mathbf{v}) = \begin{cases} 1 & \text{when } \mathbf{v} \in \mathbf{V} \\ 0 & \text{otherwise} \end{cases}$$

The preprocessing steps described in section "3-D Model Prepossessing"
are followed and Equation 25.5 is applied to the translated, rotated, and
scaled $h(\mathbf{v})$ producing the vector

$$\mathbf{u}_{\mathrm{CIT}} = [\mathrm{CIT}(\boldsymbol{\eta}_i)], \quad i = 1, \ldots, N_{\mathrm{CYL}} \tag{25.19}$$

Figure 25.3 illustrates the computation of the CIT when it is applied on
$h(\mathbf{v})$: the red dots indicate the model's vertices $\mathbf{v}_i$, the green line
segments indicate the lines $L(\boldsymbol{\eta}_i)$, which end in the points $\mathbf{P}_i$ on the surface
of the bounding unit sphere, and the yellow cylinders $\mathrm{CYL}(\boldsymbol{\eta}_i)$ indicate
the cylindrical integration area.

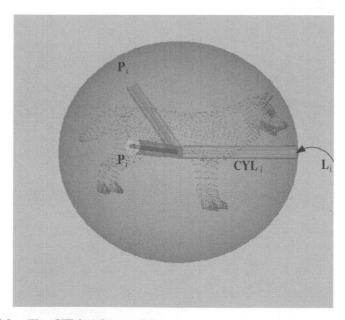

Figure 25.3.   The CIT for the model's vertices.

Selecting Regions for Watermarking

Given a 3-D model M, robustness to translation, rotation, and scaling is achieved by applying the process described in section "3-D Model Prepossessing" prior to watermark embedding and detection. The values $\text{CIT}(\eta_i), i = 1, \ldots, N_{\text{CYL}}$, of the components of the vector $\mathbf{u}_{\text{CIT}}$ are sorted in descending order and the first $K, K \in Z$, of them, which carry the most important information of the model, are selected. The proposed algorithm embeds a watermark by modifying the location of certain vertices. In order to ensure that the vectors $\mathbf{q}_1$ and $\mathbf{q}_2$ used in the model's rotation and translation will not be altered after the watermark embedding, the cylinders that contain the vertices that form these vectors are excluded from the watermarking process.

Watermark Embedding

The proposed watermarking method embeds K bits in the vertices of M. The sequence of the K bits is unique for each model and it is used as an identifier that links each model to its descriptor vector. The procedure of embedding the watermark is shown in Figure 25.4. The watermark is inserted as follows:

1. Each selected $\mathbf{u}_{\text{CIT}}(i), i = 1, \ldots, K$, corresponds to the cylinder $\text{CYL}(\eta_i)$ and furthermore to a set of vertices lying inside the cylinder $\text{CYL}(\eta_i)$. The length of the projection $l_{ij} = \mathbf{v}_{ij}^T \cdot \eta_i$ of each vector $\mathbf{v}_{ij}$ onto the line $L(\eta_i)$ is calculated. All vertices that lie in the same cylinder are sorted in descending order of their l_{ij} and the first half of them form the set $S_{i1} = \{\mathbf{v}_{ij1}\}, j = 1, \ldots, N_{S1}$; the remaining form the set $S_{i2} = \{\mathbf{v}_{ij2}\}, j = N_{S1} + 1, \ldots, N_{P_i}$ (Figure 25.5), where N_{P_i} is the total number of vertices that lie on the same cylinder.[1]

2. A watermark $B = \{b_i, i = 1, \ldots, K\}, b_i \in \{-1, 1\}$ is embedded in each vertex of set S_{i1} or S_{i2} by modifying their location, according to the following. In each set, the mean of the distances D_{ij} between each vertex $\mathbf{v}_{ij} \in \text{CYL}(\eta_i)$ and the axis $L(\eta_i)$ is calculated:

$$M_{i1} = \frac{1}{N_{S_i}} \sum_{j=1}^{N_{S_i}} D_{ij1} = \frac{1}{N_{S_i}} \sum_{j=1}^{N_{S_i}} \mathbf{v}_{ij1}^T \cdot \kappa_i, \qquad i = 1, \ldots, K \qquad (25.20)$$

where κ_i is a unit vector perpendicular to η_i ($\kappa_i^T \cdot \eta_i = 0$) directed toward the interior of the model, D_{ij1} is the distance between each vertex $\mathbf{v}_{ij1} \in \text{CYL}(\eta_i)$ and the axis $L(\eta_i)$ in S_{i1} (Figure 25.6) and N_{S_i} is

---

[1]Whenever N_{P_i} is odd, the watermarking procedure simply bypasses the vertex with minimum projection length in the set S_{i2}.

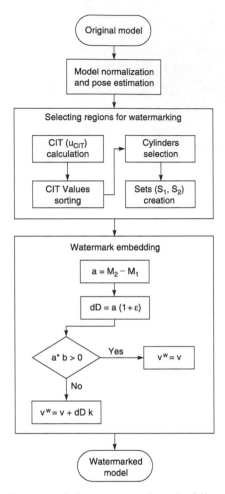

Figure 25.4. **Block diagram of the watermark embedding procedure.**

the total number of vertices in set S_{i1}. Similarly,

$$M_{i2} = \frac{1}{N_{S_i}} \sum_{j=1}^{N_{S_i}} D_{ij2} = \frac{1}{N_{S_i}} \sum_{j=1}^{N_{S_i}} \mathbf{v}_{ij2}^T \cdot \boldsymbol{\kappa}_i, \qquad i = 1, \ldots, K \qquad (25.21)$$

where D_{ij2} is the distance between each vertex $\mathbf{v}_{ij2} \in \mathrm{CYL}(\boldsymbol{\eta}_i)$ and the axis $L(\boldsymbol{\eta}_i)$ in S_{i2} (Figure 25.6) and N_{S_i} is the total number of vertices in set S_{i2}.

The difference $a_i = M_{i2} - M_{i1}$ is then calculated. If $a_i > 0$, the watermark is embedded in the vertices of the set S_{i2}, otherwise,

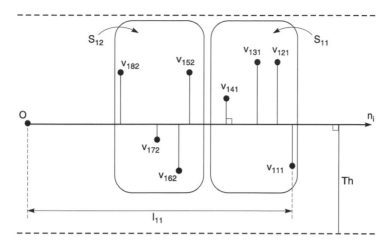

Figure 25.5. Example of the creation of two sets. For the cylinder with orientation $\eta_1, (i = 1)$, the vertices $\{v_{111}, \ldots, v_{141}\}$ with projection lengths $\{l_{11}, \ldots, l_{14}\}$ form the set S_{11}. The vertices $\{v_{152}, \ldots, v_{182}\}$ with projection lengths $\{l_{15}, \ldots, l_{18}\}$ form the set S_{12}.

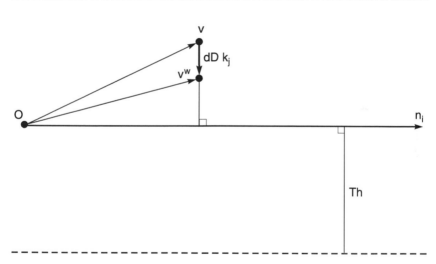

Figure 25.6. Watermark embedding.

it is embedded in those of the set S_{i1} according to the formula

$$
\mathbf{v}_{ij}^{W} = \begin{cases} \mathbf{v}_{ij} & \text{if } a_i \cdot b_i > 0 \\ \mathbf{v}_{ij} + dD_i\kappa_i & \text{if } a_i \cdot b_i < 0 \end{cases}
$$

745

where $\mathbf{v}_{ij}$ denotes the original vertex, $\mathbf{v}_{ij}^W$ denotes the watermarked vertex, and dD_i is the displacement (Figure 25.6):

$$dD_i = a_i \cdot (1 + \varepsilon) \qquad (25.22)$$

where ε is a small positive number.

After the watermark embedding, the term a_i^W has the same sign as b_i. In fact,

$$a_i = M_{i2} - M_{i1} = \frac{1}{N_{S_i}} \sum_{j=1}^{N_{S_i}} \mathbf{v}_{ij2}^T \cdot \boldsymbol{\kappa}_i - \frac{1}{N_{S_i}} \sum_{j=1}^{N_{S_i}} (\mathbf{v}_{ij1})^T \cdot \boldsymbol{\kappa}_i \qquad (25.23)$$

Let the watermark be embedded in the vertices of the set S_{i1}. Then,

$$a_i^W = M_{i2} - M_{i1}^W = \frac{1}{N_{S_i}} \sum_{j=1}^{N_{S_i}} \mathbf{v}_{ij2}^T \cdot \boldsymbol{\kappa}_i - \frac{1}{N_{S_i}} \sum_{j=1}^{N_{S_i}} \mathbf{v}_{ij1}^{W^T} \cdot \boldsymbol{\kappa}_i$$

Thus,

$$a_i^W = \begin{cases} \dfrac{1}{N_{S_i}} \displaystyle\sum_{j=1}^{N_{S_i}} \mathbf{v}_{ij2}^T \cdot \boldsymbol{\kappa}_i - \dfrac{1}{N_{S_i}} \displaystyle\sum_{j=1}^{N_{S_i}} \mathbf{v}_{ij1}^T \cdot \boldsymbol{\kappa}_i & \text{if } a_i \cdot b_i > 0 \quad (25.23a) \\[4mm] \dfrac{1}{N_{S_i}} \displaystyle\sum_{j=1}^{N_{S_i}} \mathbf{v}_{ij2}^T \cdot \boldsymbol{\kappa}_i - \dfrac{1}{N_{S_i}} \displaystyle\sum_{j=1}^{N_{S_i}} (\mathbf{v}_{ij1} + dD_i \boldsymbol{\kappa}_i)^T \cdot \boldsymbol{\kappa}_i & \text{if } a_i \cdot b_i < 0 \quad (25.23b) \end{cases}$$

Equation 25.23 and Equation 25.23a imply that $a_i = a_i^W$ for $a_i \cdot b_i > 0$. Likewise, Equation 25.23 and Equation 25.23b imply that $a_i - a_i^W = dD_i = a_i + a_i \cdot \varepsilon$ and because $\varepsilon > 0$, $a_i^W \cdot a_i < 0$ for $a_i \cdot b_i < 0$. Thus, in both cases, $a_i^W \cdot b_i > 0$. The same result is obtained in the case where the watermark is embedded in the vertices of the set S_{i2}.

Watermark Detection

The block diagram of the watermark detection procedure is depicted in Figure 25.7. Let M^d be the model, v_{ij}^d the vertices, S_{i1}^d and S_{i2}^d the sets, D_{ij}^d the distances, M_{i1}^d and M_{i2}^d the mean values of the distances, and a_i^d the difference $M_{i2}^d - M_{i1}^d$, after geometric attacks. The watermark is detected as follows:

1. The model M^d is translated, rotated, and scaled following the procedure described in section "3-D Model Prepossessing."

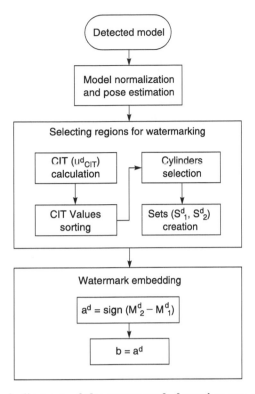

Figure 25.7. Block diagram of the watermark detection procedure.

2. The CIT vector u^d_{CIT} of the model M^d is calculated and the values of its components are sorted in descending order. The first K of the values are selected and the corresponding cylinders $\text{CYL}(\eta_i)$ are identified.

3. As in step 1 of the embedding procedure, in each selected $\text{CYL}(\eta_i)$ the two sets $S^d_{i1} = \{v^d_{ij1}\}$ and $S^d_{i2} = \{v^d_{ij1}\}$ are found according to their $l^d_{ij} = v^{d^T}_{ij} \cdot \eta_i$. The sets S^d_{i1} and S^d_{i2} are, however, identical to the sets S_{i1} and S_{i2} because the projections of their vertices onto the axis of the cylinder they belong to remain the same. Further, $M^d_{i1} = M^W_{i1}$ and $M^d_{i2} = M^W_{i2}$ and $a^d_i = a^W_i$. Thus, the watermark sequence can be easily extracted using the formula:

$$a^d_i = \text{sign}(M^d_{i2} - M^d_{i1}), \qquad i = 1, \dots, N_{P_i} \qquad (25.24)$$

and because of $a^d_i (= a^W_i)$ and b_i have the same sign, the watermark sequence b_i is easily extracted.

This procedure is repeated for all watermark bits. The extracted sequence of bits forms the identifier that links the 3-D model with its descriptors.

MATCHING ALGORITHM

Let A be the 3-D model that corresponds to the identifier extracted using the procedure described in section "Watermarking for Data Hiding." The descriptor vector $\mathbf{u}_{\mathrm{CIT}A}$ of A is found from this identifier. Also, let B be one of the database models to which A is compared. The identifier of B, extracted using the same procedure, defines the descriptor vector $\mathbf{u}_{\mathrm{CIT}B}$ of B. The similarity between A and B is calculated using

$$\text{Similarity} = \left(1 - \sum_{i=1}^{N_{\mathrm{CYL}}} \frac{|\mathbf{u}_{\mathrm{CIT}A}(i) - \mathbf{u}_{\mathrm{CIT}B}(i)|}{|\mathbf{u}_{\mathrm{CIT}A}(i) + \mathbf{u}_{\mathrm{CIT}B}(i)|/2}\right) \times 100\% \qquad (25.25)$$

The dimension of the CIT feature vector was experimentally selected to be $N_{\mathrm{CYL}} = 252$.

EXPERIMENTAL RESULTS

The proposed method was tested using two different databases. The first one, formed in Princeton University [23], consists of 907 3-D models classified into 35 main categories. Most are further classified into subcategories, forming 92 categories, in total. The second one was compiled from the Internet by us and consists of 544 3-D models from different categories.

Experimental Results for 3-D Model Retrieval

To evaluate the ability of the proposed method to discriminate between classes of objects, each 3-D model was used as a query object. Our results were compared with those of the method described in Reference 9. The retrieval performance was evaluated in terms of "precision" and "recall," where precision is the proportion of the retrieved models that are relevant to the query and recall is the proportion of relevant models in the entire database that are retrieved in the query. More precisely, precision and recall are defined as

$$\text{Precision} = \frac{N_{\mathrm{detection}}}{N_{\mathrm{detection}} + N_{\mathrm{false}}} \qquad (25.26)$$

$$\text{Recall} = \frac{N_{\mathrm{detection}}}{N_{\mathrm{detection}} + N_{\mathrm{miss}}} \qquad (25.27)$$

where $N_{\mathrm{detection}}$ is the number of relevant models retrieved, N_{false} is the number of irrelevant models retrieved, and N_{miss} is the number of relevant models not retrieved.

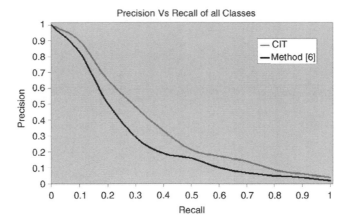

Precision Vs Recall of all Classes

Figure 25.8. Comparison of the proposed method (CIT) against the method proposed in Reference 9 in terms of precision–recall diagram using the Princeton database.

Figure 25.8 contains a numerical precision vs. recall comparison with the method in Reference 9 for the Princeton database. On average, the precision of the proposed method is 15% higher than that of the method in Reference 9.

Figure 25.9 illustrates the results produced by the proposed method in the Princeton database. The models in the first line are the query models (each belonging to a different class) and the rest are the first four retrieved models.

As discussed in Reference 23, the first database reflects primarily the function of each object and only secondarily its shape. For this reason, a new, more balanced database was compiled, supplementing the former [23] with 544 VRML models collected from the World Wide Web so as to form 13 more balanced categories: 27 animals, 17 spheroid objects, 64 conventional airplanes, 55 Delta airplanes, 54 helicopters, 48 cars, 12 motorcycles, 10 tubes, 14 couches, 42 chairs, 45 fish, 53 humans, and 103 other models. This choice reflects primarily the shape of each object and only secondarily its function. The number of the "other" models is kept high so as to better test the efficiency of the proposed method. (A large number of "other" models that do not belong to any predefined class guarantees the validity of the tests because it adds a considerable amount of "noise" in the classification procedure.) The average number of vertices and triangles of the models in the new database is 5080 and 7061, respectively.

Figure 25.10 illustrates the overall precision–recall for the new database compared with the method described in Reference 9. On average,

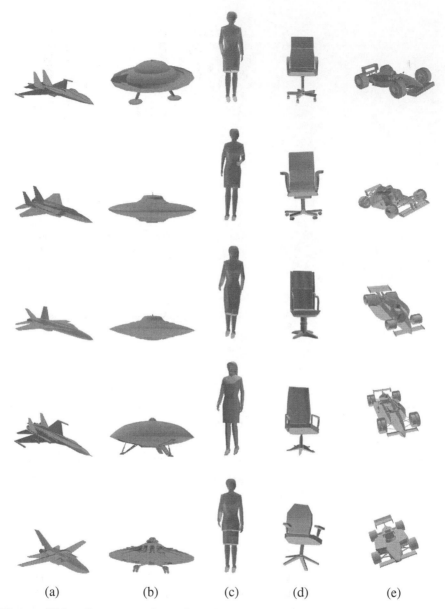

(a)　　　　(b)　　　　(c)　　　　(d)　　　　(e)

Figure 25.9. Query results using the proposed method in the Princeton database. The query models are depicted in the first horizontal line.

the precision of the proposed method is 13% higher than the method in Reference 9.

Figure 25.11 illustrates the results produced by the proposed method in the new database. The models in the first horizontal line

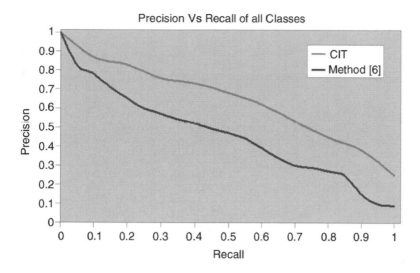

Precision Vs Recall of all Classes

Figure 25.10. **Comparison of the proposed method (CIT) against the method proposed in Reference 9 in terms of precision–recall diagram using the new database.**

are the query models and the rest are the first seven retrieved models. The similarity between the query model and the retrieved ones is obvious.

These results were obtained using a personal computer (PC) with a 2.4-MHz Pentium IV processor running Windows 2000. On average, the time needed for the extraction of the feature vectors for one 3-D model is 12 sec, whereas the time needed for the comparison of two feature vectors is 0.1 msec. Clearly, even though the time needed for the extraction of the feature vectors is relatively high, the retrieval performance is excellent.

Experimental Results for 3-D Model Watermarking

The proposed 3-D model watermarking technique for data hiding was tested using models from the above databases. It was specifically tested for the following:

1. *Robustness to geometric attacks and vertex reordering.* The geometric attacks tested were translation, rotation, and uniform scaling. Due to the preprocessing steps applied to each model prior to embedding and detecting a watermark sequence, the percentage of correct extraction, as expected, was 100% for $K = 16, 24$, and 32 bits. Similarly, because the coordinates of the vertices do not depend on their order, the percentage of correct

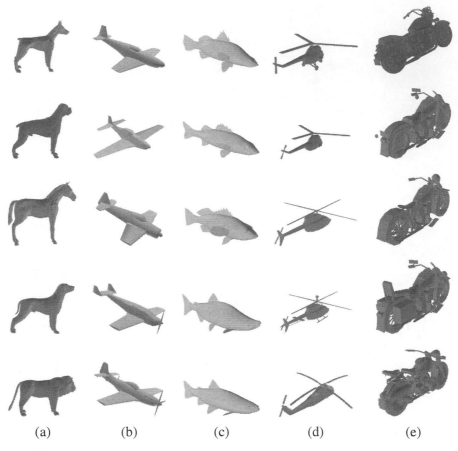

Figure 25.11. Query results using the proposed method in the new database. The query models are depicted in the first horizontal line.

extraction following a change in the order of vertices was also 100% for $K = 16, 24$, and 32 bits. It should be noted that for $K = 32$ bits, the total number of models is 10^{11}, which is sufficiently large to be representative of a large database.

2. *Imperceptibility.* In order to measure the imperceptibility of the embedded watermark, the signal-to-noise ratio (SNR) was calculated for 100 3-D models randomly selected from the new database. The following formula was used:

$$\text{SNR} = \frac{\sum_{i=1}^{N_M} x_i^2 + y_i^2 + z_i^2}{\sum_{i=1}^{N_M} (x_i - x_i^W)^2 + (y_i - y_i^W)^2 + (z_i - z_i^W)^2} \tag{25.28}$$

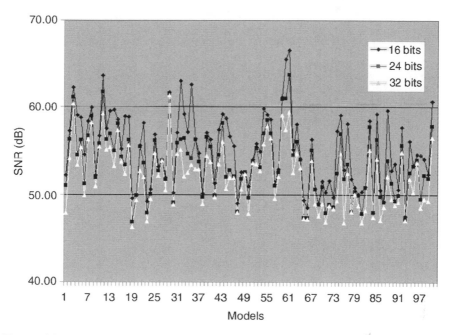

Figure 25.12. SNR measure of 3-D models.

where x_i, y_i, and z_i are the coordinates of the vertex $\mathbf{v}_i$ before the embedding of the watermark and x_i^W, y_i^W, and z_i^W are the coordinates of the same vertex after the embedding of the watermark.

Figure 25.12 illustrates the value of SNR for each model. In Figure 25.13 to Figure 25.15, the same information is shown as a histogram of the number of vertices vs. SNR values. In Figure 25.16, the percentage is plotted of the vertices that are being modified by the watermarking procedure. The original and the watermarked 3-D models are shown in Figure 25.17 for various values of K. It is obvious that the embedded watermark is imperceptible.

3. *Watermark extraction time.* Finally, the proposed method was tested in terms of time needed for the detection of the watermark sequence. The results presented in Figure 25.18 show that, on average, 0.02 sec are needed for the detection of the watermark sequence. Clearly, this is far shorter than the time needed for the descriptor vector extraction. Thus, the proposed scheme is very appropriate for use as an efficient tool for Web-based, real-time application.

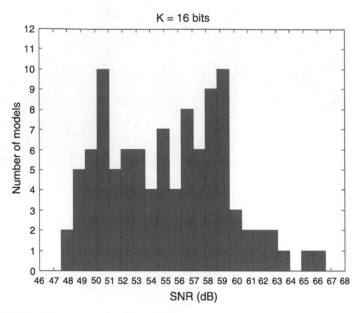

Figure 25.13. Histogram for $K = 16$ bits.

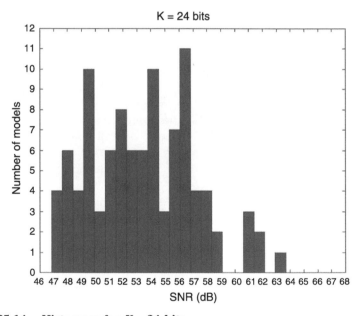

Figure 25.14. Histogram for $K = 24$ bits.

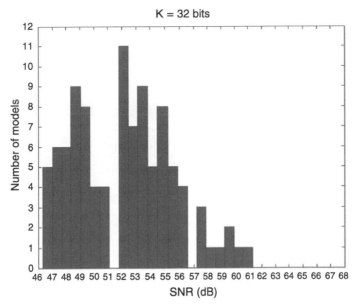

Figure 25.15. Histogram for $K = 32$ bits.

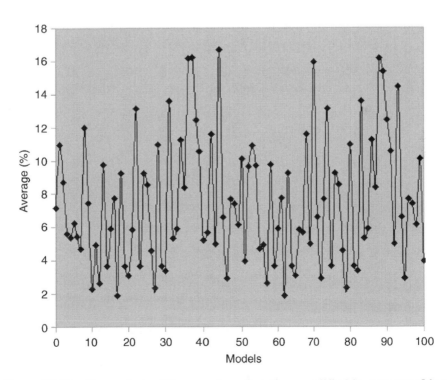

Figure 25.16. Percentage of the number of vertices modified in watermarking.

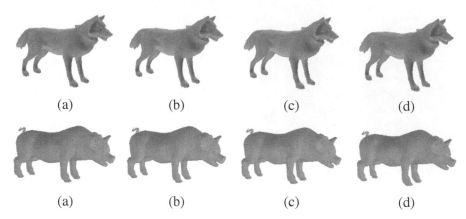

(a)   (b)   (c)   (d)

(a)   (b)   (c)   (d)

Figure 25.17. Visual results of the watermarking procedure. (a) Original model; (b) $K=16$ bits; (c) $K=24$ bits; (d) $K=32$ bits.

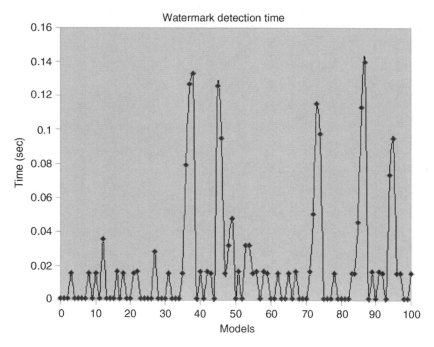

Figure 25.18. Watermark detection time.

CONCLUSIONS

A novel technique for 3-D model indexing and watermarking, based on a generalized Radon transform (GRT) was presented. The form of the GRT implemented was the cylindrical integration transform (CIT). After proper positioning and alignment of the 3-D models, the CIT was applied on (1) the model's voxels, producing the descriptor vector, and (2) the model's vertices, producing a vector that is used in the watermarking procedure. The watermarking algorithm proposed in this chapter is used in order to embed a specific model identifier in the vertices of the model by modifying their location. This identifier links the model to its descriptor vector. Every time a model is used as a query model, the watermark detection algorithm is applied in order to retrieve the corresponding descriptor vector, which can be further used in a matching algorithm. Similarity measures are then computed for the specific descriptors and introduced into a 3-D model-matching algorithm.

The overall method is characterized by properties highly desirable for efficient 3-D model search and retrieval and for 3-D model watermarking as well, for the following reasons:

- The overall method can be applied for any given model without the necessity for any preprocessing in terms of model's fixing degeneracies.
- The descriptor vectors are invariant with respect to translation, rotation, and scaling of a 3-D model.
- The complexity of the object-matching procedure is minimal, because matching involves simple comparison of vectors.
- The watermarking method is robust to geometric attacks such as translation, rotation, and uniform scaling.
- The watermarking method is robust to points reordering attack.
- The watermark is imperceptible regardless of the length of the watermark sequence.
- The extraction of the watermark sequence is very fast and accurate.

Experiments were performed using two different databases to test both content-based search and retrieval and watermarking efficiency. The results show that the proposed method can be used in a highly efficient manner.

REFERENCES

1. Koutsonanos, D., Simitopoulos, D., and Strintzis, M.G., Geometric Attack Resistant Image Watermarking for Copyright Protection, in Proceedings of IST International Conference on Digital Printing Technologies, NIP 19, New Orleans, 2003, pp. 507–510.
2. Benedens, O., Geometry-based watermarking of 3-D models, *IEEE Computer Graphics Applic.*, 19, 46–55, 1999.
3. Boulgouris, N.V., Kompatsiaris, I., Mezaris, V., Simitopoulos, D., and Strintzis, M.G., Segmentation and content-based watermarking for color image and image region indexing and retrieval, *EURASIP J. Appl. Signal Process.*, 2002(4), 420–433, 2002.

4. Zhang, C. and Chen, T., Indexing and Retrieval of 3-D Models Aided by Active Learning, in Proceedings of ACM Multimedia 2001, Ottawa, 2001, pp. 615–616.

5. Vranic, D.V. and Saupe, D., 3-D Shape Descriptor Based on 3-D Fourier Transform, in Proceedings of the EURASIP Conference on Digital Signal Processing for Multimedia Communications and Services (ECMCS2001), Budapest, 2001.

6. Kazhdan, M., Chazelle, B., Dobkin, D., Finkelstein, A., and Funkhouser, T., A reflective symmetry descriptor, in Proceedings of the European Conference on Computer Vision (ECCV2002), Lecturer Notes in Computer Science Vol. 2351, Springer-Verlag, Berlin, 2002, pp. 642–656.

7. Ohbuchi, R., Otagiri, T., Ibato, M., and Takei, T., Shape-similarity search of three-dimensional models using parameterized statistics, in Proceedings of the 10th Pacific Conference on Computer Graphics and Applications, IEEE Computer Society, 2002, pp. 265–274.

8. Osada, R., Funkhouser, T., Chazelle, B., and Dobkin, D., Shape distributions, *ACM Transactions on Graphics*, 21(4), 807–832, 2002.

9. Kolonias, I., Tzovaras, D., Malassiotis, S., and Strintzis, M.G., Fast content-based search of VRML models based on shape descriptors, *IEEE Trans. Multimedia*, in press.

10. Tangelder, J.W.H., and Veltkamp, R.C., Polyhedral Model Retrieval Using Weighted Point Sets, in Proceedings of International Conference on Shape Modeling and Applications (SMI2003), 2003.

11. Rubner, Y., Tomasi, C., and Guibas, L.J., A Metric for Distributions with Applications to Image Databases, in Proceedings of IEEE International Conference on Computer Vision, 1998, pp. 59–66.

12. Ohbucci, R., Masuda, H., and Aono, M., Watermarking three-dimensional polygonal models through geometric and topological modifications, *IEEE J. Selected Areas Commn.*, 16(4), 551–560, 1998.

13. Benedens, O., Affine Invariant Watermarks for 3-D Polygonal and NURBS Based Models, Springer Information Security, Third International Workshop, Australia, 2000, Vol. 1975, pp. 15–29.

14. Benedens, O., Watermarking of 3-D Polygon Based Models with Robustness Against Mesh Simplification, in Proceedings of SPIE: Security and Watermarking of Multimedia Contents, 1999, Vol. 3657, pp. 329–340.

15. Benedens, O., Two High Capacity Methods for Embedding Public Watermarks into 3-D Polygonal Models, in Proceedings of the Multimedia and Security Workshop at ACM Multimedia 99, Orlando, FL, 1999, pp. 95–99.

16. Benedens, O. and Busch, C., Towards blind detection of robust watermarks in polygonal models, in Proceedings of the EUROGRAPHICS'2000, Computer Graphics Forum, Vol. 19, No. 3, Blackwell, 2000, pp. C199–C208.

17. Praun, E., Hoppe, H., and Finkelstein, A., Robust mesh Watermarking, in Proceedings of SIGGRAPH 99, 1999, pp. 69–76.

18. Zhi-qiang, Y., Ip, H.H.S., and Kowk, L.F., Robust Watermarking of 3-D Polygonal Models based on Vertice Scrambling, in Proceedings of Computer Graphics International CGI'03, 2003.

19. Daras, P., Zarpalas, D., Tzovaras, D., and Strintzis, M.G., Efficient 3-D model search and retrieval using generalized Radon transforms, *IEEE Trans. Multimedia*, in press.

20. Toft, P., The Radon Transform: Theory and Implementation, Ph.D. *thesis*, Technical University of Denmark, 1996; http://www.sslug.dk/pt/PhD/.

21. Noo, F., Clack, R., and Defrise, M., Cone-beam reconstruction from general discete vertex sets using Radon rebinning algorithms, *IEEE Trans. Nuclear Sci.*, 44(3), 1309–1316, 1997.

22. Kadyrov, A. and Petrou, M., The trace transform and its applications, *IEEE Trans. Pattern Anal. Mach. Intell.*, 23(8), pp. 811–828.

23. Princeton Shape Benchmark, http://shape.cs.princeton.edu/search.html, 2003.

26
Digital Rights Management Issues for Video

Sabu Emmanuel and Mohan S. Kankanhalli

INTRODUCTION

A huge amount of digital assets involving media such as text, audio, video, and so forth are being created these days. Digital asset management involves the creation, transfer, storage, and consumption of these assets. It is chiefly in the transfer function context that the Digital Rights Management (DRM) becomes a major issue. The term DRM refers to a set of technologies and approaches that establish a trust relationship among the parties involved in a digital asset creation and transaction. The creator creates a digital asset and the owner owns the digital asset. The creators need not be owners of a digital asset, as in the case of employees creating a digital asset for their company. In this case, although the employees are the creators, the ownership of the digital asset can reside with the company. The digital asset needs to be transferred from the owner to the consumer, usually through a hierarchy of distributors. Therefore, the parties involved in the digital asset creation and transaction are creators, owners, distributors, and consumers. Each of these parties has their own rights; namely creators have *creator rights*, owners have *owner rights*, distributors have *distributor rights*, and consumers have *consumer rights*. We lay down the requirements of these rights in section "Digital Rights Requirements." A true DRM system should be able to support all of these rights. It is to be noted that there can be multiple creators jointly creating a digital asset and, likewise, multiple owners jointly owning a digital asset. There can also be a hierarchy of distributors and also distributorships allocated by geographic regions or time segments. The consumer can also be either an individual or a group of individuals or a corporate entity. Tackling DRM issues in this multiparty, multilevel scenario is a very challenging problem. We continually point out challenges in this chapter that can be the basis for further research.

The digital rights requirements vary with the kind of digital asset being transferred. Digital assets can possess different transaction values. Some can be free, whereas others can be of a high value (digital cinema, high-valued digital arts). Some can be truly invaluable, especially archival and historical photographs, audios, and videos. The creation or production cost of a digital asset is different from the transaction value. The transaction value is most often assigned by the owner (which is usually higher than the production cost). The transaction cost consists of the distribution cost and cost of setting up the DRM (DRM cost). Designing an economically viable DRM system is particularly challenging for low-valued digital assets.

The framework of DRM consists of technical, business, social, and legal components [1–4]. The business aspect of DRM involves new business models such as a freem evaluation copy for few days followed by a full permission or a downloadable stream for immediate use followed by a digital versatile disk (DVD) copy in the mail. The social angle of DRM is governed by societal norms concerning fair use, privacy management, and the education of consumers so as to inform them of the risks associated with using pirated content. The legal aspect deals with the laws of the license jurisdiction and include legislation, compliance, investigation, and enforcement. The technical DRM perspective deals with the technical standards and infrastructure, which consists of protection mechanisms, trading protocols, and rights language. In this chapter, we will discuss the business, social, legal, and technical aspects of DRM in detail.

Consider a digital commerce transaction completed over the Internet for DVDs. In such a transaction, usually the consumer and the seller are geographically far apart and possess scant knowledge of each other. To initiate the transaction, the consumer browses the Web for DVDs. In this scenario, a few typical issues are raised. Can the consumer trust that the online stores are legitimate and possess the right to distribute the copyrighted DVDs? Can the consumer be guaranteed that the payment information and order information are secure? Can the consumer trust that the digital content he is about to download is authentic, untampered material from the original author, owner, or distributor? Can the consumer trust that the content will work on his usage devices and whether he can give the content to his friend? Can the consumer trust that his privacy regarding his buying behavior is secured? The questions that the legitimate online DVD seller would ask are as follows: Can I trust the consumer that he would not create unauthorized copies and redistribute? In the event of a copyright violation, can the seller identify and trace the consumer who did it so as to bring him to justice? Without proper DRM, answers to most of these questions are negative.

Let us look at the example of a new business model where a consumer wants to buy a DVD over the Web. E-mail could straight away stream out the video to the consumer which could be watched on the personal computer (PC) or sent to the liquid crystal display (LCD) projectors of the home theater system immediately and a physical DVD could be dispatched for viewing and possession. However, the rights usage of the streamed version should be limited. The stream should be allowed to play on the machines indicated by the consumer and also may be allowed to be diverted to the systems as specified by the user. Also, the DVD should be allowed to play as and when required and on any machine. The DRM system should support these usage requirements. We discuss various usage requirements in section "Digital Asset Usage Controls."

Already, products and services are available that support DRM. However, these products are often not interoperable, do not support newer business models, and support only limited digital rights of parties. The current DRM systems are quite primitive, as they tie the content to a particular playback device. The content is not allowed to be played on any other device or to be diverted to another device. The DRM system depends on strong encryption schemes and ties the decryption to the device on which it is decrypted [5]. Often, these DRM products only support the digital rights of owners; they do not address false implication, fair use, and privacy concerns of consumers. Therefore, these DRM solutions are not widely used. Consequently, not many high-valued copyrighted or archival digital contents are available on the Web due to piracy, or recreation and reuse concerns. It must be noted that there are digital delivery channels such as satellite, cable broadcasts, and so on, other than the Internet. They, too, need to address piracy and reuse concerns. We discuss various delivery techniques in section "Digital Video Asset Management."

Although digital assets fall into various categories such as text, audio, image and video, in this chapter, we concentrate on the DRM issues for digital video. The challenges in solving DRM for digital video are multifaceted. Video files are large and they require much storage space. Therefore, to reduce the storage requirement, videos are compressed before being stored. For example, a typical 2-h, 35-mm feature film scanned at a standard high-quality resolution of 1920×1080 pixels, 1 byte per color and 24 frames per second would, in its uncompressed form, require $120 \times 1920 \times 1080 \times 3 \times 24 = 17,915,904,000$ bytes of disk space. Assuming a compression ratio of $30:1$, the disk space requirement becomes more manageable. These requirements are going to go up as the high-resolution 4000×4000-pixel scanners (Kodak Cineon and Quantel Domino) gain widespread use [6]. Therefore, compressed video would be preferred over uncompressed video. Again, the large size of the video compels the owners and distributors to broadcast the video to multiple receivers rather than to unicast to individual receivers. Broadcasting reduces the channel bandwidth requirement and also the transmission costs. Therefore, the preferred transmission technique for the digital video is broadcast. For DRM, compressed domain processing and broadcast type transmission make matters more complex. Essentially, DRM is a one-to-one contract whereby rights of both the parties are to be defined, complied, monitored, and enforced. However, broadcast is a one-to-all transmission scenario, where every receiver in the broadcast region receives the video. How to satisfy these seemingly contradictory requirements is a challenging problem for DRM in the broadcasting environment.

Past decades might have seen tremendous achievements in the creation, processing, and delivery of multimedia data. However, growth of digital commerce of these digital objects over the Internet or other digital delivery mechanisms have been somewhat slow, due to the major concerns regarding DRM.

Section "Digital Video Asset Management" to section "Digital Rights Requirements" concentrate on the environments and requirements for achieving the DRM. Particularly, section "Digital Video Asset Management" discusses digital video asset management, section "Digital Asset Usage Controls" presents digital usage controls, and section "Digital Rights Requirements" discusses digital rights requirements. Section "Business Aspect of DRM" to section "Technical Aspect of DRM" elaborate on the various components of DRM. Specifically, section "Business Aspect of DRM" discusses the business aspects, section "Social Aspect of DRM" presents social aspects, section "Legal Aspect of DRM" describes the legal aspects, and section "Technical Aspect of DRM" elaborates on the technical aspects.

DIGITAL VIDEO ASSET MANAGEMENT

After creating the digital asset, storage and delivery functions need to be addressed as part of digital asset management [2]. We discuss storage and delivery mechanisms in this section.

Storage

Digital video is usually compressed. The open compression standards from ISO/IEC, Moving Pictures Experts Group (MPEG) are MPEG-1, MPEG-2, and MPEG-4 [7]. The MPEG-1 standard is used for the video compact disk (VCD). The DVD, digital video broadcasts (digital TV transmissions), and high-definition television (HDTV) transmissions currently use the MPEG-2 standard. The MPEG-4 is a newer standard intended for low-bit-rate (wireless, mobile video applications) and high-bit-rate applications. There are other open standards such as H.261 and H.263 from ITU primarily for videoconferencing over telephone lines. Apart from the open standards, there are proprietary compression standards from Real Networks, Apple, and Microsoft, and so forth. In addition to compression, the digital assets are to be wrapped in metadata to declare the structure and composition of the digital asset, to describe the contents, to identify the contents, and to express the digital rights. We discuss the declarations, descriptions, and identifiers in section "Content Declarations, Descriptors, and Identifiers."

Digital Video Delivery

E-shops in the World Wide Web (WWW) typically offer downloading or streaming modes for delivering the video. The downloading mode employs a unicast type of transmission technique, whereas the streaming mode can use either a unicast or broadcast type of transmission. Digital TV transmissions usually use the broadcast channel. The most popular medium of digital TV transmission is terrestrial digital TV transmission. Other popular mediums are satellite and cable transmissions. Direct to home satellite transmissions are becoming more and more popular due to their ease in deployment. Stored media such as DVD and VCD are other means of distributing digital video.

Digital cinema is a system to deliver motion pictures and "cinema-quality" programs to theaters throughout the world using digital technology. The digital cinema system delivers digitized or mastered, compressed, and encrypted motion pictures to theaters using either physical media distribution (such as DVD-ROMs) or electronic transmission methods such as the Internet as well as satellite broadcast methods. Theaters receive digitized programs and store them in hard-disk storage while still encrypted and compressed. At each showing, the digitized information is retrieved via a local area network from the hard-disk storage, and is then decrypted, decompressed, and displayed using cinema-quality electronic projectors.

As we can see, digital TV, digital cinema, and Internet streaming services use the broadcasting technique, whereas the Internet downloading service uses the unicast technique for transmission. In all of the above delivery methods except in stored media such as DVD and VCD delivery for home segments, a return channel (other than the content and control signal delivery channel from the owner or distributor to the consumer) is provided for communication from the consumer to the distributors or the owner for DRM implementation. The return path need not be always online, as in the case of certain prepaid digital pay TVs for which the communication is at certain intervals of time.

DIGITAL ASSET USAGE CONTROLS

Owners often want usage controls to be associated with the digital asset. These controls essentially specify the various ways the digital asset can be used; for example, whether the digital asset can be copied or not, and if allowed to copy, the number of times that the digital asset can be copied. Currently, DRM systems are being designed with no copy, one-time copy, and many-times copy features [5,8,9]. The controls can specify whether the digital asset is allowed for usage activities such as play, copy, or reuse. If allowed, additional controls based on count, time, time period, and territory can also be placed [2,10].

The count parameter will support the service such as "pay-per-usage" (i.e., if the video is allowed to copy once, after making a copy once no further copying of the original video is allowed). A new video needs to be purchased to make another copy. The time-period parameter supports "pay-per-view-per-time period" programs (i.e., the billing can be for the period of time the video program is viewed). Another application may be to have a differential billing scheme based on the time of the day. The territory parameter may support the differential billing for different geographic regions. The reuse control specifies whether the digital asset can be reused. If reuse is allowed, new digital assets can be created from the digital asset. For example, consider the creation of a home music VCD album where the music is from different VCDs or the creation of a new VCD with new characters replacing the original characters. These controls support new business models.

DIGITAL RIGHTS REQUIREMENTS

The digital asset creation and delivery chain consists of creators, owners, distributors, and consumers. Each of these parties has their own rights and we call these the digital rights. The DRM system is expected to build trust among the parties by meeting the digital rights of each party. It must

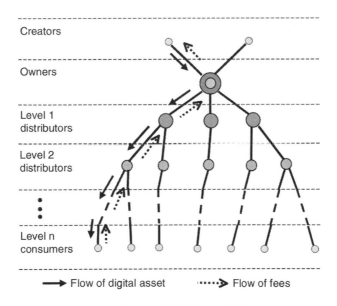

Figure 26.1. **Digital asset creation and delivery chain.**

be noted that in this chain, the created digital asset moves from the creator to the consumer through the owner and distributors. Also, the fee for the digital asset flows from consumer to the creator through owner and distributors. This can be seen in Figure 26.1. Often, the owner already pays the fee for the creation of the digital asset to the creators, as in the case of the employee–employer relationship. Therefore, the owner is the one concerned about the loss of revenue due to piracy and unauthorized usage. The owners can directly deliver the digital asset to the consumers or they may use a hierarchy of distributors. This can be seen in Figure 26.1. The hierarchy of distributors is preferred to deal with the business scalability and viability. Each of the distributors may have their own already existing consumer bases and infrastructure, which can easily be tapped into by the owners to deliver their digital asset to the consumers. When there is a hierarchy of distributors, building trust relationships between the owner and the distributor, and the distributor and the subdistributor become more complex. This is because whereas the owner is concerned about the revenue loss due to malpractices of the distributors and also of the end consumer, the distributors would be concerned about the false framing by the owners and also would be concerned about the revenue loss due to the malpractices of the subdistributors and the endconsumer. Because the consumer is at the end of the distribution chain, he would be concerned about the digital asset, its quality and integrity, and also false framing by the distributor or owner [11]. There should also be a trust relationship built between the consumer and the distributor or the owner. We now elaborate on the digital rights requirements of each party for various application scenarios.

Digital Rights of Creators

A digital asset may be jointly created by one or more creators. All participating creators would expect that their participation be provable to any party. This requires that a jointly created mark be placed on the digital asset or individually created marks be individually placed on the digital asset. In either case, the idea is to establish the right of creatorship. The mark should be robust so that no one can remove it except the concerned creator himself. The mark also should be robust against collusion removal attacks. For example, a few creators may collude and remove the creatorship information of another creator from the digital asset, which should not be allowed. Only a few researchers have addressed this problem. The marking technique should be scalable with an increase in the number of creators and the mark should not cause considerable perceptual degradation of the digital asset.

The creators may be paid royalty by the owner for every copy of the digital asset made, as in the case of publishing books. In this case, the creators would need to track the number of copies made. The typical scenario in digital movie and video creations is that the creators are the employees of the production house and, hence, no such tracking of the number of copies is needed.

Digital Rights of Owners

Video distribution can either be subscription-based distribution or free. In both of these cases, the owner would like to specify how the distributors and consumers should use the digital asset. This is primarily specified through the usage controls. Often, the copyright-protected videos are distributed using the subscription-based schemes. In these subscription-based schemes, the owners may want to associate various usage controls on viewing, copying, play back device in use, and reuse. The usage control can vary from one-time view to many-times view, no copy to many times copy, play back on dedicated device to any device, and conditional to unconditional reuse.

The free-distribution mode can fall into two categories: one that can be reused as well as modified and the other in which the asset cannot be tampered with. An example of free distribution that cannot be tampered with is archival material with historical importance. In this case, usage control needs to specify whether the digital video can be reused or not. Also, the other usage control parameters are of little importance because the distribution is free. These usage controls essentially constrain the usage of digital asset by consumers and distributors.

We list the owner's requirements for a subscription-based distribution:

- Needs to specify the usage controls
- Needs the consumer/distributor to be authenticated
- Needs the ability to trace the consumer/distributor who has violated the DRM contract
- Needs to declare the ownership of the digital asset
- Needs to distinguish consumers from non-consumers
- Needs a payment protocol to handle the fees

The authentication of consumers or distributors is needed to reliably identify the consumer or distributor. In the event of DRM contract violation, the violating consumer or distributor can be identified. This can be implemented using authentication protocols. In order to trace the DRM-contract-violating consumer or distributor and to declare the ownership of the digital asset, digital watermarking techniques can be used. The digital watermarking technique should be robust against attacks and also noninvertible to prove the ownership of the digital asset. Because the distribution is subscription based, only the consumers (who have paid the subscription fee) should get a clear video for viewing, whereas the nonconsumers should not get a clear video for viewing. This requirement is the confidentiality requirement and can be implemented using encryption techniques.

Next, we list the owner's requirements of free distribution with archival importance:

- Needs to specify the reuse policy (essentially no tampering allowed)
- Needs the consumer or distributor to be authenticated
- Needs to detect any tampering of the content
- Needs the ability to trace the consumer or distributor who has tampered with the video
- Needs to declare the ownership of the video

For tamper detection, one can employ fragile watermarks. For tracing the consumer or distributor who has tampered with the video and also to declare the ownership of the video, noninvertible robust watermarking techniques can be used. Authentication can be implemented using authentication protocols.

For free distribution with no archival importance, there need not be any rights associated, as the digital asset is free to use in whichever way the consumer desires.

Digital Rights of Distributors

It can be seen from the digital asset delivery chain that the distributors act as middlemen. The owners fall at the upper end, the distributors are at the middle part, and the consumers fall at the lower end of the delivery chain. To the owners or upstream distributors, the downstream distributors are consumers with the same or less restrictions on usage compared to end consumers. For example, usage control on viewing or playing on any device may be unimportant, whereas usage control on copying and reuse would be important. As far as trust-building measures between the downstream distributor and the owner or upstream distributor, the downstream distributor will have the following concerns for both subscription-based and free distribution with archival importance, which need to be addressed:

- Needs to address false implication by the owner or upstream distributor
- Needs the owner or upstream distributor to be authenticated
- Needs proof of ownership or distributorship of the video
- Needs the integrity of video to be proved
- Needs a payment protocol to handle the subscription fee (for subscription-based schemes only)

The above requirements apply for both subscription-based and free distributions with archival importance. The false implication concern is that often the fingerprinting marks for tracing the copyright violator is placed by the owner or upstream distributor alone [11]. It is possible that the owner or upstream distributor may knowingly or unknowingly creates a second fingerprinted copy with the same mark, which had been already used for someone else, and distributes the second copy to a second party. Later, the owner or upstream distributor can sue the first legal recipient for the copy with the second recipient. This false implication by sellers is a major concern for downstream distributors and end consumers. As a trust-building measure between the upstream distributor and downstream distributor or end consumer, the following requirements of upstream distributors are to be addressed:

- Needs to specify the usage controls (this can be a subset of what the owner has specified)
- Needs the consumer or downstream distributor to be authenticated
- Needs to trace the consumer or downstream distributor who has violated the DRM contract
- Needs to distinguish consumers or downstream distributors from nonconsumers or nondownstream distributors (for subscription-based schemes only)

- Needs to detect the tampering (for free distribution with archival importance only)
- Needs a payment protocol to handle the subscription fee (for subscription-based schemes only)

Digital Rights of Consumers

The consumers are at the lower end of the delivery chain. The following concerns of the consumers are to be addressed:

- Needs to address false implication by the owner or distributor
- Needs the owner or distributor to be authenticated
- Needs proof of ownership or distributorship of the video
- Needs the integrity of video to be proved
- Needs a payment protocol to handle the subscription fee (for subscription-based schemes only)
- Needs to address fair use and privacy concerns

The digital rights requirements such as authentication of parties, confidentiality against the nonconsumers and nondistributors, proving the integrity, ownership or distributorship of the video, and payment protocols can be implemented using cryptographic and watermarking techniques in the DRM system. However, incorporating techniques to deal with false implication concerns, DRM contract violator tracing, fair use [12,13], and privacy management [14–16] into the designed DRM is a challenging one. Some of the requirements are seemingly contradictory (e.g., authentication and privacy management). In a broadcasting environment, how to implement DRM contract violator tracing is another problem. The false implication concern can be dealt using nonrepudiation techniques and how to integrate this into the DRM system is another challenging problem. How to support the fair use requirement is a major concern. How to design invisible multiple watermarking technique for tracing DRM contract violator in a multilevel delivery chain is another challenge.

Next, we discuss various components of DRM such as business aspects, social aspects, legal aspects, and technical aspects of DRM.

BUSINESS ASPECT OF DRM

The Internet and broadcast channels (such as satellite, cable, and terrestrial) act as low-cost (compared to hardcopy transport) and quick transmission channels for the distribution of video data. These channels often transcend geographic and national boundaries connecting owners, distributors, and consumers. Various new business models leverage on the anywhere connectivity, quickness, and low-cost properties of these channels [17–19].

Some of the new business models are as follows:

- *Preview and purchase*. The preview is through quick and low-cost channels; the purchased version can be a hard copy or soft copy with fewer constraints on the usage compared to the preview version.
- *Pay-per-usage*. The usage can be viewing, copying, or reusing. As an example, the pay-per-view programs of TV channels belong to this category. In the pay-per-view TV program, the payment is for a single program feature as desired, which could be prebooked or impulsive.
- *Pay-per-usage-per-usage constrained*. The usage constraints can be count, time, time period, territory, device, and so forth. The constraint time is for supporting differential pricing scheme based on peak hour, nonpeak hour, and so forth.
- *Subscription based*. An example for this business model is pay TV.

Some of the business models make use of dedicated transmission channels (cable, satellite link, etc.) and end-rendering hardware (set-top box, etc.) and software. Some other business models make use of open networks such as the Internet and general rendering software (Microsoft's media player, Apple's quick time, etc.), and general hardware such as computers.

Designing a DRM solution for the dedicated setup is easier compared to the open and general setups, because the owner or distributor is connected to the consumer through a dedicated transmission channel and rendering hardware and software. Designing DRM solutions for this open or general setup is challenging for the following reasons. The Internet service providers (ISPs) do not belong to the owner or distributor category. ISPs merely provide bit pipes and switching mechanisms for transferring the data from one party to the other. However, they often implement caching and other temporary storages, which can cause stale or multiple copies to reach the consumers. In the DRM solutions, if we use DRM aware general-purpose third-party rendering software and hardware, the third party who is providing the software and hardware also should be part of the DRM agreements. This is because the third party can act as an intruder. For example, the third-party software and hardware can keep track of the transactions between the owner or distributor and the customer (maybe for market analysis). This problem is especially important when the client is a corporate entity. Designing a DRM-compliant computer would restrict the usefulness of the computer. To the contrary, if the copyrighted materials cannot be played or used on the computer, the charm of the computer is lost.

New revenue models are being contemplated to make DRM viable. One is to recover the cost of the copyrighted material by allowing third-party advertisers to advertise their wares along with the video. In return, they share the cost of the video. This makes the cost of video that is to be borne by the consumers nominal and makes piracy unattractive. Another model proposes making the ISPs the retailers, which makes the DRM more implementable [19]. Yet another model proposes some ticket-based system for free or low-cost digital videos. Escrow-services-based systems propose an architecture that uses economic incentives instead of tamper resistance to motivate users to keep the digital asset within the subscription community [20].

SOCIAL ASPECT OF DRM

The social angle of DRM is governed by societal norms of fair use, such as institutions that are not for profit being allowed to use the material free of cost and make any number of copies, but those institutions and organizations that run on a profit basis need to pay for it. Other fair use problems are how does the DRM support the act of giving gifts, loaning a digital video to a friend, making a backup copy for oneself, and so forth. How to support DRM in a library scenario where loaning is major function is another problem. Guaranteeing privacy regarding consumer's buying practices is another social aspect of DRM. The social aspect includes also the education of consumers such as to inform them on the risks associated with using the pirated content. Education involves disseminating proper norms on how a copyrighted material is to be used, in addition to information on the penalties associated with the breaking of a copyright agreement.

LEGAL ASPECT OF DRM

A proper legal framework is necessary for successful implementation of the (DRM) system [18,21–25]. The framework primarily involves enacting new legislation that support DRM issues such as compliance, investigation, and enforcement. The DRM compliance lays out rules for the hardware and software manufacturers, service providers, owners, distributors, and consumers — what each one of them should do in order to be DRM compliant. In the event of DRM agreement violations, how do we track the violations and prove the culpability of the violators? How do we enforce the DRM-related laws and also what are the penalties or compensations for the DRM agreement violations? These are some of the issues tackled in the legal aspects of DRM. One of the major concerns is

771

how does the legal aspect of DRM work across the national boundaries. Often different countries have their own DRM-related laws and the digital asset transactions often take place beyond the national boundaries. Another major concern is how the fair use issue is to be tackled in the DRM laws.

The United States, the European Union, and Australia have already enacted laws to support DRM. The Digital Millennium Copyright Act (DMCA), passed in 1998, is an American law implementing the 1996 World Intellectual Property Organization (WIPO) Copyright Treaty (WCT) and the WIPO Performances and Phonograms Treaty (WPPT). In 2001, the European Union enacted the European Union Copyright Directive (EUCD), which also implements the WIPO treaties. Both of these acts of law make it compulsory to outlaw any attempt to circumvent a technical copyright protection. Both also outlaw the creation and distribution of DRM circumvention tools. Whereas the DMCA allows fair use such as using digital content for research and teaching, with the exception of distance learning, the EUCD does not allow fair use. The fair use also allows the consumer to resell what he bought or make a backup copy for personal use. However, there are no clear rules defining fair use, which is usually decided by a judge on a case-by-case basis. The Australian Copyright Amendment (digital agenda) Act 2000 (DACA) is similar to the DMCA. A new act called the Consumer Broadband Digital Television Promotion Act (CBDTPA) was introduced to the U.S. Senate in 2002. The CBDTPA mandates that anyone selling, creating, or distributing "digital media devices" must include government-approved security to prevent illegal copying of protected data. The devices include CD and DVD devices and desktop PCs. This act might restrict the deployment of open-source products such as Linux and also would limit the usefulness of the desktop PC.

TECHNICAL ASPECT OF DRM

The technical aspect of DRM includes defining technical standards and infrastructure needed to support DRM in new business models, fair use, and privacy issues of the social aspects of DRM, and provability and traceability requirements of the legal aspect of DRM [2,18]. The technical infrastructure consists of the hardware and software protection mechanisms that use cryptographic and watermarking techniques, trading protocols, rights language, and content descriptors and identifiers. We further elaborate on these technical infrastructures in this section. We also discuss the standardization efforts by various bodies such as MPEG, IETF, W3C, SMPTE, and so forth and also current DRM techniques and tools in digital video broadcasts, stored media such as DVD, and digital cinema.

Cryptographic Techniques

Cryptographic techniques provide confidentiality, authentication, data integrity, and nonrepudiation functions. They are also designed for traitor tracing. However, in the DRM world, cryptographic techniques are primarily used for providing media confidentiality, trading party authentication, and trading protocol nonrepudiation. There are well-established party authentication protocols such as Kerberos, X.509, and so forth. Some of these are based on symmetric key cryptosystem, whereas others are based on public key infrastructure. They authenticate the parties involved and also, if necessary, hand over secrets such as keys for confidential communication. The confidential communication may be intended to transfer payment information and also may be the media in transaction. Free video with tracing requirement and pay video need confidentiality to be implemented. Conditional access (CA) systems, which implement confidentiality through a scrambling technique, are used in most of the subscription-based video systems [26,27]. The nonrepudiation function provides for nonrepudiation of the trading agreement. It is easy to combine the party authentication, trading agreement nonrepudiation, and payment function into a trading protocol, as they are usually one-to-one protocols and take place before the delivery of media.

Confidentiality of media against nonintended receivers is achieved by scrambling the media under the control of a scrambling key. The descrambling key is distributed only to the intended receivers. Various traitor-tracing schemes have been proposed by the researchers in cryptography [11,28]. Some of the works focus on fingerprinting the descrambling key; thus, each intended receiver possesses and uses a different descrambling key. Therefore, by looking at the descrambling key used in the pirate decoder, the traitor receiver who has leaked out the descrambling key can be identified. These schemes make certain assumptions, such as the video being of a large size, so it is easy for traitors to hand over the descrambling key, which is small in size in comparison to the descrambled video. Thus, in these schemes, traitor tracing cannot be performed from the descrambled video because all descrambled videos are identical. With today's distribution networks, such as copy and distribute over the Internet or through CDs, the above schemes would not be effective in combating piracy.

There are other cryptographic schemes that which employ fingerprinting the video for traitor tracing. Fingerprinting of video takes place while descrambling. Combining confidentiality and video fingerprinting for traitor tracing is a challenging problem due to the following reasons. Video data being large sized, broadcast transmission techniques are preferred over unicast transmission, which reduces the transmission bandwidth requirement. Thus, a single scrambled copy of video is broad-casted to the receivers. Also, video data would be compressed before transmission. The compression scheme used is often a lossy scheme. Hence, the scrambling needs to be performed after the compression and the descrambling needs to be performed before the decompression. However, if fingerprinting is done along with descrambling, the finger-prints can be considered as bit errors occurring during transit. Especially in coding schemes, which employ interframe coding techniques, the errors can propagate into other frames and cause drifts. Thus, design-ing a scrambling technique that would simultaneously perform finger-printing while descrambling with no adverse effects is a challenging problem [29].

Although it is desirable to have the fingerprinting and the descram-bling processes combined into one atomic process to force the pirate to exert more effort to get an unfingerprinted clear video for piracy, they are often implemented as two separate processes. This allows the pirate to turn off the control bits for the fingerprinting process or bypass the fingerprinting process completely and thus obtain an unfingerprinted but descrambled video [11].

Media data integrity can be ensured through digital signature schemes or by using hashing functions. However, these techniques may not survive benign processing that might happen while in transit. However, for media data integrity, watermarking is preferred because watermarking techniques can be designed to survive processing such as compression, transcoding, and so forth. For traitor tracing, water-marking techniques are preferred because the watermarks can survive various benign or hostile operations.

In a multiparty distribution network (such as owner, distributor, consumer), implementing a trading protocol consisting of party authen-tication, trading agreement nonrepudiation, and payment function is easy, as it is essentially a one-to-one protocol (between owner and distributor, distributor and subdistributor, subdistributor and consumer, etc.). However, in order to trace the traitor along the distribution channel, owner $\rightarrow$ distributor $\rightarrow$ consumer, at every level (owner $\rightarrow$ distributor one level, distributor $\rightarrow$ subdistributor another level, and subdistributor $\rightarrow$ consumer another level) of the distribution channel in order to differentiate and identify the recipient of the media in that level, the media needs to be fingerprinted. Thus, for an n-level distribution channel, the end consumer would get a media with n fingerprints in it. Multiple fingerprints can cause them to be visible and can lead to the deterioration of video quality. The multiple fingerprinting technique is a challenging problem, which needs to be addressed by the research community.

Watermarking Techniques

Watermarking techniques are usually preferred for copyright owner-ship declaration, creator or authorship declaration, copyright violation detection (fingerprinting function), copyright violation deterrence, copy control, media authentication, and media data integrity functions. They are also devised for variety of media (viz., text, digital audio, digital image, and digital video). A digital watermark is an embedded piece of information either visible or invisible (audible or inaudible, in the case of audio) [30]. In the case of visible watermarking [31,32], the watermarks are embedded in a way that is perceptible to a human viewer. Hence, the watermarks convey an immediate claim of ownership or authorship, providing credit to the owner or author, and also deter copyright violations. In the case of invisible watermarking [33,34], the watermarks are embedded in an imperceptible manner. The invisible watermarks can be fragile or robust. Fragile invisible watermarks [35] attempt to achieve data integrity (tamper-proofing). The fragile invis-ible watermarks must be invisible to the human observers, altered by the application of most common image processing techniques, and should be able to be quickly extracted by authorized persons. The extracted watermark indicates where the alterations have taken place. Sometimes, it is necessary that the marks remain detectable after legitimate tampering and fail to provide content authentication for illegitimate operations. These watermarks are referred to as semifragile watermarks. The desired properties for robust invisible watermarks are that they must be invisible to a human observer and that the watermarks should be detectable or extractable by an authorized person even after the media object is subjected to common signal processing techniques or after digital-to-analog and analog-to-digital conversions. A robust invisible watermarking technique is suitable for copyright ownership declaration, creator or authorship declaration, copyright violation detection (fingerprinting function), copyright violation deter-rence, and copy control. Many watermark algorithms have been proposed in the literature [36,37]. Some techniques modify spatial and temporal data samples, whereas others modify transform coefficients [29]. To be robust, the watermark must be placed in perceptually significant regions [33].

The input to a watermark embedding system is usually original data, watermark information, and a secret embedding key; the output of watermark embedding system is watermarked data. The watermarking system can also be classified based on the watermark detection scheme in use. If original data or watermark information is needed along with watermarked data and the secret key, the system is called a *private watermarking* system [33,37–40]. If neither the original data nor the watermark information is needed for detection, the watermarking scheme is known as a *public watermarking* scheme or a *blind watermarking* scheme [29,34]. The original data and the watermark information in certain applications are to be kept secret even at the time of detection. Therefore, in these applications, blind watermarking schemes are preferred. Sometime, during a dispute, the author has to prove the existence of watermark to a third party. However, divulging the original data or watermark information or the key may not be desirable. The *zero knowledge watermark detection* schemes [41,42] are designed for this purpose. *Asymmetric watermarking* schemes [43,44] make use of another key for the detection rather than the embedding key.

Various types of attack on watermarking techniques have been described in References 30 and 45–47. There are several software tools available for benchmark watermarking techniques (e.g., Stirmark, Checkmark, and Optimark). These tools primarily support benchmarking of image watermarking techniques, but Checkmark also supports benchmarking of video watermarking techniques.

Trading Protocols

The Internet engineering task force (IETF) trade working group is designing a trading protocol framework for Internet commerce under the title "Internet Open Trading Protocol (IOTP)" [48]. IOTP provides an interoperable framework for Internet commerce. IOTP tries to provide a virtual capability that safely replicates the real world — the paper-based, traditional, understood, accepted methods of trading, buying, selling, value exchanging that has existed for many years. The negotiation of who will be the parties to the trade, how it will be conducted, the presentment of an offer, the method of payment, the provision of a payment receipt, the delivery of goods, and the receipt of goods are the events that would be taken care of in the IOTP. IOTP messages are Extensible Markup Language (XML) documents, which are physically sent between the different organizations that are taking part in a trade. IOTP v2 (version 2) is intended to extend the IOTP v1 (version 1). IOTP v2 will provide optional authentication via standards-based XML Digital Signatures; however, neither IOTP v1 nor v2 provide any confidentiality mechanism. Both depend on the security mechanisms of payment system (such as Secure Electronic Transaction [SET], CyberCash, etc.) used in conjunction with them to secure payments, and both require the use of secure channels such as those provided by Secure Sockets Layer and Transport Layer Security (SSL/TLS) or Internet Protocol Security (IPSEC) for confidentiality. Electronic commerce frequently requires a substantial exchange of information in order to complete a purchase or other transaction, especially the first time the parties communicate. Electronic Commerce Modeling Language (ECML) provides a set of hierarchical payment-oriented data structures (in an XML syntax) that will enable wallet software to automate the task of supplying and querying the needed information to complete a purchase or other transaction. IOTP is also designed to support the use of ECML.

The World Wide Web Consortium (W3C) is also developing standards for e-commerce on the Web. W3C has actively contributed in developing an XML signature, XML encryption, XML protocol, micropayment, and platform for Privacy Preferences Project (P3P) standards [49]. The XML signature provides a mechanism for signing documents and metadata in order to establish who made the statement. XML encryption specifies a process for encrypting or decrypting digital content and an XML syntax used to represent the encrypted content and information that enables an intended recipient to decrypt it. The XML protocol is aimed at developing technologies, which allow two or more peers to communicate in a distributed environment, using XML as its encapsulation language, allowing automated negotiations. The Platform for Privacy Preferences Project (P3P) provides communication about data privacy practices between consumers and merchant sites on the Web as well as enhanced user control over the use and disclosure of personal information. The micropayment initiative specifies how to provide in a Web page all of the information necessary to initialize a micropayment and transfer this information to the wallet for processing.

The IOTP and W3C's standards for e-commerce do not address the DRM issues explicitly. However, there is a proposal for a digital Rights Management System (RMS) that is integrated into IOTP for electronic commerce applications and services [50]. It introduces a rights insertion phase and a rights verification phase for IOTP. In the proposed framework, digital watermarking plays a very important role in facilitating DRM.

Rights Languages

In order to manage the rights of the parties, the rights are to be specified in a machine understandable way [1,51–53]. For this purpose, a rights data dictionary of allowed words and allowed constructs of these words is defined. The allowed constructs are often defined using the XML scheme, which forms the rights language. One of the rights languages is eXtensible rights Markup Language (XrML) defined by ContentGuard Inc., and another is Open Digital Rights Language (ODRL) from IPR Systems Pvt Ltd. Both have their own rights data dictionary and vocabulary for the language. These languages express the terms and conditions over any content, including permissions, constraints, obligations, offers, and agreements with rights holders. XrML is designed to be used in either single-tier or multi-tier channels of distribution with the downstream rights and conditions assigned at any level. The MPEG-21 Rights Expression Language (REL) is using XrML as the base language. In addition, the OASIS (Organization for the Advancement of Structured Information Standards) rights language technical committee and the Open eBook Forum (OeBF) are using XrML as the base for their rights language specification. The Open Mobile Alliance (OMA) accepted the ODRL as the standard rights expression language for mobile content [54].

Content Declarations, Descriptors, and Identifiers

A digital content may comprise many subcontents. The makeup, structure, and organization of the content or subcontent can be specified using XML-based declaration techniques. Metadata can be used to describe the contents or subcontents (by specifying the media type, playback requirements, etc.). In order to specify the digital rights information associated with the contents or subcontents, they need to be identifiable. For example, in a video, the visual, audio, and text streams each can be considered as subcontents. Another example, an audio CD, may contain different audio tracks and each track can be considered as individual subcontents. In the first example, the types of subcontent are different; however, for the second example, the types of subcontent are the same. Several identifier codes are already defined and can be seen in Table 26.1. Except for the DOI and URI, all of the others are content-type-specific (e.g., ISBN for books, ISMN for printed music, etc.) [49,51]. MPEG-21 uses URIs to identify digital items and their parts.

TABLE 26.1. List of identifier codes and the type of content for which it is intended

Identifier Codes	Type of Content
BICI (Book Item and Component Identifier)	Book content
SICI (Serial Item and Contribution Identifier)	Serial content
DOI (Digital Object Identifier)	Any creation
URI (Uniform Resource Identifier)	Any creation
ISAN (International Standard Audiovisual Number)	Audiovisual programs
ISBN (International Standard Book Number)	Books
ISMN (International Standard Music Number)	Printed music
ISRC (International Standard Recording Code)	Sound recordings
ISSN (International Standard Serial Number)	Serials
ISWC (International Standard Musical Works Number)	Musical works
UMID (Unique Material Identifier)	Audiovisual content

MPEG and Digital Rights Management

The MPEG standards MPEG-2, MPEG-4, MPEG-7, and MPEG-21 address the DRM issues. The MPEG's specific term for DRM is "Intellectual Property Management and Protection" (IPMP) [38,55–58].

The MPEG-2 intellectual property management and protection (IPMP) standard [57] provides place holders for sending the information about whether the packet is scrambled, the information about which conditional access system (CAS) is used, control messages such as the encryption control message (ECM) and the encryption management message (EMM), and the copyright identifier of 32 bits, which identifies the type of work (like ISBN, ISSN, etc.). The IPMP specification only addresses confidentiality and watermarking. However, the IPMP-X (intellectual property management and protection – extension) includes an authentication function also along with the confidentiality and watermarking functions. The standard, however, does not specify the algorithms to be used for confidentiality, watermarking, or authentication.

The MPEG-4 IPMP "Hooks" covers identification of content, automated monitoring and tracking of creations, prevention of illegal copying, tracking object manipulation and modification history, and supporting transactions among users, media distributors, and rights holders [58]. The more recent call for proposal IPMP-X (extensions) stresses the interoperability of the solution to IPMP, renewability of the security, flexible expression of different business models or rules, protection of user privacy, user rights, and mobility of terminal and content. The IPMP–Descriptors (IPMP-Ds) and IPMP–Elementary Streams (IPMP-ESs) provide a communication mechanism between the IPMP system and the MPEG-4 terminal. IPMP-Ds are a part of the MPEG-4 object descriptors that describe how an object can be accessed and decoded. These IPMP-Ds are used to denote which of the IPMP systems was used to encrypt the object and, consequently, at the decoder, which IPMP system to be used for decrypting the object. IPMP-ESs are special elementary streams that are used to specify IPMP specific data, which can be used to deliver cryptographic keys, to deliver key update information, and to initiate rekeying protocols. The IPMP-Ds and IPMP-ESs are the inputs to the IPMP system at the MPEG-4 decoder and the IPMP system decides where to apply the IPMP control in the MPEG-4 decoder depending on the inputs. Watermarking technology is also envisaged for sending the IPMP information. There are mainly three scenarios in which the watermarking technologies are envisaged: broadcast monitoring, watermarking-based copy control, and fingerprinting. For broadcast monitoring purposes, the MPEG-4 decoder need not contain the watermark detectors, as the required monitoring can be done after rendering. The copy control systems require that the IPMP system inside the MPEG-4 decoder implements the watermark retrieval algorithm, and for fingerprinting, the IPMP system inside the MPEG-4 decoder must implement the watermark embedding algorithm.

MPEG-7 is a content description standard for content management [7]. The descriptions can be used for searching and retrieval of the content. The description consists of a description scheme (DS), descriptors (Ds), and a binary streamable representation of DSs and Ds called the binary format for MPEG-7 (BiM). Because MPEG-7 is for descriptions, the IPMP manages, protects, and authenticates these descriptions. MPEG-7 does not manage and protect the contents; however, it will accommodate and provide a mechanism for pointing to the content rights. MPEG-7 will provide the usage rules, usage history, identification of contents, and identification of contents in descriptions.

The vision of MPEG-21 is to define a multimedia framework to enable transparent and augmented use of multimedia resources across a wide range of networks and devices used by the different sectors [38,56]. MPEG-21 multimedia framework will identify and define the key elements needed to support the multimedia value and delivery chain, the relation between them, and the operations supported by these elements. There are seven key elements that have been identified for supporting inter-operability and one of them is IPMP. In the IPMP area, MPEG-21's aim is to provide a uniform framework that enables all users to express their rights and interests in agreements related to digital objects and to have assurance that those rights, interests, and agreements will be persistently and reliably managed and protected across a wide range of networks and devices. The MPEG IPMP extensions are designed so that they can be applied to any MPEG multimedia representation (MPEG-2, MPEG-4, MPEG-7, and MPEG-21).

Digital Video Broadcasts and DRM

There are three popular digital TV broadcasting standards: the American Advanced Television Systems Committee (ATSC) standard, the European Digital Video Broadcasting (DVB) standard, and the Japanese Integrated Services Digital Broadcasting–Terrestrial (ISDB-T) standard. These broadcasting standards use MPEG-2 for video coding. The pay channels of digital video broadcasts require that a confidentiality requirement be implemented against the nonsubscribers. Hence, current pay channels employ a conditional access system (CAS) for this purpose. Primarily, the CAS uses a scrambling technique for providing the confidentiality [11,22,26,27,59,60]. The broadcaster scrambles the video data using a control word (CW) and broadcasts the scrambled video data. The subscribers use the same CW to descramble the received scrambled video data to obtain clear video. The data encryption standard (DES) is used in ATSC and the digital video broadcasting common scrambling algorithm

(DVB-CSA) [59] is used in DVB. The DVB supports the simulcrypt and multicrypt standards, whereas ATSC supports only the simulcrypt standard [59,60]. The simulcrypt standard allows the coexistence of more than one CAS, simultaneously addressing different consumer bases, in one transmission. In the case of multicrypt, no program is available through more than one CAS.

The current CAS only supports confidentiality against nonsubscribers, but it does not implement any fingerprinting technique to trace the copyright violator. Because the broadcasting requires a single copy to be transmitted and the copyright violator identification requires individual subscribers copy to be unique, the fingerprinting for copyright violator identification should be performed at each subscriber end. This means that the current CASs, at the subscriber end, has to implement the fingerprinting process for copyright violator identification in addition to the descrambling process. It is more secure if the fingerprinting and descrambling processes can be combined into a single atomic process. However, it is hard to combine the existing descrambling process with fingerprinting process into a single atomic process. It is even more challenging to support confidentiality and copyright violator identification in compressed (MPEG-2) domain broadcast. The challenges are due to the quantization step, which is lossy and interframe coding, which makes use of the motion-compensated predictions. Thus, DRM in digital video broadcasts currently implements only the confidentiality requirement against nonsubscribers.

Stored Media (DVD) and DRM

The DVD Copy Control Association (DVD CCA) licenses Content Scrambling System (CSS) to the owners and manufacturers of DVD disks with video contents and manufacturers of DVD players, DVD recorders, and DVD-ROM drives [8,9,61]. CSS is a data encryption and authentication scheme intended to prevent copying video files directly from prerecorded (read-only) DVD video disks. CSS was developed primarily by Matsushita and Toshiba. The information on the DVD video disk is CSS encrypted. DVD players have CSS circuitry that decrypts the data before being decoded and displayed. Computer DVD decoder hardware and software would include a CSS decryption module, and the DVD-ROM drive would have extra firmware to exchange authentication and decryption keys with the CSS decryption module in the computer. The CSS license is extremely restrictive in an attempt to keep the CSS algorithm and keys secret. However, this CSS algorithm was broken and posted on the Internet.

Content Protection for Prerecorded Media (CPPM) is for DVD audio, which performs the same function as that of CSS . A disk may contain both CSS and CPPM content if it is a hybrid DVD video/DVD audio disk. Content protection for recordable media (CPRM) protects the exchange of entertainment content recorded on portable or removable storage media such as DVD-RAM, DVD-R, DVD-RW, and secure compact flash. The companies responsible for CPPM and CPRM are Intel, IBM, MEI, and Toshiba. The CSS, CPPM, and CPRM are not designed to protect against copying through the analog output port. However, Macrovision's DVD copy protection system supports protection against such analog copying. Macrovision's Copy Generation Management System (CGMS) can accommodate fair use copying by allowing an unlimited number of first-generation copies, but prohibiting those copies from being recorded again. The CGMS system works by storing 2 bits of CGMS copy control information (CCI) in the header of each disk sector. The CGMS is CGMS-A for analog output and CGMS-D for digital output. Verance developed a DVD audio watermarking technique (which uses CGMS CCI) to guard against the copying through audio analog output. The Watermark Review Panel (WaRP) under the DVD content protection working group (CPTWG) is working on video watermarking proposals from the Galaxy group (IBM, NEC, Pioneer, Hitachi, and Sony) and the Millennium group (Macrovision, Digimarc, and Philips) to choose one standard technique for DVD content protection. The minimum watermark payload expected is of three states (never copy, no more copying, and copy once) and an additional 2 bits for triggering analog protection system.

Digital Transmission Content Protection (DTCP) safeguards data streams transmitted over digital interfaces like IEEE-1394 and USB. DTCP is developed by Intel, Sony, Hitachi, Matsushita, and Toshiba and the Digital Transmission Licensing Administrator (DTLA) issues licenses to use DTCP. DTCP can be used to protect DVD content only if it has already been encrypted by CSS, CPRM, or CPPM. DTCP also uses 2-bit CCI to signal "copy one generation," "copy freely," "copy never," and "no more copies."

Digital Cinema and DRM

Cinemas also are not out of the compulsions of the digital age [6,62–65]. The digital cinema system delivers motion pictures that have been digitized or mastered, compressed, and encrypted to theaters using either physical media distribution (such as DVD-ROMs) or electronic transmission methods, such as via Internet and satellite multicast methods. Although today's digital cinema quality is not as good as that of film-based cinema, the digital cinema has many compelling factors in favor of it. One of the advantages is that although the quality is currently not as good, the quality of the digital cinema is preserved over the repeated shows, whereas in the film-based cinema, the quality deteriorates after multiple viewings. Another advantage is that the distribution cost is low for digital cinema, as they can be transmitted to the cinema halls through electronic transmission methods or through stored media such as DVDs, which are much more manageable than the bulkier and heavier film reels. Making copies of film reels for distributing to cinema halls is an enormous task compared to making copies in the digital domain. In addition, digital cinema can incorporate advanced theatrical experiences such as moving seats and so forth through the use of metadata.

Usually, movies are mastered on a film and are then digitized to obtain the digital version. However, with the introduction of high-definition digital camcorders, motion pictures are mastered directly in the digital domain. This digital movie is then compressed, encrypted, and delivered to the cinema halls. At the cinema hall, the digital movie is temporarily stored, decrypted, decompressed, and projected or played back. The cinema halls are equipped with a storage system and a projector system. There are two models for the secure storage and play back of digital cinema content: the broadcast server model and the data server model. In the broadcast server model, the storage system provides for temporary storage, decryption decompression, and local encryption. The projector system implements local decryption and play back. As far as security is concerned, the storage system and the projector system need to be physically secured and the communication between the projector system and the storage system need to be secured through local encryption. However, in the data server model, the storage system needs only temporary storage and the projection system implements decryption, decompression, and play back. In this model, only the projector system needs to be physically protected. The audio stream and other advanced feature control streams are routed to the relevant processor and played back after decryption and demultiplexing by the storage or play back system. The producer should employ a strong encryption scheme and the key needs to be transmitted through a secure channel to the security manager in the cinema hall. Usually for enhanced security and control, producers or distributors apply the encryption in such a way that it can be decrypted only in a designated storage or projector system and they can also control the play back schedules or revoke the play back if the storage or play back system is found compromised or the exhibitor is found aiding in piracy.

In a film-based cinema system, pirates can make a pirate copy by taking possession of film reels and digitizing the content. This is possible because the reel content is not encrypted. In the case of digital cinema, because the digital cinema content is transferred in an encrypted mode, it would be difficult for the pirates to break the encryption scheme to make a pirated copy. Another technique for making a pirated copy is by directly videographing off the screen during the show. In order to trace and control this kind of piracy in digital cinema systems, digital watermarking techniques can be employed. None of the current digital cinema systems explicitly mention the use of any schemes for tracing piracy. The current systems only provide conditional access using an encryption scheme. Thus, much work remains to be done in this area.

The Society for Motion Picture and Television Engineers (SMPTE) of the United States has constituted the DC28 digital cinema technology committee, a working group to discuss the various issues that face the full deployment of digital cinema. DC28 is divided into seven study groups: mastering, compression, conditional access, transport and delivery, audio, theater systems, and projection. The European counterpart of SMPTE is the European Digital Cinema Forum (EDCF), where study groups from different European countries share their information on digital cinema and the "technical module" is where all of the security research is being performed.

CONCLUSION

This chapter discusses the rights requirements of each party involved over the life cycle of the digital video. It also discusses various requirements to DRM stemming from asset management and usage controls needs. In addition, the requirements and constraints to DRM, business, legal, and social aspects of DRM have also presented. The technical aspect of DRM elaborates on the technologies used for obtaining DRM, techniques and tools for obtaining DRM in digital video broadcasts, stored video such as DVDs, and digital cinema, and standardization efforts by bodies such as MPEG, IETF, W3C, SMPTE, and so forth.

Most of the current DRM systems support confidentiality against nonconsumers. Some of the systems implement copy protection and control mechanisms. Some envisage the use of fingerprinting mechanisms for tracing piracy. These systems are essentially intended to mitigate the concerns of owners. Although the trading protocols and rights languages are being designed with the intention of supporting consumer concerns, the supporting cryptographic and watermarking techniques and protocols are yet to be designed. A full-fledged DRM system should be capable of supporting the rights of all parties (creators, owners, distributors, and consumers) involved in digital asset creation and transaction. DRM in its broadest sense is end-to-end management of digital trust. Therefore, tremendous research challenges abound in this vital area of information security.

REFERENCES

1. The Open Digital Rights Language, http://odrl.net/, 2002.
2. Iannella, R., *Digital rights management (DRM) architectures*, D-Lib Mag., 7(6), 2001.
3. Löytynoja, M., Seppänen, T., and Cvejic, N., *Experimental DRM Architecture Using Watermarking and PKI*, presented at First International Mobile IPR Workshop: Rights Management of Information Products on the Mobile Internet. Helsinki, 2003.

4. Mooney, S., *Interoperability: Digital rights management and the emerging ebook environment, D-Lib Mag.*, 7(1), 2001.
5. Linnartz, J.P., Depovere, G., and Kalker, T., *Philips Electronics Response to Call for Proposals Issued by the Data Hiding SubGroup Copy Protection Technical Working Group*, July 1997.
6. Peinado, M., Petitcolas, F.A.P., and Kirovski, D., *Digital rights management for digital cinema, Multimedia Syst.*, 9, 228–238, 2003.
7. The MPEG Home Page, http://www.chiariglione.org/mpeg/.
8. The Copy Protection Technical Working Group, http://www.cptwg.org/.
9. DVD The DVD Copy Control Association, http://www.dvdcca.org/.
10. Park, J. and Sandhu, R., *Towards Usage Control Models: Beyond Traditional Access Control*, in Proceedings of 7th ACM Symposium on Access Control Models and Technologies, 2002, pp 57–64.
11. Emmanuel, S. and Kankanhalli, M.S., *A digital rights management scheme for broadcast video, Multimedia Syst.*, 8(6), pp. 444–458, 2003.
12. Erickson, J.S., *Fair Use, DRM, and trusted computing, Commn. ACM*, 46(4), 34–39, 2003.
13. Fox, B.L. and LaMacchia B.A., *Encouraging recognition of fair uses in DRM systems Commn. ACM*, 46(4), 61–63, 2003.
14. Cohen, J.E., *DRM and privacy, Commn. ACM*, 46(4), 47–49, 2003.
15. Korba, L. and Kenny, S., Towards Meeting the Privacy Challenge: Adapting DRM, in Proceedings of ACM Workshop on Digital Rights Management, 2002.
16. Durfee, G. and Franklin, M., *Distribution chain security*, in Proceedings of the 7th ACM Conference on Computer and Communications Security, Athens, ACM Press, New York, 2000, pp. 63–70.
17. DRM Business Models, http://www.microsoft.com/windows/windowsmedia/wm7/drm/scenarios.aspx.
18. Liu, Q., Safavi-Naini, R., and Sheppard, N.P., *Digital Rights Management for Content Distribution*, presented at Australasian Information Security Workshop, Adelaide, 2003.
19. Sobel, L.S., DRM as an Enabler of Business Models: ISPs as Digital Retailers, presented at Digital Rights Management Conference, The Berkeley Center for Law and Technology, February, 2003.
20. Horne, B., Pinkas, B., and Sander, T., *Escrow Services and Incentives in Peer-to-Peer Networks*, presented at 3rd ACM Conference on Electronic Commerce, 2001.
21. Clark, D., *Future of Intellectual Property: How Copyright Became Controversial*, in Proceedings of the 12th Annual Conference on Computers, Freedom and Privacy, 2002.
22. Clayson, P.L., and Dallard, N.S., *Systems Issues in the Implementation of DVB Simulcrypt Conditional Access*, Presented at International Broadcasting Convention, 1997, pp. 470–475.
23. Rosnay, M.D., *Digital Right Management Systems Toward European Law: Between Copyright Protection and Access Control*, in Proceedings of the 2nd International Conference on WEB Delivering of Music, 2002.
24. Samuelson, P., *Encoding the law into digital libraries, Commn. ACM*, 41(4), 13–18, 1998.
25. Williams, J., IT Architecture meets the real (legal) world, *IT Prof.*, 3(5), 65–68, 2001.
26. Mooij, W., *Advances in Conditional Access Technology*, presented at International Broadcasting Convention, 1997, pp. 461–464.
27. Zeng, W. and Lei, S., *Efficient Frequency Domain Digital Video Scrambling for Content Access Control*, presented at ACM Multimedia '99 Proceedings, Orlando, FL, 1999.
28. Anderson, R. and Manifavas, C., *Chameleon—A New Kind of Stream Cipher*, in Proceedings of the 4th Workshop on Fast Software Encryption, Haifa, 1997.
29. Emmanuel, S. and Kankanhalli, M.S., *Copyright Protection for MPEG-2 Compressed Broadcast Video*, in Proceedings of the IEEE International Conference on Multimedia and Expo (ICME 2001), Tokyo, 2001.

30. Hartung, F. and Kutter, M., *Multimedia watermarking techniques, Proc. IEEE*, 87(7), 1079–1107, 1999.
31. Braudaway, G.W., Magerlein, K.A., and Mintzer, F., *Protecting Publicly Available Images with a Visible Image Watermark*, presented at International Conference on Image Processing, 1997, Vol. 1, pp. 524–527.
32. Meng, J. and Chang, S.F., *Embedding Visible Video Watermarks in the Compressed Domain*, presented at International Conference on Image Processing, 1998, Vol. 1, pp. 474–477.
33. Cox, I.J., Killian, J., Leighton, T. and Shamoon, T., *Secured spread spectrum watermarking for multimedia. IEEE Trans. Image Process.*, 6(12), 1673–1687, 1997.
34. Hartung, F. and Girod, B., *Watermarking of uncompressed and compressed video, Signal Process.*, 66(3), 283–301, 1998.
35. Kundur, D. and Hatzinakos, D., *Digital Watermarking Using Multiresolution Wavelet Decomposition*, in Proceedings of IEEE International Conference on Acoustics, Speech, and Signal Processing, 1998, Vol. 5, pp. 2969–2972.
36. Barni, M., Bartolini, F., Cappellini, V., and Piva, A., *A DCT domain system for robust image watermarking, Signal Process.*, 66(3), 357–372.
37. Podilchuk, C.I. and Zeng, W., *Digital image watermarking using visual models*, in Proceedings of the IS&T/SPIE Conference on Human Vision and Electronic Imaging II, San Jose, CA, USA, B.E. Rogowitz and T.N. Pappas, Eds., IS&T and SPIE, 1997, pp. 100–111.
38. Bormans, J., Gelissen, J., and Perkis, A., *MPEG-21: The 21st century multimedia framework, IEEE Signal Process. Mag.*, 20(2), 53–62, 2003.
39. Piva, A., Barni, M., Bartolini, F., and Cappellini, V., *DCT-based Watermark Recovering Without Resorting to the Uncorrupted Original Image*, presented at International Conference on Image Processing, 1997, Vol. 1, pp. 520–523.
40. Wolfgang, R.B. and Delp, E.J., *A Watermarking Technique for Digital Imagery: Further Studies*, in Proceedings of IEEE International Conference on Imaging, Systems, and Technology, Las Vegas, 1997, pp. 279–287.
41. Adelsbach, A. and Sadeghi, A.R., *Zero-Knowledge Watermark Detection and Proof of Ownership*, in Proceedings of Fourth Information Hiding Workshop, IHW2001, Pittsburgh, 2001.
42. Craver, S., *Zero knowledge watermark detection*, in *Proceedings of the Third International Workshop on Information Hiding*, Lecture Notes in Computer Science Vol. 1768, Springer-Verlag, Berlin, 2000, pp. 101–116.
43. Eggers, J.J., Su, J.K., and Girod, B., *Asymmetric watermarking schemes*, in Proceedings of Sicherheit in Mediendaten, Berlin, September 2000, Springer-Verlag, Berlin, 2000.
44. Pfitzmann, B. and Schunter, M., *Asymmetric fingerprinting (extended abstract)*, in *Advances in Cryptology — EUROCRYPT'96*, Maurer, U., Ed., Lecture Notes in Computer Science, Vol. 1070, Springer-Verlag, Berlin, 1996, pp. 84–95.
45. Langelaar, G.C., Lagendijk, R.L., and Biemond, J., *Removing Spatial Spread Spectrum Watermarks*, Presented at European Signal Processing Conference (EUSIPCO'98), Rhodes, Greece, 1998.
46. Cox, I.J. and Linnartz, J.P., *Some general methods for tampering with watermarks, IEEE J. Selected Areas Commn.*, 16(4), 587–593, 1998.
47. Craver, S., Memon, N., Yeo, B., and Yeung, M.M., *Resolving rightful ownerships with invisible watermarking techniques: Limitations, attacks and implications, IEEE J. Selected Areas Commn.*, 16(4), 573–586, 1998.
48. Internet Open Trading Protocol (trade), http://www.ietf.org/html.charters/trade-charter.html.
49. The W3C Trading Protocols, http://www.w3.org/Help/siteindex.

50. Kwok, S.H., Cheung, S.C., Wong, K.C., Tsang, K.F., Lui, S.M., and Tam, K.Y., *Integration of digital rights management into internet open trading protocol (IOTP) Dec. Support Syst.*, 34(4), 413–425, (2003).
51. The Extensible rights Markup Language, http://www.xrml.org.
52. Mulligan D. and Burstein A., *Implementing Copyright Limitations in Rights Expression*, in Languages Proceedings of ACM Workshop on Digital Rights Management, 2002.
53. Wang, X., Lao, G., DeMartini, T., Reddy, H., Nguyen, M., and Valenzuela, E., *XrML — Extensible rights Markup Language,* in Proceedings of the 2002 ACM Workshop on XML Security, 2002, pp. 71–79.
54. Messerges, T.S. and Dabbish, E.A., *Digital Rights management in a 3G Mobile Phone and Beyond*, in Proceedings of ACM Workshop on Digital Rights Management, 2003.
55. Hartung, F. and Ramme, F., *Digital rights management and watermarking of multimedia content for M-commerce applications*, *IEEE Commn. Mag.*, 38(11), 78–84, 2000.
56. ISO/IEC TC JTC1/SC 29/WG11/N 6269: Information Technology — Multimedia Framework (MPEG-21) – Part 1: Vision, Technology and Strategy.
57. ISO/IEC 13818-1: *Generic Coding of Moving Pictures and Associated Audio: Systems.* (MPEG-2 Systems).
58. ISO/IEC 14496-1: *Coding of Audiovisual Objects: Systems.* (MPEG-4 Systems).
59. Cutts, D.J., *DVB conditional access, Electron., Commn., Engi. J.*, 9(1), pp. 21–27, 1997.
60. *ATSC Standard A/70: Conditional Access System for Terrestrial Broadcast with Amendment*, http://www.atsc.org/standards.html, 2004.
61. Simitopoulos, D., Zissis, N., Georgiadis, P., Emmanouilidis, V., and Strintzis, M.G., *Encryption and watermarking for the secure distribution of copyrighted MPEG video on DVD, Multimedia Syst.*, 9(3), 217–227, 2003.
62. Byers, S., Cranor, L., Cronin, E., Kormann, D., and McDaniel, P., *Analysis of Security Vulnerabilities in the Movie Production and Distribution Process*, in Proceedings of ACM Workshop on Digital Rights Management, 2003.
63. The Motion Picture Association's Role in Digital Cinema, http://www.mpaa.org/dcinema.
64. Korris, J. and Macedonia, M., *The end of celluloid: Digital cinema emerges, Computer*, 35(4), 96–98, 2002.
65. Lubell, P.D., *A coming attraction: Digital cinema, IEEE Spectrum*, 37(3), 72–78, 2000.
66. Camp, L.J., DRM: Doesn't Really Mean Digital Copyright Management, in Proceedinss of the 9th ACM Conference on Computers and Communications Security, 2002, pp. 78–87.
67. The Digital Object Identifiers, http://www.doi.org/.

Index

Index

Index

Index

Fragile watermarks, 256
Frequency division, multibit watermarking, 244
Frequency domain
  image, 106
  video, 106
Frequency domain watermarking, 390

G

G.723.1 Speech Codec by Wu and Kuo, 2000, selective encryption algorithm for, 128
Generalized three-dimensional radon transform, 737–738
  three-dimensional models, watermarking, indexing, 733–759
Generic attack, 69–84
  on linear detectors, 71–76
  on QIM scheme, 76–80
  on quantization-based schemes, 76–83
  on quantized projection scheme, 80–83
Generic framework, for system security evaluation, 666
Geneva Convention for Protection of Producers of Phonograms Against Unauthorized Duplication of Their Phonograms, 49
Geometric attacks, robustness, 751–752
Geometric distortion
  image watermarking resistant to, 331–358
    capacity, 355
    complexity, 355
    embedding without exact inversion, 355–356
    experimental results, 349–352
    geometric distortions, 334–342
    rotation invariance, 353–355
    RST-invariant watermarking method, 342–349
    watermark system design, 345–349
    watermarking framework, 333–334
  image watermarking robust to, JPEG compression, 467–492
    message encoding, 472–475
    template detection, 479–482
    template embedding in DFT domain, 475–478
    training sequence embedding in DWT domain, 472–475
    translation registration, decoding, 482–484

watermark embedding, 472–478
watermark extraction with resynchronization, 479–484
JPEG compression, image watermarking robust to, 467–492
Granularity, 265
Graphic works, copyright, 8
Group authentication, 35
Group signatures, 620

H

Halftone image watermarking techniques, 408–413
  halftoning technique, defined, 408–410
  roles of watermarking printed for textual images, 413–414
  watermarking techniques for halftone images, 410–413
  watermarking techniques for printed textual images, 414
HCIE. See Hierarchical Chaotic Image Encryption
Health supervision, 419
Hearable frequency spectrum, audio quality layers, 129–130
Heterogeneous networks
  security in, intellectual property multimedia, 173
  streaming media over, to various devices, layout, 202
Hierarchical Chaotic Image Encryption, 152
Hierarchical key-based schemes, 31
Hierarchical key distribution trees, 31
Hierarchical node-based schemes, 31
Hierarchy of intermediaries
  watermarking multicast video with, 588
  watermarking with, 37–38
High-capacity real-time audio watermarking, perfect correlation sequence/repeated insertion, 283–310
  audio similarity measure, 293–294
  correlation properties, 287
  correlation property, 289–291
  experimental audio clip, PN binary data profiles, 297–298
  experimental results, 294–308
  perfect sequences, 286–288
  product theorem, 287–288
  proposed watermarking technique, 291–294

Index

806

Printed and bound by CPI Group (UK) Ltd, Croydon, CR0 4YY

30/10/2024

01781339-0001